MOLECULAR ACTIVITIES OF
PLANT CELLS

MOLECULAR ACTIVITIES OF PLANT CELLS

An introduction to plant biochemistry

JOHN W. ANDERSON
BAgrSc, PhD
Reader, Botany Department,
School of Biological Sciences,
La Trobe University, Bundoora,
Victoria, Australia

JOHN BEARDALL
BSc, PhD
Department of Ecology and
Evolutionary Biology,
Monash University,
Clayton, Victoria,
Australia

OXFORD

BLACKWELL SCIENTIFIC PUBLICATIONS

LONDON EDINBURGH BOSTON

MELBOURNE PARIS BERLIN VIENNA

© 1991 Blackwell Scientific Publications
Editorial offices:
Osney Mead, Oxford OX2 0EL
25 John Street, London WC1N 2BL
23 Ainslie Place, Edinburgh EH3 6AJ
3 Cambridge Center, Cambridge
  Massachusetts 02142, USA
54 University Street, Carlton
  Victoria 3053, Australia

Other Editorial Offices:
Arnette SA
2, rue Casimir-Delavigne
75006 Paris
France

Blackwell Wissenschaft
Meinekestrasse 4
D-1000 Berlin 15
Germany

Blackwell MZV
Feldgasse 13
A-1238 Wien
Austria

First published 1991

Set by Interprint Ltd, Malta
Printed and bound at
Dah Hua Printing Press Co. Ltd,
Hong Kong

British Library
Cataloguing in Publication Data

Anderson, John W.
  Molecular activities of plant cells.
  1. Plants. Cells. Molecular biology
  I. Title II. Beardall, John
  581.876

ISBN 0-632-02457-7
ISBN 0-632-02458-5 pbk

DISTRIBUTORS

Marston Book Services Ltd
PO Box 87
Oxford OX2 0DT
(*Orders*: Tel: 0865 791155
         Fax: 0865 791927
         Telex: 837515)

USA
Blackwell Scientific Publications, Inc.
3 Cambridge Center
Cambridge, MA 02142
(*Orders*: Tel: 800 759-6102)

Canada
Oxford University Press
70 Wynford Drive
Don Mills
Ontario M3C 1J9
(*Orders*: Tel: (416) 441-2941)

Australia
Blackwell Scientific Publications
(Australia) Pty Ltd
54 University Street
Carlton, Victoria 3053
(*Orders*: Tel: (03) 347-0300)

Library of Congress
Cataloging-in-Publication Data

Anderson John W. (John Warwick), 1940–
Molecular activities of plant cells: an
introduction to plant biochemistry John W.
Anderson, John Beardall.
          p.   cm.
Includes bibliographical references and index.
1. Plant molecular biology. 2. Plant
cytochemistry. 3. Growth (Plants)—
Molecular aspects.
I. Beardall, John. II. Title.
QK728.A53 1991.   581.87′6042—dc20

ISBN 0-632-02457-7 (hard)
ISBN 0-632-02458-5 (pbk)

Contents

Preface, xi

Nomenclature and Abbreviations, xiii

PART 1: GENERAL BIOCHEMISTRY, 1

1 Plant growth and biochemistry—the connection, 3

1.1 Plant growth involves the synthesis of biological matter, 3
1.2 Synthesis of biological matter requires an energy source, 4
1.3 Characteristics of autotrophs and heterotrophs, 5
1.4 Photosynthesis—definitions, 5
1.5 Major activities of producer and consumer cells, 5
1.6 Molecular traffic between producer and consumer cells, 6
1.7 Plant growth and carbon dioxide assimilation rates, 7
Further reading, 8

2 Molecules of plant cells, 9

2.1 Classes of compounds in plant matter, 9
2.2 Organic molecules of plants occur in specific isomeric forms, 9
2.3 Carbohydrates are quantitatively the most important compounds in plants, 10
2.4 Lipids are biological molecules with little affinity for water, 20
2.5 Proteins are constructed from 20 amino acids, 25
2.6 Nucleotides and nucleic acids, 30
2.7 Some phenylpropanoid compounds are important constituents of plant cell walls, 34
2.8 Secondary compounds are not essential for plant function, 34
Further reading, 35

3 Structural organization of plant cells, 36

3.1 Cellular organization is essential for cellular activity, 36
3.2 Membrane structure, 36
3.3 Ribosomes, 38
3.4 Nucleus, 40
3.5 Chloroplasts, 41
3.6 Mitochondria, 42
3.7 Microbodies, 42
3.8 Other membrane-limited organelles, 43
3.9 Cytoplasm—its inclusions and cytosol, 44
3.10 Plant cell walls, 45
3.11 The cell cycle involves non-sexual production of two daughter cells, 45
3.12 Preparation of subcellular fractions from plants, 46
Further reading, 49

4 Cellular energetics, 50

4.1 The free energy change of a chemical reaction is related to its equilibrium constant, 50
4.2 Cells have mechanisms for coupling endergonic reactions to exergonic reactions, 52
4.3 Hydrolysis of a few key compounds provides important intermediary energy sources in cells, 52
4.4 A few redox pairs are important intermediary electron acceptors and donors in cells, 54
4.5 The free energy change of a redox reaction is related to the electrode potentials of the two redox pairs, 56
4.6 Differences in electrical potential and ion concentration across a membrane cause free energy differences between the ions across the membrane, 58
4.7 Proton electrochemical gradients across membranes are important energy sources in cells, 59
4.8 A mole of pigment can absorb a mole of photons, the energy of which varies with the wavelength, 59
Further reading, 60

5 Enzymes and post-translational enzyme regulation, 61
5.1 Enzymes are protein molecules which catalyse specific biological reactions, 61
5.2 Some enzymes require the presence of cofactors for activity, 63
5.3 Kinetics of reactions catalysed by non-regulatory enzymes, 67
5.4 Inhibition of enzyme-catalysed reactions, 76
5.5 Some enzymes have a regulatory role in metabolism, 79
Further reading, 82

6 Membranes and metabolite transport, 83
6.1 Membranes exhibit variation in their composition and cellular function, 83
6.2 Membranes are responsible for transport of materials into and out of cells and organelles, 86
6.3 Movement of many materials across plant membranes involves carrier proteins, 92
6.4 'Patch clamping' is a powerful tool for studying membrane transport of ions, 95
Further reading, 96

PART 2: ENERGY-GENERATING MECHANISMS OF PLANTS, 97

7 Aerobic oxidation of sugars to carbon dioxide, 99
7.1 Respiratory rates vary between different tissues and stages of development, 99
7.2 Storage products are mobilized before they are respired, 99
7.3 Glycolysis involves the anaerobic oxidation of hexoses to pyruvate, 100
7.4 Reactions of glycolysis occur both in the cytosol and in plastids, 104
7.5 Flow of metabolites through the glycolytic sequence is regulated, 105
7.6 The tricarboxylic acid cycle oxidizes pyruvate to carbon dioxide, 108
7.7 NADH and $FADH_2$ are oxidized by oxygen via the mitochondrial electron transport chain, 114
7.8 The free energy of oxidation of NADH by oxygen is conserved by ADP phosphorylation, 118
7.9 Analysis of input and output associated with respiration, 123
7.10 Glycolysis and the TCA cycle contribute both ATP and carbon skeletons for biosynthetic pathways, 125
Further reading, 126

8 Secondary oxidative mechanisms in plants, 127
8.1 Oxidative pentose phosphate pathway, 127
8.2 Oxidation of lipids is important in germinating oil-bearing seeds, 133
8.3 The glyoxylate pathway provides a mechanism for linking β-oxidation to carbohydrate synthesis, 137
Further reading, 141

9 Light reactions of green plants, 142
9.1 Properties of solar radiation in relation to plant growth, 142
9.2 Plants absorb light by specific pigments contained in thylakoids, 142
9.3 Light is absorbed by chlorophyll and the energy is transferred to specialized chlorophyll molecules—the reaction centres, 144
9.4 Light is required for production of ATP and reducing equivalents, 146
9.5 Two light reactions are involved, 147
9.6 The photosynthetic unit, 149
9.7 Properties of the photosystems and their antenna complexes, 149
9.8 Transport of electrons from water to NADP, 150
9.9 Functional organization of thylakoid membranes, 154
9.10 Transverse heterogeneity of electron transport components results in a trans-thylakoid proton gradient, 156
9.11 The trans-thylakoid proton gradient is coupled to ADP phosphorylation, 157
9.12 The CF_1–CF_0 ATP synthetase, 158
9.13 How many protons are transported for each ATP formed? 159
9.14 Analysis of inputs and outputs and efficiency of non-cyclic electron transport, 159
9.15 Monitoring the photosystems *in vivo* by fluorescence and fluorescence quenching, 159
9.16 Regulation of energy distribution between PSII and PSI, 160
9.17 Cyclic photophosphorylation, 161
9.18 Light-dependent reduction of oxygen: the Mehler reaction and pseudocyclic photophosphorylation, 163
Further reading, 163

PART 3: ASSIMILATORY MECHANISMS IN PLANTS, 165

10 C$_3$ carbon reduction cycle and associated processes, 167

10.1 Most autotrophic organisms possess an active C$_3$ carbon reduction cycle, 167

10.2 Experimental approaches to define the pathway of carbon dioxide assimilation, 167

10.3 Incorporation of carbon dioxide into triose phosphates involves a series of chloroplast enzymes, 167

10.4 Carbon dioxide in chloroplasts is in equilibrium with a pool of bicarbonate, 169

10.5 Ribulose-1,5-P$_2$ is regenerated from triose-P, 170

10.6 Energy inputs from ATP and NADPH satisfy the energy requirements for production of carbohydrate from carbon dioxide, 172

10.7 The chloroplast envelope is impermeable to many intermediates of the C$_3$-CR cycle, 172

10.8 Triose phosphates are exported to the cytosol by the phosphate translocator, 173

10.9 Sucrose is synthesized via UDP-glucose in the cytosol, 174

10.10 Sucrose is exported from photosynthetic cells and loaded into the phloem, 175

10.11 Starch is synthesized in chloroplasts via ADP-glucose, 175

10.12 Starch is metabolized to sucrose when demand for carbon in the cytosol exceeds the rate of carbon dioxide assimilation, 177

10.13 Activities of the C$_3$-CR cycle and starch and sucrose synthesis are regulated, 177

10.14 Aspects of carbon metabolism can be monitored by chlorophyll fluorescence, 182

Further reading, 184

11 Photorespiration, 185

11.1 Light enhances carbon dioxide evolution by C$_3$ plants, 185

11.2 Photorespired carbon dioxide is derived from recently assimilated carbon dioxide and is strongly influenced by oxygen and carbon dioxide concentrations, 185

11.3 Ribulose-1,5-P$_2$ oxygenase activity accounts for several features of photorespiration, 185

11.4 Glycollate-2-P is returned to the C$_3$-CR cycle via the C$_2$ photorespiratory cycle, 187

11.5 Enzymes of the C$_2$-PR cycle and their subcellular location, 187

11.6 The C$_2$-PR cycle is associated with metabolite transport between organelles and with nitrogen cycling, 190

11.7 The C$_2$-PR cycle requires a large input of energy, 191

11.8 Why is photorespiration light-dependent?, 191

11.9 What does photorespiration achieve?, 191

11.10 Some plants have strategies for suppressing photorespiration, 193

11.11 Some algae possess alternative routes of photorespiratory glycollate metabolism, 194

Further reading, 194

12 C$_4$ mechanisms of carbon dioxide assimilation, 195

12.1 Carbon dioxide assimilation characteristics of some plants do not reflect the properties of ribulose-1,5-P$_2$ carboxylase, 195

12.2 C$_4$ plants and CAM plants assimilate carbon dioxide into oxaloacetate and other C$_4$ dicarboxylates, 195

12.3 C$_4$ plants have distinctive biochemical, physiological and anatomical characteristics, 196

12.4 The pathway of carbon dioxide assimilation in C$_4$ plants can be determined by the order in which metabolites are labelled with [^{14}C]carbon dioxide, 196

12.5 Establishing the functions of mesophyll and bundle sheath cells, 198

12.6 Leaves of C$_4$ plants have a quantitatively distinctive complement of enzymes, 199

12.7 Mechanisms for decarboxylating C$_4$ dicarboxylates in bundle sheath cells vary between species, 200

12.8 Enzymes of the C$_3$-CR cycle occur in bundle sheath cells but some are also found in mesophyll cells, 202

12.9 C$_4$ plants expend more energy in assimilating carbon dioxide than do C$_3$ plants, 202

12.10 Carbon dioxide assimilation in C$_4$ plants involves extensive intracellular and intercellular transport of metabolites, 204

12.11 The C$_4$ mechanism of carbon dioxide assimilation explains many of the physiological properties of C$_4$ plants, 204

12.12 CAM plants assimilate carbon dioxide into malate at night and reassimilate carbon dioxide into carbohydrate in the light, 205

12.13 CAM affords an explanation for many physiological and ecological features of CAM plants, 207

12.14 Phylogenetic distribution of C_4 and CAM plants and its implications for herbicide design, 207
Further reading, 208

13 Assimilation of inorganic nitrogen into amino acids, 209

13.1 Plants use inorganic nitrogen for synthesis of protein amino acids, 209
13.2 Plants reduce nitrate to ammonia, 210
13.3 Ammonia is incorporated via glutamine into glutamate by the C_5-ammonia assimilation cycle, 211
13.4 Ammonia produced by photorespiration is reassimilated via the C_5-ammonia assimilation cycle, 213
13.5 Light is an important energy source for the assimilation of inorganic nitrogen in leaves, 213
13.6 Aminotransferases catalyse transfer of the amino group of amino acids to oxo acids to form other amino acids, 214
13.7 Synthesis of some other amino acids, 215
13.8 Some organisms use gaseous nitrogen as a nitrogen source by the process of nitrogen fixation, 226
13.9 Some plants contain toxic non-protein amino acids, 229
Further reading, 229

14 Other light-coupled assimilatory and reductive mechanisms, 230

14.1 Light supports various processes in chloroplasts and protoplasts, 230
14.2 Assimilation of inorganic sulphur, 230
14.3 C_1 fragments have various origins, 233
14.4 Reducing equivalents and phosphorylation potential are exported from illuminated chloroplasts by shuttle mechanisms, 236
14.5 Illuminated chloroplasts reduce oxidized glutathione, 239
14.6 Illuminated chloroplasts synthesize fatty acids, protein and other metabolites, 241
14.7 Rates of processes associated with chloroplasts, 241
Further reading, 241

PART 4: SYNTHESIS OF NEW CELLS AND CELL STRUCTURES, 243

15 Synthesis of nitrogenous compounds from amino acids, 245

15.1 Synthesis of pyrimidines and pyrimidine nucleotides, 245
15.2 Synthesis and metabolism of purines and purine nucleotides, 249
15.3 Synthesis of chlorophyll, 252
Further reading, 256

16 Synthesis of lipids, 257

16.1 Fatty acids are synthesized from acetyl-CoA, 257
16.2 Waxes are derived from palmitate and stearate, 261
16.3 Phosphatidate is an important intermediate in the synthesis of acyl lipids other than waxes, 263
16.4 Biosynthesis of prenyllipids, 269
16.5 Membranes have a capacity for self-assembly, 274
Further reading, 274

17 Synthesis of plant cell walls, 275

17.1 Plant cell walls are important biologically and economically, 275
17.2 Molecular composition of plant cell walls, 275
17.3 Plant cells vary in the composition of their walls, 279
17.4 The orientation of cellulose microfibrils is normal to the direction of cell growth, 279
17.5 Model for the organization of the primary cell wall, 280
17.6 Structural studies indicate that microtubules and Golgi bodies are involved in cell wall synthesis, 281
17.7 Monosaccharides are incorporated into cell wall polysaccharides as their nucleoside diphosphate derivatives, 282
17.8 Cellulose microfibrils are thought to be formed by enzyme assemblies associated with the plasma membrane, 283
17.9 Matrix polysaccharides are synthesized in Golgi bodies and the endoplasmic reticulum, 286
17.10 Extensin synthesis involves post-translational oxidation of proline and attachment of sugars, 287
17.11 Lignin is synthesized from phenylalanine, 288
Further reading, 290

18 The plant genome and its replication, 291

18.1 Organization of the plant genome, 291

18.2 Replication of the plant genome, 296

Further reading, 300

19 Processes involved in protein synthesis, 301

19.1 The information required to build proteins is contained in a genetic code, 301

19.2 Transcription of DNA to RNA, 302

19.3 Translation of mRNA into protein occurs in the cytoplasm, 306

19.4 Many proteins are modified after mRNA translation, 311

19.5 Protein synthesis in mitochondria and chloroplasts, 313

Further reading, 314

20 Regulation of gene expression, 315

20.1 Control of gene expression can occur at several points, 315

20.2 Regulation of some genes occurs at transcription, 316

20.3 Control of gene expression also occurs post-transcriptionally or at translation, 318

20.4 Gene expression is regulated in response to a range of environmental and developmental signals, 320

20.5 Recombinant DNA technology is a powerful tool for investigating gene organization and function, 322

20.6 Genetic engineering: manipulation of the genetic information of plant cells, 333

Further reading, 337

21 Biogenesis of organelles, 338

21.1 Mitochondria and chloroplasts are not formed *de novo*, 338

21.2 Chloroplasts develop from proplastids, 339

21.3 Perception of the light stimulus in chloroplast development involves a number of receptors, 344

21.4 Biosynthesis of organelles involves interaction between organelles and the nucleus, 347

21.5 Regulation of the interaction between nuclear and organelle genomes, 349

Further reading, 352

Further reading, 353

Index, 357

Preface

What is a plant? After a moment's reflection most of us would probably give an answer in terms of function. The ability of plants to make biological matter from CO_2 and inorganic salts, using light as an energy source, must rate as their single-most distinguishing feature. Every other form of life is directly or indirectly dependent on this activity in plants and for this reason there is a long tradition of interest in the molecular activities of plant cells.

Recently, plant biochemistry has attracted additional interest because advances in this discipline have enhanced our appreciation of several more general biological principles. Studies of gene expression in chloroplasts, for example, have aided our understanding of the evolution of eukaryotic cells, whilst investigations of light-induced proton gradients in chloroplasts adds considerably to our knowledge of the chemiosmotic theory of ADP phosphorylation. Our comprehension of certain biochemical processes in plants also provides a theoretical basis for understanding and researching into various practical matters, such as the mechanism of action of herbicides and the production of seeds with a 'balanced' mixture of the amino acids essential for human nutrition. Modern recombinant DNA techniques have provided greater appreciation of gene regulation and may possibly lead to techniques for improving plant varieties, especially, of course, for agricultural purposes.

We would draw the reader's attention to the subtitle of this book. This is an introductory text which concentrates on those processes in plant cells which are relevant to primary productivity. In order to ensure that the text is within the capacity of student finances (and is therefore available to its chosen audience) we have had to make some concessions regarding the inclusion (or exclusion) of certain topics. Thus, it has not been possible to include discussion of a number of processes of lesser importance to plant growth and we make only passing reference to the structure, synthesis and metabolism of secondary plant products. Similarly, we have not included many details of the large number of experiments which form the basis of our knowledge of how plant cells function.

Generally speaking, students come to study plant biochemistry from two different directions. One group consists of those who are studying plants *per se*, i.e. plant scientists of all types, including those from the applied plant sciences of agriculture, horticulture, forestry and plant biochemistry. For these students we have endeavoured to interface this book with plant physiology and have included an introductory section on general biochemical principles (using examples involving plants wherever possible) so that these students should not need to supplement this book by also acquiring a general biochemical text.

The second main group of students are those who have studied, or are studying, general biochemistry. This book is equally relevant to these students and even the examples used to illustrate the general principles discussed in the first section should be of interest. However, the real interest for the general biochemist will be in learning about processes unique to plants. Although we have not made a point of comparing and contrasting particular processes in plants with those in other organisms, we hope that biochemistry students will recognize any differences as they occur and so add to their knowledge of biochemical diversity.

In preparing this book we have prevailed upon colleagues throughout the world to read and evaluate individual chapters. The comments, criticisms and suggestions we have received have been invaluable, especially in those aspects with which we are least familiar. We would like to express our gratitude to the following for the time they have spent in assisting us in this way: T. Ap Rees, M. Avron, J. Barber, H. Beevers, J.N. Burnell, G. Edwards, G.B. Fincher, J.C. Gray, J. Giovanelli, M.D. Hatch, T.J. Higgins, G.J. Kelly, A.J. Keys, P.J. Lea, C.J. Leaver, A. Marcus, D.J. Morrè, C.K. Pallaghy, R.W. Parish, J.A. Raven, K.S. Rowan, R.K. Scopes, B.A. Stone, P.K. Stumpf, M.E. Van Steveninck, A.B. Wardrop, R.E. Williamson and the anonymous referee who read the entire manuscript and brought some recent developments to our attention. In similar fashion we would like to thank Chris Beardall for her assistance in improving the clarity, syntax and expression of the entire manuscript. However, we accept full responsibility for the contents, especially the introduction of sometimes new but logical and consistent abbreviations and the adherence to SI units.

This book was prepared at La Trobe University where we have received support and encouragement with our task. We would like to record our gratitude to I.A. Staff for supervising the preparation of the electron micrographs and to

Linda Humphreys for her cheerful, supportive and very professional role in preparing the typescript—a task in which she was assisted by Linda Davidson. We are conscious that many colleagues, including various technical staff, past and present members of our research laboratories, and many others, have directly and/or indirectly contributed to, or influenced, the contents of this book. These contributions are gratefully acknowledged.

As undergraduates we recall being fascinated by the lectures we received on the metabolic activities of plants and other autotrophic organisms and we remember our period as Ph.D. students with great pleasure when, supervised by Kingsley Rowan (J.W.A.) and the late Ian Morris (J.B.), we were introduced to the rigours, the fascination and the excitement of the research laboratory. We have equally fond memories of postdoctoral positions with Sir Leslie Fowden (J.W.A.) and John Raven (J.B.). To all of these people, together with those with whom we have been associated in our employment and on study leave, we extend our gratitude. These associations have helped to shape this book and they serve, in their own small way, to illustrate that the general advancement of scientific knowledge is a communal activity involving contributions from scientists throughout the world.

This book was mostly written from our desks at home, in the weekends, the evenings and the small hours of the morning, isolated from our families in studies at the quiet end of our homes. To our wives, Pat and Chris, and our offspring, we extend our appreciation of their tolerance, support and understanding during the course of the writing. We also extend to them our sympathy and apologies for the isolation they have experienced during the prolonged gestation of this book.

J.W. Anderson
J. Beardall

Nomenclature and Abbreviations

Although some of the abbreviations used in this book are commonplace, we have introduced a number of abbreviations which we feel describe various compounds and processes more accurately than those previously in use. Furthermore, we have attempted to develop a system of abbreviations which are more logical and internally consistent than many of those sometimes used elsewhere. We also feel that a note on our use of some units and/or terminology is necessary to prevent any misinterpretation of our meaning.

Notation for specific atoms within a molecule

It is sometimes necessary to refer to a specific atom within a molecule or to an isotopic label (usually radioactive) at particular positions within a molecule. For individual carbon atoms, the notation C-1, C-2, C-n refers to carbon atoms $1-n$. For isotopically labelled materials, the notation [1-^{14}C]glucose for instance, refers to glucose labelled with the radioactive isotope ^{14}C at C-1. [U-^{14}C]Glucose refers to glucose containing ^{14}C-label at all carbon atoms (i.e. uniform labelling). [^{15}N]Glutamate refers to glutamate in which the nitrogen atom is labelled with the non-radioactive isotope ^{15}N.

When referring to the total number of carbon atoms in a molecule we have used a subscript. Thus, we refer to glycerate-3-P as a C_3 compound (with 3 carbon atoms) and pentoses as C_5 sugars. Also, [1-^{14}C]glucose can be described as a C_6 compound labelled with ^{14}C in C-1. We also refer to C_3 or C_4 plants where the first product of CO_2 assimilation is a C_3 or C_4 compound respectively.

Nomenclature for phosphorylated compounds and organic acids

We have used the abbreviation -P to indicate a phosphate group in ester linkage ($-OPHO_3^-$). Thus, for example, ribose phosphorylated at C-5 is referred to as ribose-5-P whilst glycerate with phosphate groups at C-1 and C-3 (i.e. glycerate-1,3-bisphosphate) is referred to as glycerate-1,3-P_2. As a general rule we have described organic acids in their anionic form since they are dissociated at physiological pH and occur in cells as their salts. Thus, malic acid, glutamic acid and pyruvic acid are referred to as malate, glutamate and pyruvate respectively.

Metabolic processes

Abbreviation	Process	Alternative term(s)
C_3-CR cycle	C_3 carbon reduction cycle	Reductive pentose phosphate pathway; Calvin cycle
C_2-PR cycle	C_2 photorespiratory cycle	Photorespiration; glycollate pathway
C_4-CC cycle	C_4-CO_2 concentrating cycle	C_4 pathway

Other processes are referred to by their previously used abbreviations, e.g.:

TCA cycle	Tricarboxylic acid cycle
OPP pathway	Oxidative pentose phosphate pathway

Abbreviations of compounds and physico-chemico symbols

Abbreviation	Full name
AMP, ADP, ATP	Adenosine 5′-monophosphate, -diphosphate, -triphosphate
APS	Adenosine 5′-phosphosulphate (adenosine 5′-sulphatophosphate)
CMP, CDP, CTP	Cytidine monophosphate, -diphosphate, -triphosphate
Cyt	Cytochrome
DCMU	3-(3,4-dichlorophenyl)-1,1-dimethyl urea
DNA	Deoxyribonucleic acid

Abbreviation	Full name
E'_0	Standard redox potential of a redox half reaction at pH 7
F	Faraday constant
FAD	Flavin adenine dinucleotide ($FADH_2$, reduced form)
FCCP	Carbonyl cyanide p-trifluoromethoxyphenyl hydrazone
Fd	Ferredoxin (Fd_{ox}, oxidized form; Fd_{red}, reduced form)
FMN	Flavin mononucleotide ($FMNH_2$, reduced form)
GMP, GDP, GTP	Guanosine monophosphate, -diphosphate, -triphosphate
GSH, GSSG	Glutathione (reduced form, oxidized form)
$\Delta G^{\circ\prime}$	Standard change in Gibbs free energy under standard temperature and pressure at pH 7 with all reactants at a concentration of 1 M
HRGP	Hydroxyproline-rich glycoprotein
HSCoA	Coenzyme A
IMP	Inosine 5'-monophosphate
Δ^2-IPP	Δ^2-isopentenyl pyrophosphate
Δ^3-IPP	Δ^3-isopentenyl pyrophosphate
kDa	Kilodalton (1 dalton $= 1.6605 \times 10^{-24}$ g, i.e. the mass of 1 hydrogen atom)
K_m	Michaelis–Menten constant for an enzyme-catalysed reaction
NAD	Nicotinamide adenine dinucleotide (NAD^+, oxidized form; NADH, reduced form)
NADP	Nicotinamide adenine dinucleotide phosphate ($NADP^+$, oxidized form; NADPH, reduced form)
P_{680}	Photo-oxidizable form of chlorophyll associated with the reaction centre of photosystem II
P_{700}	Photo-oxidizable form of chlorophyll associated with the reaction centre of phostosystem I
PC	Plastocyanin
PEP	Phosphoenolpyruvate
P_i	Orthophosphate
PP_i	Pyrophosphate
PQ	Plastoquinone (PQH_2, reduced form)
PRPP	5-phosphoribosyl-1-pyrophosphate
PSI	Photosystem I

Abbreviation	Full name
PSII	Photosystem II
Q_A	Immediate electron acceptor of photosystem II
q_E	Quenching of chlorophyll fluorescence attributable to trans-thylakoid proton gradient
q_Q	Quenching of chlorophyll fluorescence attributable to reduction of Q
RNA	Ribonucleic acid (mRNA, messenger RNA; rRNA, ribosomal RNA; tRNA, transfer RNA)
Td	Thioredoxin (Td_{ox}, oxidized form; Td_{red}, reduced form)
THF	Tetrahydrofolate (formylTHF, N^{10}-formylTHF; methenylTHF, N^5,N^{10}-methenylTHF; methylTHF, N^5-methylTHF; methyleneTHF, N^5,N^{10}-methyleneTHF)
TPP	Thiamine pyrophosphate
UMP, UDP, UTP	Uridine 5'-monophosphate, -diphosphate, -triphosphate
UQ	Ubiquinone

Enzymes

Generally speaking, we have referred to enzymes by their full name the first time they are used in a chapter. Thereafter, we sometimes resort to the abbreviated form given at that time. These abbreviations are also listed below. Note that they always end with the suffix '-ase' so that they can be recognized as enzymes, distinct from metabolites and other proteins. All enzyme abbreviations consist of two terms separated by a hyphen. The first of these is an abbreviation for the substrate(s) while the second refers to the type or class of enzyme (e.g. -DHase for dehydrogenase, -ATase for aminotransferase). Where relevant, abbreviations denoting nucleotide specificity are shown independently of the enzyme abbreviation, e.g. NADP M-DHase refers to NADP-specific malate dehydrogenase. We have not used full Enzyme Commission nomenclature for naming the common enzymes and refer the reader to the text by Dixon and Webb* and to the Recommendations (1978) of the Nomenclature Committee of the International Union of Biochemistry for full details.

*Dixon, M. & Webb, E.C. (1979) *Enzymes*, 3rd edn. Longman Group Ltd, Harlow.

Abbreviation	Full name
FP_2-Pase	Fructose-1,6-P_2 phosphatase
FP-PTase	Fructose-6-P: PP_i phosphotransferase
GO-ATase	Glutamate: oxaloacetate aminotransferase
GP-ATase	Glutamate: pyruvate aminotransferase
G-DHase	Glutamate dehydrogenase
M-DHase	Malate dehydrogenase
PEP-Case	Phosphoenolpyruvate carboxylase
PEP-CKase	Phosphoenolpyruvate carboxykinase
PF-Kase	Phosphofructokinase
PP-DKase	Pyruvate, phosphate dikinase
RuP_2-Case	Carboxylase activity of RuP_2-C/Oase
RuP_2-C/Oase	Ribulose-1,5-P_2 carboxylase/oxygenase
RuP_2-Oase	Oxygenase activity of RuP_2-C/Oase
SP_2-Pase	Sedoheptulose-1,7-P_2 phosphatase

The exceptions to these rules are the restriction endonucleases used in recombinant DNA technology. These enzymes are of bacterial origin and, by convention, are named after the bacterium from which they are obtained. Thus, the enzymes *Eco* RI and *Bam* H1 are isolated from *Escherichia coli* RY 13 and *Bacillus amyloliquefaciens* H respectively (see Section 20.5.1).

Lipids

Abbreviation	Full name
ACP	Acyl carrier protein
BCCP	Biotin carboxyl carrier protein
CDP-	Cytidine diphosphate derivative
CMP-	Cytidine monophosphate derivative
DGDG	Digalactosyl diglyceride
FAS	Fatty acid synthetase complex
HMG-CoA	3-Hydroxymethylglutaryl-CoA

Abbreviation	Full name
MGDG	Monogalactosyl diglyceride
MVA	Mevalonate
PA	Phosphatidate (phosphatidic acid)
PC	Phosphatidyl choline
PE	Phosphatidyl ethanolamine
PG	Phosphatidyl glycerol
PI	Phosphatidyl inositol
PS	Phosphatidyl serine
PLTP	Phospholipid transfer protein
SQDG	6-Sulphoquinovosyl diacylglycerol

Units

We have endeavoured to keep to the Système Internationale d'Unites (SI) throughout the text. We have used the term flux strictly in terms of velocity of transfer of materials per unit area as, for example, when dealing with transport processes across membranes. To describe the rate of passage of materials through a particular metabolic pathway we use the term overall flow rate (as determined experimentally) or flow capacity (a theoretical flow rate estimated from other criteria). Thus, for example, we talk about the *rate of flow* of carbon through the C_2-PR cycle but the *flux* of triose phosphate across the chloroplast envelope. When describing rates, flows and fluxes in terms of molecules we use the standard term, mole (mol). We have also used this nomenclature when describing rates and fluxes of an element (e.g. 1 mol of glucose contains 6 mol of carbon). Thus, for a CO_2 fixation rate of $200 \mu mol \ CO_2 \ mg^{-1}$ chlorophyll h^{-1}, the flow rate of carbon through the associated pathway is $200 \mu mol$ atom carbon mg^{-1} chlorophyll h^{-1}.

PART 1
GENERAL BIOCHEMISTRY

1
Plant growth and biochemistry — the connection

1.1 Plant growth involves the synthesis of biological matter

Plants require nothing more than sunlight, water, carbon dioxide (CO_2) and various mineral salts for their growth. These requirements come directly from the sun, soil, air and water of the environment. It is this very simplicity that makes plants so interesting to study. Plants, like all cellular organisms, contain a plethora of proteins, nucleic acids, carbohydrates and lipids essential for their function. Clearly, plants must be able to make these and many other compounds from simple inorganic compounds and the energy of sunlight. This results in the production of new biological matter and, under the control of various internal factors, morphological growth and development.

The constituent compounds of plants confer on them their characteristic elemental composition. Ignoring the water content of plants, carbon, hydrogen and oxygen account for approximately 90–95% of the weight of most plants. In a typical analysis, the numerical ratio of carbon, hydrogen and oxygen is 1.8:2.0:0.90 (Table 1.1). This ratio reflects the combination of carbon with hydrogen and oxygen principally as carbohydrate with the empirical formula $[CH_2O]_n$. Plant cells are bounded by a structural framework of the carbohydrate, cellulose, so that they are low in protein (and hence nitrogen and sulphur) relative to carbohydrate. Animal cells have no analogous structural framework and thus have much higher nitrogen:carbon and sulphur:carbon ratios. Rather, vertebrate animals have a structural framework of bone (rich in phosphorus and calcium). Table 1.1 emphasizes that the production of biological matter by plants centres on the incorporation of carbon from CO_2 into organic molecules, principally carbohydrates.

Despite the obvious importance of CO_2 assimilation, in practice the growth of plants is rarely limited by the availability of CO_2. Next to water, the materials which most commonly restrict growth are nitrogen, phosphorus, potassium and sulphur even though they are required in relatively small amounts (Table 1.1). Nitrogen, phosphorus and sulphur occur in essential molecules such as proteins, nucleic acids and phospholipids. Potassium is required for the maintenance of osmotic integrity. These elements are obtained from the environment as inorganic salts. Relative to the vast amounts of carbon assimilated by plants, the incorporation of inorganic

Table 1.1 Abundance of some elements found in plants (maize) and animals (man) and oxidation states of the elemental forms. From Anderson, J.W. (1980) *Bioenergetics of Autotrophs and Heterotrophs*. Arnold Ltd, Sevenoaks.

Element[†]	Elemental abundance*		Elemental forms	
	Zea mays (maize)	*Homo sapiens* (man)	Physical environment	Biological matter
Hydrogen	1705	2038	H_2O	—HCOH—, —CH_2—
Carbon	1000	1000	CO_2	—HCOH—, —CH_2—
Oxygen	765	252	O_2	—HCOH—
Nitrogen	28.6	143	NO_3^-, N_2, NH_4^+	—NH_2, —NH—
Potassium	6.5	6.0	K^+	K^+
Calcium	1.6	25.0	Ca^{2+}	Ca^{2+}
Phosphorus	1.8	21.6	PO_4^{3-}	PO_4^{3-}
Magnesium	2.0	1.4	Mg^{2+}	Mg^{2+}
Sulphur	1.5	5.2	SO_4^{2-}	—SH
Chlorine	1.1	2.8	Cl^-	Cl^-
Iron	0.4	0.05	Fe^{2+}, Fe^{3+}	Fe^{2+}, Fe^{3+}

* Abundance of elements in organisms is expressed as atoms of element relative to 1000 atoms of carbon.
[†] Most elements required in trace amounts are not included.

salts is a minor activity. Plants typically contain about 30 times more carbon than nitrogen and 500 times more carbon than phosphorus. Nevertheless, the incorporation of inorganic salts into organic molecules such as proteins, phospholipids and nucleic acids is indispensable as these compounds are essential for cellular activity and hence growth. Moreover, the synthesis of these compounds is of wide biological importance since animals are unable to manufacture this wide range of essential compounds from inorganic salts. Plants also require small amounts of many other elements for incorporation into molecules essential for growth (e.g. magnesium in chlorophyll, copper in the chloroplast protein plastocyanin, molybdenum in the enzyme nitrate reductase). Not all of the elements required for growth are incorporated in this way. Some are free ions (e.g. K^+), essential for the maintenance of cell turgor, water movement and enzyme function.

1.2 Synthesis of biological matter requires an energy source

Most of the elements found in plants occur in the same redox state in the environment. This includes potassium (free K^+ in cells), magnesium (in chlorophyll) and phosphorus (in various phosphorylated intermediates). However, carbon, nitrogen and sulphur occur naturally in oxidized forms (principally as CO_2, NO_3^- and SO_4^{2-}) but are found in reduced forms in the organic molecules of plants (Table 1.1). The reductive assimilation of CO_2, NO_3^- and SO_4^{2-} into biological matter consumes large amounts of energy. Reactions of this type are called endergonic (or endothermic) reactions and are discussed further in Section 4.1. The energy required can be appreciated subjectively when the reaction is reversed, i.e. the heat energy released from burning wood on a cold winter's night as the plant matter is oxidized to CO_2 and various oxides of nitrogen and sulphur. These are exergonic or exothermic reactions. The heat energy released represents a portion of the chemical energy required for the synthesis of biological matter by plants. Net formation of a product is only possible if the overall reaction conditions and mechanism permit it to proceed in an exergonic manner (Chapter 4). The obvious success with which plants grow (think how often you mow your lawn in spring!) implies that plants possess mechanisms of this type. The ultimate source of this energy is sunlight which is absorbed by the green pigment chlorophyll.

Three basic principles contribute to the synthesis of biological matter from CO_2 and inorganic salts. A typical synthetic reaction is the incorporation of CO_2 into hexose sugar ($C_6H_{12}O_6$) which only yields significant amounts of product if energy is supplied.

$$6CO_2 + 6H_2O + energy \rightleftharpoons C_6H_{12}O_6 + 6O_2 \tag{1.1}$$

Plants also support the endergonic assimilation of NO_3^- and SO_4^{2-} into amino acids. The first principle is that these reactions proceed in cells in a series of *small steps* so that the energy input required for any one step is not excessive. Thus CO_2 assimilation does not involve the simultaneous reaction of six molecules of CO_2 with six molecules of water as in Eqn. 1.1. Rather, CO_2 is incorporated one molecule at a time, each in turn involving a series of component reactions.

The second principle concerns *coupling* of endergonic synthetic reactions to exergonic reactions. Cells contain various compounds which, under appropriate conditions, can react with water in highly exergonic reactions. For example, adenosine triphosphate (ATP) can hydrolyse to adenosine diphosphate (ADP) and orthophosphate (Pi).

$$ATP + H_2O \rightleftharpoons ADP + P_i + energy \tag{1.2}$$

In cells, this reaction proceeds in a very controlled way. The energy released can be used to 'drive' an endergonic reaction which would otherwise give a small or negligible yield of product. Various other reactions also serve as energy sources. In plants the most important of these are the oxidation of the reduced forms of nicotinamide adenine dinucleotide phosphate (NADPH) and ferredoxin (Fd_{red}), both powerful biological reducing agents.

$$NADPH + H^+ + J \rightleftharpoons NADP^+ + JH_2 \tag{1.3}$$

$$Fd_{red} + K \rightleftharpoons Fd_{ox} + K_{red} \tag{1.4}$$

In Eqn. 1.3, J is a molecule which accepts hydrogen atoms ($H^+ + e^-$) from NADPH. In Eqn. 1.4, K is a molecule which accepts electrons from Fd_{red}.

The third principle concerns the way in which the endergonic biosynthetic reactions of cells are coupled to exergonic reactions such as the hydrolysis of ATP (Eqn. 1.2) through the action of *enzymes*. Cells contain thousands of different enzymes, each one being a particular protein which catalyses a specific chemical reaction at physiological pH and temperature. In biology, compounds used by organisms or which serve as reactants are known as substrates. Catalysis of a reaction by an enzyme entails binding of the appropriate substrate(s) to a specific reaction site on the enzyme followed by complex chemical interactions between the two which facilitate the overall reaction. The endergonic biosynthetic reaction $A \rightleftharpoons B$, as it stands, will not yield significant amounts of B. But, if coupled to the exergonic reaction $X \rightleftharpoons Y$ (analogous to the reaction in Eqn. 1.2) in the presence of an enzyme specific for A and X

$$A + X \xrightleftharpoons{\text{enzyme}} B + Y \tag{1.5}$$

significant synthesis of B occurs. However, the enzyme does not catalyse the conversion of A to B in the absence of X. Also, A and X do not interact significantly in the absence of enzyme. This demonstrates the essential role enzymes play in directing energy into biosynthetic energy-requiring reactions.

The supply of energy from ATP, NADPH and Fd_{red} for synthetic purposes raises the question of how they are formed. In plants, the necessary energy comes from sunlight; the pigments of green leaves absorb radiant energy to initiate the photochemical oxidation of a reaction centre, effecting the conversion of light energy into chemical energy.

Plants also generate ATP and reducing equivalents (in the form of reducing agents) by oxidizing carbohydrate (e.g. $C_6H_{12}O_6$) by aerobic respiration.

$$C_6H_{12}O_6 + 6O_2 \rightarrow 6CO_2 + 6H_2O + energy \qquad (1.6)$$

Respiration occurs in both photosynthetic and non-photosynthetic cells in the light and the dark. For whole plants, respiration in the dark typically consumes about 10–15% of the biological matter plants make. However, the importance of this process varies greatly—it provides the main mechanism for generating energy in root cells, germinating seeds and in photosynthetic cells in the dark.

1.3 Characteristics of autotrophs and heterotrophs

Autotrophs are organisms which use CO_2 as their sole source of carbon. By definition, they cannot use their carbon source (CO_2) to fulfil their energy requirements for synthetic processes. Plants are by far the most important autotrophs but other types of autotrophic organisms exist. Most use light as their energy source. Plants and cyanobacteria are oxygenic photoautotrophs as their growth is associated with the oxidation of H_2O to O_2. Anoxygenic photoautotrophs (all photosynthetic bacteria other than cyanobacteria) are highly intolerant of O_2. In these organisms the assimilation of CO_2 and thus growth, is associated with the oxidation of some compound other than water. These differences are attributable to differences in the photochemical mechanisms of oxygenic and anoxygenic photoautotrophs (see Section 9.5). Chemoautotrophic organisms derive energy to incorporate CO_2 by catalysing the oxidation of reduced forms of inorganic elements. These organisms (all of them bacteria) include *Thiobacillus*, which supports the aerobic oxidation of reduced forms of sulphur (e.g. $S^{2-} \rightarrow SO_4^{2-}$), and *Nitrosomonas*, which oxidizes NH_3 to NO_2^-. Chemoautotrophic bacteria are most prevalent in aerobic environments with relatively high concentrations of reduced forms of inorganic compounds (e.g. H_2S, NH_3, H_2, Fe^{2+}).

In contrast to autotrophs, heterotrophic organisms use organic molecules such as acetate, methane, fatty acids, sugars, etc. as carbon sources. Heterotrophs are directly or indirectly dependent on autotrophs for their supply of organic carbon—some is used to supply carbon skeletons for the synthesis of new biological matter while the remainder is oxidized to provide the energy to do this (i.e. respiration).

Respiration is the only mechanism for generating 'energy-rich' compounds such as ATP and NADPH; in autotrophs, respiration is a secondary energy-generating mechanism (although the non-photosynthetic tissues of plants could be regarded as exceptions).

1.4 Photosynthesis—definitions

We have intentionally been very sparing in the use of the term 'photosynthesis'; not because we find it an unsatisfactory term but because it is too imprecise to describe the assimilatory biochemical processes of plants and other organisms. Photosynthesis literally means *light-dependent synthesis of biological matter* and it is used in this context in this book. However, 'photosynthesis' is variously used to mean CO_2 assimilation or O_2 evolution, accumulation of dry weight, electron flow through the chloroplast electron transport chain, photoreduction of $NADP^+$ or light-dependent synthesis of carbohydrate. In truth, photosynthesis includes all these but it also includes the light-dependent incorporation of inorganic nitrogen and sulphur, various light-enhanced processes in the cytosol and the synthesis of protein and lipid in chloroplasts. In keeping with our definition, we have avoided using 'photosynthesis' to refer to the various partial processes of photosynthesis. Thus we have used expressions such as the C_3 and C_4 pathways of CO_2 assimilation (which are not directly dependent on light) and not C_3 and C_4 photosynthesis. This is illustrated by the chemoautotrophs which assimilate CO_2 via the C_3 pathway but grow independently of light; it would be patently incorrect to talk about C_3 photosynthesis in these organisms. Our nomenclature may be a mouthful but it is important that the names of the component processes are not ambiguous or misleading.

1.5 Major activities of producer and consumer cells

Plants contain many cells, each with their own characteristic form, structure and function. Photosynthesis is restricted to cells containing chlorophyll and other photosynthetic pigments. Chlorophyll is associated with chloroplasts around the periphery of mesophyll cells of the lamina or leaf blade. Mesophyll cells produce more biological matter than they require and are called producer cells. They require light, CO_2, water and ions for the production of biological matter. CO_2 is supplied by diffusion from the gas phase of the sub-stomatal cavity which exchanges gases with ambient air (containing 0.03% CO_2). The stomatal pores are the main sites for the evaporative loss of water and the acquisition of CO_2. Plants can control both the rate of water loss and the availability of CO_2 (and hence the rate of CO_2 assimilation) by regulating the aperture of the stomatal pores. The other elements required for the production of biological matter are taken up from soil as ions (e.g. nitrogen as NO_3^-, phosphorus as $H_2PO_4^-$, sulphur as SO_4^{2-}) by epidermal root cells and are transported to the producer cells in the xylem (see Section 1.6). In some species NO_3^- is extensively reduced and assimilated in root cells and in these species nitrogen is largely transported as amino acids and ureides rather than NO_3^-.

Cells that do not produce sufficient biological matter to fulfil their own requirements import material from producer cells and are called consumer cells. Groups of consumer cells constitute 'sinks'. Most cells, including the mesophyll cells of unexpanded leaves, behave as consumer cells during the early stages of their development while others remain net consumers (e.g. most root cells). Tissues such as meristems and the endosperm of developing seeds have very strong demands for biological matter either because of their intense rates of growth or the deposition of storage compounds (e.g. starch).

1.6 Molecular traffic between producer and consumer cells

Plants possess very effective mechanisms for the distribution of various metabolites. The simpler of the two main mechanisms is the passive transport of solutes, principally ions, in the xylem. This involves the movement of water containing dissolved ions and minor amounts of organic metabolites in response to the water potential gradient between root cells and atmospheric water vapour, primarily determined by a large water potential difference between the leaves and the air. This is referred to as evapotranspiration and the amount of water delivered to each leaf and evaporated depends primarily on the leaf area. The initial delivery of ions to the leaves follows the same pattern but the demand for these is unrelated to the amount delivered. For example, older fully formed leaves have much smaller demands for ions relative to the amount of water they transpire than young developing leaves. Plants rectify this by reabsorption and retranslocation mechanisms.

The second major mechanism for long distance transport of metabolites concerns the movement of organic compounds in the phloem by translocation. About 80–85% of the dry weight of the material transported in the phloem is sucrose in most species. Various organic and amino acids and ions represent minor components.

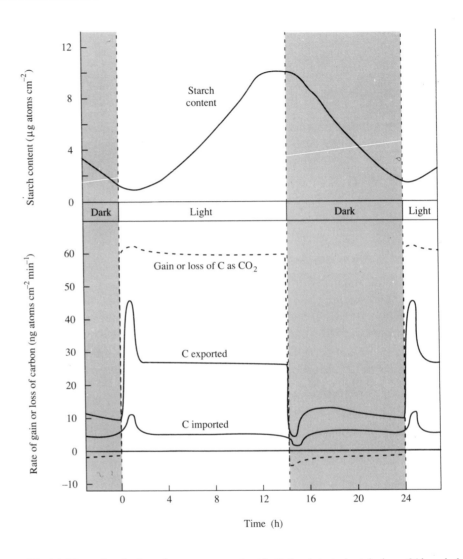

Fig. 1.1 The carbon budget of a mature sugar beet leaf of an intact plant during a 24 h period of 14 h light and 10 h dark. For about 1.5 h following illumination, approximately 80% of the newly assimilated carbon was exported, thereafter decreasing to about 50% with the onset of starch accumulation. During the dark period carbon from starch mobilization was exported at a slow rate. Small amounts of carbon were imported from other leaves throughout the 24 h period. After Fondy, B.R. & Geiger, D.R. (1982) *Plant Physiol.*, **70**, 671–6.

Xylem transport is upwardly unidirectional but phloem transport is bidirectional and generally involves transport of metabolites from producer to consumer cells. Compounds synthesized by producer cells can be exported via the phloem both downwards (e.g. to roots and underground storage tissues) and upwards (e.g. to apical buds, inflorescence and expanding leaves). Producer cells export predominantly to the nearest sink. Thus, in mature plants, lower leaves export to roots and underground storage tissues and upper leaves to expanding leaves, developing fruits, etc. In general, consumer cells have first call on newly assimilated $^{14}CO_2$; ^{14}C-label only accumulates in producer cells (mostly as starch) after the requirements of consumer cells have been satisfied (Fig. 1.1).

1.7 Plant growth and carbon dioxide assimilation rates

Growth rates are important in seeking to provide molecular explanations of plant growth. If the growth rate and the composition of plant matter (content of protein, cellulose, starch, etc.) are known, then the rate of production of the component compounds can be calculated. Biochemistry must seek to describe mechanisms which quantitatively account for the rates of processes observed in intact plants.

The amount of biological matter in a specimen is usually determined as dry weight which entails destructive sampling. Growth is measured as the change in dry weight between two sampling times. The amount of growth is related to the dry weight at the start of the sampling period. Thus, growth is frequently expressed as change in dry weight per unit time per unit of pre-existing dry matter (e.g. $g\,g^{-1}$ $week^{-1}$). This is the relative growth rate and is approximately constant in the early stages of plant growth but declines with maturity. This is partly explained by the increasing number of consumer cells relative to producer cells. Relative growth rate is expressed by two factors on the assumption that growth is related to leaf area, as leaves are the sites of producer cells. The first is the leaf area ratio which denotes the leaf area relative to the total dry weight; this declines significantly during the life of the plant. The second is the net assimilation rate and is expressed as change in dry weight per unit leaf area per time (e.g. $g\,dm^{-2}\,week^{-1}$); this also decreases, especially as the plant approaches maturity, but much more slowly.

Since the assimilation of CO_2 is by far the most important determinant of plant growth, measurements of the CO_2 assimilated per unit area of leaf per unit time provide a good indicator of net assimilation rate. This can be measured non-destructively in relatively short-term experiments during which changes in dry weight and leaf area are negligible. However, the rate of CO_2 assimilation per unit leaf area varies by as much as 100-fold between species because of differences in leaf thickness, amount of chlorophyll per unit leaf area, different CO_2 assimilation mechanisms, internal control mechanisms and inherent differences in sensitivity to light intensity, temperature, water availability, etc. Molecular expressions of growth rates (and hence CO_2 assimilation rates) are usually expressed in relation to the light (i.e. energy) absorbed. Since this is determined principally by chlorophyll, assimilatory processes are normally expressed as rates per unit of chlorophyll (e.g. $\mu mol\ CO_2\ mg^{-1}$ chlorophyll h^{-1}). For a given temperature and light intensity, this (Table 1.2) shows far less variation than the assimilation rates expressed per unit leaf area.

Table 1.2 Rates of CO_2 assimilation for plants with differing CO_2 assimilation mechanisms. After Edwards, G. & Walker, D.A. (1983) C_3, C_4: *Mechanisms, and Cellular and Environmental Regulation of, Photosynthesis*. Blackwell Scientific Publications, Oxford.

CO_2 assimilation mechanism	Temperature (°C)	CO_2 assimilation rate* ($\mu mol\ mg^{-1}$ chlorophyll h^{-1})
C_3	15–25	100–200
C_4	20–35	200–400
CAM†	15–35	30–80

* Values are for plants grown under optimum conditions and measured for CO_2 assimilation in atmospheric air containing 0.03% CO_2.
† Crassulacean acid metabolism.

A rate of $200\ \mu mol\ CO_2\ mg^{-1}$ chlorophyll h^{-1} is typical for plants grown under temperate conditions. Knowing this value provides a base against which all the component processes of CO_2 assimilation and various related processes can be assessed. For example, the enzymes associated with CO_2 assimilation should exhibit activities (corrected for the appropriate stoichiometry where necessary) consistent with the flow of $200\ \mu g$ atoms of carbon mg^{-1} chlorophyll h^{-1}. If the observed activity is well below this value then either the proposed mechanism of CO_2 assimilation is questionable or the conditions of extraction and/or assay are suspect. Knowing the rate of CO_2 assimilation allows many predictions; if the light-dependent assimilation of a molecule of CO_2 consumes three molecules of ATP then illuminated chloroplasts should be able to phosphorylate ADP at a minimum rate of $600\ \mu mol\ mg^{-1}$ chlorophyll h^{-1}. The theoretical rates of export of assimilated carbon from the chloroplasts of photosynthetic cells can also be assessed in relation to the rate of CO_2 assimilation. Further, the mean theoretical rates of assimilation of the other elements present in plants can be calculated, knowing the rate of CO_2 assimilation and the molar ratio of elements present relative to carbon (Table 1.1). Assuming that assimilation occurs entirely in leaves, the theoretical rates for inorganic nitrogen and sulphur are 5–7 and 0.3–0.4 μmol mg^{-1} chlorophyll h^{-1} respectively.

Further reading

General texts on physiology and growth: Leopold & Kriedemann (1975); Salisbury & Ross (1985).

Requirements of consumer cells: Ho (1988); Thorne (1985).

Plant requirements for elements: Läuchli & Bieleski (1983); Mengel & Kirkby (1982).

Molecular traffic in vascular tissue: Giaquinta (1983); Pate (1980); Preiss (1980) Chapter 8.

2
Molecules of plant cells

2.1 Classes of compounds in plant matter

This chapter briefly outlines the main classes of compounds found in plant cells, principally the four major classes of relatively simple organic molecules which account for about 90% of the biological matter. These compounds occur principally as components of macromolecules, some of very high molecular weight. Cellulose and amylose, for example, are macromolecules of the sugar D-glucose while proteins are macromolecules of amino acids. This chapter is also concerned with the arrangement of the basic molecules into macromolecules.

Plants also contain other compounds, collectively less than 10% of the biological matter. They include many low molecular weight compounds which act as intermediates in the synthesis of essential molecules and others concerned with the oxidation and utilization of energy, many of which are discussed in subsequent chapters. Other compounds of varying molecular weights, including commercially important ones (e.g. the constituents of rubber latex and various pharmacologically active compounds), are only found in some of the 250 000 or so plant species. Many are confined to just a few closely related plants. Excepting certain phenylpropanoid compounds, important structural components of some cell walls, most of the compounds with restricted distribution are not essential to cell structure or function. They are known as secondary compounds and are not discussed extensively in this chapter.

Some knowledge of organic chemistry is assumed but some terms referring to carbon bonding, chemical groups, asymmetric carbon atoms and stereoisomers are summarized in Section 2.2.

2.2 Organic molecules of plants occur in specific isomeric forms

The reactivity of organic molecules is largely determined by the presence of functional groups. These are important in naming and classifying the organic molecules of plants (Table 2.1) and in determining their properties and functions *in vivo*.

The four bonding positions of a tetrahedral carbon atom can carry different singly bonded substituents, asymmetrically arranged about the atom in two isomeric

Table 2.1 Some functional groups in organic molecules of plants.

Functional group	Structure*	Notation	Examples
Hydroxyl	R^1—O—H	—OH	Alcohols (e.g. sugars)
Aldehyde	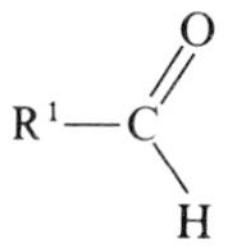 R^1—CHO structure	—CHO	Aldehydes (e.g. open chain forms of aldo-sugars)
Carbonyl	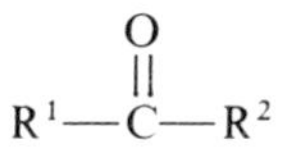 R^1—CO—R^2	—CO—	Ketones (e.g. open chain forms of keto-sugars)
Carboxyl	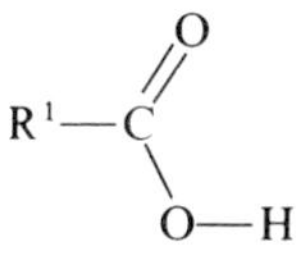 R^1—COOH structure	—COOH	Carboxylic acids (e.g. fatty acids, amino acids, dicarboxylic acids)
Amino	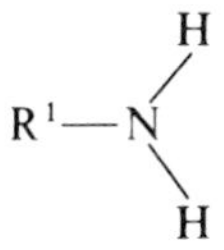 R^1—NH structure	—NH$_2$	Amines (e.g. amino acids)
Amido	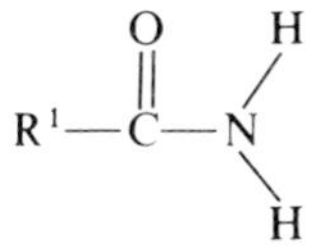 R^1—CO—NH structure	—CO—NH$_2$ (or —CO.NH$_2$)	Amides (e.g. asparagine)
Thiol	R^1—S—H	—SH	Thiols (e.g. cysteine)
Disulphide	R^1—S—S—R^2	—S—S— (or —S.S—)	Disulphides (e.g. cystine)
Ester	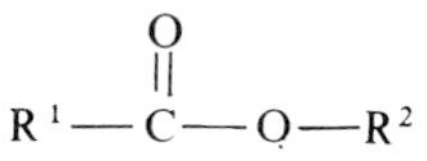 R^1—CO—O—R^2	—CO—O— (or —CO.O—)	Esters (e.g. lipids)
Ethers	R^1—O—R^2	—O—	Ethers

*R^1 and R^2 designate structures to which the functional group is attached.

forms, one the mirror image of the other. These are known as stereoisomers or enantiomers.

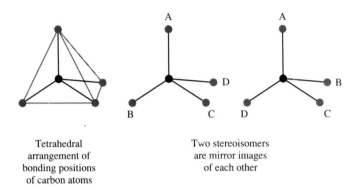

Tetrahedral
arrangement of
bonding positions
of carbon atoms

Two stereoisomers
are mirror images
of each other

The stereoisomers are identical chemically but one rotates plane-polarized light to the left, the other to the right. Many processes in cells are specific for one stereoisomer (e.g. reactions involving amino acids and sugars). Carbon atoms with only three different substituents do not form stereoisomers.

2.3 Carbohydrates are quantitatively the most important compounds in plants

Carbohydrates are polyhydroxyaldehydes or polyhydroxyketones (i.e. aldehydes or ketones with two or more hydroxy groups) or substances which yield these compounds upon acid hydrolysis. They commonly account for 60–85% of the bulk of a plant and include simple sugars (monosaccharides) and macromolecules degraded to sugars by mineral acids (the majority). Monosaccharides are not decreased to smaller units by mild acid hydrolysis and generally have the empirical formula $[CH_2O]_n$ (e.g. glucose, $C_6H_{12}O_6$). The disaccharide sucrose ($C_{12}H_{22}O_{11}$) and the polysaccharide starch are also carbohydrates because they yield monosaccharides upon treatment with acid. Carbohydrates also include various compounds closely related to sugars (e.g. deoxysugars, uronic acids, sugar alcohols, cyclitols and amino sugars).

2.3.1 Monosaccharides are the simplest carbohydrates

The simplest of the carbohydrates are the trioses with three carbon atoms. They include the two structural isomers, glyceraldehyde (an aldehyde) and dihydroxy-

acetone (a ketone). The carbon atoms of glyceraldehyde only are numbered from the carbon atom bearing the aldehyde group (—CHO), designated C-1. Glyceraldehyde has an asymmetric carbon atom (C-2, shown below in bold) but dihydroxyacetone does not. Thus glyceraldehyde has two optical isomers while dihydroxyacetone has only one. By convention, the L-isomer of glyceraldehyde has the hydroxy group at C-2 to the left of the C-1 carbon atom when viewed from the C-3 end of the molecule whereas in the D-isomer, the hydroxy group at C-2 is to the right.

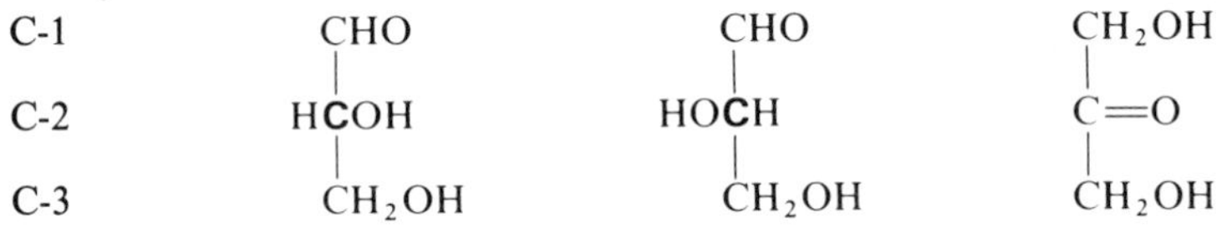

C-1	CHO	CHO	CH$_2$OH
C-2	HCOH	HOCH	C=O
C-3	CH$_2$OH	CH$_2$OH	CH$_2$OH
Carbon atoms	D-Glyceraldehyde	L-Glyceraldehyde	Dihydroxyacetone

(asymmetric carbon atoms shown in bold)

For glyceraldehyde only, a solution of the L-isomer rotates plane-polarized light to the left (laevorotatory) and the D form to the right (dextrorotatory).

A series of sugars can be formed by increasing the chain length of each of the three trioses above. Each additional carbon atom bearing —H and —OH substituents can have the configuration —(H)C(OH)— or —(HO)C(H)—, i.e. it is asymmetric, doubling the number of optical isomers. Extension of each isomer of glyceraldehyde by one carbon atom gives two tetroses with four carbon atoms (Fig. 2.1). The addition of three carbon atoms forms eight aldohexoses (with six carbon atoms) for each parent optical isomer. Monosaccharides containing up to eight carbon atoms occur in plants so that, in theory, a very large number of isomers is possible although not all of these have been found. All monosaccharides can exist in an 'open chain' or acyclic form, but sugars containing five or more carbon atoms exist predominantly in ring or cyclic forms (Section 2.3.2).

Some aspects of nomenclature and structure are important (Table 2.2). The sugars derived from glyceraldehyde, all with an aldehyde group in their acyclic form, are aldo-sugars or aldoses and the carbons are numbered as above. Sugars derived from dihydroxyacetone (a ketone) are referred to as ketoses and the carbon atoms are numbered from the carbon adjacent to the one bearing the carbonyl or keto group. The L and D configurations for any monosaccharide can be ascertained by viewing the molecule from the end with the highest carbon number: in D-sugars

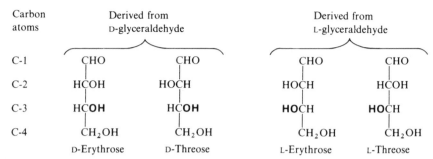

Fig. 2.1 Tetroses formed by increasing L- and D-glyceraldehyde by one carbon atom. The hydroxy group on the penultimate carbon atom (C-3) is shown in bold. Note that D- and L-erythrose are mirror images of each other and are thus enantiomers. Erythrose and threose are optical isomers but are not mirror images; compounds with this type of relationship are known as diastereoisomers.

Table 2.2 Terms used to describe monosaccharides.

Terms	Description
Triose, tetrose, etc.	Designates the number of carbon atoms in the monosaccharide (C_3, C_4, etc.)
Structural isomers	Refers to whether the sugar is an aldose (—CHO on C-1) or a ketose (—CO— on C-2)
Diastereoisomers	Refers to stereoisomers of an aldose or ketose which are not mirror images (e.g. D-glucose, D-galactose and D-mannose are all D-aldohexoses)
L- and D-isomers	Refers to the configuration of hydroxy groups on the penultimate carbon atom to which the hydroxy groups on other carbon atoms bear a fixed geometry (see text)
Epimers	Refers to a pair of sugars which differ with respect to their configuration at a single carbon atom (e.g. D-glucose and D-mannose)
Enantiomers	Refers to a pair of isomers which form mirror images (e.g. L- and D-erythrose)
Anomers	Refers to the two forms (α and β) which result from formation of a ring
Pyranose/furanose	Refers to six-membered (pyranose) or five-membered (furanose) rings formed by hexoses and pentoses

the hydroxy group on the penultimate carbon atom is on the right of the carbon atom of next lowest number; in L-sugars it is on the left. Note that the L and D series of sugars are not defined in terms of their capacity to rotate plane-polarized light. Indeed many L-sugars with chain lengths greater than that of glyceraldehyde are dextrorotatory. Similarly, many D-sugars are laevorotatory (e.g. D-fructose). A systematic list of the D-aldopentoses and D-aldohexoses is shown in Fig. 2.2 and for the D-ketopentoses and D-ketohexoses in Fig. 2.3. Similar lists of structures could be compiled for the L-monosaccharides. As seen from the structure of L- and D-erythrose (Fig. 2.1), the arrangement of the substituents is the same in L and D series; they are mirror images. Thus, the L and D forms of a sugar are stereoisomers (or enantiomers). Isomers of aldo-sugars or keto-sugars which have the same number of carbon atoms but are not enantiomers are called diastereoisomers (e.g. D-ribose, D-arabinose and D-xylose). Only some of the L- and D-monosaccharides occur in plants (e.g. allose and idose are not found). Generally, only one of the two possible enantiomers (L or D) of any one diastereoisomer occurs in plants (see legends to Figs 2.2 & 2.3) or, if both enantiomers do occur, usually one form predominates (e.g. D-glucose is quantitatively much more important than L-glucose). For this reason it is common to omit reference to the form of the enantiomer. In plants, the majority of the important monosaccharides have the D configuration; the only important exceptions are L-arabinose, L-sorbose and some deoxy sugars (Section 2.3.3).

2.3.2 Monosaccharides containing five or more carbon atoms occur mostly in ring forms

The ring or cyclic forms of the pentose and hexose monosaccharides are of great importance to biologists. Rings are formed from acyclic structures by an internal reaction between the aldehyde group at C-1 and a hydroxy group at either C-5 or C-4. Glucose most commonly forms a stable hemiacetal with an oxygen bridge between C-1 and C-5. (An aldehyde group can react with two hydroxy groups; a hemiacetal is formed by reaction with only one hydroxy group.) This results in a six-membered ring of five carbon atoms and an atom of oxygen with C-6 outside the ring (2.Ia & 2.Ic).

Formation of rings also produces a hydroxy group at C-1 known as the hemiacetal hydroxy group (outlined by a dashed line), absent in the open chain or acyclic form. An important consequence is that the ring form of a specific sugar can exist in two isomeric states—the hydroxy group extends to the right in the α-form

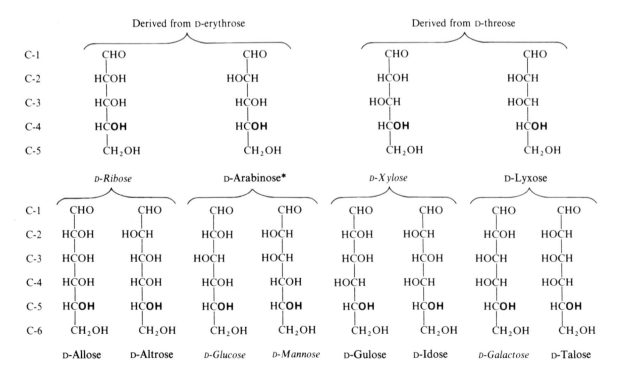

	α-D-Glucose (cyclic form) (36%)	D-Glucose (acyclic form) (1%)	β-D-Glucose (cyclic form) (63%)
C-1	HC(OH)	CHO	(HO)CH
C-2	HCOH	HCOH	HCOH
C-3	HOCH O	HOCH	HOCH O
C-4	HCOH	HCOH	HCOH
C-5	HC——	HCOH	HC——
C-6	CH_2OH	CH_2OH	CH_2OH

Carbon atoms

2.Ia   **2.Ib**   **2.Ic**

and to the left in the β-form. Two isomers which differ only in this way are known as anomers (e.g. α-D-glucose and β-D-glucose). When the α and β anomers of D-glucose are dissolved in water they exhibit changes in their specific optical rotations indicating that they form an equilibrium mixture. This interconversion of the two anomers, in which the open-chain form of D-glucose serves as an intermediate, is known as mutarotation. At equilibrium, solutions of D-glucose contain 36% of the α-anomer, 63% of the β-anomer and 1% of the open chain form. Polymers of α-D-glucose have very different properties to polymers of β-D-glucose.

The six-membered ring form of glucose is depicted as a planar hexagon (i.e. Haworth representation) in Fig. 2.4. This emphasizes the resemblance of the six-membered ring to the ring structure of pyran (2.IIa) and sugars of this type are described as pyranose forms. In Haworth representations all groups shown to the left in the straight-chain structures extend above the horizontal plane and those to

Fig. 2.2 Structures and inter-relationships of the D-aldopentoses and D-aldohexoses. The hydroxy groups on the penultimate carbon atoms which confer the D configuration, are shown in bold. The L series of aldopentoses and aldohexoses are the same in that the hydroxy groups have the same geometry relative to the hydroxy group on the penultimate carbon atom which, in L-sugars, occurs to the left. Of the D-sugars shown here, only those which are in italics are important in plants. *Of the L-aldo sugars, only L-arabinose occurs commonly in plants.

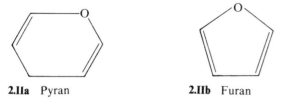

2.IIa Pyran   **2.IIb** Furan

the right extend below the plane. So, for the β-pyranose form of D-glucose the hydroxy group at C-1 is above the plane whereas for the α-form it is below the plane. Note, too, that the geometry of the hydroxy group at C-5 determines whether the glucose molecule is in the L or D configuration. Thus, for D-glucose,

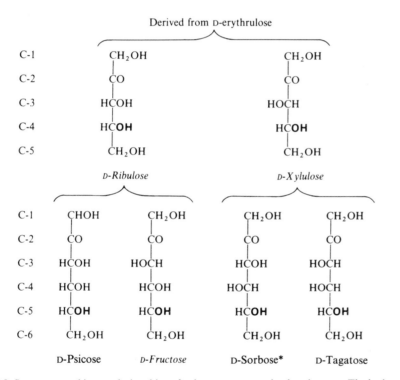

Fig. 2.3 Structures and inter-relationships of D-ketopentoses and D-ketohexoses. The hydroxy groups on the penultimate carbon atoms which confer the D configuration are shown in bold. Of the D-sugars shown here, only those which are in italics are important in plants. *Of the L-keto sugars, only L-sorbose occurs commonly in plants.

(a) Pyranose forms

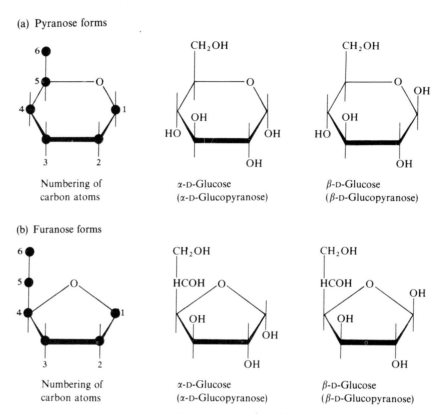

(b) Furanose forms

Fig. 2.4 Haworth representations of the (a) pyranose form (six-membered ring) and (b) furanose form (five-membered ring) of the α and β anomers of D-glucose. To aid clarity, hydrogen atoms bonded directly to carbon atoms in the ring have been omitted to emphasize the geometry of the —OH groups; this practice has also been adopted in the ensuing figures.

C-6 protrudes above the plane but in L-glucose it lies below. Although Haworth diagrams are an extremely convenient way of representing the structure of monosaccharides, in fact the real three-dimensional structure of glucopyranose is not planar but chair-shaped (2.IIIa & 2.IIIb).

As noted above, the aldehyde group of acyclic forms of aldo-sugars also reacts internally with the hydroxy group at C-4 to form a five-membered ring of four atoms of carbon and one of oxygen. For hexoses in this configuration, both C-5 and C-6 lie outside the ring. Haworth representations of these forms (shown for

α-D-Glucopyranose

2.IIIa

β-D-Glucopyranose

2.IIIb

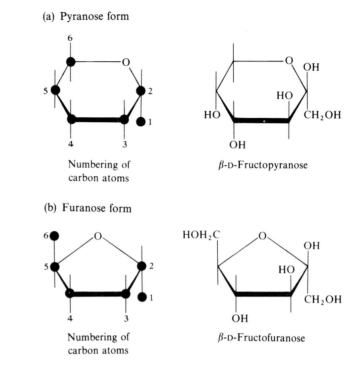

(a) Pyranose form

Numbering of
carbon atoms

β-D-Fructopyranose

(b) Furanose form

Numbering of
carbon atoms

β-D-Fructofuranose

Fig. 2.5 Haworth representation of the (a) pyranose and (b) furanose forms of β-D-fructose.

glucose in Fig. 2.4) are known as furanose forms because of their similarity to the structure of furan (2.IIb).

The aldopentoses also form six- or five-membered rings. In the latter case C-5 projects from the ring but in the former case all five carbon atoms form part of the ring.

Both aldopentoses and aldohexoses form an equilibrium between their ring and acyclic forms, although the equilibrium lies very much towards the ring structures (>99% for glucose). Nevertheless, glucose and other aldoses exhibit a variety of reactions typical of an aldehyde (e.g. quantitative reduction of Cu^{2+} to Cu^+). Thus, the ring forms of aldoses must rapidly equilibrate with their acyclic forms.

The ketohexoses form five- and six-membered cyclic structures by way of an oxygen bridge of the pyranose and furanose type between the keto group at C-2 and the hydroxy group at C-5 or C-6 to produce a hemiketal, each enantiomer giving rise to α and β anomers. For the furanose form of fructose (fructofuranose), C-1 and C-6 lie outside the ring; for the pyranose form only C-1 lies outside (Fig. 2.5). Like the aldo-sugars, the ring forms of the keto-sugars are in equilibrium with their acyclic forms, allowing quantitative reactivity of their carbonyl groups.

Some terms used to describe the monosaccharides are summarized in Table 2.2.

2.3.3 Monosaccharide derivatives are important constituents of plant cells

The most important of the monosaccharide derivatives commonly found in plants are the uronic acids (or, at physiological pH, uronates) which mainly occur in polymerized form in several pectic substances in cell walls (see Section 17.2.2). The uronates differ from the sugars in that the terminal carbon atom is oxidized to form a carboxylic acid. Uronates with six carbon atoms can form rings in the same manner as their parent compounds since the —COO⁻ group at C-6 lies outside

the ring. Some typical uronates named from the parent sugar are shown in Fig. 2.6. Uronic acids also occur as methyl esters.

The aldonic acids are formed by oxidation of the aldehyde carbon atom at C-1 and are therefore unable to form cyclic structures. Gluconic acid (or, at physiological pH, gluconate) is a common metabolite.

The aldehyde and carbonyl groups associated with the C-1 and C-2 atoms of aldo- and keto-sugars respectively, can be reduced to form a family of compounds with open chains known as sugar alcohols. Thus, mannitol, ribitol and glycerol are the corresponding sugar alcohols of mannose, ribose and glyceraldehyde respectively. Note, however, that the reduction of certain aldo- and keto-sugars can result in the formation of a common sugar alcohol (e.g. reduction of the aldehyde group of glucose and reduction of the carbonyl group of fructose both produce sorbitol).

The inositols are sugar derivatives because of their structure and origin. Like

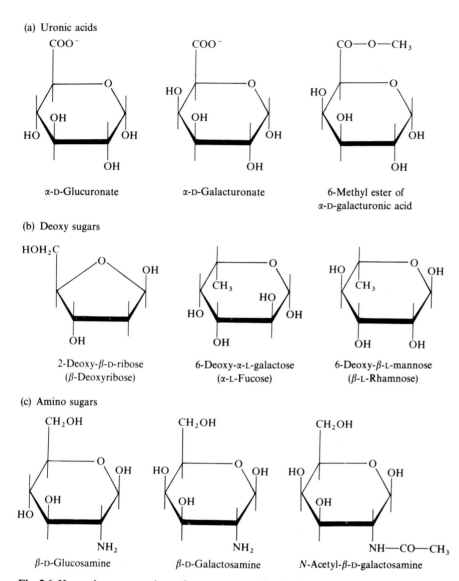

(a) Uronic acids

α-D-Glucuronate

α-D-Galacturonate

6-Methyl ester of
α-D-galacturonic acid

(b) Deoxy sugars

2-Deoxy-β-D-ribose
(β-Deoxyribose)

6-Deoxy-α-L-galactose
(α-L-Fucose)

6-Deoxy-β-L-mannose
(β-L-Rhamnose)

(c) Amino sugars

β-D-Glucosamine

β-D-Galactosamine

N-Acetyl-β-D-galactosamine

Fig. 2.6 Haworth representations of some monosaccharide derivatives which occur in plants: (a) uronic acids of aldohexoses, with the carboxylic group at C-6, and their esters (uronic acids are shown in their dissociated form); (b) deoxypentoses and deoxyhexoses; and (c) amino sugars, with the amino group at C-2, and their N-acetyl esters.

the C_6 sugar alcohols, they are hexahydric alcohols but have cyclic structures of six carbon atoms. Inositol has nine stereoisomers, seven of which occur naturally.

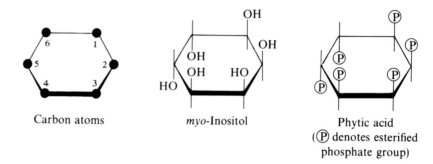

Carbon atoms

myo-Inositol

Phytic acid
(Ⓟ denotes esterified
phosphate group)

myo-Inositol and the other isomers occur both in free forms and in combination. For example *myo*-inositol is especially abundant as the hexaphosphate ester (or phytic acid) in many seeds where it usually occurs as the calcium/magnesium salt (or phytin), and commonly accounts for 50–80% of the phosphorus present. Inositol is also found in some phospholipids (see Section 2.4.4) and as a component of various other essential and secondary products.

Ascorbic acid is also related to the sugars. It is usually associated with actively metabolizing cells, especially chloroplasts of photosynthetic cells. Ascorbic acid, (*threo*-hex-2-enono-1,4-lactone) is an oxidation product of an aldohexose and is also known as vitamin C which prevents scurvy in humans.

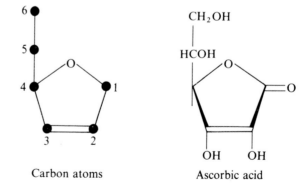

Carbon atoms

Ascorbic acid

Plants also contain monosaccharides in which hydroxy groups at specific carbon atoms are reduced (Fig. 2.6). 2-Deoxyribose is a constituent of deoxyribonucleic acid (DNA); L-rhamnose (6-deoxy-L-mannose) is a component of some pectic substances of plant cell walls; and L-fucose (6-deoxy-L-galactose) is a major component of the cell walls of brown algae.

The amino sugars found in plants are commonly aldohexoses in which the hydroxy group on C-2 is replaced by an amino group. D-Glucosamine (2-amino-2-deoxy-D-glucose) and D-galactosamine (2-amino-2-deoxy-D-galactose) are the most common, often occurring as their *N*-acetyl derivatives (Fig. 2.6) in certain macromolecules.

2.3.4 Monosaccharides can be linked together by glycosidic bonds

Sometimes two or three monosaccharides are linked together to form di- and trisaccharides; for example, sucrose is a disaccharide consisting of glucose linked to fructose. Most monosaccharides are linked together to form much larger molecules known as polysaccharides. The nature of the linkages between the various component monosaccharides is of considerable importance as it has a marked effect on the chemical and biological properties of the product.

The linking of monosaccharides is based on the principle that the hemiacetal hydroxy group of the ring forms of aldoses (C-1) or the hemiketal hydroxy group of the ring forms of ketoses (C-2) can react with the hydroxy group of another compound to form a full acetal or ketal. A simple example is the reaction between the ring form of glucose and methanol.

α-D-Glucose   Methanol   α-Methyl-D-glucoside

The glucose unit in the product is a glucosyl residue (outlined by a dashed line) and the linkage via the oxygen atom (shown in bold) which joins, in this case, the methyl group to the glucosyl residue is referred to as a glycoside (or glycosidic) bond. Clearly, the formation of the glycoside bond abolishes the hemiacetal

function of C-1 of the glucosyl residue. Thus, glucose will support the reduction of Cu^{2+} to Cu^+ but methylglucoside will not.

This principle can be extended to reactions between the ring forms of sugars and the hydroxy groups of other monosaccharides. The linkage between any two monosaccharide components must involve the C-1 hydroxy group of aldo-sugars or the group at C-2 in keto-sugars. In theory, the hemiacetal (or hemiketal) hydroxy group of one sugar molecule can undergo a reaction with any of the hydroxy groups of another sugar molecule, including one of the same enantiomer.

The glycosidic bonds between monosaccharides are named according to the molecules linked together and the carbon atoms involved. Thus, if the hemiacetal hydroxy group at C-1 of α-glucose is linked to the hydroxy group at C-4 of galactose the linkage is glucose-α (1→4)-galactose, and for β-fructose (hemiketal hydroxy group at C-2) joined to the hydroxy group at C-1 of another fructose, the linkage is fructose-β (2→1)-fructose. Repeating glycosidic linkages can be represented more simply, e.g. amylose consisting of repeated glucose-α (1→4)-glucose linkages is described as a polymer of glucose in α(1→4) linkage.

Macromolecules of carbohydrates with predominantly one type of monosaccharide are commonly named after the monosaccharide. Thus, polymers of glucose are glucans; cellulose is a β(1→4)-glucan and amylose an α(1→4)-glucan. Polymers of galactose and fructose are galactans and fructans (or fructosans) respectively. Polysaccharides consisting principally of several monosaccharides are named according to their component monosaccharides (e.g. rhamnogalacturonans consist of rhamnose and galacturonic acid).

2.3.5 Glycosidic bonds occur in many carbohydrates in plants

The disaccharide maltose (2.IV) is one of the simplest compounds with a glycosidic bond. It consists of two molecules of α-D-glucose linked to form an α(1→4) glycosidic bond with the elimination of water. The hemiacetal hydroxy group at C-1 in the right-hand glucosyl residue of maltose is not linked and therefore exhibits the characteristics of an active aldehyde group. For this reason maltose supports the reduction of certain reagents (e.g. Cu^{2+}) and, like the aldohexoses, is a reducing sugar. The C-1 group can then link with additional molecules of glucose to form tri- and tetrasaccharides, etc.

The hemiacetal hydroxy group at C-1 of aldohexose molecules can form glycosidic bonds with any of the hydroxy groups of other molecules including those of glucosyl residues already joined in glycosidic linkage. Thus, several molecules of

CH₂OH   CH₂OH

α-D-Glucose   α-D-Glucose

HOH

2.IV   α-Maltose

(glucose residue)   (fructose residue)

2.V   Sucrose

Compounds comprising 2–10 monosaccharides are called oligosaccharides. Some common plant oligosaccharides are listed in Table 2.3; note that all of them terminate with sucrose. Stachyose appears to replace sucrose as the principal form in which carbon is transported in some plants. Incidentally, various beans used in

glucose can link via C-1 to separate 'free' or 'vacant' hydroxy groups of a common glucose residue causing 'branching'. A simple example of branching at C-6 in a chain consisting of residues in $\alpha(1 \rightarrow 4)$ linkage is shown in Fig. 2.7. Branching is common in plant polysaccharides.

β-Glucose can also link through its hemiacetal hydroxy group. The disaccharide cellobiose consists of two glucose residues linked by a $\beta(1 \rightarrow 4)$ glycoside bond. Cellulose, the principal structural component of plant cell walls, is a polymer of β-glucose joined in $\beta(1 \rightarrow 4)$ linkage.

The disaccharide sucrose (2.V) is the sugar of commerce, produced principally from sugar cane and sugar beet. It consists of α-D-glucose linked via C-1 to the C-2 hemiketal hydroxy group of β-D-fructose. Since the glucose residue exists in the pyranose ring form and fructose in the furanose ring form sucrose can be defined as α-D-glucopyranosyl-(1→2)-β-D-fructofuranoside. Since the hemiacetal and hemiketal hydroxy groups of glucose and fructose respectively are involved in the glycosidic bond, sucrose shows none of the reactions of an aldehyde or carbonyl group. It does not support the reduction of Cu^{2+} (or other reagents) and is a non-reducing sugar.

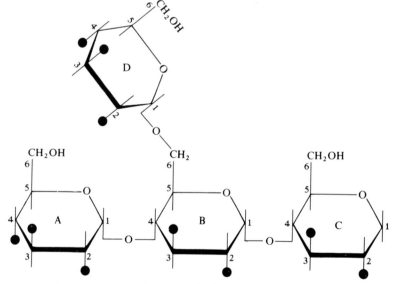

Fig. 2.7 Haworth representation of a tetrasaccharide of α-D-glucose in which three molecules (A, B and C) are joined in $\alpha(1 \rightarrow 4)$ linkage and the fourth molecule (D) is attached to glucose residue B in $\alpha(1 \rightarrow 6)$ linkage. Free hydroxy groups attached to the carbon atoms in the rings of the glucose residues are represented as solid symbols (●). Attachment of residues A, C and D to residue B gives rise to branching of glycosidic chains.

Table 2.3 Some oligosaccharides found in plants.

Compound	Number of Residues	Sequence and linkage
Sucrose	2	Glucose-(1α→2β)-fructose
Raffinose	3	Galactose-(1α→6)-glucose-(1α→2β)-fructose
Stachyose	4	Galactose-(1α→6)-galactose-(1α→6)-glucose-(1α→2β)-fructose
Verbascose	5	[Galactose-(1α→6)-galactose]$_2$(1α→6)-glucose-(1α→2β)-fructose

the food industry contain large amounts of raffinose and stachyose which humans are unable to hydrolyse into forms that can be absorbed. They pass into the lower alimentary system and are anaerobically metabolized by various micro-organisms with a good deal of gas being produced in the process!

2.3.6 Properties of polysaccharides are determined by constituent monosaccharides and their glycosidic bonds

Carbohydrates with more than about 10 monosaccharide residues are called polysaccharides. The polysaccharides in plants are divided into two broad groups according to their structure and function.

The first group comprises the storage polysaccharides of which starch is the most important. Starch is especially abundant in storage tissues such as underground organs (e.g. potato tubers) and the endosperm of seeds but is also found in the chloroplasts of photosynthetic cells where it is deposited as starch grains. Individual plant species tend to be very specific in the form, shape and size of their starch grains, a feature which can be used to trace the origin of various foodstuffs. Starch consists of two components, amylose and amylopectin. Amylose contains about 60–500 glucosyl residues in α(1→4) linkage, causing the polymer to take up a helical configuration (Fig. 2.8) which iodine can enter to form the characteristic blue starch reaction. Most starches contain 20–25% amylose though 'waxy' corn contains low levels. Amylopectin, like amylose, consists of glucosyl residues linked via α(1→4) bonds but it is distinguishable by having α(1→6) branch points. Amylopectin typically contains about one α(1→6) linkage (i.e. branch point) per 30 or so α(1→4) glycosidic bonds. These branches are, in turn, branched by further

(a) Amylose

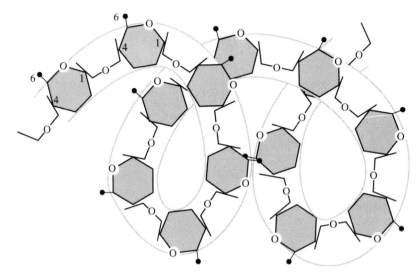

(b) Cellulose

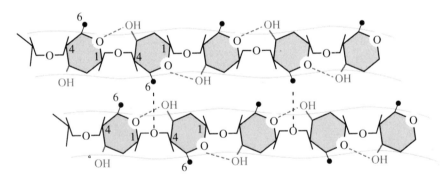

Fig. 2.8 Comparison of the structures of polymers of (a) amylose and (b) cellulose. These two polymers differ only in that amylose is a polymer of α-D-glucose joined in α(1→4) linkage whereas cellulose is a polymer of β-D-glucose joined in β(1→4) linkage. The numbering of the carbon atoms is shown for the first two glucosyl residues of each polymeric chain, C-6 being represented by the symbol ● outside the pyranose rings. Details of the hydroxy groups have been omitted except for those at C-3 in cellulose. Intramolecular hydrogen bonds are represented by grey dashed lines and intermolecular bonds by dashed lines.

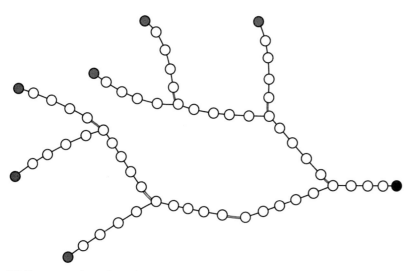

Fig. 2.9 Representation of amylopectin, a branched polymer of α-glucose. Circular symbols denote glucose residues which are joined in α(1→4) linkage (black bonds) except at branch points which involve α(1→6) linkages (grey bonds). All glucose residues are linked via their hemiacetal hydroxy groups at C-1 except for the residue shown in black. The residues shown in grey contain hydroxy groups at C-4 which are not linked to other residues.

α(1→6) linkages (Fig. 2.9) to make the whole structure rather globular. Amylopectin contains up to 5000 glucose residues. The array of glucosyl units in α(1→4) linkage between the branch points is referred to as a chain. Amylose and amylopectin have very open structures and many of the hydroxy groups within these molecules can interact with water. Amylose, for example, can be leached from starch grains with hot water leaving a residue of amylopectin.

In some plants polymers of fructose (i.e. fructans) are important storage polysaccharides. Many members of the family Asteraceae (Compositae) contain the polysaccharide inulin which has about 30 fructosyl residues in β(2→1) linkage with a terminal glucose unit — Jerusalem artichoke tubers (*Helianthus tuberosus*) are an especially rich source. Fructans containing fructose in β(2→6) linkage are also important in stems, roots and leaves of many grasses, including most cereals.

The second main group of plant polysaccharides fulfil structural functions in the walls of plant cells. Indeed, cellulose is the most abundant organic compound in the biosphere and is of prime importance in commercial products such as timber, cotton, paper, etc. Cellulose is a polymer of glucose in β(1→4) linkage (contrast

with the α(1→4) linkage of amylose). In amylose, the C-1 hydroxy group of the α-D-glucose residues are axial with the pyran rings (see Fig. 2.8a) causing coiling. However, since the C-1 hydroxy groups of the β-D-glucose residues of cellulose are not axial with the rings, adjacent glucosyl units are rotated at about 180° to each other thereby preventing coiling. This permits intramolecular bonding between adjacent glucosyl residues within the chain, involving hydrogen bonding between the C-3 hydroxy group of one residue and the oxygen atom of the pyranose ring of the next.

Intramolecular hydrogen bonding between glucose residues in β(1→4) linkage

Structurally this is stiff, straight and ribbon-like. In plant cell walls, cellulose molecules also form intermolecular hydrogen bonds between the oxygen atom of the glycosidic bond of one cellulose molecule with the C-6 hydroxy group of another (Fig. 2.8b). The hydrogen bonding *between* cellulose molecules results in the formation of highly ordered crystalline regions, not readily accessible to water. In higher plants, about 50–70 cellulose chains (each comprising up to 5000–10 000 glucosyl residues) lie parallel to form microfibrils (Section 3.10) which impart great strength and rigidity and comprise the single most important structural unit of cell walls (see Chapter 17).

Cellulose is very resistant to both enzyme-mediated hydrolysis and chemical attack (strong acids are required to hydrolyse it). This biological stability is due to the numerous hydrogen bonds between molecules. The enzyme cellulase, which attacks the β(1→4) linkages, appears to be expressed only at very low levels in most plant cells in which it is presumed to be involved in certain aspects of cell development. Other cellulases are expressed in specific cells at specific times (e.g. the abscission layer of petioles at leaf abscission). However, substantial levels of cellulase activity occur in some organisms such as gastropods, wood-rotting fungi and certain micro-organisms, the latter including many in the gastrointestinal tracts of herbivorous animals and insects (e.g. termites) for digestion of cellulose. Other naturally occurring polysaccharides also form linear chains which aggregate into bundles to form microfibrils. The most important is chitin, a polymer of

N-acetylglucosamine in $\beta(1\rightarrow4)$ linkage, which is the principal structural polysaccharide in the cell walls of many fungi. The cell walls of some green algae (e.g. *Codium* and *Acetabularia*) contain microfibrils of a polymer of $\beta(1\rightarrow4)$-mannose. Still other green algae contain cell wall microfibrils of a polymer of $\beta(1\rightarrow3)$-xylose. Plants also contain polymers of $\beta(1\rightarrow3)$-glucose but these do not attain the same degree of polymerization as cellulose and do not form microfibrils.

The matrix polysaccharides are another main group of structural polysaccharides which constitute the filling material of the cell wall between the cellulose microfibrils. Until recently matrix polysaccharides were divided into two main groups: the pectic substances (soluble in water); and the hemicelluloses (insoluble in water but soluble in dilute alkali). However, these terms are very imprecise since the material extracted consists of many components. Nowadays, these non-cellulosic polysaccharides are named according to the principal monosaccharide components. The structures of some of these are described in Chapter 17.

2.4 Lipids are biological molecules with little affinity for water

The lipids are a rather heterogeneous group of naturally occurring compounds, some of which have few common structural features, but all lipids are sparingly soluble in water; they are said to be hydrophobic ('water hating'). Conversely, most lipids exhibit considerable affinity for non-polar solvents (e.g. ether and petroleum spirit) due to the presence of one or more alkyl chains or other residues which convey aliphatic properties to at least part of the molecule. By far the most common source of alkyl chains are fatty acids of the structure

$$CH_3—(CH_2)_n—COOH$$

but other types of molecules (e.g. long chain alcohols) can also be involved. The principal reason for considering such a diverse group of compounds as a single class is that, with the exception of the storage lipids associated with some seeds, they are mostly associated with the cellular membranes. The role of lipids in the structure and function of various membranes in plant cells is discussed in Section 3.2.

2.4.1 Fatty acids are important components of most lipids

Fatty acids are the source of the alkyl components in the majority of plant lipids. The simplest are saturated straight-chain monocarboxylic acids with an even number of carbon atoms. The most common saturated fatty acids found in plant

Table 2.4 Some fatty acids associated with plants. 'Unusual' fatty acids, often found in very high concentrations in a few species (e.g. ricinoleic acid in *Ricinus communis*) are not shown.

Chain length	Saturated fatty acids		Unsaturated fatty acids	
	Symbol*	Trivial name	Symbol*	Trivial name
6	6:0	Caproic acid		
8	8:0	Caprylic acid		
10	10:0	Capric acid		
12	12:0	Lauric acid		
14	14:0	Myristic acid		
16	16:0	Palmitic acid	16:1(9c)	Palmitoleic acid
18	18:0	Stearic acid	18:1(9c)	Oleic acid
			18:2(9c, 12c)	Linoleic acid
			18:3(9c, 12c, 15c)	Linolenic acid
20	20:0	Arachidic acid	20:4(5c, 8c, 11c, 14c)	Arachidonic acid

* The symbol 18:1 (9c) denotes a fatty acid of 18 carbon atoms with a single *cis* double bond between carbon atoms 9 and 10.

Table 2.5 Abundance of the major fatty acids in commercial vegetable oils. Data refer to fatty acids present in the major vegetable oils produced in 1969–70. After Hitchcock, C. & Nichols, B.W. (1971) *Plant Lipid Biochemistry*. Academic Press, London.

Fatty Acid	Symbol	Abundance (% of total)
Lauric acid	12:0	4
Myristic acid	14:0	2
Palmitic acid	16:0	11
Stearic acid	18:0	4
Oleic acid	18:1 (9c)	34
Linoleic acid	18:2 (9c, 12c)	34
Linolenic acid	18:3 (9c, 12c, 15c)	5
Ricinoleic acid*	18:1 (9c) 8-hydroxy	1
Arachidic acid	20:0	1
Erucic acid*	22:1 (13c)	3

* Ricinoleic acid and erucic acid are unusual fatty acids present in very high concentrations in the oils of *Ricinus communis* (88% ricinoleic acid) and *Brassica campestris* (50% erucic acid). These compounds are not associated with the oils of most plants but since *R. communis* and *B. campestris* are grown very extensively, they make significant contributions to the total commercial fatty acid production.

lipids contain 16 or 18 carbon atoms although the fatty acids contained in the waxes of leaf cuticles are appreciably bigger. Some of the common saturated fatty acids are listed in Table 2.4. Usually only palmitic acid (C_{16}) and stearic acid (C_{18}) are present in significant amounts but the saturated fatty acids collectively account for only about 20% of the total fatty acid content of most plants (Table 2.5). Those with one or more double bonds (unsaturated fatty acids—see Table 2.4) account for the remaining 80%. In many fatty seeds, oleic and linoleic acid frequently account for more than 70% of the fatty acid content (>90% in sunflower). Fatty acids are usually esterified through their —COOH group to an alcohol (most commonly glycerol), forming constituent parts of lipids.

The atoms of fatty acids are numbered in sequence from the carboxyl end (C-1). By convention, the structure of a fatty acid is described by symbols which denote the number of carbon atoms, the number of double bonds and the position and nature of these bonds. Myristic acid (14 carbon atoms, no double bonds) is denoted as 14:0; linolenic acid (18 carbon atoms, three double bonds) is shown as 18:3. The position of double bonds and their configuration (*cis* or *trans*) are important with respect to the structure and shape of unsaturated fatty acids. Those with the *cis* configuration are kinked or bent and have a different conformation to the unsaturated antecedent (Fig. 2.10). Unsaturated fatty acids are described by citing the carbon atom with the lowest number involved in the double bond followed by the letter *c* or *t* for *cis* or *trans*. Thus, oleic acid (for which the systematic name is *cis*-9-octadecenoic acid) is denoted as 18:1 (9*c*). Linolenic acid is 18:3 (9*c*, 12*c*, 15*c*) indicating that it contains three *cis* double bonds located between carbon atoms 9 and 10, 12 and 13, and 15 and 16. However, since most unsaturated fatty acids exhibit the *cis* configuration, this is implied unless specified otherwise.

The introduction of a double bond into a fatty acid decreases the melting point dramatically. The melting points of stearic acid (18:0), oleic acid (18:1) and linoleic acid (18:2) are 70°C, 14°C and −12°C respectively. Hence, at room temperature, animal lipids (mostly saturated fatty acids) are typically solids (fats) and vegetable lipids are usually liquids (oils). Vegetable oils can be converted to solids by saturating a proportion of the unsaturated fatty acid residues by hydrogenation, a process used in the commercial production of margarine.

2.4.2 Glycerides are fatty acid esters of glycerol

The glycerides are fatty acid esters of glycerol. They are the chief components of the natural animal fats and vegetable oils used in commerce. They account for most

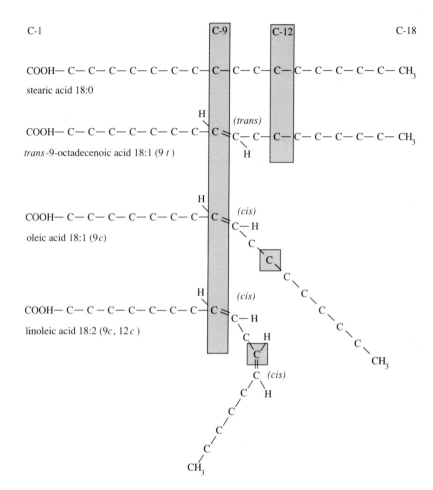

Fig. 2.10 Structures of some C_{18} fatty acids. Note that the *cis* arrangement about the double bond(s) of unsaturated fatty acids causes the molecule to bend or 'kink'.

of the lipid present in fatty storage tissues (e.g. fatty seeds). Soya bean seeds, for example, contain about 20% lipid of which 85–90% is triacylglycerol. Glycerol contains three hydroxy groups, any one of which can theoretically form an ester with a fatty acid. Glycerides in which one to three of the hydroxy groups of glycerol are esterified by fatty acids (shown as COOH—R[1,2,3] below) are referred to as mono-, di- or triglycerides or as mono-, di- or triacylglycerols.

$$CH_2OH \qquad CH_2O-CO-R^1 \qquad\qquad CH_2O-CO-R^1$$
$$|\qquad\qquad\quad | \qquad\qquad\qquad\qquad\qquad |$$
$$HOCH \qquad\quad HOCH \qquad\quad R^2-CO-O-CH$$
$$|\qquad\qquad\quad | \qquad\qquad\qquad\qquad\qquad |$$
$$CH_2OH \qquad\quad CH_2OH \qquad\qquad\quad CH_2O-CO-R^3$$

Glycerol   Monoglyceride   Triglyceride

Glycerides exhibit little diversity since they occur mostly as triglycerides and as few as three or four fatty acids occur in significant amounts in some tissues of some species (Table 2.5). Importantly, the triglycerides lack polar groups so that they have very little affinity for water. This distinguishes them and the true waxes (Section 2.4.3) from the polar lipids in which at least one group confers some affinity for water.

Since C-2 of glycerol is asymmetric, enantiomers of glycerides are possible when C-1 and C-3 are esterified with different substituents.

$$C\text{-}1 \qquad\quad CH_2OH \qquad CH_2OH \qquad\quad C\text{-}3$$
$$|\qquad\qquad\qquad |$$
$$C\text{-}2 \qquad\quad HOCH \qquad\quad HCOH \qquad\quad C\text{-}2$$
$$|\qquad\qquad\qquad |$$
$$C\text{-}3 \qquad\quad CH_2OH \qquad CH_2OH \qquad\quad C\text{-}1$$

By convention, the carbon atom appearing at the top is C-1 if the hydroxy group on the central carbon atom (C-2) is on the left in a planar projection (stereospecific numbering (*sn*)). In fact, plants synthesize only one of the two potential enantiomers of any one glyceride.

2.4.3 Waxes, cutin and other cuticular lipids

The term 'wax' is used in two senses. It is often used, incorrectly, in a general sense to refer to the cuticular lipids which collectively form the relatively hard, firm and water-resistant cuticular surfaces, especially on leaves and fruits. The cuticular lipids are especially important in preventing loss of water by evapotranspiration although they also affect the rate of exchange of CO_2 and other gases. Plants associated with dry environments have a very thick layer of cuticular lipids.

Strictly speaking, waxes are *n*-alkyl-1-anoic fatty acids, typically containing 20–28 carbon atoms, esterified to long-chain primary alcohols containing 24–28 carbon atoms. They are readily soluble in organic solvents. Cutin is another esterified cuticular lipid consisting of fatty acids with 16 or 18 carbon atoms (the latter often containing a double bond), usually bearing one to three hydroxy groups, one of which is attached to the terminal carbon atom (C-16 or C-18). The hydroxy fatty acids are presumed to be highly cross-esterified to form a very tough and water resistant product, insoluble in organic solvents. Leaf cuticles also contain simple *n*-alkane hydrocarbons, usually with an odd number of carbon atoms, 29 and 31 being the most common. Unesterified fatty acids with chain lengths longer than those in waxes and cutin are also present in the cuticle. There is considerable variation in the composition of cuticular lipids and some species have a characteristic composition.

Suberin fulfils a role similar to cutin in underground organs and is also produced in response to wounding. Like cutin, it is highly cross-linked and impermeable to water but it differs in that the main components are unsubstituted long chain fatty acids (C_{20}–C_{26}), very long chain alcohols and phenylpropanoid residues of the type described in Section 2.7.

2.4.4 Phospholipids are polar lipids consisting of fatty acid esters of an alcohol linked via phosphate to a hydrophilic side-chain

The phospholipids, like all lipids, are hydrophobic due to the presence of alkyl components but are regarded as polar lipids because they contain a group which gives them some affinity for water—part or all of which is conferred by a phosphate group. The phospholipids collectively are the predominant lipids in most membranes, the glycolipids of chloroplast membranes excepted. Unlike triglycerides, phospholipids do not occur in lipid-rich storage tissues except as components of membranes.

The phosphoglycerides differ from the triglycerides in that only the C-1 and C-2 hydroxy groups of glycerol are esterified with fatty acids. The C-3 hydroxy group is esterified with phosphate which in turn is usually esterified to one of a number of substituents. Glycerol-3-phosphate (glycerol-3-P) is regarded as the parent compound of the phosphoglycerides. The fatty acid esterified to C-1 is usually saturated but the component at C-2 is usually one of the more common unsaturated fatty acids listed in Table 2.5 (e.g. 18:1, 18:2, or 18:3) which, by nature of the *cis* configuration of the double bond(s), causes the lipid to have an open but bulky structure. The simplest of the phosphoglycerides is phosphatidic acid in

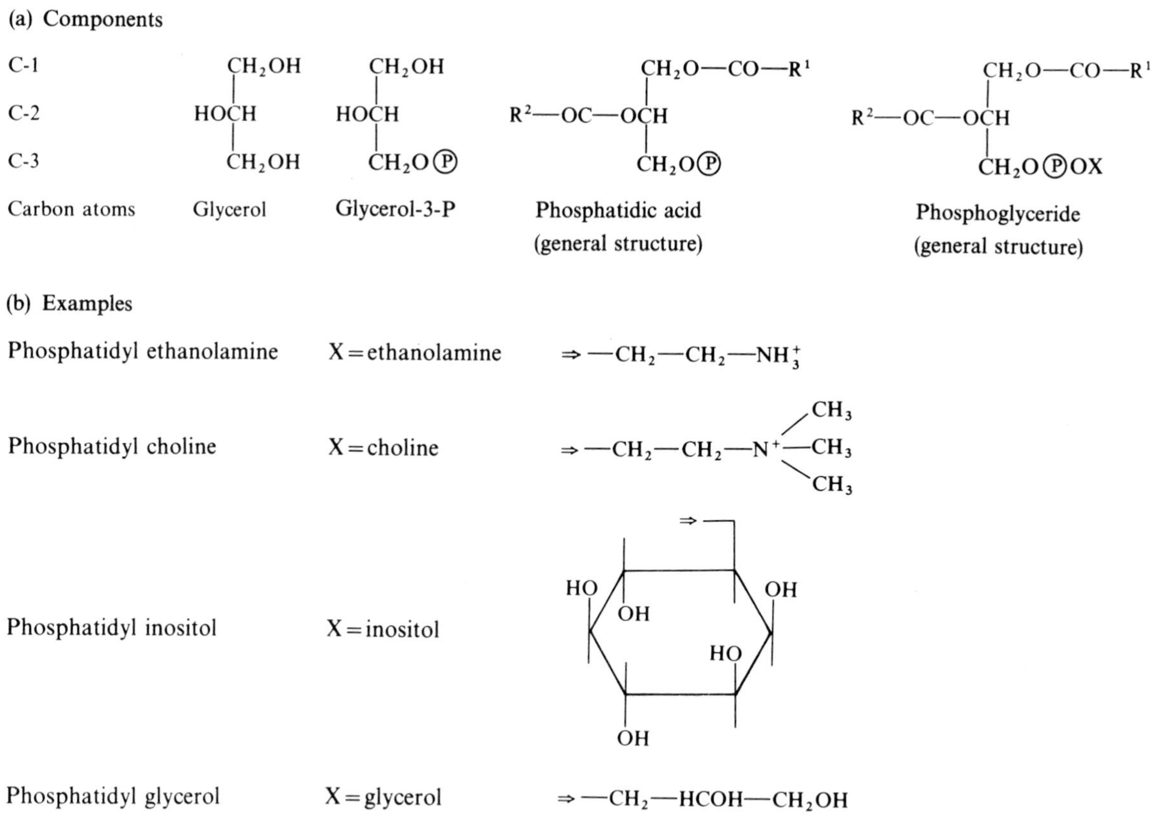

(a) Components

C-1	CH_2OH	CH_2OH	$CH_2O-CO-R^1$	$CH_2O-CO-R^1$
C-2	$HOCH$	$HOCH$	$R^2-OC-OCH$	$R^2-OC-OCH$
C-3	CH_2OH	$CH_2O\textcircled{P}$	$CH_2O\textcircled{P}$	$CH_2O\textcircled{P}OX$
Carbon atoms	Glycerol	Glycerol-3-P	Phosphatidic acid (general structure)	Phosphoglyceride (general structure)

(b) Examples

Phosphatidyl ethanolamine   X = ethanolamine $\Rightarrow -CH_2-CH_2-NH_3^+$

Phosphatidyl choline   X = choline $\Rightarrow -CH_2-CH_2-N^+\begin{smallmatrix} CH_3 \\ -CH_3 \\ CH_3 \end{smallmatrix}$

Phosphatidyl inositol   X = inositol

Phosphatidyl glycerol   X = glycerol $\Rightarrow -CH_2-HCOH-CH_2OH$

Fig. 2.11 (a) Component residues and general structure of a phosphoglyceride. (b) Examples of phosphoglycerides. For simplicity, the phosphate group is shown as $\textcircled{P}$ and R^1 and R^2 denote fatty-acyl side-chains. Arrows indicate point of attachment of X to phosphoglyceride.

Sphingosine
2.VIa $CH_3-(CH_2)_{12}-CH= CH-\overset{\overset{\displaystyle H}{|}}{C}-\overset{\overset{\displaystyle H}{|}}{\underset{\underset{\displaystyle NH_2}{|}}{\underset{\underset{\displaystyle }{|}}{C}}}-CH_2OH$ OH

Phytosphingosine
2.VIb $CH_3-(CH_2)_{12}-CH_2-\overset{\overset{\displaystyle H}{|}}{\underset{\underset{\displaystyle OH}{|}}{C}}-\overset{\overset{\displaystyle H}{|}}{\underset{\underset{\displaystyle OH}{|}}{C}}-\overset{\overset{\displaystyle H}{|}}{\underset{\underset{\displaystyle NH_2}{|}}{C}}-CH_2OH$

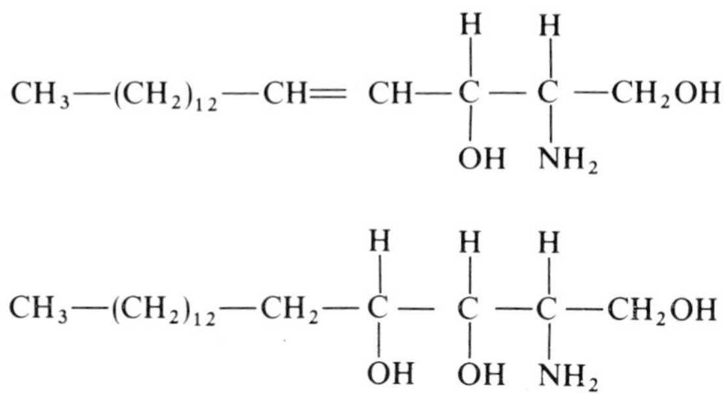

Carbon atoms   C-18...   C-5   C-4   C-3   C-2   C-1

which the C-3 hydroxy group is esterified to phosphate only (Fig. 2.11) although this is relatively uncommon in cells.

Some of the more common phosphoglycerides are shown in Fig. 2.11. They differ from phosphatidic acid only in that the phosphate group at C-3 is further esterified to one or another of various alcohols (including a second glycerol molecule) all of which contain additional polar groups (e.g. ethanolamine, choline, inositol). When phosphatidic acid is esterified to the C-3′ hydroxy group of a second molecule of glycerol then the hydroxy groups at the C-1′ and C-2′ positions are usually esterified with further fatty acids to form diphosphatidylglycerols.

Phosphosphingolipids, the second main group of phospholipids, differ from the phosphoglycerides in that sphingosine (2.VIa) or a derivative serves as the main structural alcohol in place of glycerol. In plants the principal alcohol is phytosphingosine (2.VIb). (In animals it is sphingosine and the phosphate group is esterified to ethanolamine). Phytosphingosine is esterified to phosphate through the hydroxy group at C-1 and a fatty acid is joined to phytosphingosine in amide linkage via

the amino group at C-2. In addition, the phosphate moiety at C-1 is esterified to an oligosaccharide (Fig. 2.12).

Phospholipids are amphipathic substances—they possess both hydrophobic and hydrophilic properties. In the phosphoglycerides, the acyl residues are highly hydrophobic but the phosphate group and the additional substituents esterified to it are hydrophilic (Fig. 2.13). Similarly, the phosphate group and the oligosac-

(a) Phosphosphingolipids

$$CH_3-(CH_2)_{12}-\overset{\overset{\displaystyle H}{|}}{C}-\overset{\overset{\displaystyle H}{|}}{C}-\overset{\overset{\displaystyle H}{|}}{C}-CH_2-O-\textcircled{P}-O-X$$

with OH OH NH below, CO below, R below.

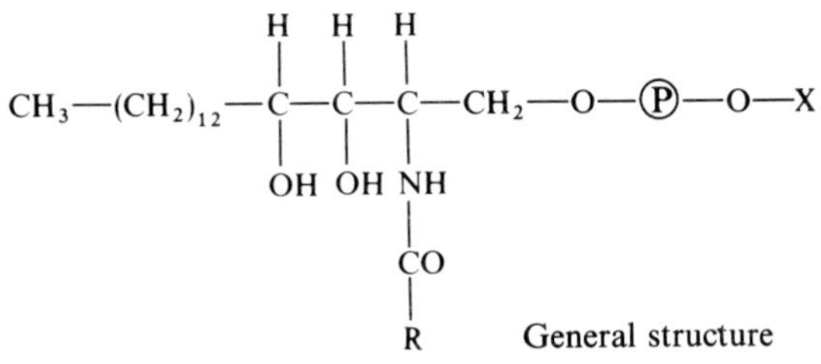

General structure

(b) Glycolipids

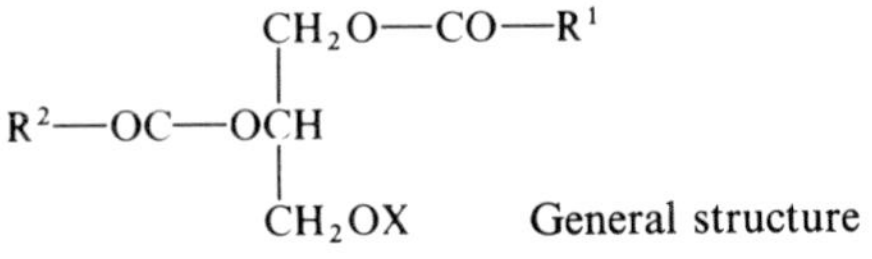

$$\begin{array}{l} CH_2O-CO-R^1 \\ R^2-OC-OCH \\ CH_2OX \end{array}$$

General structure

(c) Examples

Monogalactosyl diglyceride   X = −1[β-D-galactose]

Digalactosyl diglyceride   X = −1[β-D-galactose]6→1[α-D-galactose]

Sulpholipid   X = −1[6-sulpho-α-D-quinovose]

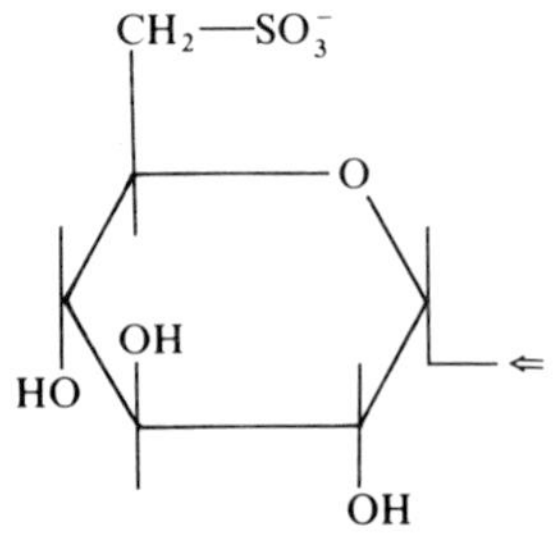

$CH_2-SO_3^-$

Fig. 2.12 (a) General structure of phosphosphingolipids, based on phytosphingosine where X denotes an oligosaccharide and R denotes a fatty acid. (b) Structures of glycolipids based on glycerol where R^1 and R^2 denote fatty acids and X denotes a mono- or oligosaccharide linked via C-1 or the position indicated by the arrow.

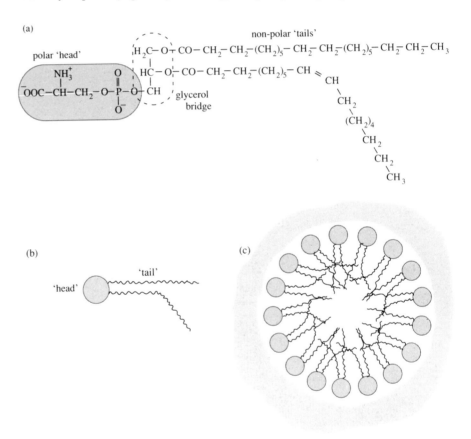

Fig. 2.13 (a) Structure of the hydrophilic and hydrophobic regions of the phospholipid phosphatidyl serine and (b) representation of the structure in models of membranes. (c) In water, phospholipid molecules arrange with their hydrophobic 'tails' inwards and their polar 'heads' directed towards the water phase to form stable spherical associations known as micelles. Other phospholipids exhibit similar characteristics.

charides attached to phosphosphingolipids are hydrophilic but the rest of the molecule is hydrophobic. Phospholipids disperse quite readily in water to form spherical aggregations (micelles) in which the hydrophobic portions aggregate, leaving the hydrophilic regions oriented outwards where they interact with water to form a stable configuration (Fig. 2.13). The amphipathic properties of the phospholipids and other polar lipids account for many of the properties of cell membranes (see Section 3.2).

2.4.5 Glycolipids are polar lipids containing one or more monosaccharides or monosaccharide derivatives

Glycolipids are polar lipids which contain one or more monosaccharides or monosaccharide derivatives. The most important are the galactosyldiglycerides (galactolipids) which are the principal lipids in the internal membrane complex of chloroplasts. The monogalactosyldiglycerides (see Fig. 2.12) are distinguishable from phosphoglycerides as the polar substituent at C-3 of glycerol is esterified to the hemiacetal hydroxy group of β-D-galactose. Digalactosyldiglycerides are similar except that the β-D-galactose residue is linked to another residue of α-D-galactose by an $\alpha(1\rightarrow6)$ glycosidic bond.

Another class of glycolipid found is the sulpholipids, named because of the presence of a sulphonate moiety. They are similar to the galactolipids except that the C-3 of glycerol is esterified to the hemiacetal hydroxy group (C-1) of 6-sulpho-α-D-quinovose (6-sulpho-6-deoxy-α-D-glucose; Fig. 2.12). This lipid is found principally in chloroplasts, typically at less than 10% of the concentration of galactolipid.

2.4.6 Terpenes and terpenoids are composed of, or derived from, isoprenoid units

Plants contain a large number of compounds with carbon skeletons comprising one or more residues of a branched C_5 repeating structure known as an isoprenoid unit (because they derive from a pathway which also gives rise to isoprene).

Isoprene   Isoprenoid unit

Terpenes consist of an integral number of isoprenoid units (or related structures) which may or may not have undergone some modification of the substituents without altering the carbon skeleton. Terpenoids are obviously derived from terpenes but have modified carbon skeletons. Most of the 10 000 or so terpenes and terpenoids characterized are secondary plant products, with cis 1,4-polyisoprene (rubber) from Hevea brasiliensis having considerable commercial importance. However, several terpenes and terpenoids have essential roles in plants. Like the lipids they are extremely hydrophobic and are usually associated with some essential function within the lipophilic region of membranes. Examples include phytol (C_{20}, of four isoprenoid units), the hydrophobic substituent of chlorophyll, and the carotenoids (C_{40}), both of which are located in the internal membrane complex of chloroplasts. Ubiquinone and plastoquinone are essential components of the electron transport chains of membranes of mitochondria (ubiquinone) and chloroplasts (plastoquinone). Several of the plant growth regulator substances (e.g. the gibberellins) are also terpenoids.

2.5 Proteins are constructed from 20 amino acids

Proteins are macromolecules of L-amino acids. The genetic information to construct a cell and carry out its functions is expressed through proteins. They are involved in many essential processes including the catalysis of chemical reactions, storage of reserve materials, transport mechanisms across membranes, regulatory phenomena, intracellular structure and energy-generating mechanisms involving electron transport, to name but a few. Proteins are heteropolymers because the basic repeating unit is always one of 20 different L-amino acids (glycine, with no optical isomers, excepted). All of these 20 (known as protein amino acids) contain nitrogen and two of them also contain sulphur. Cells of higher plants contain thousands of different proteins, each with its own characteristic complement of amino acids linked together in a very specific sequence.

Vegetative plant material contains low levels of protein due to the high amounts of structural polysaccharides (see Chapter 1). However, most seeds (especially legumes—e.g. soya bean contains about 35% protein) contain mainly storage proteins to provide a reserve of amino acids and energy for germination. Interestingly, humans, unlike plants, are unable to synthesize all 20 of the protein amino acids and are dependent on sources such as plants for a supply of nine amino acids ('essential' amino acids). Many storage proteins (and hence seeds) contain insufficient amounts of some of these for complete human nutrition (e.g. some legume seeds contain too little of the sulphur amino acids cysteine and methionine while cereals contain insufficient lysine—see Section 13.1).

2.5.1 Protein amino acids are classified according to the properties of their side-chains

Nineteen protein amino acids are carboxylic acids bearing a primary amino group; the other (proline) is a cyclic secondary amine. The amino acids differ in the composition of the side-chain attached to the carbon atom bearing the primary amino group (some have additional amino groups). Carbon atoms are always numbered from the carboxyl end (C-1). Thus, the primary amino group is attached at C-2 (also known as the α carbon atom). This carbon atom is asymmetric in all of the amino acids except glycine, giving rise to enantiomers. Since the naturally occurring C_3 amino acid serine is chemically converted to L-glyceraldehyde by procedures which do not interfere with the configuration at C-2, this establishes the nomenclature for the amino acid enantiomers.

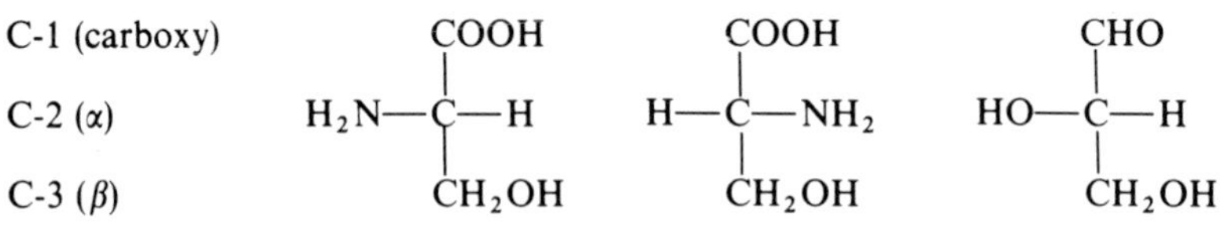

Carbon atoms			
C-1 (carboxy)	COOH	COOH	CHO
C-2 (α)	H₂N—C—H	H—C—NH₂	HO—C—H
C-3 (β)	CH₂OH	CH₂OH	CH₂OH
	L-Serine	D-Serine	L-Glyceraldehyde

For sugars the L and D refer to the position of the hydroxy group on the penultimate carbon atom whereas for the amino acids, L and D always signify the configuration of the amino group at C-2, regardless of the length of the side-chain. All of the amino acids associated with proteins (except glycine) have the L configuration.

The protein amino acids can be classified on the basis of the structure and chemical reactivity of their side-chains although protein chemists prefer to classify them according to their charge and polarity since these, together with the sequence in which they are linked together, largely determine the shape, form, reactivity and function of proteins. The amino acids are shown in Fig. 2.14 in a way which emphasizes the structure and chemical relationships of the side-chains as this is relevant to their biosynthesis. Nonetheless, classification according to polar-

ity/charge is also shown. Both the carboxyl group at C-1 and the amino group at C-2 are dissociated at physiological pH to confer both a positive and negative charge —NH_3^+ and —COO^-) to form a zwitterion. Thus, the *free* amino acids are amphoteric (i.e. they receive and donate protons) and the charged groups confer hydrophilic character. Amino acids differ in the polarity of their side-chains (see Fig. 2.14).

Some of the 20 protein amino acids listed in Fig. 2.14 undergo modification after they have been incorporated into certain specific proteins. Most notable is the formation of hydroxyproline (from proline) in certain cell wall proteins, and hydroxylysine in some algae.

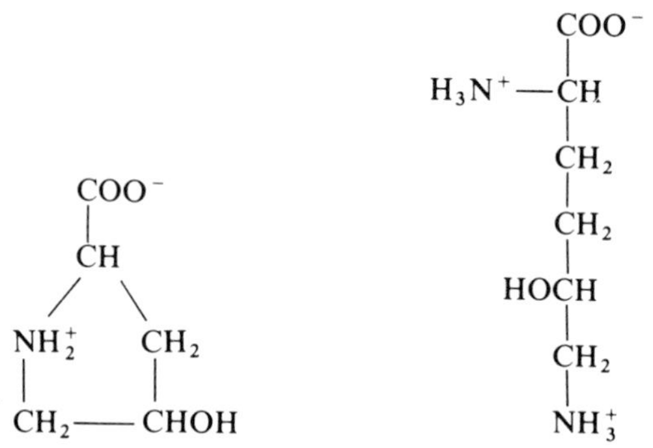

Hydroxyproline   Hydroxylysine

Although most of the amino acid content is found in proteins, a proportion of each individual amino acid occurs in an unpolymerized or 'free' form. Some of these (e.g. aspartate and glutamine) are especially important in metabolism. Under some conditions (e.g. phosphorus deficiency) the level of free amino acids can rise appreciably and in certain forms of stress (e.g. salt stress) the free proline content can rise to very high levels.

Plants also contain other L-amino acids. Some (e.g. homocysteine, homoserine, cystathionine) are essential intermediates in the synthesis of protein amino acids (see Chapters 13 and 14). Additionally, some plants contain L-amino acids, often in high concentrations, which are not incorporated into proteins or known to be essential. More than 200 of these non-protein amino acids are known. They are regarded as secondary products and some are very toxic (see Section 13.9).

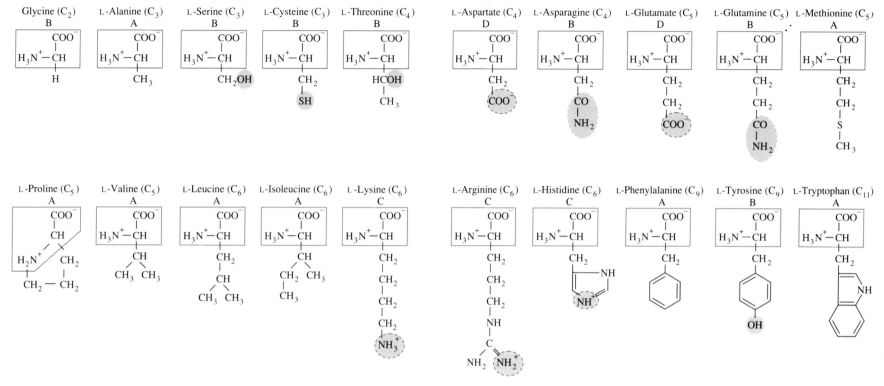

Fig. 2.14 Structures of the protein L-amino acids, arranged in order of increasing chain length to emphasize structural similarities. The carboxyl group (C-1) and the carbon atom bearing the amino group (C-2), common to all the amino acids, are boxed. (Note that proline with its ring structure is a special case and that glycine, which does not have an asymmetric carbon atom at C-2, exists only in one isomeric form.) The side-chains of amino acids indicated by: A are hydrophobic; B are polar but uncharged; C are polar with positive charge; and D are polar, with negative charge. Polar groups are shaded.

2.5.2 Protein primary structure is determined by covalent peptide and disulphide bonds

In proteins the carboxyl group at C-1 of one amino acid is linked to the amino group of C-2 of another to make a peptide. The constituent amino acids are referred to as aminoacyl residues and the link between two residues is a peptide bond. Tripeptide, tetrapeptide and pentapeptide refer to peptides containing three, four or five aminoacyl residues. Proteins are, therefore, polypeptides. By conven-

tion the aminoacyl residue at the beginning of a peptide (or protein) bears an amino group and is shown on the left (cyclic peptides excepted) while the corresponding residue at the other end bears a carboxyl group (shown on the right). The hypothetical pentapeptide shown in Fig. 2.15 illustrates these principles.

The molecular masses of proteins and some other biological macromolecules are usually expressed in kilodaltons. One dalton (Da) is the mass of one hydrogen atom, i.e. 1.6605×10^{-24} g.

The number of aminoacyl residues in a protein varies greatly. Soluble spinach

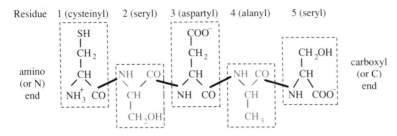

Residue   1 (cysteinyl)   2 (seryl)   3 (aspartyl)   4 (alanyl)   5 (seryl)

Fig. 2.15 A hypothetical pentapeptide consisting of an alanyl, cysteinyl, aspartyl and two seryl residues to illustrate the principles of peptide nomenclature. The bold symbols joining the aminoacyl residues denote peptide bonds.

ferredoxin contains 97 residues and has a molecular mass of about 11 kDa but proteins of more than 5000 residues (molecular mass > 500 kDa) are not uncommon. Any one protein has a fixed amino acid composition and sequence, called the primary structure, conferred by the covalent peptide bonds.

Disulphide bonds (—S—S—) form in many proteins by oxidative linking of the thiol groups (—SH) in the side-chains of two separate residues of cysteine, thereby causing cross-linking within the primary structure and some degree of folding. Oxidative linking and reductive unlinking of —S—S— bridges regulate the activity of some proteins.

2.5.3 Biologically active proteins have other levels of organization in addition to their primary structure

Denaturing a protein with an organic solvent or by heating does not interfere with the primary structure but it usually abolishes the biological activity (e.g. the catalytic activity of enzymes). It is, therefore, evident that native proteins are organized at other levels. The enzyme ribulose 1,5-bisphosphate carboxylase, for example, contains helical coils and sheets and many other levels of organization (Fig. 2.16).

The secondary structure of a protein involves hydrogen bonding between the —CO— group of an aminoacyl residue with the —NH— group of another, non-adjacent, residue. Intra-chain associations occur between —CO— and —NH— groups spaced about four residues distant, causing the peptide backbone to form a right-handed α-helix with an average of 3.6 residues per turn. The extent

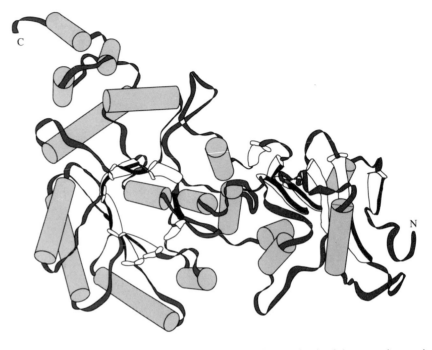

Fig. 2.16 Schematic representation of the structure of a large subunit of the enzymic protein, ribulose 1,5-bisphosphate carboxylase from the photosynthetic bacterium *Rhodospirillum rubrum*. C denotes the end of the peptide with the terminal —COOH group and N denotes the end with the terminal —NH₂ group. Cylinders represent regions in which the peptide chain forms α-helices and white regions represent β-strands involving bonding between non-adjacent parts of the main chain, thereby forming sheets. From Schneider, G., Lindqvist, Y., Brändén, C-I. & Lorimer, G. (1986) *EMBO J.*, **5**, 3409–15.

of helical coiling is influenced considerably by the nature and grouping of the side-chains of the residues in the peptide chain. Residues with non-polar side-chains (e.g. leucine, valine, alanine) are conducive to coiling but clusters of residues with side-chains of similar charge repel each other, straining the primary backbone and preventing formation of the secondary bonds which promote helical coiling (helical destabilization). Negatively charged destabilization involves side chains with —COO⁻ groups (i.e. C-5 of glutamate and C-4 of aspartate) while destabilization by positive charges involves the secondary —NH₃⁺ and —NH₂⁺— groups of lysine and histidine respectively. Clusters of residues with large bulky polar but uncharged groups in their side-chains and the rigid ring system of proline residues (in which C-2 and the nitrogen atom are incorporated into the primary backbone) also interfere with helix formation.

β-Pleated sheets are formed by hydrogen bonding between the —CO— and —NH— groups of amino acids residues in separate peptide chains (inter-chain associations) or between residues spaced well apart within a chain. This generally involves proteins with unusually high proportions of aminoacyl residues with small side-chains (glycine and alanine)—they generally form regular arrays of folded sheet-like structures.

The tertiary structure of a protein is conferred by additional internal bonds between amino acids not in immediate proximity on the covalently linked peptide chain, resulting in folding and twisting to give the protein its characteristic native state. This includes ionic bonds (e.g. —COO$^-$ with —NH$_3^+$), interaction between hydrophobic side chains (e.g. amino acids with category A side-chains—see Fig. 2.14) and hydrogen bonding (e.g. —OH with —CO—). Also, hydrophobic side-chains tend to aggregate in the interior of the protein to form hydrophobic 'domains' while residues with polar side-chains orientate outwards and interact with water. This maximizes internal bonding and binding with the solute (water), conferring a stable conformation to many proteins in solution. Some proteins, like the membrane lipids, are strongly amphipathic, especially those in membranes (see Section 3.2) in which the hydrophobic domains form strong and stable associations with the non-polar interior of the membrane.

Oligomeric proteins consist of more than one polypeptide subunit (protomer). Those with two, three or more protomers are called dimers, trimers, tetramers, etc. The quarternary structure is conferred by the binding of the protomers which may or may not be identical. Usually the catalytic activity of oligomeric enzymes with identical subunits is abolished when the oligomer is dissociated into its protomers, suggesting that the catalytic or active site is inaccessible to the substrate. Restoration of the quarternary structure causes a conformational change, exposing the active site and restoring catalytic activity.

For other dissociated oligomeric enzymes, usually those consisting of two or more different protomers, one protomer exhibits catalytic activity (the catalytic subunit) but the other(s) does not (non-catalytic subunit). The non-catalytic subunit is able to bind one or more specific compounds which inhibit or enhance the activity of the catalytic subunit. This is important for regulation of enzyme activity (see Section 5.5.2).

2.5.4 Proteins fulfil many essential functions in plants

Plants contain thousands of different proteins, each with its own characteristic function. Various examples are discussed throughout this book and it is appropriate here to classify these and indicate where further details can be found.

The most ubiquitous group of proteins in plants are the enzymes, each being a catalyst for a specific chemical reaction (Chapter 5). Since plants carry out thousands of reactions they contain thousands of different enzymes. Some proteins contain, in addition to their peptide component(s), an additional structure known as a prosthetic group which is essential for catalytic activity (Section 5.2.2).

The electron carrier proteins facilitate the transfer of electrons from one compound to another, often to other electron carrier proteins, forming a chain for the passage of electrons from donor to acceptor. These proteins support the passing of electrons from the reduced nicotinamide adenine dinucleotide (NADH) to O_2 in mitochondria (Section 7.7) and from photosystem II to photosystem I in chloroplasts (Section 9.8.3). Most (but not all) electron carrier proteins are bound to membranes and contain prosthetic groups, often containing a metal.

Various proteins are involved in the transport of specific substances across the intracellular and peripheral membranes of cells (see Section 6.3). For example, translocator proteins facilitate the transport of certain metabolites across the chloroplast envelope (Section 10.8) while others are involved in the transport of specific ions (e.g. proton pumps). An assembly of proteins (ATP synthetase) in association with the proton pumps of mitochondria and chloroplasts supports the proton-driven phosphorylation of adenosine diphosphate (see Section 7.8 & 9.11).

Some proteins serve structural roles—for example, the chlorophyll binding protein within the lipophilic regions of the inner membrane complex of chloroplasts (Section 3.2). Other proteins fulfil structural roles of a different kind. Actin consists of protomeric units with a molecular mass of about 43 kDa which, under appropriate conditions, reversibly polymerize and depolymerize. The polymerized form maintains a level of structure and order within the cytoplasm (Section 3.9). Tubulin also undergoes reversible polymerization/depolymerization. The polymerized form (arranged in microtubules) confers a level of structural organization to cells, especially during mitosis and cellulose synthesis (Sections 3.11 & 17.6 respectively).

Storage proteins in seeds are discussed above (Section 2.5). Glycoproteins, sugar residues covalently linked to a polypeptide backbone, are common components of the matrix of cell walls (Section 17.2.3). Proteoglycans consist of a main backbone of polysaccharide with a polypeptide component attached (Section 17.2.3). They are thought to be involved in various recognition phenomena such as the specificity of pollination and infection of plants by pathogenic and symbiotic organisms.

2.6 Nucleotides and nucleic acids

The nucleotides are indispensable to intermediary metabolism and the energy relations of cells. However, most of the nucleotides present in plants occur as the two macromolecules, deoxyribonucleic acid (DNA) and ribonucleic acid (RNA). DNA is a repository for genetic information while RNA is largely concerned with expressing the genetic information contained in DNA. The sequence of the component nucleotides determines the information content of strands of DNA and RNA in much the same way that the amino acid sequence determines the properties and biological function of a protein.

2.6.1 Nucleotides are purines or pyrimidines linked to ribose-5-P or deoxyribose-5-P

Nucleotides (or nucleoside phosphates) are composite molecules consisting of a pentose sugar, a base containing nitrogen and orthophosphate.

Five compounds account for most of the nitrogen bases associated with the nucleotides and nucleic acids of cells. Two of these have a purine ring structure while the other three are derivatives of pyrimidine (Fig. 2.17). The two purines adenine and guanine occur in both DNA and RNA. Adenine contains a single amino substituent at C-6, whereas guanine has an amino substituent at C-2 and a keto group at C-6. The three most important pyrimidines in plants are cytosine, thymine and uracil. Small amounts of other purines and pyrimidines can be found, some in the nucleic acids (e.g. 5-methylcytosine in DNA and 7-methylguanine in transfer-RNA) and others as intermediates in the synthesis or degradation of the common purines and pyrimidines (see Chapter 15).

The sugars associated with the nucleotides and nucleic acids are either the furanose form of β-D-ribose (β-D-ribofuranose) or its deoxy derivative, β-D-deoxyribose (2-deoxy-β-D-ribofuranose; see Fig. 2.6). The carbon atoms of the pentose are designated with a prime (e.g. C-5′) to distinguish the numbering of the pentose ring from the purine (or pyrimidine) component. The hemiacetal hydroxy group at C-1′ of ribose/deoxyribose is joined to N-9 of the purine rings of adenine or guanine or N-1 of the pyrimidine ring of cytosine, thymine or uracil to form nucleosides. Ribonucleosides and deoxyribonucleosides contain ribose and deoxyribose respectively (see Table 2.6 for nomenclature).

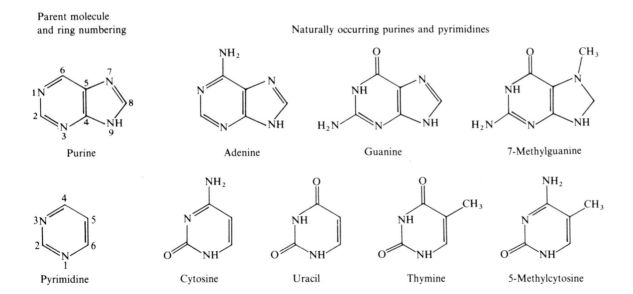

Fig. 2.17 The structure and numbering of the carbon and nitrogen atoms of purine and pyrimidine and the structures of some purine and pyrimidine derivatives. 7-Methylguanine and 5-methylcytosine are less common bases found in plant nucleic acids.

Table 2.6 Nomenclature of common nucleosides and nucleotides.

Base	Nucleoside	Nucleotide	Symbol*
Adenine	Adenosine	Adenosine 5′-monophosphate	AMP
		Adenosine 5′-diphosphate	ADP
		Adenosine 5′-triphosphate	ATP
		Adenosine 3′-monophosphate	3′-AMP
	Deoxyadenosine	Deoxyadenosine 5′-monophosphate	dAMP
		Deoxyadenosine 3′-monophosphate	3′-dAMP
Guanine	Guanosine	Guanosine 5′-monophosphate	GMP
Cytosine	Cytidine	Cytidine 5′-monophosphate	CMP
Uracil	Uridine	Uridine 5′-monophosphate	UMP
Thymine	Deoxythymidine	Deoxythymidine 5′-monophosphate	dTMP

*For nucleotides represented by symbols the phosphate group is assumed to be attached at the 5′-position of the pentose, unless specified otherwise.

One or more of the 2′-, 3′- or 5′-hydroxy groups of ribonucleosides and of the 3′- and 5′-hydroxy groups of the deoxyribonucleosides can be esterified with phosphate to form nucleotides. Nucleotides are named after the parent nucleosides (Table 2.6). The best known examples, discussed in Section 4.3, are adenosine 5′-monophosphate (AMP or adenylic acid) and its derivatives adenosine 5′-diphosphate (ADP) and adenosine 5′-triphosphate (ATP). Compounds formed by further esterification through the phosphate group of AMP are known as adenylates (e.g. aminoacyl adenylates). The adenine nucleotides are especially important in cellular metabolism due to their role in energy conservation and the use of the energy associated with the exergonic hydrolysis of ATP (see Chapter 4) in biosynthetic processes.

The phosphate esters of guanosine, cytidine, uridine and thymidine fulfil important roles in cellular metabolism in much the same way as the adenosine phosphates, although they are quantitatively less important. Nucleotides are also found phosphorylated at the 3′-position (and to a lesser extent the 2′-position). As discussed below, the nucleotides of the nucleic acids are linked via the 3′- and 5′-positions.

2.6.2 RNA and DNA are polynucleotides

Nucleic acids are polymers of 5′-mononucleotides in which the vacant 3′-hydroxy group of one nucleotide is esterified to the phosphate group at the 5′-position of

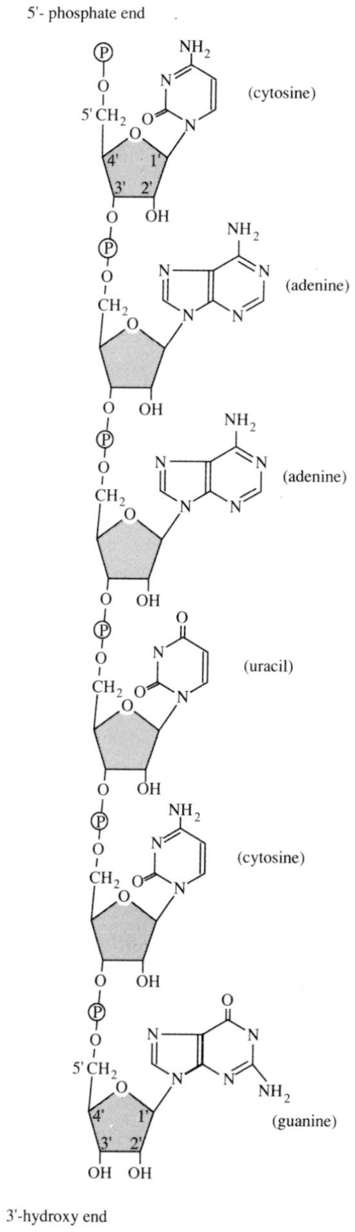

Fig. 2.18 Hypothetical strand of RNA showing the backbone consisting of ribose residues esterified together via phosphate groups. The purine and pyrimidine bases attached at C-1′ of each ribose residue and the sequence in which they occur give each RNA molecule its specific characteristics and function.

an adjoining nucleotide to form a phosphate diester bond (Fig. 2.18). This arrangement gives the nucleic acids a polymeric pentose-phosphate backbone in which the purines and pyrimidines are attached at the 1'-positions of the pentose-phosphate residues. Thus, the ribose residue at one end of a polynucleotide chain bears a free (unesterified) phosphate group at the 5'-position while the other end terminates with an unesterified 3'-hydroxy group (circular DNA excepted). Nucleic acids contain from as few as 60 nucleotides (e.g. transfer RNA) up to 10^9 nucleotides linked in this way. Since the polymeric pentose-phosphate chain is unchanged along the entire length of the polymer, the distinctive function and composition of each nucleic acid must be determined by the sequence of the bases.

2.6.3 DNA contains thymine and is double-stranded whereas RNA contains uracil and is single-stranded

Although DNA and RNA are primarily distinguished by the composition of their respective pentoses they also differ in other ways. Both DNA and RNA contain the purines, adenine and guanine, and the pyrimidine, cytosine, but DNA contains thymine and lacks uracil whereas RNA contains uracil and lacks thymine. Further, DNA contains the purine, adenine and the pyrimidine, thymine in equimolar quantities. Guanine and cytosine also occur in equimolar quantities. The double-stranded structure shown in Fig. 2.19 explains why—the purine and pyrimidine bases of one DNA-strand are linked through hydrogen bonds to specific complementary bases in the other. Hydrogen bonds form between complementary adenine and thymine residues and guanine and cytosine residues on opposite strands (Fig. 2.20). Thus, one DNA-strand is complementary to the other. This bonding causes the well-known secondary structure of a double helix of DNA in which the base pairs are directed inwards and the two strands adopt an 'anti-parallel' arrangement with the 5' end of one strand aligned with the 3' end of the other. Each turn of the helix involves 10 base pairs with a repeating distance of 3.4 nm. Dividing cells are able to synthesize exact replicas of double-stranded DNA (Chapter 18). By contrast, RNA is not double-stranded and does not necessarily contain equimolar amounts of purines and pyrimidines.

The principle of hydrogen bonding between complementary bases in each of the two strands of DNA can be exploited for analytical purposes. For example, a short synthetic segment of DNA (deoxyoligonucleotide) containing some form of 'tagging' (e.g. radioactive isotope) can bind to an appropriate complementary region in a much larger DNA molecule and so can be used to identify particular regions

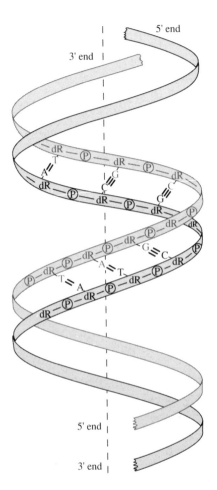

Fig. 2.19 Double helix of DNA showing two strands maintained in helical conformation by hydrogen bonds between the base pairs A–T and G–C where A, T, G and C denote adenine, thymine, guanine and cytosine respectively. Details of the hydrogen bonds involved are given in Fig. 2.20. The deoxyribose residues and phosphate groups are denoted by dR and Ⓟ respectively.

or sequences on the large DNA molecule. A similar technique can be used for RNA even though RNA does not naturally occur in double-stranded form.

In eukaryotic cells about 98% of the DNA is located in the nucleus; the remainder occurs in mitochondria and chloroplasts. The DNA from chloroplasts has a different nucleotide composition, density, and molecular weight to nuclear DNA and is known as satellite DNA. Most of the RNA in cells is associated with ribosomes present in the cytosol, chloroplasts and mitochondria.

The RNA of cells is of three principal types. About 5–10% is messenger RNA

Adenine   Thymine

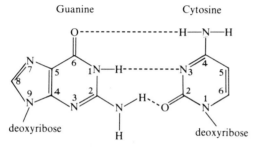

deoxyribose   deoxyribose

Guanine   Cytosine

deoxyribose   deoxyribose

Fig. 2.20 Hydrogen bonding between the base pairs adenine–thymine and guanine–cytosine in double-stranded DNA. Hydrogen bonds are shown as dashed lines.

(mRNA) which is homologous with sections of DNA (i.e. it has a base sequence complementary to a segment of a DNA strand). It differs from a single strand of complementary DNA in that ribose replaces deoxyribose and uracil replaces thymine. Messenger RNA acts as the intermediary in translating genetic information, encoded in DNA, into specific protein(s) (see Chapter 19).

Most of the RNA (60–70%) is ribosomal RNA (rRNA). It occurs in two principal forms, one of molecular weight about 1.3 million which is associated with a number of proteins in the large ribosomal subunit, and one of molecular weight about 0.7 million associated with the proteins in the small subunit (see Section 3.3). Like all RNA molecules they are single-stranded but internal cross-bonding causes some helical coiling.

The third main category is transfer RNA (tRNA), about 15% of the total RNA. Cells contain a population of different transfer RNA molecules, at least one for each amino acid. Each type of transfer RNA molecule binds a specific amino acid.

They provide the last link in translating the encoded genetic information into protein (see Section 19.3). Transfer RNA molecules are quite small, about 60–80 nucleotides, and often contain several other purines and pyrimidines (e.g. 1-methylguanine, 1-methylhypoxanthine) in addition to the four common bases. Many also contain pseudouridine which consists of uracil linked from N-1 in the pyrimidine ring to C-1′ of ribose in the usual way, but also from C-5 in the pyrimidine ring to C-1′ of a second molecule of ribose. This second ribose is not involved in the formation of the polynucleotide chain. The structure of a typical transfer RNA molecule is shown in Fig. 2.21. Although single-stranded, complementary base-pairing occurs in four separate regions and confers a clover-leaf structure on the molecules. The central unpaired region (centre 'loop') contains a sequence of three nucleotides which constitute an 'anticodon' and is recognized by

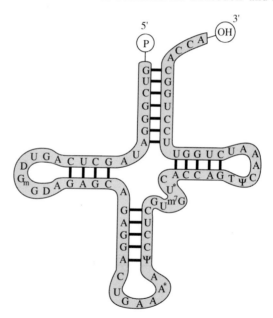

Fig. 2.21 Structure and ribonucleotide sequence of the phenylalanyl transfer RNA from the chloroplasts of *Phaseolus vulgaris*. The molecule comprises several nucleosides in addition to the four which are common to RNA. Abbreviations: A = adenosine; A* = 2-methyl thio-N^6-isopentenyl adenosine; C = cytidine; D = 5,6-dihydrouridine; G = guanosine; G_m = 2′-methylribosylguanosine; m^7G = 7-methylguanosine; T = ribothymidine; U = uridine; U* = an unknown derivative of uridine; ψ = pseudouridine. From Guillemaut, P. & Keith, G. (1977) *FEBS Lett.*, **84**, 351–6.

a complementary series of three nucleotides ('codon') on messenger RNA. This recognition is of crucial importance in ensuring the synthesis of proteins with a specific amino acid sequence, determined ultimately by the genetic information in the DNA.

2.7 Some phenylpropanoid compounds are important constituents of plant cell walls

Plants contain a vast array of phenylpropanoid compounds with an aromatic ring and a C_3 side-chain. They are mostly derived from the aromatic protein amino acids, phenylalanine and tyrosine.

Although most phenylpropanoid compounds and their derivatives are secondary products, the polymer lignin has an especially wide distribution among pteridophytes (ferns), gymnosperms and angiosperms. In many species it largely determines the strength and rigidity of the walls of particular cells. Lignin is especially abundant in conducting cells such as xylem vessels and in structural cells such as fibres with extensive secondary thickening, and often constitutes 20–30% by weight of these lignified walls (Chapter 17).

When lignin is non-oxidatively degraded it yields various *para*-hydroxy phenyl-

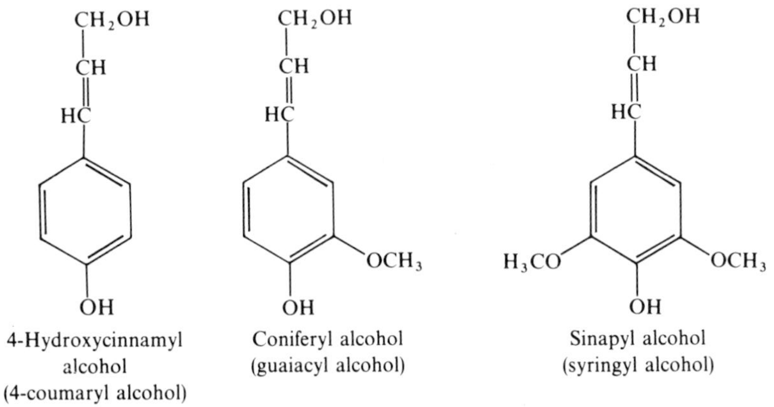

Fig. 2.22 The principal phenylpropanoid compounds obtained by non-oxidative degradation of lignin from various plants. The lignin of gymnosperms yields almost entirely coniferyl alcohol whereas angiosperms (monocotyledons and dicotyledons) yield both coniferyl alcohol and sinapyl alcohol. The lignin of monocotyledons also contains small amounts of 4-coumaryl alcohol.

propane derivatives, the composition varying between species. Some of the more important degradation products are shown in Fig. 2.22. An example of how these units are thought to be linked together in beechwood lignin is given in Section 17.2.3.

2.8 Secondary compounds are not essential for plant function

The compounds referred to so far are regarded as essential since they are indispensible to cell function. Various other molecules, including intermediates in the synthesis, degradation and regulation of vital cellular constituents, are also essential. However, most plants also contain secondary compounds, not known to be essential. More than 20 000 of these are known and some attain very high concentrations, although any one compound often occurs in just a few species, genera or families. Closely related species often contain several secondary compounds structurally related but differing between species. Specific roles have been reported for a few secondary compounds in some plants and although none of these can be regarded as essential in the metabolic sense, they are nonetheless important in ensuring biological success. For example, the flavonoid group of

Table 2.7 Examples of secondary compounds in plants.

Class	Examples
Alkaloids (heterocyclic compounds containing nitrogen)	Morphine alkaloids (e.g. morphine, codeine) Ergoline alkaloids (e.g. lysergic acid) Tropane alkaloids (e.g. hyoscyamine, cocaine) Pyridine alkaloids (e.g. nicotine) Quinoline akaloids (e.g. quinine)
Non-protein acids	Azetidine-2-carboxylic acid
Phenols	Simple phenolics (e.g. arbutin) Phenylpropanoids (e.g. coumarins) Flavonoids (e.g. anthocyanins)
Terpenoids	Monoterpenes (e.g. menthol, camphor) Triterpenes (e.g. squalene) Tetraterpenes (e.g. non-essential carotenoid pigments associated with various fruits Polyterpenes (e.g. rubber)

compounds are the principal determinants of colour in flowers, which is important for insect pollination. Also, certain toxic substances (e.g. cyanogenic glycosides) provide protection against predators, and phytoalexins, which inhibit fungal growth, are produced in response to infection by some fungal pathogens. However, it is not known why mechanisms exist for the synthesis of most of the vast array of secondary compounds. Perhaps they are vestiges of ancestral (essential?) biosynthetic pathways, which are now largely redundant.

Table 2.7 lists just a few representative examples of the 20 000 secondary compounds in plants, some of which are of considerable commercial importance. They include various pharmacologically active compounds (e.g. quinine, one of the quinoline alkaloids from *Cinchona* species, and the polyterpenes (rubber) from the latex of *Hevea*).

Further reading

Carbohydrates: Loewus & Tanner (1982); Preiss (1980); Preiss (1988); Tanner & Loewus (1981).
Lipids: Stumpf (1980); Stumpf (1987).
Amino acids and proteins: Boulter & Parthier (1982); Chothia (1984); Eisenberg (1984); Marcus (1981).
Nucleic acids: Marcus (1981); Parthier & Boulter (1982).
Enzyme classification: International Union of Biochemistry (1979).
Secondary plant products: Bell & Charlwood (1980); Conn (1981).

3
Structural organization of plant cells

3.1 Cellular organization is essential for cellular activity

The various chemical processes associated with cellular activity take place within specific parts of cells. Some of these processes are mutually antagonistic; the mesophyll cells of plant leaves, for example, produce carbohydrate from CO_2 but they also oxidize carbohydrate to CO_2. Net production would be impossible unless the two processes are physically separated in cells and independently regulated. This illustrates the importance of cell structure in understanding metabolic activity. The functional integrity of many vital cellular processes relies on their location in specific membranes or within compartments surrounded by particular membranes.

Plant cells exhibit considerable diversity in their structure and function. Some are biochemically inactive (i.e. dead) although they fulfil important physiological roles (e.g. dead xylem cells in water transport). Other specialized cells like the sieve tubes of the phloem lack a nucleus and do not exhibit the full complement of membranes dividing the cell into internal compartments. These, therefore, cannot be regarded as biochemically fully active. Parenchyma cells are the most common type of fully active cell, often about 80% of the population, and they exhibit considerable variation, some photosynthetic, others non-photosynthetic. They include mesophyll and bundle sheath cells of leaves, cortical cells of roots and stems and transfer cells. Parenchyma cells consist of a thin primary cell wall and a protoplast, delineated by the plasma membrane (or plasmalemma). The living material of the protoplast, including the plasma membrane, is known as protoplasm. Plasmodesmata are occasional protoplasmic strands of very small cross-section joining adjacent cells and which provide protoplasmic continuity (symplasm). They are important in the intercellular transport of metabolites.

The protoplasm contains bodies bounded by membranes, known as organelles (Fig. 3.1), as well as membrane structures which do not enclose a body. It also contains structures without an enclosing membrane (particulate structures). There is some confusion about naming the aqueous milieu surrounding the organelles and particulate structures due to recent changes to the definition of cytoplasm and the tendency by some biologists to use 'cytoplasm' and cytosol synonymously. Former-ly, cytoplasm was defined as the contents of the protoplast with the exception of the nucleus but nowadays most biologists do not include organelles in their definition (Table 3.1). In this book cytoplasm includes the various membrane vesicles, strands, filaments, particles and other structures (cytoplasmic inclusions) as well as the aqueous phase of the cell internal to the plasma membrane and external to the membrane-bound organelles. The aqueous phase of the cytoplasm, devoid of all particulate material, is defined as cytosol.

3.2 Membrane structure

Membranes are essential components of cells. The plasma membrane, for example, delineates a cell from its environment. It permits controlled entry of certain essential substances and prevents uncontrolled loss of intracellular metabolites. Organelles are also delineated by membranes permeable to some metabolites but essentially impermeable to others which permit controlled movement of specific metabolites between various subcellular compartments. Other membranes are involved in equally vital processes. The internal membrane complex of chloroplasts, for example, plays key structural and functional roles in light-dependent electron transport and photophosphorylation. Most cell membranes are about 8 nm in cross-section as determined by electron microscopy, although the tonoplast surrounding the vacuole is somewhat less (about 5–6 nm) and the plasma membrane is somewhat greater (about 9–10 nm).

Membranes are lysed by lipid solvents and various lipases, especially phospholipases, implying that lipids, especially phospholipids, are major constituents of membranes. Proteins are also major components. Membranes commonly comprise about 50% lipid and 50% protein although energy-transducing membranes have a higher proportion of protein (e.g. about 75–80% in mitochondria).

The general non-specific permeability characteristics of membranes are largely determined by the membrane lipids. Thus, the permeability of membranes to non-electrolytes correlates with the affinity of the non-electrolytes for lipid. The proteins associated with membranes account for the specific permeability properties

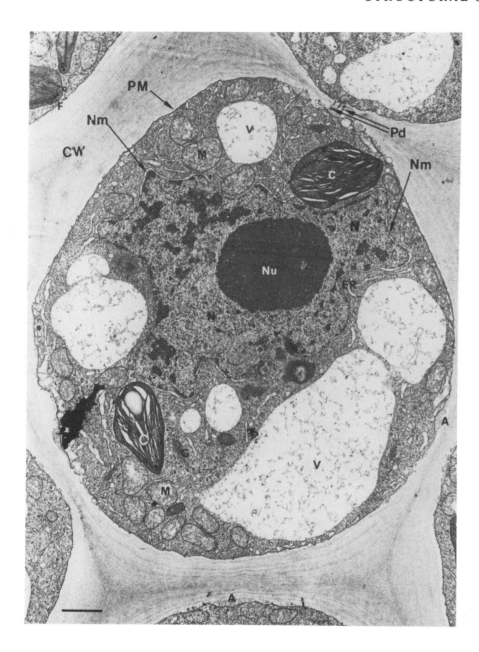

Table 3.1 Definitions of some terms used to describe plant cell components.

Term	Definition
Protoplast	Contents of a cell excluding the cell wall
Protoplasm	Living material within the protoplast of a cell
Organelles	Structures surrounded by a membrane (or pair of membranes), sometimes exhibiting a high degree of internal organization
Cytoplasm	Protoplasm lacking organelles
Cytoplasmic inclusions	Structures within the cytoplasm (e.g. vesicles, microtubules, microfilaments and ribosomes) which do not constitute a body surrounded by a membrane
Cytosol	Cytoplasm lacking cytoplasmic inclusions

towards water-soluble compounds including many electrolytes with little, if any, affinity for lipid. Specific transport proteins (different for different membranes) recognize particular compounds and facilitate their transport across the non-polar membrane.

The currently accepted model of membrane structure is the fluid mosaic model (Fig. 3.2). It proposes that membranes consist of proteins embedded in a lipid bilayer and that the proteins can migrate within the plane of the bilayer. Membranes differ in the composition of their component lipids (see Section 6.1.2) and their protein complement. Some amphipathic membrane proteins, known as integral or intrinsic proteins, contain several separate hydrophilic and hydrophobic domains resulting from long sequences of amino acids with predominantly non-polar side chains interspersed between extended sequences of amino acids with

Fig. 3.1 Electron micrograph of a collenchyma cell of hop (*Humulus lupulus*). Although the micrograph shows many of the features commonly observed in plant cells, most plant cells do not exhibit all of these features. In particular, the vacuoles of most mature cells commonly occupy about 90% of the cell volume and the nucleus is usually less prominent. Also, most photosynthetically active cells contain more chloroplasts and the cell wall is normally less well-developed in relation to the nucleus and protoplasm. The scale of the micrograph is given by the bar which represents 1 µm. Abbreviations: C = chloroplast; CW = cell wall; ER = endoplasmic reticulum; G = Golgi body; M = mitochondrion; N = nucleus; Nm = nuclear membrane; Nu = nucleolus; Pd = plasmodesmata; PM = plasma membrane; V = vacuole. From Chafe S.C. (1970) Ph.D. Thesis, La Trobe University.

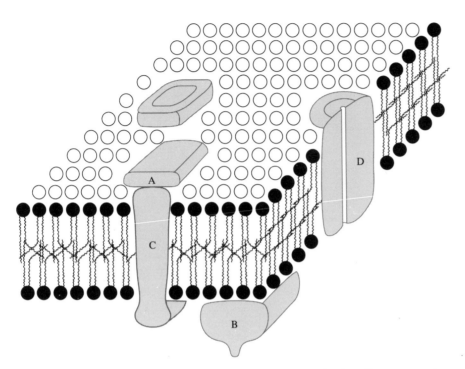

Fig. 3.2 Fluid mosaic model of membrane structure. The main sheet-like structure is a phospholipid bilayer; the circular heads (shown in black) denote polar regions of individual phospholipid molecules and the tails denote lipophilic regions. The grey structures denote proteins. Proteins A and B are extrinsic proteins (non-amphipathic) whilst C and D are intrinsic proteins (amphipathic). A is attached to the polar region of another protein (C) and B is attached to the polar heads of molecules in the phospholipid bilayer.

predominantly polar side chains. The protein is folded *in vivo* in such a way that the hydrophobic domains form strong associations with the hydrophobic 'tails' of the lipid bilayer while the hydrophilic domains associate with the hydrophilic 'heads' and the aqueous medium surrounding the membrane. These principles are illustrated in Fig. 3.3 by the light-harvesting chlorophyll-binding protein, which spans the inner membrane complex of chloroplasts.

Some membrane proteins are non-amphipathic, i.e. they exhibit predominantly hydrophilic properties at their periphery. Known as extrinsic or peripheral proteins, they are linked to the exterior hydrophilic regions and are thought to associate with the polar heads of the phospholipid bilayer or the hydrophilic regions of intrinsic proteins (Fig. 3.2).

Differences in the binding of intrinsic and extrinsic proteins to membranes can be used to extract them. Extrinsic proteins (non-amphipathic) are extracted by salts and chelating agents, which dissociate ionic bonds binding them to hydrophilic regions of phospholipids and to intrinsic proteins embedded in the bilayer but do not affect the binding of intrinsic proteins to the lipid bilayer. Ionic detergents such as sodium dodecyl sulphate extract both classes of membrane protein.

The orientation of specific active sites, binding sites, etc., of individual proteins with respect to each side of the membrane is of particular importance. Immunological techniques can be used to demonstrate that the active sites of a specific membrane protein are located on one side of a membrane. This entails isolating and purifying the protein ('target' protein) which is then injected into an animal. The animal's immune system elicits the production of antibodies which bind to specific sites on the target protein. If the antibody is mixed with a vesicle preparation of the membrane (self-sealing spheres of the original membrane) and it binds specifically to the vesicle then the active site of the target protein is exposed or located on the outer surface. If the active site is exposed only on the inner surface the antibody (being a large molecule) cannot cross the membrane to react with it. By tagging the antibody with a fluorescent dye or a heavy metal opaque to electrons, the reaction site(s), if any, can be identified. To check, inside-out vesicles (see Section 7.7.3) should give a contrasting result.

Freeze-fracture techniques for preparing material for electron microscopy cleave membranes in the centre of the lipid bilayer. The two halves are asymmetric and different patterns protrude above and below the plane of the membrane.

The various proteins within a membrane fulfil a variety of functions. Some are enzymes, others are involved in the transport of metabolites, ions and electrons and some (at least in bacteria—see Section 6.3) are both. Others again are involved in recognition phenomena or the attachment of pigments to membranes (see Section 9.7.1). Certain intrinsic proteins are thought to attach to actin filaments and microtubules which are linked to other cellular structures thereby giving the cell an internal framework (see Section 3.9).

3.3 Ribosomes

Ribosomes are small spheroid particles without a surrounding membrane, found in large numbers in plant cells. Cytoplasmic ribosomes are about 20 nm in diameter

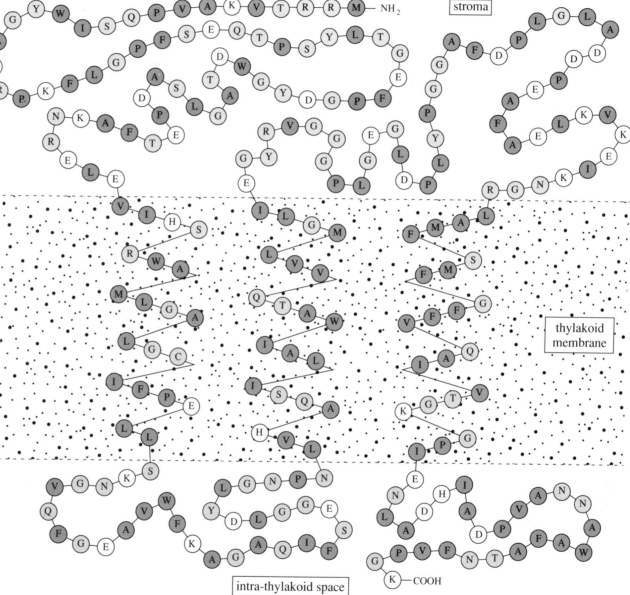

Fig. 3.3 Structure of the apoprotein of the light-harvesting chlorophyll *a/b* binding protein (LHCP) from the thylakoids of *Lemma gibba* chloroplasts. LHCP is an intrinsic protein which exhibits amphipathic properties and spans the entire membrane three times. The upper part of the protein is associated with the chloroplast stroma, the middle region with the hydrophobic regions of the thylakoid membrane and the lower region with the intra-thylakoid space. The amino acid residues shown in dark grey have hydrophobic side-chains and are especially numerous in three helical regions which bind to the hydrophobic regions of the membrane. Note that residues D, E, H and K (shown in white circles), which form charged groups and therefore destabilize helical coiling, are mostly present in regions that protrude into the stroma and intra-thylakoid space and are largely absent from the helices that span the membrane. Individual amino acids are denoted by letters as follows: A = alanine; C = cysteine; D = aspartate; E = glutamate; F = phenylalanine; G = glycine; H = histidine; I = isoleucine; K = lysine; L = leucine; M = methionine; N = asparagine; P = proline; Q = glutamine; R = arginine; S = serine; T = threonine; V = valine; W = tryptophan; Y = tyrosine. From Karlin-Neumann, G.A., Kohorn, B.D., Thornber, J.P. & Tobin, E.M. (1985) *J. Mol. Appl. Genet.*, **3**, 45–61.

and are known as 80 S ribosomes because of their rate of sedimentation in an analytical ultracentrifuge (see Section 3.12.2 for details and a definition of S, the Svedberg unit). Chloroplastic ribosomes are typically about 15 nm in diameter, and have a sedimentation constant of about 70 S like those from prokaryotic organisms. Mitochondrial ribosomes are of intermediate size (77–78 S; see Section 19.5).

In the presence of Mg^{2+}, both 80 S and 70 S ribosomes are stable nucleoprotein particles, about 50–60% protein and 40–50% ribosomal RNA (rRNA). Ribosomes dissociate into large and small subunits in the absence of Mg^{2+} but readily reassociate in 1 mM Mg^{2+}. Some data on the composition of the subunits of 80 S and 70 S ribosomes are given in Table 3.2. (Note that the sedimentation constant is not directly proportional to mass.)

Table 3.2 RNA and protein components of ribosome subunits from cytoplasm, chloroplasts and mitochondria of plants. Ribosomes, subunits and component RNAs are designated by their sedimentation coefficients.

Ribosome type	Subunits	Component RNAs	Number of component proteins
Cytoplasm 80 S	60 S	25 S, 5.8 S, 5 S	45–50
	40 S	18 S	~30
Chloroplasts 70 S	50 S	23 S, 5 S	~55
	30 S	16 S	
Mitochondria 77–78 S	50 S	26 S, 5 S	~35*
	32 S	18 S	~20*

*Data for mammalian systems.

The proteins of ribosomes are extracted by treatment with ribonuclease and dissolving the proteins in 8 M urea. When the proteins and rRNA from a ribosomal subunit are mixed the subunit self-assembles showing that the structure of the subunit, and ultimately the ribosome itself, is determined by binding between RNA and the various component proteins.

Individual ribosomes form complexes with other structures and are often attached to the cytoplasmic side of the endoplasmic reticulum (rough endoplasmic reticulum). When cells are homogenized the endoplasmic reticulum is disrupted and parts of it remain attached to the ribosomes (microsomes). The membrane fragments can be removed by treatment with the bile salt, deoxycholate, or compounds with detergent properties.

Ribosomes also form polysomes which appear in electron micrographs as short spirals. Polysomes consist of several ribosomes synthesizing protein as they travel along a strand of messenger RNA through a groove between the small and large subunits of the ribosomes. Ribonuclease abolishes polysomes and so increases the number of single ribosomes.

3.4 Nucleus

The cell nucleus is a large organelle of variable size and shape. In non-dividing cells it is commonly about 10 μm in diameter and surrounded by a double membrane, the nuclear envelope, but in dividing cells it varies in size, shape and appearance and lacks an intact membrane (see Section 3.11). The nucleus is the main centre for the control of gene expression and this is reflected in its chemical composition; RNA and DNA collectively comprise about 25% of the mass. Protein accounts for about 75% of the mass with considerable variation between species and tissues. Nuclear proteins are of two main types, acidic and basic. The basic proteins (histones) are especially numerous, although quantitatively minor, and bind to specific base sequences on template DNA coiling into compact nucleosomes (see Section 18.1.5). Histones regulate DNA transcription since dissociation of one or more from the nucleoprotein complex precedes unwinding of the DNA for transcription into messenger RNA (see Section 18.2.3). The acidic proteins are largely associated with the nucleoplasm.

Chromatin, a complex of DNA and protein, stains with various basic and DNA-specific dyes. It occurs in certain regions of nuclei and in dividing cells it thickens and is distinguishable as chromosomes. Material which does not stain in this way is called nucleoplasm. Nucleoli, small bodies within the nuclei, stain more heavily with heavy metals than the rest of the nucleus (Fig. 3.1). They are mostly protein with about 5% RNA. The large and small ribosomal subunits are assembled here using ribosomal RNA and ribosomal proteins imported from the cytoplasm (see Section 19.2.4).

The two membranes of the nuclear envelope are separated by the perinuclear space. These membranes are continuous around a series of pores about 50–100 nm in diameter which contain a central granule and a series of globular structures around an annulus. In animal cells the pores behave as though they contain an aqueous channel of about 9 nm diameter, much smaller than the diameter of the pores themselves. The pores are sites of macromolecular transport between nucleus and cytoplasm: messenger RNA from nucleus to the cytoplasm; ribosomal proteins from the cytoplasm to the nucleus; and assembled ribosomal subunits back to the cytoplasm.

The nuclear membrane disappears and reappears during mitosis (see Section 3.11) and is continuous with the endoplasmic reticulum (see Section 3.9).

3.5 Chloroplasts

Chloroplasts are typically lentiform with a diameter of about 5–10 μm and about 2–3 μm thick. They normally occur around the perimeter of photosynthetic cells and adjacent to intercellular spaces, with their discoid faces parallel to the cell walls to maximize the absorption of light. Mesophyll cells of higher plants usually contain 40–50 chloroplasts per cell although higher numbers are not uncommon (e.g. 200–300 in mature spinach leaves). Chloroplasts are bounded by a double membrane (envelope) and are highly organized internally by an extensive internal membrane system containing the pigment chlorophyll, surrounded by the soluble phase, the stroma (Fig. 3.4a) which includes 70 S ribosomes, starch grains, DNA strands and other minor components. Only isolated chloroplasts with intact envelopes support light-dependent assimilation of CO_2.

The most extensive components of the internal membrane complex are the 'double membranes' which ramify through the stroma (Fig. 3.4a) and are known as stromal thylakoids or stromal lamellae. This 'double membrane' is a single membrane folded back on itself, enclosing the intra-thylakoid space. The face bounding the intra-thylakoid space is the inner side, and that in contact with the stroma is the outer side.

The other main internal structure is a series of parallel-sided sacs (granal sacs). Each sac consists of a membrane folded back on itself (granal thylakoid) to form a flat structure about 200–300 nm across. The lumen of granal sacs, like that of stromal thylakoids, is known as the intra-thylakoid space. Granal sacs form stacks called grana which are interposed between stromal thylakoids (Figs 3.4b & 3.5), like stacks of coins sandwiched between parallel sheets of paper. Tight stacking is evident from heavier staining along the common margins of adjacent sacs (i.e. two thicknesses of membrane) compared with stromal thylakoids (Fig. 3.4b). Chlorophyll is located in both the granal and stromal thylakoids making the grana, which typically contain 20–50 granal sacs, appear intensely green.

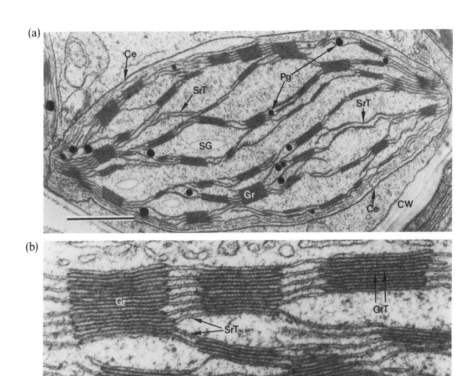

Fig. 3.4 Electron micrographs of a chloroplast from a mesophyll cell of *Digitaria sanguinalis*. (a) Entire chloroplast (bar denotes 1 μm). Note the proximity of the chloroplast to the cell wall. (b) Details of the internal membrane complex associated with several grana (bar denotes 0.25 μm) Abbreviations: Ce = chloroplast envelope; CW = cell wall; Gr = granum (of chloroplast); GrT = granal thylakoid; Pg = plastoglobuli; SG = starch grain; SrT = stromal thylakoid. Micrograph supplied by Botany Department, La Trobe University.

When isolated chloroplasts are transferred from iso-osmotic medium to a medium of higher water potential they take up water and burst ('osmotic shock'). The thylakoids can be separated from stroma by centrifugation. When these are suspended in distilled water they tend to reseal into vesicles (sub-chloroplast particles), each effectively with its own intra-thylakoid space.

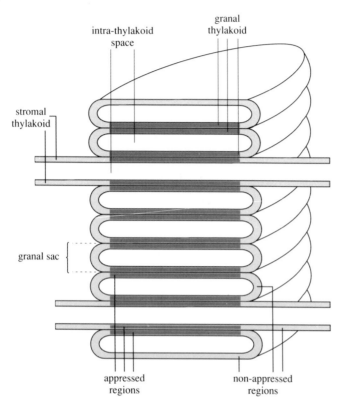

Fig. 3.5 Representation of the stromal and thylakoid membranes of the inner membrane complex of chloroplasts and their arrangement to form a granum. Appressed regions show tightly packed surfaces in grana; non-appressed regions occur in the stroma.

The closely adjoined surfaces between granal sacs are described as appressed regions; those parts bounded on one side by stroma are non-appressed (Fig. 3.5). Some structural details of thylakoids can be observed by freeze-fracturing the plane of the membrane and examining the two faces by electron microscopy. The outer appressed fracture face contains about 5000 particles μm^{-2} with diameters of about 8 nm. The inner appressed fracture face contains about 1500 particles μm^{-2} with diameters of about 14–16 nm. The two sets of particles interdigitate in the ratio 4:1. For inner non-appressed regions the fracture face has fewer particles (600 μm^{-2}) of smaller diameter than the inner appressed regions. Also, the stromal surface of these regions bears protruding particles ('knobs') about 10–12 nm in diameter

consisting of the protein complex CF_1 involved in the phosphorylation of ADP (see Chapter 9 for details). Analysis of the protein complement from appressed and non-appressed regions, which includes several protein–chlorophyll complexes and various proteins involved in light-driven electron transport, provides some details of the molecular organization. These data, together with a model of the molecular structure, are discussed in Section 9.9.

Not all photosynthetic cells have chloroplasts with the general features described above. In the NADP M-E variants of C_4 plants (see Section 12.7) the chloroplasts in bundle sheath cells lack well-formed grana and exhibit several other distinctive features (see Section 12.3).

3.6 Mitochondria

Mitochondria vary in shape from spherical (about 1 μm in diameter) to cylindrical (about 2–3 μm long and 0.5–1 μm in diameter). They aerobically oxidize pyruvate to CO_2 with the concomitant phosphorylation of ADP to ATP in processes involving the tricarboxylic acid cycle (TCA cycle) and oxidative phosphorylation. Mitochondria are bounded by a double membrane, the outer one forming a simple enclosing perimeter and the inner one having numerous inwardly directed invaginations—the cristae. The soluble phase enclosed by the inner membrane (matrix) contains soluble proteins and various inclusions. The inner and outer membranes are separated by the perimitochondrial space and the intra-cristal space separates the infolds of the cristae.

Both the inner and outer membranes are essential for oxidative phosphorylation which depends on a proton gradient between the intra-cristal space and the matrix. The various proteins involved in electron transport and oxidative phosphorylation are located in the inner membrane (see Section 7.7), while the matrix contains various enzymes (e.g. those for the TCA cycle), mitochondrial ribosomes, DNA filaments and other particulate material. Negatively stained mitochondrial membranes reveal stalked particles about 10 nm in diameter protruding from the matrix side of the inner membrane; these represent the F_1 complexes responsible for the phosphorylation of ADP.

3.7 Microbodies

Microbodies are organelles about 1 μm in diameter bounded by a single membrane. They are delicate structures and difficult to isolate. All microbodies contain oxidase(s) which produce H_2O_2 and catalase, which detoxifies H_2O_2. They support various oxidative processes but these are not coupled to the phosphorylation of ADP.

Microbodies differ in their enzyme complement. The most ubiquitous are peroxisomes, found in photosynthetic cells, usually in close association with chloroplasts. They sometimes contain crystalline inclusions but generally show little internal structure. Leaf peroxisomes metabolize glycollate (via glycollate oxidase) to glycine with the consumption of O_2 and the production and detoxification of H_2O_2. They also metabolize serine to glycerate via glycerate dehydrogenase (also known as hydroxypyruvate reductase). These are component reactions of photorespiration which occurs in the leaves of C_3 plants in the light (see Chapter 11).

Glyoxysomes are microbodies which occur in close association with lipid deposits (spherosomes) in the storage tissues of germinating seeds rich in lipid (e.g. castor bean endosperm). They appear during germination, coinciding with the period of maximum lipid consumption. Glyoxysomes contain an enzyme complement which supports the synthesis of succinate from long-chain fatty acids (see Sections 8.2 & 8.3).

3.8 Other membrane-limited organelles

3.8.1 Golgi bodies

Golgi bodies (or dictyosomes) consist of a stack of about five flattened sacs (or discs) known as cisternae, about 1 μm in diameter, each with an enveloping membrane enclosing a lumen. A reticulated mesh of tubules with smooth membranes is attached to the circumference of the cisternae and spreads out to give the whole structure a diameter of 2–3 um (Fig. 3.6). The flattened faces of the two outermost cisternae are referred to as *cis* and *trans* faces, or forming and maturing faces respectively. The *cis* face acquires membranous material from the endoplasmic reticulum while the *trans* face gives rise to spheres bounded by a membrane (Golgi vesicles), which are released into the cytoplasm. The assembly of Golgi bodies and vesicles is known as the Golgi apparatus, which adopts a characteristic appearance in sectioned material (Fig. 3.7). The membranous material of Golgi vesicles is compatible with the plasma membrane and fuses with it, supplying materials to it during cell growth. Golgi bodies are the sites for the synthesis of most of the matrix polysaccharides of cell walls (see Chapter 17) and Golgi vesicles transport these polysaccharides to the plasma membrane where they are released to the cell exterior and incorporated into the wall.

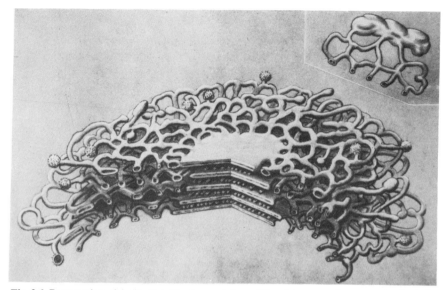

Fig. 3.6 Proposed model of a Golgi body consisting of five cisternae and a network of tubules. Inset shows a Golgi vesicle in the process of forming. From Mollenhauer, H.H. & Morré, D.J. (1966) *Annu. Rev. Plant Physiol.*, **17**, 27–46.

Golgi bodies are found throughout the cytoplasm and do not appear to be arranged in any specific way (Fig. 3.7). Dividing cells and other cells which synthesize non-cellulosic structural polysaccharides at rapid rates have very active Golgi apparatus.

3.8.2 Vacuoles

The vacuoles of plant cells are of variable size and number although there is usually only one in most mature cells. In many cells they represent up to 90% of the cell volume. Vacuoles are surrounded by a single membrane, the tonoplast, and usually have no internal structure. They store salts, various organic metabolites (e.g. sucrose) and certain end-products of metabolism with a slow turnover (e.g. tannins). As a result vacuoles have a considerable solute potential which promotes water uptake, making the cell turgid. Most of the organic compounds within vacuoles are synthesized in other metabolic compartments. Vacuoles are often regarded as being of little interest since they do not possess energy-generating mechanisms or exhibit significant biosynthetic activity. However, they do contain a range of hydrolytic

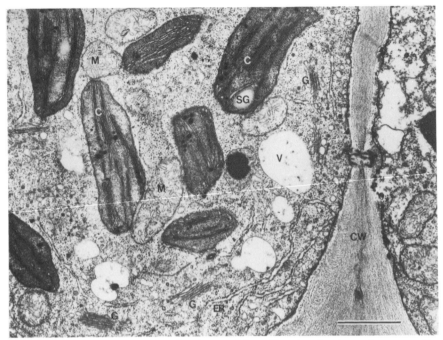

Fig. 3.7 Electron micrograph of a collenchyma cell from *Humulus*, showing various organelles and details of inclusions in the cytoplasm. The bar represents 1 μm. Abbreviations: C = chloroplast; CW = cell wall; ER = endoplasmic reticulum; G = Golgi body; M = mitochondrion; SG = starch grain; V = vacuole. From Chafe, S.C. (1970) Ph.D. Thesis, La Trobe University.

enzymes (e.g. RNAases and proteases) which promote hydrolysis of certain cellular components. Thus, if the tonoplast is ruptured, RNA, protein, etc., in the cytoplasm are attacked by these enzymes and could seriously impair cellular activity. This, together with the release of various metabolites stored in the vacuole into the cytoplasm, is relevant to plant senescence and the extraction (in active form) of enzymes and sub-cellular organelles from plant cells.

3.9 Cytoplasm—its inclusions and cytosol

Cytosol, the aqueous phase of cytoplasm, contains enzymes which support many metabolic processes (e.g. glycolysis and the pentose phosphate pathway). These

enzymes are recovered in the supernatant fraction when cells are disrupted in iso-osmotic media and centrifuged in a strong field.

When material is fixed and stained using procedures which are not known to destroy cell components (e.g. glutaraldehyde fixation and post-fixation staining with osmium tetroxide and uranyl acetate), various particles and membranous structures are observed in the cytoplasm (Fig. 3.7). In addition to ribosomes the most prominent structure is an extensively folded membrane system, the endoplasmic reticulum. This forms one large continuous membrane system and is continuous with evaginations of the nuclear envelope. The 'spaces' within the folds are interconnected to form a single continuous lumen. Some regions on the cytoplasmic surface of the endoplasmic reticulum are studded with ribosomes giving them a 'rough' appearance but other regions lack ribosomes and are 'smooth'.

The endoplasmic reticulum is the principal site for the synthesis of membrane lipids and contains most of the enzymes required (see Chapter 16). Membrane proteins are also synthesized on the ribosomes of the rough endoplasmic reticulum. Some of these proteins are transported into the lumen and some function within the lumen. Others are packaged into vesicles enclosed by fragments of endoplasmic reticulum and transported to other sites, most commonly specific organelles where membrane lipids are added to the membrane of the organelle and protein is added to the lumen. Some proteins remain within the endoplasmic reticulum while others are selectively exported to other membranes (e.g. the *cis* face of Golgi bodies and the plasma membrane).

Microtubules are tubular inclusions within the cytoplasm having variable length (up to 1000 nm) and an external diameter of about 25 nm. They consist of 13 protofilaments, each essentially a polymer of the globular protein, tubulin, itself a dimer comprised of α- and β-tubulin, although several other proteins occur in smaller amounts. Tubulin polymerizes/depolymerizes reversibly. Polymerization is enhanced by Mg^{2+} and GTP and results in some GTP hydrolysis. Polymerization to form arrays of microtubules is important *in vivo* at different stages of the cell cycle as microtubules direct the physical orientation of various components within the cytoplasm; for example, microtubules comprise the spindle fibres involved in the separation of dividing nuclei and they determine the orientation of cellulose microfibrils in cell walls (see Chapter 17).

Microfilaments are aggregations of the polymeric form of the protein, actin, which also reversibly polymerizes and depolymerizes. Small amounts of other proteins are also present. Microfilaments are about 6 nm in section but they usually occur in bundles. With microtubules they provide some form of structural framework to the cytoplasm (the cytoskeleton). Some microfilaments cause organelle movements (cytoplasmic streaming) which is presumed to involve the hydrolysis of ATP by a protein with myosin-like properties.

3.10 Plant cell walls

Plant cell walls are tough, rigid and chemically stable. They give plants and materials derived from them (e.g. timber) their strength and durability.

The structure of the wall can be considered in the sequence of formation. In a new cell the first-formed component is the cell plate consisting of non-cellulosic structural polysaccharides derived from fusion of a series of Golgi vesicles, with the fused membrane material simultaneously forming the plasma membrane. Additional non-cellulosic polysaccharides are added to this plate, external to the plasma membrane to form the middle lamella. This is constructed principally of polysaccharides with a high proportion of galacturonic acid. As the cell grows, more material is secreted and added to the middle lamella. Initially, this is mostly amorphous and non-cellulosic (e.g. pectin) but later, an increasing amount of cellulose is deposited. This primary wall does not necessarily become thicker as material is added. Primary walls, although quite strong, are not rigid due to the high pectin content, the absence of lignin and the loose packing of the cellulose microfibrils. The microfibrils of primary walls tend to be like hoops around the axis of cell elongation at the time of synthesis but distort into a loose net-like mesh as the cell elongates; they generally lack a high degree of geometrical order. If wall growth ceases at the same time as growth of the cell volume then the cell wall consists of a primary wall only. After growth of the cell volume ceases many cells deposit more material into the wall to form the secondary cell wall. The microfibrils of the secondary wall tend to be laid down in distinct layers, the microfibrils in each layer having a characteristic orientation. In many species three distinct layers can be seen. The ultimate strength of the cell results from this network of microfibrils and the deposition of matrix material, such as lignin, locking them together (see Chapter 17).

3.11 The cell cycle involves non-sexual production of two daughter cells

The cell cycle is the sequence of events involved in the non-sexual reproduction of cells. Although all cells have the potential to divide, most divisions are restricted to a few regions—the meristems. The majority of cells produced at these centres follow a specific course of development and become specialized, a matter determined by position, environment and information from other cells.

The cell cycle is divided into a series of cytological events, summarized in Table 3.3, with two main stages—interphase and mitosis. During interphase, the longer of the two stages, the nuclear envelope is readily discernable and various cytological features associated with mitosis are absent. Interphase is divided into three phases, involving growth of the cytoplasm, replication of the organelles and doubling of the DNA content with nuclear growth (Table 3.3). Throughout interphase the chromatin is not obviously confined to any specific part of the nucleus.

Table 3.3 Phases of the cell cycle and their duration in a typical meristematic cell.

Phase	Events (including cytokinesis)	Approximate duration
Interphase		
G_1	Cytoplasmic growth	
S	DNA doubles, cytoplasmic growth	12–30 h
G_2	Nuclear and cytoplasmic growth	
Mitosis		
Prophase	Nuclear membrane disappears; joined pairs of chromatids appear in nucleus	1–2 h
Metaphase	Formation of mitotic spindle; migration of paired chromatids to equatorial plane between the two poles formed by the spindle	5–15 min
Anaphase	Division of paired chromatids and their separate migration to opposite poles	2–10 min
Telophase	Chromatids disappear, nuclear membrane reappears	10–30 min

Mitosis involves the production of two daughter nuclei, each containing half of the DNA complement of the interphase cell. At the onset of mitosis (prophase) the chromatin thickens and chromatin fibrils or chromosomes appear. Each chromosome duplicates to form two identical chromatids, which remain joined together at a specific point, the centromere. The chromatids shorten and thicken and both the nucleolus and the nuclear envelope disappear. This is followed by metaphase, in which the spindle fibres, consisting of bundles of microtubules, radiate from opposite ends ('poles') of the cell. The chromosomes then migrate to the equatorial plane where they attach to one of the spindle fibres. Then follows anaphase during which each chromosome separates to release the two chromatids which give rise to two daughter chromosomes. These migrate to the opposite poles of the cell on the spindle fibres and constitute two new complete sets of genetic information. The final process, telophase, produces two daughter nuclei; it more or less involves prophase in reverse.

Cytokinesis involves division of the cytoplasm and organelles and formation of two independent plasma membranes and a cell plate between the two new cells. These processes begin during telophase and continue during the ensuing interphase period. Another array of microtubules, the phragmoplast, is produced between the two daughter nuclei. Golgi vesicles align across the equatorial plane of the phragmoplast and fuse to form the cell plate and the plasma membranes as discussed in Section 3.10.

Many molecular activities are involved in cell division. They include doubling of the DNA content during interphase, transmission of the genetic information into the two daughter cells, formation of the plasma and nuclear membranes of the daughter cells, formation of the cell plate and the new cell walls and division of the subcellular organelles.

3.12 Preparation of subcellular fractions from plants

The study of the molecular processes conducted by the subcellular structures of plant cells requires their isolation without contamination by unwanted material. This section describes briefly the theory of plant cell extraction and fractionation procedures.

3.12.1 Organelles and particles can be released from plant cells by mechanical and enzymic methods

Subcellular suspensions are prepared by extracting plant material into a buffered iso-osmotic medium. The medium controls pH (plant vacuoles are frequently very acidic) and prevents uptake of water into the organelles which would otherwise burst and release their contents. Special procedures must be adopted to deal with two important problems unique to plants: the tough cell wall; and the generation of phenolic oxidation products which inactivate many enzymes.

The wall surrounding the protoplast has to be torn open to release the organelles and other structures without rupturing them! Shearing open plant cells by harsh mechanical techniques inevitably leads to breakage of at least some of the organelles, resulting in poor yields. There are various ways of avoiding or minimizing this problem. Plant species and tissues with little fibre give higher yields of intact organelles than tough fibrous tissues. Also, very brief extraction times are used; for example, chloroplasts prepared by mechanically blending spinach leaf tissue for 2–3 s are usually 60–75% intact although only about 2% of the mesophyll cells are disrupted. Longer treatment times (e.g. 5–10 s) rupture more cells but the proportion of intact chloroplasts approaches zero. Mechanical techniques are not suitable for the extraction of cytosol since it becomes contaminated with the aqueous contents of ruptured organelles (e.g. chloroplast stroma).

One technique used very successfully to surmount this problem involves producing protoplasts prior to shearing the cell. This entails incubating the tissue for several hours with pectinase(s) and cellulase(s) to remove the cell wall and release the protoplasts into suspension (various wood-rotting fungi are excellent sources of these enzymes). The naked cells are recovered by centrifugation and gently squeezed through a mesh which is slightly smaller than the diameter of the protoplasts (30 μm is frequently used for mesophyll cells). This disrupts the plasma membrane and releases the cell contents with very few organelles ruptured.

The other factor that bedevils the extraction of organelles and some enzymes concerns their inactivation by endogenous phenolic compounds and their oxidation products (e.g. sucrose synthase from sugar cane is completely inactivated). Some species are particularly troublesome (e.g. various members of the family Solanaceae such as tobacco, potato and tomato) but the problem can arise in any tissue containing high levels of phenolic compounds due to mixing of these with an oxidative enzyme (*o*-diphenoloxidase) during extraction. In intact cells, phenolic compounds are confined to vacuoles and are spatially separated from *o*-diphenoloxidase by the tonoplast. *o*-Diphenoloxidase has a low affinity for O_2. When the cell and the tonoplast are disrupted high concentrations of O_2 are admitted into the tissue, providing ideal conditions for the production of phenolic oxidation products which inactivate the organelles and susceptible enzymes. Sometimes the oxidation products cause the extracts and organelles to turn brown (e.g. the cut surfaces of potatoes and apples). The most common procedures to control this are to add to the extracting medium reagents which bind phenolics (e.g. polyvinylpyrrolidone), scavenge O_2 (e.g. metabisulphite and some thiols), react with the phenolic oxidation products (e.g. some thiols) or inhibit *o*-diphenoloxidase (e.g.

diethyldithiocarbamate). Polyvinylpyrrolidone is available in a soluble form for preparation of particulate fractions (e.g. mitochondria) and in an insoluble form (polyvinylpolypyrrolidone) for soluble fractions. Alternatively, O_2 can be excluded by extracting under N_2.

3.12.2 Subcellular components can be fractionated by centrifugation

Extracts containing subcellular particles and organelles can be fractionated by centrifugation in several ways. The rate of sedimentation (V) of a spherical particle of radius r_p is given by the equation

$$V = \frac{2r_p^2(\rho_p - \rho_m)g}{9\eta} \qquad (3.1)$$

where ρ_p is the density of the particle, ρ_m is the density of the iso-osmotic medium, η is the viscosity of the iso-osmotic medium and g is the gravitational force (9.8 m s^{-2}).

If a tube contains particles of three different diameters, all of a similar density, the large particles will travel to the bottom of the tube more rapidly than smaller ones. After the large particles have sedimented the supernatant fraction containing the remaining particles can be poured into another tube, recentrifuged until all the middle sized particles have sedimented, and so on. This procedure, which must employ a medium with a density (ρ_m) less than that of the particles (ρ_p), is termed differential centrifugation. In order to keep centrifugation times for very small particles as short as possible, higher centrifugal fields are used to sediment these (Table 3.4).

An alternative approach (Fig. 3.8) involves carefully filling a tube with a series of solutions of gradually decreasing density to form a density gradient (i.e. ρ_m varies); usually a relatively inert solute (e.g. sorbitol) is used. The cell extract is layered onto the top of the gradient and the tube is subjected to prolonged high-speed centrifugation. The particles move at rates given by Eqn. 3.1 but, as they move down the gradient, ρ_m increases. Particles move until a position is reached where $\rho_p = \rho_m$ and so $V = 0$, i.e. they equilibrate in the gradient according to density, irrespective of diameter (isopycnic centrifugation or density gradient centrifugation). This process is more time-consuming than differential centrifugation, even in very high speed centrifuges, and less material can be processed but it gives better resolution of the particles.

Mitochondria and peroxisomes have a similar diameter (Table 3.4) and cannot

Table 3.4 Common centrifugation conditions for preparation of protoplasts and subcellular organelles.

Protoplast/ organelle	Average largest dimension	Differential centrifugation conditions	Equilibrium density (g cm^{-3})
Protoplasts (from mesophyll cells)	35 µm	$100\,g \times 5$ min	1.24
Nuclei	10 µm	$600\,g \times 5$ min	1.32
Chloroplasts	5 µm	$2500\,g \times 3$ min	1.21
Mitochondria	1.5 µm	$10\,000\,g \times 10$ min	1.18
Peroxisomes	1 µm	$10\,000\,g \times 10$ min	1.25
Ribosomes	20 nm	$150\,000\,g \times 1$ h	

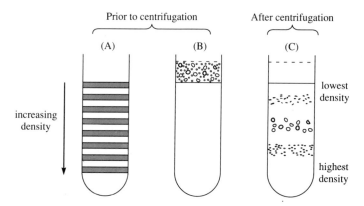

Fig. 3.8 Principles of isopycnic centrifugation. (A) A tube containing solutions of increasing density is prepared. (B) A cellular extract containing particles of different density is layered onto the top. (C) During centrifugation a particle migrates in the gravitational field until it reaches a region with the same density as itself.

be satisfactorily separated by differential centrifugation. Figure 3.8 shows their separation in spinach leaf extracts by isopycnic centrifugation. Peroxisomes, detected by the presence of glycollate oxidase, equilibrate at a density of 1.25 g cm^{-3} while mitochondria, detected by the presence of cytochrome c oxidase and isocitrate dehydrogenase, equilibrate at 1.18 g cm^{-3}. The mitochondrial fraction

48

shows little activity of the enzymes of peroxisomes and vice versa, indicating that little cross-contamination has occurred. The presence of chlorophyll at 1.21 g cm^{-3} implies that this fraction contains either chloroplasts or chloroplast fragments (i.e. thylakoids). Ribulose-1,5-P_2 carboxylase, an enzyme of the chloroplast stroma, is found in fractions with this density showing that this fraction contains intact chloroplasts. The data also show that there is very little activity of mitochondrial and chloroplast enzymes at the top of the tube (low density), where cytosolic enzymes (e.g. nitrate reductase) and soluble enzymes from broken organelles are recovered. This implies that relatively few chloroplasts and mitochondria have been disrupted. Significantly, the analysis shown in Fig. 3.9 used protoplast extracts and not mechanical blending of whole leaf tissue. Nevertheless, a considerable amount of glycollate oxidase and other peroxisome enzymes equilibrated at the top of the tube, implying that a substantial proportion of the peroxisomes have been ruptured, consistent with their fragile nature. Collectively, the data show that the activity of certain enzymes (referred to as marker enzymes) can be used to identify particular subcellular components and quantify cross-contamination by other components. Incidentally, as the data for malate dehydrogenase in Fig. 3.9 show, some enzyme activities are not restricted to specific organelles and cannot be used as markers.

The principles of centrifugation can be used analytically as well as preparatively. Analytical centrifugation is used to characterize small particles and macromolecules, especially RNA. The material for analysis is placed in a cell through which light can be passed during centrifugation and the movement of the particle/macromolecule in the centrifugal field is monitored by refractive index, since migration of the particle produces a sharp boundary. The rate of sedimentation is given by

$$s = \frac{dr}{dt} \cdot \frac{1}{\omega^2 r}$$

Fig. 3.9 Fractionation of the subcellular organelles from extracts of protoplasts from spinach leaf tissue by centrifugation on a sucrose density gradient (see Fig. 3.8 for principle). Following centrifugation, fractions were progressively removed from the tube and analysed for the activity of various enzymes: fractions are numbered from the top of the tube (lowest density) to the bottom (highest density). Shaded regions denote fractions with densities of 1.19 g cm^{-3} (mitochondria), 1.21 g cm^{-3} (chloroplasts) and 1.25 g cm^{-3} (peroxisomes). An interpretation of the data is given in the text. After Nishimura, M., Graham, D. & Akazawa, T. (1976) *Plant Physiol.*, **58**, 309–14.

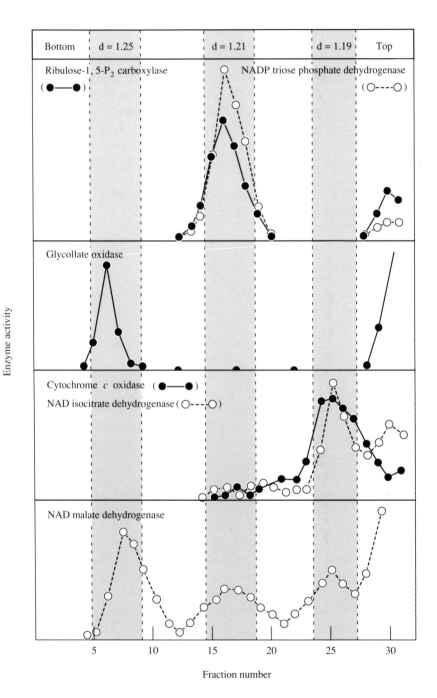

where r is the radial distance between the particle and the axis of rotation, t is time, ω is the angular velocity and, s is the sedimentation coefficient of the molecule. The sedimentation coefficient is the rate per unit centrifugal field. The basic unit adopted for this work is the Svedberg unit $(S) = 10^{-13}$ s. The value of S for a given molecule/particle is particularly influenced by molecular weight and shape. For this reason it is common to cite the property of a particle/molecule in terms of its S value in the analytical ultracentrifuge (e.g. 70 S ribosomes from chloroplasts and 80 S ribosomes from cytoplasm).

Further reading

General texts and treatises on plant cell structure: Alberts *et al.* (1989); Gunning & Steer (1975); Hall *et al.* (1981); Tolbert (1980).
Chloroplasts: Kirk & Tilney-Basset (1978); Staehlin & Arntzen (1986) Chapter 1.
Membranes: Eisenberg (1984); Finean *et al.* (1984); Robertson (1983); Robinson (1985).
Microtubules: Davies (1987) Chapter 1; Gunning & Hardham (1982).
Plant cell development: Burgess (1985).

4

Cellular energetics

Plants use solar energy to phosphorylate adenosine diphosphate (ADP) to adenosine triphosphate (ATP) and to produce reduced forms of strong biological reducing agents such as nicotinamide adenine dinucleotide phosphate (NADPH) and ferredoxin (Fd_{red}).

This chapter quantifies the energy changes of chemical reactions and establishes the relationship between energy change and yield of product to assess the importance of ATP, NADPH and Fd_{red} and other key compounds as mediators of energy transfer. Analysis shows that the energy required to phosphorylate ADP and to produce reduced forms of NADP and Fd can be readily supplied by light.

The SI system of units is used throughout this chapter, with energy expressed in joules (J). One joule is equivalent to 0.239 thermochemical calories.

4.1 The free energy change of a chemical reaction is related to its equilibrium constant

The specific internal bond configurations linking the atoms within a molecule confer it with a characteristic internal energy. The product(s) of a chemical reaction has different internal bonding and thus different internal energy to the substrate(s) so that chemical reactions cause a change in internal energy. In a closed system at constant pressure, the change in energy between substrate and product is manifested as a change in heat (ΔH) which consists of two components. One component involves a change due to decreased order (entropy, S) of the products. The change in entropy (ΔS) must be adjusted by the temperature (T) at which the reaction is carried out and increases with temperature ($T\Delta S$). The other component is the free energy change (ΔG). It can be used to perform work, including biological work. The relationships can be specified by the following equation.

$$\Delta H = T\Delta S + \Delta G$$
$$or \quad \Delta G = \Delta H - T\Delta S \tag{4.1}$$

The supply of energy from some chemical reactions to attain significant yields of product in other reactions is of great importance to biology.

Consider a reversible reaction between E and F. Regardless of their initial concentrations, these compounds will react until the ratio of the product to the substrate attains a value specific to the reaction. The reactants then undergo no further *net* interconversion and the reaction has reached equilibrium. This ratio is known as the equilibrium constant (K_{eq}). When the reaction is displaced from equilibrium in either direction it has the potential to do work until enough of E is converted to F, or F to E, to restore equilibrium i.e. the reaction is 'downhill' (Fig.

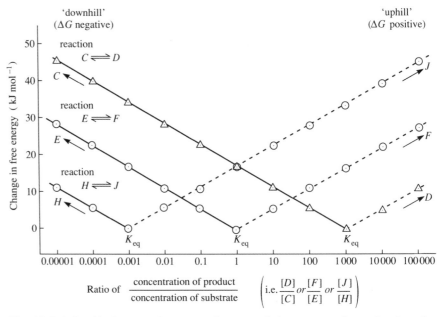

Fig. 4.1 Relationship between the energy changes and the concentration ratio of product relative to substrate (shown on a logarithmic scale) for the reactions (i) $C\rightleftharpoons D$, (ii) $E\rightleftharpoons F$ and (iii) $H\rightleftharpoons J$. The K_{eq} for reactions (i), (ii) and (iii) are 1000, 1 and 0.001 respectively. Reaction (i) can be regarded as exergonic and reaction (iii) as endergonic.

50

4.1). At equilibrium this potential is zero ($\Delta G = 0$). Once the system is at equilibrium net production of E or F is not possible without expending energy. The energy changes of the reaction (Fig. 4.1) are like the potential energy of a ball on the slopes of a valley. The higher the ball on either side (E or F) the greater its potential energy. In practice all reactions only proceed 'downhill'. However, equilibrium yields of E and F other than those predicted by the equilibrium constant can be obtained if energy is supplied to the system, i.e. the reaction is displaced 'uphill' towards E or F. In reality, the reaction is still proceeding 'downhill' (exergonic) but it helps conceptually to envisage the reaction as going 'uphill', as this gives some idea of the energy expenditure from some other source to displace the reaction. A reaction which in theory must be 'forced' to produce an equilibrium yield other than that predicted by the equilibrium constant is an endergonic or endothermic reaction. It should be re-emphasized that this is a theoretical concept only.

The free energy changes for other reactions with equilibrium constants ranging from 0.001 to 1000 are shown in Fig. 4.1. The reaction $C \rightarrow D$ proceeds with a large free energy change and gives a high yield of product (D) at equilibrium ($D/C = 1000$); even quite small amounts of C relative to D have the potential to do work (i.e. exergonic or exothermic). However, the reaction $H \rightarrow J$ yields insignificant amounts of J and undergoes a very small free energy change to attain equilibrium. A significant yield of J is only possible if large amounts of energy are supplied from an external source. Assuming that a reaction mechanism exists, then it can be seen from Fig. 4.1 that 17.1 kJ is needed to convert 50% of H to J (i.e. $J/H = 1$). The reaction $E \rightarrow F$ is readily reversible, exhibiting relatively little change in free energy for F/E ratios from 0.1 to 10.

These examples demonstrate that ΔG not only varies with the product/substrate ratio but that each reaction has its own quantitative relationship. The standard free energy change (ΔG°) has been devised to specify the free energy change of a reaction when all the reactants are present at a concentration of 1 M. However, some metabolic reactions in cells involve H^+ and if ΔG° is specified at a H^+ concentration of 1 M the pH would be 0—scarcely a physiological pH! Accordingly, it is normal biological practice to express standard free energy changes at pH 7 (i.e. $\Delta G^{\circ\prime}$). For reactions which do not involve H^+, $\Delta G^\circ = \Delta G^{\circ\prime}$. The free energy change at physiological pH under non-standard conditions is designated as $\Delta G'$.

The free energy change of a reaction ($\Delta G'$) can be calculated from

$$\Delta G' = \Delta G^{\circ\prime} + RT \log_e \frac{[\text{product}]}{[\text{substrate}]}$$

or

$$\Delta G' = \Delta G^{\circ\prime} + 2.303\, RT \log_{10} \frac{[\text{product}]}{[\text{substrate}]} \qquad (4.2)$$

where R is the gas constant (0.00831 kJ mol^{-1} K^{-1}), T is the temperature (298 K at 25°C) and [product] and [substrate] denote the relevant concentrations under the experimental conditions. $\Delta G^{\circ\prime}$ provides an indicator of the concentration of product(s) relative to the concentration of substrate(s) remaining at equilibrium. At equilibrium $\Delta G' = 0$ so, after inserting values for $\Delta G'$, R and T, Eqn. 4.2 simplifies to

$$\Delta G^{\circ\prime} = -5.7 \log_{10} \frac{[\text{product(s)}]_{eq}}{[\text{substrate(s)}]_{eq}} = -5.7 \log_{10} K_{eq} \qquad (4.3)$$

where $\Delta G^{\circ\prime}$ is expressed in kJ mol^{-1} and K_{eq} is the equilibrium constant. This expression explains the quantitative values for the reactions in Fig. 4.1.

Compare the aerobic oxidation of a hexose sugar ($C_6H_{12}O_6$) to CO_2 with the assimilation of CO_2 into hexose.

$$C_6H_{12}O_6 + 6O_2 \rightleftharpoons 6CO_2 + 6H_2O \, (\Delta G^{\circ\prime} = -2870 \text{ kJ mol}^{-1}) \qquad (4.4)$$

$$6CO_2 + 6H_2O \rightleftharpoons C_6H_{12}O_6 + 6O_2 \, (\Delta G^{\circ\prime} = +2870 \text{ kJ mol}^{-1}) \qquad (4.5)$$

The $\Delta G^{\circ\prime}$ for Eqn. 4.4 predicts that hexoses are almost completely oxidized to CO_2 at chemical equilibrium. The $\Delta G^{\circ\prime}$ value for Eqn. 4.5 tells us that, assuming a chemical mechanism exists, significant yields of hexose (about 50% conversion) can be obtained only if about 2870 kJ of chemical energy are supplied from an external source per mole of hexose synthesized. Similarly, large amounts of energy and specialized mechanisms are required to effect the reduction of NO_3^- and SO_4^{2-} to NH_3 and H_2S.

The $\Delta G^{\circ\prime}$ provides a useful index of the likely direction of a reaction and the yield of product when, and if, equilibrium is attained. However, the $\Delta G^{\circ\prime}$ value does not provide any information about the reaction rate or whether a mechanism exists. Also, the $\Delta G^{\circ\prime}$ value does not necessarily indicate the capacity of a reaction to perform work within a functioning cell, since this is determined by $\Delta G'$ and not $\Delta G^{\circ\prime}$ (see Eqn. 4.2). In addition most reactions in cells rarely come to chemical equilibrium; they tend to attain a steady state, displaced from equilibrium ('dynamic equilibrium').

4.2 Cells have mechanisms for coupling endergonic reactions to exergonic reactions

Endergonic reactions do not provide significant yields of products unless energy is supplied. Consider the hypothetical reaction

$$L \rightleftharpoons M \ (\Delta G^{\circ\prime} = +34.2 \ \text{kJ mol}^{-1}). \tag{4.6}$$

Equation 4.3 predicts that the yield of M at equilibrium will be only 1 part of M per 1 000 000 parts of L. Now consider X, which undergoes a highly exergonic reaction to form Y

$$X \rightleftharpoons Y \ (\Delta G^{\circ\prime} = -39.9 \ \text{kJ mol}^{-1}). \tag{4.7}$$

If a mechanism exists to couple the formation of M to the production of Y then the $\Delta G^{\circ\prime}$ is given by the sum of the two component reactions

$$L + X \rightleftharpoons M + Y \ (\Delta G^{\circ\prime} = -5.7 \ \text{kJ mol}^{-1}). \tag{4.8}$$

Thus,

$$K_{eq} = \frac{[M]_{eq}[Y]_{eq}}{[L]_{eq}[X]_{eq}} = 10$$

where $[M]_{eq}$, $[Y]_{eq}$, $[L]_{eq}$ and $[X]_{eq}$ represent the equilibrium concentrations of M, Y, L and X respectively. If L and X are initially supplied at a concentration of 1 000 000, the equilibrium concentrations of M and Y are about 760 000 and of L and X about 240 000. Thus, at equilibrium, the yield of M relative to L is much greater for the coupled reaction than the uncoupled reaction. In effect the half reaction $X \rightarrow Y$ provides the free energy to drive the synthesis of L to M 'uphill'. Compounds in cells exhibiting the features shown by X in Eqn. 4.8 are known as 'high energy compounds' (e.g. ATP). This term, of course, is not strictly correct as it is the displacement of the compound from the equilibrium position with its reaction product rather than the compound itself that provides the free energy. Nevertheless, 'high energy compound' is used in this book as a succinct way of expressing the concept.

4.3 Hydrolysis of a few key compounds provides important intermediary energy sources in cells

Cells contain several compounds which hydrolyse in highly exergonic reactions and act as important intermediary sources of chemical energy. These compounds rarely

react directly with free H_2O as this would achieve little useful purpose. Rather, the exergonic hydrolysis of high energy compounds is usually coupled to the synthesis of a product in which H_2O is removed from the substrate(s). The hydrolysis of a high energy compound can be written as a half reaction, remembering that the H_2O required comes from the half reaction to which it is coupled.

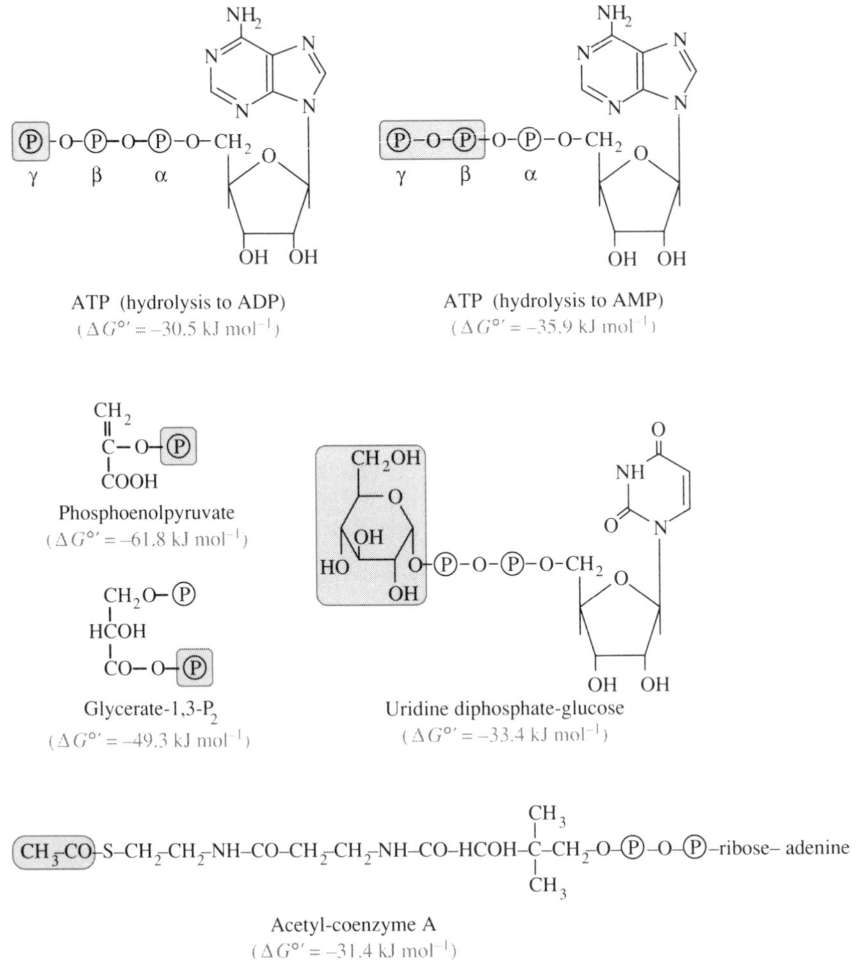

ATP (hydrolysis to ADP)
($\Delta G^{\circ\prime} = -30.5$ kJ mol^{-1})

ATP (hydrolysis to AMP)
($\Delta G^{\circ\prime} = -35.9$ kJ mol^{-1})

Phosphoenolpyruvate
($\Delta G^{\circ\prime} = -61.8$ kJ mol^{-1})

Glycerate-1,3-P_2
($\Delta G^{\circ\prime} = -49.3$ kJ mol^{-1})

Uridine diphosphate-glucose
($\Delta G^{\circ\prime} = -33.4$ kJ mol^{-1})

Acetyl-coenzyme A
($\Delta G^{\circ\prime} = -31.4$ kJ mol^{-1})

Fig. 4.2 Structures of 'high energy' compounds used in intermediary energy transfer. The groups shown in shaded boxes are released in highly exergonic hydrolytic reactions. Note the nomenclature for naming the phosphate groups of ATP (α, β, γ).

The structures of the most important 'high energy compounds' in plants are shown in Fig. 4.2. By far the most important of the half reactions are those involving the hydrolysis of ATP.

$$ATP + H_2O \rightleftharpoons ADP + P_i \quad (\Delta G^{\circ\prime} = -30.5 \text{ kJ mol}^{-1}) \tag{4.9}$$

$$ATP + H_2O \rightleftharpoons AMP + PP_i \quad (\Delta G^{\circ\prime} = -35.9 \text{ kJ mol}^{-1}). \tag{4.10}$$

A biological reaction involving the hydrolysis of ATP according to the half reaction shown in Eqn. 4.9 is the synthesis of glutamine,

Glutamate   Glutamine

where the free energy of hydrolysis of ATP is used to incorporate NH_3 into the amide group of glutamine. Hydrolysis of ATP to ADP and Pi implies removal of H_2O from the substrates: —OH from glutamate and —H from NH_3 so that H_2O does not appear in Eqn. 4.11. A further example is given by the phosphorylation of glucose in which the terminal phosphate of ATP is incorporated into the product, glucose-6-P.

$$Glucose + ATP \rightleftharpoons glucose\text{-}6\text{-}P + ADP. \tag{4.12}$$

The free energy of hydrolysis of ATP drives the reaction strongly to the right ($\Delta G^{\circ\prime} = -16.7 \text{ kJ mol}^{-1}$) to yield a product considerably more reactive than glucose (i.e. with higher free energy). Many biological reactions involve the hydrolysis of ATP to AMP (as in Eqn. 4.10) and these are important in biosynthetic pathways. They often involve attachment of the substrate to AMP to form an adenylate, for example

$$Fatty\ acid + ATP \rightleftharpoons fatty\text{-}acyl\ adenylate + PP_i. \tag{4.13}$$

The distinction between the hydrolysis of ATP to yield P_i or PP_i (Eqns 4.9 & 4.10) is important since PP_i is readily hydrolysed in another strongly exergonic reaction

$$PP_i + H_2O \rightleftharpoons 2P_i \quad (\Delta G^{\circ\prime} = -33.4 \text{ kJ mol}^{-1}). \tag{4.14}$$

The reaction obtained by summing Eqns 4.10 and 4.14

$$ATP + 2H_2O \rightleftharpoons AMP + 2P_i \quad (\Delta G^{\circ\prime} = -69.3 \text{ kJ mol}^{-1}) \tag{4.15}$$

is much more exergonic than the hydrolysis of ATP to ADP (Eqn. 4.9). However, this involves increased breakdown of ATP and much more energy must be expended to regenerate it from AMP than from ADP.

The $\Delta G^{\circ\prime}$ of hydrolysis of the β-phosphate group of ADP to AMP and P_i ($-27.2 \text{ kJ mol}^{-1}$) is similar to that for the hydrolysis of the γ-phosphate of ATP (Eqn. 4.9) but hydrolysis of ADP is rarely used in cells as a coupled energy source. Hydrolysis of the α-phosphate group of AMP to adenosine and P_i is a much less exergonic reaction ($\Delta G^{\circ\prime} = -9.2 \text{ kJ mol}^{-1}$) and is not used at all.

The tri-, di- and monoribonucleotides of cytosine, guanine and uracil participate in reactions with characteristics similar to those for ATP, ADP and AMP. The $\Delta G^{\circ\prime}$ values for the hydrolysis of guanosine triphosphate (GTP), uridine triphosphate (UTP) and cytidine triphosphate (CTP) to their corresponding diphosphates and monophosphates are similar to those shown for ATP in Eqns 4.9 and 4.10 respectively. GTP, UTP and CTP are important energy sources in cells; for example, the hydrolysis of UTP with the production of PP_i is important in the formation of nucleoside diphosphate sugars from sugar-1-phosphates.

$$UTP + glucose\text{-}1\text{-}P \rightleftharpoons UDP\text{-}glucose + PP_i. \tag{4.16}$$

Cells use the nucleoside diphosphate derivatives of sugars (see Fig. 4.2 for structure of UDP-glucose) as substrates for the formation of the glycosidic bonds found in the various plant polysaccharides and oligosaccharides.

Several other compounds which undergo highly exergonic hydrolysis reactions (typically -30 kJ mol^{-1} or more negative) also act as important intermediaries in the transfer of energy. These reactions are also coupled to the removal of H_2O from another substrate.

$$Phosphoenolpyruvate + H_2O \rightleftharpoons pyruvate + P_i$$
$$(\Delta G^{\circ\prime} = -61.8 \text{ kJ mol}^{-1}). \tag{4.17}$$

$$Glycerate\text{-}1,3\text{-}P_2 + H_2O \rightleftharpoons glycerate\text{-}3\text{-}P + P_i$$
$$(\Delta G^{\circ\prime} = -49.3 \text{ kJ mol}^{-1}). \tag{4.18}$$

$$UDP\text{-}glucose + H_2O \rightleftharpoons UDP + glucose$$
$$(\Delta G^{\circ\prime} = -33.4 \text{ kJ mol}^{-1}). \tag{4.19}$$

$$Acetyl\text{-}coenzyme\ A + H_2O \rightleftharpoons acetate + coenzyme\ A$$
$$(\Delta G^{\circ\prime} = -31.4 \text{ kJ mol}^{-1}). \tag{4.20}$$

Note that the $\Delta G^{\circ\prime}$ values for the hydrolysis of phosphoenolpyruvate (PEP) and glycerate-1,3-P_2 are much more negative than that for hydrolysis of ATP to ADP (Eqn. 4.9). Thus, in theory, hydrolysis could be coupled to phosphorylation of ADP in an exergonic reaction

$$PEP + H_2O \rightleftharpoons pyruvate + P_i \quad (\Delta G^{\circ\prime} = -61.8 \text{ kJ mol}^{-1})$$
$$ADP + P_i \rightleftharpoons ATP + H_2O \quad (\Delta G^{\circ\prime} = +30.5 \text{ kJ mol}^{-1})$$
$$PEP + ADP \rightleftharpoons pyruvate + ATP \quad (\Delta G^{\circ\prime} = -31.3 \text{ kJ mol}^{-1}). \quad (4.21)$$

Plants contain the enzyme pyruvate kinase which supports the coupled reaction (Eqn. 4.21). Synthesis of ATP in this way is referred to as substrate-level phosphorylation. An analogous set of equations can be compiled for the phosphorylation of ADP by glycerate-1,3-P_2.

$$Glycerate-1,3-P_2 + ADP \rightleftharpoons glycerate-3-P + ATP$$
$$(\Delta G^{\circ\prime} = -18.8 \text{ kJ mol}^{-1}). \quad (4.22)$$

This reaction is catalysed by the enzyme glycerate-3-P kinase and is another example of substrate-level phosphorylation. Incidentally, ATP, ADP and certain other high energy compounds are often referred to as 'cofactors'. However, ATP and ADP are true substrates subject to continual turnover.

The synthesis of nucleoside diphosphate sugars is coupled to the hydrolysis of nucleoside triphosphates (see Eqn. 4.16). Since the $\Delta G^{\circ\prime}$ of hydrolysis of the nucleoside diphosphate sugars (see Eqn. 4.19) is similar to that for hydrolysis of nucleoside triphosphates (see Eqn. 4.10) then most of the energy must be conserved in the sugar derivative (i.e. nucleoside diphosphate sugars are 'activated' sugars). They supply the free energy required for synthesis of the glycosidic bonds present in oligosaccharides and polysaccharides (see Section 10.9).

Acetyl-coenzyme A and other fatty-acyl thioesters of coenzyme A (Eqn. 4.20) are intermediates in energy transfer in plants in processes involving substrate-level phosphorylation and the synthesis and oxidation of fatty acids.

The $\Delta G^{\circ\prime}$ values for the hydrolysis of the high energy compounds (typically -30 kJ or more negative) (see Fig. 4.2) are considerably more exergonic than the hydrolysis of certain bonds in biological macromolecules, such as peptide and glycoside bonds and some bonds in metabolic intermediates (Table 4.1). Thus, in theory, the free energy of hydrolysis of ATP (or another high energy compound) can be used in a coupled exergonic reaction to synthesize the bonds in the compounds listed in Table 4.1.

Table 4.1 Standard free energy of hydrolysis ($\Delta G^{\circ\prime}$) of some compounds of biological importance.

Substrate	Products	$\Delta G^{\circ\prime}$ (kJ mol^{-1})
AMP	Adenosine + P_i	-9.2
Glycerol-3-P	Glycerol + P_i	-9.2
Glucose-6-P	Glucose + P_i	-13.8
Fructose-6-P	Fructose + P_i	-15.9
Maltose*	2 Glucose	-15.5
Lactose*	Glucose + galactose	-15.9
Glycylglycine†	2 Glycine	-9.2
Glutamine	Glutamate + NH_3	-14.2

* Maltose and lactose are typical of glycosidic bonds.
† Glycylglycine is representative of a peptide bond.

4.4 A few redox pairs are important intermediary electron acceptors and donors in cells

Oxidation/reduction reactions are extremely important in the energy relations of organisms. For example, oxidation of sugars fulfils the energy requirements of non-photosynthetic cells and the assimilation of oxidized forms of carbon (CO_2), nitrogen (NO_3^-) and sulphur (SO_4^{2-}) into organic compounds involves reduction.

An oxidation/reduction reaction moves electrons from the reducing agent or reductant (thereby oxidizing it) to the oxidizing agent or oxidant (thus reducing it). In cells, most oxidation/reduction reactions also involve proton exchange. The oxidized and reduced forms of a compound constitute a redox pair. The reduced form of one redox pair is oxidized by the oxidized form of another redox pair. Electron transfer is usually depicted as two half reactions to show the events for each of the redox pairs involved. This method of setting out an oxidation/reduction reaction (see Section 4.5) facilitates the calculation of the standard free energy change.

Plant cells support many different oxidation/reduction reactions but just a few redox pairs are common to the great majority. In some reactions the oxidized form is an electron acceptor and in others the reduced form acts as an electron donor. Thus, these compounds are continuously turned over between the oxidized and reduced forms. The nature of the important redox pairs varies greatly. Some are water soluble, others are strongly hydrophobic and migrate within the lipophilic regions of membranes, while many protein redox pairs are bound to specific membranes.

The most ubiquitous redox compounds in plants are the oxidized and reduced forms of nicotinamide adenine dinucleotide (NAD) and its phosphorylated derivative, nicotinamide adenine dinucleotide phosphate (NADP). They are water soluble and occur in free form. They consist of two ribonucleotides, one containing adenine, the other nicotinamide (strictly speaking not a nucleotide), joined in ester linkage via their phosphate groups (Fig. 4.3). NADP contains an extra phosphate group

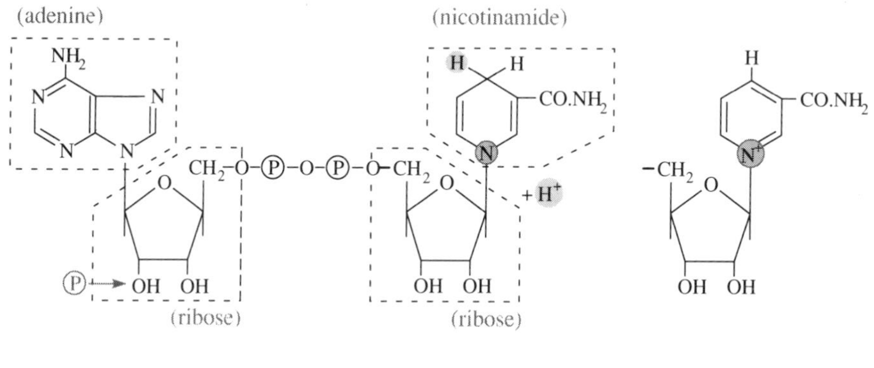

Fig. 4.3 Structures of the reduced and oxidized forms of NAD. The oxidized form (NAD$^+$) differs only with respect to the nicotinamide moiety. Reduced and oxidized forms of NADP (NADPH and NADP$^+$) differ in having a phosphate group on the hydroxy group at C-2' of the ribose residue attached to adenine.

attached to the hydroxy group at C-2' in the ribose residue attached to adenine. The redox properties of both NAD and NADP are conferred by the nicotinamide residue. NAD is used collectively for both the oxidized and reduced forms or when a distinction between the two forms is not relevant. Specifically, the oxidized form is NAD$^+$ and the reduced form is NADH or NADH$_2$ since the reduced form consists of NADH + H$^+$. Similarly, NADP$^+$ and NADPH (or NADPH$_2$) are used to denote the oxidized and reduced forms of NADP. The derivation of these terms is evident from their structures (Fig. 4.3). Like ATP and ADP, NAD(P)$^+$ and NAD(P)H are often referred to as 'cofactors' but they are true substrates subject to rapid rates of turnover. Reactions involving NADH and NADPH can be

monitored by absorbance of ultraviolet light since the dihydronicotinamide residue present in the reduced forms has an absorption maximum at 340 nm but is absent in the oxidized forms. Thus, oxidation of NADH and NADPH can be monitored by the decrease in absorbance at 340 nm and vice versa.

Flavin mononucleotide (FMN) and flavin adenine dinucleotide (FAD) differ from the nicotinamide nucleotides as they are usually firmly bound to the enzymes whose activity they support and are regarded as prosthetic groups (see Section 5.2.2). As the structures in Fig. 4.4 show, FMN is inappropriately named because

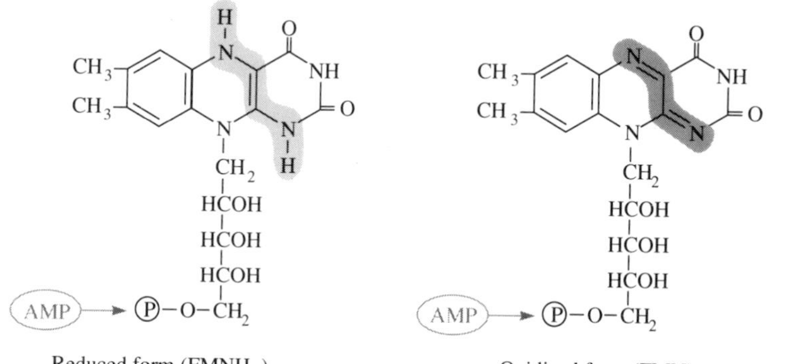

Fig. 4.4 Structures of the reduced and oxidized forms of FMN. Reduced and oxidized forms of FAD (FADH$_2$ and FAD) are as for FMNH$_2$ and FMN, respectively, except that the phosphate group of FMN is attached to AMP.

the isoalloxazine ring (which confers the redox characteristics) is not joined to the phosphate group through a true pentose. FAD consists of FMN esterified through its phosphate group to the phosphate group of AMP. The oxidized and reduced forms of the flavin nucleotides also have different spectral characteristics which can be used to monitor the interconversion between the two forms.

Ferredoxins are proteins containing iron which undergo one electron redox change and do not carry protons. Depending on the source, they contain two (or four) atoms of iron and 95–100 amino acid residues, five of which are usually cysteine and have a molecular mass of about 11 kDa. Besides the sulphur associated with the cysteinyl residues, the ferredoxins contain two (or four) additional sulphur atoms which link the iron atoms (Fig. 4.5) and are released as H$_2$S when treated with acid. For this reason, ferredoxins are referred to as

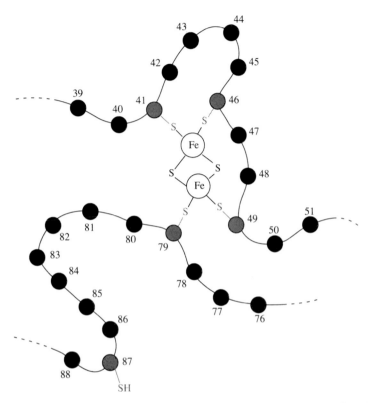

Fig. 4.5 Schematic representation of ferredoxin (2Fe–2S) from the cyanobacterium *Spirulina platensis*. The amino acids in the peptide backbone are denoted by black symbols with the cysteinyl residues and cysteinyl sulphur atoms shown in grey. Acid-labile sulphur atoms are shown in black. Of the five cysteinyl residues shown, only four are involved in linking the two atoms of iron.

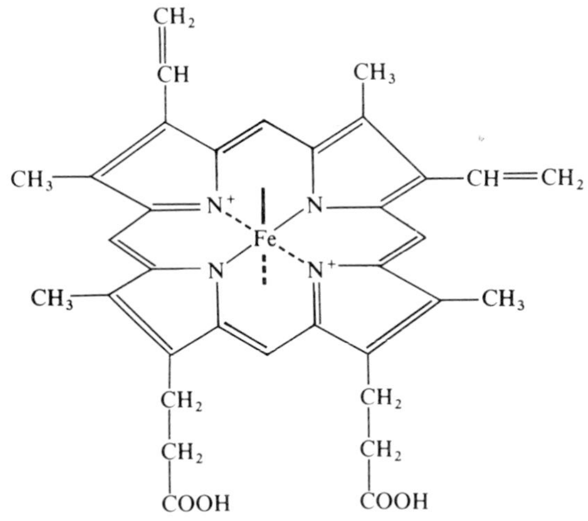

iron–sulphur proteins. Soluble ferredoxins contain a reaction centre of two atoms of iron and two acid-labile sulphur atoms with iron attached to the protein through the sulphur atoms of four cysteinyl residues (Fig. 4.5). Membrane-bound forms have four iron atoms and four atoms of acid-labile sulphur in a cubic arrangement.

Cytochromes are also redox proteins containing iron but with iron in a haem prosthetic group attached to a polypeptide (apoprotein). Haem consists of four pyrrole rings linked by methine bridges ($=$CH$—$) to form a porphyrin ring with a central atom of iron chelated through the nitrogen atoms of the pyrrole rings

Fig. 4.6 Structure of haem, a prosthetic group present in several proteins with redox properties. The iron atom has six co-ordination bonds, four of which are bound to the nitrogen atoms of the porphyrin ring. In the cytochromes, the other two sites (shown in bold) are attached to amino acid residues in the protein backbone (e.g. one is attached to the nitrogen atom in the ring of histidine and the other to the sulphur atom of cysteine in cytochrome *c*).

(Fig. 4.6). Cytochromes differ in the composition of their apoproteins. The reduced (Fe^{2+}) forms of the various cytochromes are spectrally distinct so that in aerobic cells with many different cytochromes the form and amount of each can be determined (see Section 7.7.1).

Several other important biological redox pairs are discussed elsewhere. They include the lipophilic compounds, ubiquinone, plastoquinone and various proteins containing metals (e.g. plastocyanin and nitrogenase).

4.5 The free energy change of a redox reaction is related to the electrode potentials of the two redox pairs

The free energy changes of oxidation/reduction reactions can be derived from the standard redox potentials (E_o) of redox pairs. The standard redox potential denotes the electrical potential (measured in volts) generated by a redox pair (e.g. NAD^+/NADH) relative to the H^+/H_2 redox pair (the standard hydrogen electrode) when all the reactants are at 1 M except for H_2 which is supplied at a pressure of 101.3 kPa (1 atmosphere). H^+ is present in the standard hydrogen electrode at 1 M (pH 0) but, for obvious reasons, biologists cite standard redox potentials at pH 7 (E'_o); the H^+/H_2 electrode at pH 7 has a potential of -0.42 V relative to a standard hydrogen electrode.

Table 4.2 Standard redox potentials (E'_o) at pH 7 and 25 °C of some important redox half reactions (redox pairs).

Redox pair (oxidized/reduced)	E'_o(V) (ox/red)	$\Delta G^{o'}$ of oxidation of reduced form by O_2 (kJ mol^{-1})*
$Fd_{ox} + e^- \rightarrow Fd_{red}$	-0.43	-241
$NADP^+ + 2H^+ + 2e^- \rightarrow NADPH + H^+$	-0.32	-220
$NAD^+ + 2H^+ + 2e^- \rightarrow NADH + H^+$	-0.32	-220
Glycerate-1,3-P_2 + 2H^+ + 2e^- → glyceraldehyde-3-P + P_i	-0.29	-214
$FMN + 2H^+ + 2e^- \rightarrow FMNH_2$	-0.22	-201
Pyruvate + 2H^+ + 2e^- →lactate	-0.19	-195
Oxaloacetate + 2H^+ + 2e^- →malate	-0.18	-193
Fumarate + 2H^+ + 2e^- →succinate	$+0.03$	-152
Cytochrome $b_{ox} + e^- \rightarrow$ cytochrome b_{red}	$+0.06$	-147
Cytochrome $c_{ox} + e^- \rightarrow$ cytochrome c_{red}	$+0.25$	-110
Cytochrome $f_{ox} + e^- \rightarrow$ cytochrome f_{red}	$+0.36$	-89
Cytochrome $a_{3\,ox} + e^- \rightarrow$ cytochrome $a_{3\,red}$	$+0.55$	-52
$\frac{1}{2}O_2 + 2H^+ + 2e^- \rightarrow H_2O$	$+0.82$	

*Values assume a transfer of two electrons in all cases.

The E'_o values of some important biological redox pairs are shown in Table 4.2. They can be used to predict the direction of electron flow between two redox pairs and to calculate $\Delta G^{o'}$ for the reaction. The more negative the E'_o value of a redox pair, the more easily the reduced form loses electrons (i.e. it is a more powerful reducing agent). Conversely, the more positive the E'_o value, the greater the tendency of the oxidized form to gain electrons. Thus, provided a reaction mechanism exists, the reduced forms of redox pairs high in Table 4.2 can be oxidized by the oxidized forms of those lower in the Table. Such reactions are always exergonic.

The standard free energy change at pH 7 ($\Delta G^{o'}$) of an oxidation/reduction reaction can be determined from the difference in the E'_o values ($\Delta E'_o$) between the oxidizing redox pair and the E'_o value of the reducing pair. $\Delta G^{o'}$ is then calculated from the equation

$$\Delta G^{o'} = -nF\Delta E'_o \tag{4.23}$$

where n is the number of electrons transferred from reductant to oxidant and F is the Faraday constant (the charge of one mole of electrons = 96.496 kJ mol^{-1} V^{-1}). For the oxidation of NADPH by O_2 (Eqn 4.26), the redox pair $NADP^+$/NADPH has a more negative E'_o value than the O_2/H_2O pair so that O_2 oxidizes NADPH.

$$\frac{1}{2}O_2 + 2H^+ + 2e^- \rightleftharpoons H_2O \qquad E'_o = +0.82 \text{ V} \tag{4.24}$$

$$NADPH + H^+ \rightleftharpoons NADP^+ + 2H^+ + 2e^- \qquad E'_o = -0.32 \text{ V} \tag{4.25}$$

$$NADPH + H^+ + \tfrac{1}{2}O_2 \rightleftharpoons NADP^+ + H_2O \qquad \Delta E'_o = 1.14 \text{ V}. \tag{4.26}$$

Substituting $\Delta E'_o = 1.14$ V and $n = 2$ into Eqn. 4.23, then

$$\Delta G^{o'} = -2 \times 96.5 \times 1.14 = -220 \text{ kJ mol}^{-1}.$$

The reaction is therefore strongly exergonic.

Equation 4.23 predicts that the reduced forms of redox pairs high in Table 4.2 participate in more strongly exergonic reactions with a given oxidizing agent than those lower in the Table. Examples involving oxidation by O_2 are also given. Although the reactants in cells are unlikely to be present at a concentration of 1 M, and reaction mechanisms may not exist for the oxidation by O_2 of all the compounds listed, the data nonetheless indicate the $\Delta G^{o'}$ values for the important biological reducing agents. Significantly, the $\Delta G^{o'}$ for the oxidation of NADH by O_2 (-220 kJ mol^{-1}) is far more negative than the $\Delta G^{o'}$ of hydrolysis of ATP (-30.5 kJ mol^{-1}) so that, in theory, oxidation of NADH by O_2 could provide the free energy to phosphorylate not just one ADP but several.

To determine the free energy change for the non-standard conditions prevailing in cells ($\Delta G'$), it is necessary to determine the redox potentials of the two half reactions for these conditions (E'). For each redox pair, E' is dependent on the ratio of the concentrations of the oxidized form, $[C_{ox}]$ and the reduced form, $[C_{red}]$

$$E' = E'_o + \frac{RT}{nF} \log_e \frac{[C_{ox}]}{[C_{red}]} \tag{4.27}$$

where R is the gas constant (0.00831 kJ mol^{-1} K^{-1}) and T is temperature (298 K).

Substituting values for R, T, F, and $2.303 \log_{10}$ for $\log_e$, then Eqn. 4.27 simplifies to

$$E' = E'_o + \frac{0.059}{n} \log_{10} \frac{[C_{ox}]}{[C_{red}]}. \tag{4.28}$$

By calculating E' for each of the redox pairs, $\Delta E'$ can be ascertained and the $\Delta G'$ calculated by substituting $\Delta E'$ for $\Delta E'_o$ in Eqn 4.23.

4.6 Differences in electrical potential and ion concentration across a membrane cause free energy differences between the ions across the membrane

Plants establish electrical potential differences and ion gradients, most notably proton gradients, across certain membranes. The ion electrochemical potentials resulting from membrane potential differences and ion gradients can perform chemical work. These are energy-transducing membranes.

The free energy of an ion electrochemical gradient across a cellular membrane is expressed as ΔG rather than $\Delta G'$ since the gradients frequently involve protons. Imagine two hypothetical compartments, A and B, separated by a membrane (Fig. 4.7) impermeable to the ion X^m (for H^+, $m=1$; for Ca^{2+}, $m=2$, etc.). Suppose that the electrical charges in A and B are the same and the electrical potential difference across the membrane is zero. Then the ΔG of the ion gradient (expressed in the direction A to B) is given by the difference in the free energy of X^m between B and A (Fig. 4.7a). This is dependent on the log of the ratio of the concentrations of X^m in B, $[X^m]_B$, and in A, $[X^m]_A$, according to the following expression.

$$\Delta G = 2.303 \, RT \log_{10} \frac{[X^m]_B}{[X^m]_A} = 5.7 \log_{10} \frac{[X^m]_B}{[X^m]_A}. \tag{4.29}$$

Thus, for an ion concentration gradient ΔG alters by $5.7 \, kJ$ for each 10-fold variation in the ratio of the ion concentrations. For protons this equals 1 pH unit. $\Delta G = 0$ when the ion concentrations on each side of the membrane are the same.

Now imagine that the concentrations of X^m in A and B are the same but one side of the membrane has a different electrical potential to the other side (Fig. 4.7b). This difference, known as the membrane potential ($\Delta \Psi$), is measured in volts. Cations tend to move to the side of lower electrical potential and anions more to the side of higher electrical potential. As before, the change in free energy of X^m in the direction A to B is given by the expression

$$\Delta G = -mF \Delta \Psi = 96.5 \, m \Delta \Psi. \tag{4.30}$$

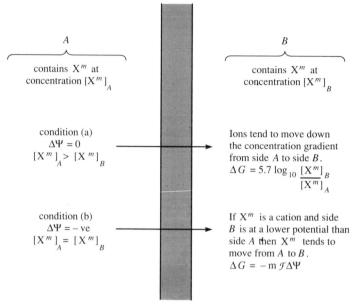

Fig. 4.7 Factors contributing to the difference in free energy of an ion (X^m) across a membrane which is impermeable to X^m. In condition (a), the two sides of the membrane have the same electrical potential ($\Delta \Psi = 0$) but the concentration of X^m in A is greater than in B ($[X^m]_A > [X^m]_B$). In condition (b), $[X^m]_A = [X^m]_B$ but a potential difference across the membrane exists. The difference in free energy of ions between the two sides of the membrane under these conditions is described in the text.

When there is both an ion concentration gradient and an electrical gradient across a membrane the total free energy difference between the ions on each side is given by the ΔG for ion concentration (Eqn. 4.29) plus the ΔG for membrane potential (Eqn. 4.30).

$$\Delta G = 2.3 \, RT \log_{10} \frac{[X^m]_B}{[X^m]_A} + (-mF \Delta \Psi) \tag{4.31}$$

$$= 5.7 \log_{10} \frac{[X^m]_B}{[X_m]_A} - 96.5 \, m \Delta \Psi \tag{4.32}$$

where ΔG is expressed in $kJ \, mol^{-1}$ and $\Delta \Psi$ in volts. Customarily, the free energy change is expressed as an ion electrochemical potential gradient ($\Delta \mu_{xm}$) in volts,

5.2 Some enzymes require the presence of cofactors for activity

Many biological reactions require other, non-protein, substances as well as the protein component of the enzyme and, of course, the substrate to support catalysis. These substances are known as cofactors. There are three basic types of cofactor; activators, coenzymes and prosthetic groups, although the distinction is sometimes vague. Table 5.2 lists some common activators and prosthetic groups and shows examples of reactions in which they are important.

Many enzymes require free metal ions for full catalytic activity. These *activators*, such as magnesium, potassium and calcium ions are not tightly bound in the enzyme but, nonetheless, they exert an effect by affecting substrate binding or by

Table 5.2 Common cofactors which act as activators or prosthetic groups in enzymes and examples of enzyme-catalysed reactions for which the cofactor is important.

Cofactor	Enzymes
Activators	
Mg^{2+}	Hexokinase; succinyl-CoA synthetase
Mn^{2+}	Isocitrate dehydrogenase
Ca^{2+}	Phospholipase
K^+	Aldolase; pyruvate kinase
Prosthetic groups	
Haem (iron porphyrin)	Cytochromes; catalase
Cu^{2+}	Cytochrome oxidase; phenol oxidases
Zn^{2+}	Carbonic anhydrase; alcohol dehydrogenase
Pyridoxal phosphate	Aminotransferases
Flavin mononucleotide (FMN)	Nitrate reductase
Flavin adenine dinucleotide (FAD)	Pyruvate dehydrogenase
Biotin	Acetyl-CoA carboxylase
Thiamine pyrophosphate (TPP)	Transketolase; pyruvate decarboxylase
Nicotinamide adenine dinucleotide (NAD) and NADP*	Glyceraldehyde-3-P dehydrogenase

*For most enzymes described in this book, NAD and NADP are substrates rather than cofactors. However, in the case of NAD-dependent glyceraldehyde-3-P dehydrogenase, one of the four NAD molecules associated with the enzyme can only be removed by rather drastic treatment suggesting that it is tightly bound as a prosthetic group. In NADP-dependent glyceraldehyde-3-P dehydrogenase, four NADP molecules are tightly bound as prosthetic groups. NAD(P) is described in Chapter 4. Coenzymes such as coenzyme A and tetrahydrofolate are, like NAD, NADP and ATP, best considered as substrates for enzymic reactions and consequently are not included in this table.

altering the conformation of the enzyme to give a more active catalytic site.

The distinction between coenzymes and prosthetic groups is less marked. Generally, both types of cofactor are complex organic compounds showing a marked specificity in the reactions in which they are involved. Certain metal ions such as Cu^{2+} or Fe^{2+} can also act as cofactors tightly bound to some enzymes. Coenzymes and prosthetic groups play a more direct and essential part in the enzymic reaction than activators by acting, for example, as carriers of specific chemical groups. Broadly speaking, a *prosthetic group* is a moiety which is a tightly bound integral part of an enzyme molecule and is distinct from a *coenzyme* such as coenzyme A or tetrahydrofolate which is free to move from one enzyme molecule to another. The enzyme without its particular prosthetic group is called an apoenzyme. It is important to note that although NAD can act both as a coenzyme (by this definition) and, in certain enzymes, as a prosthetic group (e.g. glyceraldehyde-3-P dehydrogenase) it is more correct to consider the pyridine nucleotides NAD, NADP, ATP and ADP, as well as compounds such as coenzyme A and tetrahydrofolate, as *substrates* for the particular reactions in which they are involved. These compounds are chemically changed during the enzyme-catalysed reactions in which they participate and are free to move around the cell, unlike prosthetic groups which cannot move from one enzyme molecule to another. In those enzymes in which NAD is bound tightly as a prosthetic group, the NAD is bound to enzyme subunits which do not participate in the enzyme-catalysed reaction but it is nevertheless structurally necessary for activity of the catalytic subunit.

5.2.1 Coenzymes

As noted above, the coenzymes are best regarded as substrates. The most common of these are NAD/NADH, NADP/NADPH and ATP/ADP, the structures and functions of which are discussed in Sections 4.3 and 4.4. Of the other coenzymes, coenzyme A is probably the most important. It has the structure 3'-phospho-pantoyl-β-alanyl-cyteamine-ADP (see Fig. 4.2). Coenzyme A (represented below as HSCoA) is important because it reacts through its thiol (—SH) group with a variety of acylated compounds (e.g. fatty acids) to form acylated thiol esters (acyl-S-CoA). Thus, the —SH group is referred to as the active group of coenzyme A. The acylated CoA product is regarded as a high energy compound because it hydrolyses with a large negative $\Delta G^{o'}$ (see Section 4.3). Broadly speaking, thiol esters are more reactive than oxygen esters. In cells the acyl groups of thiol esters are readily transferred to a number of acyl-acceptors in the generalized reactions shown in Eqn. 5.2.

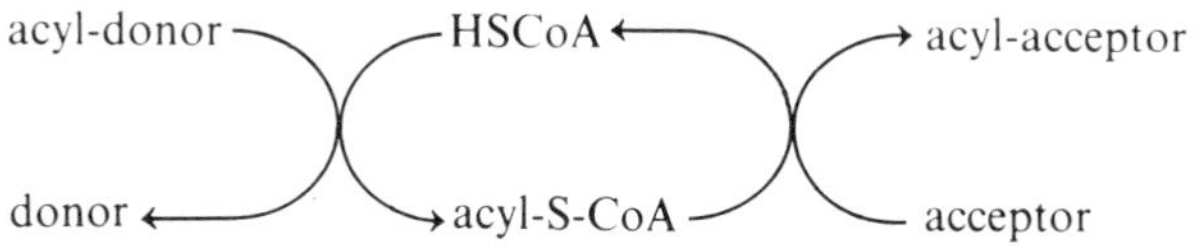

$$(5.2)$$

Coenzyme A plays an important role in a number of biological processes such as lipid biosynthesis (Chapter 16) and the TCA cycle (Chapter 7).

Tetrahydrofolate (THF) is a coenzyme involved in the transfer of C_1 fragments such as formyl, methylene or methyl groups from a donor compound to an acceptor molecule. Tetrahydrofolate has the structure shown below.

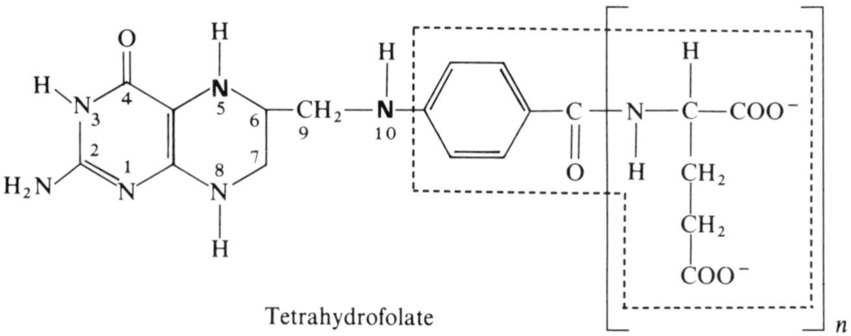

Tetrahydrofolate

Characteristically, THF and its derivatives in plants contain not just one glutamate residue (in brackets above) but five to seven. Tetrahydrofolate forms several C_1-derivatives by transfer of various C_1 moieties from a donor compound to either or both of the N atoms shown in bold at positions 5 and 10. There are four common derivatives of THF in plants,

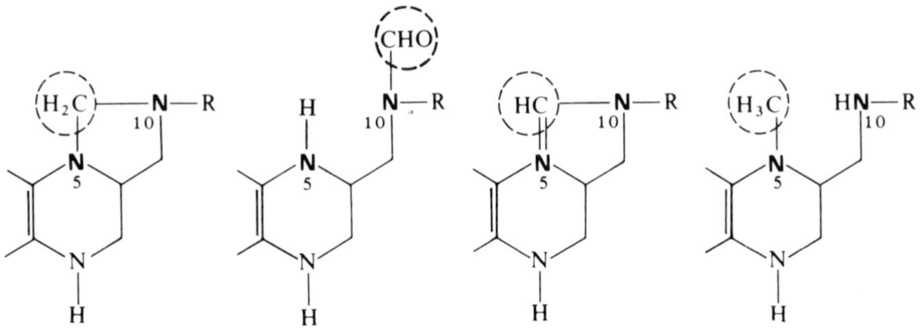

N^5, N^{10}-methyleneTHF N^{10}-formylTHF N^5, N^{10}-methenylTHF N^5-methylTHF

where R is the boxed group in the diagram of THF above. The reactions of THF and its derivatives and their biochemical origin are discussed in Section 14.3. A good example of the role of THF comes from the conversion of glycine to serine in photorespiration (see Section 11.4).

5.2.2 Prosthetic groups are cofactors which bind tightly to proteins including many enzymes

Many prosthetic groups involve metal ions chelated with a porphyrin ring. This is a tetrapyrrole consisting of four pyrrole rings joined by methine ($-CH=$) bridges. There is a whole range of porphyrins derived from this basic structure by the addition of various side groups and chelation with different metals. Cytochromes, for example, are proteins containing iron chelated with protoporphyrin IX (haem). The protoporphyrins contain two vinyl, four methyl and two propionic acid groups as ring substituents (see Fig. 4.6). Other important porphyrins are the etioporphyrins, with four methyl and four ethyl side groups, and the mesoporphyrins which have four methyl, two ethyl and two propionic acid side-chains attached in these positions. Chlorophyll (see Section 9.2.1) consists of a porphyrin ring containing magnesium.

Haem serves as a prosthetic group in a range of proteins/enzymes involved in the transfer of electrons from an electron donor (e.g. NADH) to molecular oxygen in oxidation reactions (as for the cytochromes), and in the oxidative cleavage of peroxides in reactions involving the enzymes catalase and peroxidase. Haem also occurs as a prosthetic group in the protein haemoglobin which is crucial for oxygen transport in animals and oxygen binding in root nodules (as leghaemoglobin). In functions involving oxygen binding, the iron remains in the ferrous state. However, in functions involving electron transfer (e.g. cytochromes), the iron alternates between ferrous and ferric states as electrons are transferred (see Section 4.4). For enzymes involved in peroxide cleavage, the ferric iron of haem is oxidized to a higher valance state (ferryl, Fe^{4+}).

The flavin cofactors, flavin mononucleotide (FMN) and flavin adenine dinucleotide (FAD), whose structures are shown in Fig. 4.4, are involved in a wide range of reduction/oxidation reactions in plant cells, including those catalysed by nitrate reductase and the pyruvate dehydrogenase complex. This reduction/oxidation involves the isoalloxazine ring system of the flavin prosthetic group (see Fig. 4.4).

Biotin functions in plant cells as a prosthetic group of various enzymes which catalyse reactions concerned with the transfer of carboxyl groups and incorporation of CO_2. Biotin has the structure shown below.

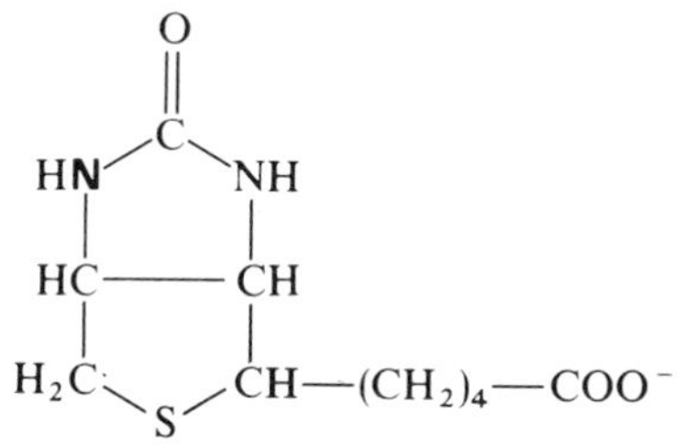

Biotin

Biotin is usually covalently bound to enzymes via an amide linkage between the ε-amino group of a lysine residue at the active site of the enzyme and the carboxyl group of the biotin to form biotinyl lysine. The enzymic reaction mechanism then involves carboxylation at the N atom shown in bold in the diagram. This carboxyl group can subsequently be transferred to an acceptor molecule. An example of the role of biotin in carboxylation reactions is afforded by acetyl-CoA carboxylase (see Section 16.1.1).

Thiamine pyrophosphate (TPP) functions as a prosthetic group in a number of enzymic reactions involving C—C bonds adjacent to a carbonyl group of keto-acids and keto-sugars.

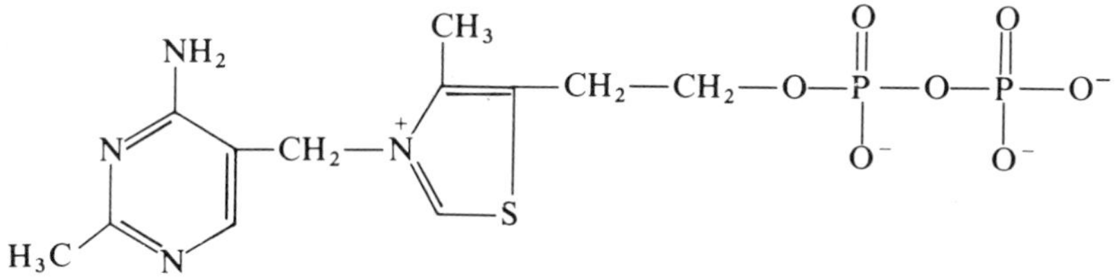

Thiamine pyrophosphate

Examples involving keto-acids include the oxidative and non-oxidative decarboxylation of pyruvate involving pyruvate dehydrogenase (Eqn. 5.3) and pyruvate decarboxylase (Eqn. 5.4) respectively.

Transketolase affords a very important example of a plant enzyme using TPP as a cofactor in reactions involving keto-sugars. It is involved in the oxidative pentose phosphate pathway and the C_3-CR cycle. Transketolase catalyses the transfer of a glycoaldehyde moiety (—CO—CH_2—OH) from a phosphorylated keto-sugar (donor) to a phosphorylated aldo-sugar (acceptor), thereby increasing the chain

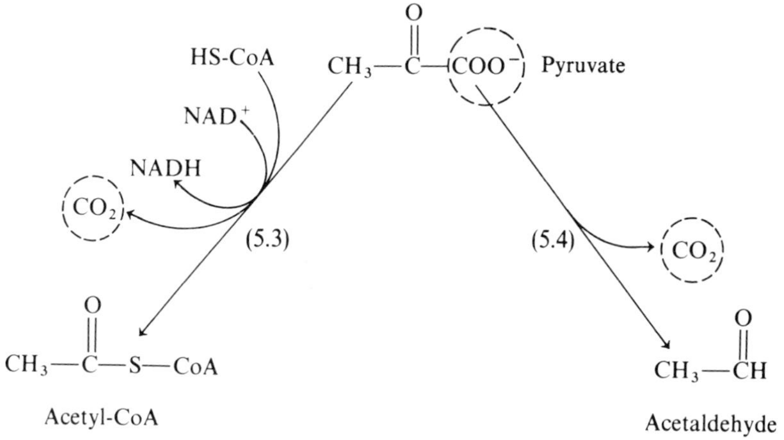

length of the acceptor by two carbon atoms. An example is shown in Eqn. 5.5.

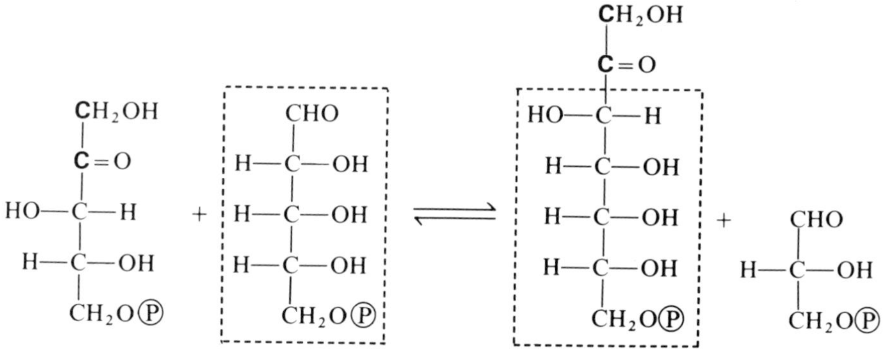

(5.3) (5.4)

Pyruvate

Acetyl-CoA

Acetaldehyde

(5.5)

The glycoaldehyde moiety binds tightly to TPP as it is transferred from donor to acceptor. Transketolase supports reactions with a range of donors and acceptors (see Table 8.1). The reaction mechanism centres on the carbon atom between the nitrogen and sulphur atoms of the thiazole ring of TPP as outlined in Fig. 5.2. Other enzymes using TPP as a prosthetic group have a similar reaction mechanism.

Pyridoxal phosphate (pyridoxal-P) is an important prosthetic group in a large number of reactions involving amino acids, particularly those catalysed by aminotransferases. It is also a prosthetic group of enzymes involved in reactions such as

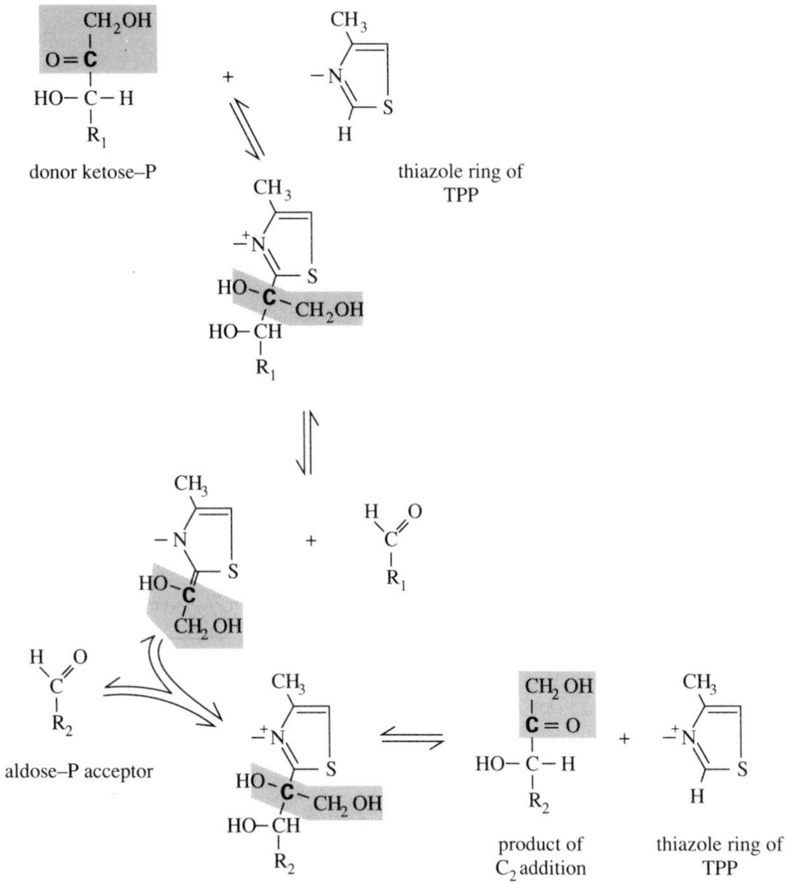

Fig. 5.2 The mechanism of action of TPP. This coenzyme participates in a number of enzymic reactions involving carbon–carbon bonds adjacent to a keto group. The example shown here is a transketolase reaction which involves the transfer of a glycoaldehyde moiety (shaded) from a donor ketose phosphate to an acceptor molecule, an aldose phosphate. The reacting carbon of the keto group is shown in bold.

the decarboxylation of aspartate, the desulphuration of homocysteine and the sulphuration of *O*-acetylserine to form cysteine.

In those enzymes in which it acts as a prosthetic group, pyridoxal-P is attached through the aldehyde group (shown in bold) to the ε-amino group of a lysine

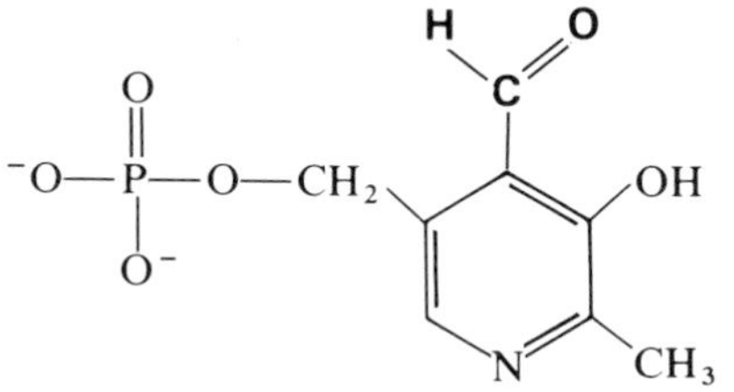

residue in the peptide chain of the enzyme. This leads to the formation of a 'Schiff' base. The pyridoxal-P moiety is then transferred from the enzyme to the amino acid substrate (Eqn. 5.6).

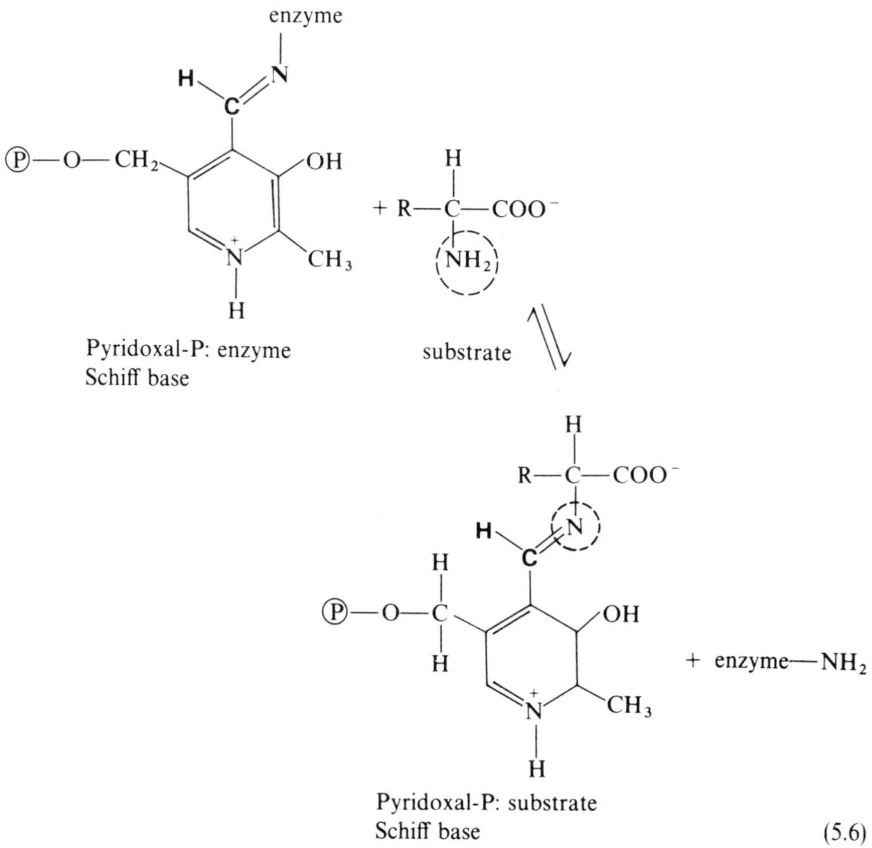

(5.6)

Various enzyme-mediated reactions involving rearrangement of the pyridoxal-P:substrate Schiff base with a second substrate (e.g. a keto-acid) result in different types of reaction products and reattachment of pyridoxal-P to the ε-amino group of the lysyl residue in the enzyme concerned. An example of a pyridoxal-P-mediated reaction involving the aminotransferases is described in Section 13.6.

It is interesting to note that many cofactors synthesized by plants contain moieties that serve as vitamins in animal nutrition. Thus THF and coenzyme A contain folic acid and pantothenic acid residues respectively, the flavin cofactors contain riboflavin and TPP contains thiamine (Vitamin B_1) whilst biotin functions as a vitamin in its own right.

5.3 Kinetics of reactions catalysed by non-regulatory enzymes

5.3.1 Kinetics of chemical reactions depend on the concentration and number of reactants

Before examining the effects of enzymes on chemical reactions, it is informative to consider the rates of the uncatalysed processes and how these depend on the concentrations of the reactants, i.e. the reaction kinetics. Chemical reactions can be classified into monomolecular, bimolecular or termolecular reactions if one, two or three molecules respectively, react to give the product(s).

Different reactions show reaction rates which depend in differing ways on the concentrations of reactants, usually referred to as substrate(s) in enzyme-catalysed reactions. This relationship between rate and concentration is described as the *order* of the reaction. For example, if a reaction proceeds at a rate which is directly proportional to the concentration of a reactant, this is termed a *first order reaction* (Fig. 5.3a). If, in the simple monomolecular reaction

$$S \longrightarrow P$$

substrate   product

the rate of formation of P is proportional to the concentration of substrate, $[S]$, the first order rate equation is given by

$$-d[S]/dt = k[S] \qquad (5.7)$$

where $-d[S]/dt$ is the rate at which the concentration of S decreases. The constant k is called the rate constant and for a first order reaction this has units of $time^{-1}$ (usually s^{-1}).

Equation 5.7 can be integrated to give

$$-\ln[S]_t = kt + \text{constant} \qquad (5.8)$$

where t denotes time and $[S]_t$ is substrate concentration at time t. At the beginning of the reaction when $t=0$ and $[S]_t=[S]_0$, kt is zero and so the constant in Eqn. 5.8 is equal to $-\ln[S]_0$. Thus

$$-\ln[S]_t = kt - \ln[S]_0.$$

This can be rearranged, and by converting to $\log_{10}$

$$\log_{10}([S]_0/[S]_t) = kt/2.303. \qquad (5.9)$$

So, for a first order reaction, a plot of $\log_{10}([S]_0/[S]_t)$ against t gives a straight line with slope equal to $k/2.303$. (Fig. 5.3d).

In *second order reactions*, the rate is proportional to the product of the concentration of two reactants in a bimolecular reaction or to the second power of the concentration of the one substrate of a monomolecular reaction. For the bimolecular reaction

$$S + S' \rightarrow P \qquad (5.10)$$

the rate of reaction, which can be specified by either the rate of consumption of S or S' or by the production of P, is given by

$$-d[S]/dt = k[S][S']. \qquad (5.11)$$

Of course, if the reaction involves the production of P from two molecules of S (i.e. $2S \rightarrow P$), then $-d[S]/dt = k[S]^2$.

For a second order reaction, k has units of $1/(\text{concentration} \times \text{time})$ (e.g. $M^{-1} s^{-1}$). Integration of Eqn. 5.11 yields two different equations depending on the relative initial concentrations of the two reactants.

If $[S]_0 \simeq [S']_0$

$$\frac{[S]_0 - [S']_0}{[S_0][S']_0} = kt. \qquad (5.12)$$

If $[S]_0 > [S']_0$

$$\log_{10}\left[\frac{[S']_0[S]_t}{[S]_0[S']_t}\right] = k\frac{([S]_0 - [S']_0)}{2.303}t \qquad (5.13)$$

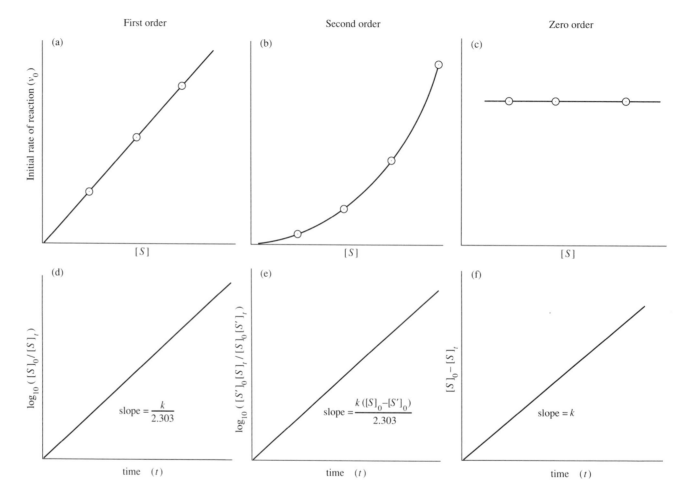

Fig. 5.3 A schematic representation of the rates of (a) first, (b) second and (c) zero order reactions as a function of substrate concentration, $[S]$. The progress curves of these reactions with time differ for the three types or reaction but these can be expressed as linear functions as shown for the (d) first, (e) second and (f) zero order reactions. Note that in d–f the ordinates reflect different functions of the concentration(s) of the substrate(s). S_0 and S_t are substrate concentrations at times 0 and t respectively, whilst k is the rate constant. For bisubstrate reactions the two substrates are denoted by S and S'.

and a plot of

$$\log_{10}\left[\frac{[S']_0[S]_t}{[S]_0[S']_t}\right]$$

against time is a straight line with

$$\text{slope} = k\,\frac{([S]_0 - [S']_0)}{2.303} \quad \text{(Fig. 5.3e)}.$$

If the rate of reaction is independent of the concentration of reactants then it is said to be zero order (Fig. 5.3c). The rate equation is given by

$$-\mathrm{d}[S]/\mathrm{d}t = k \tag{5.14}$$

which, when integrated, gives

$$[S]_0 - [S]_t = kt \tag{5.15}$$

so that a plot of $[S]_0 - [S]_t$ against t gives a straight line of slope k (Fig. 5.3f).

It is important to note that a second order reaction, as shown for example in Eqn. 5.10, can appear to be first order if one of the reactants is present at a much greater concentration than the other. In this case, the rate of reaction will seem to be proportional to the concentration of only the less concentrated reactant. This is termed a *pseudo-first-order* reaction.

The rates of many catalysed reactions depend on the concentration of catalyst, not reactant(s). In this case, the reaction is zero order with respect to the reactants.

5.3.2 Enzymes decrease the free energy of activation of chemical reactions by binding substrates at specific sites

Chemical reactions involve breakage and/or formation of one or more chemical bonds in reactant molecules or substrates. The reactions only take place if a certain fraction of the reactant molecules possess sufficient energy to enter an activated condition—the *transition state* (Fig. 5.4). In this state, the chemical bonds in the reactants are broken and there is a 50% probability that they will reform to yield the product(s) rather than the reactants. The minimum energy required to elevate the reactant molecules to the transition state is termed the activation energy. The *free energy of activation*, $\Delta G^{\ddagger}$, for a reaction is the energy required to convert all the molecules in 1 mol of the reactant(s), at a given temperature, to the transition state.

There are two ways in which the rates of chemical reactions can be accelerated. The first is, obviously, a rise in temperature, since this imparts increased thermal motion and kinetic energy to molecules, increasing the proportion entering the transition state. The second way is through the activity of a catalyst. Catalysts form a transition state complex with the reactants, which has a lower activation energy than that of the uncatalysed reaction (Fig. 5.4). When the products are formed the free catalyst is released in an unchanged form and can be recycled.

Chemical reactions can be catalysed by a range of inorganic compounds. In biological systems enzymes very effectively catalyse specific reactions, increasing rates of reaction by 10^8 to 10^{11} times the uncatalysed rate. Each enzyme has a characteristic tertiary structure and the substrate(s) for the reaction bind to highly specific regions termed active sites. One example concerns the enzyme ribulose-1,5-P_2 carboxylase/oxygenase (RuP_2-C/Oase) (Fig. 5.5). The specificity of the binding between RuP_2 and CO_2 ensures that only molecules with very similar structure will attach to the active site.

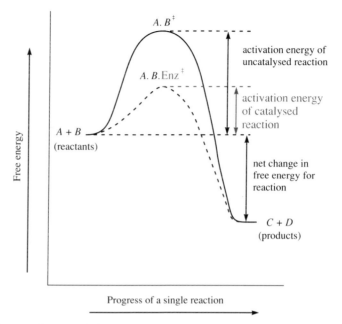

Fig. 5.4 Enzymes decrease the free energy of activation for chemical reactions. For chemical reactions to proceed reacting molecules must attain a minimum energy for the chemical bonds to be altered. The energy that those molecules have to acquire in order to react is known as the activation energy. Although enzymes do not affect net free energy change of reactions they do facilitate them by decreasing the activation energy (dashed curve). Enzymes combine with the substrate(s) to form a *transition state complex* which has a lower free energy than the transition state formed by the reactants in the uncatalysed reaction (black line), thereby permitting the catalysed reaction to proceed more readily at physiological temperatures. A and B are substrates, and C and D are products of the reaction $A + B \rightarrow C + D$. Abbreviations: Enz = enzyme; $\ddagger$ = transition state complex.

5.3.3 Enzyme-catalysed reactions show saturation with respect to their substrates

The general principles of the kinetics of chemical reactions are applicable to enzyme-catalysed processes. However, for a given set of reaction conditions and a given amount of enzyme, the rate does not increase indefinitely with substrate concentration as for non-enzymic reactions but approaches a maximum, characteristic of the reaction. This is known as substrate saturation. If the initial velocity of an enzyme-catalysed reaction (v_0) is plotted as a function of the concentration of

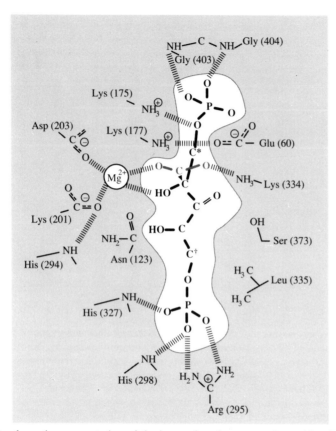

Fig. 5.5 A schematic representation of the interactions between amino acid residues at the active site of spinach ribulose-1,5-P_2 carboxylase (RuP$_2$-Case) and the C$_6$ transition state intermediate of the RuP$_2$-Case reaction. It is believed that binding between charged amino acid residues and phosphate groups of ribulose-1,5-P_2 opens up the central portion of ribulose-1,5-P_2 for attachment of CO_2 (shown in grey). The RuP$_2$-Case reaction requires the presence of Mg^{2+} which is involved in binding ribulose-1,5-P_2 and CO_2. The C atoms C* and C† denote atoms derived from C-1 and C-5 respectively of ribulose-1,5-P_2. After Andersson, I., Knight, S., Schneider, G., Lindqvist, Y., Lundqvist, T., Brändén, C-I. & Lorimer, G. (1989) *Nature*, **337**, 229–34.

the reactant or substrate, then a rectangular hyperbola results as shown in Fig. 5.6. At very low substrate concentrations (and in the presence of non-limiting enzyme levels) the reaction is first order with respect to substrate

$$v_0 = k[S] \tag{5.16}$$

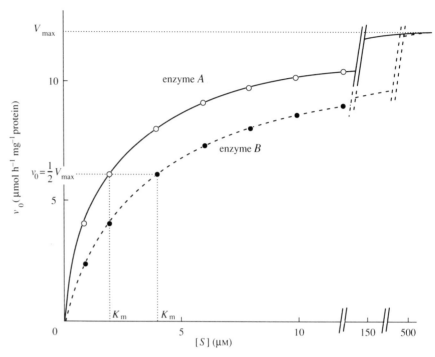

Fig. 5.6 The relationship between the initial rate of reaction (v_0) and substrate concentration [S] for a simple monomolecular reaction catalysed by a non-regulatory enzyme. The relationship takes the form of a rectangular hyperbola described by the Michaelis–Menten equation (Eqn. 5.26). Two hypothetical examples are given here (shown by the solid and dashed curves). Both cases have identical values for V_{max} (12 μmol substrate h^{-1} mg^{-1} protein) but for enzyme A, $K_m = 2$ μM, whilst for enzyme B, $K_m = 4$ μM. Enzyme B attains its maximum rate at much higher substrate concentrations than does enzyme A. V_{max} and K_m are defined in the text.

(cf. Eqn. 5.7 for non-enzymic reactions). At high concentrations of substrate, v_0 is independent of substrate concentration (i.e. substrate saturation) and the reaction is zero order (Eqn. 5.14). The rate of reaction at substrate saturation is maximal and is termed V_{max}.

5.3.4 Reaction kinetics of non-regulatory enzymes are expressed by the Michaelis–Menten equation

The saturation effect in enzyme-catalysed reactions led to the concept that enzymes react reversibly with their substrates to form an enzyme–substrate complex and that this is an essential step in enzyme catalysis. Substrate saturation implies that a given amount of enzyme has a limited number of sites at which the enzyme–substrate complex can form. This idea was developed and used by Michaelis and Menten and is best described for the simple case of a reaction involving one substrate, S. The Michaelis–Menten theory assumes that the substrate forms a reversible complex (ES) with the enzyme and that this enzyme–substrate complex breaks down in a second step to yield the free enzyme and the product (P). Each of these steps (both forward and backward reactions) has a particular rate constant (Eqns 5.17 & 5.18).

$$E + S \underset{k_2}{\overset{k_1}{\rightleftharpoons}} ES \underset{k_4}{\overset{k_3}{\rightleftharpoons}} E + P$$

$$(5.17) \quad (5.18)$$

Considering the initial rate of reaction v_0, the formation of ES from E and P will be negligible as the concentration of P will be nearly zero. As a result, k_4 can be ignored in subsequent analysis. The rate of the reaction (as measured by the formation of P) is given by

$$v_0 = k_3 [ES]. \tag{5.19}$$

The formation of ES from E and S obeys second order kinetics such that

$$d[ES]/dt = k_1 ([E_T] - [ES])[S] \tag{5.20}$$

where E_T is the total concentration of enzyme in the system, i.e. the sum of free enzyme (E) and enzyme–substrate complex (ES). ES is removed in two reactions: the formation of product (with rate constant k_3); and the reverse reaction in Eqn. 5.17 (rate constant k_2). These are both first order reactions. Thus,

$$-d[ES]/dt = k_2[ES] + k_3[ES]. \tag{5.21}$$

When the rate of the overall reaction (production of P) is constant the system is at a 'steady state'; the rate of formation of ES equals its rate of breakdown, so that the rates of the forward and reverse reactions (Eqns 5.20 & 5.21) are equal.

$$k_1([E_T] - [ES])[S] = k_2[ES] + k_3[ES]$$

$$k_1([E_T] - [ES])[S] = (k_2 + k_3)[ES]$$

$$\frac{([E_T] - [ES])[S]}{[ES]} = \frac{(k_2 + k_3)}{k_1} = K_m. \tag{5.22}$$

The term $(k_2 + k_3)/k_1$ is called the Michaelis–Menten constant, K_m. Rearranging

Eqn. 5.22 gives

$$[E_T][S] - [ES][S] = K_m[ES]$$

$$[E_T][S] = K_m[ES] + ([ES][S])$$

$$[E_T][S] = ES(K_m + [S])$$

$$[ES] = \frac{[E_T][S]}{K_m + [S]}. \tag{5.23}$$

Since the initial rate of the enzymic reaction, v_0, is given by Eqn. 5.19, this can be used with Eqn. 5.23 to give

$$v_0 = \frac{k_3[E_T][S]}{K_m + [S]}. \tag{5.24}$$

When the concentration of S is very high, the enzyme is virtually all present as enzyme–substrate complex. The maximum rate of reaction, V_{max}, is then given as

$$V_{max} = k_3[E_T]. \tag{5.25}$$

Substituting in Eqn. 5.24 we have

$$v_0 = \frac{V_{max}[S]}{K_m + [S]}. \tag{5.26}$$

This is the *Michaelis–Menten equation* and is the rate equation for a single-substrate, enzyme-catalysed reaction. As discussed in Section 5.3.6, this equation also applies to bisubstrate reactions when one substrate is at saturating concentration. It describes the relationship between initial substrate concentration, initial velocity of reaction and maximum velocity. The latter term, V_{max}, is also a function of enzyme concentration (Eqn. 5.26).

The Michaelis–Menten constant, K_m, is equal to the substrate concentration at which the initial rate of reaction, v_0, equals half the maximum rate (i.e. where $v_0 = \frac{1}{2}V_{max}$). This can be demonstrated as follows.

$$v_0 = \frac{V_{max}}{2} = \frac{V_{max}[S]}{K_m + [S]}.$$

Dividing through by V_{max} gives

$$\frac{1}{2} = \frac{[S]}{K_m + [S]}$$

$$K_m + [S] = 2[S]$$

$$\therefore K_m = [S] \text{ when } v_0 = \tfrac{1}{2} V_{max}.$$

The importance of K_m and V_{max} in describing the properties of specific enzyme-catalysed reactions will become apparent subsequently. Not only do these parameters describe the kinetic behaviour of enzymes but they also provide insights into likely *in vivo* activities and regulatory mechanisms. V_{max} describes the maximum rate at which a given amount of enzyme catalyses a reaction whilst the K_m value is indicative of the affinity of the enzyme for a substrate. Thus, an enzyme with a high affinity for its substrate attains a rate of $\tfrac{1}{2} V_{max}$ at a much lower substrate concentration than an enzyme with a low affinity, i.e. it has a lower K_m value (see Section 5.3.7, for example). K_m has a theoretical derivation (Eqn. 5.23) but, in practice, it is determined experimentally as the concentration of substrate which supports $\tfrac{1}{2} V_{max}$.

For some enzymic reactions k_2 is very large compared with k_3. In this case, K_m simplifies to k_2/k_1 and K_m is approximately equal to the dissociation constant of *ES*. This is termed K_s where

$$K_s = [E_T][S]/[ES]. \tag{5.27}$$

Although K_m and K_s are often regarded as synonymous, this is often an incorrect assumption; it is *only* true if k_2 is much larger than k_3.

5.3.5 The Michaelis–Menten equation can be rearranged to allow accurate determination of K_m and V_{max}

The Michaelis–Menten equation describes the relationship between initial reaction rate and substrate concentration. Since this relationship takes the form of a rectangular hyperbola (Eqn. 5.26, Fig. 5.6), it is sometimes difficult to get an accurate estimate of K_m and V_{max} from experimental plots of v_0 against $[S]$. But, as noted in the previous section, these parameters provide a great deal of information about the likely role of an enzyme *in vivo*. It is therefore important to obtain reliable estimates, especially of K_m. This can be achieved by algebraically transforming Eqn. 5.26 to one of several linear forms.

The simplest and most often used transformation involves taking the reciprocal of both sides of the Michaelis–Menten equation (Eqn. 5.26).

$$\frac{1}{v_0} = \frac{K_m + [S]}{V_{max}[S]}$$

This can be arranged to

$$\frac{1}{v_0} = \frac{K_m}{V_{max}} \cdot \frac{1}{[S]} + \frac{1}{V_{max}}. \tag{5.28}$$

This is the *Lineweaver–Burk equation*. Plotting $1/v_0$ against $1/[S]$ gives a straight line with slope K_m/V_{max} and an intercept on the $1/v_0$ axis of $1/V_{max}$. The intercept of the $1/[S]$ axis is equal to $-1/K_m$ (Fig. 5.7a). Such a plot can also provide information on inhibition of enzymic reactions (Section 5.4).

Lineweaver–Burk plots tend to give undue emphasis to points obtained at low substrate concentrations. Due to the use of reciprocal plots these points are likely to suffer from greater error, with less reliable estimates of kinetic parameters being obtained, than for data obtained at higher substrate levels. Thus, alternative derivations of the Michaelis–Menten equation are often preferable.

In an *Eadie–Hofstee* plot (Fig. 5.7b), both sides of Eqn. 5.28 are multiplied by V_{max} and rearranged to give

$$v_0 = V_{max} - K_m \frac{(v_0)}{[S]}. \tag{5.29}$$

Plotting v_0 against $v_0/[S]$ then gives a straight line of slope $-K_m$ with intercepts of V_{max} on the v_0 axis and V_{max}/K_m on the $v_0/[S]$ axis.

Multiplying Eqn. 5.28 through by $[S]$ yields

$$\frac{[S]}{v_0} = \frac{K_m}{V_{max}} + \frac{[S]}{V_{max}} \tag{5.30}$$

and a plot of $[S]/v_0$ against $[S]$ also gives a straight line but with slope $1/V_{max}$, an intercept on the $[S]/v_0$ axis of K_m/V_{max} and an intercept on the $[S]$ axis of $-K_m$. This is termed a *Woolf* plot (Fig. 5.7c).

Eadie–Hofstee and Woolf plots not only yield values of K_m and V_{max} but they also emphasize deviation from Michaelis–Menten kinetics which might not show up in a Lineweaver–Burk plot. However, as shown in later sections, Lineweaver–Burk plots can yield very useful information about inhibition of enzyme activity and about reaction mechanisms.

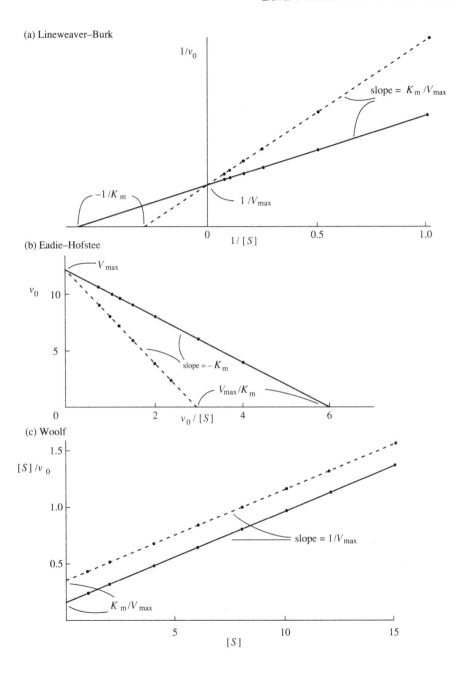

(a) Lineweaver–Burk

$1/v_0$

slope = K_m/V_{max}

$-1/K_m$

$1/V_{max}$

$1/[S]$

0   0.5   1.0

(b) Eadie–Hofstee

V_{max}

v_0 10

5

slope = $-K_m$

V_{max}/K_m

0   2 $v_0/[S]$ 4   6

(c) Woolf

1.5

$[S]/v_0$

1.0

slope = $1/V_{max}$

0.5

K_m/V_{max}

5   10   15

$[S]$

5.3.6 Enzyme catalysis of bisubstrate reactions

The foregoing equations relate to enzyme-catalysed reactions involving a single substrate. Most enzymic reactions involve two substrates. Analysis of the kinetics of a reaction such as

$$A + B \rightleftharpoons Y + Z$$

is far more complicated. There are several possible enzyme–substrate or enzyme–product complexes (EA, EB, EY, EZ) and ternary complexes such as EAB, EAY, EBZ or EYZ. However, it is relatively simple to estimate K_m and V_{max} for each of the substrates individually, since (see Section 5.3.1) bisubstrate reactions can be treated as single-substrate reactions if one of the substrates is at saturating levels. Thus, if A is at a saturating concentration, then measuring v_0 with respect to the concentration of B yields information regarding K_m and V_{max} with respect to B. These are usually written as K_m^B and V_{max}^B. Similarly, values of K_m^A and V_{max}^A can be found by repeating the process with saturating levels of B and varying the concentration of A.

Bisubstrate reactions can be either single-displacement or double-displacement reactions. These can be distinguished by their kinetics.

For *single-displacement reactions*, both substrates must be present at the enzyme active site (forming a ternary complex EAB). The binding can be *ordered*, in which case one of the substrates must bind first (although the release of products is not necessarily ordered, see Eqn. 5.31) or it can be *random* with either of the substrates binding, or either of the products being released, in a random order (Eqn. 5.33).

Ordered reactions can be differentiated from random bisubstrate reactions by studying the effect of the products on the overall reaction. In an ordered reaction, one specific substrate (known as the *leading* substrate) must be bound first (e.g. A in Eqn. 5.31). The second or *following* substrate (e.g. B in Eqn. 5.31) then binds to form a ternary complex (EAB). The enzyme catalyses the reaction to yield a ternary complex of enzyme+products X and Y. These products are then released. As

Fig. 5.7 The rectangular hyperbola obtained by plotting v_0 against $[S]$ for an enzymic reaction can be transformed into a number of linear forms which permit easier estimation of V_{max} and K_m (see text). Shown here are the data for enzymes A (black line) and B (dashed line) from Fig. 5.6 presented as (a) a double reciprocal Lineweaver–Burk plot, (b) an Eadie–Hofstee plot and (c) a Woolf plot. For simplicity units for the axes have been omitted.

shown below, release of X and Y takes place in an ordered manner, although this is not necessarily the case.

$$E \rightleftharpoons EA \rightleftharpoons EAB \rightleftharpoons EXY \rightleftharpoons EX \rightleftharpoons E \quad (5.31)$$

A typical example of such a reaction is malate dehydrogenase which catalyses the reaction

$$NAD^+ + malate \rightleftharpoons oxaloacetate + NADH + H^+. \quad (5.32)$$

In this reaction, NAD^+ is the leading substrate. This is an ordered bimolecular reaction since the last product to be released inhibits the overall reaction by competing with the leading substrate only, i.e. NADH competes with NAD^+ but not with malate.

In random bisubstrate reactions either substrate A or substrate B can combine with the free enzyme, E, giving either of the binary complexes EA or EB. The ternary complex EAB is formed when the other substrate binds to the binary complex (Eqn. 5.33).

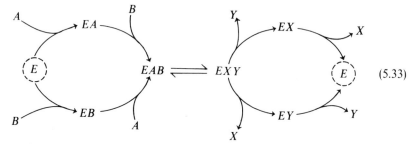

$$(5.33)$$

Release of the products X and Y is also shown here as a random process, but in some cases product release can be ordered. The rate equation for an ordered single-displacement reaction is given by Eqn. 5.34.

$$v_0 = \frac{V_{max}}{\dfrac{K_s^A K_m^B}{[A][B]} + \dfrac{K_m^A}{[A]} + \dfrac{K_m^B}{[B]} + 1} \quad (5.34)$$

This can be converted to a double reciprocal form for one of the substrates (in this case A) as

$$\frac{1}{v_0} = \frac{1}{V_{max}}\left(K_m^A + \frac{K_s^A K_m^B}{[B]}\right)\left(\frac{1}{[A]}\right) + \frac{1}{V_{max}}\left(1 + \frac{K_m^B}{[B]}\right). \quad (5.35)$$

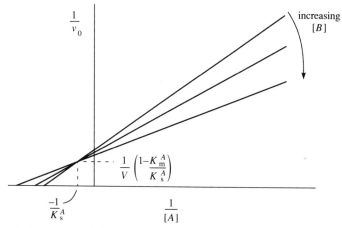

Fig. 5.8 A double reciprocal Lineweaver–Burk plot for an ordered, single-displacement, bisubstrate reaction involving two substrates A and B. Plotting $1/v_0$ against $1/[A]$ at different concentrations of B yields a series of lines with a common intersect.

A plot of $1/v_0$ against $1/[A]$ yields a straight line with a slope of

$$\frac{1}{V_{max}}\left(K_m^A + \frac{K_s^A K_m^B}{[B]}\right) \quad \text{(see Fig. 5.8).}$$

Each concentration of B gives a different slope and the intercept on the $1/v_0$ axis for each of these lines gives values of

$$\frac{1}{V_{max}}\left(1 - \frac{K_m^A}{K_s^A}\right).$$

Similarly, the intercepts on the $1/[A]$ axis give values for $-1/K_s^A$.

Some bisubstrate reactions have a *double-displacement* or *ping-pong* mechanism. In these reactions, as illustrated below, the first substrate combines with free enzyme (E) to give an enzyme–substrate complex. This dissociates to release product X leaving a modified enzyme E^*. The second substrate combines with E^* to form a second complex E^*B.

$$E \rightleftharpoons EA \rightleftharpoons E^* \rightleftharpoons E^*B \rightleftharpoons E$$

This subsequently dissociates to release free enzyme and product Y. The amino-transferase reactions discussed in Section 13.6 are an excellent example.

The general equation relating the rate of double-displacement reactions to substrate concentration for a reaction in which A is the first substrate and B the second, takes the form

$$v_0 = \frac{V_{max}}{1 + \frac{K_m^A}{[A]} + \frac{K_m^B}{[B]}}. \tag{5.36}$$

This can be linearly transformed to

$$\frac{1}{v_0} = \frac{K_m^A}{V_{max}}\left(\frac{1}{[A]}\right) + \left(1 + \frac{K_m^B}{[B]}\right)\left(\frac{1}{V_{max}}\right) \tag{5.37}$$

so that plots of $1/v_0$ against $1/[A]$ at different concentrations of B yield a set of parallel lines of slope K_m^A/V_{max} and intercepts on the $1/v_0$ axis of

$$\frac{1}{V_{max}}\left(1 + \frac{K_m^B}{[B]}\right)$$

(see Fig. 5.9).

These bisubstrate reactions can often be distinguished by their kinetic behaviour. In some instances, however, kinetic analysis is not enough and a much more detailed analysis of reaction mechanisms is necessary. For example, double-displacement reactions exhibit a very active partial reaction, with no formation of the final product, when only one of the two displacement reactions is involved. This is known as equilibrium exchange and is best explained by reference to the reactions catalysed by the aminoacyl-tRNA synthetases. These reactions have the general outline

ATP   amino   PP$_i$ tRNA   aminoacyl   AMP
      acid   -tRNA

Enzyme ———————————————————————— Enzyme.

First displacement reaction | Second displacement reaction

In these reactions PP$_i$ is in equilibrium with the substrates of the first displacement reaction. Thus, if the enzyme is incubated with amino acid, ATP and PP$_i$ (but without tRNA) and the PP$_i$ is labelled with ^{32}P, ^{32}P-label will be incorporated from [32]PP$_i$ into ATP. A single displacement reaction would require all the substrates to be present before ^{32}P-label appears in ATP.

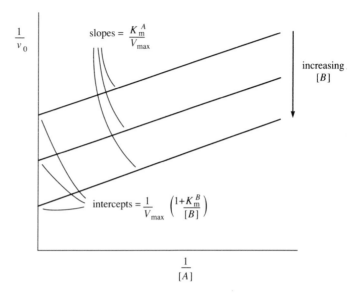

Fig. 5.9 A double reciprocal Lineweaver–Burk plot for a double displacement enzymic reaction in which A is the first substrate. Plots of $1/v_0$ against $1/[A]$ at different concentrations of the second substrate (B) yield a series of parallel lines with intercepts on the ordinate.

5.3.7 Enzymes with different kinetic properties can compete for a common substrate

Some enzyme-catalysed reactions have substrates in common. If two enzymes are competing for a single substrate then the fate of that substrate will depend on its concentration and the kinetic features (K_m and V_{max}) of the enzymes involved. This is illustrated hypothetically for enzymes A and B competing for substrate S, in Fig. 5.6. Enzyme A has a higher affinity (lower K_m) for substrate than does enzyme B. At sub-saturating substrate concentrations the rate at which substrate is consumed is greater via enzyme A than via enzyme B. So, even though it is possible to demonstrate activity of enzyme B with significantly high V_{max} values, the reaction catalysed by B does not necessarily occur at significant rates under the low substrate concentrations likely to be found within cells. Consequently, it is important to know values of K_m as well as of V_{max} when considering whether an enzyme-catalysed reaction occurs to a physiologically meaningful extent *in vivo*.

5.3.8 Rates of enzymic reactions are strongly affected by pH and temperature

The rates of all enzymic processes are affected markedly by pH and temperature, as well as by substrate concentration. Most enzymes have a noticeable pH optimum above or below which activity decreases. The optimal pH and the degree to which activity declines on either side of the optimum varies considerably between enzymes. The sensitivity of enzymes to pH depends on the acid–base behaviour of both the enzyme and substrate, since pH affects ionization of the substrate and of the various dissociable groups on the enzyme (e.g. ε-amino groups and γ-carboxyl groups of lysyl and glutamyl residues respectively) and thereby influences protein structure and/or substrate binding sites on the enzyme. These changes can affect the enzyme–substrate complex, and thus the V_{max} value. They can also modify the affinity of the enzyme for its substrate (K_m).

As stated in Section 5.3.2 an increase in temperature increases the rate of chemical reactions generally, and enzymic reactions are no exception. However, the rate of reaction only increases up to a certain temperature; further rises cause a decrease in reaction rate. This decrease is caused by a lowered stability of the enzyme as the protein becomes denatured with a loss of the higher level of structure—see Section 2.5.3. The temperature optimum for enzymic reactions is, thus, a balance between increased reaction rates and increased denaturation and inactivation of the enzyme itself. The optimum also varies with the duration of the treatment since substantial inactivation occurs even at relatively mild temperatures given prolonged incubation times.

5.4 Inhibition of enzyme-catalysed reactions

Biochemical reactions are inhibited by a range of substances which act in a number of different ways. Studies of enzyme inhibition can be of great assistance in elucidating various problems, for example, an understanding of metabolic regulation and the sequence of reactions in various metabolic pathways. They have also aided understanding of the mechanism of enzyme–substrate interactions and the nature of the active sites of enzymes. There are two main types of enzyme inhibition: reversible and irreversible. Reversible inhibition involves the establishment of an equilibrium between inhibitor and free enzyme without permanently altering or destroying the functional groups of the enzyme. It is especially important in functioning cells as it provides a means of regulating the activity of metabolic pathways. Reversible inhibition can be analysed through derivations of

Michaelis–Menten kinetics. As discussed below, it can be competitive or non-competitive, although in bisubstrate reactions there is a third type—uncompetitive inhibition. Irreversible inhibition involves the chemical alteration or destruction of various functional groups on the enzyme and is less important than reversible inhibition in regulation of the functioning of plant cells. Consequently, subsequent subsections only consider the kinetic behaviour of reversible inhibition.

5.4.1 Competitive inhibition

A competitive inhibitor acts by binding to the active site of the enzyme and usually resembles the normal substrate in some way. For example, malonate inhibits the oxidation of succinate to fumarate by succinate dehydrogenase. Malonate and succinate both have two carboxyl groups. However, whereas the C_4 dicarboxylate, succinate, is dehydrogenated by succinate dehydrogenase by removal of hydrogen atoms at C-2 and C-3 to yield fumarate, an analogous reaction with the C_3 dicarboxylate, malonate, is impossible.

In competitive inhibition, the inhibitor competes with the normal substrate for binding at the active site of the enzyme, forming an enzyme–inhibitor complex (EI) analogous to the enzyme–substrate complex:

$$E + I \rightleftharpoons EI.$$

The inhibitor is not subject to chemical modification by the enzyme in the same way as the substrate. The binding of the inhibitor to the substrate binding site can be quantified by the inhibition constant (K_i). This parameter is analogous to K_s (Eqn. 5.27) and can be defined as the dissociation constant of the enzyme–inhibitor complex

$$K_i = \frac{[E][I]}{[EI]}. \tag{5.38}$$

With inhibitor (I) and substrate (S) in competition for the same binding site on the enzyme (E), the binding of I to E decreases the amount of enzyme–substrate (ES) complex formed. Because the rate of the overall reaction is proportional to [ES] (Eqn. 5.19), it follows that, for a given substrate concentration, the rate of formation of product from S decreases with increasing concentration of I. The degree of inhibition depends on the *ratio* of the concentrations of inhibitor and substrate. Competitive inhibition is overcome, therefore, by increasing [S] and raising the [S]/[E] ratio. As a result, V_{max} is unaffected by competitive inhibition whilst K_m is increased (Fig. 5.10). A plot of $1/v_0$ against $1/[S]$ for different inhibitor concentrations gives a family of straight lines intersecting on the $1/v_0$ axis (Fig. 5.11a). K_i can then be determined since the slope of the plot for the inhibited reaction is given by

$$\left(\frac{K_m}{V_{max}}\right)\left(1+\frac{[I]}{K_i}\right).$$

The intercepts on the $1/[S]$ axis are equivalent to values of $-1/K_m(1+[I]/K_i)$.

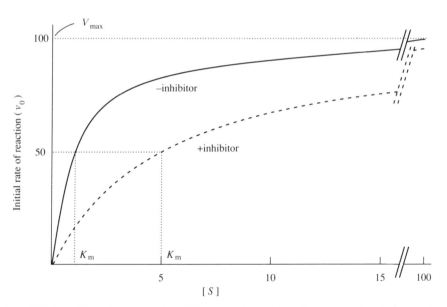

Fig. 5.10 The effect of a competitive inhibitor on the activity of an enzyme in relation to the concentration of substrate. The affinity of the enzyme for substrate is substantially decreased by the presence of the inhibitor.

Although most competitive inhibitors do not react with the enzyme, some enzymes catalyse reactions with different substrates competing for the same active site and so give rise to two different products. A good example of this is the competition between CO_2 and O_2 for the active site of the enzyme ribulose-1,5-P_2 carboxylase/oxygenase discussed in Chapter 11 (see Fig. 11.2). Each substrate competitively inhibits the activity shown by the enzyme towards the other substrate and each substrate results in the formation of different products.

A similar form of competitive inhibition can be the result of enzyme activity towards two closely related substrates (one of which is often non-physiological). In these cases, the enzyme catalyses the formation of two analagous products—that from the non-physiological compound is often toxic. An example of this is the incorporation of the amino acid analogue, fluorophenylalanine, by aminoacyl-tRNA synthetase into polypeptides in place of phenylalanine. Selenocysteine can also be incorporated into polypeptides in place of cysteine.

5.4.2 Non-competitive inhibition

In non-competitive inhibition, the inhibitor binds to the enzyme without competing with the substrate for the active site. This can either be with the free enzyme (Eqn. 5.39) or the enzyme–substrate complex (Eqn. 5.40).

$$E+I \rightleftharpoons EI \tag{5.39}$$

$$ES+I \rightleftharpoons ESI \tag{5.40}$$

There are thus two inhibitor constants which may or may not be equal:

$$K_i^{EI}=\frac{[E][I]}{[EI]} \quad \text{and} \quad K_i^{ESI}=\frac{[ES][I]}{[ESI]}.$$
$$\tag{5.41} \qquad\qquad \tag{5.42}$$

The binding between enzyme and inhibitor or enzyme–substrate and inhibitor affects the conformation and/or function of the enzyme so it less readily forms a complex with the substrate(s), and/or the complex, once formed, dissociates more slowly to yield the products. The degree of inhibition is independent of the substrate/inhibitor ratio and consequently, unlike competitive inhibition, it cannot be relieved by increasing substrate concentration. This means that V_{max} is lowered by inhibitor, however high the substrate concentration. With some inhibitors, the affinity of the enzyme for substrate is unaffected so K_m remains the same but in other cases K_m and V_{max} are changed. The effect of a non-competitive inhibitor on K_m is thus highly variable depending on the enzyme and inhibitor.

Non-competitive inhibition is easily recognized from double reciprocal plots of $1/v_0$ against $1/[S]$ in the presence of different inhibitor concentrations. Such plots have different slopes but, unlike competitive inhibition, do not have a common intercept on the $1/v_0$ axis. When extrapolated, such lines intersect somewhere to the left of the $1/v_0$ axis but, dependent on the nature of the inhibitor and interactions with enzyme and substrate, this can be above, below or on the $1/[S]$ axis, again emphasizing the variable effect upon K_m. The double reciprocal form of the rate equation for non-competitive inhibition is given by

$$\frac{1}{v_0} = \frac{K_m}{V_{max}}\left(1+\frac{[I]}{K^{EI}}\right) \times \frac{1}{[S]} + \frac{1}{V_{max}}\left(1+\frac{[I]}{K^{ESI}}\right) \qquad (5.43)$$

(assuming that the ESI complex is inactive). Where K_i^{EI} and K_i^{ESI} are equal, intersection of the lines from double reciprocal plots occurs on the $1/[S]$ axis at $-1/K_m$.

5.4.3 Uncompetitive inhibition

Uncompetitive inhibition is occasionally encountered in bisubstrate reactions. It is characterized in double reciprocal plots by a family of parallel straight lines with different intercepts on the $1/v_0$ axis, indicating that V_{max} decreases with increasing concentration of inhibitor.

5.4.4 Some applications of the kinetics of enzyme inhibition

As mentioned above, the enzyme ribulose-1,5-P_2 carboxylase/oxygenase catalyses reactions between ribulose-1,5-P_2 and both O_2 and CO_2. This is of considerable practical importance since the reaction involving O_2 inhibits the reaction involving CO_2 and so affects the plant's acquisition of carbon and hence its growth. The kinetic constants and V_{max} values can be used to predict the relative rates of the two reactions over a wide range of O_2 and CO_2 concentrations (see Section 11.3).

Inhibition kinetics can yield a considerable amount of information about the mechanism of action of certain herbicides. One example is given in Fig. 5.11 and concerns the mode of action of glyphosate. The data suggest that glyphosate binds

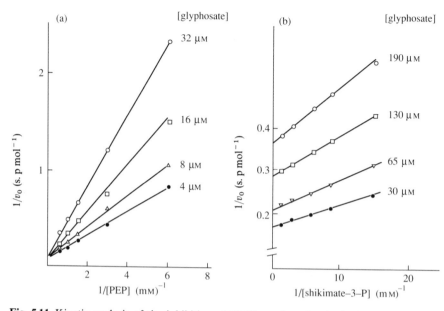

Fig. 5.11 Kinetic analysis of the inhibition of EPSP synthase by the herbicide glyphosate, showing how the site of action of glyphosate on the enzyme can be established. EPSP synthase supports the synthesis of 5-enolpyruvyl-shikimate-3-P (EPSP) from shikimate-3-P and PEP, which is an essential step in the synthesis of aromatic amino acids (see Section 13.7.4). The effect of glyphosate with respect to (a) PEP at a fixed concentration of shikimate-3-P and (b) shikimate-3-P at a fixed concentration of PEP is shown. Data are shown as double reciprocal plots and indicate that inhibition by glyphosate is (a) competitive with respect to PEP (i.e. varying glyphosate concentrations gives a family of straight lines which intersect on the $1/v_0$ axis) and (b) non-competitive with respect to shikimate-3-P (i.e. varying glyphosate concentrations gives a series of lines intersecting on the $1/[S]$ axis to the left of the $1/v_0$ axis). Since glyphosate is competitive with respect to PEP, this implies that the herbicide competes for the PEP binding site of EPSP synthase. After Amrhein, N., Holländer-Cyztko, H., Leifield, J., Schulz, A., Steinrücken, H.C. & Topp, H. (1982) *Journées internationales d'études du Groupe Polyphenols. Bulletin de Liason*, Vol. III, 21–31.

to the phosphoenolpyruvate (PEP) rather than to the shikimate-3-P binding site of 5-enolpyruvylshikimate-3-P synthase (EPSP synthase). Curiously, however, glyphosate seems to have no significant effect on other enzymes which catalyse reactions involving PEP (e.g. pyruvate kinase and PEP carboxylase). This is fortunate as many of these enzymes are important to animal metabolism!

5.5 Some enzymes have a regulatory role in metabolism

5.5.1 Processes catalysed by enzymes in cells require regulation

Cell functioning involves the combined activity of a great many biochemical reactions catalysed by an equally large number of enzymes. However, some enzymes catalyse reactions with completely opposite effects to others and would, uncontrolled, result in complete disorder. Clearly, these processes must be regulated for cells to function correctly and efficiently. Many examples of regulation of metabolic activity will be found elsewhere in this book but perhaps the most striking example of control of metabolic activity concerns glycolysis and can be observed in a range of cells including bacteria, yeast and plant and animal cells. Under anaerobic conditions cells break down sugars such as glucose by glycolysis to pyruvate and then to ethanol or lactate by fermentation. Fermentation of sugars yields only two ATP for every glucose broken down, so the rate of glucose consumption and hence glycolysis has to be high to maintain a sufficient supply of ATP for cellular function. However, on transfer to air, the pyruvate produced by glycolysis is oxidized by the tricarboxylic acid (TCA) cycle to yield more than 30 ATP per glucose. If glycolysis continued at a high rate the supply of ATP would increase about 15-fold, vastly in excess of cellular metabolic requirements. However, in air the rate of consumption of glucose by plant (and other) cells via glycolysis decreases dramatically. This is the 'Pasteur' effect and is primarily due to regulation of the activity of phosphofructokinase, one of the enzymes of glycolysis (see Section 7.5.2). This demonstrates that plant cells can regulate metabolic pathways through specific enzymes, a phenomenon known as metabolic regulation.

The formation and hydrolysis of fructose-1,6-P_2, a key intermediate in glycolysis and gluconeogenesis (see Section 7.5.2), provides an example of two antagonistic reactions involving two separate enzymes. Fructose-1,6-P_2 is formed by phosphorylation of fructose-6-P in a reaction dependent on ATP and catalysed by phosphofructokinase. However, fructose-1,6-P_2 phosphatase activity catalyses the dephosphorylation of fructose-1,6-P_2 to fructose-6-P and Pi. Since these reactions can occur in the same compartment of the cell it is theoretically possible for cycling to take place between fructose-6-P and fructose-1,6-P_2, a futile cycle which achieves nothing but consumption of ATP. There are a number of other potentially futile cycles in plant cells. The activities of the enzymes that constitute these cycles are strictly regulated to exclude futile hydrolysis of ATP.

There are many ways in which the activities of enzymes are controlled. A 'coarse' control can be achieved by varying the amount of enzyme protein present through regulation of transcription and translation (Chapter 20). Enzyme activities (and hence, cellular metabolic activities) are responsive to changes in the level of substrates or to other factors such as pH and the concentration of certain metal ions. Most importantly, however, enzyme activity, especially the activity of regulatory enzymes, can be modified by more direct effects, outlined in the following sections.

5.5.2 Activity of allosteric enzymes is modulated by binding of metabolites at non-catalytic sites

In many metabolic pathways involving a series of related enzymes, the flow of material through the pathway is controlled by the activity of just one, or occasionally a few, key enzymes. Generally, these enzymes show activity which is controlled or modulated by certain molecules (effectors) which bind non-covalently to the enzyme molecule, changing the kinetic properties of the enzyme. Enzymes modified in this way are known as allosteric enzymes and specific examples will be found throughout this book. Quite often the effector molecule is an end-product of the pathway, structurally dissimilar to the substrate but able to bind to the regulatory subunit and affect enzyme activity. Any one allosteric enzyme might have stimulatory (positive) effectors or inhibitory (negative) effectors. If the effector molecule(s) is not the enzyme substrate, the allosteric enzyme is under *heterotropic* control; if the effector molecule is the substrate, control is *homotropic*. Most allosteric enzymes show a mixture of homotropic and heterotropic control with binding sites both for substrate molecules (since all enzymes by definition have substrate binding sites) and for structurally unrelated compounds.

5.5.3 Kinetics of allosteric enzymes differ from those of non-regulatory enzymes

Allosteric enzymes do not conform to the standard Michaelis–Menten equation (see Section 5.3.4). For positive regulation of allosteric enzymes by an effector, a plot of initial velocity against substrate concentration yields a sigmoidal curve rather than a rectangular hyperbola. This is particularly true of homotropic allosteric enzymes and reflects the fact that binding of one substrate molecule can enhance the affinity of the enzyme for binding of subsequent molecules of the same substrate. This is termed *positive co-operativity*. Such sigmoid responses often mean that a small change in substrate concentration results in a far greater change in initial velocity than a hyperbolic response. The sigmoid response also spreads the 'useful' range in which activity is very sensitive to small changes in substrate concentration.

Negative co-operativity occurs when binding of one substrate *decreases* the affinity of the enzyme towards subsequent molecules of that substrate. In this case the relationship between initial velocity and substrate concentration shows a much flattened curve (Fig. 5.12). The enzyme is, thus, far less sensitive to changes in substrate concentration than non-regulatory enzymes or positively regulated enzymes.

Most allosteric enzymes respond to the binding of effector molecules by changing the affinity for the substrate without any change in the maximum rate at substrate saturation. A negative effector produces an increase in the apparent half-saturation constant without an effect on V_{max}. (Note that since allosteric

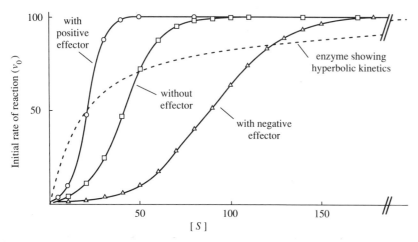

Fig. 5.12 Kinetics of an allosteric enzyme showing the effect of positive and negative effectors. Binding of an effector molecule alters the enzyme's affinity for substrate molecules and binding of additional effector molecules further alters the affinity for the substrate; the result is a sigmoid response of v_0 to substrate concentration $[S]$ (□—□). A positive effector enhances the affinity of the enzyme for substrate (○—○) whereas negative effectors (△—△) make the enzyme less sensitive to changes in substrate concentration. The effect of substrate concentration on a non-regulatory enzyme is shown for comparison (----). In the case shown here, the allosteric enzyme shows a change in activity from 10% V_{max} to 90% V_{max} if $[S]$ is changed threefold. For a similar change in activity, the non-regulatory enzyme, which obeys the Michaelis–Menten equation, requires nearly a 50-fold change in $[S]$.

enzymes do not conform to Michaelis–Menten kinetics the half-saturation constant is not K_m and 'apparent half-saturation constant', $K_{1/2}$, is preferable). Allosteric enzymes showing changes in $K_{1/2}$ but not V_{max} are sometimes called *K* enzymes. Some allosteric enzymes show the converse effect (i.e. a change in V_{max} without a change in $K_{1/2}$) and are termed *M* enzymes, but these are not common. In fact changes in V_{max} are rarely important as substrate levels within cells are usually far below saturation values.

5.5.4 *Some regulatory enzymes are converted between active and inactive forms by covalent modification of their structure*

As well as allosteric enzymes, there are other enzymes involved in regulation of metabolic processes. In these enzymes, regulation is brought about by covalent modification of enzyme structure. There are two main ways in which such changes can occur.

Several of the enzymes associated with chloroplasts exhibit light-dependent activation or deactivation. Changes from inactive to active states can be brought about *in vitro* and in the absence of light, by the thiol, dithiothreitol, suggesting that the light-induced changes in activity are associated with the oxidation state of thiol/disulphide groups within the enzymes. For those enzymes which are light-activated, the reduced (—SH) form is active whereas for enzymes which are activated in the dark the —SH form of the enzyme is inactive. It has been proposed that, in the dark, these groups exist in an oxidized state (—S—S—) but that, during light-dependent electron transport, reduced ferredoxin (Fd_{red}) can reduce the disulphide groups present in a class of low molecular weight proteins known as thioredoxins (Td). This reaction is catalysed by Fd:Td reductase. The reduced form of Td (Td_{red}) can in turn effect the reduction of oxidized forms of enzymes according to the overall scheme in Eqn. 5.44, in which the enzyme is light-activated.

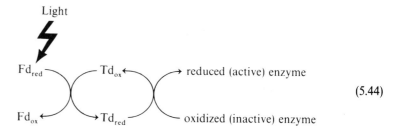

(5.44)

Specific examples of enzymes whose activity is modified in this manner are given in Section 10.13.1.

There are other types of covalent modification which affect enzyme activity. For example, some enzymes are subject to reversible phosphorylation/dephosphorylation. Depending on the enzyme, either the phosphorylated or the dephosphorylated form can be active. These changes are brought about by the activity of specific protein kinases, which catalyse the phosphorylation of specific proteins (including enzymes) at particular amino acid residues, and protein phosphatases (also specific), which cause dephosphorylation.

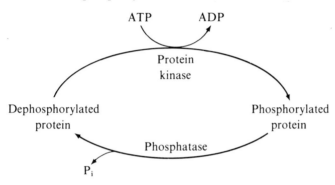

The precise mechanism of these changes is not understood. In the case of glutamate dehydrogenase from the yeast *Candida utilis*, the dephosphorylated enzyme is the more active with a four-to-10-fold increase in V_{max} and a six-fold decrease in K_m. Phosphorylation occurs at the —OH groups of seryl residues. However, in the case of yeast glycogen phosphorylase, phosphorylation appears to be at a threonyl residue.

To date, seven specific enzymes in plants are believed to be regulated by reversible phosphorylation, although not all of these involve regulation of catalytic activity. The seven are: pyruvate dehydrogenase from mitochondria; pyruvate P_i dikinase (see Section 12.6.1) and ribulose-1,5-P_2 carboxylase (RuP$_2$-Case) from chloroplasts; PEP carboxylase and fructose-6-P 2-kinase from the cytosol; hydroxymethylglutaryl-CoA reductase from the endoplasmic reticulum; and quinate:NAD$^+$ oxidoreductase. In these enzymes, phosphorylation is also associated with seryl or threonyl residues. In the case of RuP$_2$-Case, phosphorylation of the small subunit is associated with its transport across the chloroplast envelope and not with the regulation of catalytic activity. The light-harvesting chlorophyll *a/b* binding protein (LHCP), present in thylakoids, is also subject to reversible phosphorylation. In this case phosphorylation is associated with regulation of energy distribution between the two photosystems (see Section 9.16).

5.5.5 Activity of some enzymes is regulated by the level of Ca^{2+} in the cytosol

Protein phosphorylation and dephosphorylation appears to be involved in regulating the activity of certain key enzymes. There is growing evidence that calcium and the calcium binding protein, calmodulin, cause the protein kinase to act on the target enzyme.

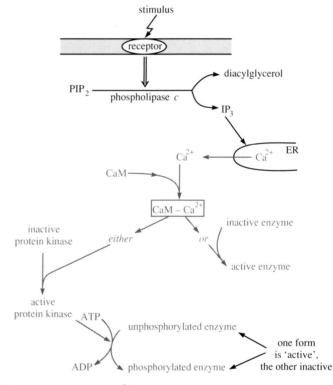

Fig. 5.13 The proposed role of Ca^{2+} in regulating some specific enzymes in plant cells in response to an extracellular stimulus. Details are given in the text. Abbreviations: CaM = calmodulin; PIP$_2$ = phosphatidyl inositol-4,5-P$_2$; IP$_3$ = inositol triphosphate; ER = endoplasmic reticulum. After Poovaiah, B.W., Reddy, A.S.N. & McFadden, J.J. (1987) *Physiol. Plantarum*, **69**, 569–73.

Some enzymes have been shown to exist in active/inactive forms due to phosphorylation/dephosphorylation which is Ca^{2+} dependent. Calcium is believed to be involved in the way illustrated in Fig. 5.13. The concentration of free Ca^{2+} in the cytosol is maintained at very low levels (typically 10^{-7} M) but increases in response to an extracellular stimulus such as light or hormones. The stimulus is believed to increase the activity of the enzyme phospholipase C which produces inositol triphosphate (IP$_3$) from phosphatidyl inositol-4,5-P$_2$ (PIP$_2$). Inositol triphosphate increases the permeability of the endoplasmic reticulum to Ca^{2+}, permitting the movement of Ca^{2+} from the lumen of the endoplasmic reticulum into the cytosol and increasing cytosolic concentration of Ca^{2+}. The Ca^{2+} binds to a Ca^{2+}-binding protein called calmodulin. Calmodulin is a protein of molecular mass of about 14.5–17 kDa and is found, for example, in maize coleoptile cells at levels in the micromolar range. The Ca^{2+}–calmodulin complex (but not free calmodulin) binds, with a high affinity, to a number of enzymes, modifying their activities. Several protein kinases are activated by binding to this complex. In Fig. 5.13 this is illustrated as a change from inactive to active protein kinase, bringing about the phosphorylation of a number of other proteins. In the case of NAD kinase, activation by calmodulin and Ca^{2+} is direct and does not proceed via phosphorylation/dephosphorylation mediated by a protein kinase.

It is believed that various primary stimuli from outside a cell (for example, light or hormones) can elicit the release of Ca^{2+} into the cytosol, causing various biochemical responses. Ca^{2+} is referred to as a secondary messenger. It is becoming increasingly apparent that Ca^{2+} is important in the regulation of a wide range of biochemical and physiological process in plants.

Further reading

Monographs and treatises: Davies (1987); Dixon & Webb (1979).
Nomenclature: International Union of Biochemistry (1979).
Kinetics: Cornish-Brown (1979).
Regulation: Ranjeva & Boudet (1987).

6
Membranes and metabolite transport

6.1 Membranes exhibit variation in their composition and cellular function

6.1.1 Membranes fulfil several essential cellular functions

As shown in Chapter 3, plant cells and the organelles they contain, are surrounded by one or more membranes. Membranes serve a number of critical functions. Firstly, and perhaps most obviously, membranes provide a relatively impermeable container for the different metabolites and enzymes necessary for cellular functioning. Membranes are able to transmit information from the external environment to the interior of the cell or organelle. They can also regulate metabolic processes by regulating transport of ions and metabolites into or out of the cell or from one cellular compartment to another. Membranes also play a crucial role in the energetics of cells by providing a matrix for the proteins involved in vectorial reactions associated with energy transduction in mitochondria (see Chapter 7) and in chloroplasts (see Chapter 9).

The capacity of living cells and organelles to regulate the concentration of solutes (metabolites) within them is of vital importance to cell growth and metabolism. Differences in the concentration of solutes between cells and their surroundings allow essential nutrients to move to areas of the tissue or plant where they are needed. Control of these differences is also very important in regulating osmotic relationships and hence the volume and shape of cells (factors crucial to, amongst other things, the functioning of stomata). This chapter is mainly concerned with the ability of membranes to provide a permeability barrier, whilst selectively transporting materials across that barrier.

6.1.2 Different membranes within the cell have different compositions

As discussed in Chapter 3, the basic structure of membranes is a fluid bilayer of lipids in which are embedded a range of different proteins. These proteins are either attached to both the hydrophobic and hydrophilic regions of the membrane (intrinsic proteins) or attached to the outermost hydrophilic regions only (peripheral proteins). However, the ratio of lipid to protein and the proportions of different lipids varies from membrane to membrane.

Separation of the various types of membrane within plant cells is not easy. Density gradient centrifugation makes it comparatively easy to isolate fractions at least enriched in a particular membrane type but such fractions are often contaminated with other membrane components. However, density gradient centrifugation in combination with aqueous polymer two-phase partition (see Fig. 9.9) is proving a powerful new technique for obtaining highly purified membrane fractions for analysis.

Since membrane functions such as transport and catalysis are mediated by proteins and each membrane has its own unique function, it is not surprising to find that different membranes have different protein contents and composition. Thylakoids contain about 40 different polypeptides making up 60–65% of the membrane, the remainder being pigments and acyl lipids. Relatively small proportions of acyl lipid are also found in the inner mitochondrial membrane which is up to 74% protein. Even within membranes there are local differences in composition. For example, within the internal membranes of higher plant chloroplasts, granal thylakoids have a protein:lipid ratio of 1.8 compared to 1.2 for stromal thylakoids. In fact, in appressed regions of thylakoids (see Fig. 3.15) only about 24% of the surface area is occupied by acyl lipids—rising to 51% in non-appressed regions. In contrast, the plasma membrane of plant cells contains only about 40–50% protein by weight whilst the tonoplast, which has even fewer associated biochemical or transport functions, has a protein composition approximately 30% lower than this.

Not only does the proportion of protein in membranes vary but the particular protein composition also differs depending on membrane function. For many, but not all, plant membranes highly specific proteins have been identified, serving as markers for the presence of the various types of membrane in subcellular preparations. Some marker proteins associated with various compartments and several membranes are given in Table 6.1. For example, endoplasmic reticulum is characterized by the presence of NADPH cytochrome c reductase whilst only Golgi bodies contain a nucleoside diphosphatase specific for inosine diphosphate (IDP-Pase). The IDP-Pase of Golgi bodies exhibits 'latency'; in its native state the enzyme is inaccessible, to a variable extent, to its substrates and generally shows lower activity than a non-latent, solubilized form. Some marker enzymes, known

Table 6.1 Examples of marker proteins and other constituents useful in identifying cell components. From Morré, D.J., Brightman, A.O. & Sandelius, A.S. (1987) In Findlay, J.B.C. & Evans, W.H. (eds) *Biological Membranes: a practical approach*, pp. 37–72. IRL Press, Oxford.

Cell component	Marker*	Specific activity[†]	Units
Endoplasmic reticulum	NADPH cytochrome *c* reductase	15.3	μmol cytochrome *c* reduced h^{-1} mg^{-1} protein
Golgi apparatus	Latent nucleoside diphosphatase with inosine-P$_2$ as substrate (pH 7.2)	47	μmol P$_i$ h^{-1} mg^{-1} protein
Plasma membrane	Glucan synthetase II	585	nmol glucose incorporated h^{-1} mg^{-1} protein
Tonoplast	NO$_3^-$ inhibited, Cl$^-$ stimulated, ATPase	0.55	μmol h^{-1} mg^{-1} protein
Mitochondria	Cytochrome oxidase	67.8	μmol O$_2$ h^{-1} mg^{-1} protein
Etioplasts	Carotenoids	80	ng mg^{-1} protein[‡]
Peroxisomes	Catalase	–	

* Markers listed here have been used for studies on etiolated hypocotyls of soybean.
[†] Specific activities of purified fractions provide values for comparison with other preparations.
[‡] This is not strictly a 'specific activity' but a 'specific ratio' of carotenoid:protein in etioplast membranes.

as *operational* markers, are not absolutely restricted to a particular membrane but are, nonetheless, more concentrated in some membranes than in others. For example, glucan synthetase II is mainly present in the plasma membrane but is also present in Golgi bodies. Some membranes are also characterized by non-protein components (e.g. chlorophyll in chloroplast lamellae).

Turning to the lipid component of membranes, it is possible to identify both similarities and striking differences between different membranes. In non-photosynthetic tissues, the phospholipid composition of various membranes is very similar (Table 6.2). For example, phosphatidyl choline and phosphatidyl ethanolamine are dominant and together comprise 68–94% of the total phospholipids in all the membranes listed in Table 6.2. In contrast, chloroplasts have no phosphatidyl ethanolamine and phosphatidyl glycerol represents more than 20% of the total phospholipid content.

Table 6.3 presents data on the major glycerolipids in plant membranes. From this some major differences in membrane composition emerge. The mitochondrial membrane lipids are almost entirely phospholipid whilst the plasma membrane and tonoplast have only about 50% phospholipid. The plasma membrane is characterized by a high proportion of neutral lipids and sterols whilst the tonoplast has both glycolipids and neutral lipids/sterols. In marked contrast to other membranes the chloroplast contains a high proportion of glycolipids. Table 6.4 demonstrates other differences. For example, the inner membrane of mitochondria is characterized by high levels of diphosphatidyl glycerol whilst the chloroplast envelope and lamellae contain large proportions of mono- and digalactosyl diglyceride respectively.

Chloroplast membranes are also characterized by the presence of the fatty acid *trans*-3-hexadecenoic acid esterified to the C-2 position of glycerol. They also contain the sulpholipid, diacylsulphoquinovosyl glycerol. Linolenic acid is the dominant fatty acid in the galactolipids and phospholipids of thylakoid lamellae, comprising more than 60% of the total fatty acids in chloroplasts.

In mung beans, up to 40% of plasma membrane lipid is free sterols, sterylglycoside and acylated sterylglycoside. Indeed the sterol:phospholipid ratio of the plasma membrane is much higher than in any other cell membrane. Ceramide monohexoside (CMH), which is present as 7% of lipids in the plasmalemma, is present at twice this level in tonoplasts.

The composition of the individual fatty-acyl residues found in the membrane lipids varies considerably, both between type of lipid and type of membrane. Some examples are given in Table 6.5. In general terms, the chloroplast lamellae are rich in unsaturated fatty acids compared to the chloroplast envelope and they contain a higher proportion of unsaturated fatty acids than any other type of membrane. Similarly, the inner mitochondrial membrane has more unsaturated fatty acids than the outer membrane. The plasma membrane and other membranes tend to have a lower percentage of unsaturated fatty acids, although these are still predominant. The fatty acid composition of particular lipids also varies between membranes. For example, phosphatidyl inositol is rich in 16:0 fatty acids in thylakoid and envelope membranes of spinach chloroplasts whilst the fatty acids associated with phosphatidyl ethanolamine and phosphatidyl choline in the mitochondrial fractions are more unsaturated.

Several types of membrane can be distinguished on the basis of their lipid

Table 6.2 Composition of the phospholipid fraction of various membranes of potato tubers. Values are given as a percentage of the total phospholipid in a particular membrane. From Mazliak, P.(1977). In Tevini, M. & Lichtenthaler, H.K. (eds) *Lipids and Lipid Polymers in Higher Plants*, pp. 48–71. Springer Verlag, New York.

			Mitochondria			
Lipid	Plasma membrane	Whole organelles	Inner membrane	Outer membrane	Peroxisomes	Microsomes*
PC	38	43	27	52.6	61	44.8
PE	47	30	29	25	19.6	32.5
DPG	–	8	19.5	12.1	–	1
PA	–	5	–	–	–	3.5
PI	15	7	–	–	4.3	16
PS	–	3	24.5	10.3	–	1
PG	–	3	–	–	15.1	1

* Membranes associated with the microsomal fraction are believed to be vesicles derived from the endoplasmic reticulum, plasmalemma and tonoplast—they are not recognized as biological entities *in vivo*.

Abbreviations: PC, phosphatidyl choline; PE, phosphatidyl ethanolaine; DPG, diphosphatidyl glycerol; PA, phosphatidic acid; PI, phosphatidyl inositol; PS, phosphatidyl serine; PG, phosphatidyl glycerol.

Table 6.3 Glycerolipids of plant membranes. Data given represent average values from different plant species. After Douce, R., Holtz, R.B. & Benson, A.A. (1973) *J. Biol. Chem.*, **248**, 7215–22; Bligny, R. & Douce, R. (1980) *Biochim. Biophys. Acta*, **617**, 254–63; Yoshida, S. & Uemura, M. (1986) *Plant Physiol.* **82**, 807–12.

	Abundance (mol % of total lipids)		
Membrane	Phospholipids	Glycolipids	Neutral lipids and sterols
Plasma membrane	49	1	50
Mitochondria			
Inner membrane	99	0	trace
Outer membrane	99	0	trace
Chloroplasts			
Envelope membranes	37	63	trace
Thylakoids	17	83	trace
Tonoplast	51	31	18

Table 6.4 Abundance of polar lipids associated with membranes of chloroplasts and mitochondria. Data represent mean values obtained from different plant species. After Douce, R., Holtz, R.B. & Benson, A.A. (1973) *J. Biol. Chem.*, **248**, 7215–22; Bligny, R. & Douce, R. (1980) *Biochim. Biophys. Acta*, **617**, 254–63.

	Abundance (% by weight of total lipids)			
	Mitochondria		Chloroplasts	
Lipid	Outer membrane	Inner membrane	Outer membrane	Inner membrane
Phospholipids				
Phosphatidyl choline	68	29	20	3
Phosphatidyl glycerol	2	1	8	9
Phosphatidyl inositol	5	2	1	1
Phosphatidyl ethanolamine	24	50	trace	0
Diphosphatidyl glycerol	n.d.	17	0	0
Glycolipids				
Monogalactosyl diglyceride	0	0	20	51
Digalactosyl diglyceride	0	0	30	26
Sulpholipid	0	0	6	7

Abbreviation: n.d., not determined.

composition: (i) the inner mitochondrial membrane, characterized by the presence of diphosphatidyl glycerol; (ii) the chloroplast lamellae, characterized by the presence of *trans*-3-hexadecenoic acid 16:1 (3*t*) as a component of lipids; and (iii) other membranes from the rest of the cell including the plasma membrane,

Table 6.5 Fatty acid composition of lipids of plant membranes. Data on chloroplast membranes from *Vicia faba* (broad bean), nuclear membrane data refer to sunflower hypocotyls and other data refer to membranes from potato tuber. After Mazliak, P. (1977). In Tevini, M. & Lichtenthaler, H.K. (eds) *Lipids and Lipid Polymers in Higher Plants*, pp. 48–74. Springer Verlag, New York.

Membrane fraction	Fatty acid (% of total fatty acid)								
	14:0	16:0	16:0	16:1Δ³ *trans*	18:0	18:1	18:2	18:3	DBI*
Chloroplasts									
Envelope	0	13.3	1.8	0	4.3	5.8	11.5	63.3	93.9
Thylakoids	0	5.9	0	1.2	1.6	3.2	5.3	82.9	263.5
Mitochondria									
Outer membranes	0	24.0	0	0	3.5	1.8	58.0	12.7	155.9
Inner membranes	0	12.5	0	0	2.3	0.8	67.4	16.8	186.0
Peroxisomes	2	22.0	0	0	2.1	4.3	56.0	13.3	156.2
Microsomes	0	31.0	0	0	7.0	0.5	49.5	12.0	135.5
Nuclear membranes	0	25.5	trace	0	3.0	12.1	54.2	3.9	132.2
Plasma membrane	0	26.9	0	0	6.2	8.7	26.0	27.8	144.1

*Double Bond Index (DBI) represents the number of double bonds in every 100 fatty-acyl residues and is calculated as: DBI = (1 × % fatty acids with one double bond) + (2 × % fatty acids with two double bonds) + (3 × % fatty acids with three double bonds).

tonoplast and microsomal membranes with significant proportions of neutral lipids and sterols.

6.1.3 Differences in membrane composition are reflected by membrane properties

Membranes differ qualitatively in their protein composition, which reflects the specialized functions of particular membranes, e.g. the proteins involved in electron transport in the inner mitochondrial membrane or in chloroplast thylakoids. Variations in lipid composition also confer differences in physical properties.

In an elementary membrane and in artificial lipid bilayers, increasing the proportion of unsaturated fatty acids results in a greater degree of fluidity of the membrane. The high proportion of unsaturated fatty acids in the lipid bilayer of choloroplast lamellae and mitochondrial membranes means that certain proteins (as well as lipophilic substances such as plastoquinone) can move laterally within the membrane (see Section 9.8). In fact, the thylakoids from higher plants are extremely fluid with a viscosity of only 0.034 Pa.s at 25 °C. Rat liver mitochondria

have an even lower value of 0.029 Pa.s at 35 °C but, in contrast, the plasma membranes of human erythrocytes have a viscosity of 0.042 Pa.s at 35 °C. (To put these figures into context, water, 50% glycerol in water and 1.89 M sucrose have viscosities of 0.001, 0.006 and 0.020 Pa.s respectively at 20 °C.)

The relative stability and impermeability of plasma membranes is increased partly by the presence of sterols (which interact with phospholipids), and partly by the higher proportion of saturated fatty acids in the constituent phospholipids. Although thylakoids have an extremely high level of unsaturated fatty acids, they need to have a very low permeability to establish the pH gradient necessary to drive ATP synthesis (see Section 9.11). Stability and impermeability of thylakoids is a consequence of the high levels of galactolipids, presumably as a result of strong interactions between the polar monosaccharide components of galactolipid molecules.

Digalactosyl diglyceride is a typical bilayer-forming lipid. However, monogalactosyl diacylglycerol has unusual properties which are believed to confer on this lipid an important role in protein packaging. The presence of large integral proteins are believed to cause deformation of the planar lipid bilayer. Monogalactosyl diacylglycerol adjacent to such a protein can induce a local concavity in the membrane and act to stabilize and seal the bilayer. Lipid–protein interactions in membranes are important in other respects. For example, monogalactosyl diacylglycerol enhances ATP synthetase activity of thylakoids. Other thylakoid acyl lipids restore full activity of a number of processes of light-dependent electron transport to chloroplast preparations depleted of lipid. In mitochondria, too, lipids modify protein function. Figure 6.1 shows that removal of lipids from cauliflower mitochondria causes activity of NADH cytochrome *c* reductase to decrease by about 66%. Addition of mitochondrial lipids (as liposomes) to preparations depleted in lipids restores, and even increases, activity. The NADH dehydrogenase activity of mitochondria responds in a similar manner.

6.2 Membranes are responsible for transport of materials into and out of cells and organelles

6.2.1 Permeability of membranes to different substances reflects the presence of the lipid bilayer

Since lipids form the majority of the surface area of most (although not all) membranes, it is perhaps not surprising that the permeability properties of membranes closely resemble those of lipid bilayers. Membranes are, with some

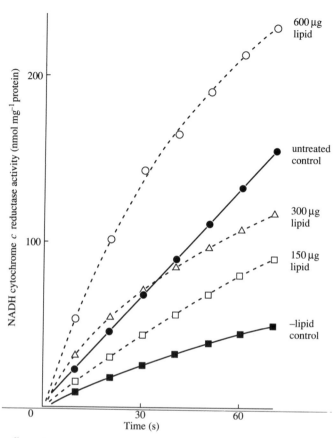

Fig. 6.1 The effect of mitochondrial lipids on the activity of NADH cytochrome c reductase in mitochondria of cauliflower. Cauliflower mitochondria (containing 28 µg protein) were extracted for 10 min in 90% acetone. This treatment (■—■) removed lipids from the mitochondria and decreased enzyme activity to about 30% of that found with unextracted, control, mitochondria (●—●). Incubation of lipid-extracted mitochondria with 150 µg, 300 µg or 600 µg of mitochondrial phospholipids restored, and even stimulated, enzyme activity. After Mazliak, P. (1977) In Trevini, M. & Lichtenthaler, H.K. (eds) *Lipids and Lipid Polymers in Higher Plants*, pp. 48–74. Springer Verlag, New York.

notable exceptions, relatively impermeable to polar and large non-polar molecules. For example, propan-1-ol and urea both have a molecular weight of about 60. However, propanol is less polar than urea and permeates the cell membrane of the alga *Nitella mucronata* some 10^4 times faster.

In general terms, the permeability of membranes to materials is related to the partition coefficient of those materials between lipid and water. This is illustrated for the plasma membrane of *N. mucronata* in Fig. 6.2. The more readily a compound dissolves in oil ($\cong$ lipid), that is, the greater the lipid:water partition coefficient, the greater the permeability of the plasma membrane to that compound.

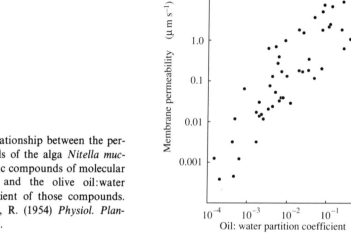

Fig. 6.2 The relationship between the permeability of cells of the alga *Nitella mucronata* to organic compounds of molecular weight 59–234 and the olive oil:water partition coefficient of those compounds. After Collander, R. (1954) *Physiol. Plantarum*, 7, 420–45.

Water is an exception to this rule. Water is a polar molecule, yet it can permeate lipid bilayers (artificial or natural membranes) much faster than most of the solutes it usually contains in cells (Table 6.6). This transfer of water was long assumed to occur via the various proteins in membranes but since water also rapidly permeates artificial lipid bilayers this seems unlikely. In many cases, the passage of water through the lipid bilayer is now thought to be through the formation of small open cavities filled with water or non-electrolyte molecules. These cavities are caused by conformational changes in the hydrophobic lipid 'tails'. The cavities move across the membrane transferring water molecules one by one. In some cases, however, the reported permeability of natural plant membranes to water is so high that it has been postulated that pores might account for the values obtained.

Gases such as CO_2, N_2 and O_2 readily permeate membrane lipid bilayers (Table 6.6). This is generally advantageous as it allows, for example, ready transfer of CO_2 to the site of CO_2 assimilation in chloroplasts and of O_2 for respiration in mitochondria. A high permeability of membranes to CO_2 would, however, be disadvantageous to operation of the 'CO_2-concentrating mechanism' of microalgae (Section 11.10), which it would effectively short circuit.

Table 6.6 also shows that membranes are less permeable to charged ionic

Table 6.6 Permeability coefficient (P_s) of lipid bilayers to selected solutes. After Raven, J.A. (1984). *Energetics and Transport in Aquatic Plants*. Alan R. Liss Inc., New York.

Solute*		Permeability coefficient[†] ($\mu m\ s^{-1}$)
Organic acids		
Maleic acid	– undissociated	4×10^{-1}
	– dissociated	4×10^{-5}
Salicylic acid	– undissociated	1×10^{3}–7.7×10^{3}
	– dissociated	1×10^{-3}
Indoleacetic acid	– undissociated	3.4×10^{1}
	– dissociated	1×10^{-5}
Organic bases		
Histamine	– unassociated	3.5×10^{-1}
	– associated	~ 0
Tryptamine	– unassociated	1.8×10^{3}
	– associated	~ 0
Organic non-electrolytes		
Urea		6×10^{-3}–40×10^{-3}
Glycerol		45×10^{-3}–54×10^{-3}
Acetamide		2.4×10^{-1}
Glucose, sorbitol and mannitol		1×10^{-6}
Inorganic non-electrolytes and weak acids		
H_2O		2.36
HCl		3×10^{4}
H_3BO_3		1.4×10^{-2}
H_4SiO_4		1×10^{-4}
CO_2		3.5×10^{3}
Inorganic ions		
H^+		5×10^{-6}–100×10^{-6}
OH^-		1.8×10^{-5}
$Li^+, Na^+, K^+, Rb^+, Cs^+$		2×10^{-6}
$Cl^-, NO_3^-, SCN^-, HCO_3^-$		1.5×10^{-12}

*In cases where the solute can exist in dissociated/undissociated forms (for acids) or associated/unassociated forms (for bases), permeability coefficients for both forms are given.
[†]The permeability coefficient is a measure of the distance moved by a solute molecule through a membrane per unit time. In some cases a range of values are available in the literature—this variation is indicated in the table.

species (e.g. Na^+, K^+, H^+) than to non-electrolytes, i.e. ions are less permeant. Gradients in proton concentration across specific membranes are vital to energy transduction (see Chapters 7 & 9). The low membrane permeability to protons permits these proton gradients. Charged organic acids or bases also have a low capacity to cross lipid bilayers whilst undissociated forms permeate much more

readily. This has practical considerations, since the formation of charged species of organic acids and bases depends on the pH. Thus, when CO_2 dissolves in water it exists in a complex equilibrium of different forms (Eqns. 6.1–6.3).

$$CO_2 + H_2O \rightleftharpoons H_2CO_3 \rightleftharpoons HCO_3^- + H^+$$
$$(6.1) \qquad (6.2)$$
$$CO_2 + OH^- \rightleftharpoons HCO_3^-$$
$$(6.3)$$

At alkaline pH, HCO_3^- (and at very high pH, CO_3^{2-}) is the dominant ion. This is far less permeant than CO_2. Thus, many aquatic plants which grow at pH values greater than 8.0 have developed mechanisms to use HCO_3^- ions to supply the inorganic carbon needed for growth.

The herbicide and synthetic auxin 2,4-dichlorophenoxyacetic acid (2,4-D) has an acetic acid group. At low pH, the molecule is largely uncharged and far more permeant through lipid bilayers than the ionized form which is dominant at neutral or high pH. Leaves or plant cells in tissue culture therefore absorb 2,4-D far more effectively at acidic pH values. Similar arguments apply to the natural plant growth substance indoleacetic acid (IAA). On the other hand, ammonia diffuses more rapidly through membranes at neutral or alkaline pH as the molecule is uncharged under these conditions. As shown in Section 6.2.3, this differential permeability of membranes to certain charged or uncharged molecules can be used to determine pH gradients across membranes.

6.2.2 *Some solutes cross membranes by passive diffusion*

Many solutes cross membranes by dissolving in the lipid portion of the membrane, diffusing to the opposite lipid–water interface and redissolving in the aqueous phase on the other side of the membrane. This passive diffusion is increased if the solute readily dissolves in the lipid phase. Thus, passively permeating substances tend to be non-polar (Fig. 6.2) and also of relatively small size. A number of substances important to plant growth such as CO_2, H_2O, O_2 and H_3BO_3 cross membranes in this way.

The net transport of uncharged solutes by passive diffusion across a membrane is expressed as a flux, i.e. amount per unit area per unit time. This can be calculated from Eqn. 6.4

$$J_s = P_s([S]_A - [S]_B) \qquad (6.4)$$

where, for the diffusing solute S, J_s is the net flux ($mol\ cm^{-2}\ s^{-1}$), P_s is the permeability coefficient (cm s^{-1}) and $[S]_A$ and $[S]_B$ are the concentrations of S (mol cm^{-3}) on the two sides of the membrane. It should be emphasized that J_s represents a *net* flux of S across the membrane. Even if $[S]_A \gg [S]_B$, so that the net flow of S is from side A to side B, some solute molecules will pass from side B to side A (Fig. 6.3). Net flux of S will continue until $[S]_A = [S]_B$ and $J_s = 0$. Consequently J_s is proportional to the concentration gradient of S between side A and side B, $([S]_A - [S]_B)$. If $[S]_B$ is constant, J_s is proportional to $[S]_A$.

The driving force for diffusion is the gradient in the free energy of S on either side of the membrane (ΔG_s), related to $[S]_A$ and $[S]_B$ by

$$\Delta G_s = - RT \log_e([S]_A/[S]_B) \qquad (6.5)$$

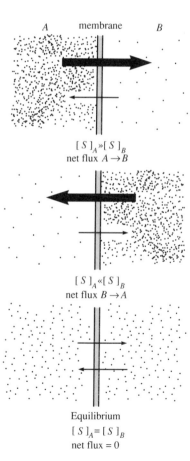

Fig. 6.3 A schematic representation of the diffusion of molecules across a membrane. The net flux of solute depends on the relative concentration of solute ($[S]$) on either side of the membrane. At equilibrium the flux is the same in both directions.

A membrane B

$[S]_A \gg [S]_B$
net flux $A \to B$

$[S]_A \ll [S]_B$
net flux $B \to A$

Equilibrium
$[S]_A = [S]_B$
net flux $= 0$

where ΔG_s is the change in free energy, R is the gas constant and T is the absolute temperature. Net flux of S proceeds with a continual decrease in free energy to equilibrium, i.e. $[S]_A = [S]_B$ and $\Delta G_s = 0$.

The permeability coefficient, P_s, describes the rate of movement of solute molecules across a membrane and is expressed in units of μm s^{-1}. It is a complex parameter corresponding to D_s/l where l is the length of the diffusion path (i.e. the thickness of the membrane) and D_s is diffusivity of the solute species (i.e. the ability of the solute to diffuse within the membrane). The term D_s takes into account the partition coefficient between aqueous and lipid phases as well as diffusivity of the solute within the membrane.

For charged solutes (ions), the driving force for transport involves an electrical term, in addition to the concentration term in Eqn. 6.5. For ions, ΔG_s is given by the expression

$$\Delta G_s = - RT \log_e([S]_A/[S]_B) + mF\Delta\psi_{A,B} \qquad (6.6)$$

where m is the charge on the ionic solute, F is the Faraday constant and $\Delta\psi_{A,B}$ is the difference in electrical potential across the membrane (the membrane potential). Membrane potential is related to the flux J_s of the charged solute across the membrane as

$$J_s = - P_s \left[\frac{(mF\Delta\psi_{A,B}/RT)}{1 - \exp(zF\Delta\psi_{A,B}/RT)} \right] ([S]_A - [S]_B \exp(mF\Delta\psi_{A,B}/RT)). \qquad (6.7)$$

However, as shown in Table 6.6, values of P_s for most inorganic ions in lipid bilayer membranes are very low ($\sim 10^{-6}$ μm s^{-1}). The mobilities of positively and negatively charged ions are very seldom equal and, as a result, a *diffusion potential* is established as ions diffuse at different rates across a membrane. If, for example, K$^+$ ions from a solution of KCl diffuse faster than the Cl$^-$ ions, then an electrical potential difference ($\Delta\psi_{A,B}$) will be set up between one side of the membrane and the other as the K$^+$ ions 'outpace' the Cl$^-$ ions. The relationship between $\Delta\psi_{A,B}$ and the ion concentrations is then

$$\Delta\psi_{A,B} = \frac{RT}{F} \log_e \left[\frac{P_{K^+}[K^+]_A}{P_{K^+}[K^+]_B} + \frac{P_{Cl^-}[Cl^-]_B}{P_{Cl^-}[Cl^-]_A} \right] \qquad (6.8)$$

where P_{K^+} and P_{Cl^-} are the permeability coefficients for K$^+$ and Cl$^-$ respectively, and the subscripts A and B denote the concentrations on either side of the membrane.

6.2.3 Passive diffusion of some materials across membranes can be used to determine transmembrane electrical potential differences and internal pH of cells and organelles

Although for most ions the permeability coefficient, P_s, is extremely low there are some ions which have a delocalized charge in lipid (or lipoprotein) membranes and these ions are consequently soluble in lipids and have much higher P_s values. These 'lipid-soluble' cations, such as triphenylmethylphosphonium (TPhMP$^+$) and tetraphenylphosphonium (TPhP$^+$), or anions, such as thiocyanate (SCN$^-$), come to equilibrium across cell membranes relatively rapidly and can be used to determine the electrical potential difference between two phases separated by a membrane. At equilibrium there will be no net flux of the anion or cation and so ΔG_s in Eqn. 6.6 will be zero. Setting $\Delta G_s = 0$ and rearranging, then

$$\Delta \psi_{A,B} = -\frac{RT}{mF}\ln([S]_A/[S]_B) \qquad (6.9)$$

which is the Nernst equation (see Section 4.6). The equilibrium concentrations $[S]_A$ and $[S]_B$ can easily be measured if radioactively labelled lipophilic ions are used. These values are then substituted into Eqn. 6.9 and used to calculate $\Delta \psi_{A,B}$. (Results obtained using lipophilic cations such as TPhP$^+$ must be interpreted carefully, however, particularly if used with intact cells, as at least some of the uptake could be attributable to transport via a thiamine translocator. Furthermore, mitochondria accumulate TPhP$^+$ and related ions to concentrations of 10^4–10^6 times that outside the organelle.)

Similarly, the distribution of weak acids or bases can be used to determine pH gradients across membranes. An uncharged (electroneutral), permeant solute (e.g. CO_2 or glycerol) is unaffected by $\Delta \psi_{A,B}$ and net movement across the membrane ceases at equilibrium when the concentration gradient is zero. Some weak acids and bases with a pK_a of 3–11 can permeate in the uncharged form, whilst the ionized form is impermeant even if it is present at much higher concentrations (Table 6.6). The Henderson–Hasselbach equation (Fig. 6.4) predicts that if a pH gradient exists across a membrane then, at equilibrium, the concentration of the ionized species will be different on either side of the membrane. Since at equilibrium the concentration of the uncharged species will be the same on either side of the membrane, it follows that measurements of the concentration of the uncharged plus charged species on either side can be used to calculate the pH gradient (Fig. 6.4).

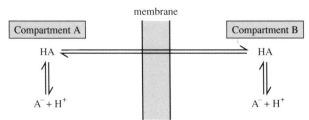

The Henderson–Hasselbach equation relates the dissociation of the weak acid to pH as

$$pH_A = pK_A + \log\left(\frac{[A^-]_A}{[HA]_A}\right) \qquad\qquad pH_B = pK_B + \log\left(\frac{[A^-]_B}{[HA]_B}\right)$$

which can be rewritten as

$$[A^-]_A = [HA]_A (10^{\,pH_A - pK_A}) \qquad\qquad [A^-]_B = [HA]_B (10^{\,pH_B - pK_B})$$

and at equilibrium

$$[HA]_A = [HA]_B \; .$$

The ratio of the total concentration of weak acid (HA+A$^-$) on either side of the membrane is given by

$$\frac{\text{total concentration on side } A}{\text{total concentration on side } B} = \frac{[S]_A}{[S]_B} = \frac{([HA]_A + [A^-]_A)}{([HA]_B + [A^-]_B)} = \frac{[HA]_A + [HA]_A (10^{\,pH_A - pK_A})}{[HA]_B + [HA]_B (10^{\,pH_B - pK_B})}$$

$$= \frac{1 + (10^{\,pH_A - pK_A})}{1 + (10^{\,pH_B - pK_B})}$$

Fig. 6.4 Use of the equilibrium distribution of a weak acid to determine the pH gradient across a membrane. Only the undissociated, uncharged species (HA) can cross the membrane; permeability to the ionized form (A$^-$) is essentially zero. The proportion of the acid that is dissociated is given by the Henderson–Hasselbach equation. Since this is pH dependent, and at equilibrium the concentration of the undissociated form will be the same on either side of the membrane, it follows that if there is a pH gradient across the membrane then the total concentration (dissociated and undissociated) of the weak acid will be different on either side of the membrane. Using radioactively labelled weak acids or bases such as dimethyloxazolidinedione (DMO) or methylammonium, $[S]_A/[S]_B$ can be measured experimentally and used to calculate the pH of one compartment if the pK values and pH of the other compartment are known. It is possible to calculate the average internal pH of a cell or organelle from the external pH, the internal and external pK_a values and the distribution of the acid or base between the inside and outside of the cell/organelle when the system comes to equilibrium.

6.2.4 Ionophores modify the permeability of membranes to some ions

A number of bacteria and fungi have evolved, as a mechanism for protection against attack by other microbes, the ability to synthesize molecules called ionophores, which make the membranes of other organisms permeable to various ions. Passage of these ions disturbs the membrane properties and cellular functions of the affected organisms and they die. Both natural and synthetic ionophores have proved extremely useful in studies of membrane transport and energy-generating mechanisms in plants and other organisms.

Ionophores can be divided into several groups.

(1) Some ionophores transfer cations and their associated charge but not protons. The naturally occurring 'transport antibiotics' valinomycin and gramicidin are included in this category although they act very differently. Valinomycin is a mobile carrier and a depsipeptide (it consists of alternating hydroxy and amino acids), comprising L-lactate, L-valine, D-hydroxyisovalerate and D-valine. These four residues are repeated three times around a ring with their hydrocarbon groups oriented towards the outside. Valinomycin binds various univalent cations with greatest specificity towards K^+ ions, although it also binds to Na^+, Cs^+, Rb^+ or NH_4^+. (The ability to bind and transport Na^+ is, however, at least 10^4 times less than that for K^+!). When K^+ interacts with valinomycin, the ion loses its water of hydration and binds to the six oxygen atoms located towards the centre of the doughnut-like molecule. In this state, the K^+ is effectively covered in a lipophilic coat which permeates rapidly across the membrane. On the other side of the membrane, water competes with valinomycin for K^+ and hydrated K^+ is released into the aqueous solution. Other compounds such as nonactin and monactin transport K^+ in like manner but with a decreased selectivity.

Gramicidin is an open chain polypeptide which forms a transient channel through membranes. Gramicidin A consists of four residues each of valine, leucine and tryptophan together with two residues of alanine and one of glycine. Four of these amino acids belong to category A, as defined in Fig. 2.14, and cause gramicidin A to form a hollow helix about 0.2 nm in diameter. The hydrophobic side chains of the amino acids are oriented to the outside of the cylinder, which thus readily associates with the lipid bilayer. A dimer of two helices forms a cylindrical channel through which cations can pass but, unlike valinomycin, with little specificity.

(2) Certain ionophores such as nigericin carry protons across membranes in exchange for other cations without a net transfer of charge. The nigericin group of compounds consists of a series of cyclic ethers linked together to form polyethers of molecular mass 1–2 kDa, with a carboxyl group at one end of the molecule and one or two hydroxyl groups at the other end. At physiological pH values the carboxyl group is dissociated and forms a hydrogen bond with a hydroxyl group at the other end of the molecule, forming a ring structure analogous to valinomycin which can bind with a univalent cation. Since the ionophore has a dissociated carboxyl group with a negative charge, the ring formed with a univalent metal ion has no net charge (Fig. 6.5). However, the protonated form of nigericin is also mobile in membranes, so that when the complexed form gives up a metal ion on one side of the membrane, it can accept a proton for transport in the opposite direction. Nigericin in effect catalyses an electroneutral exchange of K^+ for H^+.

(3) The final group of ionophores considered here results in the transfer of protons but with a net change in charge. These are the proton translocators or uncouplers used so often in studies of energy-generating mechanisms (see Chapters 7 & 9). This groups of compounds, of which carbonylcyanide-4-trifluoromethoxyphenylhydrazone (FCCP) is typical, has, in anionic form, a system of delocalized electrons in a π orbital.

This makes the molecule lipid-soluble even when charged. As a result, at concentrations between 10^{-9} and 10^{-5} M, FCCP and related compounds transport protons across membranes and catalyse the net unidirectional transport of protons (Fig. 6.5). This greatly increases the conductance of membranes to protons and short circuits the proton gradients of mitochondria and chloroplasts which are necessary for ATP synthesis, i.e. it uncouples the generation of proton gradients (proton motive force) from ATP synthesis (see Sections 7.8 & 9.11).

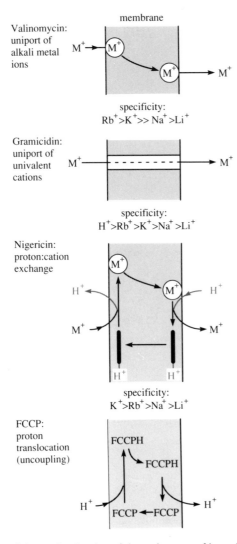

Fig. 6.5 A summary of the mode of action of the main types of ionophore. Valinomycin and gramicidin bring about uniport of alkali metal ions whilst nigericin causes exchange of metal ions for protons. Uncouplers such as carbonylcyanide-4-trifluoromethoxyphenylhydrazone (FCCP) mediate proton transport across membranes. Cation specificity for each of the ionophores is indicated. Note that, in all cases, the ionophores only facilitate passive ion movement *down* ion gradients which have been independently established by other active processes.

6.3 Movement of many materials across plant membranes involves carrier proteins

Most of the molecules required for physiological processes are required at rates higher than can be satisfied by diffusion. Supply of these materials requires mediated transport processes.

6.3.1 Mediated transport involves protein carriers in membranes

The membrane-bound transport systems involved in the movement of materials across membranes have been extensively studied in animal and bacterial cells but are only now being identified and characterized from plant cells. These carriers (or permeases, also known as translocators) are proteins with a high specificity for the compounds they transport. For example, the phosphate translocator of the inner membrane of the chloroplast envelope is a 29 kDa protein and a highly specific transporter of P_i, triose-P and glycerate-3-P (see Section 10.8). This protein can be isolated and reconstituted into liposomes which then show the transport properties of the phosphate translocator of the intact chloroplast membrane.

The proteins involved in mediated transport show a number of similarities to enzymes. Firstly, dissimilar compounds are not transported by the same carrier, which suggests a specificity between carrier and transported molecule. Also, structural analogues compete for binding sites on the protein and inhibit transport. For example, the transport of glycollate across the chloroplast envelope is carrier-mediated. In experiments investigating the uptake of [1-^{14}C]glycollate by pea chloroplasts, of the many compounds tested only D-glycerate, glyoxylate and D-lactate inhibited glycollate uptake significantly. L-Lactate, however, was not inhibitory, suggesting stereospecificity (i.e. specificity to one stereoisomer) between carrier and transported molecule. The inhibitory compounds are all two- or three-carbon molecules with similar structures.

$$
\begin{array}{cccc}
\text{CH}_2\text{OH} & \text{CHO} & \begin{array}{c}\text{CH}_3\\ |\\ \text{HO--C--H}\\ |\end{array} & \begin{array}{c}\text{CH}_2\text{OH}\\ |\\ \text{HO--C--H}\\ |\end{array}\\
| & | & & \\
\text{COO}^- & \text{COO}^- & \text{COO}^- & \text{COO}^-\\
\text{Glycolate} & \text{Glyoxylate} & \text{D-Lactate} & \text{D-Glycerate}
\end{array}
$$

The second similarity between enzymes and permeases is the kinetics of the transport process. Carrier-mediated transport shows saturation with respect to uptake and the flux (J_s) of a solute into or out of cells can be described by the

Michaelis–Menten relationship

$$J_s = \frac{J_s\max[S]}{K_m + [S]}.$$ (6.10)

It should be pointed out, however, that although the relationship between rate of transport and substrate concentration bears a formal resemblance to Michaelis–Menten kinetics, the actual process can be very complicated and deviations from true Michaelis–Menten behaviour are common.

The details of the physical process by which carrier proteins transport materials are not yet clear. A number of possibilities have been suggested.

(1) Diffusion of the carrier protein from one side of the membrane to the other. The carrier is envisaged as binding solute molecules on one side of the membrane, diffusing through the lipid bilayer and releasing the solute molecule on the opposite side.

(2) Rotational movement of the carrier protein. The carrier, seen as a large protein molecule spanning the membrane, picks up a solute molecule on one side of the membrane, rotates, and releases the solute on the other side ('flip-flop'). Proteins can move in the plane of the membrane, but the 'flip-flop' movement is less likely because the hydrophilic bonds at the exterior faces of the membrane would have to be broken in order to allow the protein to rotate through the lipid bilayer.

(3) The carrier protein accepts a solute molecule on one side of the membrane and then undergoes a conformational change releasing the solute on the opposite side of the membrane. This type of mechanism is sometimes referred to as a 'gated pore'. It has been proposed for the ADP/ATP translocator of mitochondria (see Section 7.8.4) and might also occur with the chloroplast P_i translocator.

Furthermore, some ions can cross membranes by means of 'gated channels'. Proteins spanning the membrane are believed to be in continual motion changing through many different conformational states, some of which open a channel across the membrane through which ions diffuse at very high rates (10^3–10^4 times that of carrier-mediated transport). Each type of protein which forms channels shows specificity to a particular ion. The opening or closing of channels can be regulated by stimuli such as light or Ca^{2+} concentration and is referred to as channel 'gating'.

6.3.2 Movement of a molecule or ion in one direction can be linked to the transport of another

Often the transport of one solute molecule on a carrier is linked to the movement of a second molecule. In fact such linkage is common and can take a number of forms (Fig. 6.6).

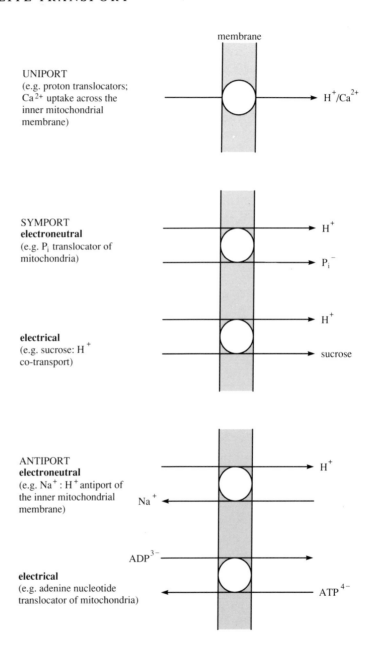

Fig. 6.6 Modes of transport of materials across membranes.

Uniport involves the movement of only a single substance across the membrane via the carrier. The movement of Ca^{2+} across the inner mitochondrial membrane and the facilitated transport of protons across membranes by uncouplers such as FCCP are examples.

For some carrier mechanisms, the movement of one substance is inextricably linked to transfer of another, i.e. the two processes are coupled. If such coupling involves the movement of two solutes in the same direction the process is termed *co-transport* or *symport*. The transport of two solutes in opposite directions is called *antiport* (Fig. 6.6).

Transport is described as electroneutral if the transported solute has no charge or if the movement of an ion is balanced by co-transport of an oppositely charged ion or antiport of a similarly charged species. If solute transport is not balanced in this way, the transport process will lead to transfer of electrical charge. Such an example is the sucrose–proton co-transport system described in Section 10.10.

6.3.3 *Many solutes are actively transported across membranes*

Uniport, symport or antiport can proceed without the direct input of metabolic energy. For neutral solutes transport proceeds until the solute concentration gradient across the membrane is zero (cf. Eqn. 6.4). For charged molecules, the driving forces for solute/ion transport are the gradients in electrochemical potential described by Eqns 6.5 and 6.6. For univalent cations, Eqn. 6.6 predicts that net flux from one side of a membrane (source) to the other (sink) proceeds if the electrochemical potential of the sink is 58 mV more negative than that of the source, until the solute concentration on the sink side reaches 10 times that of the source. For univalent anions, net flux will occur provided that the sink solute concentration does not exceed one tenth that of the source.

Plants accumulate many neutral solutes against their concentration gradients. Sucrose, for example, is loaded into the phloem to a concentration far in excess of that found at the site of sucrose formation in mesophyll cells. Similarly, many ions are found in cells at concentrations much greater than those predicted by the Nernst equation (Eqn. 6.9).

Active transport against concentration (or electrochemical potential) gradients requires expenditure of energy and can occur in a number of ways. If the driving force for active transport is an exergonic biochemical reaction, such as ATP hydrolysis, it is said to be *primary active transport*. Alternatively, solute transport can be linked to the exergonic transmembrane flux of a second solute, this being known as *secondary active transport*. In either case, the overall transport process occurs with a decrease in free energy. The primary active transport processes in plants generally involve only a few inorganic ions. There is good evidence that ATP hydrolysis drives the primary transport of Ca^{2+} and, in the case of some marine algae and salt glands of halophytic higher plants, Cl^-, but the major primary active transport process appears to involve an ATP-powered H^+ uniport. Such systems are known as proton-pumping ATP phosphatases, proton pumps or H^+-ATPases.

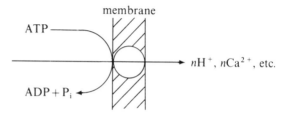

H^+-ATPases are located at the plasma membrane and tonoplast, as are the sites of Ca^{2+}-ATPase activity.

Primary active uniport of ions is electrogenic, i.e. it results in the transfer of charge across the membrane. In the case of H^+ and Ca^{2+} pumps, transport is from the cytoplasmic phase with a relatively high protein content and diversity, to phases with relatively low protein content and diversity such as the vacuole and external medium. This electrogenic ion transport undoubtedly contributes to the transmembrane electrical potential although the diffusion potential (Section 6.2.2) is also important. Generally speaking, H^+-ATPases have a stoichiometry of ATP hydrolysed to H^+ transported (ATP:H^+) of 1:1 compared to values of 1:3 found for the ATP synthetases of mitochondria and chloroplasts (see Sections 7.8 & 9.12).

Electrogenic transport of the ion X^{m+} establishes an electrochemical potential gradient $\Delta\tilde{\mu}_{X^{m+}}$, analogous to the change in free energy when one mole of X^{m+} is transported from phase A, with a concentration of $[X^{m+}]_A$, to phase B, of concentration $[X^{m+}]_B$ (see Eqn. 6.6).

$$\Delta\tilde{\mu}_{X^{m+}} = m\Delta\psi - \frac{2.3\,RT}{F}\log_{10}\frac{([X^{m+}]_B)}{([X^{m+}]_A)}. \quad (6.11)$$

For protons, $[X^{m+}]_A$ and $[X^{m+}]_B$ are $[H^+]_A$ and $[H^+]_B$, and the factor $\log([H^+]_B/[H^+]_A)$ is equal to the pH gradient, ΔpH (Eqn. 6.12).

$$\Delta\tilde{\mu}_{H^+} = \Delta\psi - \frac{2.3RT}{F}\Delta\text{pH}. \quad (6.12)$$

The function $\Delta \tilde{\mu}_{H^+}$ is referred to as the *proton motive force* which can be used *inter alia* to drive the transport of other solutes uphill against their electrochemical potential gradients.

In contrast to the few ions involved in primary active transport processes, the materials transported by secondary active transport are legion and include other ions such as K^+ or Na^+, sugars and amino acids. This secondary active transport can involve uniport, antiport or symport as shown in Fig. 6.6. Thus, the primary active transport of H^+ generates a proton motive force. Protons are then driven, by this proton motive force, in the opposite direction in various processes linked to electroneutral antiport of cations, symport of anions, or $1 H^+:1$ neutral solute symport. Incidentally, it is impossible to differentiate between H^+ and OH^- linked co-transport processes; to all intents and purposes a H^+:cation antiport is the same as an OH^-:cation symport. In addition, it is important to note that although secondary active transport processes can result in the movement of ions or neutral solutes against their electrochemical potential differences, such fluxes are in fact coupled to the '*downhill*' movement of H^+ and in this sense they are not 'active' processes at all!

6.4 'Patch clamping' is a powerful tool for studying membrane transport of ions

Most studies of membrane transport of ions have, in the past, relied heavily on the use of microelectrodes or radioactive tracers. However, microelectrode techniques can only be used with relatively large cells and also they only provide an 'overall picture' of electrochemical potentials between the external environment and the cell interior or between major compartments (organelles) within the cell.

The technique known as 'patch clamping' offers much finer resolution of ion transport processes. A fine glass electrode (a hollow micropipette with a tip diameter of approximately 1 μm) is brought into contact with the outer membrane of a protoplast or isolated organelle suspended in a suitable medium. Slight suction is applied to give a tight seal between membrane and electrode to form a 'cell-attached' configuration. Depending on the process to be studied, the protoplast is treated in different ways. The pipette can be withdrawn from the protoplast or organelle in such a way that a small piece of membrane remains across the tip of the pipette. The membrane fragment or 'patch' has the inside surface towards the bathing medium and this is termed an 'inside-out' patch. Alternatively, in the cell-attached configuration, the plasma membrane inside the boundaries of the pipette

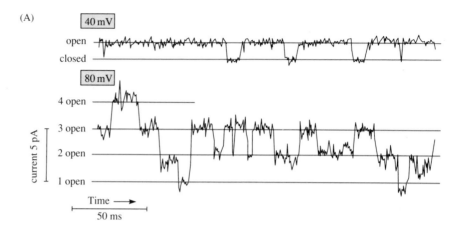

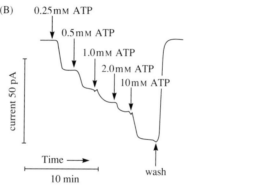

Fig. 6.7 (A) The use of a right-side out patch from a protoplast of the legume *Samanea* to detect currents due to opening of ion channels. Currents were detected when the voltage between the inside of the pipette and the bathing medium was held constant (clamped) at 40 mV or 80 mV. Transient increases in current indicate opening of ion channels. Positive (upward) fluctuations indicate the outward flow of cations or inward flow of anions. In the 40 mV trace there are two levels of current corresponding to the channel being open or closed. When the voltage across the patch is increased to 80 mV, current fluctuates between four different levels (the zero current corresponding to closed channels is not shown). These levels are multiples of the current through a single open channel indicating that up to four channels are open at a given time. (B) Direct measurement of the current generated by the H^+-ATPase in a vacuole of a mesophyll cell. Increasing the concentration of ATP in the bathing solution caused an increase in the electrogenic current from 0 to 65 pA. Removal of ATP by washing caused the current to revert to zero. After Satter, R.L. & Moran, N. (1988) *Physiol. Plantarum*, **72**, 816–20; Hedrich, R., Flügge, U.I. & Fernandez, J.M. (1986) *FEBS Lett.*, **204**, 228–32.

can be ruptured chemically. This leaves a whole protoplast or organelle attached to the electrode in a 'whole-cell' configuration. If the electrode is then pulled away from the protoplast, the membrane breaks and a portion outside the electrode seals the tip. This membrane fragment faces 'right-side out'. 'Inside-out' and 'right-side out' patches can be used to investigate ion flow through individual ion channels. A voltage is applied across the membrane patch, that is between the interior of the electrode and the bathing medium. When ion channels open in response to a stimulus, ion flow causes a small current in the picoampere range. This is detected by an electronic circuit used to keep the voltage constant, 'clamped' by generating an equal but opposing current across the membrane. This generated current is monitored and used to detect the flow of ions across a membrane patch (Fig. 6.7).

Ion transport through the single or few ion channels present in a small patch can be very rapid and is easily detectable, although carrier-mediated active or passive transport is too slow to cause a significant current. Nonetheless, active ion pumps such as the H^+-ATPase have been investigated using whole protoplasts or organelles attached to the electrode (Fig. 6.7). Patch clamping, therefore, offers a versatile means of studying transport processes and their regulation in plants.

Further reading

Monographs and treatises: Finean *et al.* (1984); Robertson (1983).
Techniques: Findlay & Evans (1987).
Membrane composition: Murphy (1986).
Transport: Davies (1987) Chapter 2; Reinhold & Kaplan (1984); Spanswick (1987).
Energetics: Harold (1986); Nicholls (1982).

PART 2
ENERGY-GENERATING
MECHANISMS OF PLANTS

7
Aerobic oxidation of sugars to carbon dioxide

The light-harvesting reactions of plants use the energy of solar radiation to form the ATP and reductant necessary to power the assimilation of inorganic carbon to form sugars. These sugars are converted to various products within cells. This chapter is concerned with the controlled oxidation of carbon skeletons 'locked up' in carbohydrates and other reserve materials by respiration, releasing energy, as ATP and reductant, for use in the maintenance and growth of plant tissues.

7.1 Respiratory rates vary between different tissues and stages of development

The biosynthetic activities of cells and their associated demand for energy vary between different tissues and different stages in plant development and there is, therefore, considerable variation in rates of respiration found in tissues. Table 7.1 gives some typical rates recorded from plants and other organisms. Dormant and

Table 7.1 Respiration rates of various species and tissues. After Bidwell, R.G.S. (1979), *Plant Physiology*, 2nd edn. Collier Macmillan, New York.

Species/Tissue	Respiration rate (μmol O_2 h^{-1} g^{-1} fresh wt)
Barley	
seed (non-germinating)	0.003
Wheat	
seedling	65
5-day-old leaf	22
13-day-old leaf	8
Potato	
whole plant	5
tuber	0.3
Apple	
developing fruit	10
ripened fruit	0.5
Humans	
resting	10
running	200

inactive tissues use far less energy, and show lower respiratory rates, than actively growing tissues. Mature leaves often show lower levels of respiration than actively growing younger leaves and matured fruit show very little respiration in comparison to ripening fruit. Tissues such as woody stems, with a large proportion of dead or non-metabolizing cells, tend to have lower respiration rates than tissues with a high proportion of very active cells, such as cambial tissue. Thus, overall respiratory rates reflect the physiological and metabolic activities of the plant or tissue. Respiration is the sole means of generating energy in non-photosynthetic cells but it is also important in photosynthetic cells.

7.2 Storage products are mobilized before they are respired

Although plant tissues oxidize a range of other substrates such as protein and lipids, respiration in most plants is dominated by the oxidation of carbohydrate via the processes of glycolysis and the tricarboxylic acid (TCA) cycle. The starting point for respiration is usually the hexose sugars, which comprise the basic components of the disaccharide sucrose and the polysaccharide starch, respectively the principal transport and storage forms of carbohydrate.

In many tissues carbon is stored in reserve material and is not immediately available as hexose for respiration. These storage materials must first be converted to soluble, readily metabolized compounds, i.e. they must be 'mobilized'. In some cases, lipid is the reserve material and its mobilization is discussed in Section 8.2. However, the most common storage material is starch, stored in chloroplasts (in photosynthetic tissue) or amyloplasts (in non-photosynthetic tissue). Usually this is mobilized to sucrose via starch phosphorylase and other enzymes as described in Section 10.12.

Sucrose can be hydrolysed to its constituent hexoses (glucose and fructose) by invertase (Eqn. 7.1).

$$\text{Sucrose} + H_2O \rightleftharpoons \text{D-glucose} + \text{D-fructose}. \qquad (7.1)$$

Alternatively, sucrose can be broken down to UDP-glucose and fructose via sucrose synthase (Eqn. 7.2).

$$\text{Sucrose} + \text{UDP} \rightleftharpoons \text{fructose} + \text{UDP-glucose}. \tag{7.2}$$

UDP-glucose is subsequently converted to glucose-1-P plus UTP via UDP-glucose pyrophosphorylase (Eqn. 7.3). Glucose-1-P can in turn be rearranged to glucose-6-P by glucose-P mutase (Eqn. 7.4).

$$\text{UDP-glucose} \xrightarrow[\text{PP}_i \quad \text{UTP}]{} \text{glucose-1-P} \longrightarrow \text{glucose-6-P}$$

$$(7.3) \qquad\qquad\qquad\qquad (7.4)$$

In some tissues (e.g. cereal seeds) the breakdown of starch involves α- and β-amylases (and not starch phosphorylase) which catalyse the hydrolysis of the two components amylose and amylopectin. α-Amylase hydrolyses the $\alpha(1\rightarrow4)$ linkages of amylose at random to produce a mixture of glucose and maltose but it does not hydrolyse the maltose to its two constituent glucose moieties. β-Amylase also hydrolyses $\alpha(1\rightarrow4)$ linkages of amylose, but it attacks only the penultimate $\alpha(1\rightarrow4)$ linkages from the ends of the molecule, releasing maltose. Both α- and β-amylases attack amylopectin but they cannot hydrolyse the $\alpha(1\rightarrow6)$ linkages at the branch points (see Section 2.3.6, Fig. 2.19). As a result, the end-products of β-amylase action on amylopectin are maltose and a highly branched core called a 'limit dextrin' because it represents the limit of β-amylase action. Complete hydrolysis of the amylopectin is possible because germinating seeds also possess oligo-1,6-glucosidase which attacks the $\alpha(1\rightarrow6)$ linkages in the limit dextrin, permitting further breakdown by the amylases to yield maltose and glucose. The $\alpha(1\rightarrow4)$ linkages in maltose are hydrolysed by α-glucosidase to glucose. Thus, the combined action of the amylases and glucosidases converts starch to glucose.

In some species the main carbohydrate reserve is not starch but fructans. These are polymers of β-D-fructofuranose and are found in two forms; the inulin type in which the fructofuranose residues are joined by $\beta(2\rightarrow1)$ glycosidic linkages, and the levan type (sometimes termed phlein) in which the joining glycosidic linkages are $\beta(2\rightarrow6)$. Both types have a terminal unit of glucose (see Section 2.3.6). Inulin-type fructans are found in underground storage organs of members of the Asteraceae (Compositae) such as dahlias and Jerusalem artichokes. The levan type are found as temporary storage reserves in leaves, stems and roots of monocotyledons such as Poaceae (Gramineae). Fructans are hydrolysed by β-fructofuranosidases, specific for the $\beta(2\rightarrow1)$ or $\beta(2\rightarrow6)$ linkages. Jerusalem artichokes, for example, contain a β-fructofuranosidase which breaks down inulin by successively cleaving off fructose units, leaving a residual dimer of sucrose unhydrolysed.

Maize kernels contain, in addition to starch, a glycogen-like polysaccharide called phytoglycogen. This compound, synthesized and stored in the cytosol, shows a greater degree of branching than the amylopectin of starch and approximately 10% of the glycosidic bonds are $\alpha(1\rightarrow6)$ compared to 5% in amylopectin (see Section 2.3.6). In addition to α- and β-amylases, maize seeds contain an additional hydrolase, isoamylase, which hydrolyses the $\alpha(1\rightarrow6)$ linkages of phytoglycogen.

Very little is known about the degradation of the other storage materials found in plants, such as the galactomannans of legume seeds or the glucomannans found in the tubers of some plants such as orchids.

7.3 Glycolysis involves the anaerobic oxidation of hexoses to pyruvate

The first step in the respiration of hexose sugars, produced by the breakdown of reserve materials, is a series of reactions termed glycolysis. These reactions concern the degradation of hexose sugars via a series of hexose phosphates and triose phosphates to pyruvate, a process which occurs in both aerobic and anaerobic conditions but does not require oxygen (Fig. 7.1).

7.3.1 Initial steps in glycolysis represent 'priming reactions' in which hexose sugars are converted to triose phosphate

Glycolysis can be divided into two distinct phases, the first of which involves 'priming reactions' which convert hexose sugars to triose phosphate. The first steps in the 'priming' phase entail the phosphorylation of glucose or other hexose sugars and their rearrangement to fructose-1,6-bisphosphate (fructose-1,6-P_2). Hence, ATP is used in the phosphorylation of glucose in a reaction catalysed by hexokinase (Fig. 7.1) and requiring a divalent cation such as Mg^{2+}, which forms a complex with ATP before ATP participates as a substrate in the reaction. Glucose is phosphorylated at C-6 to form glucose-6-P (Eqn. 7.5).

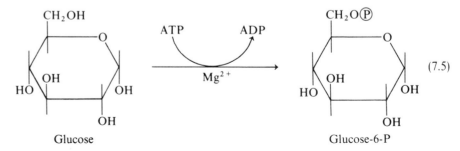

(7.5)

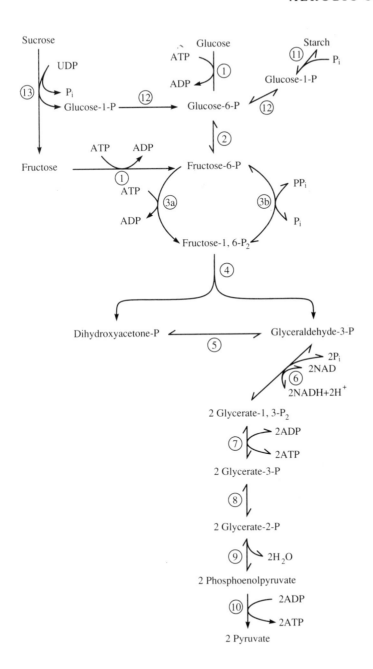

Hexokinases also catalyse the conversion of other hexoses such as D-mannose and D-fructose to their respective 6-phosphates. The phosphorylated sugars have a much lower standard free energy of hydrolysis than ATP, and the reactions are therefore essentially irreversible. For the glucose hexokinase reaction of Eqn. 7.5 for example, $\Delta G^{\circ\prime}$ is $-16.7\,\mathrm{kJ\,mol^{-1}}$.

In most cases carbon enters glycolysis as glucose-6-P produced from sucrose by the reactions described in Eqns 7.2–7.4. The sucrose, in turn, is derived (see Section 10.12) from triose-P and starch synthesized in chloroplasts. In other cases where starch is mobilized directly to glucose-1-P via starch phosphorylase, glucose-6-P is formed via glucose-P mutase activity.

Glucose-6-P is converted to fructose-6-P by a reversible reaction ($\Delta G^{\circ\prime} = -1.7\,\mathrm{kJ\,mol^{-1}}$) catalysed by hexose-P isomerase (Eqn. 7.6).

$$(7.6)$$

Glucose-6-P   Fructose-6-P

A similar enzyme converts mannose-6-P to fructose-6-P, although this is only found in some plants.

Fructose-6-P is then phosphorylated to yield fructose-1,6-P_2. There are two enzymes in plants which can carry out this reaction. Phosphofructokinase (PF-Kase)

Fig. 7.1 Reactions of glycolysis. Although the glycolytic sequence is represented here as starting from glucose, carbon can enter the pathway at any of the steps shown. Under aerobic conditions, the pyruvate produced by glycolysis is further oxidized to CO_2 via the TCA cycle. Under anaerobic conditions, pyruvate can be metabolized in fermentative metabolism to ethanol or to organic acids such as lactate. Enzymes of the glycolytic sequence are: (1) hexokinase; (2) hexose-P isomerase; (3a) phosphofructokinase; (3b) fructose-6-P:PP$_i$ phosphotransferase; (4) aldolase; (5) triose-P isomerase; (6) glyceraldehyde-3-P dehydrogenase; (7) glycerate-3-P kinase; (8) glycerate-P mutase; (9) enolase; (10) pyruvate kinase. If carbon enters the pathway as hexose-P (from starch or sucrose) the enzymes involved are: (11) starch phosphorylase; (12) glucose-P mutase; (13) UDP-glucose pyrophosphorylase.

uses ATP to phosphorylate fructose-6-P in an essentially irreversible reaction ($\Delta G^{\circ\prime} = -14.2 \text{ kJ mol}^{-1}$); see Eqn. 7.7.

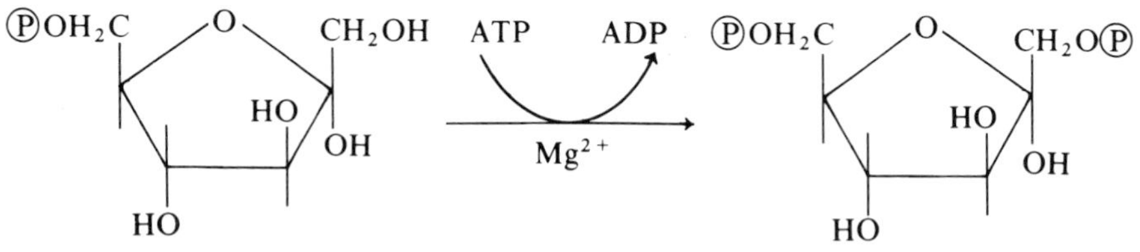

$$\text{Fructose-6-P} \qquad\qquad\qquad\qquad\qquad\qquad \text{Fructose-1,6-P}_2 \quad (7.7)$$

However, the cytosol also contains fructose-6-P:PP$_i$ phosphotransferase (FP-PTase) which phosphorylates fructose-6-P using pyrophosphate in place of ATP (Eqn. 7.8).

$$\text{Fructose-6-P} + \text{PP}_i \rightleftharpoons \text{Fructose-1,6-P}_2 + \text{P}_i \qquad (7.8)$$

This reaction occurs with a relatively small change in free energy ($\Delta G^{\circ\prime} = -2.9 \text{ kJ mol}^{-1}$) and is thought to operate principally in the reverse direction to supply PP$_i$ for sucrose breakdown via UDP-glucose pyrophosphorylase (see Section 7.5.2). FP-PTase has been found only in higher plants, some algae and bryophytes and the photoautotrophic bacterium *Rhodospirillum*. It is not present in animal cells. Generally speaking, non-photosynthetic plant tissues have a greater proportion of FP-PTase activity. PF-Kase is one of the key regulatory enzymes of glycolysis although the precise role of FP-PTase is uncertain (see Section 7.5.2).

The next step in the 'priming phase' of glycolysis is the conversion of one molecule of fructose-1,6-P$_2$ to two molecules of glyceraldehyde-3-P, in two reactions. Initially, fructose-1,6-P$_2$ is cleaved to the two triose phosphates, dihydroxyacetone phosphate and glyceraldehyde-3-P (Eqn. 7.9) in a reaction catalysed by aldolase. Although this reaction has a $\Delta G^{\circ\prime}$ of $+29.9 \text{ kJ mol}^{-1}$, it proceeds readily in the forward direction given normal intracellular concentrations of the reactants.

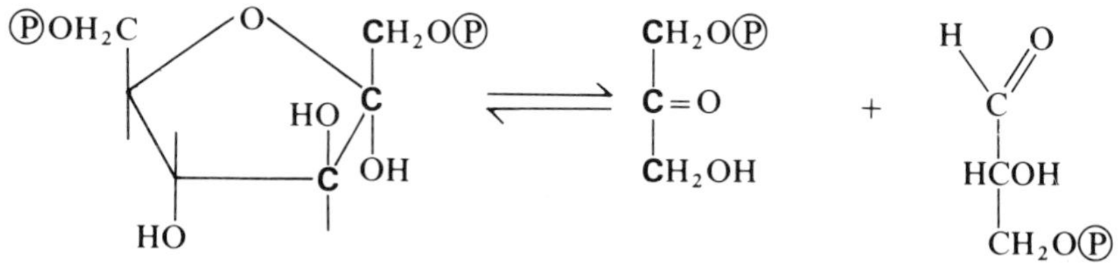

$$\text{Fructose-1,6-P}_2 \qquad\qquad \text{Dihydroxyacetone-P} \quad \text{Glyceraldehyde-3-P}$$

Dihydroxyacetone-P and glyceraldehyde-3-P are interconverted in a reversible reaction ($\Delta G^{\circ\prime} = +9.5 \text{ kJ mol}^{-1}$) catalysed by triose phosphate isomerase (Eqn. 7.10). At equilibrium dihydroxyacetone-P forms more than 90% of the two triose phosphates, but since glyceraldehyde-3-P is, *in vivo*, continuously consumed in ensuing reactions (see below), dihydroxyacetone-P is effectively converted to glyceraldehyde-3-P.

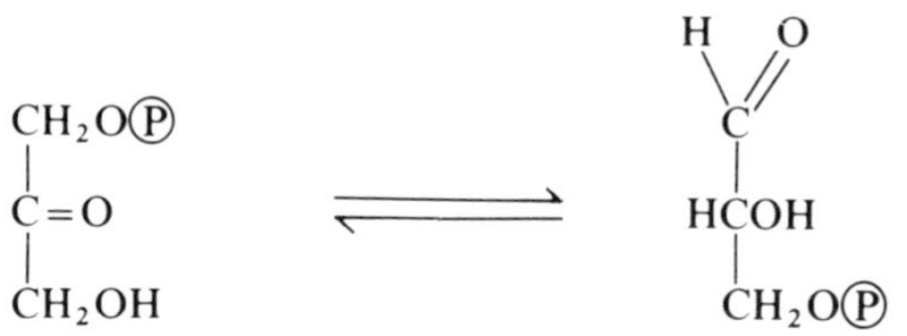

$$\text{Dihydroxyacetone-P} \qquad\qquad \text{Glyceraldehyde-3-P}$$

The reaction sequence given by Eqns 7.5–7.10 results in the conversion of one molecule of hexose sugar to two molecules of triose phosphate with the expenditure of two molecules of ATP, although only one ATP is consumed if the starch and/or sucrose mobilization yields glucose-1-P (Fig. 7.1). This ATP is recouped in the subsequent phase of glycolysis.

7.3.2 Glyceraldehyde-3-P is oxidized to pyruvate in a series of energy-conserving reactions yielding NADH and ATP

In the second phase of glycolysis, NADH and ATP are generated in a series of energy-conserving steps. The first step involves the oxidation of glyceraldehyde-3-P by NAD$^+$ to form the 'high energy' compound glycerate-1,3-P$_2$ in a reaction catalysed by NAD-dependent glyceraldehyde-3-P dehydrogenase. The large decrease in standard free energy associated with the oxidation of glyceraldehyde-3-P is used to support the incorporation of P$_i$ to form glycerate-1,3-P$_2$ (Eqn. 7.11).

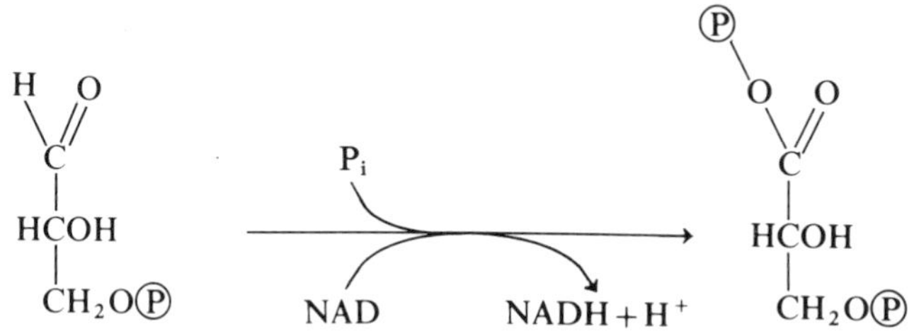

$$\text{Glyceraldehyde-3-P} \qquad\qquad\qquad\qquad \text{Glycerate-1,3-P}_2$$

This reaction is freely reversible ($\Delta G^{\circ\prime} = +6.3\,\text{kJ mol}^{-1}$) and effectively conserves the free energy of oxidation of glyceraldehyde-3-P to glycerate-3-P by NAD^+ in the high energy acyl-P group of glycerate-1,3-P_2 (see Section 4.3).

The ensuing reactions are principally concerned with the phosphorylation of ADP by substrate phosphorylation. The first reaction involves glycerate-1,3-P_2 (Eqn. 7.12).

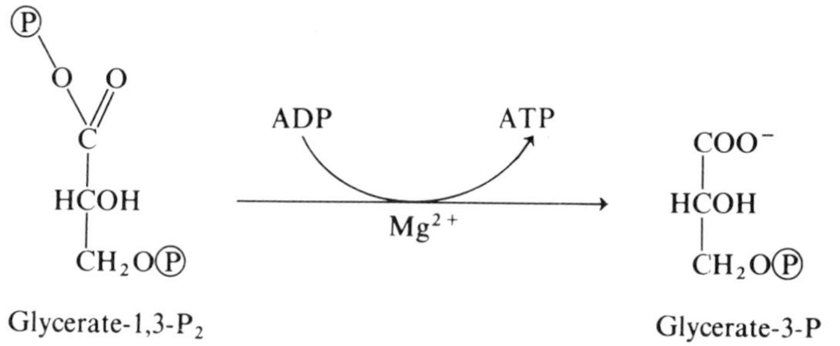

(7.12)

This reaction, catalysed by glycerate-3-P kinase, lies strongly towards ATP formation ($\Delta G^{\circ\prime} = -18.5\,\text{kJ mol}^{-1}$). The product, glycerate-3-P, equilibrates with glycerate-2-P in a reaction catalysed by glycerate-P mutase (Eqn. 7.13), and enolase subsequently catalyses the dehydration of glycerate-2-P to PEP (Eqn. 7.14). Both reactions are freely reversible with $\Delta G^{\circ\prime}$ values of 4.4 and 1.8 kJ mol^{-1}, respectively.

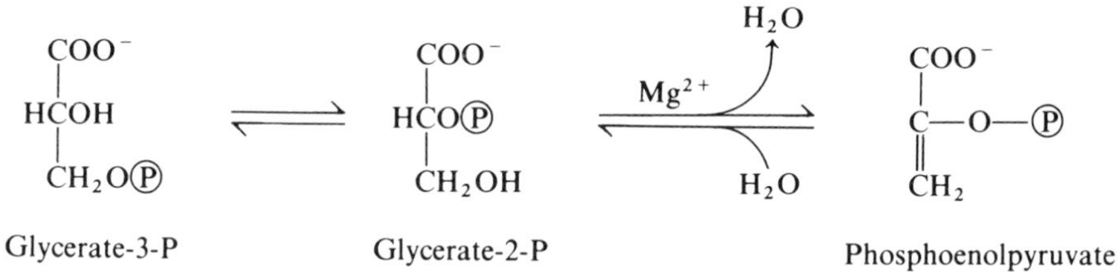

(7.13)   (7.14)

The phosphate group of PEP has a very negative standard free energy of hydrolysis of $-61.9\,\text{kJ mol}^{-1}$, more than sufficient to support the phosphorylation of ADP ($\Delta G^{\circ\prime} \doteq +30.5\,\text{kJ mol}^{-1}$) in the final reaction of glycolysis by which phosphoenolpyruvate is converted to pyruvate (Eqn. 7.15). This reaction is catalysed by pyruvate kinase and is essentially irreversible.

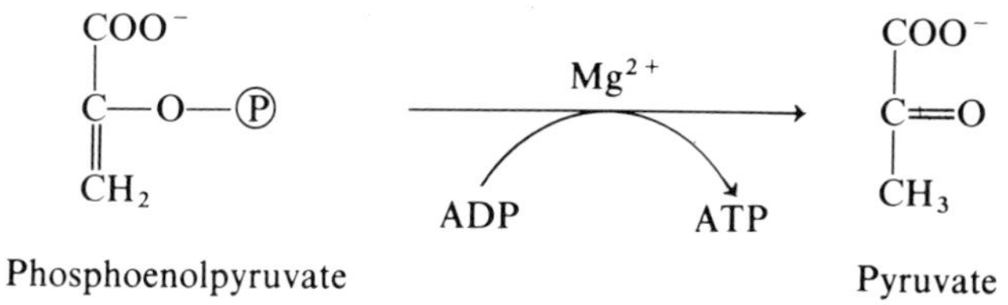

(7.15)

The priming phase of glycolysis (Section 7.3.1) results in the formation of two molecules of glyceraldehyde-3-P per molecule of glucose. Oxidation of the two molecules of glyceraldehyde-3-P supports the phosphorylation of four ADP to four ATP. With the consumption of two ATP per hexose in phase 1 (Section 7.3.1), glycolysis results in the net phosphorylation of two ADP for every molecule of hexose oxidized. Alternatively, if glucose-1-P is the initial substrate, three ADP (net) are phosphorylated. In addition, the oxidation of one hexose molecule results in the reduction of two NAD^+.

7.3.3 The fate of pyruvate formed by glycolysis is dependent on the availability of oxygen

Under aerobic conditions the NADH formed in the oxidation of hexose to pyruvate is reoxidized by the mitochondrial electron transport chain (Section 7.7). This ensures the continued supply of NAD^+ for glycolysis.

Under anaerobic conditions lack of oxygen prevents reoxidation of NADH by mitochondrial electron transport using O_2 as a terminal electron acceptor. Plant cells are not often exposed to anaerobic conditions but occasionally, for example with waterlogging of roots, this does occur. Using such circumstances there are two major routes by which pyruvate can be metabolized to regenerate NAD^+ from NADH and thereby permit continued operation of glycolysis. The first of these involves the formation of acetaldehyde by the decarboxylation of pyruvate and the subsequent reduction of acetaldehyde to ethanol in reactions (Eqns 7.16 & 7.17) catalysed by pyruvate decarboxylase and alcohol dehydrogenase respectively.

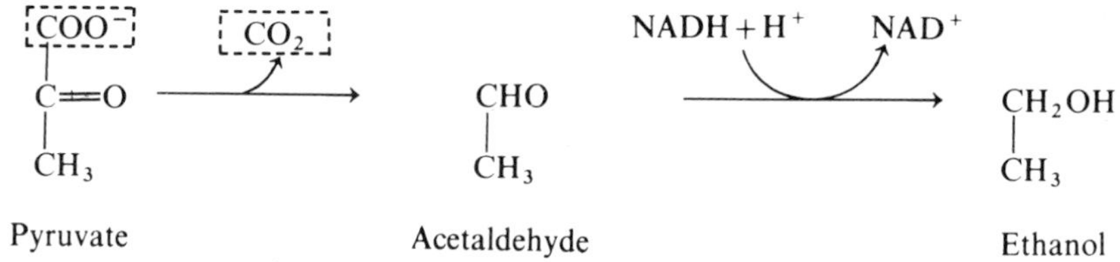

(7.16)   (7.17)

This process has interested humans for thousands of years and is, of course, the basis of the winemaking and brewing industries!

Alternatively, pyruvate is initially reduced to lactate in a reaction catalysed by lactate dehydrogenase (Eqn. 7.18).

$$
\begin{array}{ccc}
COO^- & \xrightarrow{\quad NADH+H^+ \quad NAD^+ \quad} & COO^- \\
| & & | \\
C{=}O & \longrightarrow & HCOH \\
| & & | \\
CH_3 & & CH_3 \\
\text{Pyruvate} & & \text{Lactate}
\end{array}
\qquad (7.18)
$$

Usually, however, lactate formation proceeds for only about 20 min, as the increasing lactate concentration causes the cytosolic pH to drop, switching off lactate dehydrogenase, and allowing ethanol production to commence.

Under anaerobic conditions the net yield of glycolysis from glucose to lactate or ethanol is only two ATP per glucose. This yield is low compared to the six ATP generated per hexose if the two NADH formed in glycolysis are oxidized by the mitochondrial electron transport chain (Section 7.9). The relative yield is even lower when the comparison is made to the 33.33 ATP generated when pyruvate is oxidized by the TCA cycle (see Sections 7.6 & 7.9).

7.4 Reactions of glycolysis occur both in the cytosol and in plastids

There is convincing evidence for the occurrence of all the steps of glycolysis in the cytosol of germinating castor bean seeds and the cytosol of leaf tissue, although the location of the various hexose kinases is disputed. Activity of these enzymes with glucose and fructose is found in the cytosol but much of the hexokinase activity of pea leaves is believed to be associated with the outer mitochondrial membrane. In pea stems, it has been suggested that *all* hexokinase activity is associated with the outer mitochondrial membrane.

A number of the enzymes involved in glycolysis occur in plant cells as different isoenzymes (see Section 5.1.2), one of which is located in chloroplasts. Taken with the evidence that chloroplasts readily convert $\alpha(1\rightarrow4)$glucans (such as the amylose component of starch) to glycerate-3-P it seems very likely that the glycolytic sequence as far as glycerate-3-P is present in these organelles. Less certain is whether the remaining steps to pyruvate occur in chloroplasts. For example,

purified pea chloroplasts contain enolase and pyruvate kinase, but the level of glycerate-P-mutase is very low and this could arise from contamination by cytosol. The available evidence suggests that the complete glycolytic sequence occurs in chloroplasts of spinach and in the plastids involved in fatty acid synthesis in developing seeds of plants such as castor bean. In preparations of isolated cauliflower bud plastids, glycerate-P-mutase activity is present but 80% of the activity is only detectable when the plastids are broken, i.e. the activity is 'latent'. This 'latency' suggests that the enzyme is really present within the plastids and is not due to contamination from cytosol although the activities are very low (Table 7.2). Latency is also exhibited by all other enzymes of the glycolytic pathway in cauliflower bud plastids (Table 7.2), each represented by a particular isoenzyme. Activities of the plastid enzymes are generally high, ranging, for example, from 20–70% of total activity (plastid plus cytosol) in developing endosperm tissue of castor beans.

One major difference between the glycolytic pathways in the cytosol and

Table 7.2 Activities and latency* of enzymes involved in glycolysis in isolated, intact, plastids from cauliflower buds. After Journet, E-P. & Douce, R. (1985) *Plant Physiol.*, **79**, 458–67.

Enzyme	Activity (nmol min^{-1} mg^{-1} protein)	Latency (%)
Hexokinase	40	93
Glucose-P mutase	300	91
Hexose-P isomerase	1100	90
Phosphofructokinase	70	90
Fructose-6-P: PP$_i$ phosphotransferase	0	—
Aldolase	190	90
Triose-P isomerase	1600	65
Glyceraldehyde-3-P dehydrogenase	190 (NAD specific) 30 (NADP specific)	95 95
Glycerate-3-P kinase	900	90
Glycerate-P mutase	0–30	80
Enolase	85	45
Pyruvate kinase	100	75
Pyruvate dyhydrogenase	20	90

*Latency is the latent activity 'released' when an organelle is ruptured. It is thus the total activity of a preparation of ruptured organelles (T) less the activity obtained with intact organelles (I). The % latent activity is then $100 \times (T-I)/T$.

plastids is that PF-Kase is found in both the cytosol and plastids of germinating bean (*Phaseolus vulgaris*) seeds but FP-PTase is found only in the cytosol.

7.5 Flow of metabolites through the glycolytic sequence is regulated

7.5.1 Primary regulation of glycolysis is at the stage of reactions catalysed by phosphofructokinase and pyruvate kinase

Our understanding of the control of glycolysis in plants is imperfect. The major sites of regulation appear to be the reactions catalysed by phosphofructokinase (PF-Kase; enzyme 3a in Fig. 7.1) and pyruvate kinase (enzyme 10 in Fig. 7.1). This is supported by a number of observations.

(1)   Of the 10 reactions involved in glycolysis, in cells only the reactions catalysed by PF-Kase, hexokinase and pyruvate kinase have ratios of product:substrate (mass-action ratios) very much lower than their equilibrium constants. This suggests that reactions catalysed by these enzymes are displaced from equilibrium and are irreversible under physiological conditions. Measurements of reactants and products and the calculated mass-action ratios for a number of tissues suggest that all other reactions in the sequence are close to equilibrium (Table 7.3).

(2)   The rate of glycolysis by organs or tissues can be perturbed by transferring them from a gas phase of nitrogen to one of air, by exposure to ethylene or, for tissue slices, by immersion in salt solutions. Measurements of changes in the concentrations of glycolytic intermediates are consistent with regulation at the PF-Kase and pyruvate kinase steps. Thus, in cells of *Acer pseudoplatanus*, transfer from air to nitrogen, which causes an acceleration of glucose consumption by

glycolysis (see Section 7.5.2), results in increased concentrations of fructose-1,6-P_2 and pyruvate (the products of the PF-Kase and pyruvate kinase catalysed reactions respectively), and in decreased concentrations of glucose-6-P and fructose-6-P. Similarly, in carrot tissue, changing the gas phase induces a three-fold increase in the concentration of fructose-1,6-P_2 and a 60% increase in pyruvate, with smaller decreases in concentrations of PEP and glycerate-3-P.

(3)   Both PF-Kase and pyruvate kinase are sensitive to a range of positive and negative effectors (Section 7.5.2) and have strategic positions in metabolism. The former acts as a control point between glycolysis and gluconeogenesis (see Section 7.5.2) and the latter governs the flow of carbon into the TCA cycle.

7.5.2 Regulation of pyruvate kinase and phosphofructokinase

Under certain physiological conditions many of the reactions of glycolysis can proceed in the reverse direction to support the synthesis of sugar. For example, PEP, derived from acetyl-CoA, can be metabolized in a number of steps to glucose-6-P which is used in sucrose synthesis (see Fig. 10.5). This 'reversal' of glycolysis is known as gluconeogenesis and is summarized in Table 7.4.

The reactions of glycolysis and gluconeogenesis differ most notably in the reactions catalysed by hexokinase (Eqn. 7.5), PF-Kase (Eqn. 7.7) and pyruvate kinase (Eqn. 7.15), which are irreversible under normal physiological conditions. The formation of hexose-P from PEP via gluconeogenesis requires two other enzymes in addition to those involved in the reversible reactions of glycolysis. The

Table 7.3 Comparison between equilibrium constants (K_{eq}) of the reactions of glycolysis and measured mass-action ratios for these reactions in various tissues. From Turner, J.F. & Turner, D.H. (1980). In Stumpf, P.K. & Conn, E.D. (eds) *The Biochemistry of Plants*, vol. 2, pp. 279–316. Academic Press, New York.

		Mass-action ratio					
Reaction	Apparent equilibrium constant	Geminating barley	Castor bean endosperm	Carrot discs	*Arum maculatum* spadix	Grape berries	*Acer pseudoplatanus* cells
Hexokinase	4700	0.0004	–	0.0007	–	–	–
Glucose-P mutase	17.2	89	–	–	12.6	–	–
Hexose-P isomerase	0.42	0.54	0.4	0.22	0.18	0.27	0.20
Phosphofructokinase	1050	0.08	0.09	0.013	0.69	0.042	0.086
Triose-P isomerase	0.041	0.30	–	0.23	–	–	–
Glycerate-P mutase	0.15	–	–	–	0.10	–	–
Enolase	3.7	–	–	–	1.87	–	–
Pyruvate kinase	11000	–	1.7	9.25	11.3	1.62	5.7

$\Delta G^{\circ\prime}$ (KJ mol^{-1}) in glycolytic direction	Reaction				$\Delta G^{\circ\prime}$ (KJ mol^{-1}) in gluconeogenic direction
−16.7	Glucose + ATP	→	glucose-6-P + ADP		*
+1.7	Glucose-6-P	⇌	fructose-6-P		−1.7
−14.2	Fructose-6-P + ATP	→	fructose-1,6-P$_2$ + ADP		*
−2.9	Fructose-6-P + PP$_i$	⇌	fructose-1,6-P$_2$ + P$_i$		+2.9
*	Fructose-6-P + P$_i$	←	fructose-1,6-P$_2$ + H$_2$O		−16.3
+29.9	Fructose-1,6-P$_2$	⇌	glyceraldehyde-3-P + dihydroxyacetone-P		−29.9
+9.5	Dihydroxyacetone-P	⇌	glyceraldehyde-3-P		−9.5
+6.3	Glyceraldehyde-3-P + NAD$^+$ + P$_i$	⇌	glycerate-1,3-P$_2$ + NADH + H$^+$		−6.3
−18.9	Glycerate-1,3-P$_2$ + ADP	⇌	glycerate-3-P + ATP		+18.9
+4.4	Glycerate-3-P	⇌	glycerate-2-P		−4.4
+1.8	Glycerate-2-P	⇌	phosphoenolpyruvate + H$_2$O		−1.8
−31.4	Phosphoenolpyruvate + ADP	→	pyruvate + ATP		*
*	Phosphoenolpyruvate + ADP + CO$_2$	←	oxaloacetate + ATP		+4.2

Table 7.4 Comparison of reactions and $\Delta G^{\circ\prime}$ values of glycolysis and gluconeogenesis. Reactions of glycolysis proceed from left to right and those of gluconeogenesis from right to left.

best-known gluconeogenic process in plants is concerned with the conversion of storage triglycerides to sucrose in germinating seeds, which contain lipid as storage reserves (see Chapter 8). In such plants (e.g. *Ricinus communis* or castor bean), fatty acids are converted via succinate to the C$_4$ dicarboxylate, oxaloacetate. Carbon from oxaloacetate enters the gluconeogenic pathway via the reaction catalysed by PEP carboxykinase (Eqn. 7.19).

$$ATP + oxaloacetate \rightleftharpoons ADP + CO_2 + PEP \qquad (7.19)$$

PEP is subsequently metabolized, via a number of steps (Table 7.4), to fructose-1,6-P$_2$. Because the reaction catalysed by PF-Kase is irreversible, dephosphorylation of fructose-1,6-P$_2$ requires an alternative enzyme. There are two such enzymes in plant cells, namely fructose-1,6-P$_2$ phosphatase (FP$_2$-Pase; Eqn. 7.20) and fructose-6-P:PP$_i$ phosphotransferase (FP-PTase; Eqn. 7.8).

$$Fructose-1,6-P_2 + H_2O \rightleftharpoons fructose-6-P + P_i \qquad (7.20)$$
$$\Delta G^{\circ\prime} = -16.3 \text{ kJ mol}^{-1}$$

Fructose-6-P is metabolized to sucrose via the mechanism described in Section 10.9. All of the enzymes of gluconeogenesis occur in the cytosol.

Since all the enzymes participating in gluconeogenesis and glycolysis are cytosolic (and in many cases involve the same enzymes operating in opposing directions), there must be control over these enzymes to direct carbon flow in one direction or the other, depending on metabolic need. Inhibition of pyruvate kinase, for example, is necessary during gluconeogenesis to prevent simultaneous breakdown of PEP and its formation via PEP carboxykinase. In addition, the concomitant synthesis and dephosphorylation of fructose-1,6-P$_2$ is a futile cycle (see Section 5.5.1). Simultaneous operation of PF-Kase and FP$_2$-Pase would simply lead to loss of energy conserved in ATP.

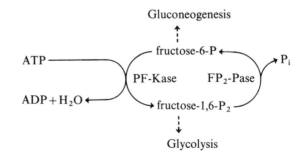

A scheme for the regulation of glycolysis is shown in Fig. 7.2. Pyruvate kinase is inhibited by ATP whilst ADP stimulates activity. The enzyme is also inhibited by citrate and stimulated by the concentration of divalent and monovalent cations such as K^+ and Mg^{2+}, although Ca^{2+} is inhibitory. The regulation of pyruvate kinase by ATP and citrate can be observed when tissue is moved from air to nitrogen; levels of ATP and citrate (formed in the TCA cycle, Section 7.6) decrease and inhibition is relieved. A decrease in concentration of ADP results in lowered pyruvate kinase activity and hence, accumulation of PEP. Since the enolase and the glycerate-P mutase reactions (Eqns 7.9 & 7.8, respectively) are reversible, inhibition of pyruvate kinase also results in an increase in the concentrations of glycerate-2-P and glycerate-3-P. Both these and PEP inhibit activity of PF-Kase.

The regulation of the reaction(s) converting fructose-6-P to fructose-1,6-P_2 is complex. As shown above, regulation of PF-Kase can be achieved through effects of the concentration of various metabolites on pyruvate kinase activity. In addition, the activity of FP$_2$-Pase is strongly inhibited by fructose-2,6-P_2 (not to be confused with fructose-1,6-P_2), synthesized from fructose-6-P via the action of fructose-6-P 2-kinase. Fructose-2,6-P_2 is dephosphorylated to fructose-6-P via fructose-2,6-P_2 phosphatase which, like the kinase, is in the cytosol. The concentration of fructose-2,6-P_2 is controlled by the relative activities of the appropriate kinase and phosphatase. High concentrations of triose phosphate stimulate dephosphorylation of fructose-2,6-P_2 and, together with glycerate-3-P, inhibit fructose-6-P 2-kinase. Conversely, the kinase is stimulated by P_i and fructose-6-P—high concentrations of these compounds thus cause an increase in fructose-2,6-P_2 concentration and thereby inhibit FP$_2$-Pase activity. As a result, carbon flows in the direction of glycolysis. If, however, triose phosphate and glycerate-3-P concentrations are high, the inhibition of FP$_2$-Pase will be relieved, PF-Kase activity will be inhibited and carbon will flow in the direction of gluconeogenesis. The reversible phosphorylation of fructose-6-P is thus a crucial step in controlling the direction of carbon flow between glycolysis and gluconeogenesis. The synthesis and role of fructose-2,6-P_2 in regulation of sucrose degradation and biosynthesis in the cytosol are discussed further in Section 10.13.5.

The role of FP-PTase in glycolysis and gluconeogenesis is unclear. Although the activity of this enzyme is strongly stimulated by fructose-2,6-P_2, the activities in both directions, namely phosphorylation of fructose-6-P and dephosphorylation of fructose-1,6-P_2, are stimulated by about 10–15-fold. FP-PTase activities, however, are greatest in tissues which import and consume sucrose at rapid rates, e.g. sinks such as very young leaves and roots. Possibly the role of FP-PTase in these

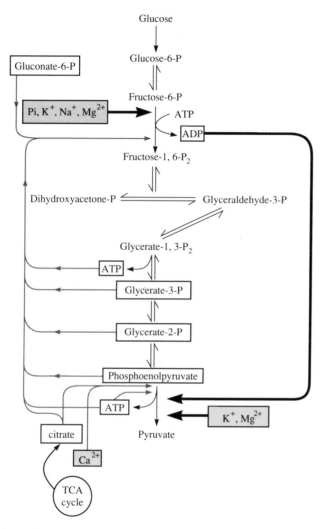

Fig. 7.2 A scheme for the regulation of glycolysis. Stimulation of enzyme activity is indicated by bold lines while inhibition is shown by grey lines. In general, several reaction products late in the pathway stimulate the amount of glucose entering the pathway via PF-Kase while the substrates of the PF-Kase reaction stimulate the pyruvate kinase reaction. (Fructose-6-P:PP$_i$ phosphotransferase is not represented here.) After Turner, J.F. & Turner, D.H. (1980) In Davies, D.D. (ed.) *The Biochemisty of Plants*, Vol. 2, pp. 487–523. Academic Press, London.

tissues is to generate PP_i for the metabolism of UDP-glucose catalysed by UDP-glucose pyrophosphorylase (Eqn. 7.3), thereby stimulating the cleavage of sucrose, via sucrose synthase, to UDP-glucose and fructose (Fig. 7.3). This represents a mechanism for generating two molecules of hexose-P (fructose-6-P) from one sucrose molecule with the expenditure of only one ATP. This is twice as efficient, in terms of ATP use, as sucrose cleavage via invertase (Eqn. 7.1) and subsequent phosphorylation of hexose via hexokinase (Eqn. 7.5).

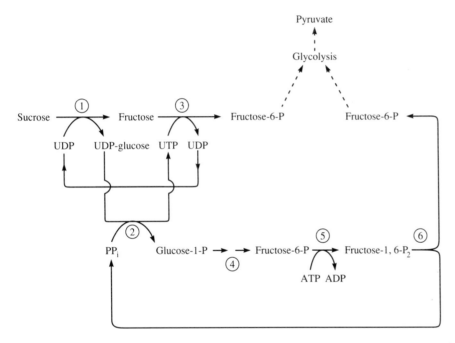

Fig. 7.3 A possible role for fructose-6-P:PP_i phosphotransferase (FP-PTase) in the cytosol of cells importing and using sucrose. The net reaction for fructose-6-P formation from sucrose is

$$\text{sucrose} + \text{ATP} + P_i \rightarrow 2 \text{ fructose-6-P} + \text{ADP}.$$

Fructose-6-P can be converted to pyruvate via subsequent reactions of glycolysis. Enzymes catalysing the reactions are: (1) sucrose synthase; (2) UDP-glucose pyrophosphorylase; (3) fructokinase; (4) glucose-P mutase and hexose-P isomerase; (5) phosphofructokinase; (6) fructose-6-P:PP_i phosphotransferase. After Black, C.C. Mustardy, L., Sung, S.S., Kormanik, P.P., Xu, D.-P. & Paz, N. (1987) *Physiol. Plantarum*, **69**, 387–94.

Plants possess an alternative mechanism, known as the oxidative pentose phosphate (OPP) pathway, for the partial oxidation of glucose in the cytosol. Although the flow of glucose through the OPP pathway is quantitatively less important than through glycolysis, it must be stressed that the two pathways act together in the breakdown of carbohydrate in plant cells. The interaction of the OPP pathway with glycolysis is discussed in Section 8.1.

7.6 The tricarboxylic acid cycle oxidizes pyruvate to carbon dioxide

Glycolysis results in the formation of two molecules of pyruvate, the net phosphorylation of two molecules of ADP and the reduction of two molecules of NAD^+ for every hexose molecule oxidized (Section 7.3). Under aerobic conditions plants oxidize pyruvate still further to three CO_2, involving the reduction of NAD^+ to form even more NADH, which is then oxidized by O_2. The large free energy change is associated with oxidation of NADH used in the phosphorylation of ADP. The oxidative metabolism of pyruvate to CO_2 is known as the tricarboxylic acid (TCA) cycle.

7.6.1 The pathway and enzymes involved

Pyruvate is oxidized to CO_2 in the mitochondria of aerobic tissues. Most of the enzymes involved are located in the matrix of these organelles, although at least one is associated with the inner mitochondrial membrane.

Experimental observations by Krebs and others demonstrated that the rate of oxidation of the C_3 glycolytic product, pyruvate, was promoted by added C_4, C_5 and C_6 compounds, suggesting that these were involved in pyruvate oxidation by O_2. These observations were reconciled in the pathway known originally as the Krebs citric acid cycle, now termed the TCA cycle (Fig. 7.4). In outline, pyruvate is oxidized to CO_2 and a C_2 derivative, acetyl-coenzyme A (acetyl-CoA), which then combines with a C_4 acceptor (oxaloacetate) to yield a C_6 product (citrate). Oxidative reactions remove a further two molecules of CO_2 and regenerate oxaloacetate, thereby comprising a cycle which is accompanied by the production of NADH, $FADH_2$ and ATP.

The first step really lies outside the 'cycle' and comprises the oxidation of pyruvate to acetyl-CoA and CO_2, catalysed by pyruvate dehydrogenase. Pyruvate dehydrogenase is, in fact, a complex of three enzymes, each catalysing a separate reaction. The first of these enzymes is pyruvate decarboxylase which catalyses the

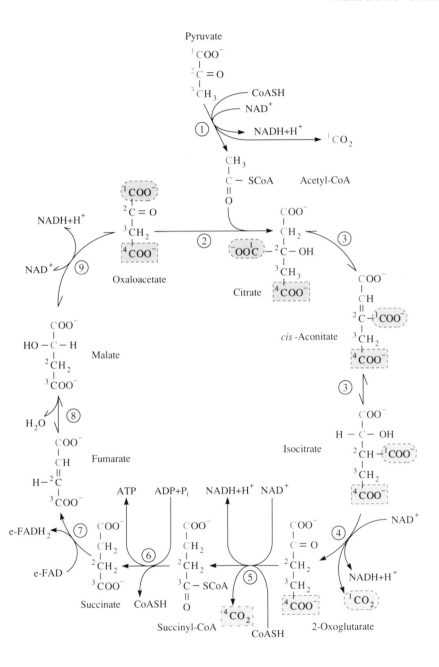

decarboxylation of pyruvate in a reaction involving thiamine pyrophosphate (TPP) and Mg^{2+}, to yield CO_2 (derived from the carboxyl group of pyruvate) and α-hydroxyethyl thiamine pyrophosphate (Eqn. 7.21).

$$\begin{array}{ccc} \mathbf{CH_3} & & \mathbf{CH_3} \\ | & Mg^{2+} & | \\ \mathbf{C}{=}O \ + \ TPP \ \longrightarrow & & H{-}\mathbf{C}{-}OH \\ | & & | \\ \overline{|\,COO^-\,|} & \overline{|\,CO_2\,|} & TPP \end{array} \qquad (7.21)$$

Pyruvate   α-Hydroxyethyl TPP

The hydroxyethyl moeity (shown by carbon atoms in bold) is then transferred to oxidized lipoate, which is bound by amide linkage to lysine residues of lipoate acetyl transferase (Eqn. 7.22). Lipoate functions as a coenzyme in a number of reactions; the oxidized form is a cyclic disulphide, readily reduced by hydroxyethyl TPP to form the acetyl thioester of dihydrolipoate. The acetyl group is eventually transferred to coenzyme A (HSCoA), producing acetyl-CoA and reduced lipoate (dihydrolipoate) in another reaction also catalysed by lipoate acetyl transferase (Eqn. 7.23).

Fig. 7.4 The TCA cycle, which involves oxidative decarboxylation of pyruvate by NAD^+ to form NADH and acetyl-CoA and the subsequent oxidation of the C_2 (acetyl) moiety of acetyl-CoA to two CO_2 with the concomitant generation of a further three NADH, one $FADH_2$ and one ATP per turn of the cycle. Some intermediates of the cycle are used as carbon skeletons for the synthesis of other compounds, a process which potentially could lead to cessation of the cycle. Enzymes are: (1) pyruvate dehydrogenase complex; (2) citrate synthase; (3) aconitase; (4) isocitrate dehydrogenase; (5) 2-oxoglutarate dehydrogenase; (6) succinate thiokinase; (7) succinate dehydrogenase; (8) fumarase; (9) malate dehydrogenase. In some cyanobacteria enzyme (5) is missing and the cycle is incomplete. As a result, acetyl-CoA is only partly oxidized to CO_2 and the incomplete cycle generates far less reductant than the complete pathway shown here. Also indicated on the figure are the fates of individual carbon atoms of pyruvate (grey) and oxaloacetate (black), and of C-1 and C-4 of oxaloacetate (shaded). After one turn of the cycle C-2 and C-3 of pyruvate end up in malate and become C-3 and C-4 of oxaloacetate. Abbreviation: e-FAD=enzyme–FAD complex.

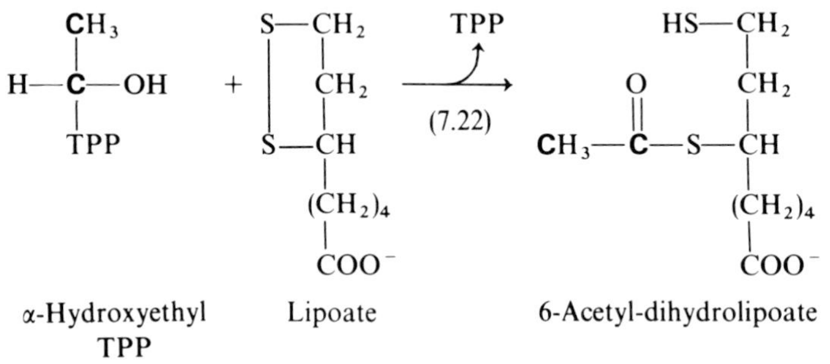

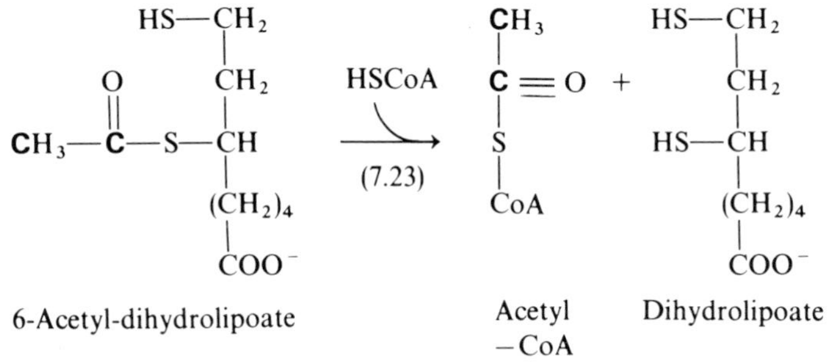

α-Hydroxyethyl TPP   Lipoate   6-Acetyl-dihydrolipoate   (7.22)

6-Acetyl-dihydrolipoate   Acetyl—CoA   Dihydrolipoate   (7.23)

Oxidized lipoate is regenerated in a reaction catalysed by dihydrolipoate dehydrogenase. This enzyme contains tightly bound FAD (e–FAD in Eqn. 7.24), which is an intermediate electron acceptor with NAD^+ as the final oxidant.

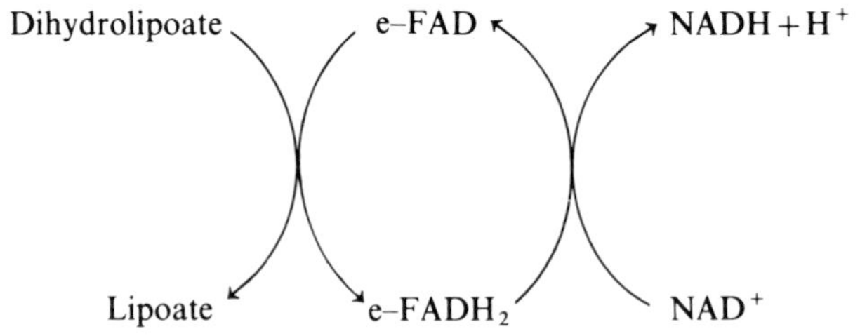

(7.24)

The overall $\Delta G^{\circ\prime}$ for this series of reactions (Eqns 7.21–7.24), summarized in Eqn. 7.25, is $-9.4 \, kJ \, mol^{-1}$.

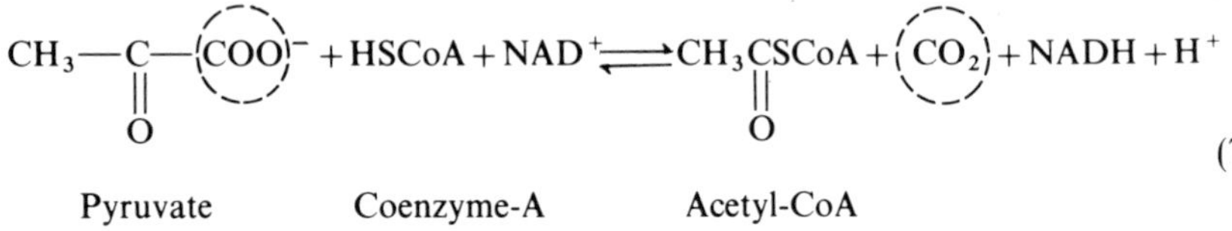

(7.25)

Pyruvate   Coenzyme-A   Acetyl-CoA

In the first reaction of the cyclic phase of the TCA cycle the acetyl group of acetyl-CoA is added, via citrate synthase, to oxaloacetate to form citrate (Eqn. 7.26). This reaction has a large negative change in free energy ($\Delta G^{\circ\prime} = -32.2 \, kJ \, mol^{-1}$), making the formation of citrate essentially irreversible.

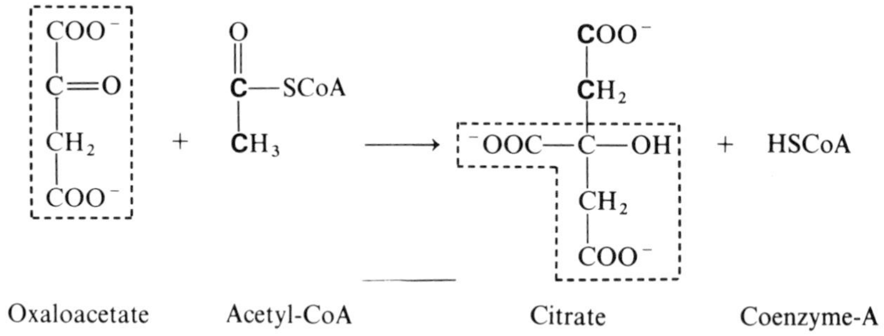

(7.26)

Oxaloacetate   Acetyl-CoA   Citrate   Coenzyme-A

The reaction catalysed by citrate synthase is also of interest as this enzyme supports the synthesis of fluorocitrate from fluoroacetyl-CoA. Fluoroacetate is not in itself toxic to living organisms but fluoroacetyl-CoA can substitute for acetyl-CoA in Eqn. 7.26 forming fluorocitrate, a potent inhibitor of the next enzyme in the cycle—aconitase. Interestingly, several plants synthesize and accumulate high concentrations of fluoroacetate, e.g. *Dichapetalum*, *Gastrolobium* and *Acacia georginae*, and are consequently very toxic to livestock.

Citrate is rearranged to isocitrate via a dehydration and a subsequent rehydration catalysed by aconitase, with *cis*-aconitate as the dehydrated intermediate (Eqn. 7.27).

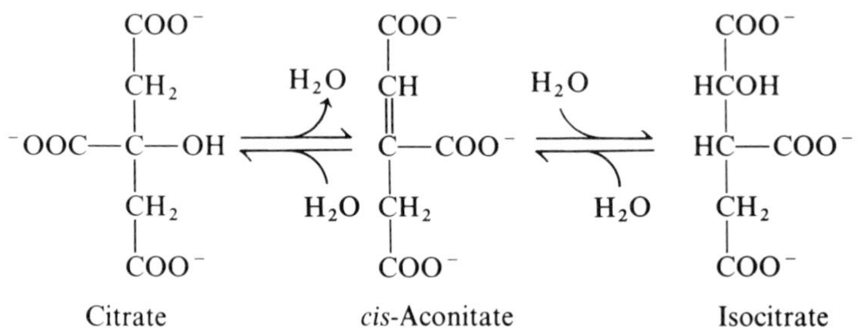

(7.27)

Citrate   *cis*-Aconitate   Isocitrate

Although the equilibrium position of the aconitase reaction favours citrate formation, isocitrate is removed rapidly in the next step of the cycle, thus pulling the reaction in the direction of isocitrate formation.

The ensuing reaction, catalysed by isocitrate dehydrogenase, involves the oxidative decarboxylation of isocitrate to 2-oxoglutarate using NAD^+ as oxidant. The reaction requires Mg^{2+} or Mn^{2+} (Eqn. 7.28), unusual for dehydrogenases.

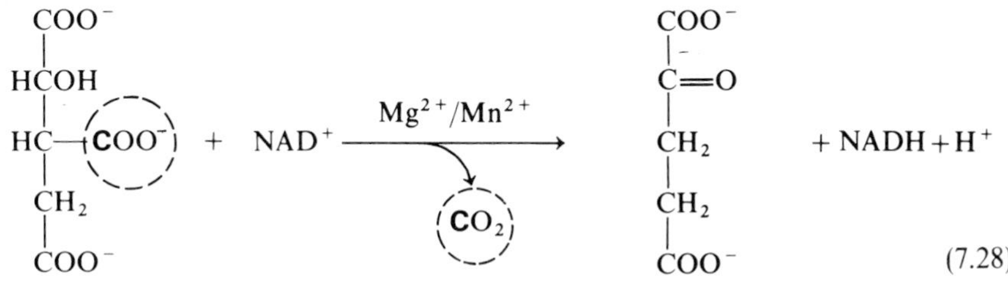

$$
\begin{array}{c}
COO^- \\
| \\
HCOH \\
| \\
HC-(COO^-) \\
| \\
CH_2 \\
| \\
COO^-
\end{array}
+ NAD^+
\xrightarrow{Mg^{2+}/Mn^{2+}}
(CO_2)
\begin{array}{c}
COO^- \\
|^- \\
C=O \\
| \\
CH_2 \\
| \\
CH_2 \\
| \\
COO^-
\end{array}
+ NADH + H^+
\qquad (7.28)
$$

Isocitrate   2-Oxoglutarate

The loss of the β-carboxyl group as CO_2 is an exergonic process and the reaction has a large decrease in free energy ($\Delta G^{\circ\prime} = -20.9 \text{ kJ mol}^{-1}$).

The C_5 molecule, 2-oxoglutarate, is also oxidized and decarboxylated. The enzyme which catalyses the conversion of 2-oxoglutarate to succinyl-CoA is 2-oxoglutarate dehydrogenase but, like pyruvate dehydrogenase, this is really a complex of enzymes. The reaction itself shows other analogies with those supported by the pyruvate dehydrogenase complex; TPP and lipoate are involved (cf. Eqns 7.21–7.24), although the product is succinyl-CoA instead of acetyl-CoA (Eqn. 7.29).

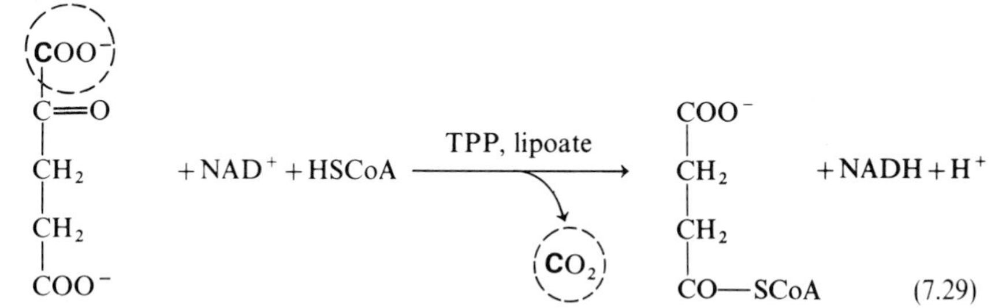

$$
\begin{array}{c}
(COO^-) \\
| \\
C=O \\
| \\
CH_2 \\
| \\
CH_2 \\
| \\
COO^-
\end{array}
+ NAD^+ + HSCoA
\xrightarrow{TPP, \text{ lipoate}}
(CO_2)
\begin{array}{c}
COO^- \\
| \\
CH_2 \\
| \\
CH_2 \\
| \\
CO-SCoA
\end{array}
+ NADH + H^+
\qquad (7.29)
$$

2-Oxoglutarate   Succinyl-CoA

This reaction has a strongly negative $\Delta G^{\circ\prime}$ of -33 kJ mol^{-1} and is irreversible under physiological conditions. As with acetyl-CoA, the thioester bond of succinyl-CoA hydrolyses with a $\Delta G^{\circ\prime}$ similar to that of ATP hydrolysis. This is used to phosphorylate ADP (i.e. substrate-level phosphorylation) in a reaction catalysed by succinate thiokinase (Eqn. 7.30). This reaction proceeds with a standard free energy change of only -2.9 kJ mol^{-1} and is freely reversible.

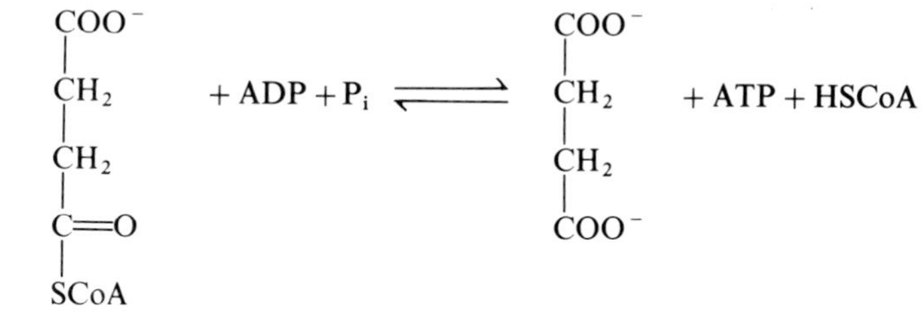

$$
\begin{array}{c}
COO^- \\
| \\
CH_2 \\
| \\
CH_2 \\
| \\
C=O \\
| \\
SCoA
\end{array}
+ ADP + P_i
\rightleftharpoons
\begin{array}{c}
COO^- \\
| \\
CH_2 \\
| \\
CH_2 \\
| \\
COO^-
\end{array}
+ ATP + HSCoA
\qquad (7.30)
$$

Succinyl-CoA   Succinate

Succinate is dehydrogenated to fumarate by succinate dehydrogenase, which is tightly bound to the inner mitochondrial membrane. This enzyme is a flavoprotein and also contains proteins with bound iron and sulphur that participate in electron transport (Fe–S centres; see Section 7.7.1). The FAD is an integral part of the enzyme (e–FAD in Eqn. 7.31) and accepts hydrogen from succinate as shown in Eqn. 7.31.

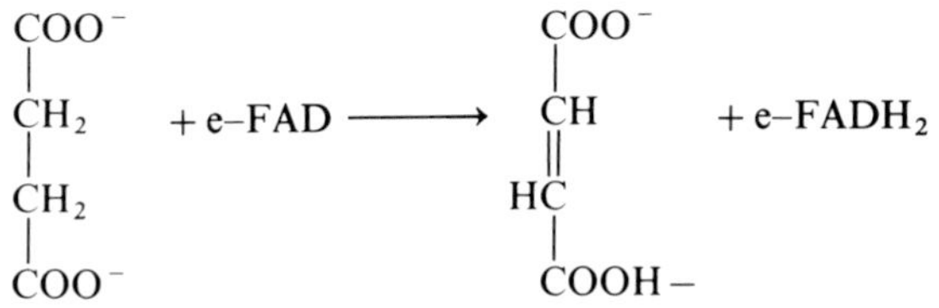

$$
\begin{array}{c}
COO^- \\
| \\
CH_2 \\
| \\
CH_2 \\
| \\
COO^-
\end{array}
+ e\text{-FAD}
\longrightarrow
\begin{array}{c}
COO^- \\
| \\
CH \\
|| \\
HC \\
| \\
COOH-
\end{array}
+ e\text{-FADH}_2
\qquad (7.31)
$$

Succinate   Fumarate

The $FADH_2$ formed in this reaction is reoxidized by the electron transport chain, this being linked to the generation of ATP by oxidative phosphorylation (Section 7.8). In this respect, succinate dehydrogenase is part of both the electron transport chain and the TCA cycle.

The TCA cycle is completed by the hydration of fumarate to malate in a reaction catalysed by fumarase, and by the oxidation of malate to oxaloacetate by malate dehydrogenase (Eqns 7.32 & 7.33 respectively).

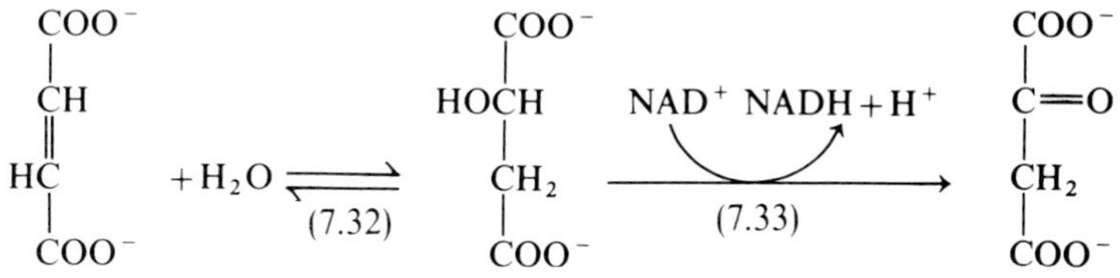

$$
\begin{array}{c}
COO^- \\
| \\
CH \\
|| \\
HC \\
| \\
COO^-
\end{array}
+ H_2O
\underset{(7.32)}{\rightleftharpoons}
\begin{array}{c}
COO^- \\
| \\
HOCH \\
| \\
CH_2 \\
| \\
COO^-
\end{array}
\xrightarrow[(7.33)]{NAD^+ \quad NADH+H^+}
\begin{array}{c}
COO^- \\
| \\
C=O \\
| \\
CH_2 \\
| \\
COO^-
\end{array}
$$

Fumarate   Malate   Oxaloacetate

The formation of malate from fumarate proceeds with virtually no change in free energy ($\Delta G^{\circ\prime} \simeq 0$). The $\Delta G^{\circ\prime}$ of the malate to oxaloacetate conversion is strongly positive ($\Delta G^{\circ\prime} = +29.7 \, \text{kJ mol}^{-1}$), but proceeds towards oxaloacetate formation *in vivo* because oxaloacetate is coupled to the highly exergonic reaction catalysed by citrate synthase (Eqn. 7.26) as the TCA cycle continues, while NADH is oxidized by the mitochondrial electron transport chain (see Section 7.7).

7.6.2 *Evidence for the TCA cycle in plants*

There is little doubt that a functional TCA cycle exists in all plants; the exception among O_2-evolving photoautotrophs appears to be the cyanobacteria which lack 2-oxoglutarate dehydrogenase and, therefore, have an incomplete cycle.

Evidence for operation of a complete cycle in plants comes from a variety of approaches. All the intermediates and enzymes have been isolated from tissues. Studies with various inhibitors are consistent with the cycle. For example, malonate inhibits succinate dehydrogenase and pyruvate oxidation (Section 5.4.1) so that the TCA cycle ceases to operate.

Trace amounts of TCA cycle intermediates support the oxidation of pyruvate by plant mitochondria. For example, castor bean mitochondria exhibit little O_2 uptake ($\equiv$ pyruvate oxidation) when supplied with pyruvate in the absence of intermediates of the TCA cycle; similarly, low rates are obtained when trace amounts of the intermediates are supplied without pyruvate. However, when both pyruvate and intermediates are supplied, the rates of O_2 uptake are greatly accelerated.

Labelling studies are also consistent with the operation of the cycle. Figure 7.4 gives some indication of the expected distribution of label from acetyl-CoA into TCA cycle intermediates. It predicts that the rate of labelling in CO_2 is greater from the C-1 of pyruvate than from C-2, and greater from C-2 than from C-3. Release of CO_2 from C-2 and C-3 should be slower as additional turns of the cycle are required for these carbon atoms to be rearranged to positions where they can be oxidized to CO_2 (see Fig. 7.4). These predictions are borne out by experimental observations. It is worth noting that the eventual recovery of CO_2 from the C-2 and C-3 positions is very rarely 100% because the cycle provides carbon skeletons for other synthetic purposes in cells (Section 7.10).

7.6.3 *A number of metabolites are transported across mitochondrial membranes*

The outer membrane of mitochondrial (and chloroplast) envelopes contains a protein, porine, which forms pores in the membranes rendering them permeable to molecules of molecular mass up to 100 kDa. The inner membranes of mitochondria and chloroplasts, however, are impermeable to most metabolites and transport across these requires various transport systems. Mitochondrial membranes contain polypeptides which transport a range of compounds from the cytosol to the mitochondrial matrix and vice versa. This is evidenced by the fact that isolated mitochondria oxidize various intermediates of the TCA cycle when these are added to the suspending medium. However, mitochondria also require other metabolites such as phosphate and ADP to support their activities, implying that these enter the matrix. Further, mitochondria export compounds such as ATP and amino acids such as glutamate. Polypeptides catalysing these transport processes are called transporters or translocators (see Section 6.3) and some are represented in Fig. 7.5. Thus, pyruvate enters mitochondria in exchange for hydroxyl ions, i.e. it enters via a pyruvate:hydroxyl antiport. Malate, oxaloacetate and succinate are all transported by a dicarboxylate transporter, whilst citrate and isocitrate cross mitochondrial membranes via a tricarboxylate transport system. The efflux of citrate from mitochondria is particularly important as it inhibits phosphofructokinase and pyruvate kinase, key enzymes in glycolysis (Section 7.5.2). Glutamate is exchanged for 2-oxoglutarate, malate or citrate, suggesting that transport of glutamate might not proceed by any one specific carrier. The movement of ADP and ATP into and out of mitochondria is dealt with separately (Section 7.8.4). Inorganic phosphate is also necessary for the exchange of metabolites via the dicarboxylate and adenosine diphosphate transporters; it is also exchanged for hydroxyl ions via the phosphate transporter.

7.6.4 *The TCA cycle is regulated by concentrations of various metabolites*

There are a number of sites at which the TCA cycle is regulated. The cyclic phase of the pathway is dependent on an input of carbon, usually as acetyl-CoA derived from pyruvate or from fatty acids, although amino acids and other organic acids can be used. Control can be exercised by restricting either the supply or rate of transport of these substrates into the mitochondrion. The mechanisms of metabolite transfer in mitochondria have been thoroughly studied (see Section 7.6.3) and are important in the interchange of metabolites between mitochondria and other compartments of the cell because they determine which substrates are oxidized and which components are removed for biosynthesis. However, these processes are probably not important in determining the overall rate of the TCA cycle (hence O_2 uptake) in plant tissues.

Although operation of a biochemical pathway can be regulated by altering the

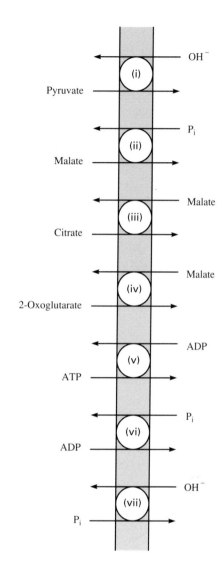

Fig. 7.5 Transport of some metabolites across the inner mitochondrial membrane. A range of transporters have been detected in isolated plant mitochondria and these carry out a number of exchange–diffusion or antiport processes (see Section 6.3.2). The transporters are (i) pyruvate, (ii) dicarboxylate, (iii) tricarboxylate, (iv) 2-oxoglutarate, (v) adenine nucleotide, (vi) adenosine diphosphate and (vii) phosphate. After Wiskich, J.J. (1980) In Davies, D.D. (ed.) *Biochemistry of Plants*, Vol. 2, pp. 243–78. Academic Press, London.

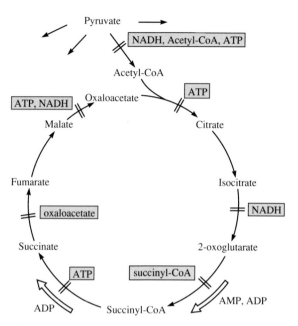

Fig. 7.6 A scheme showing the sites of regulation of the TCA cycle. Inhibition is shown by ⇥ and stimulation by ⇒. After Wiskich, J.J. (1980) In Davies, D.D. (ed.) *Biochemistry of Plants*, Vol. 2, pp. 243–78. Academic Press, London.

level and maximum capacity of a rate-limiting enzyme in the pathway (i.e. 'coarse control' based on transcriptional or translational control of gene expression—see Chapters 5 & 20), this is unlikely in the case of the TCA cycle. The maximum activity of all the enzymes involved is far greater than that needed to account for measured *in vivo* rates of respiration.

A number of the enzymes involved in the reactions of the TCA cycle are, however, regulated by various metabolites (Fig. 7.6). Pyruvate can be metabolized via several reactions other than that involving pyruvate dehydrogenase. Enzymes involved in the divergences of metabolic routes are frequently the sites of regulation. This, together with the irreversibility of the pyruvate dehydrogenase reaction, makes it likely that the pyruvate dehydrogenase complex is a site of regulation of the TCA cycle. The enzyme is inhibited by its end-products, acetyl-CoA, (but only by relatively high concentrations) and NADH. Pyruvate dehydrogenase can exist in activated and deactivated forms, associated with protein phosphorylation (see Section 5.5.4). The non-phosphorylated form of the enzyme is active but phosphorylation brings about rapid deactivation. ATP inactivates pyruvate dehydrogenase by virtue of its stimulatory effect on pyruvate dehydrogenase kinase—an enzyme which catalyses the phosphorylation of pyruvate dehydrogenase into its inactive form. Conversely, ADP inhibits pyruvate dehydrogenase kinase and thereby prevents the phosphorylation and deactivation of pyruvate dehydrogenase.

The reactions catalysed by the 2-oxoglutarate dehydrogenase complex and citrate synthase are also irreversible. The activities of both of these enzymes in mitochondria are regulated by adenine nucleotide levels; 2-oxoglutarate dehydrogenase activity is stimulated (especially by AMP), whilst citrate synthase is inhibited by elevated levels of ATP. 2-Oxoglutarate dehydrogenase is also inhibited by succinyl-CoA, an intermediate of the cycle. The inhibition of citrate synthase by ATP has been explained as a secondary effect of increased ATP levels causing feedback inhibition on succinate thiokinase (see below), resulting in increased concentrations of succinyl-CoA (which inhibits 2-oxoglutarate dehydrogenase).

Succinate thiokinase catalyses the freely reversible substrate-level phosphorylation of ADP using the high free energy of hydrolysis of the thioester bond in succinyl-CoA. The formation of succinate is thus tightly coupled to the relative concentrations of ADP and ATP. High ADP concentrations stimulate succinate formation, whereas high ATP concentrations inhibit the reaction.

Thus, the concentrations of adenine nucleotides strongly affect the activities of pyruvate dehydrogenase, citrate synthase, 2-oxoglutarate dehydrogenase and succinate thiokinase and are, therefore, particularly important in controlling the flow of carbon through the TCA cycle.

Succinate dehydrogenase also exists in activated and deactivated states with activation brought about by a number of metabolites, including succinyl-CoA, reduced ubiquinone, ATP, various metabolites such as succinate and formate, certain anions and low pH. However, although the activity of the enzyme can certainly be influenced by such treatments, their physiological significance with respect to regulation of the TCA cycle within mitochondria is not fully understood.

A number of the enzymes of the cycle, including pyruvate dehydrogenase and isocitrate dehydrogenase, are subject to allosteric inhibition (see Chapter 5) by increased levels of NADH. Malate dehydrogenase catalyses the conversion of malate to oxaloacetate and is inhibited both by NADH and oxaloacetate.

Although the activity of many of the enzymes of the TCA cycle can be altered *in vitro* by changes in the concentrations of metabolites, the significance of this *in vivo* is uncertain. It does appear, however, that the major determining factors are likely to be the concentrations of ATP and ADP as well as NADH. Changes in the intra-mitochondrial concentrations of ADP and ATP, brought about by oxidative phosphorylation (Section 7.8.1), can thereby control carbon flow through certain reactions of the pathway. Since ADP is a substrate for enzymes which bring about substrate-level phosphorylation (e.g. succinyl-CoA synthetase), changes in ADP concentration would also, via electron transport and oxidative phosphorylation,

affect NADH/NAD ratios (which would affect, in turn, activities of enzymes such as isocitrate dehydrogenase); this would afford a means of 'fine control' of the cycle. Citrate indirectly exerts a powerful regulatory effect on glycolysis and therefore, the amount of pyruvate being supplied to the cycle. It must be stressed, however, that as yet there is no full understanding of how the TCA cycle is regulated in plants.

7.7 NADH and FADH$_2$ are oxidized by oxygen via the mitochondrial electron transport chain

Under aerobic conditions, the NADH formed during the glycolytic oxidation of hexose and the oxidation of pyruvate by the TCA cycle is ultimately oxidized by O$_2$. FADH$_2$ produced in the TCA cycle by the reaction catalysed by the succinate dehydrogenase complex is also oxidized by O$_2$. These oxidations are accompanied by a large change in free energy, some of which is normally conserved by phosphorylating ADP in the process of oxidative phosphorylation. The reduction of O$_2$ to water involves a series of carriers of electrons and/or protons in the mitochondrial electron transport chain.

7.7.1 Components of the mitochondrial electron transport chain are arranged in discrete complexes within the inner mitochondrial membrane

Various components of the respiratory electron transport chain of mitochondria are found in or on the inner mitochondrial membrane and are arranged in discrete units or complexes, each comprising a number of different proteins. In plant mitochondria four such complexes are involved in the reactions that transport electrons from endogenous NADH and FADH$_2$ to O$_2$, thereby effecting their oxidation. Two of these complexes are common to the oxidation of both NADH and FADH$_2$. The components involved (Fig. 7.7) are as follows:

(1) Complex I, NADH dehydrogenase, catalyses the reduction of ubiquinone (UQ) by NADH. UQ is a hydrophobic compound of low molecular weight which can exist in reduced (UQH$_2$) and oxidized (more correctly termed ubiquinol) forms. It is substituted 1,4-benzoquinone with a polyprenyl side-chain of 9–10 isoprenoid units and is involved in the transfer of protons and electrons from Complexes I and II to Complex III. UQ is freely soluble in lipids and, as a result, can diffuse readily in the membrane allowing it to act as a mobile redox carrier between various components of the electron transport chain. The NADH-oxidizing site of Complex I is on the matrix side of the inner mitochondrial membrane, so

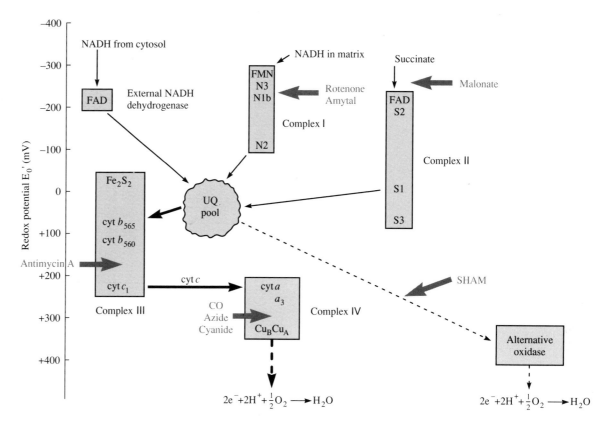

Fig. 7.7 The electron transport chain in higher plant mitochondria. Components of the chain are grouped into their presumed complexes. Also indicated are the sites of action of some inhibitors commonly used in investigations of respiratory electron transport (grey arrows) and the approximate standard redox potential (E'_o) of the components. Abbreviations: N1b, N2, N3 = iron–sulphur protein complexes of Complex I; S1, S2, S3 = iron–sulphur protein complexes of Complex II; Cu_A, Cu_B = copper atoms associated with cytochrome oxidase in Complex IV; Fe_2S_2 = Rieske iron–sulphur protein centre; UQ = ubiquinone.

that only NADH within the mitochondrial matrix is oxidized. The complex is made up from at least 26 different polypeptides with a combined molecular mass of around 850 kDa. Only some of these polypeptides, however, are involved in the redox reactions of electron transport. Flavin mononucleotide (FMN) is an integral part of the complex and can accept two protons and two electrons from NADH and H^+ to yield the reduced form, $FMNH_2$ (Eqn. 7.34).

$$NADH + H^+ + complex–FMN \rightarrow complex–FMNH_2 + NAD^+ \qquad (7.34)$$

Complex I also contains at least three iron–sulphur (Fe–S) proteins in which, unlike cytochromes, the iron is not bound to a tetrapyrrole but is acid-labile (see Fig. 4.5). These can undergo a redox change from Fe^{2+} to Fe^{3+} and vice versa. They are termed N1b, N2 and N3 and are involved in the passage of electrons from

enzyme-bound $FMNH_2$ to UQ, yielding UQH_2. Unlike FMN, the Fe–S proteins cannot accept protons, only electrons (Eqn. 7.35—the designations ox and red indicate the oxidized and reduced forms respectively of the Fe–S proteins).

$$(7.35)$$

Thus, the sequence of electron transfer within Complex I is

$$NADH \rightarrow FMN \rightarrow N3 \rightarrow N1b \rightarrow N2.$$

Electrons are transferred from N2 to UQ which at the same time accepts protons to yield UQH_2.

(2) Complex II, succinate dehydrogenase, spans the inner mitochondrial membrane (Fig. 7.8). There are two major subunits, Fp and Ip, with molecular masses of 65–67 kDa and 26–30 kDa respectively. The Fp (flavo-iron–sulphur protein) subunit contains flavin adenine nucleotide (FAD) and two iron–sulphur proteins which each contain two iron and two sulphur atoms per molecule, and are hence termed 2Fe–2S proteins. These proteins are called centres S1 and S2. The Ip subunit is another iron–sulphur protein with a centre containing four atoms of iron and four atoms of sulphur per molecule (4Fe–4S protein) and is termed S3. Complex II mediates the transfer of electrons from $FADH_2$, produced in the succinate dehydrogenase reaction, to S3 via S2 and S1.

(3) Complex III is the cytochrome b/c_1 complex. Cytochromes are proteins containing iron in a porphyrin group. They exist in reduced and oxidized forms but do not have enzymic activity. Plant mitochondria contain a number of these, distinguished from each other by the absorption spectra of their reduced forms, which have a well-defined peak (the α peak) not present in the oxidized state. Each different type of cytochrome has a characteristic wavelength for the α peak, used in the nomenclature of the b cytochromes. Complex III contains cytochrome b_{560} (α peak at 560 nm), cytochrome b_{565} and cytochrome c_1, together with an iron–sulphur centre containing two atoms of iron and two atoms of sulphur (usually termed the Rieske centre). Complex III accepts electrons from UQH_2 and passes them to cytochrome c_1 in the sequence

$$UQH_2 \rightarrow cyt\ b_{565} \rightarrow cyt\ b_{560} \rightarrow cyt\ c_1.$$

Electrons are transferred from cytochrome c_1 of Complex III to Complex IV via the hydrophilic redox component, cytochrome c.

(4) Complex IV, the cytochrome oxidase complex, spans the inner mitochondrial membrane and catalyses the transfer of electrons from cytochrome c to O_2.

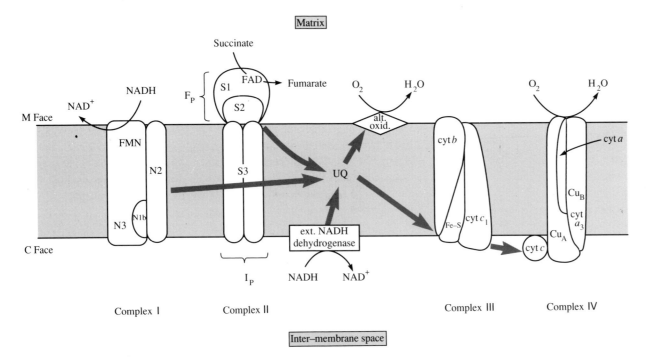

Fig. 7.8 A schematic representation of the distribution of redox components within the inner mitochondrial membrane. The bold grey arrows indicate the general pathway of electron flow from Complexes I or II via the mobile pool of ubiquinone (UQ) to Complexes III and IV. Abbreviations: alt. oxid. = 'alternative oxidase'; ext. NADH dehydrogenase = external NADH dehydrogenase located on the C face of the inner mitochondrial membrane; N1, N2, N3 and S1, S2, S3 = iron–sulphur protein centres of Complex I and Complex II respectively; Cu_A, Cu_B = copper atoms associated with cytochrome oxidase; cyt = cytochrome; Fe–S = iron–sulphur protein of Complex III (Rieske centre).

Complex IV purified from sweet potato comprises at least five major polypeptides. It contains two copper atoms (designated Cu_A and Cu_B) associated with the cytochrome oxidase enzyme itself and two haem moieties, one associated with cytochrome a and the other with cytochrome a_3. The final reaction, catalysed by cytochrome oxidase, involves the formation of water from O_2 (the normal terminal electron acceptor) using electrons from cytochrome a_3 and protons. The sequence of electron transfer is thus from cytochromes to oxygen with the final step catalysed by the cytochrome oxidase enzyme.

$$\text{cyt } a \rightarrow \text{cyt } a_3 \xrightarrow[2e^-]{\text{cytochrome oxidase}} \begin{array}{l} \frac{1}{2}O_2 + 2H^+ \\ H_2O \end{array}$$

Plant mitochondria contain, in addition to the above, two other complexes.
(5) External NADH dehydrogenase. The NADH dehydrogenase of Complex I can only oxidize endogenous NADH produced within the matrix. Plant mitochondria possess another NADH dehydrogenase, located on the external face of the inner membrane, which oxidizes NADH from the cytosol. Oxidation of cytosolic NADH is thereby coupled to reduction of UQ within the membrane.
(6) Alternative oxidase. In the mitochondria of many higher plants it appears that UQH_2 can be oxidized by O_2 via an alternative mechanism not involving the cytochrome oxidase complex (Eqn. 7.36).

$$2UQH_2 + O_2 \rightarrow UQ + 2H_2O \qquad (7.36)$$

This process does not transfer protons from the matrix to the inter-membrane space (cf. Section 7.8) and, consequently, is not coupled to the phosphorylation of ADP. The reaction is catalysed by an enzyme (the 'alternative oxidase') but this has not yet been clearly identified, let alone purified and characterized. Unlike cytochrome oxidase however, the alternative oxidase is not inhibited by cyanide but it is inhibited by substituted hydroxamic acids.

7.7.2 Electrons are transferred between components of the electron transport chain in a specific sequence

The overall reaction involving the mitochondrial electron transport chain is the oxidation of NADH by oxygen (Eqn. 7.37).

$$NADH + H^+ + \tfrac{1}{2}O_2 \rightarrow NAD^+ + H_2O \qquad (7.37)$$

This is accompanied by a very large negative change in free energy. If this is to be conserved by the phosphorylation of ADP, the redox reactions must proceed in an ordered sequence to avoid 'short circuiting' electron transport. This sequence is reflected in the redox potential of the components shown in Fig. 7.7.

Evidence for this linear series of reactions comes from several sources. First, there is the redox potential of the carriers. Electrons can move 'downhill' along the free energy gradient between NADH and O_2 with electrons moving successively from NADH to the redox component with the next lowest redox potential and so on until the electron is transferred to O_2. Because there are a number of steps involved, each of the individual redox reactions involves a fairly small change in redox potential and hence, a relatively small change in free energy (see Section 4.5).

Secondly, the judicious use of inhibitors can cast a great deal of light on the sequence of the reactions involved. Each of the electron carriers has two reaction sites. For example, in the sequence

$$NADH \rightarrow A \rightarrow B \rightarrow C \rightarrow D \rightarrow E \rightarrow O_2$$

carrier B has one site for the reaction involving the oxidation of A and another site at which C is reduced. Thus, an inhibitor of the reduction of C by B does not affect the oxidation of component A by B. For example, it is possible to raise antibodies to the two sites of cytochrome c, one antibody inhibiting the reduction of oxidized cytochrome c by the succinate–cytochrome c reductase complex but having no effect on the oxidation of cytochrome c by the cytochrome oxidase complex, and the other antibody having the opposite effect. Not all of the redox components of the respiratory electron transfer chain are so amenable to such antibody-based techniques. However, there are a number of non-protein inhibitors which can be used to great effect. Some examples are given in Fig. 7.7. They can be used to 'rough out' the reaction sequence of the electron transfer carriers and reactions. By way of illustration, consider the above sequence for electron transfer from NADH to O_2. Adding one inhibitor might reveal that components from NADH to D are reduced (shown in bold) whilst E is oxidized.

$$\textbf{NADH} \rightarrow \textbf{A} \rightarrow \textbf{B} \rightarrow \textbf{C} \rightarrow \textbf{D} \nrightarrow E \rightarrow O_2.$$

With a second inhibitor, however, perhaps only NADH, A and B are reduced.

$$\textbf{NADH} \rightarrow \textbf{A} \rightarrow \textbf{B} \nrightarrow C \rightarrow D \rightarrow E \rightarrow O_2.$$

This suggests that NADH, A and B come before C, D and E in the electron transfer sequence. By repeating this sort of experiment with different inhibitors, the whole sequence can be determined.

Thirdly, the fine details of a sequence can be discovered by examining the kinetics of redox changes of the different components following a pulse of reductant (such as NADH) or oxidant (usually O_2). Thus, if plant mitochondria are supplied with succinate to generate the reduced forms of the intermediate components and given a pulse of O_2, then the rate of oxidation of each component is determined by its position in the sequence; thus, cytochrome a_3 is oxidized first, followed by (in order) cytochromes a, c_{552} and c_{550} and succinate last. These components are very rapidly oxidized with a half time for oxidation of about 10 ms. The other redox components such as cytochrome b_{556}, UQ and the flavoproteins all have considerably longer half times of oxidation of around 300 ms. The sequence of carriers determined from this sort of experiment is consistent with Fig. 7.7. Data from NADH pulse experiments are more complex to interpret but, again, the kinetics of the changes observed support the scheme in Fig. 7.7.

7.7.3 Redox components are arranged in a specific way in the inner mitochondrial membrane

The various electron transport components described in the preceding section are arranged in an ordered fashion in the inner mitochondrial membrane. Some components span the membrane, others are located on the inner or matrix (M) face whilst others are situated towards the inter-membrane space on the cytosolic (C) side of the membrane. Our knowledge of the location of redox components comes largely from work on animal mitochondria and is based mainly on histochemical analysis and the use of macromolecular 'probes' which react with particular membrane components without passing through the membrane.

Much of the work in this area has been carried out using sub-mitochondrial particles (SMPs), formed when mitochondria are ruptured by ultrasonication and the fragments thus formed reseal. When this is carried out with high salt concentrations, the ions tend to cancel out the repulsion between adjacent C faces, making the M face of the inner membrane more hydrophilic and causing it to appear on the outside of the fragments or SMPs (Fig. 7.9). Such particles are termed 'inside-out'. Alternatively, SMPs can be prepared under low salt conditions in which case, due to repulsion between charged adjacent C faces, the particles generated are 'right-side out'. These SMPs can then be tested with, for example, antibodies specific to the various proteinaceous redox components. Antibodies to cytochrome c do not react with inside-out SMPs, suggesting that cytochrome c is

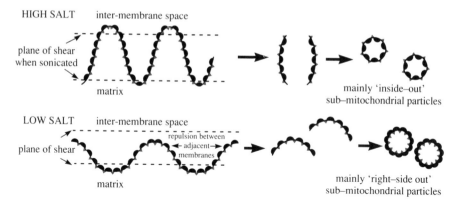

Fig. 7.9 The formation of sub-mitochondrial particles from the folded inner mitochondrial membrane. These particles have been used extensively in studies of the location of electron transport components. 'Inside-out' sub-mitochondrial particles are also useful in that, unlike intact mitochondria, the substrate binding sites for the respiratory chain and the ATP synthetase are located on the outside.

not on the M face, but they *do* inhibit respiration if the SMPs are prepared in the presence of antibody when the C face is exposed, confirming that cytochrome c is located at the C face. In a similar manner, cytochrome a has also been shown to be located on the C face, whereas cytochrome a_3 and succinate dehydrogenase are exposed to the mitochondrial matrix. A further component of the mitochondrial inner membrane, the ATP synthetase, is arranged with its active component, the F_1 constituent, on the M face (see Section 7.8).

Figure 7.8 shows a schematic representation of the organization of the various components of the electron transport chain in the inner membrane of plant mitochondria. Each of the four complexes has components exposed at either the C or M faces. As explained in Section 7.8.2, this arrangement of redox components results in proton transport from the matrix to the inter-membrane space in association with electron transport.

7.8 The free energy of oxidation of NADH by oxygen is conserved by ADP phosphorylation

7.8.1 Electron flow down the electron transport chain is coupled to ADP phosphorylation

The transport of electrons from NADH to O_2 normally occurs with the concomitant phosphorylation of ADP. Since one process does not normally occur without

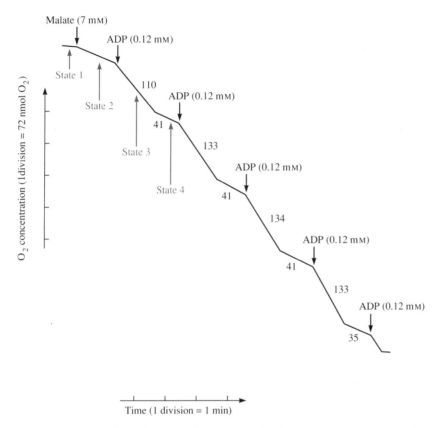

Fig. 7.10 An oxygen electrode trace of oxygen uptake by respiring mitochondria from cauliflower. Values beside the trace denote rates of O_2 consumption in nmol O_2 mg^{-1} protein min^{-1}. See text for explanation of States 1–4. After Wiskich, J.J. & Dry, I.B. (1985) In Douce, R. & Day, D.D. (eds) *Encyclopedia of Plant Physiology*, Vol. 18, pp. 281–313. Elsevier, Amsterdam.

the other, the two processes are said to be tightly coupled as illustrated in Fig. 7.10, which shows O_2 consumption by isolated mitochondria from cauliflower. The mitochondria show a slow uptake of O_2 in the absence of substrate, a condition known as State 1. When substrate is added (e.g. malate) the rate of O_2 uptake increases and the mitochondria are said to be in State 2. When ADP is added there is a very marked increase in the rate of O_2 consumption (State 3) which declines when all the ADP is phosphorylated to ATP (State 4). The ratio of the O_2 uptake rate of State 3 relative to State 4 is used as an index of the functional capacity of

isolated mitochondria. This is termed the respiratory control ratio and, for plant mitochondria, is of the order of 2–10. The important feature is that O_2 consumption is tightly linked to the phosphorylation of ADP. The supply of ADP can thus limit rates of electron transport and oxidation of TCA cycle intermediates.

7.8.2 Electron transport causes proton translocation and the generation of a 'proton motive force'

Current understanding of the way in which energy, made available by the oxidation of substrates or by light absorption, can be used to drive the synthesis of ATP from ADP or establish ion gradients across membranes, is based on the chemiosmotic hypothesis by Peter Mitchell. Although, over the last 20 years or so, a fierce debate has raged over this and alternative concepts, his hypothesis has held up well and has gained (almost) universal acceptance.

The central tenet of the hypothesis is that the transfer of electrons down an electron transport chain generates a proton electrochemical potential gradient or proton motive force across the membrane in which the reactions occur. The proton motive force is made up of both membrane potential ($\Delta\psi$, charge) and pH gradient (ΔpH, proton concentration) (see Section 4.7) and is generated by the active transfer of protons across the membrane. The free energy associated with the return of protons down the proton electrochemical gradient is used to phosphorylate ADP to ATP. Since such membranes convert the energy of a proton electrochemical gradient to a 'high energy' compound, ATP, they are known as energy-transducing membranes.

Proton transport can be demonstrated by following the extrusion of H$^+$ ions by isolated mitochondria into the incubation medium. With a suitable intermediate of the TCA cycle present, which can be oxidized within the mitochondria to generate NADH, injection of a small pulse of O_2 results in a rapid acidification of the external medium as the respiratory electron transport chain functions for 2–3 s. Acidification reaches a maximum and then declines as O_2 is used up and protons diffuse slowly back across the membrane. Addition of an uncoupler, such as FCCP (see Section 6.2.4), immediately accelerates the decline as protons rapidly move down the concentration gradient.

The active transport of protons from the matrix of the mitochondria to the inter-membrane space is possible because of the nature of the various components of the electron transport chain and their location in the membrane. Thus, proton transport comes about as electrons are transferred from the reduced form of a carrier bearing both protons and electrons (e.g. UQH$_2$) to a carrier of electrons only, like (e.g. cytochrome c, reduced form cyt c (Fe^{2+})). The 'spare' protons resulting from electron transfer in the absence of an associated proton transfer are extruded at the outer face of the inner membrane into the inter-membrane space. Transfer of electrons from a carrier of electrons to one that carries protons as well as electrons causes the 'pick up' of protons from the matrix. Each of these transfers is termed, perhaps misleadingly, a 'redox loop' and has a stoichiometry of two protons for every two electrons. Thus, protons and electrons are transferred at the M face of the membrane to FMN in Complex I (see Fig. 7.12). At the C face, electrons are transferred to the Fe–S centre N3 and thence through N1b and N2 back to the M face. Since Fe–S centres of Complex I do not accept protons from $FMNH_2$ these are released into the inter-membrane space. UQ accepts two protons from the matrix (and two electrons from N2) to give UQH_2. UQH_2 migrates to the C face and gives up two electrons to the Fe–S centre and cytochrome c_1 of Complex II. These components only accept electrons and the two protons from UQH_2 are released into the inter-membrane space. The electrons are then transferred back to the M face via the b cytochromes to the cytochrome oxidase complex (Complex IV) where they are used in the reduction of O_2 to H_2O.

In the simplest scheme, 'redox loops' are associated with FMN in Complex I and with ubiquinone with a stoichiometry of *four* protons for every two electrons moving from NADH to O_2. Most observations, however, suggest at least *six* protons are transported into the inter-membrane space from the matrix for every two electrons entering the chain from NADH, and four protons for every two electrons from succinate. One solution to this problem is the incorporation of a proton motive Q-cycle in the model.

Also proposed by Mitchell, the concept of the Q-cycle (Fig. 7.11) derives from the observation that the inhibitor of electron transport, antimycin A, enhances the reduction of b cytochromes and prevents that of cytochrome c, when electrons are entering Complex III from ubiquinone. Addition of the electron acceptor ferricyanide to mitochondrial preparations not only causes the oxidation of cytochrome c, as might be expected, but also the reduction of the b cytochromes. Mitchell proposed that ubiquinone was important in these reactions and undergoes the redox changes shown in Fig. 7.11. Thus, on the matrix side of the membrane, a semi-quinone form, $UQ\dot{H}$, picks up an electron from the transport chain and a proton from the matrix to form UQH_2. At the C face of the membrane the proton is released to yield the semi-quinone and the electron is passed on to cytochromes b_{556} and b_{562}. UQH formed at the C face by reduction of cytochrome b_{556} loses another proton to the cytosol and yields the UQ required for continuation of the cycle, as well as passing on the electron for transfer further down the electron

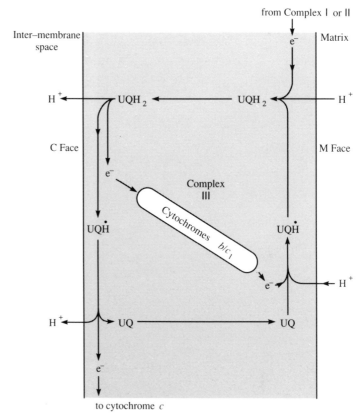

Fig. 7.11 A proposal for a 'proton motive Q-cycle' in the inner membrane of mitochondria. Oxidized ubiquinone (UQ) accepts a proton from the matrix and an electron from the cytochrome b/c_1 complex (Complex III) to form the semi-quinone $UQ\dot{H}$. The semi-quinone accepts another proton from the matrix and a second electron from Complex I or II to yield the fully reduced form, UQH_2. This compound migrates within the membrane to the C face bounding the inter-membrane space where it releases an electron to Complex III and a proton to the inter-membrane space. The UQH thus formed releases a second proton to the inter-membrane space and the remaining electron is passed on via cytochrome c to Complex IV. UQ is recycled. Thus for every electron entering the Q cycle from Complex I (or II), two protons are transported from the matrix to the inter-membrane space.

transport chain. The electron donated to cytochromes $b_{556/562}$ is redirected back into UQ which picks up a second proton and is reduced to yield the semi-quinone. Thus, the redox reactions of ubiquinone generate four protons for every two electrons passing down the transport chain.

If mitochondria are treated with the inhibitor *N*-ethylmaleimide which blocks the H^+/P_i symporter, then $H^+/2e^-$ ratios of eight protons for every two electrons from succinate are observed. It has been suggested that the extra protons are supplied by cytochrome oxidase acting as a 'conformational pump', i.e. the passage of electrons along the electron transport chain causes a conformational change in the cytochrome oxidase molecule such that protons are transported across the membrane.

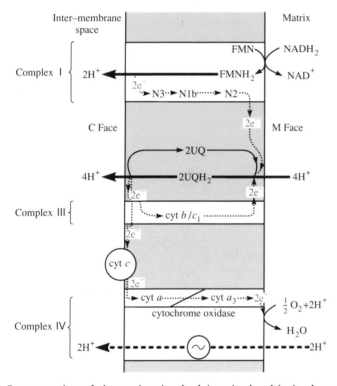

Fig. 7.12 Representation of the carriers involved in mitochondria in the movement of electrons from NADH down the mitochondrial electron transport chain to O_2 and the associated transport of protons (proton translocation) from the mitochondrial matrix to the inter-membrane space. Proton translocation is associated with: (i) the transfer of electrons from endogenous NADH to UQ; (ii) a proton motive Q-cycle; and (iii) a cytochrome oxidase proton pump shown as ⊖. Movement of electrons only are shown by dotted lines whilst combined H^+/e^- transport is shown by the bold black arrows. The bold dashed arrow indicates the conformational proton pump associated with cytochrome oxidase.

The model shown in Fig. 7.12 contains a redox loop associated with Complex I ($2H^+/2e^-$), a proton motive Q-cycle ($4H^+/2e^-$) and a cytochrome oxidase conformational pump ($2H^+/2e^-$). These, acting together, yield a stoichiometry of eight protons translocated for every two electrons moving down the electron transport chain.

7.8.3 The proton motive force drives ADP phosphorylation

Measurements of the proton motive force and its components ($\Delta\psi$ and ΔpH; see Sections 4.6, 4.7 & 6.3.3) suggest that in plant mitochondria this has a value of 150–180 mV. A look at the stoichiometry between H^+ transport and ATP formation can determine whether this is sufficiently large to account for the phosphorylation of ADP by P_i.

In inside-out sub-mitochondrial particles, ATP synthetase works in reverse, i.e. it catalyses the hydrolysis of ATP to ADP and P_i. Nevertheless, inside-out sub-mitochondrial particles in which electron transport is inhibited by the absence of substrate generate a measurable proton gradient when supplied with a known amount of ATP. This allows calculation of the H^+/ATP ratio. No significant proton translocation is observed when the ATP synthetase is blocked by oligomycin (Section 7.8.4) or when the proton transporting ionophore carbonyl cyanide *m*-chlorophenyl hydrazone (CCCP—see Section 6.2.4) is added.

Alternatively, it is possible to analyse the thermodynamics of ADP phosphorylation compared to the free energy associated with the proton gradient. As discussed in Section 4.7, the proton motive force ($\Delta\tilde{\mu}_{H^+}$) between two sides of a membrane, A and B, is given by

$$\Delta\tilde{\mu}_{H^+} = \Delta\psi - \frac{2.3RT}{F}\log\frac{[H^+]_B}{[H^+]_A} \qquad (7.38)$$

where F is the Faraday constant, $\Delta\psi$ is the electrical potential gradient (V), R is the gas constant and T is temperature (K). Thus, the observed values for $\Delta\tilde{\mu}_{H^+}$ of, say -180 mV, are equivalent to $(-180 \times F)$ or -17.4 kJ mol^{-1}. A typical value of $\Delta G'$ for the phosphorylation of ADP at the concentrations of P_i, ADP and ATP present in organelles is around 57 kJ mol^{-1}. Therefore, on theoretical grounds, the synthesis of one mol of ATP requires that 57/17.4 or 3.28 mol of H^+ are transported from matrix to inter-membrane space and then return via the ATP synthetase (see below).

Most values obtained from a range of systems (plant, mammalian and bacterial) suggest a minimum requirement of $3H^+/ATP$.

7.8.4 ATP synthetase of mitochondria

ADP is phosphorylated to ATP as protons are driven, by the $\Delta\tilde{\mu}_{H^+}$, through the macromolecular complex of the inner mitochondrial membrane known as ATP synthetase. This can be seen in electron micrographs of mitochondrial membranes negatively stained with phosphotungstate. The complexes appear as 'knobs' projecting into the matrix. Similar structures occur in thylakoids of chloroplasts, which also contain an ATP synthetase complex, where the 'knobs' are directed into the stroma. In both organelles, ADP is phosphorylated on the side of the membrane where the 'knobs' appear.

The ATP synthetase of mitochondria contains at least nine different polypeptides, several as more than one copy. The complex comprises two major components. One of these, F_1 (the 'knob'), contains the catalytic function. The second component, F_0, comprises hydrophobic proteins embedded in the membrane and serves as a channel through which protons are conducted to the site of ATP formation by F_1 (Fig. 7.13). F_1 and F_0 can be separated from one another by treatment of the complex with urea, various chelating agents or by exposing mitochondrial membranes to low ionic strength solutions. These effectively dissociate F_1 from the membrane.

F_1 has an overall molecular mass of about 360 kDa and comprises five distinct polypeptides into which it readily dissociates at low temperatures. These polypeptides are the α-, β-, γ-, δ- and ε-subunits with molecular masses of 56 kDa, 53 kDa, 33 kDa, 16 kDa and 11 kDa respectively. Their stoichiometry in the complete F_1 component is believed to be $\alpha_3\beta_3\gamma\delta\varepsilon_3$ where the subscripts indicate the number of each subunit present. Various functions have been ascribed to the subunits. Since certain analogues of ATP bind specifically to the β-subunit, this is believed to contain the catalytic site for the phosphorylation of ADP. The δ-subunit binds F_1 to F_0.

Although the isolated F_1 component hydrolyses ATP rapidly, both F_1 and F_0 are required for ADP phosphorylation. Sub-mitochondrial particles treated with, for example, urea to remove F_1 support proton flow but particles treated in this way do not support the phosphorylation of ADP. Similarly oligomycin, which binds to the F_0 complex and prevents proton flow, inhibits the phosphorylation of ADP. Binding of F_1 to F_0 requires an 'oligomycin-sensitivity-conferring-protein' (OSCP). This has a molecular mass of 18 kDa and is only present in the mitochondrial ATP synthetase, not in the chloroplast counterpart (see Section 9.12).

It is not yet known precisely how the ATP synthetase functions, either in the hydrolytic or synthetic direction. It appears likely that the proton motive force induces changes in the three-dimensional structure of F_1 (conformational changes) which alter the relative affinities of F_1 for the substrates and products. In the model shown in Fig. 7.14, the formation of ATP at the catalytic site involves a far lower change in free energy than the eventual release of ATP from the enzyme. Certainly, applying a proton motive force alters the conformation of the chloroplast equivalent of F_1, the CF_1 component. In mitochondria, changes in the fluorescence of the inhibitor aureovertin, which binds to F_1, have been detected and interpreted in terms of a conformational change of the F_1 component induced by a proton motive force.

Studies of the exchange of ^{18}O between $H_2{}^{18}O$ and phosphate (HPO_4^{2-}) according to the reaction

$$ADP + HPO_4^{2-} \rightleftharpoons ATP + H_2O \tag{7.39}$$

have been used to monitor the processes involved in ATP synthesis at the catalytic

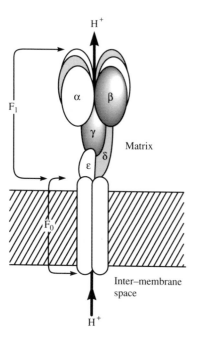

Fig. 7.13 Representation of the subunit composition and arrangement of the polypeptides of the mitochondrial ATP synthetase.

Fig. 7.14 A model for the action of the mitochondrial ATP synthetase (F_1). In this model F_1 has two conformations: in one conformation (light grey) the affinity of the binding sites for $ADP + P_i$ and for protons is low; in the other conformation (dark grey) the affinity of the binding sites for their substrates is high. (a) $ADP + P_i$ bind to the catalytic site of the low affinity form. (b) Protons at high concentration (i.e. low pH) bind to the low affinity proton binding sites, thereby initiating a conformational change in the F_1 complex to a form which has high affinity for protons. (c) As a consequence of the conformational change, the affinity of the catalytic site for ADP and P_i is also increased. (d) The phosphorylation of ADP proceeds at the catalytic site with a free energy change which is low relative to that of the conformational change. (e) If the concentration of protons in the matrix or M phase is low (high pH) then, even given a high affinity of protons for their binding site, protons dissociate from the F_1 complex. (f) When this happens, the complex reverts to the low affinity form and the product (ATP) is released. After Nicholls, D.G. (1982) *Bioenergetics.* Academic Press, London.

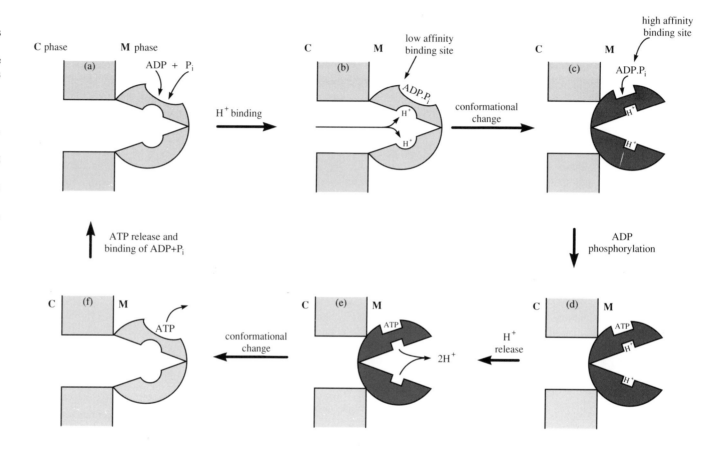

site, without ATP being released (Fig. 7.15). The exchange of ^{18}O requires ADP, is inhibited by oligomycin, and proceeds at low values of the proton motive force, suggesting that the synthesis of ATP is not the major energy-demanding process.

The ATP synthesized is released into the matrix of the mitochondria. If, however, this ATP is used to drive biosynthetic processes in the cytosol, it must be exported. An adenine nucleotide translocator (Fig. 7.16) takes ADP across the inner mitochondrial membrane in exchange for ATP. The phosphate for ADP phosphorylation enters the mitochondria by a proton:P_i symport (see Section 6.3.3). Thus, for every molecule of ATP synthesized, one of the three protons

required is actually necessary, not for the operation of the ATP synthetase itself, but for phosphate transport.

7.9 Analysis of input and output associated with respiration

The inputs and outputs of glycolysis and the TCA cycle can now be compared. The overall oxidation of glucose to CO_2 and water occurs with a change of standard free energy of $-2870 \, kJ \, mol^{-1}$ glucose.

During glycolysis conversion of one molecule of a hexose sugar results in the

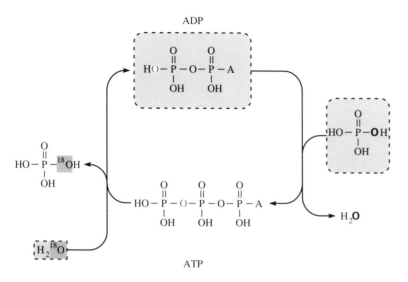

Fig. 7.15 Theory of ^{18}O isotopic equilibrium exchange between $H_2^{18}O$ and $HP_2O_4^-$. Isotope exchange detects the formation of very small amounts of product, even if these remain bound to the enzyme. In the scheme shown here, $H_2^{18}O$, ADP and P_i (shaded) are supplied to sub-mitochondrial particles. During the subsequent equilibrium reaction between $ADP + P_i$ and ATP, ^{18}O from $H_2^{18}O$ is transferred to P_i. This exchange occurs without the release of ADP from the catalytic site of F_1 into solution. Note that the oxygen (shown in grey) connecting the β and γ phosphate groups of ATP is derived from ADP and not from P_i. Abbreviation: A = adenosine.

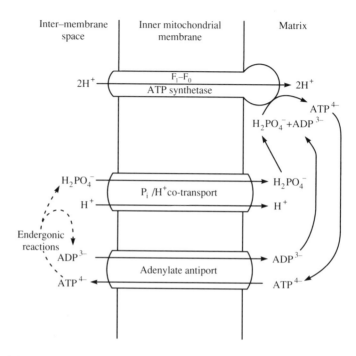

Fig. 7.16 Some features of the adenine transporter in mitochondria. The proton gradient established by respiratory electron transport is used to synthesize ATP via the F_1-F_0 ATP synthetase of the inner mitochondrial membrane. Two protons are required for the synthesis of one ATP within the matrix whilst the transport of adenylate and phosphate across the membrane consumes a further proton per ATP synthesized. Overall, three protons are transported for every external $ADP + P_i$ converted to ATP in the matrix. If $H^+/2e^- = 8$ from the oxidation of NADH by O_2, then 2.67 ATP are formed for every two electrons moving from NADH to O_2 as shown in Fig. 7.12. After Raven, J.A. & Beardall, J. (1981) *Can. Bull. Fish. Aquat. Sci.*, **210**, 55–82. Published by Fisheries and Oceans Canada and reproduced with the permission of the Minister of Supply and Services Canada, 1990.

formation of two pyruvate, together with the generation of two ATP by substrate phosphorylation and two NADH. This NADH can enter the electron transport chain via the external NADH dehydrogenase complex of the inner mitochondrial membrane. The two pyruvate can be fully oxidized to yield six CO_2. In this case, an additional eight NADH and two $FADH_2$ are generated in the matrix of the mitochondria.

Taking the simplest case, in which six protons are transported for every two electrons passing down the electron transport chain from endogenous NADH to H_2O ($H^+/2e^- = 6$) and a ratio of two protons required for every ADP phosphorylated ($H^+/ATP = 2$), then three ATP are produced for each endogenous NADH oxidized by O_2. Two molecules of ATP are produced per $FADH_2$. Also two ATP are formed per molecule of exogenous (cytosolic) NADH oxidized (Table

7.5a). This yields a total of 36 ATP per hexose oxidized to CO_2. If the free energy of phosphorylation of ADP under physiological conditions is $+57\,kJ\,mol^{-1}$, the energy conserved as ATP is $36 \times 57 = 2052\,kJ\,mol^{-1}$ and the energy in glucose is being conserved with an efficiency of 2052/2870 = 71%.

With the more likely values of $H^+/2e^- = 8$ for endogenous NADH and $H^+/ATP = 3$, yielding a ratio of $ATP/2e^- = 2.67$, then 33.33 molecules of ATP are produced per hexose oxidized, equivalent to an efficiency of 66% (Table 7.5b).

Table 7.5 Analysis of inputs and outputs in plant respiration.

	Protons translocated	ATP synthesized
(a) Case 1*†		
Glycolysis		
Substrate phosphorylation	–	2
2 NADH	8	4
TCA cycle		
Substrate phosphorylation	–	2
8 NADH	48	24
2 FADH$_2$	8	4
(b) Case 2†		
Glycolysis		
Substrate phosphorylation	–	2
2 NADH	12	4
TCA cycle		
Substrate phosphorylation	–	2
8 NADH	64	21.33
2 FADH$_2$	12	4

*All values are expressed per mole of glucose.

†Case 1 and Case 2 differ in the stoichiometry of ATP/2e$^-$ as described in the text.

7.10 Glycolysis and the TCA cycle contribute both ATP and carbon skeletons for biosynthetic pathways

7.10.1 Glycolysis and the TCA cycle serve three main functions

As shown above, the reactions of glycolysis and the TCA cycle result in the formation of ATP and the generation of reductant in the form of NADH and FADH$_2$, in turn used to generate ATP. The reactions involved also support the metabolism of triose phosphate in many biosynthetic processes in the cell. These biosyntheses are driven by the ATP produced in both glycolysis and the TCA cycle and by reductant generated by the oxidative pentose phosphate (OPP) pathway (see Chapter 8). In non-photosynthetic cells respiration is the only means of generating ATP or reductant for the synthesis of new biological matter from imported material.

Acetyl-CoA is a major source of carbon skeletons for fatty acid (and hence lipid) biosynthesis. The TCA cycle is an important source of carbon skeletons for the biosynthesis of many amino acids and pyrimidines whilst triose phosphate is used in the synthesis of glycerol and thereby lipids (see Chapter 16). Certain glycolytic intermediates can also serve as C sources for various other biochemical reactions,

e.g. PEP for the shikimate pathway. NADPH and pentose sugars required for nucleic acid synthesis and erythrose for the formation of shikimate required in lignin and quinone biosynthesis are derived from hexoses via the OPP pathway (see Chapter 8).

7.10.2 Intense biosynthetic activity can deplete the TCA cycle of intermediates, which are replenished by anaplerotic reactions

As described above, the TCA cycle recycles one molecule of oxaloacetate for every molecule of pyruvate entering the cycle. However, the TCA cycle is the source of carbon skeletons for many growth processes. If, as a result, the pool of intermediates in the cycle undergoes a net loss, oxaloacetate will not be regenerated. In this event, acetyl-CoA would accumulate and pyruvate oxidation and respiratory metabolism would cease.

However, there are mechanisms by which the intermediates of the TCA cycle can be replenished to allow its continued operation. In higher plant cells oxaloacetate is formed from the carboxylation of phosphoenolpyruvate (Eqn. 7.40) in an essentially irreversible reaction ($\Delta G^{\circ\prime} = -27.2 \text{ kJ mol}^{-1}$).

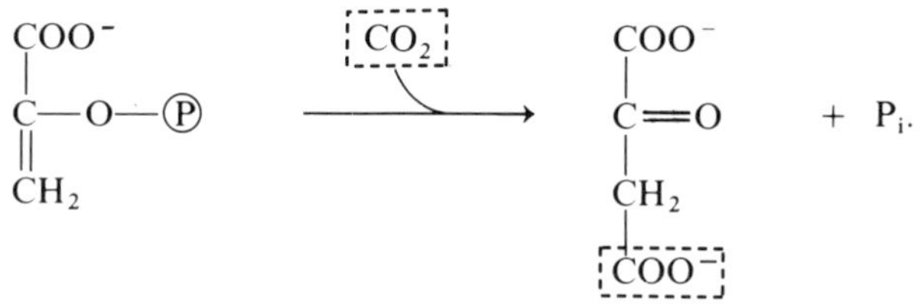

$$\text{(7.40)}$$

Phosphoenolpyruvate   Oxaloacetate

Alternatively, pyruvate can be carboxylated to form malate

$$\text{Pyruvate} + CO_2 + NADPH + H^+ \rightleftharpoons \text{malate} + NADP^+ \qquad \text{(7.41)}$$

in a reversible reaction ($\Delta G^{\circ\prime} = 1.5 \text{ kJ mol}^{-1}$) catalysed by malic enzyme. In animal tissues and some algae pyruvate carboxylase also serves this function (Eqn. 7.42).

$$\text{Pyruvate} + CO_2 + ATP + H_2O \xrightarrow{Mn^{2+}} \text{oxaloacetate} + ADP + P_i \qquad \text{(7.42)}$$

$$\Delta G^{\circ\prime} = 2.1 \text{ kJ mol}^{-1}$$

Pyruvate carboxylase is an allosteric enzyme. The rate of formation of oxaloacetate is virtually zero in the absence of acetyl-CoA, a very strong positive modulator of pyruvate carboxylase activity. So, if acetyl-CoA levels rise, activity is stimulated and more oxaloacetate formed to accept acetyl-CoA into the TCA cycle. Although of importance in animal and algal cells, the role of this enzyme in cells of higher plants is less certain.

Reactions responsible for 'topping up' the regeneration of oxaloacetate in this way are termed anaplerotic reactions. These reactions are crucial if the TCA cycle is to continue to function as a generator of NADH for electron transport and oxidative phosphorylation and as a source of the carbon skeletons for the synthesis of other metabolites for growth.

Further reading

Monographs and treatises: Davies (1980); Davies (1987); Douce & Day (1985).
Electron transport and oxidative phosphorylation: Nicholls (1982).

8
Secondary oxidative mechanisms in plants

Carbohydrate is the major respiratory substrate in plants. It is also the principal product of CO_2 assimilation and the main form in which carbon is transported around the plant. The principal route for oxidation of carbohydrate is glycolysis followed by entry of pyruvate into the TCA cycle. Carbohydrate can, however, be oxidized by alternative routes in plant cells and some reaction sequences result in the oxidation of certain other substrates. These secondary oxidative mechanisms, which can be regarded as respiratory processes, are the subject of this chapter.

8.1 Oxidative pentose phosphate pathway

Arguably the most important of the pathways which are alternatives or supplements to glycolysis is the oxidative pentose phosphate (OPP) pathway. This can be seen as an adjunct to glycolysis—hence it was originally called the hexose monophosphate shunt. A general outline of the sequences involved and their relationship to glycolysis is shown in Fig. 8.1. The reaction sequence can be divided into two parts, an oxidative phase from glucose-6-P to the pentose derivative, ribulose-5-P, and a non-oxidative phase from ribulose-5-P to hexose phosphate (hexose-P) and triose phosphate (triose-P).

8.1.1 The initial phase of the pathway involves oxidation of glucose-6-P to ribulose-5-P

The pathway commences with phosphorylation of glucose to glucose-6-P, as described for glycolysis (see Eqn. 7.5), followed by oxidation of glucose-6-P in a reaction catalysed by glucose-6-P dehydrogenase (Eqn. 8.1). However, as for glycolysis, much of the hexose entering the OPP pathway does so as hexose-P rather than as the unphosphorylated form (cf. Section 7.3.1). The reaction catalysed by glucose-6-P dehydrogenase is freely reversible ($\Delta G^{\circ\prime} = -0.42 \, \text{kJ mol}^{-1}$).

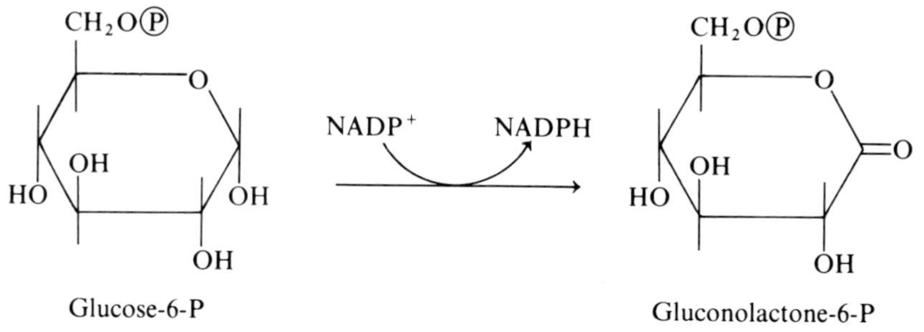

Glucose-6-P   Gluconolactone-6-P

(8.1)

The enzyme shows a specificity for NADP rather than for NAD. The lactone formed is hydrolysed to gluconate-6-P. Although the non-enzymic hydrolysis of gluconolactone-6-P can occur spontaneously, under physiological conditions formation of gluconate-6-P is catalysed by gluconate-6-P lactonase in a reaction requiring Mg^{2+} (Eqn. 8.2). The equilibrium strongly favours gluconate-6-P ($\Delta G^{\circ\prime} = -20.9 \, \text{kJ mol}^{-1}$) and the reaction is essentially irreversible.

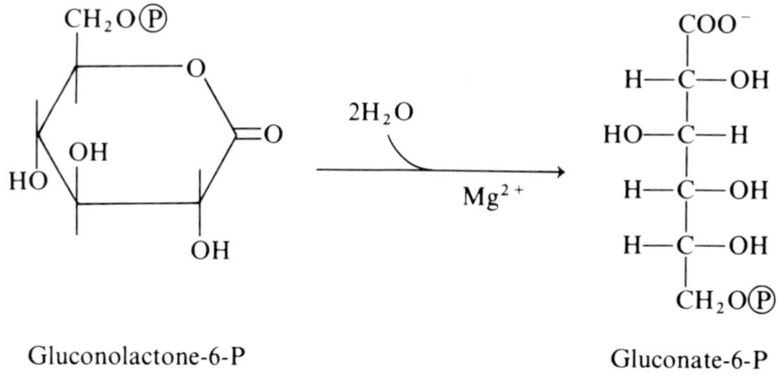

Gluconolactone-6-P   Gluconate-6-P

(8.2)

Gluconate-6-P is then decarboxylated to ribulose-5-P by gluconate-6-P dehydrogenase. This is an oxidative step, as is that catalysed by glucose-6-P

127

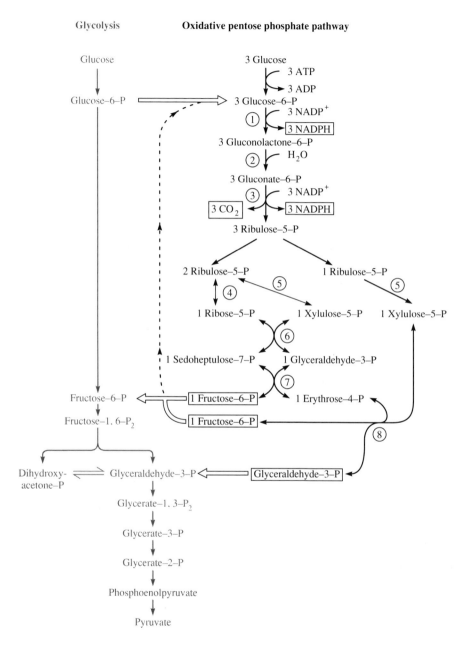

dehydrogenase, and is also NADP specific (Eqn. 8.3). The reaction is, however, irreversible.

$$(8.3)$$

Gluconate-6-P   Ribulose-5-P

During this sequence (Eqns 8.1–8.3), the carbon lost during decarboxylation is C-1 of the original glucose (indicated in bold in Eqn. 8.3). Hence, when [1-^{14}C]glucose is metabolized via the OPP pathway, the pentose phosphate produced is unlabelled and $^{14}CO_2$ is evolved. The reaction sequence also predicts that [2-^{14}C]glucose gives rise to pentose labelled in the C-1 position.

8.1.2 Ribulose-5-P is converted to hexose phosphate and triose phosphate

The next steps in the pathway are a series of interconversions and rearrangements of sugar phosphates. These reactions entail little change in free energy and are readily reversible. The first two steps involve rearrangements of ribulose-5-P to form two other pentose phosphates. One reaction involves isomerization of ribulose-5-P to ribose-5-P by ribose-5-P isomerase (Eqn. 8.4). The other step is catalysed by an epimerase and inverts the configuration of the hydroxyl group of

Fig. 8.1 Interaction between glycolysis and the oxidative pentose phosphate (OPP) pathway. Enzymes involved in the OPP pathway are: (1) glucose-6-P dehydrogenase, (2) gluconate-6-P lactonase, (3) gluconate-6-P dehydrogenase, (4) ribose-5-P isomerase, (5) ribulose-5-P 3-epimerase, (6) transketolase, (7) transaldolase, (8) transketolase. Note that glucose-6-P is a branch point for glycolysis and the OPP pathway and that carbon from glucose-6-P can enter glycolysis (grey) at the level of fructose-6-P or triose-P. Stoichiometries are only given for the OPP pathway. Some recycling of fructose-6-P can occur—this is represented by the dashed line.

C-3 (shown in bold in Eqn. 8.5) of ribulose-5-P to yield xylulose-5-P. Both reactions are freely reversible.

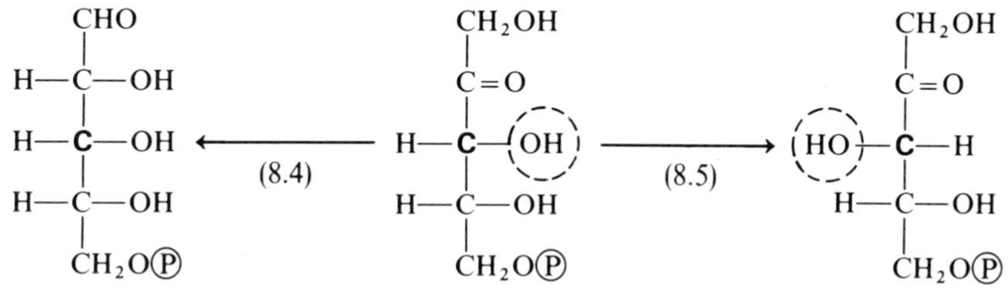

$$\begin{array}{ccc}
\text{CHO} & \text{CH}_2\text{OH} & \text{CH}_2\text{OH} \\
| & | & | \\
\text{H—C—OH} & \text{C=O} & \text{C=O} \\
| & | & | \\
\text{H—C—OH} \xleftarrow{(8.4)} & \text{H—C—(OH)} \xrightarrow{(8.5)} & \text{(HO)—C—H} \\
| & | & | \\
\text{H—C—OH} & \text{H—C—OH} & \text{H—C—OH} \\
| & | & | \\
\text{CH}_2\text{O}\text{P} & \text{CH}_2\text{O}\text{P} & \text{CH}_2\text{O}\text{P} \\
\text{Ribose-5-P} & \text{Ribulose-5-P} & \text{Xylulose-5-P}
\end{array}$$

In the next step, the enzyme transketolase catalyses the transfer of a two carbon moiety from a ketose donor to an aldehyde acceptor molecule. The donors are characteristically keto-sugar-phosphates of five or more carbon atoms, while the acceptors are invariably aldo-sugar-phosphates of shorter chain length (Table 8.1). The transketolase reaction of particular relevance to the oxidative pentose phosphate pathway is shown in Eqn. 8.6 and involves the transfer of a C_2, glycoaldehyde unit (shown in bold) from xylulose-5-P to ribose-5-P resulting in the production of a C_7 keto-sugar-phosphate, sedoheptulose-7-P, and a C_3 aldo-

Table 8.1 Substrates and products of transketolase reactions. Only some combinations are shown here—*any* acceptor can react with *any* donor.

C_2 donor	Acceptor	Products of transketolase reaction
D-xylulose-5-P	D-erythrose-4-P	D-fructose-6-P, D-glyceraldehyde-3-P
D-fructose-6-P	D-glyceraldehyde-3-P	D-xylulose-5-P, D-erythrose-4-P
D-sedoheptulose-7-P	D-glyceraldehyde-P	D-ribose-5-P, D-xylulose-5-P
D-xylulose-5-P	D-ribose-5-P	D-sedoheptulose-7-P, D-glyceraldehyde-3-P

N.B. Since the transketolase reactions are readily reversible, the product of a C_2 addition can also act as a C_2 donor.

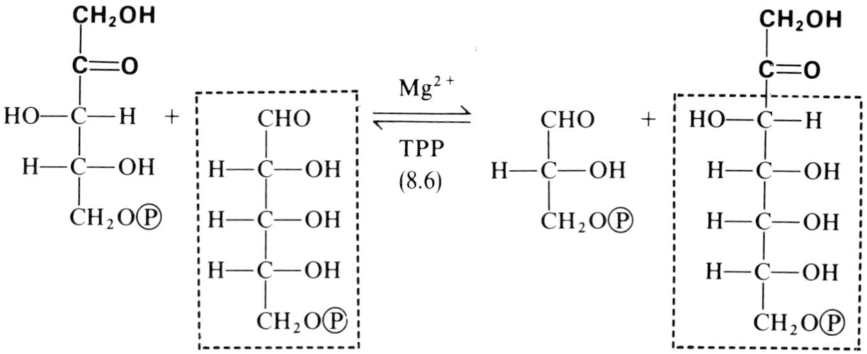

sugar-phosphate, glyceraldehyde-3-P. The reaction requires Mg^{2+} and thiamine pyrophosphate (TPP) (see Section 5.2.2) as a cofactor.

$$\begin{array}{ccccc}
\textbf{CH}_2\textbf{OH} & & & \text{CHO} & \textbf{CH}_2\textbf{OH} \\
| & & & | & | \\
\textbf{C=O} & \text{CHO} & \xrightarrow[\text{TPP}]{Mg^{2+}} & \text{H—C—OH} & \text{HO—C—H} \\
| & | & (8.6) & | & | \\
\text{HO—C—H} + \text{H—C—OH} & & & \text{CH}_2\text{O}\text{P} & \text{H—C—OH} \\
| & | & & & | \\
\text{H—C—OH} & \text{H—C—OH} & & & \text{H—C—OH} \\
| & | & & & | \\
\text{CH}_2\text{O}\text{P} & \text{H—C—OH} & & & \text{CH}_2\text{O}\text{P} \\
& | & & & \\
& \text{CH}_2\text{O}\text{P} & & &
\end{array}$$

Xylulose-5-P   Ribose-5-P   Glyceraldehyde-3-P   Sedoheptulose-7-P

In the next step, carbon atoms C-1 to C-3 of sedoheptulose-7-P are transferred to glyceraldehyde-3-P. This effectively is a transfer of a dihydroxyacetone moiety and results in the production of the C_6 product, fructose-6-P, and the C_4 compound, erythrose-4-P. The reaction is readily reversible and is catalysed by transaldolase (Eqn. 8.7). Unlike the transketolase reaction, no coenzyme is required.

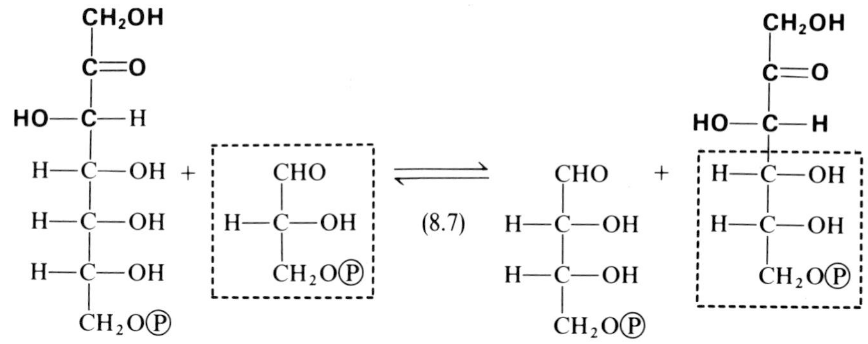

$$\begin{array}{ccccc}
\textbf{CH}_2\textbf{OH} & & & \text{CHO} & \textbf{CH}_2\textbf{OH} \\
| & & & | & | \\
\textbf{C=O} & \text{CHO} & \rightleftharpoons & \text{H—C—OH} & \textbf{C=O} \\
| & | & (8.7) & | & | \\
\text{HO—C—H} + \text{H—C—OH} & & & \text{H—C—OH} & \text{HO—C—H} \\
| & | & & | & | \\
\text{H—C—OH} & \text{CH}_2\text{O}\text{P} & & \text{CH}_2\text{O}\text{P} & \text{H—C—OH} \\
| & & & & | \\
\text{H—C—OH} & & & & \text{H—C—OH} \\
| & & & & | \\
\text{H—C—OH} & & & & \text{CH}_2\text{O}\text{P} \\
| & & & & \\
\text{CH}_2\text{O}\text{P} & & & &
\end{array}$$

Sedoheptulose-7-P   Glyceraldehyde-3-P   Erythrose-4-P   Fructose-6-P

As shown in Table 8.1, erythrose-4-P (Eqn. 8.7) can be an acceptor for a C_2-fragment from xylulose-5-P (formed in Eqn. 8.5) in a transketolase-catalysed reaction, to yield a further molecule of fructose-6-P and one of glyceraldehyde-3-P (Eqn. 8.8).

Erythrose-4-P   Xylulose-5-P   Fructose-6-P   Glyceraldehyde-3-P

The fructose-6-P and glyceraldehyde-3-P thus formed can enter the glycolytic sequence as shown in Fig. 8.1.

8.1.3 Products and stoichiometry of the OPP pathway

The products and stoichiometry of the OPP pathway are illustrated in Fig. 8.1. From a pool of three molecules of glucose-6-P entering the OPP pathway, these are oxidized to three molecules each of the pentose phosphate ribulose-5-P and CO_2, with the generation of six NADPH. Subsequently, the three molecules of ribulose-5-P are used to regenerate two molecules of a hexose phosphate, fructose-6-P, with the concomitant formation of glyceraldehyde-3-P. Both fructose-6-P and glyceraldehyde-3-P are glycolytic intermediates (Fig. 8.1). In this event, the overall process acts as a unidirectional pathway and can be written as

$$3\text{-Hexose-P} + 6\ NADP^+ \longrightarrow 2\ \text{hexose-P} + \text{glyceraldehyde-3-P} + 3\ CO_2 + 6\ NADPH. \tag{8.9}$$

Alternately, the fructose-6-P formed by the OPP pathway can be converted to glucose-6-P via hexose-P isomerase (see Section 7.3.1) and recycled via the OPP pathway. If the fructose-6-P is recycled to glucose-6-P, then for a pool of three molecules of glucose-6-P the *net* reaction for each turn of the cycle is

$$\text{Glucose-6-P} + 6\ NADP^+ \rightarrow \text{glyceraldehyde-3-P} + 3\ CO_2 + 6\ NADPH \tag{8.10}$$

The first part of the OPP pathway (Phase I) is, then, the oxidative stage and consists of those steps from glucose-6-P to ribulose-5-P where the reactions are essentially irreversible under normal physiological conditions. The second part (Phase II) is the subsequent non-oxidative portion and can be considered a regenerative phase.

In summary, for every three molecules of pentose phosphate entering the non-oxidative Phase II, two molecules of hexose phosphate and one of triose phosphate are formed.

8.1.4 The OPP pathway is present in both the cytosol and chloroplasts of photosynthetic cells

The enzymes of the OPP pathway in plants have not been extensively investigated. Nevertheless, many studies have shown that the cytosol of both non-photosynthetic and photosynthetic cells contains the complete sequence of enzymes involved in both Phases I and II of the pathway. At least some of the enzymes are present as two isoenzymes (Section 5.1.2), one in the cytosol and one in the plastids. For photosynthetic cells, certain of the enzymes involved in the OPP pathway in the cytosol are also involved in the C_3-CR assimilation pathway in chloroplasts (see Chapter 10). In such cells there are two isoenzymes of glucose-6-P dehydrogenase, 6-phosphogluconate dehydrogenase and ribose-5-P isomerase. Data on the subcellular location of transketolase isoenzymes in leaves is inconclusive and neither transaldolase nor ribulose-P 3-epimerase have been examined in this respect. However, it is thought that all the enzymes of the OPP pathway are present in both cytosol and chloroplasts of photosynthetic cells. In pea shoots up to 40% of the activity of the two dehydrogenases is found in the chloroplasts. For non-photosynthetic cells, information is even more sparse.

The enzymes and intermediates of the OPP pathway are present in a number of plant tissues and metabolism of intermediates such as erythrose, ribose or sedoheptulose is consistent with the scheme shown in Fig. 8.1. Importantly, the labelling pattern of the various carbon atoms of intermediates of the pathway following administration of [1-^{14}C]glucose, [2-^{14}C]glucose, etc., is consistent with the predicted labelling pattern (Fig. 8.2), which also strongly supports the role of transketolase and transaldolase in the pathway.

Glycolysis and the OPP pathway share a number of intermediates and a close relationship exists between the two, as shown in Fig. 8.1. In the cytosol, the bulk of the triose and hexose phosphates produced by the OPP pathway are converted to pyruvate by the reactions of glycolysis. There is a limited amount of recycling of fructose-6-P through the OPP pathway, although recycling of glyceraldehyde-3-P is generally considered unlikely.

Although there is ample evidence that enzymes of the OPP pathway are present in plastids of both photosynthetic tissue (e.g. pea and spinach leaves) and of non-photosynthetic tissue (e.g. cauliflower buds), there are few studies to show the flow of carbon from glucose via intermediates of the pathway. Nonetheless, in pea chloroplasts, the pattern of CO_2 release from glucose labelled in specific carbon atoms is consistent with considerable activity of the pathway. Furthermore, isolated spinach chloroplasts convert glucose-6-P to gluconate-6-P and pentose monophosphates. This is stimulated when the flow of carbon via glycolysis is inhibited. Thus it seems likely that chloroplasts and other plastids possess a functionally active OPP pathway. However, due to the regulatory properties of glucose-6-P dehydrogenase (Section 8.1.6), the OPP pathway in chloroplasts is only active in the dark. Furthermore, in chloroplasts, in contrast to the cytosol, the C-2 and C-6 carbon atoms of glucose are converted to CO_2, suggesting appreciable recycling of both fructose-6-P and glyceraldehyde-3-P.

The pathway fulfills a number of functions in plants. Firstly it generates NADPH as a reductant for biosynthetic processes in the cytosol. Although NADPH is generated in the light through electron transport in chloroplasts, it is

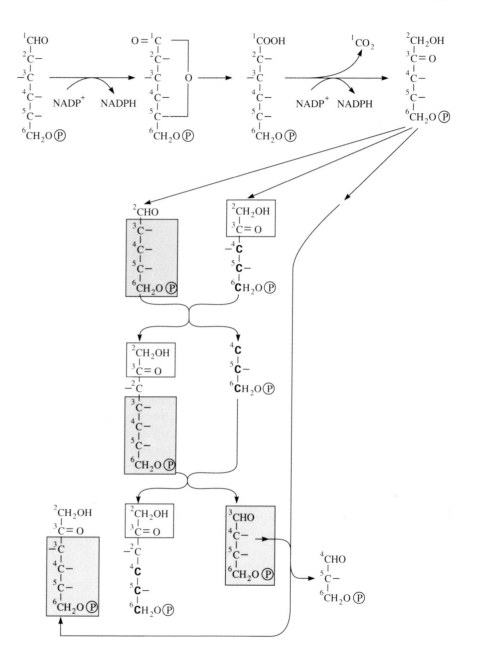

Fig. 8.2 The fate of the individual carbon atoms of hexose phosphate in the OPP pathway. Numbers refer to the carbon atoms of the original hexose and bars on C atoms denote the attachment of hydroxyl groups. Note the loss of C-1 from hexose in the conversion of hexose to pentose. The C transfer reactions involve an initial pool of 3 mol of ribulose-5-P formed from 3 mol of hexose-P.

not freely available to the cytosol, nor is it formed in the dark. Nevertheless, NADPH is required in non-photosynthetic tissues, such as roots and germinating seeds, and in the cytosol of photosynthetic cells (in the light and the dark), for processes such as lipid biosynthesis and the formation of deoxyribonucleotides. Significantly, oxidation of NADPH generated by the OPP pathway is not coupled to phosphorylation of ADP. Secondly, the pathway generates ribose-5-P for nucleotide and nucleic acid synthesis. Thirdly, the erythrose-4-P formed is used in the synthesis of shikimate, an intermediate in the formation of aromatic rings (e.g. in lignin in plant cell walls). It has also been suggested that, in chloroplasts only, the cycle is important to maintain adequate levels of reduced sulphydryl compounds (e.g. reduced glutathione) as the NADPH generated is the primary reductant for oxidized glutathione; the reduced form of glutathione fulfills several important cellular functions (see Section 14.5).

8.1.5 Is the pathway a cycle?

In the foregoing discussion, the reactions have been mainly represented as a unidirectional pathway (Eqn. 8.9) in which three molecules of a hexose sugar (glucose) undergo a series of reactions, eventually yielding two molecules of another hexose (fructose-6-P) and one molecule of triose (glyceraldehyde-3-P). In theory, however, the fructose-6-P can be recycled to glucose-6-P. The extent of such recycling is limited and variable between tissues and species. Whether or not the pathway operates as a cycle might depend on the cellular requirements for NADPH. Where this requirement is high, as in cells undergoing massive lipid synthesis, NADPH will be formed in the oxidative phase (Phase I) with any excess pentose phosphate converted to fructose-6-P via Phase II and thence possibly recycled. Where the demand is for pentose phosphates for biosynthesis but without a concomitant demand for NADPH, these could be supplied via Phase II operating in the reverse direction using hexose-P and triose-P from glycolysis.

8.1.6 Regulation of the OPP pathway is achieved via glucose-6-P dehydrogenase

As shown in Fig. 8.1, the major branch point between glycolysis and the OPP pathway is at glucose-6-P. Whereas most reactions of the OPP pathway are close to equilibrium, the steps catalysed by glucose-6-P dehydrogenase and lactonase have mass-action ratios far from equilibrium. As discussed for some glycolytic enzymes in Section 7.5.1, these reactions are, therefore, likely to be controlling

points in the pathway. There is some evidence that glucose-6-P dehydrogenase is under coarse control with levels varying, for example, with age of tissue. This enzyme is also under fine control *in vivo* through the inhibitory effect of NADPH. Fine control of glucose-6-P dehydrogenase activity is important in regulating the flow of carbon through the OPP pathway because in most plant tissues the activities of the other enzymes of the pathway are far greater than needed to account for the observed *in vivo* rates of carbon flow through the pathway.

In spinach leaf chloroplasts, the activity of glucose-6-P dehydrogenase is strongly inhibited by high ratios of NADPH/NADP$^+$. This suggests that, *in vivo*, the enzyme is inhibited in the light since the light reactions of the chloroplast generate NADPH. The extent of regulation by the NADPH/NADP$^+$ ratio is further dependent on the concentration of ribulose-1,5-P$_2$, with high levels of this compound increasing inhibition (Fig. 8.3). This again is consistent with inhibition of the enzyme in the light. In pea leaves, both chloroplastic and cytoplasmic forms

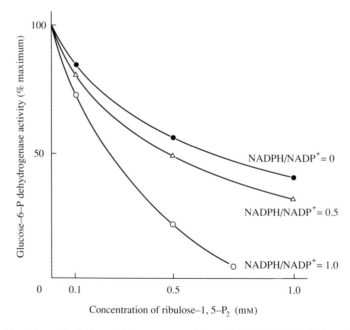

Fig. 8.3 The effect of ribulose-1,5-P$_2$ concentration and NADPH/NADP$^+$ ratio on the activity of glucose-6-P dehydrogenase from spinach chloroplasts. After Lendzian, K. & Bassham, J.A. (1975) *Biochim. Biophys. Acta*, **396**, 260–75.

of glucose-6-P dehydrogenase are activated in the dark (low $NADPH/NADP^+$ ratio). This activation would allow chloroplasts to generate pentoses and NADPH for maintenance when these are not being generated via the C_3-CR cycle or the light reactions of the chloroplast respectively.

8.1.7 Rate of carbon flow through the pathway

The proportion of glucose metabolized via the OPP pathway relative to glycolysis in plant cells has been much debated. Glucose entering the pentose phosphate pathway rapidly loses the C-1 atom as CO_2. If, however, glucose is respired via glycolysis and the TCA cycle, carbon atoms C-1 and C-6 both become C-3 of pyruvate (Section 7.6.2) and are the last to be released as CO_2, being lost at *equal rates*. Therefore, much stock has been placed, in the past, on the relative rates of release of C-1 and C-6 as CO_2 from glucose in respiring tissues, a process which can be examined by supplying tissues with glucose specifically labelled with ^{14}C at C-1 or C-6. A ratio for C-6:C-1 of less than unity suggests that some carbon is flowing through the OPP pathway. A ratio of unity implies, but does not prove, that glucose is being used entirely by glycolysis and the TCA cycle. However, because of fructose-6-P and glyceraldehyde-3-P recycling and concurrent operation of other metabolic pathways which release CO_2, yields of $^{14}CO_2$ from C-1 and C-6 labelled glucose are unreliable as quantitative indicators of the relative activities of glycolysis and the OPP pathway. However, measurements of yields of $^{14}CO_2$ and labelling patterns of various intermediates (see Fig. 8.2) suggest that 5–15% of respiratory glucose metabolism in plant cells occurs via the OPP pathway. Although data is scarce this is unlikely to exceed 30% on further investigation.

The values cited above are overall figures for whole cells. Within chloroplasts, at least those of pea and spinach, the OPP pathway contributes considerably more to the total respiration of glucose. Some 35–56% of glucose-6-P metabolized in pea chloroplasts enters the OPP pathway.

8.2 Oxidation of lipids is important in germinating oil-bearing seeds

The respiratory processes described so far concern the oxidation of carbohydrate. Starch and sucrose are, in general terms, the usual substrates of respiration but lipids and, possibly, proteins can also fulfil this role. The storage of lipid is rare in most non-storage tissues of growing plants but does occur in the fruits of some, such as avocados and olives. Lipids *are* stored, however, in the seeds of many species where they are found primarily in the endosperm, cotyledons and, less markedly, in the embryonic axis. In terms of energy stored per unit volume, lipids are very efficient compared to carbohydrate and lipids are the primary storage material in most small seeds, providing the energy required for development and establishment of the seedling. The converse, however, is not always true—the large seeds of castor bean and sunflower are notable for their high lipid content! Fats and oils are poorly soluble in water. Lipids are not, therefore, readily translocated to growing shoots and roots, but are converted to more soluble, and hence more mobile, compounds before being used in growth processes.

8.2.1 Storage lipids are hydrolysed to fatty acids and glycerol

Most storage lipids are triglycerides (i.e. esters of glycerol and fatty acids) and are found as oil bodies (oleosomes, sometimes termed spherosomes) in seeds and fruits. The first step in the breakdown of lipid involves the action of lipases, enzymes which catalyse the hydrolytic breakage of the ester bond between the glycerol and fatty acid moieties. This results in the liberation of the free fatty acids and glycerol.

A range of such enzymes with acid, neutral or alkaline pH optima have been identified, although not all might be present in any one species or tissue. In castor beans, acid lipase activity is closely associated with the accumulation of lipid in maturing endosperm, whilst the degradation of the lipid as the seeds germinate is associated with an increase in activity of alkaline lipase. In seedlings of some species such as corn or cotton, alkaline lipase activity appears in the oleosome membrane during germination. In castor bean and other seeds the lipase is an integral part of the membrane surrounding the glyoxysome where oxidation of fatty acids occurs. This poses a question regarding the transfer of the (water-insoluble) lipid from the oleosomes to the glyoxysomes. The answer might lie in the frequently observed direct contact between oleosomes and glyoxysomes.

8.2.2 Glycerol is oxidized to dihydroxyacetone-P

The glycerol released by lipase action is phosphorylated in the cytosol to glycerol-3-P by glycerol kinase (Eqn. 8.11). ATP is consumed in this reaction. Glycerol-3-P is oxidized to dihydroxyacetone-P in a mitochondrial reaction catalysed by glycerol-3-P dehydrogenase and resulting in the reduction of NAD^+ (Eqn. 8.12).

$$\begin{array}{ccc}
CH_2OH & ATP\quad ADP & CH_2OH \\
| & & | \\
HOCH & \longrightarrow & HOCH \\
| & & | \\
CH_2OH & (8.11) & CH_2O\textcircled{P} \qquad NAD^+ \\
\text{Glycerol} & & \text{Glycerol-3-P}
\end{array}$$

(8.12)

$$\longrightarrow NADH$$

$$\begin{array}{c}
CH_2OH \\
| \\
C=O \\
| \\
CH_2O\textcircled{P}
\end{array}$$

Dihydroxyacetone-P

The dihydroxyacetone-P can be either converted to hexose sugars by gluconeogenesis or catabolized via the glycolytic sequence to pyruvate, which, in turn, can enter the TCA cycle where its oxidation to CO_2 is associated with the generation of ATP (see Chapter 7).

8.2.3 Fatty acids containing an even number of carbon atoms are broken down to acetyl-CoA by β-oxidation

The most important catabolic fate of fatty acids released by the hydrolysis of triglycerides is to enter the β-oxidation pathway. Oxidation of the fatty acid is preceded by a reaction with coenzyme A (HSCoA) to form the thioester, fatty acyl-coenzyme A (fatty acyl-CoA). This is known as fatty acid activation since it involves the synthesis of a high energy bond which makes subsequent reactions energetically favourable. This reaction is catalysed by fatty acyl-CoA synthetase (also known as fatty acid thiokinase) (Eqn. 8.13).

$$R-CH_2-CH_2-COOH + HSCoA + ATP$$

Fatty acid

$$R-CH_2-CH_2-\underset{\underset{O}{\parallel}}{C}-S-CoA + AMP + PP_i$$

Fatty acyl-CoA

(8.13)

The energetics of this reaction are important since the formation of fatty acyl-CoA from coenzyme A and free fatty acid involves the synthesis of a compound with a high energy bond (i.e. the product hydrolyses with $\Delta G^{\circ\prime} = -32.2\,kJ\,mol^{-1}$). The formation of the coenzyme A derivative, however, is coupled to ATP hydrolysis ($\Delta G^{\circ\prime} = -35.9\,kJ\,mol^{-1}$) so that the $\Delta G^{\circ\prime}$ for the reaction (Eqn. 8.13) is $-3.7\,kJ\,mol^{-1}$. Further, the PP_i produced is subject to hydrolysis by pyrophosphatase. Coupling this reaction to that shown in Eqn. 8.13, the net $\Delta G^{\circ\prime}$ is $-37\,kJ\,mol^{-1}$, making the overall reaction energetically favourable. Also, the regeneration of ATP from the AMP produced in this reaction entails the expenditure of ATP to yield two molecules of ADP in a reaction catalysed by adenylate kinase. The ADP formed can be reconverted to ATP by oxidative or substrate-level phosphorylation. This gives a net consumption of two ATP per fatty acid activated.

Activation of fatty acids occurs outside the glyoxysome as fatty acyl-CoA synthetase is associated with the microsomal fraction of the cell. Unlike the transfer of activated fatty acids across the mitochondrial membrane of animal cells, transport of fatty acyl-CoA across the plant glyoxysomal membrane does not involve its conversion to a derivative of the carrier molecule, carnitine.

Once activated, fatty acids are degraded in glyoxysomes by a series of reactions summarized in Fig. 8.4.

(1) Oxidation (strictly speaking a dehydrogenation) at the C-2 and C-1 atoms to yield *trans*-2-enoyl-CoA (a *trans*-2,3-dehydro-fatty acyl-CoA). This is catalysed by acyl-CoA oxidase, a flavoprotein containing FAD (FAD–FP), and during the reaction FAD is reduced to $FADH_2$ (Eqn. 8.14). In fatty seeds, the intrinsic energy in $FADH_2$ is wasted as the hydrogen atoms combine with O_2 to yield H_2O_2 (subsequently detoxified by catalase to form $\frac{1}{2}O_2$ and H_2O) without the concomitant phosphorylation of ADP.

C-3   C-2   C-1

$$R-CH_2 \underset{}{-} CH_2 - \underset{\underset{O}{\parallel}}{C} - S-CoA \qquad FAD-FP \qquad H_2O_2 \longrightarrow H_2O + \tfrac{1}{2}O_2$$

Fatty acyl-CoA

$$R-\underset{\underset{H}{|}}{C}=\underset{\underset{H}{|}}{C}-\underset{\underset{O}{\parallel}}{C}-S-CoA \qquad FADH_2-FP \qquad O_2$$

2-Enoyl-CoA

(8.14)

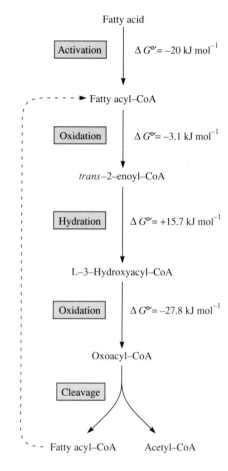

Fig. 8.4 A summary of the processes leading to the formation of acetyl-CoA from fatty acids with an even number of carbon atoms. Details of reactions are given in Eqns 8.13–8.17. The overall $\Delta G^{\circ\prime}$ is -35.7 kJ mol^{-1}, indicating that the reaction sequence is energetically favourable. The fatty acyl-CoA, shortened by two carbon atoms after the removal of the C$_2$ acetyl-CoA moiety, is recycled as shown by the dashed line.

This results in the net uptake of one molecule of O$_2$ for the dehydrogenation of two molecules of fatty acyl-CoA.

(2) The *trans*-2-enoyl-CoA produced is hydrated, in a stereospecific reaction, by enoyl-CoA hydratase (sometimes termed crotonase) to form the L-isomer of β-hydroxy-fatty acyl-CoA (Eqn. 8.15).

$$R—C \doublebond C—C—S—CoA + H_2O \longrightarrow R—C—CH_2—C—S—CoA \qquad (8.15)$$

(3) The next stage is an oxidation catalysed by an NAD-linked β-hydroxyacyl-CoA dehydrogenase (Eqn. 8.16). This enzyme acts only on the L-isomer of β-hydroxy-fatty acyl-CoA.

$$R—C—CH_2—C—S—CoA + NAD^+ \rightarrow R—C—CH_2—C—S—CoA + NADH + H^+ \qquad (8.16)$$

(4) The product of this reaction is 3-oxofatty acyl-CoA which is subsequently cleaved, in the presence of coenzyme A, at the 2,3 bond by acetyl-CoA acyl transferase (or thiolase) to form acetyl-CoA and a new fatty acyl-CoA two carbon atoms shorter than the original molecule (Eqn. 8.17).

$$R—C—CH_2—C—S—CoA + HSCoA \longrightarrow R—C—S—CoA + CH_3—C—S—CoA \qquad (8.17)$$

(5) The shortened fatty acid (R—CO—S—CoA) is subject to further oxidation by the fatty acyl-CoA oxidase (Eqn. 8.14) and the whole sequence (Eqns 8.14–8.17) repeats, with two carbon atoms being removed for each turn through the sequence until two molecules of acetyl-CoA are produced in the final reaction of the sequence. Thus, for example, a C$_{16}$ fatty acid would yield eight molecules of acetyl-CoA.

8.2.4 Oxidation of fatty acids with an odd number of carbon atoms

Although the initial stages of oxidation of fatty acids with an odd number of carbon atoms are similar to those described above, the final reaction of β-oxidation gives rise to one molecule of acetyl-CoA and one molecule of propionyl-CoA (Eqn. 8.18).

$$CH_3—CH_2—CH_2—CH_2—C—S—CoA + HSCoA$$

Pentanyl-CoA

$$CH_3—CH_2—C—S—CoA + CH_3—C—S—CoA$$

Propionyl-CoA   Acetyl-CoA   (8.18)

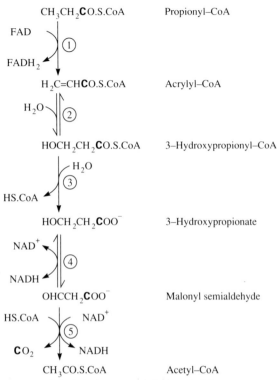

Fig. 8.5 Oxidation of propionyl-CoA to acetyl-CoA and CO_2. Enzymes are: (1) acyl-CoA oxidase; (2) enoyl-CoA hydratase; (3) 3-hydroxypropionyl-CoA hydrolase; (4) 3-hydroxypropionate dehydrogenase; (5) malonyl semialdehyde dehydrogenase.

Propionyl-CoA is dehydrogenated in the glyoxysomes to form acrylyl-CoA (Fig. 8.5). Following addition of water, involving a hydrolase, further oxidation occurs via a NAD-linked dehydrogenase to yield malonyl semialdehyde, NADH and coenzyme A. The malonyl semialdehyde and coenzyme A are converted to acetyl-CoA and CO_2 in an oxidative decarboxylation reaction, involving the reduction of NAD^+ catalysed by malonyl semialdehyde dehydrogenase.

8.2.5 *Oxidation of unsaturated fatty acids requires the involvement of two additional enzymes*

Before unsaturated fatty acids enter the β-oxidation pathway, two further enzymes are required. Figure 8.6 illustrates the oxidation of linoleic acid (an 18:2 fatty acid with

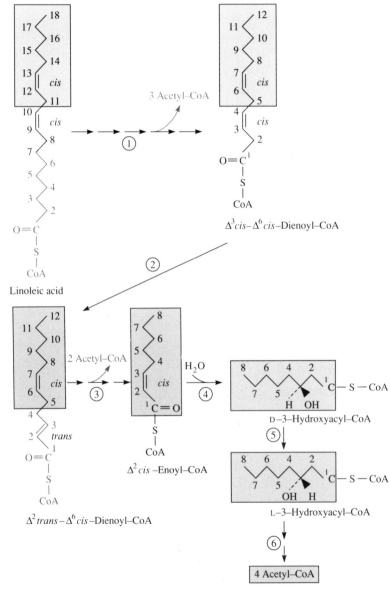

Fig. 8.6 Oxidation of the unsaturated fatty acid, linoleic acid ($C_{18:2}$). Reaction processes and enzymes: (1) β-oxidation; (2) enoyl-CoA isomerase; (3) β-oxidation; (4) hydration; (5) 3-hydroxyacyl-CoA epimerase; (6) β-oxidation.

double bonds at C-9 and C-12). Initially, oxidation of linoleic acid proceeds as described in Section 8.2.3 and three molecules of acetyl-CoA are released. However, this leaves a C_{12} coenzyme A derivative with a *cis* double bond at C-3. Since the normal substrate of the enoyl-CoA hydratase reaction (Eqn. 8.15) is the *trans*-2-enoyl isomer, the *cis*-3-enoyl product is converted to the *trans*-2-enoyl form by enoyl-CoA isomerase. Next, two further molecules of acetyl-CoA are removed to form a C_8 residue with a *cis* double bond at C-2, a feature which derives from the *cis* configuration at C-12 in the starting molecule, linoleic acid. When the C_8 product in its C-2 *cis* configuration is hydrated, D-3-hydroxyacyl-CoA is formed. Unlike the L-isomer, the D-form is not a substrate for the next enzyme in the β-oxidation sequence (3-hydroxyacyl-CoA-dehydrogenase). Plants contain the enzyme 3-hydroxyacyl-CoA epimerase which converts the D-isomer to the L-isomer so that the C_8 residue can then be processed via the β-oxidation sequence to release further molecules of acetyl-CoA.

8.2.6 Enzymes of β-oxidation are located in microbodies

In animal cells, β-oxidation reactions occur in mitochondria. However, when cells of germinating seeds of castor beans and other plants are disrupted and subjected to density gradient centrifugation, the various reactions of β-oxidation are found

Table 8.2 Activities of enzymes of β-oxidation and of marker enzymes associated with microbodies and mitochondrial fractions of *Helianthus tuberosum* (Jerusalem artichoke) tubers. From Macey, M.J.K. & Stumpf, P.K. (1982). *Plant Science Letters*, **28**, 207–12.

Enzyme	Activity (nmol min^{-1} fraction^{-1})	
	Mitochondria	Microbodies
Thiolase	149	327
Enoyl-CoA hydratase	130	305
Hydroxyacyl-CoA dehydrogenase	71	226
Catalase	44*	84
Cytochrome oxidase	26 260	650

*High levels of catalase in the mitochondrial fraction suggest that there is substantial contamination of this fraction by microbodies. This contamination is sufficient to account for all the activity of thiolase, enoyl-CoA hydratase and hydroxyacyl-CoA dehydrogenase apparently associated with mitochondria. Despite the cross-contamination between fractions, microbodies contain much higher activity of the three enzymes of β-oxidation and of catalase while virtually all of the cytochrome oxidase, a mitochondrial marker enzyme, is associated with mitochondria.

to be in the fraction of density between 1.24–1.26 g cm^{-3}. This fraction contains the membrane-bound particles termed glyoxysomes. These are microbodies unique to germinating fatty seeds and are distinguishable by their enzyme complement (Sections 3.7 & 8.3.2).

Until recently, β-oxidation in plant tissues other than those of lipid-rich seeds was thought to occur in mitochondria. Careful studies, however, indicate that even in non-fatty tissues the enzymes of β-oxidation are found in microbodies (Table 8.2). However, the microbodies from leaves have most of the enzymes involved in the C_2-PR cycle in photorespiration (see Section 11.4) as well, and are clearly distinguishable from those of germinating fatty seeds. It is concluded that the enzymes of β-oxidation in leaves, like those associated with glycollate metabolism, are associated with peroxisomes.

Lipids and their constituent fatty acids can be oxidized in other ways (α-oxidation, ω-oxidation, lipoxygenase activity) but such reactions are not important in generating energy for plant growth and will not be dealt with here.

8.3 The glyoxylate pathway provides a mechanism for linking β-oxidation to carbohydrate synthesis

The acetyl-CoA formed by the β-oxidation of fatty acids in microbodies can be processed by two routes. One involves the direct or indirect export of acetyl-CoA to mitochondria where it could either be oxidized to CO_2 via the TCA cycle or it could act as a supply of carbon skeletons for the synthesis of other compounds such as amino acids. Alternatively, the acetyl-CoA can be incorporated into succinate by the glyoxylate cycle in the glyoxysomes of germinating fatty seeds. The succinate formed acts as a substrate for the synthesis of sucrose in the cytosol.

8.3.1 Elucidation of the pathway and enzymes involved

The glyoxylate pathway was originally elucidated by Kornberg and Krebs in the mid-1950s as a result of studies of micro-organisms grown on acetate as their sole source of carbon. Growth of organisms on acetate implies that they can synthesize sugars from C_2 compounds. Cells grown on acetate contain two additional enzymes, malate synthase and isocitrate lyase. These two enzymes together with the enzymes of the TCA cycle provide a means, known as the glyoxylate pathway, for the production of C_4 acids from acetate. The C_4 acids produced can be readily converted to various C_3 glycolytic intermediates, from which sugars can be synthesized by gluconeogenesis (Section 7.5.2).

Malate synthase catalyses the condensation of acetyl-CoA and glyoxylate to yield malate (Eqn. 8.19) while isocitrate lyase catalyses the cleavage of isocitrate into glyoxylate and succinate (Eqn. 8.20).

$$\begin{array}{ccc} \boxed{\begin{matrix} CO\!-\!SCoA \\ | \\ CH_3 \end{matrix}} + \begin{matrix} H\!-\!C\!=\!O \\ | \\ COO^- \end{matrix} + OH^- \longrightarrow & \boxed{\begin{matrix} COO^- \\ | \\ CH_2 \end{matrix}} + HSCoA \\ & H\!-\!C\!-\!OH \\ & | \\ & COO^- \end{array} \tag{8.19}$$

Acetyl-CoA   Glyoxylate   Malate

$$\begin{matrix} COO^- \\ | \\ H\!-\!C\!-\!OH \\ \boxed{\begin{matrix} ^-OOC\!-\!C\!-\!H \\ | \\ CH_2 \\ | \\ COO^- \end{matrix}} \end{matrix} \longrightarrow \begin{matrix} COO^- \\ | \\ H\!-\!C\!=\!O \end{matrix} + \boxed{\begin{matrix} ^-OOC\!-\!CH_2 \\ | \\ CH_2 \\ | \\ COO^- \end{matrix}} \tag{8.20}$$

Isocitrate   Glyoxylate   Succinate

Thus, acetate taken up by micro-organisms is converted to acetyl-CoA. Some of this acetate is incorporated into C_4 acids via Eqns 8.19 and 8.20, avoiding the oxidative decarboxylation steps of the TCA cycle (Fig. 8.7). Succinate is subsequently converted to glycolytic intermediates which are used to synthesise sugars and other compounds required for growth.

In fatty tissues such as the endosperm of castor bean seeds, acetyl-CoA formed by β-oxidation of fatty acids from lipids is somehow converted to sucrose. Beevers and his associates found that the tissues of seedlings which use lipid as a storage reserve contain isocitrate lyase and malate synthase, providing the basis of a reaction sequence, involving the glyoxylate pathway and gluconeogenesis, to achieve this conversion. Figure 8.8 shows the marked correlation between the disappearance of lipid and the appearance of sugars in germinating fatty seeds. The appearance of isocitrate lyase is very clearly associated with the disappearance of lipid, suggesting that this enzyme plays an active role in lipid metabolism. These observations are consistent with lipid being the source of carbon for carbohydrate synthesis.

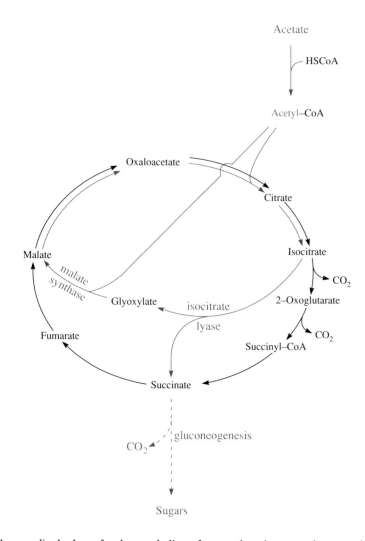

Fig. 8.7 A generalized scheme for the metabolism of acetate by micro-organisms growing on this compound. Acetate taken up by the cells is activated to acetyl-CoA and then enters the glyoxylate pathway (shown in grey) to form malate and citrate via the activity of malate synthase and citrate synthase respectively. The reaction catalysed by isocitrate lyase is important in bypassing the oxidative decarboxylations of the TCA cycle (shown in black). The acetate moieties incorporated into C_4 acids can be used in the synthesis of sugars by gluconeogenesis and sucrose synthesis.

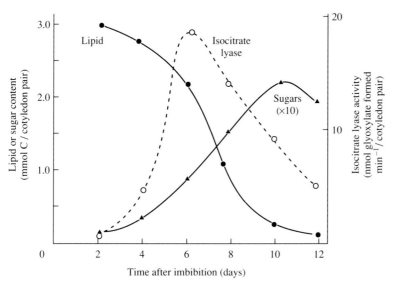

Fig. 8.8 Changes in lipid and sugar content, and activity of isocitrate lyase in germinating pumpkin seeds after imbibition. Seeds were germinated in the dark so that no CO_2 assimilation occurred.

The sequence of reactions for the synthesis of sucrose from lipid is shown in Fig. 8.9. In the sequence involving the glyoxylate pathway proposed for glyoxysomes, succinate is regarded as the net product (Eqn. 8.21) serving as the C_4 dicarboxylic acid for the production of C_3 glycolytic intermediates in the cytosol.

$$2\ \text{Acetyl-CoA} + \text{NAD}^+ \longrightarrow \text{succinate} + \text{NADH} + \text{H}^+ \qquad (8.21)$$

Fig. 8.9 Summary of the reaction sequences involved in the conversion of fatty acids to sucrose in germinating fatty seeds. β-Oxidation and the glyoxylate pathway occur in the glyoxysome and the succinate formed is transported to the mitochondrion. Here, malate is formed and transported to the cytosol where its oxidation yields NADH. Oxaloacetate is produced and metabolized to PEP with the loss of CO_2. Ultimately triose phosphates and hexose phosphates are produced via gluconeogenesis (see Section 7.5). In this way, four acetyl-CoA give rise to one hexose and two CO_2. Enzymes and reaction sequences: (1) β-oxidation; (2) citrate synthase; (3) aconitase; (4) isocitrate lyase; (5) malate synthase; (6) malate dehydrogenase; (7) succinate dehydrogenase; (8) fumarase; (9) malate dehydrogenase; (10) PEP carboxykinase; (11) gluconeogenesis; (12) sucrose synthesis (see Section 10.9). After Nishimura, M. & Beevers, H. (1979) *Plant Physiol.* **64**, 31–7.

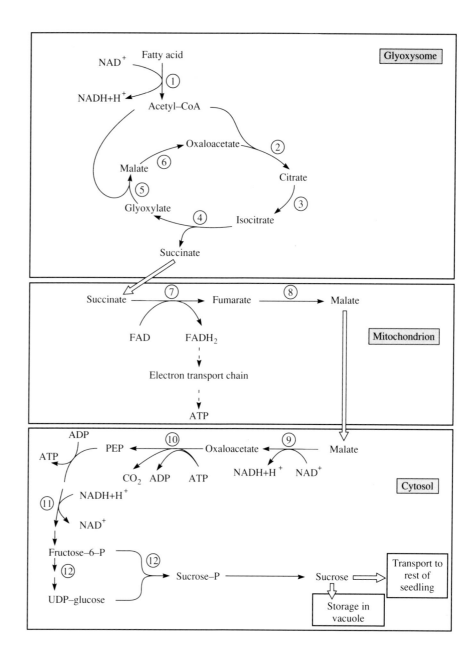

Succinate production is sustained by the formation of glyoxylate and oxaloacetate which act as acceptor molecules for acetyl-CoA. Oxaloacetate is formed by oxidation of the malate produced by malate synthase activity (Eqn. 8.19) in a reaction catalysed by malate dehydrogenase in glyoxysomes.

8.3.2 Enzymes of the glyoxylate pathway are located in glyoxysomes

All the enzymes involved in the glyoxylate cycle are associated with the glyoxysome fraction of storage tissue of fatty seeds (Table 8.3). More than 80% of malate synthase and isocitrate lyase activities, unique to the glyoxylate pathway, are found in this fraction. Malate dehydrogenase and citrate synthase are located in both mitochondrial and glyoxysomal fractions, as might be expected from their dual roles in this pathway and the TCA cycle (Fig. 8.8). Fumarase and succinate dehydrogenase, associated solely with the TCA cycle, are found only in the mitochondria.

Table 8.3 Location of some enzymes associated with β-oxidation, the glyoxylate cycle and the TCA cycle in subcellular fractions of germinating castor bean seeds. From Cooper, T.G. & Beevers, H. (1969) *J. Biol. Chem.*, **244**, 3507–13 and 3514–20.

	Enzyme activity (μmol g^{-1} fresh wt h^{-1})		
Enzyme	Mitochondria	Glyoxysomes	Proplastids
β-Oxidation of palmitoyl-CoA	0	11.4	2.8
Thiolase	0	22.9	6.3
Catalase	258	3360	1039
Isocitrate lyase	8	207	9
Malate synthase	46	790	27
Citrate synthase	218	74	2
Malate dehydrogenase	10 700	4475	311
Succinate dehydrogenase	65	0	0
Fumarase	394	2	2

8.3.3 Energetics of β-oxidation and the glyoxylate pathway

The standard free energy changes associated with the reactions of β-oxidation for a saturated, even-numbered fatty acid are shown in Fig. 8.4. The overall $\Delta G^{\circ\prime}$ for the steps leading to the shortening of the fatty acid by a two carbon fragment is

-35.2 kJ mol^{-1}, so the whole sequence is energetically favourable. In glyoxysomes, one molecule of NADH is formed per molecule of acetyl-CoA produced by the cycle.

Overall, the β-oxidation of a fatty acid of chain length C_{n+1}, containing $n+1$ carbon atoms, can be expressed as in Eqn. 8.22.

$$C_{n+1} + ATP + \tfrac{n}{2}NAD + \tfrac{n}{2}FAD + \tfrac{n}{2}H_2O + (\tfrac{n+1}{2}HSCoA)$$
$$\downarrow \qquad\qquad (8.22)$$
$$\tfrac{(n+1)}{2}\text{acetyl-CoA} + AMP + PP_i + \tfrac{n}{2}NADH + \tfrac{n}{2}FADH.$$

Glyoxysomes lack membrane-bound electron transport chains as found in mitochondria (Section 7.7) or chloroplasts (Section 9.8). The FADH synthesized by β-oxidation cannot, therefore, be used to generate ATP but is instead oxidized as shown in Section 8.2.3. The NADH, however, is exported, indirectly, to the mitochondria and used in ATP generation by oxidative phosphorylation. Reducing equivalents associated with NADH are believed to be transported from glyoxysomes to mitochondria by a malate–aspartate shuttle (see Section 14.4 for a discussion of shuttles).

8.3.4 Succinate, the end-product of the glyoxylate pathway, is converted to sucrose

Glyoxysomes lack fumarase and succinate dehydrogenase so further metabolism of succinate is not possible in this organelle. Therefore, succinate is regarded as the end-product of fatty acid metabolism in the glyoxysome. However, as mentioned above, most of the carbon associated with lipid in fatty seeds is converted, on germination, to sugars. It has been proposed that succinate is exported to the mitochondria where, in reactions catalysed by succinate dehydrogenase, fumarase and malate dehydrogenase (see Chapter 7), it is oxidized to oxaloacetate. Thus, two molecules of oxaloacetate are formed from four molecules of acetyl-CoA without loss of carbon.

The sugars in germinating fatty seeds are formed from oxaloacetate. In a reversible reaction requiring ATP, oxaloacetate is decarboxylated to phosphoenolpyruvate (PEP) by the enzyme PEP carboxykinase (PEP-CKase) (Eqn. 8.23).

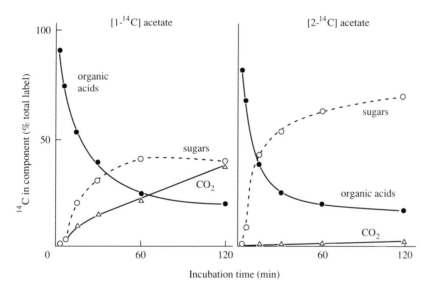

$$\underset{\text{Oxaloacetate}}{\begin{array}{c} (\text{COO}^-) \\ | \\ \text{CH}_2 \\ | \\ \text{C}{=}\text{O} \\ | \\ \text{COO}^- \end{array}} + \text{ATP} \rightleftharpoons \underset{\text{PEP}}{\begin{array}{c} \text{CH}_2 \\ \| \\ \text{C}{-}\text{O}\textcircled{P} \\ | \\ \text{COO}^- \end{array}} + (\text{CO}_2) + \text{ADP} \qquad (8.23)$$

Figure 8.9 implies that for every two molecules of oxaloacetate formed from four molecules of acetyl-CoA, two molecules of CO_2 are released in the conversion of oxaloacetate to PEP which, in its turn, is converted to sucrose via the reactions of reverse glycolysis (gluconeogenesis—See Section 7.5.2) and sucrose synthesis (see

Fig. 8.10 The time course for distribution of ^{14}C among various components, obtained when slices of endosperm tissue of castor bean were incubated with [1-^{14}C]acetate or [2-^{14}C] acetate. Incorporation of ^{14}C into organic acids, sugars and CO_2 is shown. Note that this labelling pattern is consistent with the pathway shown in Fig. 8.9 and the theoretical labelling pattern (Fig. 8.11) predicted by the pathway. From Canvin, D.T. & Beevers, H. (1961) *J. Biol. Chem.*, **736**, 988–95.

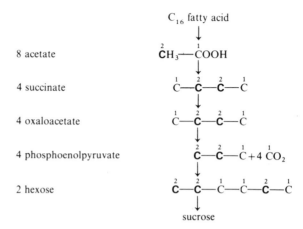

Fig. 8.11 Theoretical fate of the C-1 and C-2 atoms of acetyl-CoA during the production of sucrose from fatty acids during germination of fatty seeds according to the scheme shown in Fig. 8.9. Numbers refer to the carbon atoms of the acetate formed by β-oxidation of the fatty acid. Note that in this sequence the CO_2 lost during the decarboxylation of oxaloacetate is derived exclusively from C-1 of acetate.

Section 10.9). In theory, three-quarters of the carbon present in the lipid is conserved in its transformation to sugar (Fig. 8.11).

This sequence for converting acetyl-CoA to sugars is supported by the experiments shown in Fig. 8.10. The reaction scheme (Fig. 8.11) predicts that half the total carbon derived from the C-1 atom of acetate is released as CO_2 through the action of PEP-CKase, whilst the C-2 atom and the other half of the carbon from C-1 are transferred to PEP and then hexose. When [2-^{14}C]acetate was supplied to castor bean endosperm tissue, virtually no $^{14}CO_2$ was evolved and the level of labelling in sucrose was twice that obtained using [1-^{14}C]acetate. These data demonstrate that the assimilation of acetyl-CoA into oxaloacetate, PEP and carbohydrate involves the glyoxylate cycle and PEP-CKase activity. Without the cycle, all the carbon of acetyl-CoA would be lost via the oxidative decarboxylations of the TCA cycle and no net synthesis of sucrose would be possible.

Further reading

Monographs and treatises: Davies (1987); Douce & Day (1985).
β-Oxidation: Stumpf (1987) Chapter 2; Tolbert (1980) Chapter 9.

9

Light reactions of green plants

The synthesis of new living material requires an input of energy. In plants and other photoautotrophic organisms this energy is derived from sunlight and is used to make organic molecules from inorganic materials. Consequently the reactions by which plants harvest the energy of sunlight are crucial to plant growth. Indeed, most other organisms depend on these higher energy state organic molecules, making the light reactions of green plants not only of central importance to plant growth but of vital importance to all life on this planet.

9.1 Properties of solar radiation in relation to plant growth

Solar energy arriving at the earth's surface, and available to photoautotrophs, comprises wavelengths in the range $300 - \sim 1100$ nm. The light used to drive syntheses in green plants, photosynthetically active radiation (PAR), is within the range of visible light from 400–700 nm, but longer wavelengths up to 1010 nm are used by certain photoautotrophic bacteria.

In plant science, light is usually expressed in terms of photon flux (with units of μEinsteins $m^{-2} s^{-1}$) or irradiance. Since an Einstein is defined as a mol of photons (see Section 4.8) and, as described later, a pigment molecule absorbs one quantum at a time, photon flux can also be expressed as μmol photons $m^{-2} s^{-1}$. Irradiance is a measure of flux in terms of the energy of the radiation and has units of $W m^{-2}$. At sea level, full sunlight has a photon flux of PAR of approximately $2200 \, \mu$mol $m^{-2} s^{-1}$ or an irradiance of $500 \, W m^{-2}$.

9.2 Plants absorb light by specific pigments contained in thylakoids

Light is absorbed in plants by specific pigments (or aggregates of pigments) of which the chlorophylls are the most important. A range of chlorophylls are found, although higher plants have only chlorophyll a and chlorophyll b. The other major classes of light-absorbing pigments found in plants are the carotenoids and, in some algae and the cyanobacteria, the phycobilins. The distribution of the most important of these pigments is given in Table 9.1.

These pigments, and the various protein components also associated with the utilization of light, are found on or within the thylakoids of chloroplasts (see Section 3.5).

Organisms	Chl a	Chl b	Chl c_1	Chl c_2	Phycobilins	β-carotene	Major xanthophylls
Cyanobacteria	+	−	−	−	+	+	Myxoxanthin
Red algae	+	−	−	−	+	+	Lutein
Cryptomonad algae	+	−	−	+	+	trace	Alloxanthin
Dinoflagellates	+	−	−	+	−	+	Peridinin
Brown algae, diatoms	+	−	+	+	−	+	Fucoxanthin
Chrysophytes (golden algae)	+	−	+	+	−	+	Fucoxanthin
Green algae, higher plants	+	+	−	−	−	+	Lutein

Table 9.1 Distribution of chlorophylls (chl), carotenoids and phycobilins in cyanobacteria and major groups of algae and higher plants. After Prezelin, B. (1981) *Can. Bull. Fish. Aquat. Sci.*, **210**, 1–43. Published by Fisheries and Oceans Canada and reproduced with the permission of the Minister of Supply and Services Canada, 1990.

+ = Pigment present; − = pigment absent.

9.2.1 Chlorophylls

The basic structure of chlorophyll *a* is shown below. Like the haem found in cytochromes (see Fig. 4.8), chlorophylls are porphyrins with a tetrapyrrole head. Attached to the tetrapyrrole head is a long phytol tail (shown in bold) which makes the chlorophylls *in vitro* strongly hydrophobic.

Chlorophyll *a*

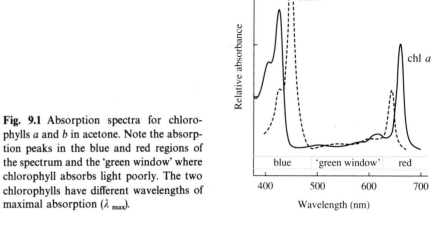

Fig. 9.1 Absorption spectra for chlorophylls *a* and *b* in acetone. Note the absorption peaks in the blue and red regions of the spectrum and the 'green window' where chlorophyll absorbs light poorly. The two chlorophylls have different wavelengths of maximal absorption (λ_{max}).

The chlorophylls differ from haem as they have magnesium in place of iron and different substituents attached to the tetrapyrrole rings A–D. Chlorophyll *b* differs from chlorophyll *a*, for example, in that it contains an aldehyde group (—CHO) in place of the methyl substituent (—CH₃) in ring B.

Chlorophylls absorb strongly in the red (absorption peak at 670 nm–680 nm) and blue (absorption peak at 435 nm–455 nm) regions of the visible spectrum. The reflection or transmittance of the intermediate green light, which is not absorbed, gives the familiar green colour to plants and to solutions of chlorophyll. Different chlorophylls each have slightly different absorption maxima. For example, chloro-

phyll *a* in acetone absorbs maximally at 432 and 663 nm, whilst in the same solvent chlorophyll *b* has peaks at 453 and 643 nm (Fig. 9.1).

Chlorophyll absorbs light because of the structure of the porphyrin ring at the 'head' of the molecule. When organic molecules absorb photons the energy is used to promote an electron from one orbital to another. In particular, if the molecule has conjugated double bonds, photons can cause an electron in a π orbital to move to a higher energy orbital, π^*; the electron becomes 'excited'. The electron in the π^* orbital has an opposite spin to its corresponding paired electron in the π orbital, and the chlorophyll molecule, in this condition, is in a singlet state. The greater the number of double bonds in the molecule, the lower the energy required to shift orbitals and, since the energy of light decreases with wavelength (see Section 4.8), the longer the wavelength that can promote the shift. Plant pigments generally have seven or more conjugated double bonds and the energy in visible light is sufficient to power the movement of electrons from the π to π^* orbitals. Chlorophylls are particularly suited as the porphyrin ring contains a number of conjugated double bonds and, in essence, is a planar ring surrounded by a dense cloud of electrons in the π orbitals. The substitution of different side groups onto the ring (as in the different types of chlorophyll) modifies the electronic state and hence the absorption characteristics.

A number of different forms of chlorophyll *a* can be distinguished *in vivo*, each with a slightly different absorption maximum. Analysis of absorbance spectra of thylakoid membranes at low temperature (77 K) indicates that there are at least four *in vivo* forms of chlorophyll *a* with absorption maxima at 662, 670, 677 and 684 nm. This results from chlorophyll *a* binding to a series of different membrane proteins in thylakoids, each of which causes a spectral shift of the absorption maximum of 5–40 nm towards the red end of the visible spectrum.

9.2.2 Carotenoids and other pigments

Carotenoids absorb light of approximately 400–495 nm. Some of the light energy absorbed by β-carotene and lutein (up to 40%) can be transferred to chlorophyll although these carotenoids also help protect against photo-oxidative damage (Section 9.3.). In some algae, carotenoids such as fucoxanthin (found in the Phaeophyceae or brown algae) and peridinin (found in the Dinophyceae or dinoflagellates) absorb light very efficiently and channel the energy into chlorophyll *a*. Absorbance by particular pigments can be shown by measuring the rates of O_2 evolution or CO_2 assimilation as a function of the wavelength of incident light. These give rise to action spectra (Fig. 9.2). Thus, in chloroplasts of *Beta vulgaris*, where chlorophylls *a* and *b* are the major light-harvesting pigments, the action spectrum closely matches the absorption spectrum for these chlorophylls. In the case of the dinoflagellate *Glenodinium*, with a high level of peridinin, the action spectrum clearly reflects the importance of this carotenoid in light-harvesting.

In the red and cryptomonad algae and the cyanobacteria, the phycobilins play a major role in the collection of light energy and its transfer to chlorophyll *a*. Phycobilins are water-soluble, comprising a tetrapyrrole chromophore attached to a protein, and are organised into discrete structures (phycobilisomes) on the outer surface of the thylakoid membrane. There are a number of different biliproteins which show absorption maxima between 500 and 650 nm.

9.3 Light is absorbed by chlorophyll and the energy is transferred to specialized chlorophyll molecules—the reaction centres

Depending on the wavelength of the light absorbed, chlorophyll can be raised to different excited states. A photon of blue light with a wavelength of, say, 438 nm has an energy content of 273 kJ mol^{-1}, sufficient to raise an electron to a high energy (but unstable) level termed the second singlet state. Within 10^{-13} s this

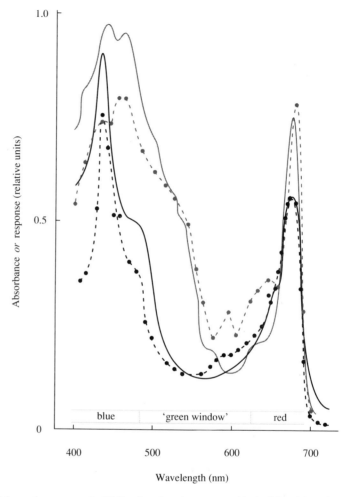

Fig. 9.2 Absorption spectra (solid lines) and action spectra (dashed lines) for photosynthesis in isolated chloroplasts from *Beta vulgaris* (black) and *Glenodinium* (grey). The action spectrum for *Glenodinium* was determined by O_2 evolution whilst that for *B. vulgaris* was measured by the reduction of dichlorophenolindophenol. Note the increased absorption in the 'green window' of *Glenodinium* due to absorption of light by the carotenoid, peridinin; that this pigment is efficient in energy transfer is shown by the close correlation between action and absorption spectra. After Chen, S.L. (1982) *Plant Physiol.*, **27**, 35–48; and Prézelin, B.B., Ley, A.C. & Haxo, F.T. (1976) *Planta*, **130**, 251–6.

second singlet decays into an excited state at a slightly lower energy level, the first singlet, and energy is released as heat. Red light of 680 nm has an energy content of only 176 kJ mol^{-1} which will elevate an electron in chlorophyll only to the first singlet (Fig. 9.3a).

The energy of the first singlet state can be dispersed in a number of ways (Fig. 9.3.).

(1) The molecule reverts to its ground state by losing all its excitation energy as heat, generated by collision or vibration.

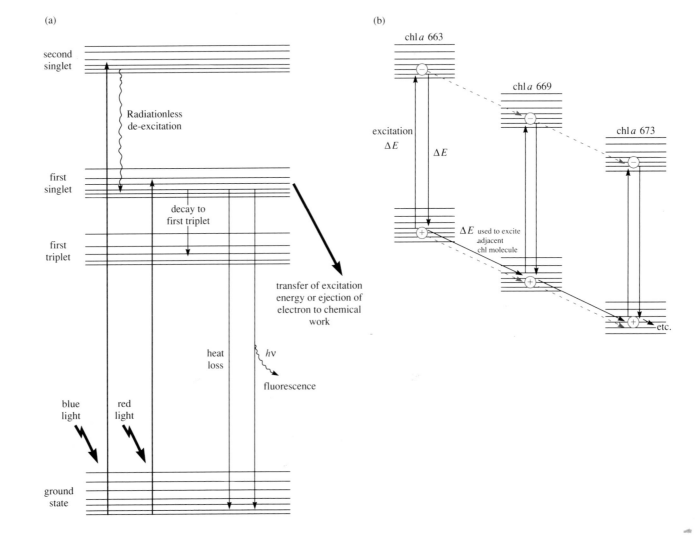

Fig. 9.3 The excitation states of chlorophyll and the fate of the energy of excited electrons. The figure is set out as an energy diagram with the lowest energy state at the bottom and the highest at the top. (a) Electrons can be raised to different excited states. Excitation energy can be lost as fluorescence or can be transferred to an adjoining chlorophyll molecule. (b) A schematic representation of excitation energy transfer between chlorophyll molecules (chl). Solid lines represent the sequential excitation/de-excitation of resonance transfer, whilst the dashed lines indicate the movement of an electron and its corresponding 'hole' by random walk. Note that energy transfer occurs only from chlorophyll molecules to others with longer absorption maxima as shown by the numbers. Different absorption maxima are attributed to binding of chl a to a series of different proteins. In (b), the energy levels represent the ground state and first singlet state of chlorophyll.

(2) Only part of the excitation energy is lost and the molecule reverts, not to its ground state, but to an excited state of lower energy called the first triplet state. This is relatively stable, lasting about 0.1 ms, and is thus potentially more reactive than the singlet state but does not occur frequently. This is just as well as, in this first triplet state, chlorophyll can convert oxygen to its first singlet state which is very reactive and causes damage to many cell components. Carotenoids minimize the risk of this damage by converting triplet chlorophyll and singlet oxygen to their ground states.

(3) Excited chlorophyll reverts to its ground state by losing energy by the emission of light of lower energy (i.e. longer wavelength) than that absorbed normally within about 10^{-9} s of excitation. This is fluorescence and is characteristically red with a maximum emission for chlorophyll a in $vivo$ at 686 nm, regardless of whether excitation is by blue or by red light. The absence of fluorescence (referred to as fluorescence quenching) is a sign that energy is being dissipated in some other way. In $vivo$ fluorescence is normally minor as most of the energy of the excited chlorophyll molecule is used instead to do work. Incidentally, fluorescence and quenching can indicate a great deal about the transfer of energy between pigments.

(4) Most importantly, an excited chlorophyll molecule can pass on its excitation energy to another molecule of chlorophyll. Excitation energy transfer is believed to occur largely through resonance transfer (sometimes termed inductive resonance). Put simply, as an electron in an excited molecule falls to the ground state, its energy is used to lift an electron in an adjoining molecule to the excited state (Fig. 9.3b). The precise mechanism is very complex and depends partly on the distance between molecules; between chlorophylls transfer is almost 100% efficient if the distance is 2.5 nm, decreasing to about 50% at 7.5 nm.

Alternatively, energy transfer can take place when an electron takes what is termed a 'random walk' through densely packed pigment molecules. The electron is thought to move from the orbital of one molecule to that of another in a similar excitation state accompanied by a corresponding movement of electrons in the ground state orbitals (Fig. 9.3). Transfer of energy is not entirely random, however, as energy can only be transferred to pigments with equal or longer wavelength absorption maxima. This is consistent with the occurence of various forms of chlorophyll a in $vivo$. Excitation energy is eventually transmitted to reaction centres (see below).

(5) Most significantly, an excited chlorophyll molecule can return to its ground state by doing work in a chemical reaction, normally involving the loss of an electron to an acceptor molecule. This involves very specialized chlorophyll molecules within membrane-bound complexes termed reaction centres (see Section 9.7). A small proportion of the chlorophyll molecules, known as P_{680} and P_{700} fulfill this function.

9.4 Light is required for production of ATP and reducing equivalents

Our present concept of the way(s) in which plants use light energy to drive carbon assimilation and growth owes much to the microbiologist C.B. van Niel who began to study photoautotrophic processes in the 1920s. He noted that light-driven CO_2 assimilation in green sulphur bacteria could be represented as

$$n\,CO_2 + 2n\,H_2S \xrightarrow{\text{Light}} [CH_2O]_n + n\,H_2O + 2n\,S \qquad (9.1)$$

where $[CH_2O]_n$ represents reduced carbon at the level of carbohydrate. Similarly, for the analogous process in higher plants

$$n\,CO_2 + 2n\,H_2O \xrightarrow{\text{Light}} [CH_2O]_n + n\,H_2O + n\,O_2. \qquad (9.2)$$

This led van Niel to suggest that water acted as a substrate which gave rise to protons, electrons and O_2 and that the reductant [H] ($H^+ + e^-$) could be used in a subsequent reaction to reduce CO_2 as shown below.

$$2H_2O \xrightarrow{\text{Light}} O_2 + 4H^+ + 4e^- \qquad (9.3)$$

$$4H^+ + 4e^- + CO_2 \longrightarrow [CH_2O] + H_2O \qquad (9.4)$$

One implication of van Niel's theory is that O_2 evolved by plants in the light is derived from water. This idea was confirmed in the early 1940s, when experiments showed evolution of $^{18}O_2$ from water labelled with the heavy isotope of oxygen, ^{18}O (Eqn. 9.5).

$$C^{16}O_2 + 2H_2^{18}O \longrightarrow [CH_2^{16}O] + H_2^{16}O + {}^{18}O_2. \qquad (9.5)$$

Although the original experiments did not take into account the numerous exchange reactions that occur between CO_2 and H_2O through the formation and dissociation of carbonic acid (H_2CO_3), van Niel's concept is still accepted.

The model devised by van Niel also implied that the reduction of CO_2 can be separated from the evolution of O_2 (Eqns 9.3 & 9.4). In 1939, Hill worked on suspensions of isolated chloroplasts and showed that they support light-dependent evolution of O_2 provided that a suitable electron acceptor (an oxidizing agent such as ferricyanide) was added. These chloroplasts did not reduce CO_2 to carbohydrate and thus confirmed van Niel's model. Electron acceptors which support O_2 evolution in this way are 'Hill oxidants' or 'Hill reagents' and the process is termed the 'Hill reaction'. The 'natural' Hill reagent is $NADP^+$.

In 1954, Arnon, Allen and Whatley showed that, in the presence of certain cofactors, illuminated chloroplast preparations supported the phosphorylation of ADP to ATP without the involvement of O_2 either as substrate or donor. Since no 'external' electron donor or acceptor was required, it was concluded that reactions taking place in the light produced both the electron donor and acceptor as illustrated in Fig. 9.4a. This is the basis of cyclic photophosphorylation (see Section 9.17) and was seen as involving photo-oxidation of chlorophyll and the transfer of electrons from oxidized chlorophyll to a cofactor (X), followed by their return via

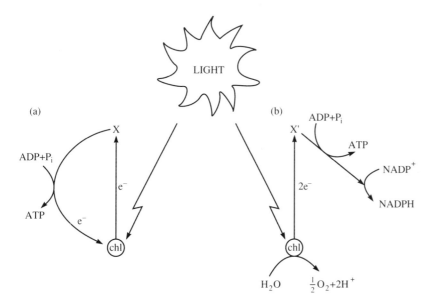

Fig. 9.4 A simple representation of (a) cyclic and (b) non-cyclic photophosphorylation. In cyclic photophosphorylation chlorophyll acts as both donor and acceptor for electrons. In non-cyclic photophosphorylation, light energy results in the oxidation of water and the raising of electrons to a strongly negative redox potential. The downhill flow of electrons is coupled to ADP phosphorylation but $NADP^+$ acts as the terminal electron acceptor.

a series of carriers to the oxidized chlorophyll; this transfer of electrons is linked to the phosphorylation of ADP.

Ruben earlier postulated that CO_2 assimilation required ATP and a reductant [H]. The ATP was to provide energy for both a carboxylation reaction and conversion of the carboxyl group to a more easily reduced carboxyl phosphate. He further suggested that the light generated reductant [H] would have the same redox potential as the $NADP^+/NADPH$ couple. This hypothesis was supported by the work of Benson and Calvin in establishing the C_3 carbon reduction cycle (Chapter 10) and the demonstration, again by Arnon and his colleagues, of the use of NADPH and ATP in carbon assimilation by intact chloroplasts.

The final piece of this biochemical jigsaw puzzle came when Arnon's group discovered that the photochemical reduction of $NADP^+$ by the 'Hill reaction' could be coupled to the production of ATP (Eqn. 9.6).

$$NADP^+ + ADP + P_i + H_2O \xrightarrow{\text{Light}} NADPH + H^+ + ATP + \tfrac{1}{2}O_2 \quad (9.6)$$

At the time it was supposed that absorbed light raised electrons, derived from water, to a potential sufficient to allow the 'downhill' flow of electrons to NADP and that ATP generation came about by coupling phosphorylation of ADP to free energy changes during this electron transport (Fig. 9.4b).

Thus, there are two forms of light-driven phosphorylation of ADP. In the first, electron flow is cyclic beginning with an electron donor and finishing with the oxidized form of that donor as the electron acceptor. This is not linked to O_2 evolution or the net production of available reducing equivalents. The second form is non-cyclic photophosphorylation in which formation of ATP is coupled to the transport of electrons from water to NADP. This is associated with O_2 evolution *and* photoreduction of $NADP^+$ (Eqn. 9.6). These processes are examined in more detail in following sections of this chapter.

9.5 Two light reactions are involved

In higher plants, the action spectrum for O_2 evolution (or CO_2 assimilation) reflects the absorption spectrum for chlorophyll, indicating the major role of chlorophyll in light harvesting. However, measuring the number of mol of O_2 evolved per mol of quanta *absorbed*, i.e. the *quantum yield*, gives a different pattern. With the exception of a noticeable dip centred around 480–490 nm (a reflection of the relatively poor efficiency of energy transfer from carotenoids, with absorption maxima around 480 nm, to chlorophylls) the quantum yield is more or less constant at wavelengths below 680 nm. However, there is a very marked drop in quantum yield as the wavelength is increased beyond 680 nm, even though such wavelengths are still absorbed by chlorophyll. This dramatic decline in efficiency is termed the *red drop*. If, however, 'long' wavelength red light (>680 nm) is supplemented with weak background light of shorter wavelengths, then the mixture of wavelengths has a synergistic effect called enhancement; this means that the mixed wavelength light supports a faster rate of O_2 evolution than the sum of rates supported by light of the different wavelengths supplied separately.

This suggests that there are two separate, yet interconnected, systems for harvesting light energy. Photosystem I (PSI) can use light from a range of wavelengths including 'long' wavelength red light (>680 nm) whilst photosystem II (PSII) can only use wavelengths of less than 680 nm or so. For maximum efficiency both systems must function co-operatively.

The two systems co-operate so that electrons from water are progressively elevated in energy, first by one photosystem and then by the other. PSII withdraws electrons from H_2O and lifts them from a redox potential equal to, or more positive, than $+0.82$ V (redox potential of the H_2O/O_2 couple) to a potential of about -0.6 V. From here, electrons can move 'downhill' to PSI where a further input of light lifts them to a potential of around -0.7 V. Downhill transfer of electrons to $NADP^+$ can then proceed.

Although co-operative, the photosystems have different functions which can be observed using inhibitors such as 3-(3′,4′-dichlorophenyl)-1,1-dimethyl urea (DCMU). This compound interferes with electron transport from one photosystem to the other and thus inhibits O_2 evolution and the photoreduction of $NADP^+$. However, if an alternative electron donor such as ascorbate together with a catalyst such as dichlorophenolindophenol (DCPIP), which can donate electrons directly to PSI, thereby bypassing PSII, is also added, the light-dependent reduction of $NADP^+$ is restored. Reduction of $NADP^+$ is, then, dependent on PSI activity and the PSI reaction alone does not evolve O_2. Evolution of O_2 must, therefore, be associated with PSII. Studies of the quantum yield of $NADP^+$ reduction show a marked 'red drop' at wavelengths >690 nm. However, if ascorbate and DCPIP are supplied, and the contribution of electrons by PSII inhibited with DCMU, then there is a marked *increase* in quantum yield at wavelengths >680 nm, thus demonstrating that light of >680 nm can be used when PSII is bypassed and illustrating the role of PSI in $NADP^+$ reduction. The separate functions of O_2 evolution and $NADP^+$ reduction associated with non-cyclic electron transport can

be represented as

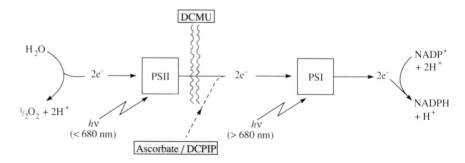

DCMU does not, however, block cyclic photophosphorylation illustrating that that process is independent of PSII activity.

Contrast this situation with the anoxygenic photosynthetic bacteria which generate ATP and reducing equivalents without evolving O_2. The green sulphur bacteria (Chlorobiaceae) have only one photosystem, analagous to PSI of plants. The reaction centre involves a specific protein-bound form of bacteriochlorophyll (P_{890}) with a redox potential of around $+0.47$ V. When this absorbs a photon, P_{890} becomes excited and effects the reduction of an electron acceptor, with a sufficiently negative redox potential (~ -0.55 V) to reduce oxidized ferredoxin and thence generate NADH. In the purple sulphur bacteria (Rhodospirillaceae), the case is more complex. Here the primary electron acceptor has a redox potential much less negative than NAD and so it cannot be used directly in NAD reduction. Instead, electrons return from the primary acceptor, via quinones and cytochromes b and c, to oxidized bacteriochlorophyll. This cyclic electron transport generates ATP which is used to power the endergonic movement of electrons from substrates such as succinate to NAD^+. Oxidized bacteriochlorophyll is not a sufficiently strong oxidizing agent to withdraw electrons from water in an exergonic reaction. In the anaerobic photosynthetic bacteria, compounds such as H_2S and H_2 can serve this function. For example, in *Rhodopseudomonas sphaeroides*, with reaction centre bacteriochlorophyll P_{890}, this can be represented as in Eqns 9.7 and 9.8 below.

$$2\,P_{890}^+ + H_2S \longrightarrow 2\,P_{890} + S^\circ + 2H^+ \qquad (9.7)$$

$$2\,P_{890}^+ + H_2 \longrightarrow 2\,P_{890} + 2H^+ \qquad (9.8)$$

This explains the requirement for compounds such as H_2 and H_2S for growth of anoxygenic photosynthetic bacteria.

9.6 The photosynthetic unit

In Section 9.3 an explanation was given for the transfer of energy from chlorophyll to a reaction centre. The number of chlorophyll molecules serving each reaction centre can be determined by supplying suspensions of an alga such as *Chlorella* with a series of variably spaced, very brief ($\sim 10\,\mu s$) and intense flashes to fully saturate the light-absorbing system. The maximal yield per flash is achieved with a time interval between flashes of about 40 ms. This then is the time taken for the light-harvesting system to complete its cycle after one flash and be ready to make use of the next. About one molecule of O_2 is evolved per flash for every 2400 molecules of chlorophyll. If the energy of the flash (equivalent to the number of quanta) is decreased whilst still at 40 ms intervals, then, a maximum of one O_2 is evolved for every eight quanta of light absorbed. This suggests that approximately 2400 molecules of chlorophyll co-operate in such a way that eight quanta absorbed by any of the 2400 molecules channel enough energy to the reaction centre to cause the evolution of one molecule of O_2. Excitation of additional chlorophyll molecules does not lead to any further energy transfer and, hence, no additional evolution of O_2. Thus, one quantum absorbed by one chlorophyll molecule in every 300 produces one photochemical event at the reaction centre, but eight such events (from eight quanta absorbed) are needed for evolution of one molecule of O_2. Each reaction centre can be visualized as being served by an arrangement or 'antenna' of roughly 300 chlorophyll molecules which channel the energy from an excited chlorophyll molecule to a reaction centre of this 'photosynthetic unit' where photochemical oxidation and hence electron transport is initiated.

9.7 Properties of the photosystems and their antenna complexes

9.7.1 Thylakoid pigments are bound in pigment–protein complexes

The two photosystems discussed above can be partially removed from the thylakoids by a number of techniques, such as mechanical disruption or treatment with mild detergents such as Triton X-100 or digitonin. The systems and their major components can then be separated by density gradient centrifugation or by polyacrylamide gel electrophoresis. All the chlorophyll, β-carotene and the major carotenoids are non-covalently bound to specific proteins, which are, seemingly, involved in the organization of the various pigments so as to optimize the efficiency of transfer of energy from the light-harvesting pigments to the reaction centres.

Although a range of different pigment–protein complexes can be isolated, studies suggest that higher plant thylakoids contain three major ones: a core of pigment–protein, tightly bound to the reaction centres; a less closely associated complex; and a major light-harvesting pigment–protein complex (LHCP) which is easily removed. These pigment–protein complexes act as the antenna to trap light and transfer its energy to the reaction centres.

9.7.2 Composition of Photosystem II

The two photosystems discussed in Section 9.5 are each an assembly of a reaction centre associated with various antenna pigments and proteins involved in light-harvesting and electron transport. The reaction centre of PSII is a specialized chlorophyll–protein dimer termed P_{680}. The core complex surrounding the reaction centre contains about 10–15% of the total chlorophyll. It has about 60 chlorophyll a molecules per P_{680} together with some bound β-carotene (but no chlorophyll b).

Most of the chlorophyll in thylakoids is contained in a major light-harvesting chlorophyll–protein complex (LHCP) (see Fig. 3.3 for structure) surrounding and acting as an antenna for PSII. A special form known as LHCP 2 is associated with PSII and contains about 40–60% of the total chlorophyll of PSII. The complex contains two major polypeptides of molecular mass 26kDa and 28kDa, has a low β-carotene content but a high xanthophyll content and a chlorophyll a/b ratio of about 1.5:1. For each P_{680} there are about 240 chlorophyll molecules in LHCP 2; about 144 chlorophyll a and 96 chlorophyll b.

Also associated with PSII are the primary electron acceptor pheophytin a, and quinones, termed Q_A and Q_B, bound to proteins. In addition, PSII contains three main polypeptides involved in the oxidation of water. These extend into the intrathylakoid space and one, of molecular mass 33 kDa, contains two tightly bound manganese atoms. Two other manganese atoms are essential for optimal O_2 evolution. A cytochrome of high redox potential, cytochrome b_{559}, is also associated with PSII.

9.7.3. Composition of Photosystem I

PSI in higher plants contains chlorophyll a, β-carotene and a small amount of chlorophyll b, all non-covalently bound to proteins. The reaction centre of PSI is a specialized chlorophyll–protein dimer called P_{700}. The antenna complex consists of around 110–160 chlorophyll molecules for every P_{700}, with about 100 in an outer shell and 50 or so in an inner region (which are less readily removed by detergent treatment). PSI has a chlorophyll a/b ratio of $\leqslant 6:1$ and it is thought that it has a light-harvesting protein–chlorophyll complex similar to LHCP 2, termed LHCP 1. PSI also contains the primary acceptor for electrons from P_{700}—a closely associated specialized chlorophyll. There is also a component known as P_{430}, the identity of which has not been proven but it seems to be associated with an iron–sulphur (Fe–S) protein. In all, PSI contains seven polypeptides including at least two Fe–S centres.

9.8 Transport of electrons from water to NADP

In addition to the components associated with the photosystems, the transfer of electrons from water to NADP involves a number of other compounds; firstly, in the transfer of electrons between PSI and PSII; and secondly, in the reduction of $NADP^+$ associated with electron transport from PSI. The sequence in which the various components operate is the Z scheme (Fig. 9.5). The reactions involved are discussed below.

Fig. 9.5 The Z scheme for non-cyclic electron transport from the primary electron donor, water, to the terminal electron acceptor, $NADP^+$. The position of various electron carriers is given in relation to their redox potentials. Transfer of two electrons from H_2O to $NADP^+$ requires the absorption of four quanta (two by each photosystem) and results in the generation of $\frac{1}{2}O_2$ and one NADPH. ATP generation is associated with electron transfer between PSII and PSI. Also shown on the figure are the sites of action of some commonly used inhibitors of electron transport (grey bars) and electron acceptors or donors (dashed arrows) used to investigate the electron transport chain. For example, in the presence of DCMU, O_2 evolution from PSII ceases as electron transport from Q_A to Q_B is inhibited. In the presence of a suitable PSI acceptor and donor, however, rates of electron transport through PSI can be determined. Such measurements can be made by measuring the reduction of $NADP^+$ by change of absorbance at 340 nm or the O_2 consumption by using methyl viologen as PSI acceptor in place of $NADP^+$. Similar arguments apply to measurement of electron flow through PSII. Abbreviations: [AB] = Fe–S proteins; DCMU = 3-(3,4-dichlorophenyl)-1,1-dimethyl urea; duroquinol = tetramethyl-p-benzoquinone ('TMQ'); DBMIB = 2,5-dibromo-3-methyl-6-isopropyl-p-benzoquinone; Fd-OR = ferredoxin:$NADP^+$ oxidoreductase; PSI denors = electron donors to PSI, e.g. reduced phenylene diamine and DCPIPH$_2$ (reduced 2,6-dichlorophenolindophenol); PSII donors = electron donors to PSII, e.g. semi-carbazide and diphenyl carbazide; PSI acceptors = acceptors of electrons from PSI, e.g. ferricyanide, $NADP^+$ and methyl viologen; PSII acceptors = acceptors of electrons from PSII, e.g. benzoquinones, phenylenediamines and DAD (diaminodurane).

9.8.1 Charge separation at the reaction centres

When a photon of light is absorbed by the antenna complex of PSI or PSII it is eventually channelled to the respective reaction centre oxidizing P_{700} (PSI) or P_{680} (PSII) (Eqns 9.9 & 9.10).

$$P_{700} \xrightarrow{h\nu} P_{700}^* \xrightarrow{Acc_1} P_{700}^+ + Acc_1^- \tag{9.9}$$

$$P_{680} \xrightarrow{h\nu} P_{680}^* \xrightarrow{Acc_2} P_{680}^+ + Acc_2^- \tag{9.10}$$

P_{680} and P_{700} are raised to the excited first singlet state (shown by *). The

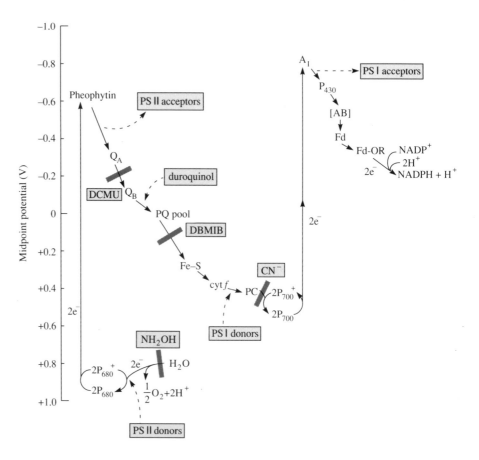

midpoint potentials of the excited states are very negative and can bring about the oxidation of the appropriate primary electron acceptors (shown as Acc_1 and Acc_2). Within 200 ps, the electrons are transferred to secondary acceptors.

9.8.2 Reduction of $NADP^+$

The electron acceptor of PSI has remained obscure for some time. At one time it was thought to be P_{430} but recent evidence suggests that the primary acceptor is, in fact, a chlorophyll component tightly bound to P_{700} (A_1 in Fig. 9.5). A_1 is a very strong reductant with a midpoint potential of between -0.8 and -0.9 V. The electron ejected from P_{700} during charge separation is passed on to A_1 and thence to the Fe–S protein, P_{430}, reducing the Fe^{3+} of this component to Fe^{2+}. Similar redox changes occur in the Fe–S proteins A and B ([AB] in Fig. 9.5) and subsequent electron transfer to ferredoxin (a soluble iron–sulphur protein) causes a similar reduction of Fe^{3+} associated with this molecule. Reduced ferredoxin reduces $NADP^+$ in the stroma in a reaction catalysed by ferredoxin:$NADP^+$ oxidoreductase (Eqn. 9.11).

$$NADP^+ + 2Fd_{red} + 2H^+ \rightarrow NADPH + 2Fd_{ox} + H^+ \qquad (9.11)$$

Two electrons are required to reduce every molecule of $NADP^+$. These come from the absorption of two photons by PSI.

9.8.3 Electron transfer from PSII to PSI

The P_{700}^+ formed in the primary charge separation (Eqn. 9.9) has to revert to P_{700} for the next photochemical event to occur. The electron required comes from the light reaction of PSII. The electron ejected by the photo-oxidation of P_{680} is trapped by pheophytin a. This modified chlorophyll a, which lacks magnesium, has a redox potential more negative than the secondary acceptor Q_A ($E_0' = -0.3$ V). From Q_A, electrons are passed down a series of electron carriers of increasingly more positive redox potential to P_{700}^+. These electron carriers can exist in a reduced or oxidized form as they receive or lose electrons.

The sequence in which these components participate in electron transfer has been determined in a number of ways. In the absence of coupling to some exergonic reaction (see Section 4.2), electron transfer must be from compounds with a more negative redox potential to those with a more positive redox potential. Also, the redox state of thylakoid components such as cytochrome f or plastoquinone (PQ)

differs according to which photosystem is excited. Hence, far red light, which excites PSI but not PSII, causes photo-oxidation of P_{700}, thus causing oxidation of cytochrome f and PQ as electrons are drained into P_{700}^+, whilst excitation of PSII by shorter wavelengths results in their reduction. It must be concluded that both cytochrome f and PQ are involved in electron transport between PSI and PSII. Judicious use of inhibitors also provides a clue as to the sequence of electron transfer. For example, DCMU blocks electron transfer from Q_A to another bound quinone, Q_B, and inhibits the reduction of both plastoquinone and cytochrome f when PSII is excited. DBMIB, however, only inhibits the reduction of cytochrome f, showing that plastoquinone is closer than cytochrome f to PSII. The sites of action of a number of other inhibitors are shown in Fig. 9.5. The kinetics of redox changes have also been used to determine the sequence of electron transfer. From Q_B, electrons are transferred to plastoquinone to form the reduced form PQH_2. This carrier is closely associated with PSII and has a redox potential of zero. Plastoquinone is relatively abundant compared to the other carriers with one present for every 10 chlorophyll molecules. It is also very mobile in the membrane which is important in the generation of ATP (Section 9.11). The reduction of PQ requires not only two electrons donated from Q_B, but also two protons from the stroma (Eqn. 9.12).

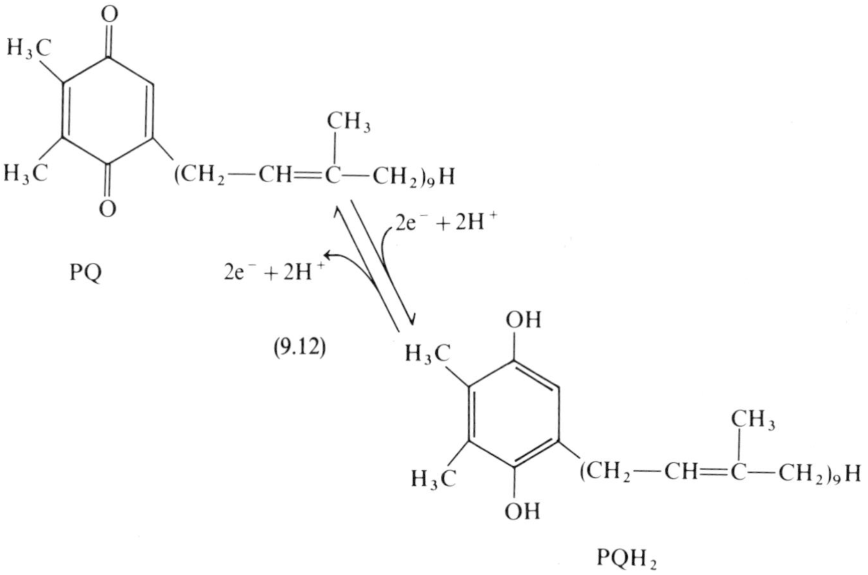

$$(9.12)$$

From reduced plastoquinone, electrons are transferred to another mobile carrier, plastocyanin (PC). This is a protein, containing copper, with a redox potential of $+0.37$ V and is present at similar levels to the other protein electron carriers of thylakoids, i.e. about one molecule for every 400 molecules of chlorophyll. Reduction of plastocyanin is achieved by the conversion of Cu^{2+} to Cu^{+}.

Reduction of plastocyanin by PQH_2 does not proceed directly but does so via a protein complex which spans the thylakoid membrane and contains one molecule of cytochrome f, one molecule of cytochrome b_{563} (with two haem groups, one on the inner edge of the membrane and one on the outer edge) and an Fe–S protein together with some bound plastoquinone. There are some five polypeptides associated with the complex. In essence, the complex acts as a plastoquinol: plastocyanin oxidoreductase (Eqn. 9.13).

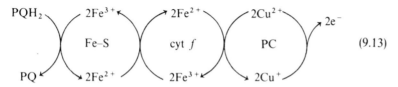

$$(9.13)$$

Finally P_{700}^{+} returns to its reduced state by accepting electrons from reduced plastocyanin.

9.8.4 Generation of electrons from the oxidation of water

P_{680}^{+} generated in the light reaction of PSII must return to P_{680} before another photoact can take place to release electrons for transport to NADP. Reduction of P_{680}^{+} is achieved by the withdrawal of an electron from water in a series of reactions involving the Mn–protein complex of PSII. This complex, possibly associated with cytochrome b_{559}, is present at a level equivalent to about five to eight atoms of manganese for every 400 molecules of chlorophyll. At least two manganese atoms are required at the catalytic site of the complex and a further two atoms are required for optimal activity.

The liberation of one molecule of O_2 in vivo requires the absorption of eight quanta of light (see Section 9.6); four by PSII (involved in O_2 evolution) and four by PSI (involved in $NADP^+$ reduction). Since the production of one O_2 involves the removal of four electrons from two H_2O molecules this implies that each electron removed requires the absorption of one quantum. Two lines of evidence suggest that four quanta absorbed by PSII support O_2 evolution.

Firstly, the rate of O_2 evolution (R) observed in weak light is not proportional to photon flux. Rather there is a fourth power dependence such that if the photon flux (I) is small, R is proportional to I^4. Secondly, the amount of O_2 evolved by illuminated chloroplasts, subjected to a series of short flashes of light following a period of darkness, is dependent on the flash numbers (Fig. 9.6): maxima are

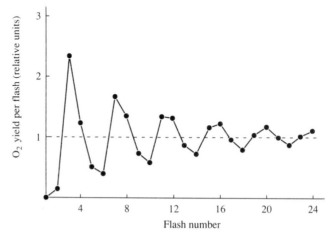

Fig. 9.6 Oxygen yield as a function of flash number. Spinach chloroplasts were pre-incubated in the dark and then exposed to saturating flashes of light of 2 μs in duration and 0.3 s apart. After many flashes the yield reaches an average value (dashed line). After Forbush, B., Kok, B. & McGoild, M. (1971) *Photochem. Photobiol.*, **14**, 307–21.

obtained after flashes 3, 7, 11, etc. and minimal rates obtained from flashes 1, 5, 9, etc. This oscillation is damped as the flash number increases until an average yield is obtained for every flash. These observations are interpreted in terms of there being four sequential states of the water splitting reaction, termed S_0, S_1, S_2 and S_3. An additional state, S_4, is unstable and its decay results in the release of O_2. The 'cycle of S states' is shown in Fig. 9.7 and corresponds to successive redox states of the manganese atoms in the Mn–protein of PSII. A photoact at the PSII reaction centre results in the removal of one electron from S_0 converting it to S_1. A second photoact converts S_1 and S_2 and so on. S_0 to S_4 are thus consecutively higher oxidation states of the chemical system. A complete cycle involves four photoacts and, in the cycle, two molecules of H_2O are oxidized to yield four protons and four electrons with the liberation of one molecule of O_2 (Eqn. 9.14).

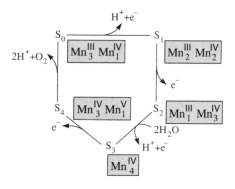

Fig. 9.7 The cycle of S states. Conversion from one of these states to the next requires input of a photon. Postulated oxidation states (and the number of Mn atoms in that state) are given in shaded boxes. After Bruding, G.W. & Crabtree, R.H. (1986) *Proc. Natl. Acad. Sci. USA*, **83**, 4586–8.

$$2H_2O + 4h\nu \rightarrow 4H^+ + O_2 + 4e^- \qquad (9.14)$$

After a dark adaptation period, 75% of S states are in S_1. Therefore, the first maximum of O_2 yield is achieved at flash number 3, i.e. after three photoacts have advanced the system from S_1 to S_2 to S_3 to S_4. Further cycling requires four photoacts for every O_2 evolved. After a while, however, double hits and misses at the reaction centres advance or retard some of the population of S states and the population becomes unsynchronized with approximately 25% in each of the states S_0, S_1, S_2 or S_3; this accounts for the fact that O_2 yield per flash approaches an average value as flash number increases.

9.8.5 *Some herbicides are effective because of their interaction with light-dependent electron transport*

A number of compounds which interfere with light-dependent electron transport are, as a consequence, potent herbicides. For example, DCMU (marketed as diuron) and other substituted ureas such as 3-(*p*-chlorophenyl)-1,1-dimethyl urea, CMU (marketed as monuron) interact with PSII, specifically blocking electron transport between the bound quinones Q_A and Q_B.

DCMU

CMU

The triazines also inhibit electron transport between these quinones and plastoquinone. The simplest examples are the *s*-triazines, atrazine (2-chloro-4-(ethylamino)-6-(isopropylamino)-*s*-triazine) and simazine (2-chloro-4,6-*bis*(ethylamino)-*s*-triazine) shown below.

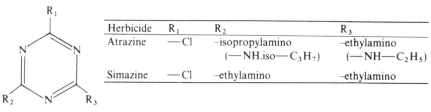

Herbicide	R_1	R_2	R_3
Atrazine	—Cl	–isopropylamino ($-NH.iso-C_3H_7$)	–ethylamino ($-NH-C_2H_5$)
Simazine	—Cl	–ethylamino	–ethylamino

General structure

In both the triazines and the substituted ureas, light stimulates the effectiveness of the herbicide. This is thought to be due to an inability of the light-harvesting system to dissipate absorbed light energy via electron transport. Although some energy is dissipated via the carotenoids (Section 9.3), eventually they become overloaded and photo-oxidative damage ensues. Some species of the alga *Dunaliella*, which have a large amount of carotenoids, are particularly resistant to substituted urea or triazine herbicides.

The triazines and substituted ureas bind reversibly to thylakoids but not to a single, unique, binding site. Rather, there is a more general 'binding domain' which contains binding sites for a whole range of herbicides. It is believed that electron transfer is blocked by the removal of plastoquinone from its binding site, due either to replacement by the inhibitor molecules or indirectly through allosteric effects.

The other class of herbicides which affect the light reactions are the bipyridyliums, such as 1,1-dimethyl-4,4′-bipyridilium dichloride (methyl viologen or paraquat). They interact with electron flow around PSI by 'draining off' electrons which would otherwise be used for the reduction of ferredoxin. Electrons from PSI instead reduce the bipyridylium to give the free radical (Fig. 9.8). This free radical is unstable and undergoes auto-oxidation, generating the superoxide radical O_2^- which spontaneously yields H_2O_2. The superoxide radical and H_2O_2 combine to yield an OH^- ion, O_2 and, importantly, the hydroxyl radical, $OH^\cdot$. The hydroxyl radical is an extremely powerful oxidant which can cause considerable damage to cell components, particularly membrane lipids. Although the enzyme superoxide dismutase (see Chapter 14) can convert the superoxide radical to H_2O_2, in turn acted on by the ascorbate/glutathione pathway (see Section 14.5), this system becomes

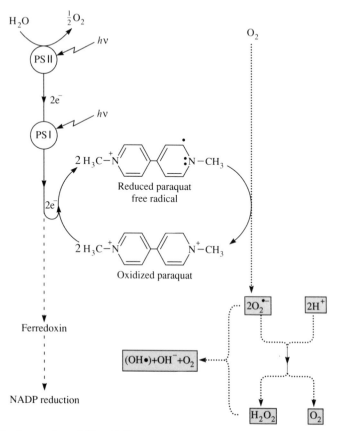

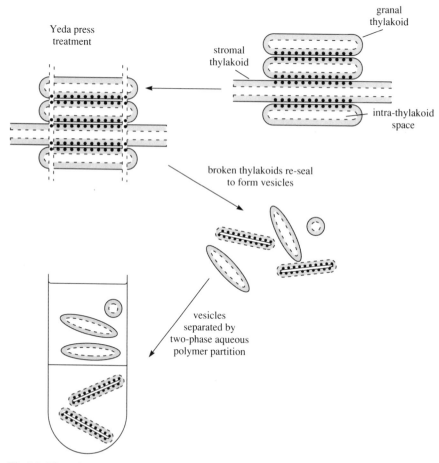

in stacks between a complex folded membrane system of stromal thylakoids to form grana (see Fig. 3.5). Some thylakoids are therefore appressed and some are non-appressed. The latter include all regions where the outer surface of the membrane is exposed to the stroma and also include the outer periphery of granal sacs and the stroma lamellae (Fig. 9.9).

The outer membrane surfaces of appressed and non-appressed regions are different but the inner surfaces are similar since they adjoin the intra-thylakoid

Fig. 9.8 The interaction of bipyridylium herbicides with electron transport. Bipyridyliums such as methyl viologen (paraquat) accept electrons from the primary acceptor of PSI before they are transferred to ferredoxin.

rapidly overloaded in the presence of bipyridylium herbicides causing cell damage and death.

9.9 Functional organization of thylakoid membranes

9.9.1 Lateral distribution of thylakoid components

The basic structure of chloroplasts has been described in Chapter 3. In higher plants, chloroplasts contain granal thylakoids, comprising granal sacs, interposed

Fig. 9.9 The origin and separation of 'inside-out' and 'right-side out' vesicles from thylakoids. After Anderson, J.M. & Andersson, B. (1982) *Trends Biochem. Sci.* **7**, 288–92.

space. These differences permit separation of appressed and non-appressed thylakoid membranes. By breaking thylakoids in a device such as a Yeda Press, which shears thylakoid sacs at their ends, and letting the membrane fragments reseal, vesicles can be obtained representative of the two types. Vesicles derived from non-appressed regions tend to reseal the 'right-side out' whilst those from appressed regions, generally reseal 'inside-out' (Fig. 9.9). Since the surface properties of these two types of vesicle are different, they can be separated using aqueous polymer two-phase partition (e.g. an upper layer of polyethylene glycol and a lower layer of dextran). Typically the 'right-side out' vesicles are found in the upper phase whilst those from appressed regions segregate into the lower phase (Fig. 9.9).

These techniques have permitted the investigation of the location of components involved in light harvesting (Fig. 9.10). Most (80–90%) of the PSII core and associated LHCP 2 is in the appressed regions, in direct contrast to PSI (including LHCP 1) which occurs in the non-appressed, stroma-exposed regions (Table 9.2). The ATP synthetase (Section 9.12) is found in non-appressed regions whilst the cytochrome b/f complex is believed to be at the junction between appressed and non-appressed membranes, the fret region. This distribution is important for the energy-transducing process.

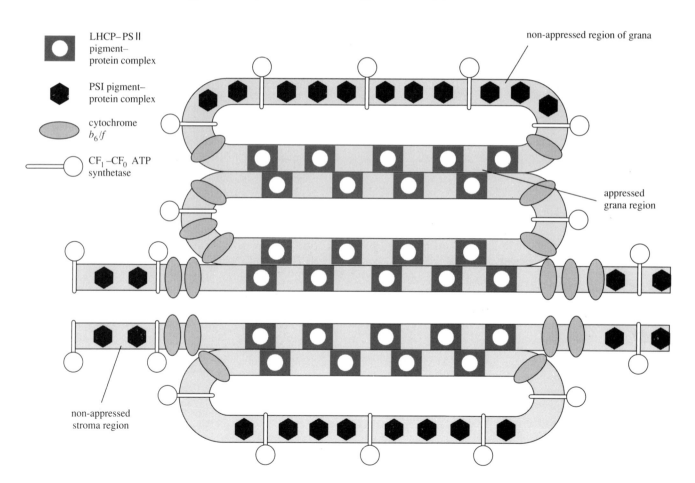

LHCP–PS II pigment–protein complex

PSI pigment–protein complex

cytochrome b_6/f

CF_1–CF_0 ATP synthetase

non-appressed region of grana

appressed grana region

non-appressed stroma region

Fig. 9.10 Lateral heterogeneity in thylakoids, showing differences in the distribution of major thylakoid complexes between appressed and non-appressed regions. After Barber, J. (1983) *Plant, Cell and Environment*, **6**, 311–22.

Thylakoid fraction	Distribution of chlorophyll (% total)			chl/cyt f ratio	ATP synthetase (arbitrary units)
	PSI core complex	PSII core complex	Light harvesting complex (LHCP)		
Unfractionated thylakoids	30	13	52	460	4.2
Grana stacks	20	13	60	–	3.4
Grana partitions (inside-out vesicles)	10	14	65	470	1.2
Stroma thylakoids	69	5	17	485	11.7

Table 9.2 Distribution of major thylakoid components in unfractionated thylakoids and in various thylakoid fractions prepared by Yeda Press treatment and aqueous polymer two-phase partition. From Anderson, J.M. & Andersson, B. (1982). *Trends Biochem. Sci.*, 7, 288–92.

(1) Because of the distribution of PSII and PSI, the notion of discrete electron transport chains is no longer valid. This is further supported by observations that the ratio of PSI to PSII assemblies is rarely one and there is not necessarily a fixed stoichiometry between the various components of electron transport from H_2O to NADP.

(2) Following on from **(1)**, there must be a mechanism for electron transport between the two, spatially separated, photosystems. The most obvious choice for a lateral shuttle within the membranes is plastoquinone which, relative to the other components, has a long half time for reoxidation (20 ms). The high content of unsaturated acyl lipid in thylakoids make such membranes fairly fluid, so it is possible for PQ/PQH_2 to diffuse through and link PSII with PSI. The large amounts of PQ in chloroplasts, its low molecular mass and lipophilic nature, are consistent with this proposal. Since the cyt b_6/f complex is not distributed evenly throughout the thylakoids, electrons must in some way be transferred from the fret region to the PSI complexes. Here plastocyanin (PC) may also act as a mobile lateral shuttle for electrons within the intra-thylakoid space.

(3) If PSI and PSII are not in close contact with each other the transfer of excess light energy from PSII to PSI, the 'spillover effect', must be very small. Nonetheless, the distribution of light energy between the photosystems *is* regulated (Section 9.16).

9.9.2 Transverse heterogeneity of thylakoid membranes

In addition to the marked lateral heterogeneity in distribution, the energy-transducing components of thylakoids also show a marked transverse heterogene-

ity across the membrane. This information comes from examination of right-side out and inside-out membrane vesicles and use of other techniques; three examples are given below.

(1) Trypsin treatment effectively inhibits PSII activity in both inside-out vesicles and control thylakoids. However, adding the electron donor diphenylcarbazide, which bypasses the water splitting reaction, restores PSII activity only in inside-out vesicles. The conclusion to be reached is that the water splitting reaction is on the inside surface of membranes adjacent to the intra-thylakoid space (ITS).

(2) Purified reduced plastocyanin reduces P_{700}^+ rapidly, only when added to inside-out vesicles. Immunological studies show that plastocyanin is released from inside-out but not right-side out vesicles, suggesting that it is located in the ITS and that P_{700} is present towards the inner side of the membrane.

(3) Treatment of either right-side out or inside-out vesicles with an antibody to LHCP 2 induces heavy agglutination, implying that it spans the membrane. A schematic representation of the transverse distribution of electron transfer components in thylakoids is given in Fig. 9.11.

9.10 Transverse heterogeneity of electron transport components results in a trans-thylakoid proton gradient

As a result of the transverse heterogeneity in the distribution of electron carriers, the movement of electrons along the electron transport chain causes a trans-thylakoid proton gradient. The water splitting reaction releases four protons from two H_2O molecules into the ITS. In addition, as electrons are added to plastoquinone this molecule picks up two more protons from the stroma, transports them across

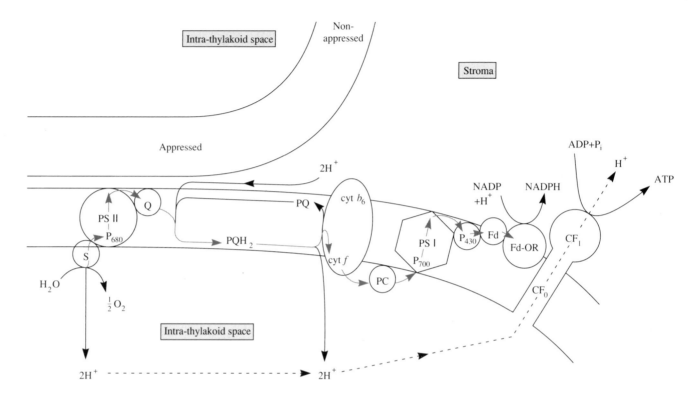

Fig. 9.11 Transverse heterogeneity in the distribution of thylakoid complexes. The diagram shows how the various components interact to bring about the transfer of electrons from water to NADP and concurrently pump protons from the stroma into the intra-thylakoid space, thus generating the proton motive force to drive ATP synthesis via the CF_0–CF_1 complex. The path of electrons is shown in grey. (No attempt is made to balance the stoichiometry of proton transport, electron flow and ATP and NADPH synthesis.) Abbreviations: S = water oxidizing Mn–protein; PQ/QH_2 = oxidized and reduced plastoquinone; PC = plastocyanin; Fd = ferredoxin; Fd-OR = ferredoxin:NADP oxidoreductase.

the membrane and releases them into the ITS as PQH_2 is oxidized. Thus, there are two sites generating the release of protons into the ITS.

The end result of the absorption of eight photons of light (four by PSI and four by PSII) is the transfer of four electrons from two molecules of H_2O (liberating one molecule of O_2) to two molecules of $NADP^+$ to form two molecules of NADPH. This electron transport results in the transfer of eight protons from the stroma to the ITS (Fig. 9.11). The pH gradient thus established between the stroma and the ITS is used to drive the formation of ATP.

9.11 The trans-thylakoid proton gradient is coupled to ADP phosphorylation

The magnitude of the proton movement and the ensuing pH gradient can be measured fairly easily. The change in pH of a weakly buffered thylakoid suspension allows measurements of proton uptake into the ITS, whilst the internal pH of that space can be measured by examining the distribution of a weak base such as methylamine (see Section 6.2.3).

The pH of the stroma of illuminated chloroplasts is about 8.0 and the pH gradient (ΔpH) between the stroma and the ITS is 3.0–3.5 units. ATP formation can be shown to depend on this ΔpH. Firstly, there is a clear relationship between light-induced proton uptake by chloroplasts, presumably the result of the transfer of protons from the stroma to the ITS, and ATP formation by those chloroplasts in a subsequent period of darkness. Secondly, if thylakoids are suspended in a solution of succinate at pH 4.0 and allowed to equilibrate in the absence of light, succinate permeates the thylakoids and the ITS soon reaches this pH. If the external pH is then quickly raised to 8.0, generating a ΔpH of 4.0, ATP is formed in the presence of ADP and P_i. This ATP formation is large and is independent of light and electron transport—only the presence of ADP, P_i and ΔpH is necessary.

Thirdly, compounds which dissipate the pH gradient inhibit ADP phosphorylation but not electron transport. The action of some of these compounds has been described in Sections 6.2.4 and 7.8.2.

The relationship between ADP phosphorylation and ΔpH can be investigated by means of energy-dependent reverse reactions. The ability of the ATP synthetase to catalyse the *hydrolysis* of ATP (ATP-phosphatase activity) is not usually expressed and the enzyme must first be activated by prior illumination in the presence of dithiols, which reduce a disulphide group on the γ subunit (see Section 9.12). In illuminated thylakoid vesicles treated with dithiothreitol, ATP hydrolysis is coupled to transport of protons from the suspending medium ($\equiv$ stroma) *into* the ITS. This results in a ΔpH across thylakoids of about 3.7 units, which is very similar in magnitude to the gradient formed by forward electron transport. With the ATP phosphatase activated, ATP drives reverse electron transport, oxidizing cytochrome f and reducing Q. Such electron flow is sensitive to uncouplers and depends upon the addition of an electron donor.

It should be noted that the transport of protons from the stroma to the ITS results in an electrical potential difference across the membrane. However, proton influx to the ITS is followed rapidly by Cl^- influx and Mg^{2+} efflux, decreasing the membrane potential component of the proton motive force (see Section 4.6) to about 10 mV, with the ITS positively charged relative to the stroma. Chemiosmotic coupling of the proton motive force to ATP synthesis is explained in Section 7.8.

9.12 The CF_1-CF_0 ATP synthetase

The way in which the pH gradient is used to drive ADP phosphorylation is thought to be similar to that of mitochondria (see Section 7.8). The proton motive force ($\Delta\tilde{\mu}_{H^+}$) across the membrane drives protons through an ATP synthetase in the membrane. The energy of the proton motive force is conserved in the phosphorylation of ADP. Although the basic mechanism of ATP synthesis is the same, the ATP synthetase of chloroplasts differs from that in mitochondria in that ATP phosphatase activity requires activation (see above).

Treatment of chloroplasts with ethylenediamine tetra-acetic acid (EDTA) releases a protein factor termed coupling factor (CF_1). Thylakoids from which CF_1 has been removed exhibit light-dependent electron transport but not photophosphorylation. This can be restored by adding back CF_1. Treatment of free CF_1 by brief heating, trypsin digestion or reduction by sulphydryl reagents gives rise to a highly active Ca^{2+} dependent ATP phosphatase activity, again reinforcing the notion that CF_1 has a role in phosphorylation. When CF_1 is removed, thylakoids

are rendered permeable to protons, due to the exposure of a hydrophobic second protein factor (CF_0) which acts as a channel for the passage of protons from the ITS into the stroma. The total coupling factor for the proton-mediated phosphorylation of ADP is thus referred to as CF_1-CF_0 or the chloroplast ATP synthetase (Fig. 9.12). The structure and subunit organization of CF_0 is still uncertain. It probably has four major polypeptides of molecular mass 15, 12.5, 8 and 20 kDa, termed subunits I, II, III and IV respectively. Six polypeptides of

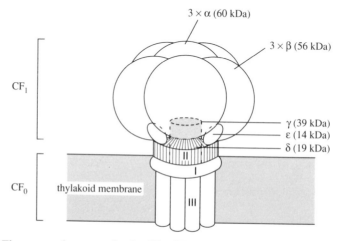

Fig. 9.12 The proposed structure for the CF_1-CF_0 complex which constitutes the chloroplast ATP synthetase. CF_0 acts as a proton channel whilst the CF_1 complex contains ATP phosphatase activity. CF_0 is shown consisting of subunits I–III. It also contains an additional subunit (IV) but how this is arranged in the complex is not known. After Gounaris, K., Barber, J. & Harwood, J.L. (1986) *Biochem. J.*, **237**, 313–28.

Table 9.3 Stoichiometry, molecular mass and probable function of the subunits of the CF_1 complex. After Clayton, B.J. (1980). *Photosynthesis: physical mechanisms and chemical patterns.* IUPAP Biophysics Series, Vol. 4. Cambridge University Press, Cambridge.

Subunit	Molecular mass (kDa)	Stoichiometry	Probable function
α	59	3	Regulatory
β	56	3	Active site of ATP synthesis
γ	37	1	Transfer of H^+ to β subunit
δ	17.5	1	Attachment of CF_1 to membrane
ε	13	2	Attachment of CF_1 to membrane; inhibition of ATPase activity?

subunit III form the proton-conducting channel and subunit II regulates the organization of subunit III. Subunit I may have a function in binding CF_0 to CF_1 whilst the function of subunit IV is uncertain.

CF_1 represents nearly 10% of the thylakoid protein and consists of five subunits, α, β, γ, δ and ε (Fig. 9.12) with a probable stoichiometry of $\alpha_3\beta_3\gamma\delta\varepsilon_2$, where the subscripts indicate the number of each subunit present. The function ascribed to each subunit is summarized in Table 9.3.

9.13 How many protons are transported for each ATP formed?

This question has kept a number of plant biochemists busy for a long time! The chemiosmotic hypothesis predicts an efflux of two protons for every ATP synthesized but measured values for the H^+/ATP ratio range from 2–4. Following the arguments outlined for ATP generation in mitochondria (see Section 7.8), it is suggested that the more correct value is H^+/ATP = 3.

A common value for $\Delta\tilde{\mu}_{H^+}$ across thylakoid membranes is 210 mV, equivalent to a $\Delta G'$ of 20.26 kJ mol^{-1}. The formation of ATP at physiological concentrations of ADP, P_i and ATP has a $\Delta G'$ (*not* $\Delta G^{\circ\prime}$) of around 57 kJ mol^{-1}. Thus, the efflux of three protons ($3 \times 20.26 = 60.78$ kJ) is sufficient to drive the phosphorylation of ADP.

9.14 Analysis of inputs and outputs and efficiency of non-cyclic electron transport

It has been shown that the light energy absorbed by chlorophyll is used in non-cyclic electron transport to form NADPH and ATP.

For the sake of argument consider light of 680 nm, one mol of which has an energy of 176 kJ. Under physiological conditions, $\Delta G'$ values for the oxidation of NADPH by O_2 and the hydrolysis of ATP to ADP and P_i are -218 kJ mol^{-1} and -57 kJ mol^{-1} respectively. Thus the following analysis can be made.

Input = 8 mol quanta at 176 kJ mol^{-1}	1408 kJ
Output = 2 mol NADPH at 218 kJ mol^{-1}	436 kJ
(derived from absorption of 8 mol quanta)	
If H/ATP = 3,	
then 8/3 ATP at 57 kJ mol^{-1}	152 kJ
Total output	588 kJ

Therefore efficiency of energy conversion = 588/1408 = 42%.

9.15 Monitoring the photosystems *in vivo* by fluorescence and fluorescence quenching

The use of fluorescence to examine the host of interconnected processes occurring during light harvesting and electron transport has increased dramatically in recent years. Energy can be lost by emission of light through fluorescence when an electron in an excited chlorophyll molecule falls back to the ground state orbital (Section 9.3), losing energy without doing any useful work. Thus, if PSII is prevented from transferring electrons to PSI by treatment with DCMU, then fluorescence increases. Low temperatures (usually those of liquid nitrogen, 77 K) are often used to study fluorescence because they prevent changes in membrane state which increase fluorescence from PSI.

9.15.1 Energy transfer between pigments

Fluorescence studies have been used to provide evidence for the presence of different chlorophyll–protein associations (chl–proteins), or of biliproteins, and on the transfer of energy between them. Much information about the complexes present in a sample are obtained from fluorescence excitation spectra in which emission fluorescence intensity is measured as a function of the wavelength of the exciting light. Examining the magnitude of emission, at the wavelength of maximum emission characteristic of a pigment, as a function of excitation wavelength can help to identify other pigments which transfer energy to the fluorescing pigment.

9.15.2 Fluorescence induction curves give information on PSII activity and its interaction with PSI

Fluorescence induction curves are measurements of the variation in fluorescence following illumination. They provide information regarding the activity of PSII and its interaction with PSI and CO_2 fixation. Two types of fluorescence induction curves are commonly examined. The first relates to the primary photochemistry associated with PSII. Electron transfer from the secondary electron acceptor of PSII, Q_A, to the next component of the electron transport chain, Q_B, is blocked at low temperatures (77 K) or in the presence of DCMU at room temperature. Illumination of dark-adapted cells or chloroplasts under such conditions allows investigation of the kinetics of reduction of Q_A (Fig. 9.13). Immediately the light is switched on, fluorescence rises very rapidly to what is termed the minimum fluorescence level F_o, Q_A is then fully oxidized and all the PSII reaction centres can

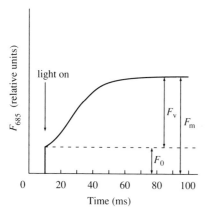

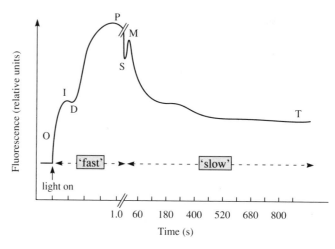

Fig. 9.13 The kinetics of induction of fluorescence at 685 nm (F_{685}) following illumination of dark-adapted, broken pea chloroplasts. The experiment was carried out at room temperature in the presence of DCMU. After Hipkins, M.F. & Baker, N.R. (1986) In Hipkins, M.F. & Baker, N.R. (eds) *Photosynthesis: energy transduction: a practical approach*, pp. 51–102. IRL Press, Eynsham.

Fig. 9.14 A stylized fluorescence induction curve during constant illumination following a period of dark adaptation. The significance of the transitions O–T are discussed in the text.

accept energy. From F_o, fluorescence rises relatively slowly (50–60 ms) to a maximum value, F_m, the rate depending on the photon flux used. F_m is reached when Q_A is fully reduced and no more reaction centres can accept energy. The difference in F_o and F_m is termed the variable fluorescence F_v and the ratio F_v/F_m is proportional to the quantum yield of the primary photochemistry of PSII.

The second form of fluorescence induction curves commonly used are Kautsky curves which relate to the variations in fluorescence when dark-adapted leaf tissue or algal cells are brightly illuminated at room temperature. The distinct phases which can be identified are designated by the letters O, I, D, P, S, M, T (Fig. 9.14). The fluorescence originates from PSII but each phase reflects various features, including the role of CO_2 fixation, as follows.

(1) A rise from the origin (O), when Q_A is fully oxidized, to an inflection point (I) as Q_A becomes reduced by electrons from PSII.

(2) Some tissues or cells show a decrease in fluorescence from I to a dip (D) as Q_A is oxidized by electron transfer to Q_B.

(3) As the plastoquinone pool becomes fully reduced and reduced Q_A cannot pass on any more electrons, there is a major rise in fluorescence to a peak (P).

(4) This is followed by a series of slower changes from P to a terminal (T), essentially steady level of fluorescence. The changes represented by P S M T are complex and are related to the commencement of CO_2 fixation in the tissue or cells (see Section 10.14). The factors affecting fluorescence in this stage of induction are consumption of NADPH by CO_2 fixation, which leads to increased rates of non-cyclic electron transport and oxidation of Q_A, and changes in the size of the

proton gradient between the stroma and ITS. The mechanism involved in the latter effect is unknown.

The time course for these changes is tremendously variable between species and conditions. Generally, the O I D P sequence is considered a fast process with I D P taking about 1 s to reach completion. The 'slow' fluorescence changes from P to T take longer; P to S takes 5–10 s, S to M about 30 s and M to T about 90 s (Fig. 9.14).

9.16 Regulation of energy distribution between PSII and PSI

As described earlier, the two photosystems act together in light harvesting and its conversion to biochemically useful forms of energy. However, PSI and PSII do not have identical absorption spectra and can be unequally excited by light of given wavelengths (Fig. 9.15). For the two photosystems to co-operate for maximum efficiency of energy transduction, the distribution of energy should be maintained at an optimal balance. This should be when the distribution of excitation energy and hence light absorption by the two systems is equal. This occurs at around 685 nm (Fig. 9.15). However, observations show that the quantum yield is more or less unchanged from 570–685 nm, so there must be some mechanism to regulate the

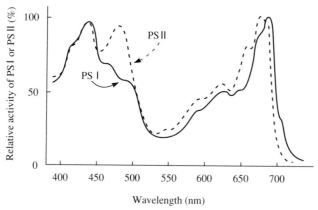

Fig. 9.15 Action spectra for PSI and PSII in *Chlorella*. The relative efficiency of incident light quanta to excite PSI (dashed line) and PSII (solid line) are given as a function of wavelength. After Reid, A. (1972) In Forti, G., Avron, M. & Melandri, A. (eds) *Proc, 2nd Internat. Congr. Photosynthesis Research*, pp. 763–72. Dr. Junk, The Hague.

distribution of excitation energy between PSII and PSI. If the available light is preferentially absorbed by PSII, the imbalance is corrected by allowing some excitation energy to be transferred to PSI—a condition known as State 2. However, if the absorbed light preferentially excites PSI (e.g. light of wavelengths >690–700 nm), then energy redistribution is minimal and the system is said to be in State 1. Changes between the two states (State 1–State 2 transitions) optimize the efficiency of energy transduction.

State 1–State 2 transitions are most commonly monitored using chlorophyll fluorescence. A fairly typical experiment using a suspension of *Chlorella* is shown in Fig. 9.16. Cells were initially illuminated with long wavelength red light (706 nm, known as 'far-red') for about 7 min to induce State 1. Significant levels of fluorescence are absent during this period, i.e. PSII is not overexcited. When the far-red light is replaced with blue-green light ($\lambda_{max} = 540$ nm), there is a large rise in fluorescence implying that much of the light trapped by PSII pigments is not used efficiently. The cells are still in State 1. After some minutes the fluorescence intensity decreases, indicating a better balance as the cells move from State 1 to State 2 and begin to redistribute excitation energy from PSII to PSI. Supplying far-red light again (together with blue-green light) reverses the situation and the cells move back to State 1, as is evidenced by the second rise in fluorescence signal when the far-red light is removed and the PQ pool is again over-reduced.

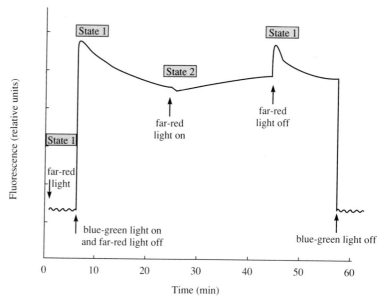

Fig. 9.16 State 1–State 2 transitions in the unicellular alga *Chlorella*. See text for explanation. After Hodges, M. & Barber, J. (1983) *Plant Physiol.* **72**, 1119–22.

This regulation of energy distribution between the two photosystems is mediated by the phosphorylation and dephosphorylation of LHCP 2 which causes a migration of these chlorophyll–protein complexes from appressed to non-appressed regions of the thylakoid membranes and vice versa. Migration of LHCP 2 to non-appressed regions causes a change in the number of pigment molecules serving each photosystem and hence, the distribution of light energy.

9.17 Cyclic photophosphorylation

9.17.1 Cyclic electron transport

The foregoing discussion of electron transport has been concerned with non-cyclic electron flow from H_2O to NADPH, linked to ATP formation. Cyclic photophosphorylation does not involve either the formation of NADPH or the splitting of water to produce O_2 (see Section 9.4, Fig. 9.4) and is associated with electron flow around PSI. Reduced ferredoxin generated by excitation of PSI, instead of transferring electrons to NADP, reduces cytochrome b_{665}. This, in turn, supplies electrons for the reduction of PQ. From PQH_2, electrons are passed back down the electron transport chain to allow P_{700}^+ to return to its ground state (P_{700}). It appears likely that cyclic electron transport involves a proton motive Q-cycle, included in the representation given in Fig. 9.17. In this case two protons are transported for every electron transferred to P_{700}^+ (instead of one proton for every electron, generated in the absence of a Q-cycle). These protons are used to drive ATP synthesis as described in Sections 9.11 and 9.12.

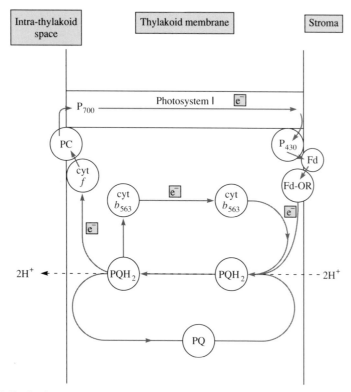

Fig. 9.17 Cyclic electron transport. The scheme shows the distribution of electron carriers in the thylakoid membrane and, as shown, involves a proton motive Q-cycle. One electron for the reduction of PQ comes from PSI and the other from the cyclic transfer of electrons involving cytochrome b_{563}. Thus, for every electron from PSI, two protons are transferred from the stroma to the intra-thylakoid space. After Raven, J.A. (1980) In Gooday, G.W., Lloyd, D. & Trinci, A.P.J. (eds) *Symp. Soc. Gen. Microbiol.*, **30**, 181–205.

9.17.2 Inputs and outputs of cyclic photophosphorylation

With $H^+/e^- = 2$ and $H^+/ATP = 3$ for ATP synthetase the following analysis can be made.

Input = 1 photon at 680 nm   176 kJ mol^{-1}
Output = $H^+/e^- = 2$, $H^+/ATP = 3$

Transport of 1 electron therefore yields
    2/3 ATP at -57 kJ mol^{-1} 38 kJ mol^{-1}
(under physiological conditions)

Therefore efficiency of energy conversion = 38/176 = 22%.

9.17.3 How important is cyclic photophosphorylation in vivo?

There has been a great deal of debate as to whether or not cyclic photophosphorylation occurs *in vivo* concurrently with non-cyclic photophosphorylation. It is certainly possible to demonstrate experimentally the maintenance of ATP levels, the sustained activity of ATP-requiring processes and various redox reactions when non-cyclic and pseudocyclic (Section 9.18) photophosphorylation are inhibited. Estimates of rates of ATP generation by cyclic photophosphorylation give values of up to 200 μmol ATP mg^{-1} chlorophyll h^{-1}, an order of magnitude less than estimates for the capacity of non-cyclic photophosphorylation.

It should be noted that the assimilation of CO_2 by C_3 plants requires three molecules of ATP and two molecules of NADPH per CO_2 fixed into carbohydrate, i.e. an ATP:NADPH ratio of 1.5:1 (see Chapter 10). This is higher than the ratio of 1.33:1 achieved from non-cyclic electron transport. Furthermore, the metabolic activities of chloroplasts (and hence, plant growth) involve a number of other ATP-requiring processes for phenomena such as ion transport, NH_3 assimilation/reassimilation, NO_2^- reduction, SO_4^{2-} assimilation and protein synthesis (see Section 14.6). It is possible that this 'extra' ATP requirement could be satisfied by cyclic photophosphorylation. It has been suggested that NADPH regulates the system so that if the utilization of NADPH (and ATP) by CO_2 assimilation is decreased, its accumulation would result in an increased cyclic flow of electrons around PSI, thus leading to generation of ATP without the formation of more NADPH.

It is also possible that additional ATP could be supplied through the process of pseudocyclic photophosphorylation.

9.18 Light-dependent reduction of oxygen: the Mehler reaction and pseudocyclic photophosphorylation

Cyclic electron transport involves the cycling of electrons around PSI. There is no involvement of PSII in this process and therefore no O_2 evolution. However, if O_2 evolution occurred through PSII activity, but was balanced by O_2 uptake, then there would still be no net O_2 exchange. It is possible to demonstrate that plants can take up O_2 in the light by using the stable isotope ^{18}O in the form $^{18}O_2$. There are three main processes contributing to this uptake; the persistence of mitochondrial respiration in the light, photorespiration (see Chapter 11), and the process known as the Mehler reaction. The Mehler reaction is a variation on the Hill reaction in which electrons are passed from Fd_{red}, produced by PSI and PSII activity, to O_2. Oxygen acts as the final electron acceptor and is thereby reduced to superoxide (O_2^-). In the presence of the enzyme superoxide dismutase, superoxide is converted to H_2O_2 which is reduced to water by the ascorbate/glutathione pathway using reducing equivalents derived from the light reactions (see Fig. 14.7). Since equal amounts of O_2 are consumed and produced, there is no net O_2 evolution in this system and it is termed pseudocyclic electron transport (Fig. 9.18).

Pseudocyclic electron transport can be linked to ATP synthesis in the same way and with the same stoichiometries as non-cyclic electron transport. Rates of up to $20\,\mu mol\,ATP\,mg^{-1}$ chlorophyll h^{-1} have been observed, although these estimates might well be on the conservative side.

Further reading

Monographs and treatises: Lawlor (1987); Staehlin & Arntzen (1986).
Techniques: Hipkins & Baker (1986).
Thylakoid composition: Anderson (1986); Murphy (1986); Stumpf (1987).
Electron transport and photophosphorylation: Harold (1986); Nicholls (1982).
Reaction centres: Glazer & Melis (1987).
Regulation of light reactions: Fork & Satoh (1986).
Herbicides: Fedtke (1982).

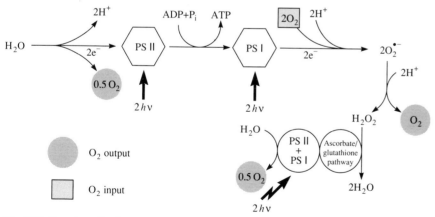

Fig. 9.18 Pseudocyclic electron transport. This is not a true cyclic process but one in which the evolution of O_2 by PSII is masked by an equal net consumption associated with O_2 reduction.

PART 3
ASSIMILATORY MECHANISMS
IN PLANTS

10
C_3 carbon reduction cycle
and associated processes

10.1 Most autotrophic organisms possess an active C_3 carbon reduction cycle

The C_3 carbon reduction cycle (C_3-CR cycle) is concerned with the net production of triose phosphates (triose-P) from CO_2. It takes its name from the C_3 compounds glycerate-3-P, the first metabolite containing newly assimilated CO_2, and glyceraldehyde-3-P and dihydroxyacetone-P, the first metabolites in which the carbon is reduced. The pathway is also known as the photosynthetic carbon reduction cycle, the reductive pentose phosphate pathway (several pentose phosphates are intermediates) and the Calvin or Calvin–Benson cycle, after the workers who discovered it. However, these names neither describe the pathway nor distinguish it from other carbon reduction mechanisms nor describe it appropriately in non-photosynthetic bacteria.

The acquisition of carbon by the C_3-CR cycle involves the reaction of CO_2 with the C_5 acceptor, ribulose 1,5-bisphosphate (ribulose-1,5-P_2) to form two molecules of glycerate-3-P. Sustained production of glycerate-3-P, and of triose-P derived from it, is only possible if some of the triose-P is used to regenerate ribulose-1,5-P_2, forming a cycle. This regeneration is, therefore, an intrinsic part of CO_2 assimilation, in much the same way as the regeneration of oxaloacetate by the tricarboxylic acid cycle is necessary for sustaining pyruvate oxidation.

An active C_3-CR cycle distinguishes almost all autotrophic organisms. The cycle is present in all green plants, chemoautotrophic bacteria, oxygenic photosynthetic bacteria (cyanobacteria) and most anoxygenic photosynthetic bacteria (excepting *Chlorobium*). About 98% of green plants assimilate CO_2 exclusively by the C_3-CR cycle and are known as C_3 plants. Some plants initially assimilate CO_2 through specialized mechanisms but even in these the CO_2 is subsequently released internally and reassimilated by the C_3-CR cycle (see Chapter 12).

In short-term experiments with C_3 plants, lasting less than a few minutes, carbon from newly assimilated CO_2 is found distributed through the intermediates of the C_3-CR cycle. In experiments of longer duration, newly assimilated carbon is mostly in sucrose and polysaccharides (mostly starch). These processes are also described in this chapter.

10.2 Experimental approaches to define the pathway of carbon dioxide assimilation

The pathway of CO_2 assimilation can be elucidated by supplying illuminated plants with air containing $^{14}CO_2$ and taking leaf samples at short intervals thereafter. In very short-term experiments (typically less than 2 s) with C_3 plants, virtually all of the ^{14}C-label is found in glycerate-3-P. For sampling times of about 3–5 s, label is also detected in triose-P and hexose-P. With still longer times various other sugar phosphates become labelled. This shows that CO_2 is initially incorporated into glycerate-3-P, that triose-P and hexose-P are derived from it and that the other sugar phosphates are derived from triose-P and hexose-P.

Virtually all the ^{14}C-label incorporated into glycerate-3-P in short-term experiments is found in the carboxyl carbon atom (C-1). This demonstrates that $^{14}CO_2$ is incorporated one molecule at a time and that the other two carbon atoms must come from a pre-existing (unlabelled) compound. Incubating leaf *extracts* with various compounds in the presence of $^{14}CO_2$ shows that the only compound which supports the synthesis of glycerate-3-P labelled in C-1 is the C_5 phosphorylated sugar, ribulose-1,5-P_2. Thus, incorporation of CO_2 must involve carboxylative cleavage of ribulose-1,5-P_2 to form two molecules of glycerate-3-P, only one of which contains ^{14}C-label.

10.3 Incorporation of carbon dioxide into triose phosphates involves a series of chloroplast enzymes

Isolated chloroplasts support the assimilation of CO_2 into glycerate-3-P and triose-P in the light without any other additions except low concentrations of phosphate. Extracts prepared by lysing isolated intact chloroplasts by osmotic shock and centrifuging to remove membrane material containing chlorophyll, also assimilate CO_2 into sugar phosphates if ribulose-1,5-P_2, ATP and NADPH are supplied. This shows that all the enzymes necessary for CO_2 assimilation occur in the stroma and that ATP and NADPH (normally produced by illuminated thylakoids) are consumed.

The assimilation of CO_2 into glycerate-3-P and its reduction to glyceraldehyde-3-P, the first of the triose phosphates is as follows.

$$\text{Ribulose-1,5-P}_2 + CO_2 + H_2O \rightleftharpoons 2 \text{ glycerate-3-P} \tag{10.1}$$

$$2 \text{ Glycerate-3-P} + 2 \text{ ATP} \rightleftharpoons 2 \text{ glycerate-1,3-P}_2 + 2 \text{ ADP} \tag{10.2}$$

$$2 \text{ Glycerate-1,3-P}_2 + 2 \text{ NADPH} + 2 \text{ H}^+ \rightleftharpoons$$
$$2 \text{ glyceraldehyde-3-P} + 2 \text{ NADP}^+ + 2 \text{ P}_i \tag{10.3}$$

Virtually all the massive amount of carbon assimilated (estimated at 200 000 million tonnes year^{-1}) proceeds via this route, making it the most active of all pathways. The reactions given by Eqns 10.2 and 10.3 are as important as Eqn. 10.1, since the energy state of the CO_2 assimilated into glycerate-3-P is not raised until the carboxyl group is phosphorylated and reduced.

10.3.1 Ribulose-1,5-P_2 carboxylase catalyses carbon dioxide incorporation and also exhibits oxygenase activity

Ribulose-1,5-P_2 carboxylase (RuP$_2$-Case) catalyses the incorporation of CO_2 into the carboxyl group of glycerate-3-P via an unstable intermediate as follows.

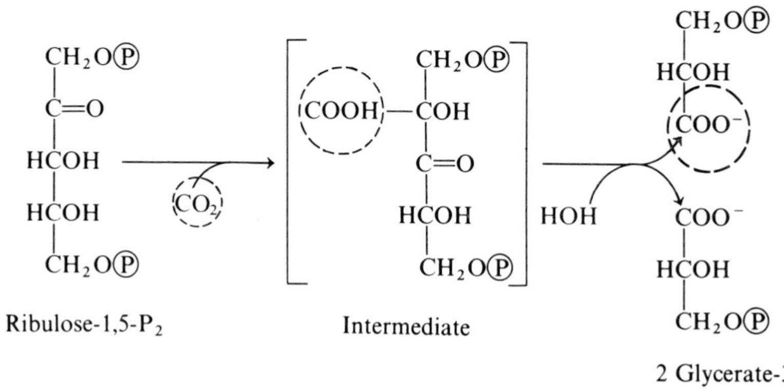

$$(10.4)$$

The active substrate is CO_2 rather than HCO_3^-. In plants, RuP$_2$-Case occurs as an oligomer (molecular mass 550 kDa) comprising eight identical small subunits (molecular mass 15 kDa) and eight large subunits (molecular mass 55 kDa). The equilibrium for the reaction lies strongly towards glycerate-3-P ($\Delta G^{\circ\prime} = -35 \text{ kJ mol}^{-1}$). The activated form of the enzyme (see Section 10.13.2)

has a high affinity for CO_2 with a K_m (CO_2) of approximately 12 μM, similar to the solubility of CO_2 in water in equilibrium with air (0.03% CO_2, 25 °C) of approximately 11 μM, so that *in vivo* RuP$_2$-Case operates well below maximum activity. The enzyme has a low turnover number (i.e. per molecule it catalyses product formation at comparatively low rates). Plants compensate by producing large amounts of RuP$_2$-Case, making it the most abundant protein in the world. Details of the interaction of RuP$_2$-Case with its substrates are given in Section 5.3.2. 2-Carboxyarabinitol-1,5-P_2 is an analogue of the intermediate shown in Eqn. 10.4 (i.e. it binds to the enzyme but does not dissociate into two equal products). Since binding of the analogue prevents enzyme turnover, this very usefully reveals details of the binding sites.

RuP$_2$-Case supports an oxidative reaction with ribulose-1,5-P_2 (as well as the carboxylative reaction—Eqn. 10.4) involving cleavage to glycerate-3-P and glycollate-2-P.

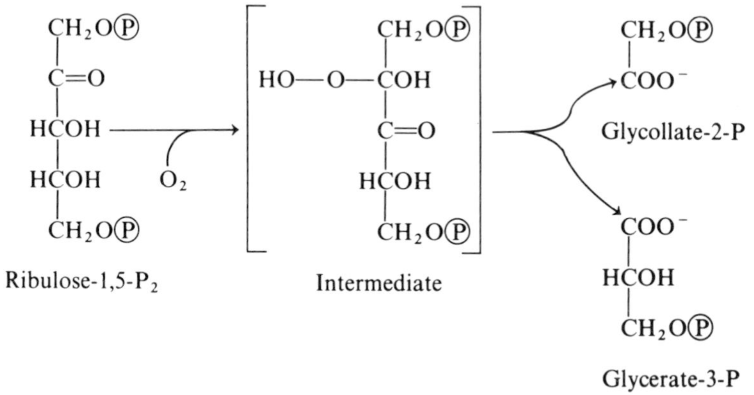

$$(10.5)$$

This is known as ribulose-1,5-P_2 oxygenase activity (RuP$_2$-Oase). There is a problem naming the enzyme since the one enzyme supports both RuP$_2$-Oase activity and RuP$_2$-Case activity. It is correctly described as ribulose-1,5-P_2 carboxylase/oxygenase (RuP$_2$-C/Oase) and is widely known by the acronym RuBisCO or Rubisco. On theoretical grounds the enzyme is a carboxylase since it has much greater affinity for CO_2 than for O_2 but since the concentration of O_2 in air (21%) is so much greater than that of CO_2 (0.03%), the activity with O_2 is of considerable practical importance.

Quantitatively, the rates of the oxygenase and carboxylase activities are determined by the concentrations of CO_2 and O_2 relative to the V_{max} and K_m

values for the two substrates. Under physiological conditions, oxygenase activity is about 20–30% of carboxylase activity. The metabolism of such a relatively large proportion of ribulose-1,5-P_2 via RuP_2-Oase activity results in a large amount of glycollate-2-P which is not an intermediate of the C_3-CR cycle (see Chapter 11).

10.3.2 Other enzymes

The phosphorylation of glycerate-3-P to glycerate-1,3-P_2 (Eqn. 10.2) in chloroplasts is catalysed by glycerate-3-P kinase which is also involved in glycolysis in the cytosol (see Section 7.3.2) but the glycolytic reaction proceeds towards glycerate-3-P ($\Delta G^{\circ'} = -18.8\,kJ\,mol^{-1}$) and that in chloroplasts runs contrary ($\Delta G^{\circ'} = +18.8\,kJ\,mol^{-1}$). Significant yields of glycerate-1,3-P_2 in chloroplasts are only possible if the concentrations of the substrates (ATP and glycerate-3-P) are high relative to the products. At the alkaline pH prevailing in the stroma of illuminated chloroplasts, glycerate-3-P is maintained as the trivalent anion which cannot pass through the chloroplast envelope, resulting in relatively high steady state concentrations in the stroma (3–5 mM). Conversely, the concentration of glycerate-1,3-P_2 in chloroplasts is extremely low due to its removal in the ensuing reaction.

Chloroplast glycerate-3-P kinase is electrophoretically distinguishable from its cytosolic counterpart. The two activities are subject to independent genetic control and are expressed independently of each other, e.g. the cytosolic enzyme, but not the chloroplast enzyme, is expressed in etiolated tissue prior to greening.

The reduction of glycerate-1,3-P_2 to glyceraldehyde-3-P by NADPH (Eqn. 10.3) in chloroplasts is catalysed by NADP-specific glyceraldehyde-3-P dehydrogenase and accounts for the all-important step of carbon reduction. The NADP specificity distinguishes it from its NAD-specific glycolytic counterpart in the cytosol. The equilibrium for the reaction lies towards glyceraldehyde-3-P ($\Delta G^{\circ'} = -6.3\,kJ\,mol^{-1}$), promoting phosphorylation of glycerate-3-P (Eqn. 10.2), as does the very high activity of NADP glyceraldehyde-3-P dehydrogenase and its very high affinity for glycerate-1-3-P_2 ($K_m \simeq 1\,\mu M$) and NADPH ($K_m = 4\,\mu M$).

NADP glyceraldehyde-3-P dehydrogenase shows sigmoidal kinetics with respect to NADPH and activity is modified by ATP but this is not thought to be physiologically important due to its very high activity relative to other enzymes of the C_3-CR cycle.

The isomerization of glyceraldehyde-3-P to dihydroxyacetone-P (another triose-P) is catalysed by triose-P isomerase in a readily reversible reaction ($\Delta G^{\circ'} = -7.5\,kJ\,mol^{-1}$).

$$
\begin{array}{ccc}
\text{CHO} & & \text{CH}_2\text{OH} \\
| & & | \\
\text{HCOH} & \rightleftharpoons & \text{C}=\text{O} \\
| & & | \\
\text{CH}_2\text{O}\,\text{Ⓟ} & & \text{CH}_2\text{O}\,\text{Ⓟ}
\end{array} \tag{10.6}
$$

$$\text{Glyceraldehyde-3-P} \qquad\qquad \text{Dihydroxyacetone-P}$$

This is an important adjunct to the previous reactions since about 40% of the glyceraldehyde-3-P produced via Eqns 10.1–10.3 is isomerized. Dihydroxyacetone-P condenses with glyceraldehyde-3-P and erythrose-4-P to form C_6 and C_7 sugar bisphosphates.

10.3.3 Reductive metabolism of glycerate-3-P is light-coupled

Isolated intact chloroplasts readily reduce glycerate-3-P in the light with the evolution of $\frac{1}{2} O_2$ since the production of NADPH is linked to light-driven electron transport.

$$ \tag{10.7}$$

Since CO_2 reacts with ribulose-1,5-P_2 to form two molecules of glycerate-3-P (Eqn. 10.1), illuminated chloroplasts also assimilate CO_2 with the evolution of an equimolar amount of O_2. If production of glycerate-3-P ceases (e.g. in the absence of CO_2 or in the presence of an appropriate inhibitor), O_2 evolution ceases, demonstrating that light-dependent electron flow and oxidation of NADPH by glycerate-1,3-P_2 in vivo are interdependent.

10.4 Carbon dioxide in chloroplasts is in equilibrium with a pool of bicarbonate

Mesophyll cells must have mechanisms for the acquisition of CO_2 commensurate with the rate of CO_2 assimilation. At the alkaline pH of illuminated chloroplasts ($\sim$ pH 8), 98% of the inorganic carbon occurs as HCO_3^- and only 2% as CO_2, the substrate for RuP_2-Case. There appears to be no evidence for a HCO_3^- transport mechanism in the chloroplast envelope of terrestrial plants. Indeed, HCO_3^- causes shrinking of chloroplast suspensions (i.e. loss of water from the chloroplasts to the external solution) implying that the envelope is impermeable to this ion. Thus, inorganic carbon must enter chloroplasts as CO_2 and equilibrate rapidly with a large internal pool of HCO_3^-, catalysed by carbonic anhydrase (Fig. 10.1); there is rapid re-equilibration if the internal CO_2/HCO_3^- ratio is perturbed (e.g. by an enhanced rate of CO_2 assimilation or a decreased rate of CO_2 entry). Unicellular algae grown at low concentrations of inorganic carbon are a special case with an active HCO_3^- pump (see Section 11.10).

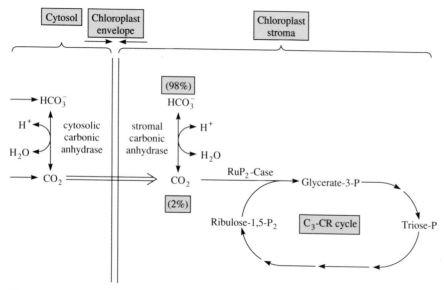

Fig. 10.1 Proposed scheme for the entry of inorganic carbon from the cytosol into chloroplasts of mesophyll cells in terrestrial C_3 plants. The chloroplast envelope is permeable to CO_2, but not to HCO_3^-. Abbreviation: RuP_2-Case = ribulose-1,5-P_2 carboxylase.

10.5 Ribulose-1,5-P_2 is regenerated from triose-P

Sustained assimilation of CO_2 into glycerate-3-P requires a constant supply of ribulose-1,5-P_2. Since five pre-existing carbon atoms are required to support the assimilation of one carbon as CO_2 (Eqn. 10.4), then five-sixths of the carbon of

glycerate-3-P must be recycled to maintain a steady state concentration of ribulose-1,5-P_2.

When plants are supplied with $^{14}CO_2$ for periods longer than 1–2 s, various C_4, C_5, C_6 and C_7 sugar phosphates (e.g. fructose-6-P, fructose-1,6-P_2, xylulose-5-P), including ribulose-1,5-P_2, become labelled. The ^{14}C-label is found predominantly in specific atoms. For example, carbon atoms C-3 and C-4 of fructose-1,6-P_2 become equally labelled but C-1, C-2, C-5 and C-6 have little label. This pattern is explained by the formation of fructose-1,6-P_2 from two molecules of triose-P labelled in C-1 only. Details of the pathway can be confirmed by monitoring the metabolism of intermediates labelled in specific carbon atoms and examining chloroplast enzymes which support the intermediate reactions.

The main features involved in regenerating ribulose-1,5-P_2 from glycerate (i.e. the C_3-CR cycle) are summarized in Fig. 10.2. It shows the fate of six CO_2 fed into a pre-existing pool of six ribulose-1,5-P_2 (C_5) producing 12 triose-P (C_3), 10 of

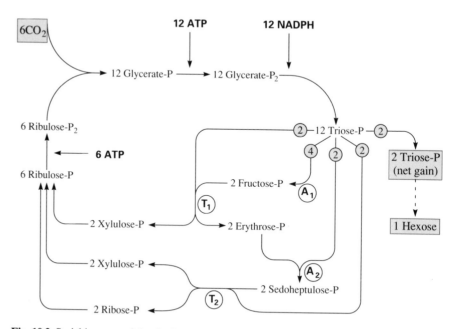

Fig. 10.2 Stoichiometry of the C_3-CR cycle. Two triose-P (or one hexose) are produced from six CO_2 with the expenditure of 18 ATP and 12 NADPH. Reactions A_1 and A_2 are catalysed by aldolase, and T_1 and T_2 are catalysed by transketolase.

which are used to regenerate the initial pool of six ribulose-1,5-P$_2$ and leaving a net gain of two triose-P. Of the 10 triose-P used for regeneration, four are used to make two fructose-6-P (C$_6$) by a sequence involving aldolase (reaction A$_1$). The two terminal carbon atoms of each fructose-6-P are transferred to each of a further two triose-P to form two pentose-P (xylulose-5-P) and two tetrose-P (reaction T$_1$), catalysed by transketolase. The two tetrose-P (erythrose-4-P) react with two more triose-P in a reaction sequence analogous to reaction A$_1$ except that two sugar phosphates containing seven carbon atoms (sedoheptulose-7-P) are formed (reaction A$_2$). Finally, the two terminal carbon atoms of sedoheptulose-7-P are transferred to the remaining two triose-P (reaction T$_2$), analogous to reaction T$_1$, catalysed by transketolase. This forms four further molecules of pentose-P (two ribose-5-P and two xylulose-5-P). The net result is that 10 triose-P are converted to six pentose-P (four xylulose-5-P and two ribose-5-P). These are rearranged internally to ribulose-5-P and phosphorylated by ATP to ribulose-1,5-P$_2$.

Aldolase and transketolase account for all the rearrangements of carbon atoms between the sugar phosphates in the regeneration of ribulose-1,5-P$_2$ from triose-P. Chloroplast aldolase is analogous to the glycolytic enzyme (see Section 7.3.1) but the glycolytic enzyme supports the cleavage of fructose-1,6-P$_2$ and illuminated chloroplasts catalyse the synthesis of fructose-1,6-P$_2$ from glyceraldehyde-3-P and dihydroxyacetone-P ($\Delta G^{\circ\prime} = -23\,\mathrm{kJ\,mol^{-1}}$). Importantly, the chloroplast enzyme also catalyses a reaction using erythrose-4-P in lieu of glyceraldehyde-3-P as the aldo-sugar, producing sedoheptulose-1,7-P$_2$, the higher homologue of fructose-1,6-P$_2$.

Transketolase catalyses the transfer of a C$_2$ fragment of a keto-sugar-phosphate to the C-1 carbon atom of an aldo-sugar-phosphate (see Table 8.1). In chloroplasts the most important transketolase reactions are the transfer of a C$_2$ fragment from fructose-6-P and sedoheptulose-7-P to glyceraldehyde-3-P.

$$\text{Fructose-6-P} + \text{glyceraldehyde-3-P} \rightleftharpoons \text{erythrose-4-P} + \text{xylulose-5-P} \qquad (10.9)$$

$$\text{Sedoheptulose-7-P} + \text{glyceraldehyde-3-P} \rightleftharpoons \text{ribulose-5-P} + \text{xylulose-5-P} \qquad (10.10)$$

Further details of the intermediates and enzymes of the C$_3$-CR cycle are shown in Fig. 10.3. The phosphatases, fructose-1,6-P$_2$ phosphatase (FP$_2$-Pase) (Eqn.

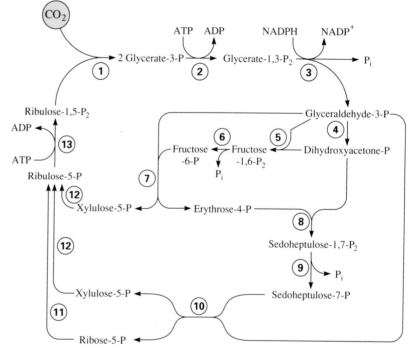

Fig. 10.3 Reactions and enzymes of the C$_3$-CR cycle. Enzymes are: (1) ribulose-1,5-P$_2$ carboxylase; (2) glycerate-3-P kinase; (3) NADP glyceraldehyde-3-P dehydrogenase; (4) triose-P isomerase; (5) aldolase; (6) fructose-1,6-P$_2$ phosphatase; (7) transketolase; (8) aldolase; (9) sedoheptulose-1,7-P$_2$ phosphatase; (10) transketolase; (11) ribose-5-P isomerase; (12) ribulose-5-P 3-epimerase; (13) ribulose-5-P kinase. The initial carboxylation reaction gives rise to two molecules of glycerate-3-P but the stoichiometries of the other reactions are not shown (see Fig. 10.2 for an example).

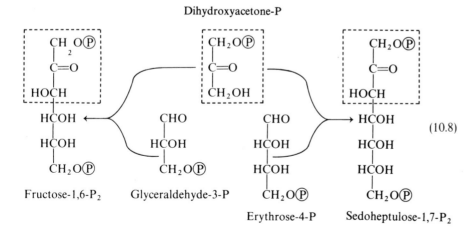

$$(10.8)$$

10.11) and sedoheptulose-1,7-P_2 phosphatase (SP_2-Pase) (Eqn. 10.12) have relatively low activity in chloroplasts.

$$\text{Fructose-1,6-}P_2 + H_2O \rightleftharpoons \text{fructose-6-P} + P_i \qquad (10.11)$$

$$\text{Sedoheptulose-1,7-}P_2 + H_2O \rightleftharpoons \text{sedoheptulose-7-P} + P_i \qquad (10.12)$$

The equilibrium for both reactions lies strongly to the right ($\Delta G^{\circ\prime} = -16.7\,\text{kJ mol}^{-1}$). Two separate and specific enzymes hydrolyse these substrates whereas a single enzyme, aldolase, supports the synthesis of the two bisphosphates (see Eqn. 10.8).

Ribose-5-P isomerase and ribulose-5-P 3-epimerase catalyse various internal rearrangements of pentose phosphates in chloroplasts. The former enzyme supports the isomerization of ribose-5-P to its corresponding keto isomer (Eqn. 10.13). The latter catalyses the epimerization of xylulose-5-P (Eqn. 10.14). Both reactions produce ribulose-5-P.

The final reaction of the pathway involves phosphorylation of ribulose-5-P by ATP to form ribulose-1,5-P_2 ($\Delta G^{\circ\prime} = -21.7\,\text{kJ mol}^{-1}$) catalysed by ribulose-5-P kinase.

$$\text{Ribulose-5-P} + \text{ATP} \rightleftharpoons \text{ribulose-1,5-}P_2 + \text{ADP} \qquad (10.15)$$

10.6 Energy inputs from ATP and NADPH satisfy the energy requirements for production of carbohydrate from carbon dioxide

Evaluation of the energy inputs required to assimilate CO_2 must include those needed to regenerate the initial pool of ribulose-1,5-P_2 as well as those directly involved in the net gain of triose-P. The energy required for net assimilation of six

CO_2 is shown in Fig. 10.2. The initial pool of six ribulose-1,5-P_2 and six CO_2 result in the production of 12 triose-P with the consumption of 12 ATP and 12 NADPH. A further six ATP are used to regenerate six ribulose-1,5-P_2 from 10 triose-P. Thus, the assimilation of six CO_2 results in the net production of two triose-P (or one hexose-P) with the expenditure of 12 NADPH and 18 ATP and the evolution of six O_2 with no net change in the ribulose-1,5-P_2 pool. The free energy input for 12 NADPH is $12 \times -220 = -2540\,\text{kJ}$, and for 18 ATP is $18 \times -30.5 = -549\,\text{kJ}$, totalling $-3189\,\text{kJ}$. Since the $\Delta G^{\circ\prime}$ for the synthesis of hexose from CO_2 and H_2O is $+2870\,\text{kJ mol}^{-1}$, the net $\Delta G^{\circ\prime}$ for the synthesis of hexose by the mechanism in Fig. 10.3 is $-319\,\text{kJ mol}^{-1}$ which ensures a significant amount of product at chemical equilibrium.

10.7 The chloroplast envelope is impermeable to many intermediates of the C_3-CR cycle

The assimilation of CO_2 and metabolism of carbon intermediates of the C_3-CR cycle in illuminated chloroplasts can be monitored by O_2 evolution (Eqn. 10.7). When CO_2 alone is supplied, the rate of O_2 evolution and CO_2 assimilation gradually increase (lag phase) before a steady rate is attained (see Fig. 10.4). The lag phase is due to a very small pool of intermediates of the C_3-CR cycle, especially ribulose-1,5-P_2, following isolation of the chloroplasts. When an experiment is initiated, the concentrations of the intermediates gradually rise with each turn of the cycle until steady state concentrations are achieved. The gradually increasing rates of O_2 evolution and CO_2 assimilation reflect the changes in the pools of intermediates (autocatalysis).

The lag phase can be shortened considerably by adding dihydroxyacetone-P (Fig. 10.4) which equilibrates rapidly with the other intermediates within the chloroplast. In theory, ribulose-1,5-P_2 or any of the intermediates of the C_3-CR cycle should also diminish the lag phase but, in practice, several compounds (e.g. ribulose-1,5-P_2 and fructose-1,6-P_2) are inactive while others are not very effective. However, all the intermediates decrease the lag phase in a 'reconstituted' chloroplast system. Isolated chloroplasts are disrupted by osmotic shock and the stroma and thylakoids are separated by centrifugation These are then recombined and supplemented with various soluble components (e.g. soluble ferredoxin and NADP), since their concentrations are diluted greatly (commonly about 100-fold) by lysis. Without any other additions, reconstituted systems do not support CO_2 incorporation. However, any one of the intermediates of the C_3-CR cycle (including ribulose-1,5-P_2 and fructose-1,6-P_2) immediately effect O_2 evolution

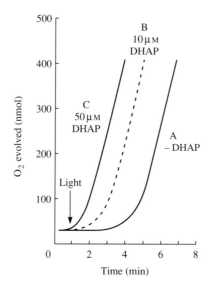

Fig. 10.4 Lag phase of CO_2-dependent O_2 evolution (A) by isolated spinach chloroplasts and the effect of very low concentrations of dihydroxyacetone-P (B = 10 µM; C = 50 µM) on the duration of the lag phase. CO_2 was supplied as 10 mM $NaHCO_3$ to all reaction mixtures. Abbreviation: DHAP = dihydroxyacetone-P. From Leegood, R.C. & Walker, D.A. (1980) *Arch. Biochem. Biophys.*, **200**, 575–82.

Table 10.1 Rates of metabolite transport across the envelope of isolated spinach chloroplasts. Different values are obtained under different experimental conditions and, therefore, cannot be strictly compared. After Edwards, G. & Walker, D.A. (1983). C_3, C_4: *Mechanisms, and Cellular and Environmental Regulation, of Photosynthesis.* Blackwell Scientific Publications, Oxford.

Metabolite	Conditions	Rate*(µmol mg^{-1} chlorophyll h^{-1})
P$_i$	4 °C, dark	57
Dihydroxyacetone-P	4 °C, dark	51
Glyceraldehyde-3-P	4 °C, dark	41
Glycerate-3-P	4 °C, dark	36
L-Aspartate	4 °C, dark	31
2-Oxoglutarate	4 °C, dark	26
L-Malate	4 °C, dark	19
L-Glutamate	4 °C, dark	8
† Glycerate	2 °C, light	13
Glycollate	0 °C, dark	30
D-Xylose	20 °C, dark	9.3
D-Mannose	20 °C, dark	8.4
L-Arabinose	20 °C, dark	7.7
D-Glucose	20 °C, dark	7.4
D-Ribose	20 °C, dark	6.0
‡ATP/ADP	20 °C, dark	<5
Ribose-5-P	20 °C, dark	1.7
PP$_i$	20 °C, dark	1.3
Fructose-6-P	20 °C, dark	0.3
Glucose-1-P	20 °C, dark	0.1
Ribulose,1,5-P$_2$	–	0
§ Fructose-1,6-P$_2$	–	0
Sedoheptulose-1,7-P$_2$	–	0
NADPH	–	0
Ferredoxin	–	0

* Rates of some rapidly transported compounds were determined at low temperature and, therefore, rates at 20°C would be considerably faster.
† Uptake of glycerate is reportedly light-dependent.
‡ The rate for chloroplasts from pea seedlings is considerably faster.
§ Definitive determinations for transport of fructose-1,6-P$_2$ are not possible since it is rapidly converted to triose-P by aldolase.

and CO_2 assimilation. It can, therefore, be concluded that the intermediates ineffective with intact chloroplasts do not freely cross the envelope. Direct measurements of the uptake of labelled intermediates support this conclusion.

CO_2 is assimilated at about 200 µmol mg^{-1} chlorophyll h^{-1} in C_3 plants (see Section 1.7). The very low rates of transport (Table 10.1) of many of the intermediates of the C_3-CR cycle are very important to maintain and control a pool of ribulose-1,5-P$_2$ within the chloroplast. The data show that only triose phosphates and glycerate-3-P (containing three carbon atoms) can transport carbon across the chloroplast envelope at rates approximating the carbon assimilation rate.

10.8 Triose phosphates are exported to the cytosol by the phosphate translocator

Isolated intact *chloroplasts* incorporate $^{14}CO_2$ into intermediates of the C_3-CR cycle but not into sucrose, even in long-term experiments. However, illuminated mesophyll *protoplasts* begin to accumulate ^{14}C-label in sucrose after about 2 min, and after 10 min or so it contains about 60% of the ^{14}C-label. Intact leaves also assimilate most of their carbon into sucrose, especially in the first hour or two after sunrise. All the ^{14}C-labelled sucrose and the enzymes involved in sucrose synthesis occur in the cytosol, not in chloroplasts. Since the incorporation of CO_2 into sucrose in intact cells is light-dependent, then CO_2 must be assimilated in chloro-

plasts by the C_3-CR cycle and a reduction product exported and used for sucrose synthesis in the cytosol. Since most of the newly assimilated CO_2 is rapidly exported as sucrose, reduced carbon must be transported out of chloroplasts at a rate approximating that of CO_2 assimilation. Of the compounds listed in Table 10.1, only the triose phosphates fulfil all the necessary criteria.

Transport of P_i, the two triose phosphates and glycerate-3-P across the chloroplast envelope is conducted by the phosphate translocator, a membrane protein which supports a strict 1:1 antiport exchange. For example, external P_i and external glycerate-3-P can exchange with an equimolar amount of internal triose-P. Since a single transport mechanism is involved, P_i, glycerate-3-P and the triose phosphates compete for sites on the translocator. Thus, depending on concentration, one metabolite can inhibit the transport of another in the same direction, although at pH 8 in the stroma of illuminated chloroplasts, glycerate-3-P exists principally as the trivalent anion which is not exported to any great extent. Several other phosphorylated C_3 metabolites are also transported on the phosphate translocator (e.g. phosphoenolpyruvate) but these are of minor importance *in vivo* in C_3 plants.

Since the C_3 skeleton of triose-P is used in the cytosol principally for sucrose synthesis, P_i, not glycerate-3-P, is usually exchanged. In view of the strict antiport exchange by the phosphate translocator, the availability of P_i in the cytosol controls the export of triose-P from chloroplasts. Very low concentrations of P_i promote CO_2 assimilation but external concentrations greater than 0.3 mM enhance export so much that regeneration of ribulose-1,5-P_2, and hence CO_2 assimilation, is inhibited. As sucrose synthesis from triose-P produces an equimolar amount of P_i (see next Section), sucrose synthesis is controlled by the concentration of P_i in the cytosol through its effect on the export of triose-P. This, in turn, affects the incorporation of triose-P into starch in chloroplasts (see Section 10.13.6).

10.9 Sucrose is synthesized via UDP-glucose in the cytosol

Sucrose is synthesized in the cytosol from four molecules of triose-P (Fig. 10.5). The initial steps involving the synthesis of two molecules of fructose-6-P are identical to those in the chloroplast (see Fig. 10.3) except that cytosolic forms of aldolase and FP_2-Pase are involved. One fructose-6-P is internally rearranged by hexose-P isomerase and glucose-P mutase to form glucose-1-P.

Glucose-1-P is prepared for joining to the other fructose-6-P by a reaction which expends UTP in one of a series of related reactions catalysed by nucleoside diphosphate sugar pyrophosphorylases. A sugar-1-phosphate (P–S) reacts with a nucleoside triphosphate (N–P–P–P). Pyrophosphate (P–P) is cleaved from the nucleotide and the resulting nucleoside monophosphate is joined to the phosphate of the sugar, thereby forming an internal pyrophosphate bond.

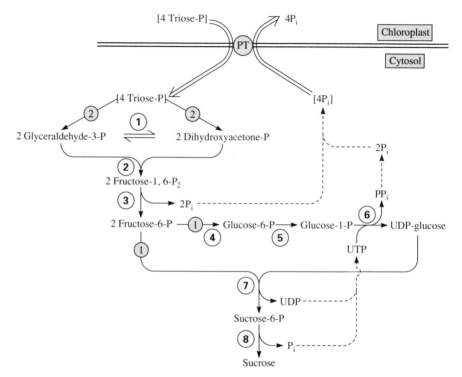

Fig. 10.5 Pathway for the synthesis of sucrose from triose-P in the cytosol of photosynthetic cells. The reaction sequence consumes four triose-P and one UTP with the net production of four P_i and one sucrose. Enzymes are: (1) triose-P isomerase; (2) aldolase; (3) fructose-1,6-P_2 phosphatase; (4) hexose-P isomerase; (5) glucose-P mutase; (6) UDP-glucose pyrophosphorylase; (7) sucrose-P synthetase; (8) sucrose-P phosphatase. Triose-P is imported into the cytosol from the chloroplast via the phosphate translocator (PT) in exchange for an equimolar amount of P_i.

$$N-\overset{\bullet}{P}-\overset{\circ}{P}-\overset{\triangle}{P} \; + \; \overset{\blacksquare}{P}-S \; \rightleftharpoons \; N-\overset{\bullet}{P}-\overset{\blacksquare}{P}-S \; + \; \overset{\circ}{P}-\overset{\triangle}{P} \qquad (10.16)$$

The resulting nucleoside diphosphate sugars are effectively 'activated' or 'high energy' forms of the sugar (see section 4.3).

Sucrose synthesis involves the cytosolic enzyme UDP-glucose pyrophosphorylase.

$$\text{Glucose-1-P} + \text{UTP} \rightleftharpoons \text{UDP-glucose} + PP_i \qquad (10.17)$$

The specificity of this enzyme for UTP distinguishes sucrose synthesis in the cytosol from the activation of glucose-1-P by ATP for starch synthesis in chloroplasts (see Section 10.11).

Sucrose, a disaccharide, is formed by reaction of fructose-6-P with UDP-glucose, catalysed by sucrose-P synthetase.

$$\text{UDP-glucose} + \text{fructose-6-P} \rightleftharpoons \text{sucrose-6-P} + \text{UDP} \qquad (10.18)$$

Fructose-6-P condenses via the hydroxy group at C-2 to C-1 of the glucosyl moiety of UDP-glucose, eliminating UDP but retaining the phosphate group at C-6 on the fructosyl residue. The equilibrium lies strongly towards sucrose-P ($\Delta G^{\circ\prime} = -20\,\text{kJ mol}^{-1}$). The enzyme exhibits sigmoidal kinetics with respect to fructose-6-P and UDP-glucose making it especially sensitive to changes in substrate concentration. The production of free sucrose is catalysed by sucrose-P phosphatase (Fig. 10.5).

Sucrose can also be synthesized directly in a reaction between UDP-glucose and fructose, catalysed by sucrose synthase.

$$\text{UDP-glucose} + \text{fructose} \rightleftharpoons \text{sucrose} + \text{UDP} \qquad (10.19)$$

However, the general view is that this enzyme is concerned with catabolism rather than with sucrose synthesis. The reaction shown in Eqn. 10.19 is freely reversible ($\Delta G^{\circ\prime}$ estimated at $4.2\,\text{kJ mol}^{-1}$), whereas the reaction sequence formed by the last two reactions in Fig. 10.5 is exergonic. Further, when illuminated photosynthetic cells are supplied with $^{14}\text{CO}_2$, the fructosyl and glucosyl moieties of sucrose become equally labelled showing that they are derived from a common source (i.e. fructose-6-P—Fig. 10.5). Also, sucrose synthase is more active in sugar storage tissue than in photosynthetic tissue.

10.10 Sucrose is exported from photosynthetic cells and loaded into the phloem

Most of the sucrose synthesized in the cytosol of photosynthesizing cells is exported to other tissues. Indeed, significant retention of carbon in these cells occurs only when the demands of other tissues for sucrose have been fulfilled (see Fig. 1.1).

The export of sucrose from photosynthetic cells involves phloem loading, in which sucrose is moved from the mesophyll cells (commonly with 10–50 mM sucrose) against a concentration gradient into phloem sieve tubes and companion cells containing up to 800 mM sucrose. Similar movement of sucrose also occurs from the cytosol across the tonoplast into the vacuoles of certain cells, especially

the sugar storage cells in stems of sugar cane (*Saccharum* spp.) and the swollen hypocotyls of various beets (*Beta vulgaris* cultivars).

Experiments on the uptake of [^{14}C]sucrose, labelled in only the glucosyl or fructosyl half of the molecule, have shown that sucrose is not readily permeable to most membranes and is not modified during uptake, thus ruling out hypotheses that it is degraded to its component monosaccharides and subsequently resynthesized. The current hypothesis, known as the sucrose/proton co-transport model, proposes that the plasma membrane of the sieve tube or an associated phloem cell contains an ATP-dependent proton pump which pumps protons out of the cell; the free energy of the resulting electrochemical gradient actively transports sucrose into the phloem. Supporting this, the uptake of [^{14}C]sucrose by perfused vascular tissue of corn leaves increases the pH of the perfused solution (i.e. uptake of protons as well as sugar into the tissue). This pH response is specific for sucrose. Also, sucrose transport is decreased if the transmembrane proton gradient is lowered by raising the external pH, and increased if the external pH is decreased. However, complete agreement on the co-transport mechanism has not been reached since the observed rates of sucrose transport in experimental systems are too slow to explain the rates of phloem loading *in vivo*.

10.11 Starch is synthesized in chloroplasts via ADP-glucose

Most triose-P is exported to the cytosol for sucrose synthesis but some is used for starch synthesis within the chloroplast. The relative amounts channelled into sucrose and starch vary greatly. Some species (e.g. tobacco) make a massive quantity of starch in their leaves while others make none. Normally, little triose-P is directed into starch for the first few hours after sunrise but the proportion increases throughout the day as sucrose synthesis decreases (see Fig. 1.1). Plants deficient in certain nutrients accumulate starch, presumably due to impaired sucrose synthesis in the cytosol or decreased demand for carbohydrate in consumer cells. Starch accumulation in leaves is a secondary route of triose-P metabolism, generally only significant when the rate of CO_2 assimilation exceeds the production and export of sucrose.

The pathway of starch synthesis from triose-P in chloroplasts (Fig. 10.6) has much in common with sucrose synthesis in the cytosol. The initial steps involve the production of glucose-1-P via fructose-1,6-P$_2$ and fructose-6-P as for sucrose synthesis except that a chloroplast complement of enzymes is involved. The important distinction is that glucose-1-P is activated by ATP (not UTP) in the presence of ADP-glucose pyrophosphorylase, a specific chloroplast enzyme.

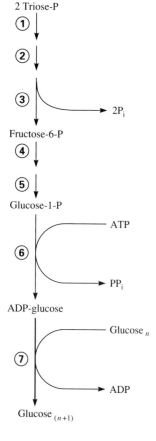

Free C-4 end   Free C-1 end

G—G—G—G—G—G—G—G—G—G—G—G—G—G   (G$_n$)

ADP—G$^\bullet$

ADP

G$^\bullet$—G—G—G—G—G—G—G—G—G—G—G—G—G—G   (G$_{n+1}$)

(10.21)

Fig. 10.6 Pathway for the incorporation of triose phosphate into starch in chloroplasts. Triose-P is metabolized to glucose-1-P as described for sucrose synthesis and incorporated into starch via ADP-glucose. Starch is denoted as glucose$_n$. Enzymes are (1)–(5) see Fig. 10.5; (6) ADP-glucose pyrophosphorylase; (7) ADP-glucose starch transglucosylase.

The equilibrium lies strongly towards chain elongation ($\Delta G^{\circ\prime} = -13.8\,\text{kJ mol}^{-1}$). This is repeated many times to make a polymer. The enzyme supports reactions between ADP-glucose and a primer of two or more glucose residues in $\alpha(1\rightarrow4)$ linkage but does not catalyse a reaction with glucose.

The synthesis of amylopectin proceeds as for amylose (Fig. 10.6) except $\alpha(1\rightarrow6)$ branch points are formed. Amylo(1,4$\rightarrow$1,6) transglucosylase ('branching' enzyme) rearranges specific $\alpha(1\rightarrow4)$ glucosyl linkages to $\alpha(1\rightarrow6)$ linkages. The number of $\alpha(1\rightarrow4)$ linkages between each $\alpha(1\rightarrow6)$ branch point tends to be characteristic for various species but the factors determining the sites of transglucosylase action are not understood.

An alternative mechanism for the synthesis of starch involves the reaction catalysed by starch phosphorylase. It successively attaches the glucosyl moieties of glucose-1-P (G$^\bullet$1P), eliminating P$_i$ from glucose-1-P and attaching the glucosyl residue from its C-1 hydroxy group to the C-4 hydroxy group of the terminal glucosyl residue of a primer of $\alpha(1\rightarrow4)$glucose (G$_n$).

$$G^\bullet 1P + G_n \rightleftharpoons G^\bullet{-}G_n + P_i \qquad (10.22)$$

This reaction is freely reversible ($\Delta G^{\circ\prime} = -2.1\,\text{kJ mol}^{-1}$) but the extremely high ratio of P$_i$ to glucose *in vivo* means the enzyme must catalyse starch degradation rather than starch synthesis.

$$ATP + \text{glucose-1-P} \rightleftharpoons \text{ADP-glucose} + PP_i \qquad (10.20)$$

The final step in the synthesis of the amylose of starch involves successive additions of glucosyl residues (G) from ADP-glucose (ADP-G$^\bullet$) to the terminal glucosyl residue of a pre-existing oligomer (primer) of glucose (G$_n$) in $\alpha(1\rightarrow4)$ linkage, catalysed by ADP-glucose starch transglucosylase. ADP-glucose is cleaved at the bond linking C-1 of glucose to ADP and the resulting glucosyl residue is attached from C-1 to the free C-4 hydroxy group of the pre-existing glucose oligomer, lengthening it by one glucose residue and eliminating ADP.

10.12 Starch is metabolized to sucrose when demand for carbon in the cytosol exceeds the rate of carbon dioxide assimilation

Carbon incorporated into starch in chloroplasts is not normally retained for an extended period. When export of triose-P to the cytosol cannot be sustained directly by CO_2 assimilation (e.g. at night), starch is mobilized and exported (see Fig. 1.1). In general, the production of triose-P from starch is promoted by conditions which decrease the ATP/ADP ratio, usually associated with a low level of triose-P and an elevated concentration of P_i.

Chloroplasts have a unique complement of enzymes to degrade starch. They lack the Ca^{2+}-dependent α-amylase and β-amylase of other compartments and cells but have a Ca^{2+}-independent endoamylase which cleaves $\alpha(1\rightarrow4)$ linkages between $\alpha(1\rightarrow6)$ branch points in amylopectin. They also contain an enzyme which attacks $\alpha(1\rightarrow6)$ linkages to yield relatively short chains of $\alpha(1\rightarrow4)$glucose and a very active starch phosphorylase which, at high concentrations of P_i, produces glucose-1-P from these shortened chains of $\alpha(1\rightarrow4)$glucose (G_n).

$$G_n + P_i \rightleftharpoons G_{n-1} + \text{glucose-1-P} \qquad (10.23)$$

Glucose-6-P is metabolized to triose-P (Fig. 10.7) and exported to the cytosol, presumably via the phosphate translocator although carbon could be exported as free hexose since the rate of carbon mobilization from starch at night is similar to that of hexose transport.

10.13 Activities of the C₃-CR cycle and starch and sucrose synthesis are regulated

The light-dependent production of ATP and NADPH, the reductive assimilation of CO_2, the regeneration of ribulose-1,5-P_2 and starch and sucrose synthesis are interlinked and interdependent. *In vivo*, these processes must be co-ordinated. For example, triose-P can be used for starch and/or sucrose but five-sixths of it must be recycled to ribulose-1,5-P_2 to maintain production. Control systems must exist to prevent futile cycles, e.g. hydrolysis of ATP with no net change in the concentration of fructose phosphates through phosphofructokinase (PF-Kase) and FP₂-Pase (see Section 5.5.1). Some regulatory mechanisms involving carbon metabolism in photosynthetic cells are discussed below.

10.13.1 Some enzymes undergo modulation of activity mediated by thioredoxin when leaf tissue is illuminated and darkened

The *in vivo* activity of several chloroplast enzymes, including some not involved in CO_2 assimilation, is influenced by light. In some instances this is attributable to light-induced alterations in the cellular environment, such as changes in the concentrations of certain ions (e.g. Mg^{2+}, H^+) and metabolites. Enzymes which respond in this way have the same activity when extracted from light- or dark-treated tissue and assayed in standard incubations *in vitro*. However, the activity of some enzymes (i.e. light-modulated enzymes) changes depending on the illumination conditions prior to extraction (Table 10.2). Most light-modulated enzymes are activated by light but at least one, glucose-6-P dehydrogenase, is inactivated.

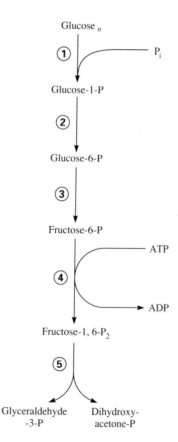

Fig. 10.7 Pathway for the mobilization of starch in chloroplasts to the triose phosphates, glyceraldehyde-3-P and dihydroxyacetone-P. Starch is shown as glucose$_n$. Enzymes are: (1) starch phosphorylase; (2) glucose-P mutase; (3) hexose-P isomerase; (4) phosphofructokinase; (5) aldolase.

Table 10.2 Some light-modulated enzymes. All the enzymes listed are light activated except glucose-6-P dehydrogenase which is light inactivated. For most light-activated enzymes the activity of light-treated tissue is about twice the activity of dark-treated tissue, when measured *in vitro* under optimum conditions. However, activation of fructose-1,6-P_2 phosphatase is considerably greater (about five fold) and NADP malate dehydrogenase is essentially inactive in dark-treated tissue. Larger differences in activity occur *in vivo* due to changes in the concentration of other parameters (e.g. pH, Mg^{2+} and various metabolites).

Enzyme	Pathway
Fructose-1,6-P_2 phosphatase	C_3-CR cycle
Sedoheptulose-1,7-P_2 phosphatase	C_3-CR cycle
NADP glyceraldehyde-3-P dehydrogenase	C_3-CR cycle
Ribulose-5-P kinase	C_3-CR cycle
NADP malate dehydrogenase	C_3-CR cycle
Glutamine synthetase	C_5-ammonia assimilation cycle
Phenylalanine ammonia lyase	Phenylpropanoid metabolism
Adenosine phosphosulphate sulphotransferase	Sulphate assimilation
Glucose-6-P dehydrogenase	OPP pathway

The activating effect on light-modulated enzymes can be simulated *in vitro* by adding various compounds (e.g. dithiothreitol) to extracts of dark-treated tissue to reduce disulphide bonds, suggesting that modulation involves oxidation/reduction of thiol groups within the enzyme with one form catalytically less active than the other form (see Section 5.5.4). With the exception of NADP-specific malate dehydrogenase, the light-modulated changes are quantitative rather than qualitative. Enzymes which are more active in the light (i.e. 'light-activated') are more active in the reduced form; light-inactivated enzymes are more active when oxidized. There is general agreement that mechanisms of the type shown in Fig. 10.8 account for light-modulation *in vivo*. The scheme proposes that Fd_{red}, formed by light-dependent electron transport, reduces a set of thioredoxins (Td), low molecular mass proteins which reduce the enzyme, modulating its activity. Reduction of Td by Fd_{red} involves Fd:Td reductase, a stromal enzyme. At least two Td molecules occur in chloroplasts, one more effective in activating NADP malate dehydrogenase than the other light-activated enzymes. The mechanism for oxidation of Td, and thus the light-modulated enzymes, is not understood.

Several enzymes of the C_3-CR cycle are light-modulated (Table 10.2). For

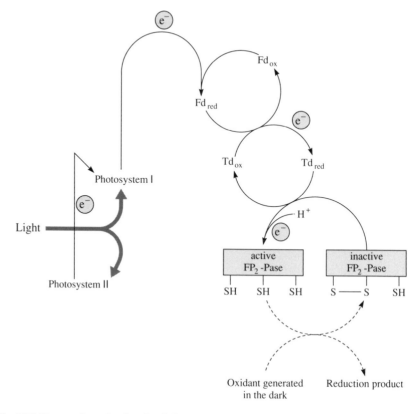

Fig. 10.8 Proposed mechanism for light activation of the light-modulated enzyme fructose-1,6-P_2 phosphatase (FP_2-Pase). The inactive form is thought to be formed by oxidation of thiol groups on the active form by an oxidant formed in the dark (dashed arrows). Abbreviations: Fd_{red} and Fd_{ox} = reduced and oxidized forms of ferredoxin; Td_{red} and Td_{ox} = reduced and oxidized forms of thioredoxin.

NADP glyceraldehyde-3-P dehydrogenase, one of the most active enzymes, the difference between the active (light) and less active (dark) form is quite small. The physiological significance of this is not clear. Ribulose-5-P kinase also shows quite small changes in extractable activity upon illumination of dark-treated plants. The activity in dark-treated tissue is still quite high but *in vivo* activity might be only 4% of this due to inhibitory effects of light-induced changes in pH and various metabolites (especially gluconate-6-P). Thus ribulose-5-P kinase is effectively shut down in darkened tissue. FP_2-Pase is one of the least active enzymes in the C_3-CR cycle and its activity in the dark is quite low so that light-modulation has a direct effect on carbon flow through the C_3-CR cycle. Dark modulation of FP_2-Pase decreases the probability of fructose-1,6-P_2 hydrolysis (and hence a futile cycle) during the production of triose-P from starch at night.

10.13.2 Light-mediated ion movements affect the activity of some enzymes of the C_3-CR cycle

Light-coupled electron transport promotes the movement of protons from the stroma into the intra-thylakoid space, decreasing the pH by about 3.5 units whilst increasing the stroma from about pH 7 to pH 8 (see Section 9.10). This promotes the movement of counter ions (most notably Mg^{2+}) from the intra-thylakoid space into the stroma. Upon illumination increased pH and higher Mg^{2+} concentrations in the stroma enhance the *in vivo* activities of several chloroplast enzymes, especially FP_2-Pase.

RuP₂-Case exists in active form (high affinity) and in inactive form (low affinity). Activation of the inactive form *in vitro* involves binding of Mg^{2+} and carbamylation by CO_2 of a specific lysine residue near the active site. However, the significance *in vivo* is unclear since changes in Mg^{2+} concentrations in the stroma of illuminated chloroplasts do not support substantial interconversion of the high and low affinity forms. Ribulose-1,5-P_2 inhibits activation. The significance of other observations concerning possible regulation *in vivo* is also unclear. Such observations include: the presence of an ATP-requiring activase to catalyse carbamylation of the lysine residue in the presence of ribulose-1,5-P_2 (absent in *Arabidopsis* mutants requiring high CO_2); formation of the inhibitor 2-carboxy-arabinitol-1-P (not to be confused with 2-carboxyarabinitol-1,5-P_2 discussed in Section 10.3.1) in the dark in some species; and reactivation of the enzyme by the activase in the light in the presence of this inhibitor.

10.13.3 Levels of some metabolites alter in response to light

Light promotes phosphorylation of ADP, reduction of $NADP^+$ and formation of new steady state pools of various intermediates of the C_3-CR cycle (e.g. increased concentration of ribulose-1,5-P_2). Light also indirectly affects the concentration of P_i. As discussed below, changes in the concentrations of these and other meta-

bolites are important in regulating the pathways of starch synthesis, starch degradation and the export of triose-P to the cytosol for sucrose synthesis.

10.13.4 Certain enzymes of the C_3-CR cycle and starch metabolism are regulated by changes in concentration of specific metabolites

The activities of several enzymes of carbon metabolism are especially sensitive to changes in the concentration of specific chloroplast metabolites, so controlling the flow of carbon. ADP-glucose pyrophosphorylase (Eqn. 10.20), the rate-limiting enzyme of starch synthesis, is strongly promoted by glycerate-3-P (positive effector) and inhibited by P_i (negative effector) although neither metabolite participates in the relevant reaction (Fig. 10.9). Glycerate-3-P and P_i attach to separate regulatory binding sites on the enzyme, causing modification of the catalytic binding site for ATP and, to a lesser extent, glucose-1-P. Clearly, factors affecting the concentration of P_i and glycerate-3-P in chloroplasts also affect the rate of starch synthesis; factors include concentration of P_i, in turn influenced by photophosphorylation of

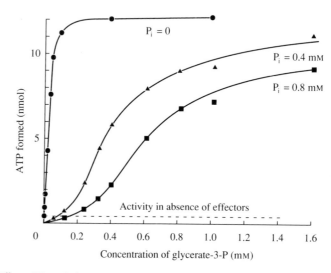

Fig. 10.9 Effect of P_i and glycerate-3-P on the activity of the regulatory enzyme ADP-glucose pyrophosphorylase from maize leaf tissue. In the experiment shown here, enzyme activity was determined by ATP synthesis (i.e. Eqn. 10.20 operating in the reverse direction). From Ghosh, H.P. & Preiss, J. (1966) *J. Biol. Chem.*, **241**, 4491–504.

ADP (P_i consumption), export of triose-P (P_i uptake), production of P_i in the cytosol and reduction of glycerate-1,3-P_2 (Eqn. 10.3), to name but a few.

10.13.5 *Fructose-2,6-P_2 controls carbon metabolism in the cytosol of photosynthetic cells*

FP$_2$-Pase and sucrose-P synthetase in the cytosol are important regulatory enzymes of sucrose synthesis. Increasing evidence shows that fructose-2,6-P_2 (not

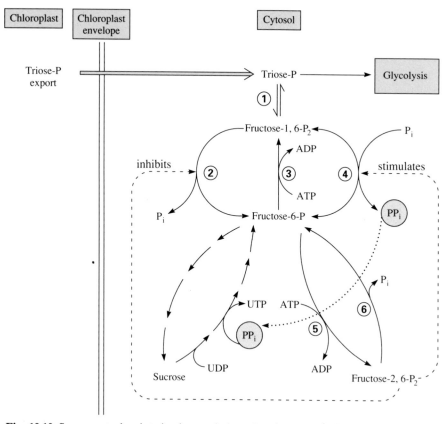

Fig. 10.10 Some control points in the regulation of carbon metabolism in the cytosol of photosynthetic cells. Enzymes are: (1) aldolase; (2) fructose-1,6-P_2 phosphatase; (3) phosphofructokinase; (4) fructose-6-P:PP$_i$ phosphotransferase; (5) fructose-6-P 2-kinase; (6) fructose-2,6-P_2 phosphatase.

to be confused with fructose-1,6-P_2) is a potent inhibitor of FP$_2$-Pase (Fig. 10.10). It is synthesized from fructose-6-P by fructose-6-P 2-kinase, a cytosolic enzyme.

$$\text{Fructose-6-P} + \text{ATP} \rightleftharpoons \text{fructose-2,6-}P_2 + \text{ADP} \qquad (10.24)$$

The product can be hydrolysed to fructose-6-P by a specific cytosolic enzyme, fructose-2,6-P_2 phosphatase.

$$\text{Fructose-2,6-}P_2 + \text{H}_2\text{O} \rightleftharpoons \text{fructose-6-P} + P_i \qquad (10.25)$$

The concentration of fructose-2,6-P_2 in the cytosol of photosynthetic tissue is inversely correlated with the rate of sucrose synthesis, suggesting that it is also a powerful negative effector of sucrose synthesis *in vivo*. The kinase is inhibited by physiological concentrations of triose-P and glycerate-3-P but is stimulated by P_i and fructose-6-P, whereas the phosphatase is stimulated by triose-P. Thus, an increase in P_i and/or fructose-6-P in the cytosol raises the concentration of fructose-2,6-P_2 and inhibits sucrose synthesis (Fig. 10.10). Conversely, an increase in triose-P in the cytosol promotes degradation of fructose-2,6-P_2 and, thus, synthesis of sucrose. Importantly, the concentration of fructose-2,6-P_2 responds to small and subtle changes in P_i, fructose-6-P and triose-P levels in the cytosol, permitting amplification of the 'signals' given. Control over cytosolic FP$_2$-Pase (and hence production of P_i) provides control over the export of triose-P from chloroplasts since triose-P is exchanged for P_i.

Sucrose-P synthetase is also regulated but little is known about the control of its activity *in vivo*.

The cytosol of photosynthetic cells contains an enzyme complement which supports glycolysis as well as sucrose synthesis. Thus, imported triose-P and certain intermediates of sucrose synthesis could be oxidized via glycolysis (Fig. 10.10). So, how is the futile flow of carbon and wastage of energy controlled? During sucrose synthesis, triose-P and fructose-1,6-P_2 must be diverted into fructose-6-P and rephosphorylation of fructose-6-P to fructose-1,6-P_2 via fructose-6-P:PP$_i$ phosphotransferase must be inhibited. The latter enzyme, like FP$_2$-Pase, is strongly influenced by fructose-2,6-P_2 but is stimulated rather than inhibited. Thus high concentrations of P_i and fructose-6-P in the cytosol promote production of fructose-2,6-P_2 and enhance the activity of fructose-6-P:PP$_i$ phosphotransferase. However, the ensuing consequences are not clear. The most popular postulate is that phosphotransferase acts as a source of PP$_i$ to support sucrose mobilization for the production of glucose-1-P from UDP-glucose, formed from sucrose by sucrose synthase activity (see Fig. 7.3 for details).

10.13.6 Concentration of P_i co-ordinates control of carbon metabolism in photosynthetic cells

The major routes of triose-P metabolism and the important regulatory steps are shown in Fig. 10.11. Sustained synthesis of sucrose is associated with high

FP$_2$-Pase activity in the cytosol and the production of P$_i$. Various regulatory enzymes in the chloroplast respond to the resulting increase in P$_i$. Sedoheptulose-1,7-P$_2$ phosphatase is stimulated, enhancing production of pentose phosphates and hence CO$_2$ assimilation. P$_i$ from the cytosol inhibits ADP-glucose pyrophosphorylase, shutting down the flow of triose-P into starch, and preventing loss of

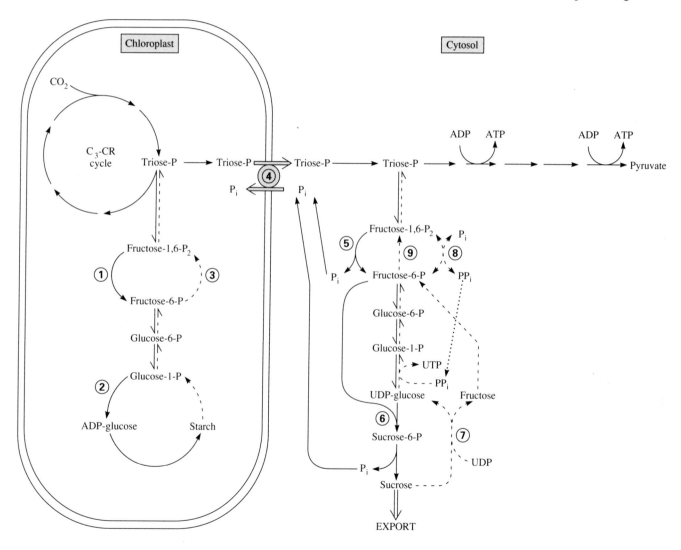

Fig. 10.11 Summary of the main routes of carbon flow in a mesophyll cell of a C$_3$ plant and the important control points for regulation of carbon flow. Enzymes and control points are: (1) Chloroplast fructose-1,6-P$_2$ phosphatase; (2) ADP-glucose pyrophosphorylase; (3) chloroplast phosphofructokinase; (4) phosphate translocator; (5) cytosolic fructose-1,6-P$_2$ phosphatase; (6) sucrose-P synthetase; (7) sucrose synthase; (8) fructose-6-P:PP$_i$ phosphotransferase; (9) cytosolic phosphofructokinase.

carbon from the C_3-CR cycle. Conversely, if production of P_i in the cytosol declines (e.g. as a result of decreased sucrose synthesis), then counter-exchange of cytosolic P_i with triose-P decreases, resulting in decreased availability of ATP within the chloroplast and diminished rates of glycerate-3-P metabolism. The elevated concentration of glycerate-3-P and decreased concentration of P_i stimulate ADP-glucose pyrophosphorylase (Fig. 10.11), diverting triose-P into starch. However, CO_2 assimilation is diminished since the metabolism of glycerate-3-P is limited by the availability of ATP. Thus, factors which affect the concentration of P_i in the cytosol have a profound effect on carbon metabolism in the chloroplast.

The flow of carbon from starch to triose-P and sucrose in the dark also involves the co-ordinated control of several processes. The factors initiating starch degradation in the chloroplast at night are not understood but the production of fructose-1,6-P_2 involves enhanced activity of phosphofructokinase. Conversely, the activity of chloroplast FP_2-Pase is diminished in the dark due to decreases in the pH and Mg^{2+} concentration in the stroma and dark-inactivation. Without these controls fructose-6-P would engage in a futile cycle with net hydrolysis of ATP.

10.14 Aspects of carbon metabolism can be monitored by chlorophyll fluorescence

10.14.1 Chlorophyll fluorescence is quenched by processes associated with carbon dioxide assimilation

Chlorophyll fluorescence increases when the excitation energy of chlorophyll associated with photosystem II (PSII) cannot be passed on to do useful biochemical work (see Section 9.15). Thus, constraints on CO_2 fixation stimulate fluorescence and vice versa (fluorescence quenching).

The two major sources of fluorescence quenching in plants are photochemical and non-photochemical. Photochemical quenching occurs when absorbed light effects a photochemical event, i.e. photochemical oxidation of $P_{680}(q_Q)$ causing reduction of the oxidized form of Q_A, the first stable acceptor of electrons from PSII. Normally, electrons are passed on from reduced Q_A to NADP and eventually intermediates of the C_3-CR cycle, and little energy is lost as fluorescence since the electrons are constantly drained, which ensures that a proportion of Q_A is oxidized. If Q_A is fully reduced, however, electrons cannot be accepted from P_{680} and fluorescence increases (or to put it another way, quenching is 'relaxed'). Other distinguishable quenching mechanisms include quenching by oxidized plastoquinone (q_p), damage to reaction centres from excess light absorption (q_I) and that

associated with transitions in the state of the reaction centres (q_T, see Section 9.16).

The major non-photochemical quenching of chlorophyll fluorescence results from formation of a proton gradient across the thylakoid membrane between the intra-thylakoid space and the stroma which changes the structure of the thylakoids. Then, a proportion of the absorbed light energy is lost as heat rather than fluorescence, causing fluorescence quenching (q_E). Importantly, some of the energy from photochemical quenching q_Q is expended to establish q_E.

10.14.2 The slow phase of fluorescence induction curves is associated with changes in q_E and q_Q quenching

Re-illumination of a leaf at low light intensity following a period of prolonged darkness results in autocatalysis (see Section 10.7) and fluorescence changes (Section 9.15.2). The initial 'fast' changes are concerned with primary photochemical events, but the slower phase represented by the SMT transient (see Fig. 9.14) concurs with the autocatalytic phase and the onset of steady state carbon assimilation. Immediately after illumination Q_A is rapidly reduced but the pool of C_3-CR cycle intermediates is very low and there is little demand for NADPH, so Q_A is not immediately reoxidized. Thus, immediately following this very rapid reduction, q_Q quenching is negligible and fluorescence rises to a peak, P (Fig. 10.12). Next, a proton gradient begins to form and fluorescence declines to a low level, (see Fig. 10.12, dashed line) due to q_Q quenching and, as the proton gradient is established, q_E quenching. At low light intensity the proton gradient (and hence q_E quenching) becomes fully established before steady state CO_2 assimilation (and its demand for ATP), so there is little tendency for the gradient to discharge and absorbed light energy is discharged as heat (i.e. q_E quenching). However, as the rate of CO_2 assimilation accelerates, the proton gradient partially discharges, q_E temporarily decreases and fluorescence increases to a peak, M. The decrease in the proton gradient accompanying the consumption of ATP and reducing equivalents stimulates electron transport and re-establishes the proton gradient and q_E quenching. This further decreases fluorescence which eventually reaches a low steady state (T) level. At high light intensities the M peak is less evident than in Fig. 10.12 due to the shorter autocatalytic phase.

The fluorescence of a leaf maintained in darkness for a short period (1-2 min) and then re-illuminated at low irradiance is shown by the solid line in Fig. 10.12. Fluorescence rises to a peak (P) and then declines. Thereafter Q_A reoxidizes, q_Q and

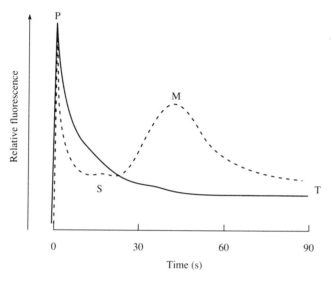

Fig. 10.12 Schematic representation of the 'slow' transients in chlorophyll fluorescence following re-illumination of dark-preincubated leaves at low light intensity. Leaves were preincubated in the dark for an extended period (dashed line) or for a short period of approximately 2 min (solid line). See text for interpretation.

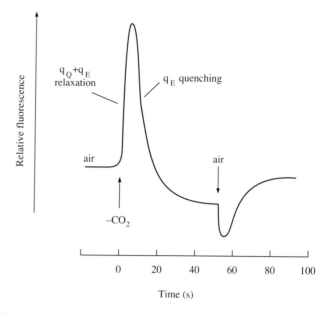

Fig. 10.13 Transients in chlorophyll fluorescence during perturbation of steady state CO_2 assimilation. See text for explanation of transients in terms of q_E and q_Q quenching which accompany changes in carbon metabolism in the C_3-CR cycle.

q_E quenching increase and fluorescence drops to the low steady terminal value (T). There is no autocatalysis and no transitory relaxation of q_E (M) due to the large pool of intermediates of the C_3-CR cycle.

10.14.3 Changes in fluorescence following perturbation of steady state conditions reflect changes in carbon metabolism

Under steady state conditions CO_2 assimilation is constant and chlorophyll *a* fluorescence remains at a relatively low steady value. However, if CO_2 assimilation is perturbed, the fluorescence signal also changes. For example, when a leaf is transferred from air to CO_2-free air fluorescence rises rapidly before decreasing equally rapidly to a lower level (Fig. 10.13). Restoration of the normal air supply causes a further drop in fluorescence, followed by a rise to the original value. The fluorescence changes reflect adjustments within the C_3-CR cycle. As the leaf is moved from air to CO_2-free air, carboxylation of ribulose-1,5-P_2 ceases. Conse-

quently, ATP consumption and oxidation of NADPH by the C_3-CR cycle stops, oxidation of the reduced form of Q_A stops, q_Q quenching is relaxed and the fluorescence rises. ATP consumption continues for a short time after carboxylation ceases as long as glycerate-3-P and ribulose-5-P are available. This discharges the proton gradient, resulting in short-term loss of q_E quenching which compounds the rise in fluorescence due to relaxation of q_Q quenching. Following cessation of ATP and NADPH consumption and the resultant high level of fluorescence, q_E quenching increases as a proton gradient is established in association with a switch to increased cyclic or pseudocyclic electron transport.

When the gas phase surrounding a leaf is changed from CO_2-free air back to normal air the fluorescence changes are reversed. After a brief decrease in fluorescence (as non-cyclic electron transport is re-introduced) electron transport decreases, resulting in a higher steady state fluorescence as the proton gradient decreases (hence less q_E quenching) following the onset of CO_2 assimilation.

10.14.4 Oscillations in fluorescence and oxygen evolution reflect mechanisms regulating carbon dioxide assimilation

Perturbation of steady state conditions for the assimilation of CO_2 causes oscillations in the rate of O_2 evolution or CO_2 uptake, which are accompanied by reciprocal oscillations in fluorescence. These oscillations appear to be related to regulatory mechanisms involving, particularly, the [ATP]/[ADP] ratio.

Glycerate-1,3-P_2 is formed from glycerate-3-P in a reaction with an unfavourable standard free energy change (Eqn. 10.2) and is especially sensitive to changes in the [ATP]/[ADP] ratio, with high ratios enhancing the reaction (Section 10.3.2). Consumption of ATP discharges the proton gradient and stimulates electron transport. Thus, for a given rate of carbon assimilation there must be a balance in the [ATP]/[ADP] ratio which favours electron transport and one which favours phosphorylation of glycerate-3-P. If this steady state condition is disturbed by, say, an increase in CO_2 concentration, transient increases in glycerate-3-P formation occur as the ribulose-1,5-P_2 pool is depleted. This causes a pulse of high metabolite concentrations to move through the C_3-CR cycle causing a disturbance in the [ATP]/[ADP] ratio.

The oscillations in O_2 evolution and fluorescence result from 'over-compensation' by the mechanisms concerned in establishing a new [ATP]/[ADP] ratio. High cytosolic P_i levels stimulate phosphorylation of ADP and the movement of triose phosphates from the stroma to the cytosol (Section 10.8) which dampens fluorescence oscillations.

Although much needs to be learnt about these oscillations, clearly variations in fluorescence are closely connected with carbon metabolism and its regulation.

Further reading

Monographs and treatises: Edwards & Walker (1983); Gibbs & Latzko (1979); Hatch & Boardman (1981); Hatch & Boardman (1987); Lawlor (1987).
Techniques and procedures: Walker (1987).
Chloroplast/cytosol metabolite exchange: Davies (1987) Chapter 2.
Phloem loading: Giaquinta (1983); Preiss (1980) Chapter 8; Preiss (1988) Chapter 2.
Properties of ribulose-1,5-P_2 carboxylase: Lorimer (1981).
Investigations with mutants: Somerville (1986).

11

Photorespiration

11.1 Light enhances carbon dioxide evolution by C_3 plants

Dark (or mitochondrial) respiration is the aerobic oxidation of sugars to CO_2 by plants in the dark. Photorespiration is the difference between the rate of CO_2 evolution (or O_2 consumption) in the light and mitochondrial respiration in the dark. Thus, photorespiration is the *light-dependent* production of CO_2 with the associated consumption of O_2.

Answering the question of whether photorespiration exists is, experimentally, not easy since the evolution of CO_2 and consumption of O_2 by respiration is masked by the much more active process of light-dependent CO_2 assimilation and O_2 evolution. One method is to place plants in the light in CO_2-free air and measure CO_2 evolution; for C_3 plants the rate is more than twice that in the dark but some of the CO_2 evolved is reassimilated, thus giving an underestimate of photorespiration.

Mitochondrial respiration, in contrast, proceeds at about the same rate in both the light and dark. For most C_3 plants the photorespiratory rate is typically two to three times the rate of mitochondrial respiration.

Studies of CO_2 exchange by detached leaves enclosed in a transparent airtight chamber provide further evidence for photorespiration. When the leaf is illuminated it begins to assimilate CO_2 but net assimilation (i.e. the difference between CO_2 assimilation and respiration) soon declines to zero as the leaf equilibrates with the residual CO_2 in the chamber. When the light is turned off, however, CO_2 is initially evolved at a rapid rate before quickly declining to a slower rate which is maintained for the remainder of the dark period (i.e. mitochondrial respiration). The short 'burst' of CO_2 evolution is accompanied by a correspondingly short 'burst' of O_2 consumption. This 'post-illumination burst' is attributed to photorespiration and the rates obtained are often used to measure the rate of photorespiration. It occurs because the metabolic processes which enhance CO_2 evolution in the light do not stop as quickly as the assimilation of CO_2 when C_3 plants are placed in the dark. The rate of CO_2 evolution during the post-illumination burst is influenced by factors that affect the rate of photorespiration.

11.2 Photorespired carbon dioxide is derived from recently assimilated carbon dioxide and is strongly influenced by oxygen and carbon dioxide concentrations

When plants are exposed to $^{14}CO_2$ for 10–20 min in the light they incorporate ^{14}C-label into the intermediates of the C_3-CR cycle as well as sucrose and other products. If the plants are then transferred to CO_2-free air and placed in the dark, the specific radioactivity of the CO_2 evolved is much less than the products (e.g. sucrose and hexose phosphates) containing newly assimilated ^{14}C. This indicates that, at least in the short-term, these are not major sources of the CO_2 evolved in dark respiration. However, when the plant is re-illuminated, the specific radioactivity of the CO_2 evolved is much greater than in the dark. Thus, the CO_2 evolved by photorespiration must be derived from a metabolite containing newly assimilated CO_2.

The rate of CO_2 assimilation, and hence growth, increases with the concentration of CO_2 but O_2 is inhibitory (Fig. 11.1). Photorespiration, on the other hand, is promoted by O_2 and inhibited by CO_2. Increasing the O_2 concentration also increases the CO_2 concentration at which evolution balances assimilation (CO_2 compensation point).

Production of the C_2 compound, glycollate in illuminated tissue is also strongly influenced by CO_2 and O_2. Conditions which promote photorespiration (high O_2, low CO_2) enhance the production of glycollate, whereas conditions which support CO_2 assimilation (high CO_2, low O_2) decrease glycollate production.

11.3 Ribulose-1,5-P_2 oxygenase activity accounts for several features of photorespiration

The antagonism between O_2 and CO_2 with respect to photorespiration, CO_2 assimilation and glycollate production is explained by the effect of O_2 and CO_2 on the oxygenase and carboxylase activities of ribulose-1,5-P_2 carboxylase/oxygenase (RuP$_2$-C/Oase) which (see Section 10.3.1) supports both the oxidative cleavage of ribulose-1,5-P_2 to glycollate-2-P and glycerate-3-P (Eqn. 11.1) and the carboxylative assimilation of CO_2 into glycerate-3-P (Eqn. 11.2).

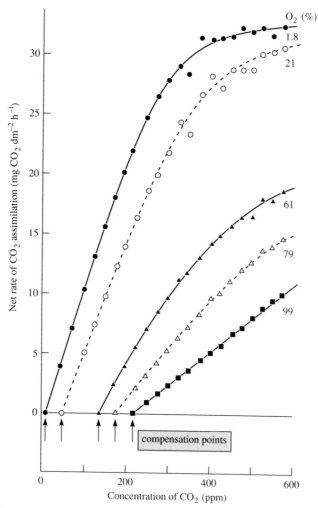

Fig. 11.1 Effect of O_2 and CO_2 concentration on the net rate of CO_2 exchange at 30°C by detached leaves of wheat, a typical C_3 plant. From Jolliffe P.A. & Tregunna E.B. (1973) *Can. J. Bot.*, **51**, 841–53.

$$Ribulose-1,5-P_2 + O_2 \rightleftharpoons glycerate-3-P + glycollate-2-P \qquad (11.1)$$

$$Ribulose-1,5-P_2 + CO_2 \rightleftharpoons 2 \ glycerate-3-P \qquad (11.2)$$

The relative rates of the two reactions are determined by the affinity of the enzyme for CO_2 and O_2, the V_{max} values of the carboxylase and oxygenase activities and the concentrations of CO_2 and O_2. Thus, carboxylase activity is inhibited competitively by O_2 (Fig. 11.2a) and the inhibitor constant (K_i) for O_2 is similar to the K_m value of oxygenase activity for O_2. Conversely, CO_2 competitively inhibits oxygen-

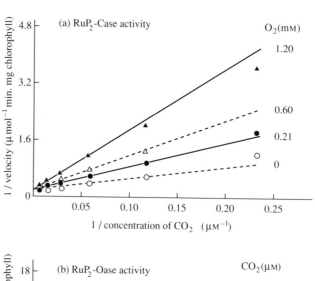

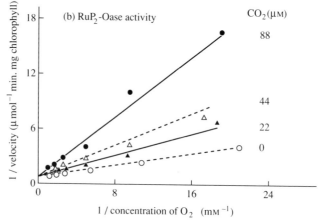

Fig. 11.2 Kinetic analysis of the effect of O_2 and CO_2 on ribulose-1,5-P_2 carboxylase activity and ribulose-1,5-P_2 oxygenase activity in crude extracts of spinach leaf tissue. Data are shown as double reciprocal plots of activity with respect to the appropriate substrate at various levels of the competing compound. For this experiment, carboxylase $K_m(CO_2) = 17.5 \ \mu M$, $K_i(O_2) = 354 \ \mu M$; for oxygenase $K_m(O_2) = 196 \ \mu M$, $K_i(CO_2) = 19.5 \ \mu M$. From Badger M.R. & Andrews T.J. (1974) *Biochem. Biophys. Res. Commun.*, **60**, 204–10.

ase activity (Fig. 11.2b) and the K_i for CO_2 is similar to the K_m value of carboxylase activity for CO_2.

Purified RuP_2-C/Oase has much greater affinity for CO_2 ($K_m = 12\ \mu M$) than for O_2 ($K_m = 250\ \mu M$) making the ratio of the two activities much more sensitive to changes in CO_2 concentration than to O_2 concentration. This causes much greater sensitivity of CO_2 assimilation and photorespiration to CO_2 than to O_2 (see Fig. 11.1). The O_2 and CO_2 antagonism of Eqns 11.1 and 11.2 also explains the effect of O_2 and CO_2 on the synthesis of glycollate, formed by hydrolysis of glycollate-2-P in the presence of a specific phosphatase.

In most C_3 plants under normal atmospheric conditions about 70% of the ribulose-1,5-P_2 is metabolized via carboxylase activity and 30% by oxygenase activity. This proportion can be calculated for concentrations of O_2 and CO_2 other that those found in air by the expression

$$\frac{v(\text{Ru-O})}{v(\text{Ru-C})} = \frac{V_{max}(\text{Ru-O})}{V_{max}(\text{Ru-C})} \cdot \frac{K_m(CO_2,\text{Ru-C})}{K_m(O_2,\text{Ru-O})} \cdot \frac{[O_2]}{[CO_2]} \qquad (11.3)$$

where $v(\text{Ru-O})$ and $v(\text{Ru-C})$ are the rates of the oxygenase and carboxylase activities at concentration $[O_2]$ and $[CO_2]$, $V_{max}(\text{Ru-O})$ and $V_{max}(\text{Ru-C})$ are the maximum activity rates and $K_m(CO_2, \text{Ru-C})$ and $K_m(O_2, \text{Ru-O})$ are the respective Michaelis constants for CO_2 as a substrate of carboxylase and O_2 as a substrate for oxygenase.

In C_3 plants photorespiration increases proportionately more than CO_2 assimilation as the temperature rises because the proportion of ribulose-1,5-P_2 metabolized via oxygenase increases. Part of the reason is that the solubility of CO_2 decreases proportionately more than that of O_2 with increasing temperatures. The higher proportion of ribulose-1,5-P_2 metabolized via oxygenase activity at higher temperatures decreases the energy efficiency of C_3 plants, since the regeneration of ribulose-1,5-P_2 from the increased amounts of glycollate-2-P formed incurs a heavy energy penalty (see Section 11.7). This is reflected in a marked decrease in the quantum yield (i.e. decreased growth per unit of light absorbed) with increasing temperature and an increase in the CO_2 compensation point.

11.4 Glycollate-2-P is returned to the C_3-CR cycle via the C_2 photorespiratory cycle

C_3 plants and C_3 mesophyll protoplasts incorporate, especially at high O_2 concentrations (i.e. enhanced photorespiration), ^{14}C-label from $^{14}CO_2$ into the C_2

compounds glycollate and glycine, the C_3 compounds serine and glycerate, and intermediates of the C_3-CR cycle. Compounds which block the oxidation of glycollate to glyoxylate (e.g. 2-pyridylhydroxymethanesulphonate, an inhibitor of glycollate oxidase) enhance accumulation of $[^{14}C]$glycollate but inhibit the incorporation of ^{14}C-label into the other C_2 and C_3 compounds.

2-Pyridylhydroxymethanesulphonate

Such compounds also inhibit CO_2 evolution in the light, suggesting that both the photorespired CO_2 and the C_2 and C_3 compounds above are derived from glycollate. Theory also predicts that the evolved CO_2 must be derived from glycollate-2-P since glycerate-3-P, the other product of oxygenase activity, is readily metabolized via the C_3-CR cycle in illuminated cells.

CO_2 is derived from glycollate-2-P via the reactions of the C_2 photorespiratory cycle (C_2-PR cycle), shown in Fig. 11.3, which metabolizes two molecules of glycollate-2-P to CO_2 and glycerate-3-P which is returned to the C_3-CR cycle. Glycollate, produced from glycollate-2-P, is oxidized to glyoxylate by molecular O_2 (further contributing to the O_2 consumption of photorespiration) and is transaminated to glycine. Two molecules of glycine give rise to serine in a complex reaction involving the production of CO_2 and NH_3, thereby accounting for the CO_2 produced by photorespiration. The amino group of serine is removed by transamination to form hydroxypyruvate which is reduced to glycerate. Finally, glycerate is phosphorylated and returned to the C_3-CR cycle.

11.5 Enzymes of the C_2-PR cycle and their subcellular location

Ribulose-1,5-P_2 is metabolized via oxygenase activity *in vivo* at about 30% of the rate of the carboxylase reaction. Since CO_2 is assimilated at about $200\ \mu mol\ mg^{-1}$ chlorophyll h^{-1}, the rate of glycollate-2-P production is about $60\ \mu mol\ mg^{-1}$ chlorophyll h^{-1}. The enzymes of the C_2-PR cycle must have rates *in vivo* consistent with this.

Isolated chloroplasts incorporate large amounts of CO_2 into glycollate but they do not produce glycine, serine and hydroxypyruvate or evolve CO_2. Isolated mesophyll protoplasts, however, photorespire and incorporate CO_2 into all the intermediates of the C_2-PR cycle shown in Fig. 11.3.

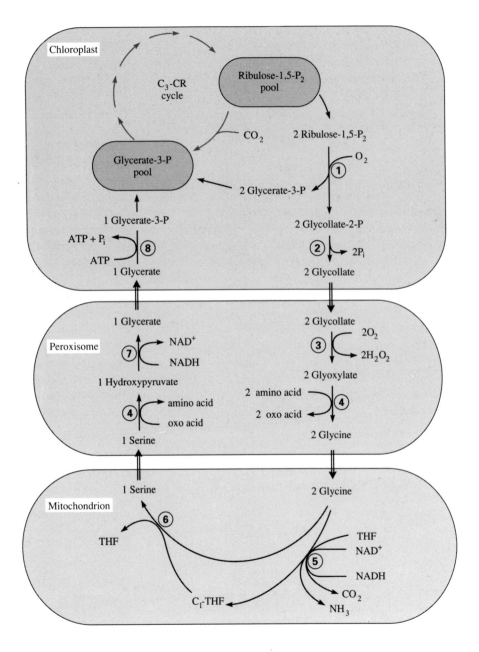

Isolated peroxisomes from mesophyll cells catalyse the production of glycine from glycollate and consume O_2 at rates consistent with the production of glycollate-2-P by illuminated chloroplasts. They contain the enzymes glycollate oxidase (Eqn. 11.4), inhibited by 2-pyridylhydroxymethanesulphonate, and glycine-specific aminotransferase (Eqn. 11.5), inhibited by amino-oxyacetate. Since photorespiration is also sensitive to these inhibitors, it is concluded that glycollate is metabolized to glycine in peroxisomes *in vivo* as follows.

$$\begin{array}{ccccc}
CH_2OH & \xrightarrow{\hspace{2cm}} & CHO & \xrightarrow{\hspace{2cm}} & CH_2 \cdot NH_3^+ \\
| & & | & & | \\
COO^- & O_2 \quad H_2O_2 & COO^- & R-NH_3^+ \quad R{=}O & COO^-
\end{array}$$

Glycollate   Glyoxylate   Glycine

(11.4)   (11.5)

Since RuP_2-Oase and glycollate oxidase both have relatively low affinities for O_2 ($K_m = 250\,\mu M$ and $170\,\mu M$ respectively), they are readily affected by changes in O_2 concentration in air (21%, equivalent to a $250\,\mu M$ solution). This distinguishes these enzymes (and hence photorespiration) from O_2 consumption via cytochrome oxidase (dark respiration) which has a much higher affinity for O_2 ($K_m = 0.5\,\mu M$) and is only sensitive to O_2 at concentrations less than 2%.

It is uncertain which amino acids donate amino groups to glyoxylate to form glycine in peroxisomes. Peroxisomes contain aminotransferases which catalyse the formation of glycine from glyoxylate with glutamate, serine, aspartate and alanine forming 2-oxoglutarate, hydroxypyruvate, oxaloacetate and pyruvate respectively. However, the reaction with glutamate (Eqn. 11.6) is thought to account for half of the amino groups incorporated into glycine in peroxisomes.

$$Glyoxylate + glutamate \rightleftharpoons glycine + 2\text{-}oxoglutarate \qquad (11.6)$$

Fig. 11.3 C_2-PR cycle for the metabolism of glycollate-2-P, derived from ribulose-1,5-P_2 through the action of RuP_2-Oase activity. For every two molecules of ribulose-1,5-P_2 oxidized, three glycerate-3-P are formed; two glycerate-3-P result directly from RuP_2-Oase activity, whilst the third is formed from two glycollate-2-P with the loss of one CO_2. Enzymes are: (1) RuP_2-Oase; (2) glycollate-2-P phosphatase; (3) glycollate oxidase; (4) aminotransferase; (5) glycine dehydrogenase; (6) serine hydroxymethyltransferase; (7) glycerate dehydrogenase; (8) glycerate kinase. Abbreviations: THF = tetrahydrofolate; C_1-THF = N^5,N^{10}-methylene THF.

Isolated mitochondria from green leaves of C_3 plants oxidatively decarboxylate glycine to CO_2, NH_3 and the methylene derivative of tetrahydrofolate.

$$
\begin{array}{c}
CH_2 \cdot NH_3^+ \quad NAD^+ \quad\quad NADH+H^+ \quad CO_2 + NH_3 \\
| \\
COO^- \\
\text{tetrahydrofolate} \quad\quad N^5,N^{10}\text{-methylene} \\
\text{tetrahydrofolate}
\end{array}
\tag{11.7}
$$

Glycine

This reaction, catalysed by glycine dehydrogenase (also known as glycine decarboxylase), accounts for the photorespiratory production of CO_2. Interestingly, leaf mitochondria from C_3 plants oxidize glycine, malate and succinate at similar rates but mitochondria from other tissues do not support significant glycine oxidation. The methylene tetrahydrofolate produced serves as the C_1 source for the hydroxymethyl group of serine in a reaction which consumes another molecule of glycine.

$$
\begin{array}{c}
H_2O \quad\quad\quad\quad CH_2OH \\
CH_2 \cdot NH_3^+ \quad\quad\quad\quad\quad\quad\quad CH \cdot NH_3^+ \\
| \quad\quad\quad\quad\quad\quad\quad\quad\quad\quad\quad\quad\quad | \\
COO^- \quad\quad\quad\quad\quad\quad\quad\quad\quad COO^- \\
N^5,N^{10}\text{-methylene} \quad\quad \text{tetrahydrofolate}
\end{array}
\tag{11.8}
$$

Glycine \quad\quad\quad\quad\quad\quad\quad\quad\quad\quad\quad Serine

This reaction is catalysed by serine hydroxymethyl transferase, another mitochondrial enzyme.

Summing Eqns 11.7 and 11.8 shows that leaf mitochondria support the production of serine from two molecules of glycine with the equimolar production of CO_2, NH_3 and NADH, i.e. about 30–40 μmol mg^{-1} chlorophyll h^{-1}. Since this rate of NH_3 production exceeds the rate of assimilation of exogenous inorganic nitrogen (about 5–7 μmol mg^{-1} chlorophyll h^{-1}), then NH_3 must be reassimilated to prevent net nitrogen loss. NH_3 is re-incorporated into glutamate in chloroplasts (see Section 13.5), so this glutamate must account for the transamination (Eqn. 11.6) of one of the two glyoxylates in peroxisomes (Fig. 11.3).

Serine is transported from mitochondria into peroxisomes where the amino group is transferred to an oxo acid, forming hydroxypyruvate (Eqn. 11.9).

$$
\begin{array}{c}
CH_2OH \quad\quad\quad\quad\quad\quad\quad CH_2OH \\
| \quad\quad\quad\quad\quad\quad\quad\quad\quad\quad\quad | \\
CH \cdot NH_3^+ \quad\quad\quad\quad\quad\quad\quad C=O \\
| \quad\quad\quad\quad\quad\quad\quad\quad\quad\quad\quad | \\
COO^- \quad R=O \quad R-NH_3^+ \quad COO^-
\end{array}
\tag{11.9}
$$

Serine \quad\quad\quad\quad\quad\quad Hydroxy-pyruvate

$$
\begin{array}{c}
\text{NADH} \\
\text{NAD}^+
\end{array}
\tag{11.10}
$$

$$
\begin{array}{c}
CH_2O\text{-}P \quad\quad\quad\quad\quad\quad CH_2OH \\
| \quad\quad\quad\quad\quad\quad\quad\quad\quad\quad\quad | \\
HCOH \quad\quad\quad\quad\quad\quad\quad\quad HCOH \\
| \quad\quad\quad\quad\quad\quad\quad\quad\quad\quad\quad | \\
COO^- \quad\quad\quad\quad\quad\quad\quad\quad COO^- \\
\quad\quad ADP \quad ATP
\end{array}
\tag{11.11}
$$

Glycerate-3-P \quad\quad\quad\quad\quad\quad Glycerate

In theory, glyoxylate could be the oxo acid with serine effectively providing the amino-N to produce the second of the two molecules of glycine required for serine synthesis in mitochondria.

$$\text{Glyoxylate} + \text{serine} \rightleftharpoons \text{glycine} + \text{hydroxypyruvate} \tag{11.12}$$

Peroxisomes contain several aminotransferases and other oxo acids could also be involved. However, the isolation of barley and *Arabidopsis* mutants deficient in serine:glyoxylate aminotransferase (Eqn. 11.12) indicates that this enzyme is especially important, since the mutants exhibit normal rates of CO_2 assimilation only under conditions which do not support photorespiration (e.g. high concentrations of CO_2). Nonetheless, serine:glyoxylate aminotransferase also exhibits serine:pyruvate and asparagine:glyoxylate aminotransferase activities so that other oxo acids could be involved. Glycerate dehydrogenase (or, in earlier literature, hydroxypyruvate reductase), which catalyses the reduction of hydroxypyruvate to glycerate (Eqn. 11.10), also occurs in peroxisomes. The source of the reducing equivalents for the production of NADH in peroxisomes is uncertain; they could be derived from the oxidation of glycine (Eqn. 11.7) in mitochondria or from illuminated chloroplasts via a shuttle mechanism (see Section 14.4). Incidentally, it is also possible that the NADH produced during glycine oxidation could be used in oxidative phosphorylation by leaf mitochondria.

The final step of the C_2-PR cycle involves transport of glycerate into the chloroplast (see Table 10.1) where it is phosphorylated to glycerate-3-P, catalysed by glycerate kinase (Eqn. 11.11).

11.6 The C_2-PR cycle is associated with metabolite transport between organelles and with nitrogen cycling

The photorespiratory pathway (Fig. 11.3) involves extensive transport of intermediates between chloroplasts, peroxisomes and mitochondria. This, and the associated transfer of glutamate and 2-oxoglutarate (which forms part of a cycle for transporting reassimilated NH_3 from chloroplasts to peroxisomes), is shown in Fig. 11.4. The overall process involving production of NH_3 by mitochondria, reassimilation of NH_3 into glutamate in chloroplasts (see Chapter 13 for details) and transport of glutamate to peroxisomes for incorporation of the amino group into glycine is known as the photorespiratory nitrogen cycle, an essential ancillary part of photorespiration.

Studies of *Arabidopsis* and barley mutants with defects in the chloroplast enzymes, glutamine synthetase and Fd-specific glutamate synthase, which catalyse the assimilation of NH_3 via glutamine into glutamate (Fig. 11.4, see Chapter 13 for details), demonstrate that the sustained activity of the C_2-PR cycle is dependent on the photorespiratory nitrogen cycle. These mutants grow in gas mixtures containing low concentrations (e.g. 1%) of O_2 but not in air (21% O_2). Since these conditions do not support significant RuP$_2$-Oase activity or photorespiration they are photorespiratory mutants. Incidentally, the survival of these mutants without photorespiration demonstrates that photorespiration *per se* is not essential for plant growth. In short-term experiments the rate of CO_2 assimilation by the barley mutants decreases strongly when they are transferred from 1% O_2 to air, whereas an ordinary cultivar quickly establishes a slightly lower rate in air. Over a period of 3 h after transfer to air, mutants attain very high levels of free NH_3, not detected

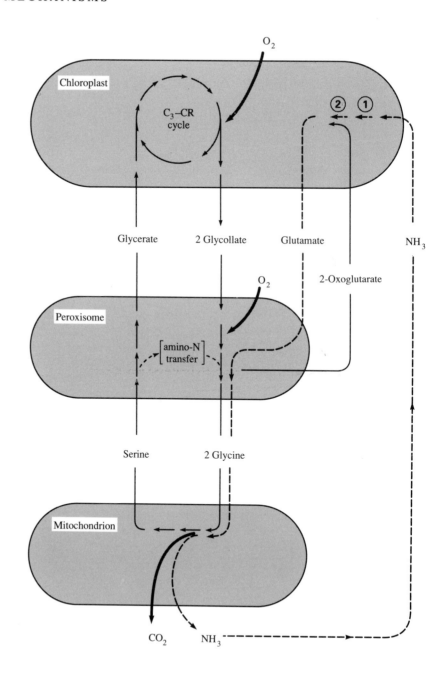

Fig. 11.4 Transport of metabolites associated with the C_2-PR cycle (solid line) and the photorespiratory nitrogen cycle (dashed line) between subcellular organelles of mesophyll cells. Note that glutamate supplies only one of the two equivalents of α-amino-N required by the peroxisome; the other is thought to be supplied by serine. Processes (1) and (2) involve the consumption of ATP and Fd$_{red}$.

in the non-mutant cultivar. The data demonstrate that the photorespiratory nitrogen cycle is essential for supporting the growth of C_3 plants in air. Indeed, the toxicity of the herbicide phosphinothricin (also known as glufosinate) also involves inhibition of glutamine synthetase activity and hence the photorespiratory nitrogen cycle.

According to Fig. 11.4 glycollate and glycine must be transported across membranes at rates similar to production of glycollate-2-P (about 60–80 μmol mg^{-1} chlorophyll h^{-1}). All the other transport processes in Fig. 11.4, including those associated with the photorespiratory nitrogen cycle, proceed at half this rate.

Little is known about the membrane transport of photorespiratory metabolites in leaf mitochondria and peroxisomes except that isolated mesophyll mitochondria oxidize exogenous glycine at rates commensurate with those of production of glycollate-2-P in chloroplasts. Carbon is returned to the C_3-CR cycle in the chloroplast as glycerate which is actively taken up by a light-stimulated mechanism specific to it. Since glycerate uptake is sensitive to proton ionophores and chloroplasts actively establish a proton gradient across the envelope in the light, uptake may involve co-transport of the glycerate anion with protons. The suggestion of a glycollate/glycerate antiport exchange mechanism at the envelope has not been greatly supported. Glycollate export may involve passive diffusion down a concentration gradient.

There is little direct evidence about which amino acid transports reassimilated nitrogen from chloroplasts to the peroxisome but it is probably glutamate (counter-exchanging with 2-oxoglutarate) or aspartate (counter-exchanging with oxaloacetate).

11.7 The C_2-PR cycle requires a large input of energy

The energy expended in metabolizing ribulose-1,5-P_2 via RuP$_2$-Oase activity is much greater than commonly recognized. Consider the metabolism of 10 ribulose-1,5-P_2 by oxygenase activity (Fig. 11.5). This produces 10 glycollate-2-P and 10 glycerate-3-P. Reductive metabolism of *all* the 10 glycerate-3-P to six ribulose-1-5P_2 via the C_3-CR cycle requires 16 ATP and 10 NADPH. The 10 glycollate-2-P is metabolized to five CO_2, five NH_3 and five glycerate. (It is assumed that the five NADH required to reduce hydroxypyruvate to glycerate are derived from glycine oxidation.) The metabolism of five glycerate to three ribulose-1,5-P_2 requires 13 ATP and five NADPH (Fig. 11.5). In addition, the five NH_3 produced from glycine

must be reassimilated to maintain the cycle. This requires a further five ATP and 10 Fd$_{red}$ (for mechanism see Chapter 13). For comparison, regard the 10 Fd$_{red}$ as equivalent to five NADPH.

Figure 11.5 shows that for 10 ribulose-1,5-P_2 metabolized, only nine are recovered (the balance of the carbon being associated with five CO_2) and that this net loss of ribulose-1,5-P_2 consumes 34 ATP and 20 NADPH. The data in Table 11.1 show that the energy requirements for the metabolism of ribulose-1,5-P_2 via the C_2-PR cycle (with net loss of carbon) are even greater than via the C_3-CR cycle (with net gain of carbon).

Alternatively, consider the energy required to metabolize ribulose-1,5-P_2 via the C_2-PR cycle on the basis that the carbon lost from the ribulose-1,5-P_2 pool is replenished through the activity of RuP$_2$-Case and the C_3-CR cycle. For the example given in Fig. 11.5, this involves the additional reaction of five CO_2 with five ribulose-1,5-P_2 to form 10 glycerate-3-P. Metabolism to six ribulose-1,5-P_2 (i.e. a net gain of one) involves the expenditure of 16 ATP and 10 NADPH. This, plus the energy inputs shown in Fig. 11.5 (34 ATP and 20 NADPH), sums to 50 ATP and 30 NADPH with no net change in the pool of ribulose-1,5-P_2 (Table 11.1).

11.8 Why is photorespiration light-dependent?

Photorespiration involves light-dependent evolution of CO_2. The schemes in Figs. 11.3 and 11.4 primarily concern the regeneration of glycerate-3-P from glycollate-2-P and do not readily explain why, but light is required to produce NADPH and ATP to maintain the level of ribulose-1,5-P_2 (the source of photorespired CO_2) in chloroplasts. Without light, the ribulose-1,5-P_2 pool is not replenished; RuP$_2$-Oase activity ceases quickly but the processes which evolve CO_2 via glycine dehydrogenase continue for some minutes until the pools of intermediates, especially glycine, fall. This explains the high ·initial rates or 'bursts' of CO_2 evolution associated with light/dark transitions.

11.9 What does photorespiration achieve?

Compared to the metabolism of ribulose-1-5-P_2 via the C_3-CR cycle, the C_2-PR cycle results in the net loss of carbon and consumes even more energy! Yet, C_3 plants metabolize a considerable portion of their ribulose-1,5-P_2 in this way. Why then should plants engage in this sort of activity?

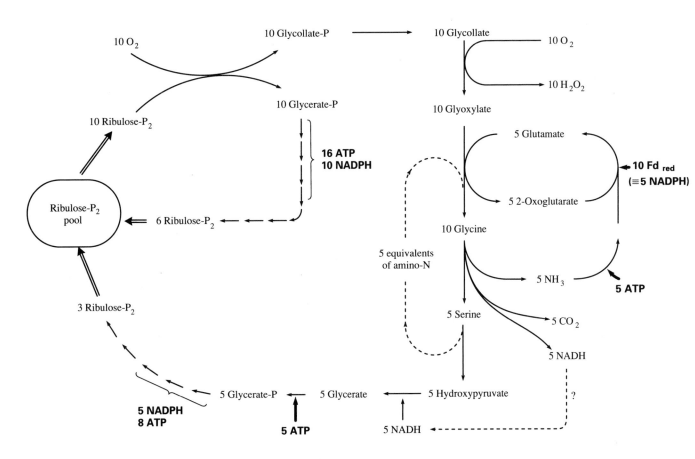

Fig. 11.5 A theoretical analysis of the stoichiometry of the C_2-PR cycle. Although 20 O_2 are consumed through the action of RuP_2-Oase and glycollate oxidase activities, the net O_2 consumption is only five due to the production of (i) 10 O_2 by the light reactions to form reducing equivalents for regenerating ribulose-1,5-P_2 from glycerate-3-P and reassimilating NH_3 and (ii) five O_2 by catalase activity. Thus, CO_2 evolution and O_2 consumption by the C_2-PR cycle are equimolar. The scheme as shown is self-sustaining with respect to nitrogen via the photorespiratory nitrogen cycle.

The most widely accepted hypothesis is that the oxygenase activity associated with RuP_2-C/Oase, and hence the sensitivity of the carboxylase function to O_2 (see Fig. 11.2), is an inevitable consequence of the carboxylation mechanism (i.e. the ribulose-1,5-P_2–enzyme complex is inherently sensitive to attack by both CO_2 and O_2). In keeping with this, every RuP_2-Case so far examined supports both oxygenase and carboxylase activity and exhibits O_2/CO_2 competition similar to the spinach enzyme (see Fig. 11.2). This includes enzymes from a wide range of photosynthetic organisms, (anaerobic and oxygenic bacteria, and C_3, C_4 and CAM plants) and chemoautotrophic organisms (e.g. *Thiobacillus* and *Nitrobacter*).

Since oxygenase activity is invariably associated with RuP_2-Case, then logically the C_2-PR cycle (Fig. 11.5) is a mechanism for retrieving carbon from the products formed by oxygenase activity (glycerate-3-P and glycollate-2-P) and returning it to the ribulose-1,5-P_2 pool (90% recovery) (see Fig. 11.5). Under natural conditions about 70% of the ribulose-1,5-P_2 is metabolized via carboxylase activity and 30% via oxygenase activity. Thus, for 100 mol of ribulose-1,5-P_2, 70 mol of carbon (as CO_2) are assimilated. If a mechanism did not exist to recover the carbon in the 30% subjected to oxygenase activity, five carbon atoms $\times$ 30 mol = 150 mol of carbon would be lost, compared with 70 mol gained by carboxylase activity; this would give the plant a net carbon deficit.

Table 11.1 Comparison of the CO_2 and O_2 inputs and outputs and the energy requirements when 10 ribulose-1,5-P_2 are metabolized via the C_3-CR cycle and C_2-PR cycle.

	C_3-CR cycle	C_2-PR cycle*	
		Photorespired CO_2 not reassimilated	Photorespired CO_2 reassimilated
CO_2 exchange	10 CO_2 assimilated	5 CO_2 evolved	No net change
O_2 exchange[†]	No net change	20 O_2 consumed	20 O_2 consumed
Net change in metabolites	3.3 triose-P synthesized	1 ribulose-1, 5-P_2 consumed	No net change
ATP input	30	34	50
NADPH input	20	20	30

* Values for the C_2-PR cycle are shown both corrected and uncorrected for reassimilation of photorespired CO_2 into ribulose-1,5-P_2.
[†] This disregards the O_2 evolved in the production of light-generated reducing equivalents for the 2 cycles and the O_2 formed from H_2O_2 by catalase activity.

11.10 Some plants have strategies for suppressing photorespiration

In theory, plants could diminish the negative effects of metabolizing ribulose-1,5-P_2 by oxygenase activity if the concentration of CO_2 in the cell was enhanced (see Eqn. 11.3). Several groups of plants have specialized mechanisms for concentrating CO_2 at the site of carboxylation, which diminish the expression of oxygenase activity and so suppress photorespiration. The most important of these are the C_4 plants described in detail in the next chapter. C_4 plants do not exhibit measurable photorespiration and have negligible CO_2 compensation points. The mesophyll cells of C_4 plants lack RuP_2-Case but contain a specialized complement of enzymes, insensitive to O_2 and which incorporate CO_2 (as HCO_3^-) into C_4 dicarboxylates. These are transported to specialized bundle sheath parenchyma cells surrounding the vascular bundle, with mechanisms for decarboxylating the dicarboxylate(s) and releasing CO_2. These cells contain a fully active complement of enzymes of the C_3-CR cycle including RuP_2-Case. Thus, the CO_2 incorporating mechanism of the mesophyll cells acts as a means of supplying CO_2 at higher concentrations to bundle sheath cells where RuP_2-Case is located.

When certain unicellular algae are grown for prolonged periods at various concentrations of inorganic carbon (CO_2 plus HCO_3^-), those grown at low concentrations exhibit a much higher affinity for inorganic carbon as measured in *short-term* experiments than do cells grown at high concentrations (Fig. 11.6). In other words, cells adapted to low carbon levels develop a mechanism for using low concentrations of inorganic carbon very effectively compared to cells adapted to high carbon; they have a low sensitivity to O_2, exhibit little photorespiratory activity and have very low CO_2 compensation points, similar to C_4 plants.

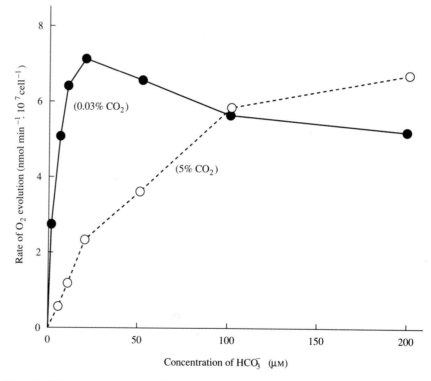

Fig. 11.6 Effect of growing cells of the green alga *Chlamydomonas reinhardtii* at high or low concentrations of CO_2, on the short-term kinetics of CO_2 assimilation. Cells were grown in a medium sparged with a gas mixture containing 5% CO_2 or air (0.03% CO_2). The rate of CO_2 assimilation by cells raised under these conditions was monitored by O_2 evolution in short-term experiments at various levels of HCO_3^- at pH 7, 20°C and saturating levels of light. From Berry J., Boynton J., Kaplan A. & Badger M. (1976) *Carnegie Inst. Washington Yearb.*, **75**, 423–32.

Cells adapted to low carbon accumulate inorganic carbon to concentrations well in excess of the external solution (up to 40-fold for CO_2). This accounts for the low photorespiratory activity and low CO_2 compensation points of these cells. However, unlike C_4 plants, the accumulation of inorganic carbon is attributed to an active uptake mechanism which does not involve a carboxylation reaction. These cells also have greatly enhanced carbonic anhydrase activity, possibly associated with the periplasmic space between the plasma membrane and the cell wall, although its function in carbon uptake is unclear.

11.11 Some algae possess alternative routes of photorespiratory glycollate metabolism

In most algae glycollate is oxidized to glyoxylate via glycollate dehydrogenase on the inner mitochondrial membrane.

$$\text{Glycollate} + NAD^+ \rightleftharpoons \text{glyoxylate} + NADH + H^+ \tag{11.13}$$

In theory, this yields two ATP per glycollate oxidized and is energetically less wasteful than the mechanism in terrestrial plants involving glycollate oxidase (Eqn. 11.4).

Some green algae (Chlorophyceae) and some cyanobacteria have an alternative mechanism for producing glycerate from glyoxylate which involves formation of tartronyl semialdehyde from two molecules of glyoxylate with the evolution of CO_2 in a reaction catalysed by glyoxylate carboligase (Eqn. 11.14). Tartronyl semialdehyde is reduced to glycerate in a second reaction catalysed by tartronyl semialdehyde reductase (Eqn. 11.15).

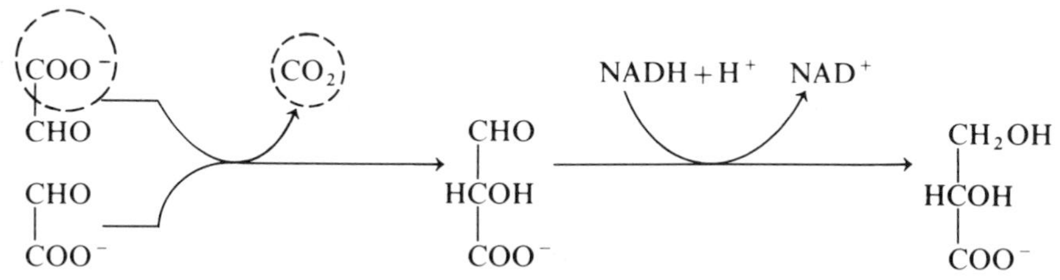

2 Glyoxylate   Tartronyl semialdehyde   Glycerate

(11.14)   (11.15)

Since this mechanism does not involve glycine or serine no NH_3 is produced. In theory, this should be energetically more favourable as no energy is expended reassimilating NH_3.

Further reading

General texts on photorespiration: Davies (1980) Chapter 12; Edwards & Walker (1983) Chapter 13; Gibbs & Latzko (1979) Chapters II.25–II.28; Hatch & Boardman (1981) Chapter 8.
Properties of ribulose-1,5-P_2 oxygenase: Lorimer (1981).
Photorespiratory nitrogen cycle and NH_3 assimilation: Singh et al. (1985).
Properties of peroxisomes: Tolbert (1980) Chapter 9.
Investigations with photorespiratory mutants: Somerville (1986).

12
C_4 mechanisms of carbon dioxide assimilation

12.1 Carbon dioxide assimilation characteristics of some plants do not reflect the properties of ribulose-1,5-P_2 carboxylase

Many of the characteristics of CO_2 assimilation of C_3 plants, such as their responsiveness to CO_2, O_2 and temperature, are largely determined by the properties of ribulose-1,5-P_2 carboxylase (RuP$_2$-Case). The gas exchange characteristics of C_4 plants, however, are not consistent in the same way; they differ in their sensitivity to CO_2, O_2, light and temperature and do not exhibit photorespiration in air, as they initially incorporate CO_2 into various C_4 dicarboxylates rather than into glycerate-3-P.

Although the estimated total number of C_4 plants is only about 1% of the 230 000 or so higher plant species, they are not mere curiosities since some of the world's most productive crop and pasture plants are C_4 grasses (e.g. sugar cane, maize, the sorghums, various millets and *Paspalum*). Moreover, the eight weeds regarded as the most troublesome in the world are C_4 plants (the grasses *Cynodon dactylon, Echinochloa colonum, E. crus-galli, Eleusine indica, Imperata cylindrica, Panicum maximum* and *Sorghum halepense* and the sedge *Cyperus rotundus*). C_4

plants are therefore of great agricultural and economic significance.

This chapter describes CO_2 assimilation in C_4 plants and in another group called CAM plants which assimilate CO_2 by Crassulacean Acid Metabolism (CAM). CAM plants assimilate CO_2 by a pathway that has much in common with that of C_4 plants.

12.2 C_4 plants and CAM plants assimilate carbon dioxide into oxaloacetate and other C_4 dicarboxylates

All C_4 and CAM plants have several common features (Fig. 12.1). Phosphoenolpyruvate (PEP), a C_3 compound, is the primary acceptor, reacting with CO_2 to form the C_4 dicarboxylate, oxaloacetate, in a reaction catalysed by PEP carboxylase (PEP-Case). In C_4 plants this occurs during the day but in CAM plants it takes place at night. In both C_3 and CAM plants, oxaloacetate is rapidly metabolized to one or more of the two related C_4 dicarboxylates, malate and aspartate, making these compounds the first most readily detected products. Malate and/or aspartate are subsequently decarboxylated, releasing CO_2 within

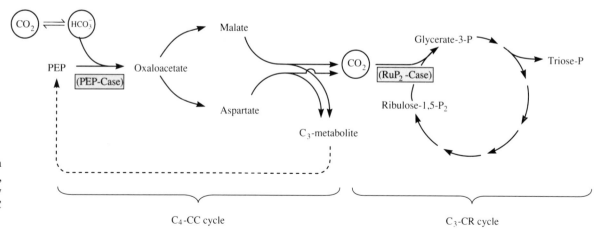

Fig. 12.1 Summary of the main features of CO_2 assimilation common to C_4 and CAM plants. Note that CO_2 is, in effect, assimilated twice; the first time as HCO$_3^-$ via PEP-Case activity and subsequently via RuP$_2$-Case activity. Abbreviation: C_4-CC cycle = CO_2 concentrating cycle.

the leaf. This CO_2 is then reassimilated in the light into glycerate-3-P and other intermediates of the C_3-CR cycle.

12.3 C_4 plants have distinctive biochemical, physiological and anatomical characteristics

By definition, C_4 plants initially assimilate CO_2 in the light into oxaloacetate, catalysed by PEP-Case. Most C_4 plants are indigenous to the tropics and warm temperate zones with high light intensity and high temperatures and often a seasonal period of low rainfall. They typically have higher photosynthetic rates and growth rates than C_3 plants under these conditions.

C_4 plants are distinguished from C_3 plants in other ways. C_4 plants do not exhibit measurable photorespiration at O_2 concentrations less than 60% and the CO_2 assimilation rates of C_4 plants are optimal at much higher light intensities than those of C_3 plants. C_4 monocotyledons also have a characteristic leaf anatomy. Most C_3 monocotyledons (especially the festucoid grasses) have vascular bundles surrounded by two layers of cells, the bundle sheath. The inner layer, known as the mesotome, consists of relatively small thick-walled cells with few, if any, chloroplasts. The outer layer consists of thin-walled parenchyma cells with chloroplasts considerably smaller than those of mesophyll cells. The vascular bundles of C_4 monocotyledons may or may not possess an achlorophyllous mesotome but are surrounded by a distinctive sheath of parenchyma cells with many chloroplasts, often larger than those in mesophyll cells. In most C_4 plants the mesophyll cells are radially arranged around the bundle sheath so that each cell has a common boundary with a bundle sheath parenchyma cell. This arrangement is known as 'Kranz anatomy' (Fig. 12.2) and occurs very rarely in C_3 monocotyledons. Virtually all the starch in the leaves of C_4 monocotyledons occurs in the bundle sheath parenchyma cells. Cytologically, the bundle sheath cells of C_4 plants appear to be very active and their presence around the vascular bundle suggests that they are involved with some metabolic activity related to it. In some C_4 plants the thylakoids of bundle sheath chloroplasts are not tightly packed into grana as in the mesophyll cells (i.e. chloroplasts exhibit structural dimorphism) (Fig. 12.3).

12.4 The pathway of carbon dioxide assimilation in C_4 plants can be determined by the order in which metabolites are labelled with [^{14}C]carbon dioxide

The pathway of CO_2 assimilation can be elucidated by supplying $^{14}CO_2$ to illuminated leaf segments for a few seconds (pulse) and rapidly replacing it with

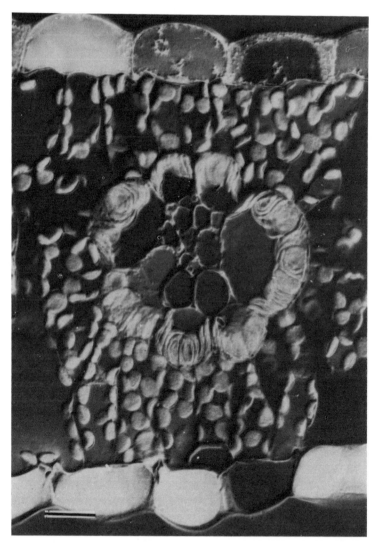

Fig. 12.2 Light micrograph of a transverse section of a leaf of *Zea mays* (maize), a typical C_4 grass showing Kranz anatomy, photographed with Nomarski optics. Mesophyll cells contain small chloroplasts arranged around the cell periphery. These cells are radially arranged around a distinctive sheath of parenchyma cells which, in turn, surround the cells of the vascular bundle. The bundle sheath comprises a layer of cells containing many large chloroplasts. The bar represents 20 µm. Micrograph supplied by Botany Department, La Trobe University.

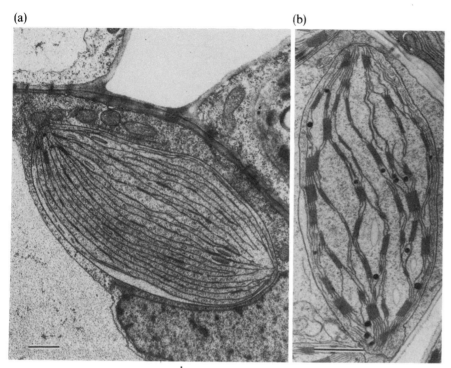

Fig. 12.3 Fine structure of the chloroplasts from (a) a bundle sheath cell and (b) a mesophyll cell of the C₄ grass *Digitaria sanguinalis*, a NADP M-E variant. Note the absence of well-defined grana in the bundle sheath chloroplast which has an incompetent PSII. Details of the fine structure for a mesophyll chloroplast are given in Fig. 3.4. The bar represents 1 μm in both micrographs. Micrograph supplied by Botany Department, La Trobe University.

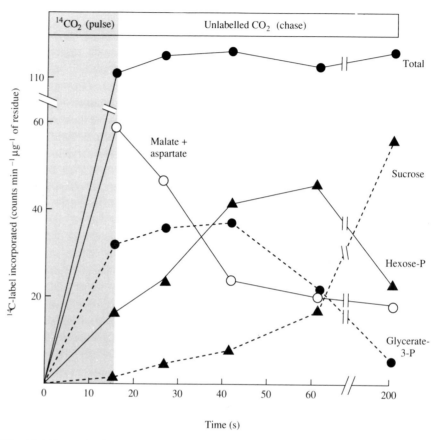

Fig. 12.4 Determination of the pathway of CO_2 assimilation in illuminated leaf tissue of sugar cane (*Saccharum officinarum*) in a pulse–chase experiment. Leaf segments were supplied with air containing $^{14}CO_2$ for 15 s (pulse) which was then replaced with air containing unlabelled CO_2 (chase). Leaf segments were withdrawn at various times during the chase and the amount of ^{14}C-label associated with each metabolite was determined. After Hatch M.D. & Slack C.R. (1966) *Biochem. J.* **101**, 103–11.

unlabelled CO_2 (chase). Leaf samples are taken at the beginning of the chase and at various times thereafter and the ^{14}C-label in individual compounds in each sample is determined. During the pulse, ^{14}C-label is incorporated into very short-term products, usually only one or two; other compounds labelled during the chase must be derived from those formed during the pulse. The results of one such experiment with sugar cane (Fig. 12.4) show that most of the radioactivity is in malate and aspartate at the end of a 15 s pulse and there is also some in glycerate-3-P. During the chase, the label in the dicarboxylates declines and that in glycerate-3-P increases during the next 30 s, before declining. Hexose-P shows a

similar pattern except that it contains very little label at the beginning of the chase which increases for up to 60 s, before decreasing. Label continues to accumulate in sucrose after 200 s. The data demonstrate that CO_2 is incorporated into the C₄ dicarboxylate prior to glycerate-3-P, which passes the label to hexose-P and eventually to sucrose. More detailed experiments show that labelling of oxaloacetate (which needs to be trapped as its phenylhydrazone) precedes labelling of aspartate and malate.

In very short-term experiments, the ^{14}C-label in the three C_4 dicarboxylates (oxaloacetate, malate and aspartate) is primarily in the carboxyl carbon atom at C-4. The label initially appearing in glycerate-3-P in C_4 plants is in the carboxyl carbon atom at C-1. This specifically shows that C-1 of glycerate-3-P is derived from C-4 of the C_4 dicarboxylates. The labelling of glycerate-3-P is identical to that of C_3 plants (see Eqn. 10.4), demonstrating that the C-4 carboxyl group of malate and/or aspartate is released as CO_2 and reassimilated via RuP_2-Case to form glycerate-3-P (see Fig. 12.1). The subsequent labelling of hexose-P (Fig. 12.4) and other intermediates is consistent with metabolism of glycerate-3-P via the C_3-CR cycle.

12.5 Establishing the functions of mesophyll and bundle sheath cells

The location of the enzymes of CO_2 assimilation in the mesophyll and bundle sheath cells is important in understanding CO_2 assimilation in C_4 plants. Methods for separating the two types of cells take advantage of the fact that bundle sheath cells have thicker and tougher walls. The most successful technique involves treating leaf tissue with fungal enzymes to degrade the cell walls and produce protoplasts. The mesophyll cells with thinner walls are released first, recovered by centrifugation, disrupted by forcing them through $20\,\mu m$ mesh (the cells are larger than this) and the chloroplasts recovered by centrifugation. Similar procedures using longer digestion times for the preparation of uniform suspensions of bundle sheath protoplasts work with only a few species (e.g. the grass *Panicum miliaceum*). Generally, strands of cells containing vascular tissue with bundle sheath cells are used after removing mesophyll cells.

Mesophyll protoplasts and intact chloroplasts from C_3 plants readily assimilate CO_2 in the light. However, mesophyll protoplasts from C_4 plants only assimilate CO_2 in the presence of the C_3 compounds pyruvate or alanine into the C_4 dicarboxylates oxaloacetate, malate and/or aspartate and not into intermediates of the C_3-CR cycle. This implies that, *in vivo*, mesophyll cells require a supply of C_3 compounds from other cells to sustain CO_2 assimilation. However, *mesophyll chloroplasts* do not support CO_2 assimilation, even in the presence of pyruvate or alanine because PEP-Case is located in the cytosol of mesophyll cells (Fig. 12.5).

Illuminated bundle sheath cells generally catalyse CO_2 assimilation without any additional compounds (For reasons described later, one of the C_4 variants is an

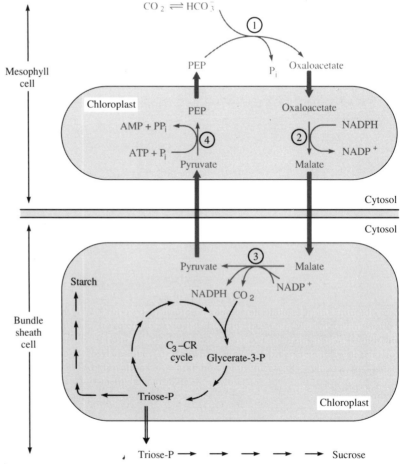

Fig. 12.5 Pathway of CO_2 assimilation in a NADP M-E variant showing the cellular and subcellular location of the enzymes and transport processes involved. Reactions which comprise the C_4-CC cycle are shown in grey. Enzymes of the C_4-CC cycle are: (1) PEP carboxylase; (2) NADP malate dehydrogenase; (3) NADP malic enzyme; (4) pyruvate, phosphate dikinase.

important exception.) CO_2 assimilation by these cells is inhibited by glyceraldehyde, an inhibitor of the C_3-CR cycle, and the products are identical to those of the C_3-CR cycle in C_3 plants.

Experiments have shown that $^{14}CO_2$ incorporation into dicarboxylates in mesophyll cells precedes that of the C_3-CR cycle in bundle sheath cells where the C-4 atom of malate and/or aspartate is decarboxylated and reassimilated by the cycle. For example, illuminated bundle sheath cells incorporate ^{14}C-label from [4-^{14}C]malate and [4-^{14}C]aspartate into intermediates of the C_3-CR cycle at rates commensurate with those for CO_2 assimilation by whole plants (up to the equivalent of 300 µmol mg^{-1} chlorophyll h^{-1}). The integration of these processes in the most common type of C_4 plant (NADP M-E variant) is shown in Fig. 12.5.

12.6 Leaves of C_4 plants have a quantitatively distinctive complement of enzymes

12.6.1 Pyruvate, phosphate dikinase

This enzyme catalyses the phosphorylation of pyruvate in an unusual reaction involving the hydrolysis of ATP (depicted as APPP in the equation below) to AMP (AP) and the phosphorylation of P_i to PP_i, giving rise to the name dikinase.

$$\text{Pyruvate} + P_i + ATP \rightleftharpoons PEP + AMP + PP_i \tag{12.1}$$

In C_4 plants pyruvate, phosphate dikinase (PP-DKase) is located mainly in the chloroplasts of mesophyll cells but is also found in photosynthetic tissues of various C_3 plants although, compared to C_4 plants, the activity is extremely low.

In illuminated C_4 plants, the extractable activity of PP-DKase is barely sufficient to produce PEP at rates commensurate with CO_2 assimilation but the activity of dark-treated leaves is less than 10% of this. Light-activation (and dark-inactivation) of PP-DKase in maize involves a regulatory protein (RE$_{PP\text{-}DKase}$). In vitro, in the presence of ATP and P_i or PEP, PP-DKase exists in a catalytically active form, phosphorylated on a histidyl residue at the catalytic site (PP-DKase-P). However, in the presence of ADP, RE$_{PP\text{-}DKase}$ catalyses the phosphorylation of PP-DKase-P on a threonyl residue to form PP-DKase-P$_2$ which is catalytically inactive (Eqn. 12.3). In vitro, this form is produced in the dark. Light-activation involves dephosphorylation of the phosphothreonyl residue in the presence of P_i. AMP and PP_i or pyruvate can dephosphorylate PP-DKase-P to PP-DKase (Eqn. 12.2). This form is also catalytically active and is not vulnerable to ADP-mediated inactivation by RE$_{PP\text{-}DKase}$.

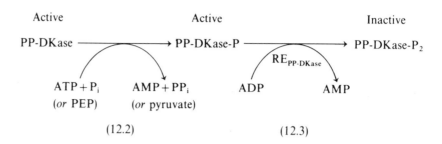

$$(12.2) \qquad\qquad (12.3)$$

The concentration of pyruvate might be important in maintaining the level of active enzyme in the light.

12.6.2 PEP carboxylase

PEP carboxylase (PEP-Case) is the primary enzyme of CO_2 assimilation in C_4 plants. It catalyses the carboxylation of PEP to form oxaloacetate containing the newly assimilated carbon in the C-4 carboxyl group.

$$\tag{12.4}$$

PEP   Oxaloacetate

This reaction is essentially irreversible due to the highly exergonic hydrolysis of PEP (see Section 4.3). In C_4 plants, PEP-Case is located in the cytosol of mesophyll cells and requires Mg^{2+} for activity. Unlike RuP$_2$-Case, the substrate is HCO_3^-, not CO_2. PEP-Case has a high affinity for HCO_3^-; the K_m is about 0.2–0.4 mM, equivalent to a dissolved CO_2 concentration of 4–8 µM at pH 8. The level of PEP-Case in C_4 plants is usually well in excess of the rate of CO_2 assimilation in vivo and is not rate-limiting. Nevertheless, the enzyme exhibits sigmoidal kinetics with respect to PEP with malate and aspartate as negative effectors and glucose-6-P a positive effector. These compounds could be important in regulating CO_2 incorporation in vivo if PEP is limiting but there is little data at present.

PEP-Case, unlike RuP$_2$-Case, does not react with O_2 and is not sensitive to it.

These features in association with the C_4 decarboxylating mechanism in bundle sheath cells largely explain why CO_2 assimilation by C_4 plants is insensitive to O_2, why they do not photorespire and why they have a higher efficiency at high temperatures (see Section 12.11). C_3 plants also have PEP-Case but the activity is typically less than 5% of C_4 plants.

12.6.3 Other enzymes

NADP-specific malate dehydrogenase (NADP M-DHase) catalyses the reduction of oxaloacetate using NADPH as reductant. It occurs in high activity in C_4 variants in which malate is the principal short-term product of CO_2 assimilation and is located in the chloroplasts of mesophyll cells.

$$\text{Oxaloacetate} + \text{NADPH} + \text{H}^+ \rightleftharpoons \text{malate} + \text{NADP}^+ \qquad (12.5)$$

The equilibrium lies strongly towards malate, partly explaining the very small pool of oxaloacetate during steady state CO_2 assimilation. The extractable activity of NADP M-DHase is strongly dependent on light and is light-modulated via the ferredoxin/thioredoxin system described in Section 10.13.1.

C_4 plants, especially variants with aspartate as the major C_4 dicarboxylate, contain several highly active aminotransferases (see Section 13.6 for properties). They include glutamate:oxaloacetate aminotransferase (GO-ATase) and glutamate:pyruvate aminotransferase (GP-ATase) which catalyse the production of aspartate (Eqn. 12.6) and alanine (Eqn. 12.7) respectively.

$$\text{Glutamate} + \text{oxaloacetate} \rightleftharpoons \text{aspartate} + \text{2-oxoglutarate} \qquad (12.6)$$

$$\text{Glutamate} + \text{pyruvate} \rightleftharpoons \text{alanine} + \text{2-oxoglutarate} \qquad (12.7)$$

These reactions are readily reversible so their direction depends on steady state conditions *in vivo*. GO-ATase and GP-ATase occur in both mesophyll and bundle sheath cells and catalyse Eqns 12.6 and 12.7 in opposite directions in the two cell types (see Fig. 12.6). For example, GO-ATase in mesophyll chloroplasts supports production of aspartate but in bundle sheath cells produces oxaloacetate.

12.7 Mechanisms for decarboxylating C_4 dicarboxylates in bundle sheath cells vary between species

C_4 plants differ in the mechanisms they use to release CO_2 from malate and aspartate in bundle sheath cells. This gives rise to three C_4 variants which take their name from the principal decarboxylating mechanism they employ.

The simplest mechanism, found in the majority of C_4 plants (NADP M-E variants), involves oxidative decarboxylation of malate, catalysed by NADP-specific malic enzyme.

$$(12.8)$$

Malate   Pyruvate

Oxalate inhibits this enzyme, making it useful for studying the role of NADP-specific malic enzyme in CO_2 assimilation, since the alternative decarboxylation mechanisms discussed below are insensitive to oxalate. NADP-specific malic enzyme is located in the chloroplasts of bundle sheath cells (Fig. 12.5).

In the other two variants, aspartate is the principal dicarboxylate. In one of these (PEP-CK variants), aspartate is transaminated in bundle sheath cells to form oxaloacetate (Eqn. 12.9) which is decarboxylated to PEP in the presence of ATP (Eqn. 12.10), catalysed by PEP carboxykinase (PEP-CKase) in the cytosol.

$$\text{Aspartate} + \text{2-oxoglutarate} \rightleftharpoons \text{oxaloacetate} + \text{glutamate} \qquad (12.9)$$

$$(12.10)$$

Oxaloacetate   PEP

PEP-CKase is sensitive to the inhibitor 3-mercaptopicolinic acid.

The other type of aspartate variant typically has large and prominent mitochondria in the bundle sheath cells. In these (NAD M-E variants), aspartate is metabolized in bundle sheath mitochondria via a series of reactions involving GO-ATase (Eqn. 12.11), NAD malate dehydrogenase (Eqn. 12.12) and NAD malic enzyme (Eqn. 12.13).

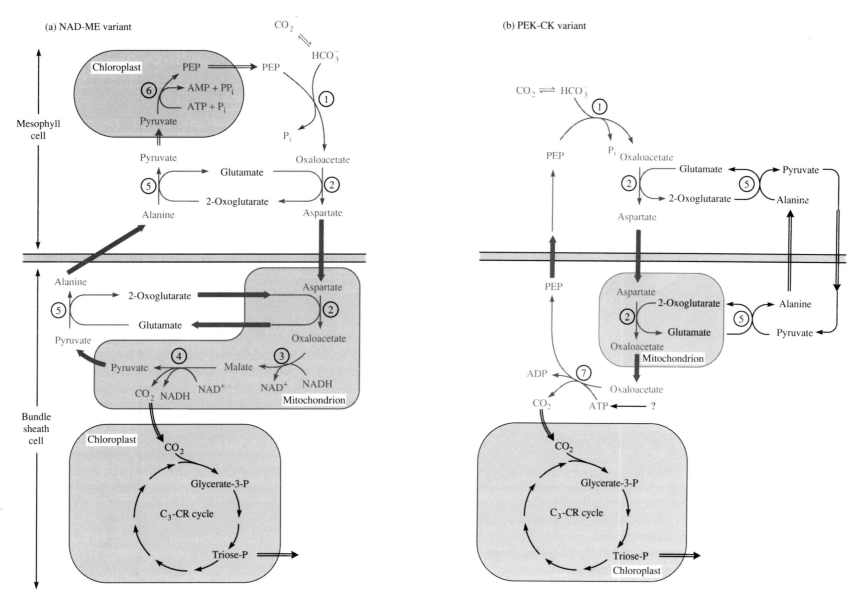

Fig. 12.6 Pathways of CO_2 assimilation in aspartate-transporting variants of C_4 plants and the cellular and subcellular location of the enzymes and processes involved. Unshaded regions within the bundle sheath and mesophyll cells represent the cytosol. Reactions which comprise the C_4-CC cycle are shown in grey. Enzymes are: (1) PEP carboxylase; (2) glutamate:oxaloacetate aminotransferase; (3) NAD malate dehydrogenase; (4) NAD malic enzyme; (5) glutamate:pyruvate aminotransferase; (6) pyruvate, phosphate dikinase; (7) PEP carboxykinase.

$$\text{Aspartate} + \text{2-oxoglutarate} \rightleftharpoons \text{oxaloacetate} + \text{glutamate} \qquad (12.11)$$

$$\text{Oxaloacetate} + \text{NADH} + \text{H}^+ \rightleftharpoons \text{malate} + \text{NAD}^+ \qquad (12.12)$$

$$\text{Malate} + \text{NAD}^+ \rightleftharpoons \text{pyruvate} + \text{NADH} + \text{H}^+ + \text{CO}_2 \qquad (12.13)$$

The pyruvate produced is transaminated to alanine in the cytosol (Fig. 12.6).

The continued assimilation of CO_2 into C_4 dicarboxylates in mesophyll cells necessitates recycling of the C_3 decarboxylation products (pyruvate, alanine and PEP) produced in bundle sheath cells to the mesophyll. The pyruvate produced in NADP M-E variants is recycled via PP-DKase activity in mesophyll chloroplasts (Fig. 12.5). Alanine, formed by NAD M-E variants is transaminated with 2-oxoglutarate in mesophyll cells to form pyruvate and converted to PEP via PP-DKase activity (Fig. 12.6). PEP produced by PEP-CK variants returns directly to the mesophyll, possibly in association with an alanine/pyruvate exchange in order to maintain nitrogen balance with aspartate (Fig. 12.6).

12.8 Enzymes of the C_3-CR cycle occur in bundle sheath cells but some are also found in mesophyll cells

All the RuP_2-Case activity of leaf tissue in C_4 plants is located in bundle sheath cells. The RuP_2-Cases from bundle sheath cells of C_4 plants and the mesophyll cells of C_3 plants are very similar and both types also support RuP_2-Oase activity. Also, they do not differ in their relative carboxylase/oxygenase activities and the properties of the oxygenase from C_3 and C_4 plants are indistinguishable.

Bundle sheath cells also contain all the other enzymes of the C_3-CR cycle. Some, like RuP_2-Case, are restricted to bundle sheath cells (e.g. ribulose-5-P kinase) and others are present in low activity. However, the enzymes involved in the metabolism of the C_3 intermediates of the C_3-CR cycle (NADP glyceraldehyde-3-P dehydrogenase and glycerate-3-P kinase) are found in approximately equal activity in mesophyll and bundle sheath cells in all the C_4 variants. It is therefore assumed that about 50% of the glycerate-3-P formed in bundle sheath cells is reduced to glyceraldehyde-3-P in each of the two types of cells.

At least in maize, the sucrose synthesis enzymes UDP-glucose pyrophosphorylase and sucrose-P synthetase are predominantly located in mesophyll cells. It is not known whether this distribution is common to C_4 plants.

Starch synthesis in C_4 plants occurs predominantly in bundle sheath cells although some is accumulated in mesophyll chloroplasts under prolonged periods of high light intensity. The starch synthesizing enzymes ADP-glucose pyrophos-phorylase the ADP-glucose transglucosylase are found mainly in bundle sheath chloroplasts.

12.9 C_4 plants expend more energy assimilating carbon dioxide than do C_3 plants

12.9.1 NADP M-E variants

The chloroplasts of bundle sheath cells of NADP M-E variants lack well-developed granal stacks (see Fig. 12.3) indicating that they are photochemically deficient. They have a normal photosystem I activity but are seriously deficient in photosystem II so that their capacity to produce NADPH and ATP by non-cyclic photophosphorylation is impaired. ATP is formed in bundle sheath chloroplasts by cyclic photophosphorylation. About 50% of the glycerate-3-P produced in bundle sheath cells is presumed to be exported and reduced in mesophyll cells but bundle sheath chloroplasts lack photochemically generated NADPH to effect the reduction of the remaining glycerate-3-P. The NADPH required is formed during oxidative decarboxylation of malate by NADP malic enzyme producing one NADPH per CO_2 (Fig. 12.7). Each CO_2 reassimilated gives rise to two glycerate-3-P, one of which is transported to mesophyll cells. Decarboxylation of malate provides the requisite NADPH to reduce the remaining glycerate-3-P in bundle sheath chloroplasts. These reducing equivalents, derived from malate, originate in the mesophyll, so these cells are the sole source of reducing equivalents for glycerate-3-P metabolism in NADP M-E variants. This explains why isolated bundle sheath chloroplasts cannot incorporate CO_2 without added malate.

For each CO_2 assimilated, two ATP are required in the bundle sheath; one each for the phosphorylation of glycerate-3-P and ribose-5-P (Fig. 12.7). In mesophyll cells three ATP are required; one for the phosphorylation of glycerate-3-P and two for the generation of PEP from pyruvate. The total input of five ATP and two NADPH per CO_2 is two ATP greater than that for the C_3-CR cycle (Table 12.1). Thus, considering nothing else, C_4 plants require a greater energy input per CO_2 assimilated than do C_3 plants. The situation *in vivo*, however, is more complex due to the energy demands associated with photorespiration in C_3 plants (see Section 11.7).

12.9.2 PEP-CK variants

Judging by the enzyme complement of mesophyll and bundle sheath cells, PEP-CK variants also metabolize glycerate-3-P to glyceraldehyde-3-P in approximately equal amounts in the two cell types. Thus, per CO$_2$ incorporated, one ATP and one NADPH are required in each cell for the reduction of glycerate-3-P to triose-P. The bundle sheath cells of PEP-CK variants have a competent photosystem II to supply this ATP and NADPH. PEP-CKase activity requires a second ATP in the bundle sheath and a third ATP to phosphorylate ribulose-5-P (Fig. 12.6). Neglect-ing the possible operation of PP-DKase to produce some of the PEP in PEP-CK variants, the total energy input of four ATP and two NADPH per CO$_2$ assimilated is one ATP more than that required by the C$_3$-CR cycle (Table 12.1). However, several aspects of metabolism in bundle sheath cells remain to be resolved.

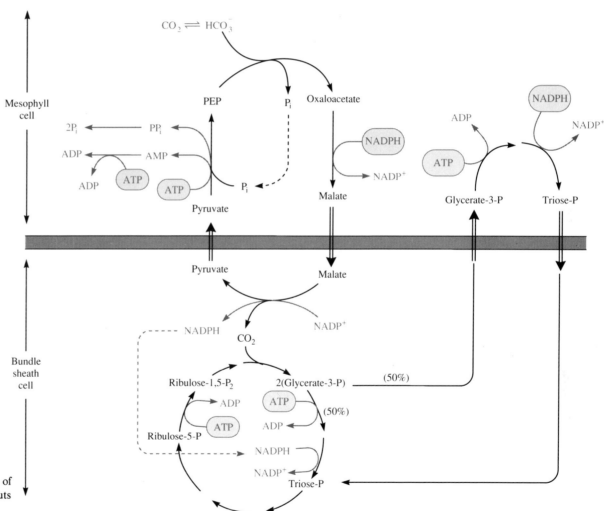

Fig. 12.7 Energy inputs for the assimilation of a molecule of CO$_2$ in a NADP M-E variant. Ultimate sources of the inputs are shaded.

Table 12.1 Energy inputs for the reductive assimilation of one CO_2 molecule in C_3, C_4 and CAM plants. Values shown are not corrected for the energy expended in recycling products formed by ribulose-1,5-P_2 oxygenase activity via the C_2-PR cycle; This is important in C_3 plants, especially at high temperatures.

| Plant type and variant | Energy inputs per CO_2 assimilated | | | | | |
| | Mesophyll | | Bundle Sheath | | Total | |
	ATP	NADPH	ATP	NADPH	ATP	NADPH
C_3	3	2			3	2
C_4:NADP M-E	3	2	2	0	5	2
C_4:NAD M-E	3	1	2	1*	5	2
C_4:PEP-CK	1	1	3	1	4	2
CAM:NADP M-E	6.5	2			6.5	2
CAM:PEP-CK	5.5	2			5.5	2

* There is some evidence to suggest that a portion of the NADPH consumed in the bundle sheath cells of NAD M-E variants is derived from mesophyll cells.

12.9.3 NAD M-E variants

The energy requirements for this variant are similar to those for the NADP M-E type, although aspartate rather than malate is the principal dicarboxylate. As before, metabolism of glycerate-3-P to glyceraldehyde-3-P accounts for one ATP and one NADPH in each cell type; production of PEP from pyruvate via PP-DKase in the mesophyll requires an additional two ATP and regeneration of ribulose-1,5-P_2 a further one ATP. Unlike NADP M-E variants, the bundle sheath chloroplasts of NAD M-E variants have a functional photosystem II to provide a substantial part of the NADPH for glycerate-3-P, although other evidence indicates that mesophyll chloroplasts contribute to this process. The energy inputs are five ATP and two NADPH per CO_2 assimilated (Table 12.1).

12.10 Carbon dioxide assimilation in C_4 plants involves extensive intracellular and intercellular transport of metabolites

The assimilation of CO_2 in C_4 plants requires the transport of various metabolites between mesophyll and bundle sheath cells and across the membranes of chloroplasts and mitochondria at rates approximating that of CO_2 assimilation. Thus, in NADP M-E variants PEP is transported from mesophyll chloroplasts into the cytosol, oxaloacetate from the cytosol into chloroplasts and malate from mesophyll chloroplasts into bundle sheath cells at rates in excess of $200\,\mu mol\,mg^{-1}$ chlorophyll h^{-1}.

Mesophyll chloroplasts from C_4 plants, unlike C_3 plants, contain a light-coupled electrogenic carrier for pyruvate. Mesophyll chloroplasts also contain a phosphate translocator similar to that of C_3 plants (see Section 10.8). However, the translocator from C_3 plants has a low affinity for PEP, but that in C_4 plants has a high affinity and fulfills two important functions. One concerns the exchange of cytosolic P_i (a product of PEP-Case activity) with PEP synthesized in chloroplasts by PP-DKase activity. The other is the exchange of glycerate-3-P (imported from bundle sheath cells) with triose-P produced by reduction of glycerate-3-P in mesophyll chloroplasts. Mesophyll chloroplasts also contain a dicarboxylate translocator, presumably to link the inward transport of oxaloacetate with the outward movement of malate and/or aspartate.

Little is known about carrier mechanisms of chloroplasts and mitochondria of bundle sheath cells as clean, uniform preparations are difficult to obtain.

The transport of metabolites between mesophyll and bundle sheath cells is no less important than the intracellular transport between metabolic compartments and is presumed to involve passive diffusion via symplastic connections in the plasmodesmata. This raises many questions about rates and the ability of cells to delineate their perimeter and control their activities. Alternatively, some form of selective transport through plasmodesmata connections could be involved.

12.11 The C_4 mechanism of carbon dioxide assimilation explains many of the physiological properties of C_4 plants

Since C_4 plants expend more energy per CO_2 assimilated than C_3 plants (Table 12.1) it would be expected that C_4 plants are less efficient in their use of energy (and hence light) and have lower growth rates. At temperatures less than about $22\,°C$, C_3 plants do assimilate CO_2 faster but at higher temperatures, C_4 plants have higher growth and CO_2 assimilation rates and use light more efficiently (Fig. 12.8). To add to the paradox, C_4 plants contain less RuP_2-Case activity per unit of chlorophyll. How then are the faster growth rates and higher efficiencies attained?

The answer lies principally in the relative expressions of the oxygenase and carboxylase activities associated with RuP_2-C/Oase. In C_3 plants the carboxylase activity accounts for nearly all CO_2 assimilation. At high temperatures various factors combine to increase the proportion of ribulose-1,5-P_2 subject to RuP_2-Oase

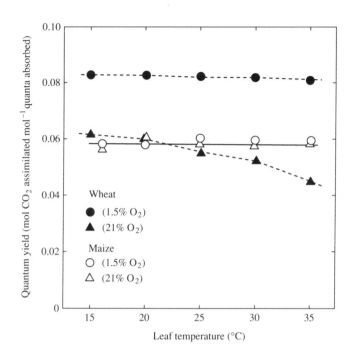

Fig. 12.8 Effect of temperature and the concentration of O$_2$ on the quantum efficiency of net CO$_2$ assimilation in wheat (*Triticum aestivum*), a typical C$_3$ plant and maize (*Zea mays*), a typical C$_4$ plant. Quantum efficiency is a measure of growth in relation to the amount of light energy absorbed. From Ku, S-B. & Edwards, G.E. (1978) *Planta*, **140**, 1–6.

activity (see Section 11.3). The regeneration of ribulose-1,5-P$_2$ requires the consumption of five ATP and three NADPH per ribulose-1,5-P$_2$ oxidized via RuP$_2$-Oase (see Section 11.7). An increase with temperature in the proportion metabolized via oxygenase causes a decrease in the efficiency with which C$_3$ plants use the absorbed light energy (Fig. 12.8). C$_4$ plants, however, initially assimilate CO$_2$ in mesophyll cells via an O$_2$-insensitive system involving PEP-Case. The subsequent release of CO$_2$ in the bundle sheath establishes a high concentration of CO$_2$ at the site of RuP$_2$-Case, suppressing RuP$_2$-Oase activity and making the photosynthetic efficiency of C$_4$ plants insensitive to temperature and O$_2$ (Fig. 12.8). Thus, at high temperatures, C$_4$ plants perform better than C$_3$ plants. The high concentrations of CO$_2$ in bundle sheath cells means that the RuP$_2$-C/Oase of C$_4$

plants works more efficiently as a carboxylase than in C$_3$ plants. This is reflected in the lower levels of RuP$_2$-C/Oase per unit of chlorophyll in C$_4$ plants.

The mechanism in C$_4$ plants involving incorporation of CO$_2$ into dicarboxylates, their subsequent decarboxylation and recycling of the C$_3$ product acts as a C$_4$-CO$_2$ concentrating cycle (C$_4$-CC cycle). Although the C$_4$-CC cycle incurs an energy penalty (Table 12.1), this is less than that of regenerating ribulose-1,5-P$_2$ from the products of RuP$_2$-Oase activity in C$_3$ plants, which increases greatly with temperature.

Concentration of CO$_2$ in bundle sheath cells via the C$_4$-CC cycle is relevant to other physiological characteristics of C$_4$ plants. At intracellular concentrations of CO$_2$ just below the CO$_2$ compensation point, C$_3$ plants undergo net carbon loss but at these concentrations, C$_4$ plants with their negligible CO$_2$ compensation points exhibit net CO$_2$ assimilation and hence growth.

To support high growth rates in air, C$_3$ plants need to maintain high concentrations of CO$_2$ within the leaf by having a large number of stomata per unit area of leaf surface, resulting in greater loss of water by evapotranspiration. This makes C$_3$ plants less efficient in their use of restricted water availability than C$_4$ plants.

12.12 CAM plants assimilate carbon dioxide into malate at night and reassimilate carbon dioxide into carbohydrate in the light

CAM plants incorporate CO$_2$ by Crassulacean Acid Metabolism, common to members of the family Crassulaceae. CAM also occurs in many other plants with thick fleshy leaves (e.g. cacti) and many orchids. CAM plants are distinguished, particularly under conditions of water stress, by assimilating their CO$_2$ at night, 90% of which accumulates as malate in large vacuoles and causes a large increase in acidity. To facilitate this the stomata of CAM plants open at night. HCO$_3^-$, formed from CO$_2$, reacts with PEP to form oxaloacetate which is rapidly reduced to malate, catalysed by PEP-Case and NAD-specific malate dehydrogenase (NAD M-DHase) respectively (Fig. 12.9). During an ensuing light period, the stomata close and the pH of the leaves increases as malate is incorporated into starch. Malate is decarboxylated by one of several processes to produce a C$_3$ derivative, releasing free CO$_2$ within the leaf. The decarboxylation mechanisms are similar to those in the bundle sheath cells of C$_4$ plants except that malate is always the initial substrate. They include reactions catalysed by (a) NADP malic

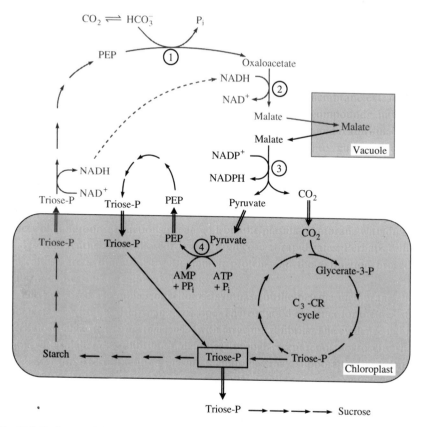

Fig. 12.9 Pathway of CO_2 assimilation in a typical CAM plant. Processes in grey concern the aquisition of CO_2 from the atmosphere and its incorporation into malate at night. Processes in black occur by day and involve the decarboxylation of malate and reassimilation of CO_2 via the C_3-CR cycle using light as an energy source. The decarboxylation mechanism shown here is characteristic of a NADP M-E variant. Enzymes associated with the C_4-CC cycle are: (1) PEP carboxylase; (2) NAD malate dehydrogenase; (3) NADP malic enzyme; (4) pyruvate, phosphate dikinase.

enzyme or NAD malic enzyme (Eqns 12.14 & 12.15) and a sequence (b) involving NAD M-DHase (Eqn. 12.16) coupled to PEP-CKase (Eqn. 12.17).

(a) $\text{Malate} + NADP^+ \rightleftharpoons \text{pyruvate} + NADPH + H^+ + CO_2$ (12.14)

$\text{Malate} + NAD^+ \rightleftharpoons \text{pyruvate} + NADH + H^+ + CO_2$ (12.15)

(b) $\text{Malate} + NAD^+ \xrightleftharpoons[\text{}]{\text{NAD M-DHase}} \text{oxaloacetate} + NADH + H^+$ (12.16)

$\text{Oxaloacetate} + ATP \xrightleftharpoons[\text{}]{\text{PEP-CKase}} PEP + ADP + CO_2$ (12.17)

CAM variants, like the C_4 variants, are named after their decarboxylating mechanism. The CO_2 released is reassimilated by the reaction catalysed by RuP_2-Case and reductively metabolized in the light via the C_3-CR cycle. The C_4 carboxylation and decarboxylation mechanisms in C_4 plants are separated in space and CO_2 assimilation is continuous, whereas in CAM plants they are separated in time and CO_2 assimilation is discontinuous.

CAM plants must have large reserves of starch to generate PEP and NAD(P)H to support CO_2 assimilation and oxaloacetate reduction in the dark. The PEP is not available for reuse until malate is decarboxylated in the subsequent light period.

Recycling of C_3 compounds formed upon decarboxylation of malate in the light (Eqns 12.14–12.17) to maintain the carbon balance is an essential feature of CAM. Pyruvate produced in NADP M-E variants is transformed to PEP via PP-DKase activity. Compared to C_4 plants, the level of PP-DKase in these CAM variants is very low, in keeping with the low rates of CO_2 assimilation. PEP formed directly during decarboxylation (Eqn. 12.17) or by phosphorylation of pyruvate (Eqn. 12.1) is metabolized to starch in the light.

The decarboxylation of oxaloacetate via PEP-CKase during the day must be accompanied by a mechanism which prevents carboxylation of PEP by PEP-Case by a futile carboxylation/decarboxylation cycle leading to net hydrolysis of ATP.

Control is most likely exercised by metabolic regulation of enzyme activity since PEP-Case, PEP-CKase and NADP malic enzyme are all cytosolic enzymes in CAM plants.

Mechanisms must also exist to initiate malate synthesis in the dark, malate decarboxylation in the light and carboxylation of ribulose-1,5-P_2 (and not PEP) in the light. Little is known about these.

Some CAM plants alter their mechanism of CO_2 assimilation under certain environmental conditions. For example *Mesembryanthemum crystallinum* incorporates CO_2 predominantly via the C_3-CR cycle (CO_2 uptake by day) during periods of water sufficiency but reverts to CAM (CO_2 uptake by night) if it is water stressed (e.g. watered with a high salt solution). In other species the pattern is controlled by the length of the photoperiod or diurnal fluctuations in temperature.

The assimilation of one molecule of CO_2 in CAM plants requires two NADPH and three ATP via the C_3-CR cycle. An additional three and a half ATP (net) are required by NADP M-E variants and two and a half ATP (net) by PEP-CK variants. Thus, CAM plants incur an even greater energy penalty than C_4 plants (Table 12.1).

12.13 CAM affords an explanation for many physiological and ecological features of CAM plants

CAM plants are very well adapted to dry and semi-arid environments because they have a very high water-use efficiency but very slow growth rates. This is clearly related to their method of CO_2 assimilation which involves acquisition of CO_2 at night when the ambient air is cool and the water potential gradient between the leaf and the atmosphere is much lower than during the searing heat of the day; this allows relatively little water loss through their stomata.

CAM plants use the carbon they acquire from the atmosphere very efficiently. They do not lose CO_2 by photorespiration and CO_2 produced by mitochondrial respiration during the night is reassimilated. However, their more efficient use of water incurs an energy penalty.

The slow growth rates of CAM plants arise from the discontinuous nature of CO_2 metabolism as CO_2 is incorporated into malate at night and reductively assimilated in the light. This accounts for the massive pool of C_4 dicarboxylates in CAM plants, some 100-fold greater than in C_4 plants, and the much longer turnover halftimes ($t_{0.5}$) for the C_4 dicarboxylates; for C_4 plants with a continuous CO_2 assimilation mechanism and a small pool of C_4 dicarboxylates $t_{0.5} = 5$ s, but for CAM plants, $t_{0.5} = 3$ h.

12.14 Phylogenetic distribution of C_4 and CAM plants and its implications for herbicide design

C_4 and CAM plants are found principally in just a few families (Table 12.2). More than 60% of the currently known C_4 plants are grasses (family Poaceae) and a

Table 12.2 Phylogenetic distribution of C_4 and CAM plants. For comparison with the 943 C_4 species and 453 CAM species listed below, over 370 families containing about 230 000 species are recognized for the angiosperms alone. From Moore, P. (1981) *New Scientist*, **89**, 394–7.

Class	Family	Number of species C_4	Number of species CAM
Angiosperms			
Dicotyledons	Aizoaceae	11	39
	Amaranthaceae	56	–
	Asclepiadaceae	1	12
	Asteraceae	39	10
	Cactaceae	–	94
	Chenopodiaceae	104	–
	Crassulaceae	–	95
	Euphorbiaceae	75	30
	Nyctaginaceae	18	–
	Portulacaceae	8	4
	Zygophyllaceae	11	–
	Other families (15)	13	18
Monocotyledons	Agavaceae	–	14
	Bromeliaceae	–	66
	Cyperaceae	31	–
	Liliaceae	1	24
	Orchidaceae	–	44
	Poaceae	575	–
Gymnosperms	Welwitschiaceae	–	1
Pteridophytes	Polypodiaceae	–	2

further five families (Chenopodiaceae, Euphorbiaceae, Amaranthaceae, Asteraceae and Cyperaceae) account for another 32%. Although no single family listed in Table 12.2 consists entirely of C_4 members, it is not uncommon to find that all the members of a particular subfamily, tribe or genus have C_4-CO_2 assimilation mechanisms. Within the Poaceae, for example, all members of the subfamily Eragrastoideae are C_4 and most members of the subfamily Panicoideae are C_4, while all the festucoid grasses are C_3.

CAM plants are also restricted to a few families; members of the Crassulaceae, Cactaceae, Bromeliaceae, Orchidaceae and Aizoaceae account for more than 80% of known CAM plants (Table 12.2).

Many C_4 plants are very important weeds (Section 12.1) and their restriction (particularly the C_4 variants) to particular species, tribes or families should make it possible to develop herbicides specific to them. For example, the sedge *Cyperus rotundus*, a NADP M-E C_4 variant, is an important weed of rice, a C_3 plant. In theory, it should be possible to design a herbicide to inhibit a vital component of the C_4-CC cycle (e.g. PEP-Case or PP-DKase?) of the C_4 weed without impairing the C_3-CR cycle of rice, provided that there are no other side-effects. By reference to the C_4-CC cycles of the three C_4 variants (see Figs 12.5 & 12.6), it might even be possible to design a compound to inhibit the C_4-CC cycle of one type of variant but not that of the others; this would allow development of herbicides towards certain C_4 weeds in particular C_4 crops. These ideas are currently attracting considerable research activity.

Further reading

Monographs and treatises: Edwards & Walker (1983); Gibbs & Latzko (1979); Hatch & Boardman (1981); Hatch & Boardman (1987); Lawlor (1987).
Physiology and ecology of C_4 and CAM plants: Lange et al. (1982) Chapter 15.

Assimilation of inorganic nitrogen
into amino acids

13.1 Plants use inorganic nitrogen for synthesis of protein amino acids

The element nitrogen (N) is a constituent of amino acids, proteins, pyrimidines, purines, nucleic acids and many other essential compounds, where it occurs in oxidation state -3 (as in NH_3). Nitrogen is usually obtained as water-soluble salts, mostly as nitrate (NO_3^-) from soil. Since NO_3^- contains nitrogen in oxidation state $+5$, assimilation of nitrogen as NO_3^- involves an eight electron reduction with a large input of energy.

$$HNO_3 + H_2O \rightleftharpoons NH_3 + 2O_2 \; (\Delta G^{\circ\prime} = 348 \, kJ \, mol^{-1}) \tag{13.1}$$

The incorporation of nitrogen as NO_3^- into organic nitrogen compounds (N-compounds) is traditionally divided into: (i) assimilatory NO_3^- reduction (reduction of NO_3^- to NH_3); (ii) assimilation of NH_3 into glutamine and glutamate via the $C_5 - NH_3$ assimilation cycle; and (iii) amino acid anabolism (synthesis) using glutamate and/or glutamine as the source of amino or amide groups. Here, a distinction must be made between assimilatory NO_3^- reduction catalysed by plants and dissimilatory NO_3^- reduction by certain anaerobic bacteria (e.g. *Paracoccus denitrificans*) where NO_3^- is a terminal electron acceptor in much the same way as aerobic organisms use O_2.

Plants readily use NH_4^+ as a nitrogen source when this is available (e.g. $(NH_4)_2SO_4$ fertilizer) and bypass assimilatory NO_3^- reduction. However, in most aerobic soils, NH_4^+ is not important due to the oxidative activities of chemoautotrophic bacteria (e.g. *Nitrosomonas* and *Nitrobacter*).

Humans cannot use inorganic nitrogen as a nitrogen source so the human diet must contain sufficient organic nitrogen (mostly as protein) to maintain nitrogen balance. Humans can synthesize some of the 20 protein amino acids from various organic nitrogen sources but there are nine amino acids which cannot be made, or made in sufficient amounts, for human needs; these essential amino acids are listed in Table 13.1. In some regions of the world plants are the only significant source of food. Ideally, plant foodstuffs should contain the essential amino acids in proportions similar to those required for human nutrition. Unfortunately most

Table 13.1 Essential amino acids required in the human diet and their concentrations in whole hen eggs, a cereal (wholemeal wheat flour) and a legume (soya bean flour). In addition to the nine amino acids listed, arginine is considered to be essential for infants and growing children. After Paul, A.A. & Southgate, D.A.T. (1978) In *McCance and Widdowson's The Composition of Foods*, 4th edn. H.M.S.O., London. Reproduced with the permission of the Controller of Her Majesty's Stationery Office.

Amino acid	Amino acid relative to total nitrogen (mg amino acid g^{-1} nitrogen)		
	Egg*	Cereal	Legume
Leucine	520	420	490
Valine	470	280	300
Lysine[†]	390	150	400
Phenylalanine	320	280	310
Threonine	320	170	240
Isoleucine	350	210	280
Methionine[†]	200	100	80
Histidine	150	130	160
Tryptophan	110	70	80

*These values are regarded as similar to those required to maintain an adult human in nitrogen equilibrium and permit maximum use of all the component amino acids.

[†] Note that values for lysine in the cereal and for methionine in the legume are low compared to those in egg.

cereals, especially maize, are low in lysine while most legumes (rich sources of protein) are low in the sulphur amino acids (Table 13.1). Thus, the synthesis of the essential amino acids in plants and the amino acid composition of plant foodstuffs is important to human nutrition.

Herbicides are used extensively to control the growth of unwanted plants which diminish crop yields. Several herbicides strongly inhibit the synthesis of some essential amino acids. In theory, compounds specifically with these characteristics are not toxic to humans.

Gaseous N_2 comprises about 80% by volume of the air in the atmosphere.

Various free-living prokaryotic organisms, including certain soil organisms (e.g. *Klebsiella* and *Frankia*) and the anaerobic photoautotrophic bacteria can use N_2 as a nitrogen source. Some cyanobacteria (blue-green algae) can use N_2 when stressed for a source of soluble nitrogen. Eukaryotic plants *per se* cannot use N_2. However, a few plants, mostly legumes, can establish symbioses with specialized prokaryotic organisms which reduce N_2 to NH_3, a process known as N_2 fixation, and they use this NH_3 as a nitrogen source for plant growth.

This chapter describes the passage of nitrogen from inorganic NO_3^- into some of the amino acids, and the process of N_2 fixation in legume–*Rhizobium* symbioses.

13.2 Plants reduce nitrate to ammonia

13.2.1 Nitrate uptake

Plants acquire NO_3^- from soil via an active uptake mechanism present in root parenchyma cells. Measuring NO_3^- uptake is beset with problems, since a suitable radioactive isotope of nitrogen is not readily available (e.g. ^{13}N has a half-life of only 10 min and ^{15}N is not radioactive). Furthermore, when NO_3^- is supplied to cells it is metabolized and effluxed, as equilibria are established with pre-existing pools. However, chlorate (ClO_3^-) is an alternative substrate of the NO_3^- uptake mechanism, so that ClO_3^- containing the radioactive isotope ^{36}Cl can be used to study nitrate uptake properties. Of course, ClO_3^- inhibits NO_3^- uptake and vice versa but ClO_3^- is not assimilated. ClO_3^- uptake into roots is extremely sensitive to the inhibitor FCCP, an uncoupler of energy-dependent ion transport (see Section 6.2.4).

13.2.2 The site of nitrate assimilation varies between species

In some species, especially crop plants such as wheat and maize grown under favourable conditions, about 70–80% of the nitrogen in the xylem fluid is NO_3^-. This is transported to leaves where it is reduced and incorporated into glutamine using light as the energy source. Indeed, most of the processes associated with NO_3^- reduction in leaves occur in chloroplasts and form an important part of photosynthesis. In other species, however, most of the nitrogen in bleeding sap is found as amino acids (mostly asparagine and glutamine) or ureides (e.g. allantoin) indicating that NO_3^- is reduced and assimilated in root tissue and the associated energy requirements come from dark respiration. However, substantial amounts of NO_3^- can be detected in the xylem sap of these plants at high concentrations of external NO_3^-.

13.2.3 Nitrate is reduced to nitrite

NO_3^- is reduced to nitrite (NO_2^-) by NADH in a reaction involving the transfer of two electrons, catalysed by NO_3^- reductase, a cytosolic enzyme.

$$NADH + H^+ + NO_3^- \rightleftharpoons NAD^+ + NO_2^- + H_2O \qquad (13.2)$$

NO_3^- reductase has a rather low affinity for NO_3^- ($K_m = 0.2$ mM) but a high affinity for NADH ($K_m = 10$ μM). The enzyme is not stable and must be extracted in the presence of a thiol and stored with an inert, or 'carrier', protein to retard loss of activity.

NO_3^- reductase is a complex enzyme containing flavin, haem (possibly as cytochrome *b*) and the transition metal molybdenum (Mo). The enzyme can use a variety of non-biological reducing agents in place of NADH (e.g. reduced viologen dyes and reduced flavins) to reduce NO_3^- (i.e. NO_3^- reductase activity). These activities and the overall reaction using NADH are inhibited by azide and cyanide. The enzyme also catalyses the reduction by NADH of various alternative substrates (e.g. exogenous oxidized cytochrome *c*) in the absence of NO_3^- (diaphorase or dehydrogenase activity). The passage of electrons from NADH to NO_3^- and the various partial reactions are as follows.

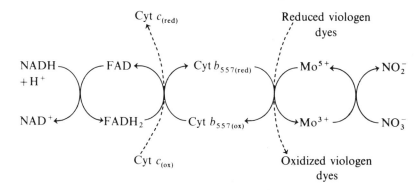

Without molybdenum, plants are unable to utilize NO_3^- as a nitrogen source for growth. Tungsten, another transition metal, inhibits the growth response of plants to molybdenum since it is incorporated into the NO_3^- reductase apoprotein in place of molybdenum to form an inactive enzyme.

NO_3^- reductase is one of the few enzymes in plants which is regulated by an exogenous substrate at the level of gene expression. Plants supplied with NH_4^+ lack NO_3^- reductase. NO_3^- induces the synthesis of NO_3^- reductase in virtually all tissues (see Chapter 20). In general, NH_4^+, a later product of NO_3^- assimilation, represses the induction of NO_3^- reductase in roots and tissue cultures and in cultures of green algae, *Lemna* and yeast but is apparently ineffective on intact leaves of higher plants.

13.2.4 Nitrite is reduced to ammonia

NO_2^- does not normally accumulate to significant concentrations, but is reduced to NH_3 in a reaction involving a six-electron transfer catalysed by NO_2^- reductase and using reduced ferredoxin (Fd_{red}) as electron donor.

$$HNO_2 + 6Fd_{red} + 6H^+ \rightleftharpoons NH_3 + 2H_2O + 6Fd_{ox} \qquad (13.3)$$

NO_2^- reductase, unlike NO_3^- reductase, has an extremely high affinity for its nitrogen substrate; K_m values for NO_2^- as low as $1\,\mu M$ have been reported. Furthermore, the potential NO_2^- reductase activity of plant tissues is usually 5–20 times greater than the rate at which NO_3^- is reduced to NO_2^- *in vivo*, reflecting the high toxicity of NO_2^-, which can react with the prosthetic group of haem proteins and the amino groups of amino acids, purines and pyrimidines.

In photosynthetic tissue NO_2^- reductase is located in chloroplasts where the formation of Fd_{red} and hence NO_2^- reduction is light-dependent (see Section 13.5). The enzyme in non-photosynthetic tissue is supposedly associated with proplastids but the mechanism for producing Fd_{red} is not clear. NO_2^- reductases from fungi use NADPH rather than Fd_{red} as electron donor.

Measurements of plant NO_2^- reductase *in vitro* commonly use reduced viologen dyes as reductant; the reduced forms are pigmented whereas the oxidized forms are colourless. NO_2^- reductase contains three atoms of iron per molecule of apoenzyme, two as labile iron–sulphur (Fe–S) centres and one in a haem compound known as sirohaem. Reduction of NO_2^- is thought to begin with reduction of sirohaem by Fd_{red} which then reacts with NO_2^- to form a NO_2^-–sirohaem complex. This initiates the reduction, by Fd_{red}, of the Fe–S centre which promotes electron transfer to the NO_2^-–sirohaem complex, progressively reducing it to NH_3.

NO_2^- reductase is an inducible enzyme but shows much smaller responses than NO_3^- reductase. There are differences between species as to whether the inducer is NO_3^- or NO_2^-.

In NADP M-E variants of C_4 plants the chloroplasts of mesophyll cells, but not those of bundle sheath cells, are photochemically fully active (see Section 12.9) and are the main sites of NO_2^- reductase activity.

13.3 Ammonia is incorporated via glutamine into glutamate by the C_5-ammonia assimilation cycle

Studies of $^{15}NH_3$ incorporation, over 25 years ago, revealed that the first major products of NH_3 assimilation were the amide and amino groups of glutamine and the amino group of glutamate. It was assumed for a long time that the reaction involving the reductive incorporation of NH_3 into 2-oxoglutarate (α-ketoglutarate) to form glutamate, catalysed by NAD-specific glutamate dehydrogenase (NAD G-DHase), present in high activity in mitochondria, was the principal mechanism for NH_3 assimilation in plants (Eqn. 13.4).

2-Oxoglutarate   Glutamate   Glutamine

$$(13.4) \qquad\qquad (13.5)$$

Plants also have NADP-specific glutamate dehydrogenase (NADP G-DHase) activity, located principally in chloroplast thylakoids.

$$\text{2-Oxoglutarate} + NH_3 + NADPH + H^+ \rightleftharpoons \text{glutamate} + NADP^+ \qquad (13.6)$$

This enzyme has a low affinity for NH_3 (K_m about $5\,mM$) and the reaction, like that catalysed by NAD G-DHase, is freely reversible. Glutamine synthetase which catalyses the amidation of glutamate in the presence of ATP (Eqn. 13.5) also occurs in plants. This reaction, coupled to Eqn. 13.4, could account for the incorporation of NH_3 into glutamate and glutamine.

An alternative sequence for the incorporation of NH_3 into glutamate involves the reactions catalysed by glutamine synthetase (Eqn. 13.7) and glutamate synthase (Eqn. 13.8).

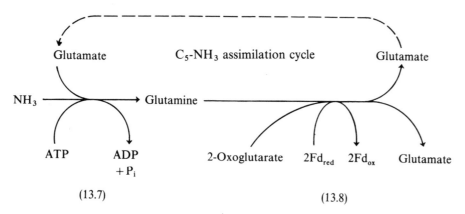

(13.7)   (13.8)

Here, NH_3 is initially incorporated into the amide group of glutamine. This is reductively transferred to 2-oxoglutarate, forming two molecules of glutamate (Fig. 13.1). The two reactions constitute a cycle which consumes one glutamate (a C_5 compound), one NH_3 and one 2-oxoglutarate (also C_5) and produces two glutamate, one of which feeds back to sustain the cycle (i.e. a C_5-NH_3 assimilation cycle). The net result is production of one glutamate from NH_3 and 2-oxoglutarate with the expenditure of one ATP and two Fd_{red}. The glutamate synthase of leaves is Fd-specific and located in chloroplasts but NADPH-linked glutamate synthase has also been reported in non-photosynthetic tissue.

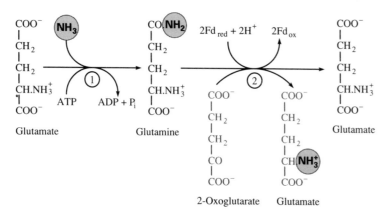

Fig. 13.1 Assimilation of NH_3 via the reaction sequence involving the enzymes glutamine synthetase (1) and glutamate synthase (2).

The incorporation of NH_3 via the reactions of Eqns 13.7 and 13.8 is similar to that of Eqns 13.4 and 13.6, as both result in net production of glutamate from NH_3 and 2-oxoglutarate. Mechanistically they differ as glutamine is not an intermediate in the production of glutamate by G-DHase activity. The NH_3 labelling patterns of glutamate and glutamine also differ; in the G-DHase sequence, NH_3 is incorporated into both the amino- and amide-N atoms of glutamine but in the C_5-NH_3 assimilation cycle NH_3 is initially incorporated into the amide-N atom of glutamine.

Currently the C_5-NH_3 assimilation cycle is held to be the major mechanism for NH_3 assimilation in leaves, since $^{15}NH_3$ is incorporated principally into the amide-N atom of glutamine in short-term experiments and the amino group of glutamate in long-term experiments. Also, the incorporation of NH_3 into the amino group of glutamate is blocked by the inhibitors methionine sulphoximine (an inhibitor of glutamine synthetase) and azaserine (an inhibitor of glutamate synthase), which have little or no effect on NAD- and NADP-specific G-DHase.

$$
\begin{array}{cccc}
O{=}C{-}NH_2 & \begin{array}{c}CH_3\\|\\O{=}S{=}NH\end{array} & COO^- & \begin{array}{c}CH_2{-}N{=}NH\\|\\CO\end{array} \\
| & | & | & | \\
CH_2 & CH_2 & CH_2 & O \\
| & | & | & | \\
CH_2 & CH_2 & CH_2 & CH_2 \\
| & | & | & | \\
CH.NH_3^+ & CH.NH_3^+ & CH.NH_3^+ & CH.NH_3^+ \\
| & | & | & | \\
COO^- & COO^- & COO^- & COO^- \\
\text{Glutamine} & \text{Methionine} & \text{Glutamate} & \text{Azaserine} \\
 & \text{sulphoximine} & &
\end{array}
$$

There are theoretical reasons too for preferring the C_5-NH_3 assimilation cycle, as incorporation of NH_3 via the cycle involves the expenditure of ATP making it essentially irreversible. It also uses Fd_{red}, a more powerful reducing agent than NADPH. Glutamine synthetase also has a much higher affinity for NH_3 ($K_m = 5$–20 μM) which would serve to keep the intracellular concentration of NH_3 low, as compared to a K_m of about 5 mM for NAD G-DHase. In chloroplasts this is important since high concentrations of NH_3 cause uncoupling of electron transport from photophosphorylation. Interestingly, the toxicity of the herbicide phosphinothricin or glufosinate (homoalanine 4-(methyl)phosphinite) is attributed to NH_3 toxicity due to inhibition of glutamine synthetase.

13.4 Ammonia produced by photorespiration is reassimilated via the C_5-ammonia assimilation cycle

Photorespiration in the leaves of C_3 plants results in the oxidation of glycine with the production of equimolar amounts of CO_2 and NH_3 at about 30–40 μmol mg^{-1} chlorophyll h^{-1} (Section 11.5).

$$2\,\text{Glycine} + NAD^+ + H_2O \rightleftharpoons \text{serine} + CO_2 + NH_3 + NADH + H^+ \qquad (13.9)$$

The NH_3 produced must be reassimilated at a similar rate to prevent rapid loss of nitrogen. When [^{15}N]glycine is fed to illuminated leaves, very little $^{15}NH_3$ is detected but label is incorporated into serine, glutamine and glutamate. However, in the presence of methionine sulphoximine ^{15}N-label accumulates in serine and NH_3, demonstrating that glutamine synthetase (and hence the $C_5 - NH_3$ assimilation cycle) is involved in assimilating NH_3 derived from glycine. Further evidence involving the use of barley mutants defective in glutamine synthetase and glutamate synthase is given in Section 11.6.

13.5 Light is an important energy source for the assimilation of inorganic nitrogen in leaves

Plants which transport nitrogen in xylem sap principally as NO_3^- assimilate inorganic nitrogen into amino acids predominantly in photosynthetic tissue. The assimilation of nitrogen as NO_3^- into glutamate requires a large input of energy involving the expenditure of one NADH (2e$^-$) for reduction of NO_3^- to NO_2^-, six Fd$_{red}$ (6e$^-$) for reduction of NO_2^- to NH_3 and two Fd$_{red}$ (2e$^-$) and one ATP for the C_5-NH_3 assimilation cycle.

In leaf tissue, all stages of the reductive assimilation of NO_3^- are either directly dependent on, or indirectly enhanced by, light. Even the reduction of NO_3^- to NO_2^- in the cytosol is stimulated by light and is further enhanced by additions of malate and other C_4 dicarboxylates as well as triose phosphates and their precursors which readily permeate the chloroplast envelope (see Section 10.7). This is attributed to oxidation of malate and glyceraldehyde-3-P via NAD-specific malate dehydrogenase and glyceraldehyde-3-P dehydrogenase in the cytosol to produce NADH for NO_3^- reduction. The oxidation products (oxaloacetate and glycerate-3-P) return to the chloroplast and are reduced in processes coupled to light-dependent electron transport. This restores the level of malate and triose-P for recycling (see Section 14.4).

In photosynthetic cells, the enzymes of NO_2^- reduction and NH_3 assimilation are located in chloroplasts. Isolated chloroplasts in the light support the reduction of NO_2^- to NH_3 with the evolution of one and a half O_2 per NO_2^- reduced, consistent with derivation of the necessary reducing equivalents (6e$^-$) from H_2O via the light reactions (Fig. 13.2, reaction (1)). Similarly, illuminated chloroplasts, which contain an endogenous pool of glutamate, support the assimilation of NH_3 into glutamate in the presence of 2-oxoglutarate with the evolution of a half O_2 per NH_3 assimilated. This is sensitive to the glutamine synthetase inhibitor, methionine sulphoximine, but in pea chloroplasts O_2 evolution can be re-initiated with glutamine. These data are consistent with the C_5-NH_3 assimilation cycle (Fig. 13.2, reactions (2) and (3)) with ATP and Fd$_{red}$ supplied by light-driven electron transport. Thus, the processes shown in Fig. 13.2 constitute the dark reactions of photosynthetic nitrogen assimilation, analogous to the dark reactions of photosynthetic CO_2 assimilation.

The theoretical rate of exogenous inorganic nitrogen assimilation needed to

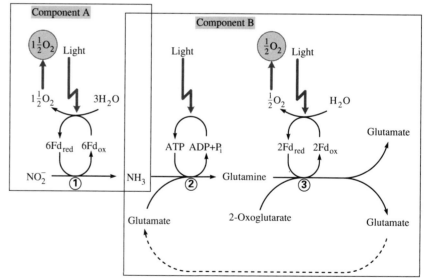

Fig. 13.2 Theory of O_2 evolution associated with the assimilation of inorganic NO_2^- into glutamate by illuminated chloroplasts. Component A shows light-dependent reduction of NO_2^- to NH_3, catalysed by NO_2^- reductase (1). Component B shows light-dependent assimilation of NH_3 into glutamate, catalysed by glutamine synthetase (2) and glutamate synthase (3). Component B is self-sustaining with respect to the glutamate required for (2).

support plant growth is 5–7 μmol mg^{-1} chlorophyll h^{-1} (see Section 1.7) and the rate of NH$_3$ production via photorespiration is about 30–40 μmol mg^{-1} chlorophyll h^{-1}. The NH$_3$-assimilating capacity of illuminated chloroplasts approaches the sum of these rates indicating that the C$_5$-NH$_3$ assimilation cycle is a major metabolic activity.

Although all C$_3$ plants have an active light-coupled C$_5$-NH$_3$ assimilation cycle to reassimilate photorespiratory NH$_3$, in some non-leguminous species nitrogen occurs in xylem sap principally as one or more of asparagine, glutamine or several other amino acids (i.e. NO$_3^-$ is assimilated in roots).

The activity of the C$_5$-NH$_3$ assimilation cycle is indirectly co-ordinated with NH$_3$ production via the C$_2$-PR cycle and NO$_2^-$ reduction in the light, which are all linked to light through their demands for carbon substrates and/or the supply of ATP and Fd$_{red}$. Subtle changes in pH and Mg^{2+} concentration in the stroma (see Section 10.13.2), maximize the activity of enzymes of the NH$_3$ cycle (especially glutamine synthetase).

13.6 Aminotransferases catalyse transfer of the amino group of amino acids to oxo acids to form other amino acids

13.6.1 Principle and examples

Glutamate and glutamine are the direct or indirect sources of nitrogen for the synthesis of all other nitrogen-containing compounds in plants, including the other 18 protein amino acids. The simplest enzyme mechanism for the transfer of nitrogen involves aminotransferases, also known as transaminases. They catalyse the transfer of an amino (—NH$_2$) group together with a proton and electron pair from an amino donor (usually an amino acid) to the carbonyl carbon atom of an amino acceptor (usually an oxo acid) as typified by glutamate:oxaloacetate aminotransferase (GO-ATase).

$$
\begin{array}{c}
\text{COO}^- \\
|\\
\text{CH}_2 \\
|\\
\text{CH}_2 \\
|\\
\text{CH.NH}_3^+ \\
|\\
\text{COO}^-
\end{array}
\; + \;
\begin{array}{c}
\text{COO}^- \\
|\\
\text{CH}_2 \\
|\\
\text{CO} \\
|\\
\text{COO}^-
\end{array}
\;\rightleftharpoons\;
\begin{array}{c}
\text{COO}^- \\
|\\
\text{CH}_2 \\
|\\
\text{CH}_2 \\
|\\
\text{CO} \\
|\\
\text{COO}^-
\end{array}
\; + \;
\begin{array}{c}
\text{COO}^- \\
|\\
\text{CH}_2 \\
|\\
\text{CH}_2 \\
|\\
\text{CH.NH}_3^+ \\
|\\
\text{COO}^-
\end{array}
$$

Glutamate   Oxaloacetate   2-Oxoglutarate   Aspartate

(13.10)

In this reaction the amino group of glutamate is transferred to oxaloacetate to form the new amino acid, aspartate. Like most reactions catalysed by aminotransferases, it is freely reversible.

Plants contain various aminotransferases which differ in their substrate specificity and subcellular location. Since glutamate is the first product containing newly assimilated nitrogen in the amino group at C-2 (α-carbon atom), it is the primary amino donor for the synthesis of the other amino acids. Several of the more common amino acids are synthesized by direct transfer of the amino group of glutamate to oxo acids (derived from other pathways) in the presence of appropriate aminotransferases (Eqns 13.11–13.13).

$$\text{Oxaloacetate} + \text{glutamate} \rightleftharpoons \text{aspartate} + \text{2-oxoglutarate} \qquad (13.11)$$

$$\text{Pyruvate} + \text{glutamate} \rightleftharpoons \text{alanine} + \text{2-oxoglutarate} \qquad (13.12)$$

$$\text{Glyoxylate} + \text{glutamate} \rightleftharpoons \text{glycine} + \text{2-oxoglutarate} \qquad (13.13)$$

The amino acids produced in this way are used to make other amino acids. For example, plants contain an aminotransferase which catalyses the transfer of the amino group of aspartate (synthesized as in Eqn. 13.10) to pyruvate to form alanine.

$$\text{Aspartate} + \text{pyruvate} \rightleftharpoons \text{oxaloacetate} + \text{alanine} \qquad (13.14)$$

The oxo acids used in the production of aspartate, alanine and glycine (Eqns 13.11–13.14) are readily available from other metabolic processes unlike the oxo acids of most other protein amino acids. Various other strategies are used to effect the synthesis of these amino acids. Some are formed by modification of other amino acids without involvement of an aminotransferase (e.g. threonine from aspartate) but transamination is usually involved. This can involve the synthesis of an oxo acid precursor via a highly specialized and often complex pathway, with the amino group added by transamination at the very last step, e.g. valine (Eqn. 13.15).

$$\text{2-Oxoisovalerate} + \text{amino acid} \rightleftharpoons \text{valine} + \text{oxo acid} \qquad (13.15)$$

However, the amino group is not always introduced at the last step as instanced by the synthesis of arogenate, an intermediate in the synthesis of phenylalanine and tyrosine.

$$\text{Prephenate} + \text{amino acid} \rightleftharpoons \text{arogenate} + \text{oxo acid} \qquad (13.16)$$

Many other transamination reactions are cited in this chapter and to these must be added the various transamination reactions involving the synthesis of glycine and serine (at least one of which is essentially irreversible) associated with photorespiration (see Section 11.5). Transamination reactions are also involved in the formation of many other nitrogen-containing products and intermediates (e.g. purines and pyrimidines—see Chapter 15).

The large number of reactions catalysed by aminotransferases and their importance to such a large number of pathways raises the matter of their substrate specificity. There are separate enzymes for each of the reactions shown in Eqns 13.11–13.14 but it is not clear whether they are the only reactions catalysed by each enzyme. For example, when GO-ATase is purified 600–fold from *Phaseolus vulgaris* it does not use pyruvate or glyoxylate as oxo acids but supports transfer of the amino group of phenylalanine to 2-oxoglutarate at 10% of the rate for aspartate and even lower for other aromatic amino acids. This suggests either that GO-ATase exhibits low but significant activity with aromatic amino acids or that even at 600-fold purification, the GO-ATase preparation is contaminated with an aromatic aminotransferase. Relatively few aminotransferases have been extensively purified and shown to be free from competing transaminase activities. Further, at least some aminotransferases from plants occur in multiple forms (isoenzymes).

13.6.2 Aminotransferases contain pyridoxal phosphate

All plant aminotransferases contain the cofactor pyridoxal phosphate (pyridoxal-P) which is essential for activity. It usually remains bound during enzyme purification but various treatments (e.g. with carbonyl reagents) can remove or destroy pyridoxal-P making the enzyme inactive. Activity is restored by addition of pyridoxal-P. Pyridoxal-P is bound to a lysyl residue in the apoprotein (see Section 5.2.2) and in this form reacts with a free amino acid (substrate) to produce pyridoxamine-P and the corresponding free oxo acid product. The reaction then occurs in reverse order; pyridoxamine-P reacts with another oxo acid (substrate) to form an amino acid (product) and regenerate pyridoxal-P.

The requirement for pyridoxal-P renders aminotransferases sensitive to several inhibitors. Amino-oxyacetate (NH_2—O—CH_2—COO^-), a carbonyl binding reagent structurally similar to the amino acid substrates, has a particularly strong affinity for the substrate binding site. It is useful for investigating processes involving transamination (e.g. photorespiration), especially as it does not inhibit the enzymes of the C_5-NH_3 assimilation cycle, although several other enzymes besides aminotransferases contain pyridoxal-P (e.g. cysteine synthase) and are also sensitive

to amino-oxyacetate. Other carbonyl binding agents such as hydroxylamine and semicarbazide also inhibit aminotransferases.

13.7 Synthesis of some other amino acids

The general routes of amino acid synthesis are outlined in Fig. 13.3. The four oxo acids, glyoxylate, pyruvate, hydroxypyruvate and oxaloacetate, together with glutamate provide the main structural components for the synthesis of 16 of the 20 protein amino acids. The carbon frameworks of the remaining four amino acids are mostly formed from ribose-5-P, erythrose-4-P and PEP; nitrogen is introduced usually by transamination or amide group transfer. The synthesis of most of the protein amino acids is discussed with particular emphasis on examples which illustrate important principles, in particular the regulation of branched pathways and the synthesis of amino acids essential for human nutrition.

13.7.1 The amide group of asparagine is derived from glutamine

Asparagine and glutamine are the most important compounds for transporting nitrogen in the majority of plants (see also Section 13.8.3). They are well suited for this as they exchange rapidly between xylem and phloem, have a high nitrogen:carbon ratio and effectively transport NH_3 in a non-toxic form. The synthesis and movement of these forms of nitrogen are especially important during seed maturation when various storage and functional proteins are synthesized; this process is reversed during seed germination. In the cotyledons of germinating *Lupinus albus* seeds the level of protein nitrogen falls by over 50% in 14 days; over 85% of the protein nitrogen lost is incorporated into asparagine and transported to the growing axes.

Despite its obvious importance, the mechanism of asparagine synthesis *in vivo* remains unclear largely because of the results of radioactive tracer experiments using likely precursors of asparagine, and from studies of enzyme activity. Asparagine is not formed by direct incorporation of NH_3 into the C_4-acceptor aspartate (as per glutamine synthesis using the C_5-acceptor glutamate). For example, in peas supplied with $^{15}NO_3^-$, ^{15}N-label was rapidly incorporated into NH_3 and glutamine but not into asparagine even though considerable net synthesis of asparagine occurs during the period. Although aspartate, in theory, would seem to be the most likely precursor, experiments with labelled aspartate have not proved very informative, perhaps because the amino group of exogenous aspartate is lost by transamination and the resulting oxaloacetate is metabolized via the TCA cycle.

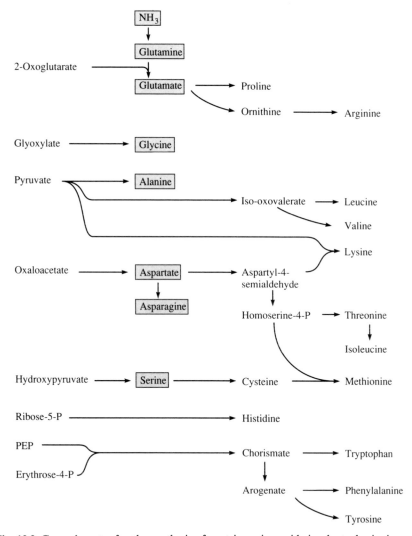

Fig. 13.3 General routes for the synthesis of protein amino acids in plants, beginning with carbon skeletons from intermediary metabolism (column 1). Glutamate, formed by incorporation of NH_3 via glutamine, is the source of amino-N for the transamination of several oxo acids supplied by intermediary metabolism to form the amino acids shown in grey boxes (column 2); the synthesis of asparagine is a special case involving amide-N transfer from glutamine.

However, a study of $^{15}NH_3$ assimilation in *Lemna* indicates that, at least in this species, about 30% (more under certain conditions) of the $^{15}NH_3$ assimilated passes through the asparagine pool and labelling occurs by transfer of the amide group of glutamine to aspartate involving asparagine synthetase.

$$(13.17)$$

Asparagine synthetase also catalyses a reaction with NH_3 in place of glutamine.

$$NH_3 + aspartate \xrightarrow[ATP \quad AMP+PP_i]{} asparagine \qquad (13.18)$$

However, the affinity for NH_3 is about 20-fold less than for glutamine and the V_{max} value is also considerably less, indicating that glutamine is the preferred substrate. The enzyme is present in germinating seeds but is notoriously unstable which might account for the very low activity in plant extracts relative to the rate of asparagine synthesis *in vivo*—hence the uncertainty about the role of this enzyme in asparagine synthesis.

An alternative mechanism involving the incorporation of cyanide into cysteine to form β-cyanoalanine and addition of water to produce asparagine is considered unlikely since many plants do not produce significant amounts of cyanide.

13.7.2 *Threonine is synthesized from aspartate but there are several branch points*

Threonine synthesis (Fig. 13.4) begins with the phosphorylation of aspartate at C-4, catalysed by aspartate kinase. Aspartyl-4-P is reduced by NADPH with the loss of P_i to aspartyl-4-semialdehyde which is further reduced by NADPH to homoserine. Homoserine is phosphorylated by ATP, catalysed by homoserine kinase. The final

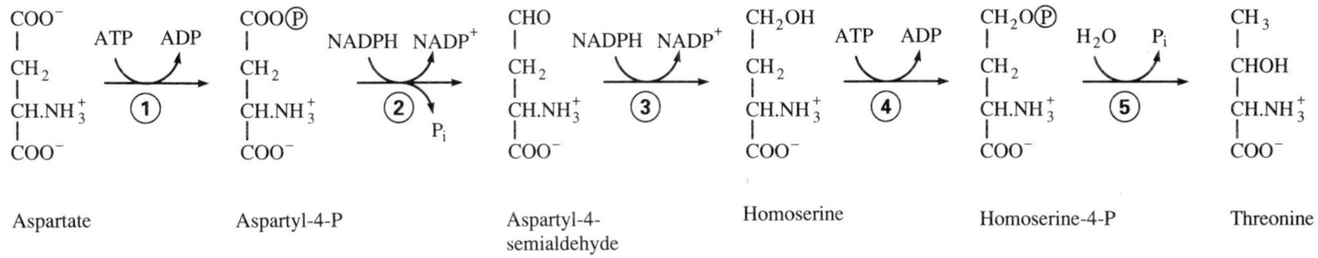

Fig. 13.4 Pathway of threonine synthesis from aspartate in plants. Enzymes are: (1) aspartate kinase; (2) aspartyl-4-semialdehyde dehydrogenase; (3) homoserine dehydrogenase; (4) homoserine kinase; (5) threonine synthase.

COO⁻	COO(P)	CHO	CH₂OH	CH₂O(P)	CH₃

$$\text{Aspartate} \xrightarrow[\text{ADP}]{\text{ATP}} (1) \text{Aspartyl-4-P} \xrightarrow[\text{NADP}^+ / P_i]{\text{NADPH}} (2) \text{Aspartyl-4-semialdehyde} \xrightarrow[\text{NADP}^+]{\text{NADPH}} (3) \text{Homoserine} \xrightarrow[\text{ADP}]{\text{ATP}} (4) \text{Homoserine-4-P} \xrightarrow[\text{P}_i]{\text{H}_2\text{O}} (5) \text{Threonine}$$

step, catalysed by threonine synthase, involves hydrolysis of homoserine-4-P with rearrangement of the oxygen atom to form a hydroxy group at C-3.

This short pathway has two branch points (Fig. 13.5): one occurs at aspartyl-4-semialdehyde, the beginning of the lysine synthetic pathway; the other occurs at homoserine-4-P which provides the α-aminobutyryl donor for the synthesis of cystathionine, an important intermediate in methionine biosynthesis (see Section 14.2.3). Aspartate kinase exhibits sigmoidal kinetics characteristic of a regulatory enzyme. Plants contain two types of isoenzyme, one inhibited by lysine (barley contains two of this type) and the other by threonine. Lysine inhibition of the lysine-sensitive enzyme is compounded by S-adenosylmethionine, a derivative of methionine (see Section 14.2.4). For example, in the absence of S-adenosylmethionine, the barley enzyme is inhibited 50% by 340 μM lysine but with 0.1 mM S-adenosylmethionine present only 48 μM lysine is required. The levels of the two isoenzymes vary with the stage of development; the lysine-sensitive type is generally more active in growing cells. Thus, lysine, threonine and S-adenosylmethionine exert powerful control over aspartate metabolism.

Lysine or threonine supplied to plants inhibits aspartate metabolism by inhibiting only the relevant isoenzyme of aspartate kinase, but addition of both inhibits both isoenzymes. This inhibits the synthesis of lysine, threonine and methionine. While the requirement for lysine and threonine can be met from the amino acids supplied, the demand for methionine can not and this results in diminished growth. Growth of the first leaf of cultured barley embryos is mildly inhibited by lysine in the absence of threonine; threonine without lysine has essentially no effect (Fig. 13.6) but with lysine it is extremely inhibitory. The inhibitory effects of lysine and threonine can be substantially reversed by homocysteine and homoserine (intermediates between aspartyl-4-P and methionine) and by methionine itself.

Homoserine dehydrogenase is positioned in the threonine pathway after the lysine branch point. It is inhibited by threonine but not by lysine, methionine or isoleucine, the latter a product of threonine metabolism (Fig. 13.5). In theory, inhibition of homoserine dehydrogenase by threonine could control the flow of aspartyl-4-semialdehyde into threonine and methionine. Conversely, dihydropicolinate synthase, the first enzyme of the lysine branch (Fig. 13.5), is inhibited by lysine, thus controlling the flow of aspartyl-4-semialdehyde into lysine. Homoserine kinase is positioned immediately prior to the threonine/methionine branch point. Isoleucine and valine reportedly inhibit this enzyme in peas.

Threonine synthase, which catalyses the production of threonine from homoserine-4-P, is enhanced up to 20 times by S-adenosylmethionine with maximum activation at very low concentrations (60 μM). Thus, an increase in methionine/S-adenosylmethionine diverts homoserine-4-P into threonine/isoleucine rather than cystathionine. Overproduction of threonine is controlled through its inhibitory effect on earlier enzymes of the pathway (Fig. 13.5).

The lysine pathway (Fig. 13.7) diverges from the threonine pathway at aspartyl-4-semialdehyde (Fig. 13.5). 2,3-Dihydropicolinate is formed in a condensation reaction with pyruvate, catalysed by the regulatory enzyme dihydropicolinate synthase. The succeeding stages are presumed to involve a further six enzymes but only some have been demonstrated in plants.

Almost all the aspartate kinase activity in leaves is in chloroplasts as is that of several enzymes of the lysine pathway (e.g. dihydropicolinate synthase) and the threonine pathway (e.g. threonine synthase and homoserine kinase). Light promotes the synthesis of homoserine, lysine and threonine from aspartate, demonstrating that the energy requirements come from light-dependent electron transport. Separate isoenzymes have been reported in chloroplasts and cytosol for homoserine dehydrogenase.

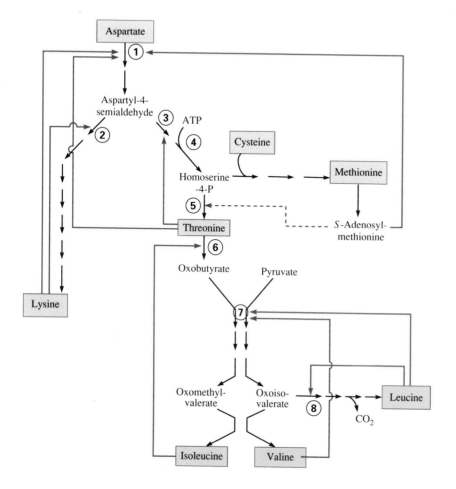

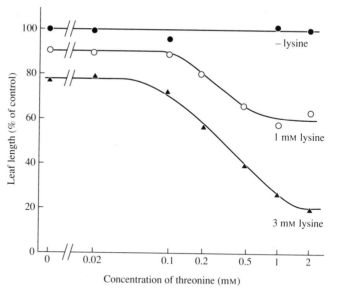

Fig. 13.6 Inhibition of growth of the first leaf of excised barley embryos by threonine and lysine. Embryos were dissected from seeds and grown aseptically for 7 days on an agar medium containing inorganic salts and sucrose at 25°C with 16 h light per day. Agar medium was supplemented with threonine and/or lysine at the concentrations shown. Leaf growth is expressed relative to a control lacking additions of threonine or lysine. From Bright, S.W.J. Wood, E.A & Miflin, B.J. (1978) *Planta*, **139**, 113–7.

13.7.3 Synthesis of the branched chain amino acids

The synthesis of valine, leucine and isoleucine have several common features (Fig. 13.5). The formation of isoleucine begins with the deamination of threonine to 2-oxobutyrate (Fig. 13.8) but this is the only reaction unique to isoleucine synthesis. The four subsequent reactions are catalysed by enzymes which also act on the lower homologues (i.e. they contain one less carbon atom) of the intermediates of the isoleucine pathway, beginning with pyruvate in lieu of 2-oxobutyrate and ending with valine in place of isoleucine. Thus, although the synthesis of isoleucine and valine are shown as independent pathways, only a single set of enzymes is involved so that there are two competing substrates at each stage of the pathway. The reaction sequence (Fig. 13.8) entails adding a C_2 unit from hydroxyethyl thiamine pyrophosphate (hydroxyethyl-TPP) to form the acetohydroxy acids of

Fig. 13.5 Summary of the amino acids synthesized from aspartate, showing the main sites of regulation. Each black arrow in the scheme denotes an enzyme, although only some of these have important regulatory functions. Solid grey arrows indicate the inhibitory effect of a compound on a specific enzyme. Details of the regulation of methionine synthesis from homoserine-4-P are not shown. Note that *S*-adenosylmethionine does not affect aspartate kinase directly but enhances lysine inhibition of the enzyme. The dashed grey arrow indicates the stimulatory effect of *S*-adenosylmethionine on threonine synthase. Enzymes of regulatory importance are: (1) aspartate kinase; (2) dihydropicolinate synthase; (3) homoserine dehydrogenase; (4) homoserine kinase; (5) threonine synthase; (6) threonine dehydratase; (7) acetohydroxy acid synthase; (8) isopropylmalate synthase.

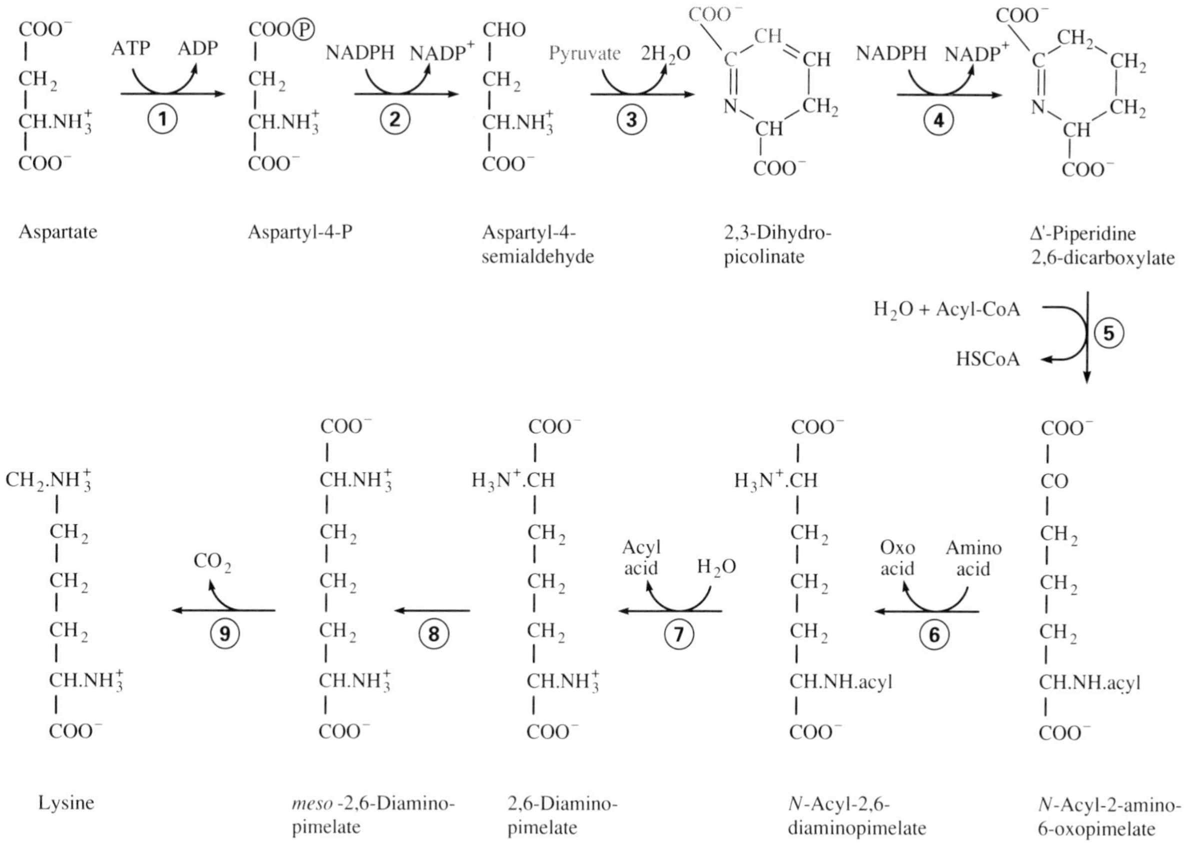

Fig. 13.7 Pathway of lysine synthesis from aspartate. Enzymes are: (1) aspartate kinase; (2) aspartyl-4-semialdehyde dehydrogenase; (3) dihydropicolinate synthase; (4) dihydropicolinate reductase; (5) Δ'-piperidine dicarboxylate acylase; (6) acyldiaminopimelate aminotransferase, (7) acyldiaminopimelate deacylase; (8) diaminopimelate epimerase; (9) diaminopimelate decarboxylase.

Top row intermediates: Aspartate → Aspartyl-4-P → Aspartyl-4-semialdehyde → 2,3-Dihydropicolinate → Δ'-Piperidine 2,6-dicarboxylate

Bottom row intermediates: Lysine ← meso-2,6-Diaminopimelate ← 2,6-Diaminopimelate ← N-Acyl-2,6-diaminopimelate ← N-Acyl-2-amino-6-oxopimelate

butyrate and pyruvate (acetolactate). The products are reductively isomerized and water is removed to form 2-oxo-3-methylvalerate and 2-oxoisovalerate, the oxo acid precursors of isoleucine and valine respectively. The amino acids are formed in the final reaction by transamination. Alternatively, 2-oxoisovalerate (but not its higher homologue) condenses with acetyl-CoA, the product undergoing isomerization to form 3-carboxy-2-hydroxyisocaproate. This intermediate is oxidatively decarboxylated to form 2-oxocaproate and transaminated to yield leucine.

Threonine dehydratase, which catalyses the production of 2-oxobutyrate from threonine, is inhibited by isoleucine, inhibiting its synthesis and decreasing the pool of 2-oxobutyrate, and thereby promoting the production of leucine and valine from pyruvate. Isopropylmalate synthase, the first enzyme of the leucine branch, is inhibited by leucine, diverting 2-oxoisovalerate into valine. Exogenous valine and leucine act singly and co-operatively to inhibit acetohydroxy acid synthase. Thus valine and/or leucine decrease(s) the synthesis of all the branched chain amino acids, especially isoleucine, inhibiting plant growth. This can be reversed by isoleucine (Fig. 13.5).

Since isoleucine is derived from threonine, which is one of several amino acids derived from aspartate, this makes the regulation of branched chain amino acid synthesis interactive with the control of lysine, threonine and methionine (summarized in Fig. 13.5).

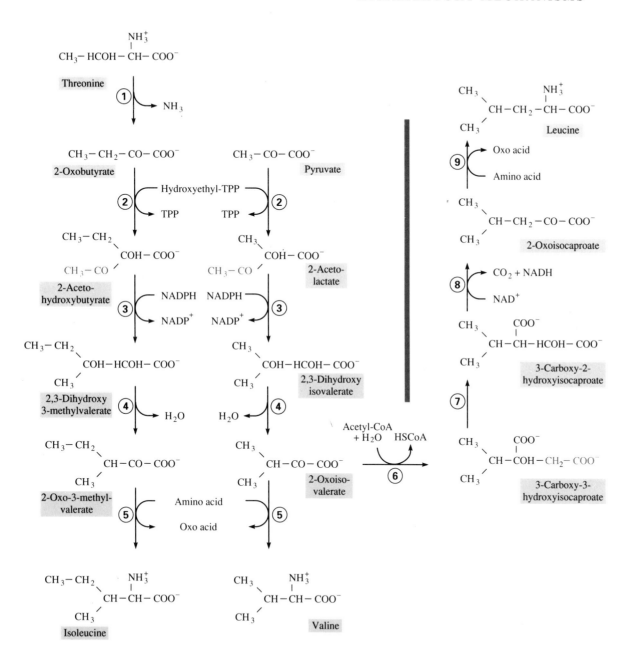

Fig. 13.8 Synthesis of the branched chain amino acids, leucine, valine and isoleucine. Enzymes are: (1) threonine dehydratase; (2) aminohydroxy acid synthase; (3)acetohydroxy acid reductoisomerase; (4) dihydroxyacid dehydratase; (5) aminotransferase; (6) isopropylmalate synthase; (7) isopropylmalate isomerase; (8) β-isopropylmalate dehydrogenase; (9) aminotransferase. Hydroxyethyl-TPP, which acts as a substrate in reaction (2), is formed by decarboxylation of pyruvate in the presence of pyruvate decarboxylase. Abbreviation: TPP=thiamine pyrophosphate.

Isolated chloroplasts assimilate [^{14}C]aspartate into isoleucine and [^{14}C]pyruvate into valine, leucine and isoleucine in light-enhanced reactions. (Note that valine and leucine are not derived from aspartate and that pyruvate also acts as a precursor of hydroxyethyl-TPP.) This affords yet another demonstration of the importance of illuminated chloroplasts in amino acid synthesis.

13.7.4 Aromatic amino acids are synthesized by a common pathway which is also an important source of secondary plant products

The aromatic amino acids phenylalanine, tyrosine and tryptophan (plus various intermediates of the biosynthetic pathway) are starting materials for the synthesis of a vast array of other compounds (Fig. 13.9) which can comprise up to 60% of the final dry weight of some plants. The flow of carbon through the aromatic (or shikimate) pathway is second only to the synthesis and turnover of carbohydrate. One of the major products is lignin. Most of the other products are also regarded as secondary compounds; they include coumarins, flavonoids and various alkaloids.

The aromatic amino acids are synthesized from PEP and erythrose-4-P via a common pathway (Fig. 13.9) as far as chorismate, involving seven enzymes (Fig. 13.10). Chorismate is the first of two branch points, one leading to anthranilate and tryptophan, the other to prephenate and arogenate (Fig. 13.11). A second branch

point occurs at arogenate, the immediate precursor of both phenylalanine and tyrosine.

Cytosol and chloroplasts contain independent complements of enzymes for the synthesis of the aromatic amino acids. The pathway in chloroplasts appears to be regulated whilst that in cytosol is not, which fuels suspicion that the latter is concerned with the production of secondary metabolites. For example, *Nicotiana silvestris* contains two forms of the enzyme DAHP synthase which catalyses the synthesis of 7-phospho-2-oxo-3-deoxyheptanoate, commonly known as 3-deoxy-arabinoheptulosonate-7-P (DAHP). One form is found in chloroplasts and is activated by Mn^{2+}; the other form is cytosolic and is activated by Co^{2+} (or Mg^{2+}). The chloroplast enzyme is extremely sensitive to arogenate (a later product of the pathway) but the cytosolic one is not. *N. silvestris* also contains two isoenzymes of chorismate mutase; again, one form is found in chloroplasts and, as discussed below, is sensitive to the end-products of the pathway whereas the other, cytosolic, form is not sensitive.

In chloroplasts, aromatic amino acid synthesis is controlled principally at five strategic points (Fig. 13.12). DAHP synthase is inhibited by arogenate so that depletion of arogenate stimulates the flow of PEP and erythrose-4-P into the pathway. Chorismate mutase is inhibited by the end-products of prephenate metabolism, phenylalanine and tyrosine but is activated by tryptophan, the end-product of the other branch of chorismate metabolism, alleviating the inhibitory

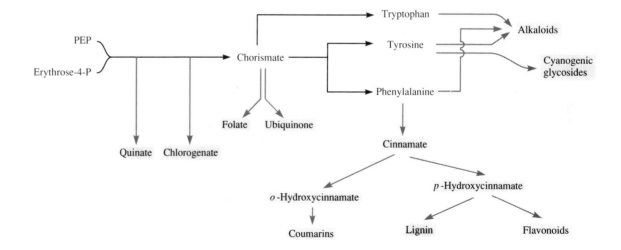

Fig. 13.9 Summary of the synthesis and fates of the aromatic amino acids and intermediates in plants. The main pathway is shown in black and compounds derived from it are shown in grey.

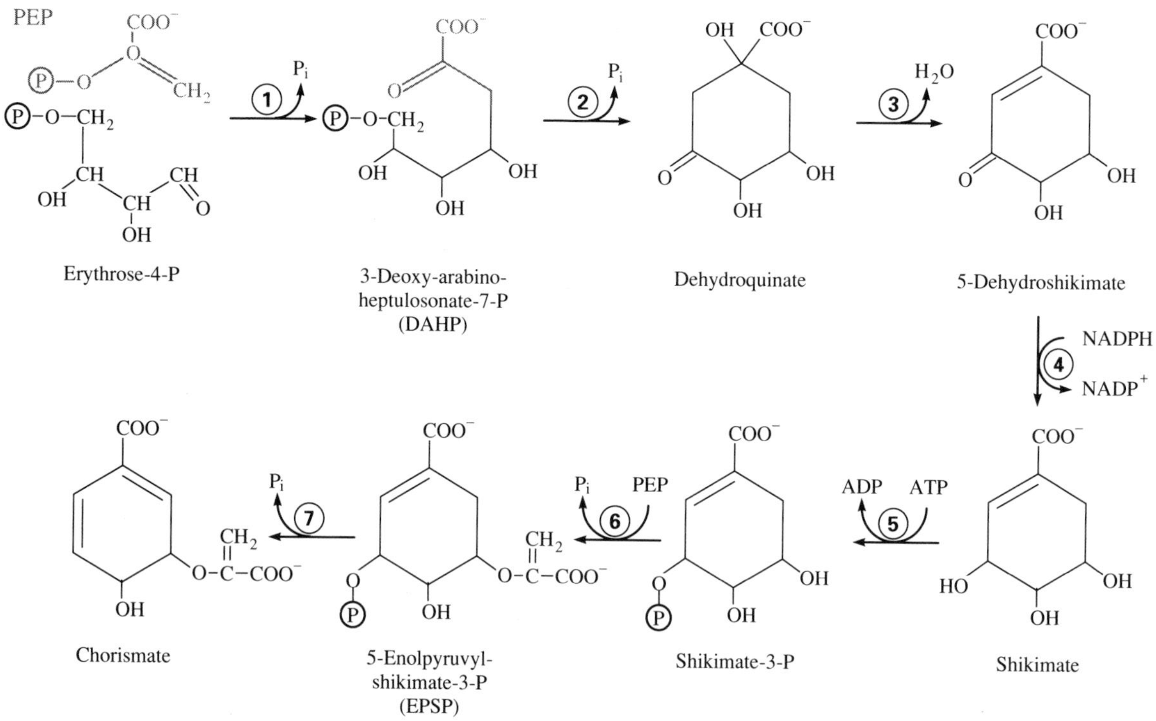

Fig. 13.10 Pathway for the synthesis of aromatic amino acids—Part I: synthesis of chorismate from PEP and erythrose-4-P. Enzymes are: (1) DAHP synthase; (2) 3-dehydroquinate dehydratase; (3) dehydroquinate dehydrogenase; (4) shikimate dehydrogenase; (5) shikimate kinase; (6) EPSP synthase; (7) chorismate synthase. Abbreviations: DAHP = 3-deoxy-arabinoheptulosonate-7-P; EPSP = 5-enolpyruvyl-shikimate-3-P.

effects of tyrosine and phenylalanine. The other three control points involve end-product inhibition: anthranilate synthase is highly sensitive to the eventual end-product tryptophan, and arogenate dehydratase and arogenate dehydrogenase are inhibited by their immediate reaction products, phenylalanine and tyrosine respectively. Thus, the end products tyrosine, phenylalanine and tryptophan control their synthesis through the two enzymes of chorismate metabolism and the two enzymes of arogenate metabolism while the flow of carbon into the pathway is regulated by arogenate through its effect on DAHP synthase.

13.7.5 Other amino acids

Serine and glycine are synthesized via the C_2-PR cycle (Section 11.4). Formation of the sulphur amino acids cysteine and methionine is discussed in Section 14.2. Note, however, that methionine is a product of both aspartate metabolism and cysteine metabolism and that methionine synthesis is integrated with the formation of other amino acids derived from aspartate (Fig. 13.5).

The major route for proline synthesis is thought to involve successive reductions of glutamate and expenditure of ATP (Fig. 13.13). An alternative mechanism involving deamination of ornithine is less important and could be used for proline catabolism. Many plants accumulate proline during drought or salt stress, sometimes accounting for as much as 50% of the free amino acids; this declines when the stress is alleviated. The precursor of the proline accumulated in stressed plants may vary between species; in stressed barley glutamate seems to be the precursor, in beans it may be arginine. Accumulation of proline might result from loss of control over synthesis. In bacteria, glutamate kinase is sensitive to proline. The same might happen in barley (Fig. 13.13).

The origin of the carbon and nitrogen atoms in arginine and histidine is shown in Fig. 13.14. In some plants arginine is an important nitrogen transport compound and is synthesized at rapid rates. The synthesis of carbamyl phosphate (carbamyl-P) by the first enzyme of the pathway (Fig. 13.15) is important since this provides an alternative way to transfer newly assimilated NH_3 from the amide group of glutamine into arginine and various other nitrogen-containing molecules (e.g. pyrimidine synthesis—see Section 15.1). The reaction is catalysed by carbamyl-P synthetase (designated below as Enz).

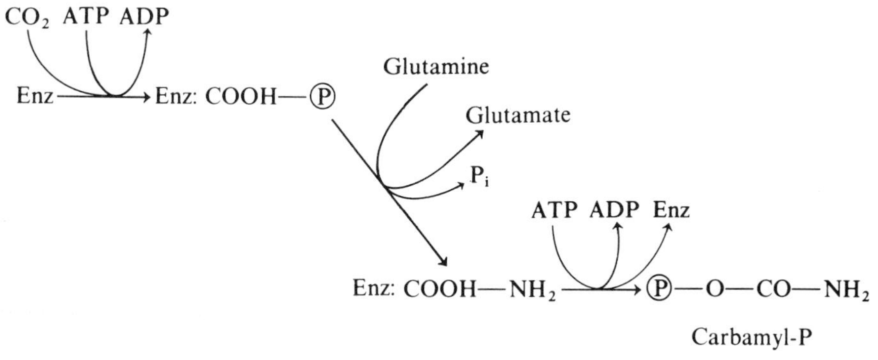

$$CO_2 \; ATP \; ADP$$

$$Enz \longrightarrow Enz: COOH \text{—} \textcircled{P}$$

Glutamine

Glutamate

P_i

ATP ADP Enz

$$Enz: COOH \text{—} NH_2 \longrightarrow \textcircled{P} \text{—} O \text{—} CO \text{—} NH_2$$

Carbamyl-P

(13.19)

Carbamyl-P reacts with ornithine (Fig. 13.15), synthesized from two molecules of glutamate. The final atom of nitrogen for the synthesis of arginine is derived from aspartate with the expenditure of ATP.

13.7.6 Some herbicides inhibit amino acid synthesis

Plants must synthesize all 20 protein amino acids for growth. If a compound specifically inhibits the synthesis of an amino acid essential for humans (see Table 13.1), then it is selectively phytotoxic. Several herbicides with very low animal toxicity are highly toxic to plants and many micro-organisms because they block

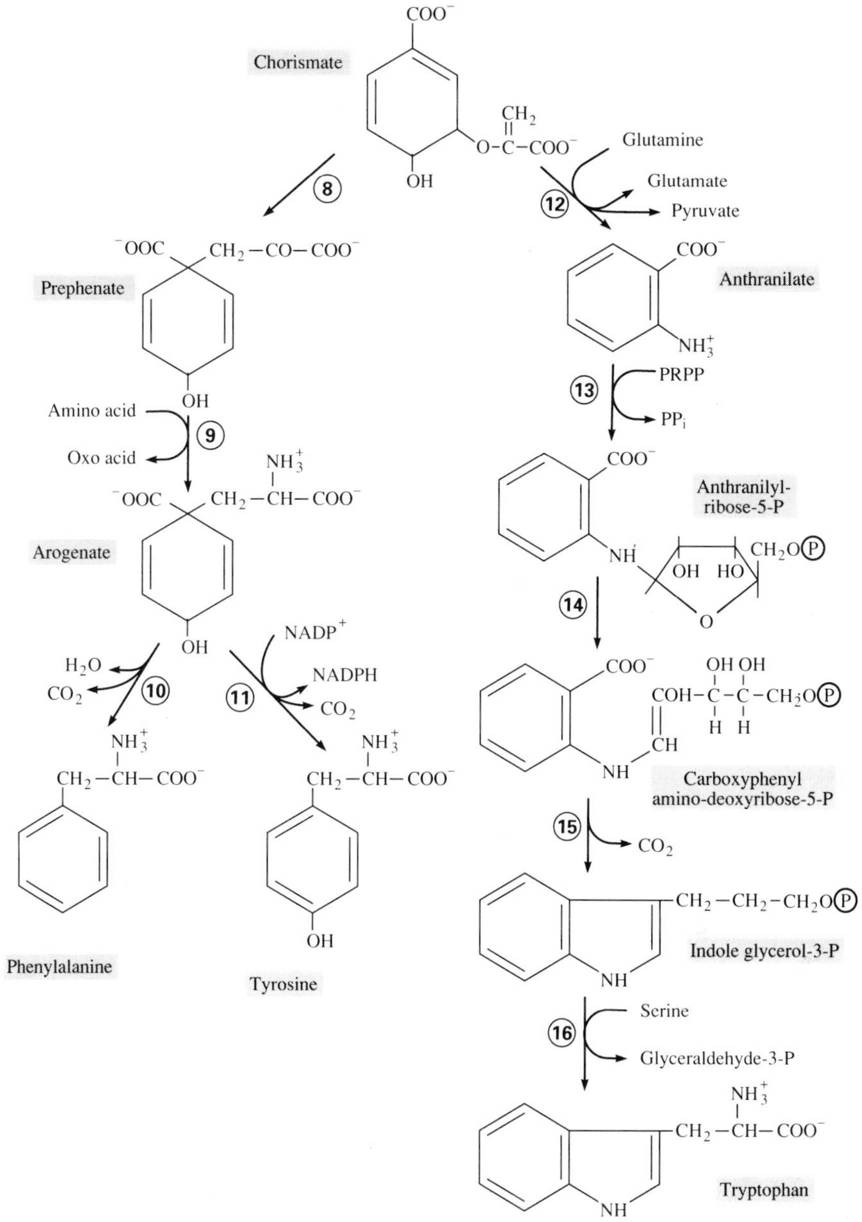

Fig. 13.11 Pathway for the synthesis of aromatic amino acids—Part II: formation of phenylalanine, tyrosine and tryptophan from chorismate. Enzymes are: (8) chorismate mutase; (9) glutamate:prephenate aminotransferase; (10) arogenate dehydratase; (11) NADP-specific arogenate dehydrogenase; (12) anthranilate synthase; (13) anthranilate phosphoribosyltransferase; (14) anthranilate phosphoribosylisomerase; (15) indole glycerol-P synthase; (16) tryptophan synthase.

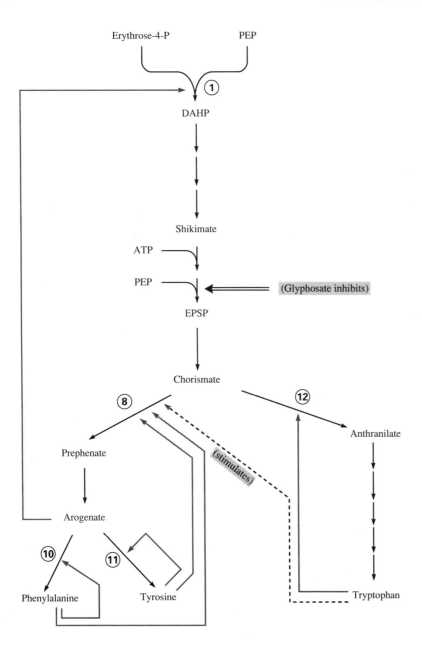

the biosynthesis of amino acids which animals are unable to make. For example, N-phosphonomethylglycine (glyphosate)

$$^-OOC-CH_2-NH-CH_2-\overset{\overset{\displaystyle O}{\|}}{\underset{\underset{\displaystyle O^-}{|}}{P}}-O^-$$

blocks the synthesis of the aromatic amino acids and promotes accumulation of shikimate (up to 10% of the dry weight in cultured cells of *Galium mollugo*). The usual control over DAHP synthase by arogenate (Fig. 13.12) is not exercised due to a block near shikimate so that carbon is constantly supplied into the pathway. Glyphosate is a powerful inhibitor of 5-enolpyruvyl-shikimate-3-P (EPSP) synthase, one of three enzymes involved in the production of chorismate from shikimate (Fig. 13.10) which catalyses the condensation of shikimate-3-P with PEP.

$$\text{Shikimate-3-P} + \text{PEP} \rightleftharpoons \text{EPSP} + \text{P}_i \qquad (13.20)$$

Glyphosate inhibition of the enzyme is competitive with respect to PEP but non-competitive with respect to shikimate-3-P, so it must bind to the PEP binding site. (Details of the inhibition kinetics are given in Section 5.4) Glyphosate is relatively specific to the aromatic amino acid biosynthetic pathway with little effect on other enzymes of PEP metabolism (e.g. PEP carboxylase, pyruvate kinase, PEP carboxykinase). Glyphosate toxicity in plants can be alleviated by the three aromatic amino acids. For example, in cultures of carrot cells treated with glyphosate, tyrosine and phenylalanine added individually have some effect in overcoming toxicity but cells supplemented with all three aromatic amino acids grow as well as cultures lacking glyphosate.

Fig. 13.12 Regulation of the synthesis of the aromatic amino acids from PEP and erythrose-4-P. Each black arrow denotes an enzyme of the pathway. Solid grey arrows indicate an inhibitory effect of a metabolite on specific enzymes of the pathway. The dashed arrow denotes the stimulatory effect of tryptophan on chorismate mutase. Enzyme numbering is consistent with that shown in Figs 13.10 and 13.11. Sites of regulation are: (1) DAHP synthase; (8) chorismate mutase; (10) arogenate dehydratase; (11) arogenate dehydrogenase; (12) anthranilate synthase. The site of inhibition (i.e. EPSP synthase) by the herbicide glyphosate is also shown. Abbreviations: DAHP = 3-deoxy-arabinoheptulosonate-7-P; EPSP = 5-enolpyruvyl-shikimate-3-P.

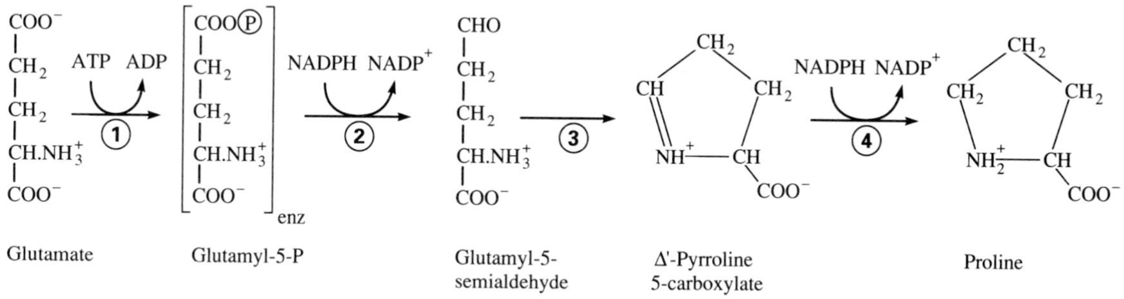

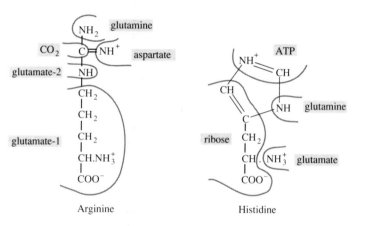

Fig. 13.13 Proposed pathway for the synthesis of proline in plants. Enzymes are: (1) glutamate kinase; (2) glutamyl-5-semialdehyde dehydrogenase; (3) non-enzymic; (4) proline dehydrogenase.

Glutamate   Glutamyl-5-P   Glutamyl-5-semialdehyde   Δ'-Pyrroline 5-carboxylate   Proline

Fig. 13.14 Structures of the protein amino acids arginine and histidine showing the origin of the carbon and nitrogen atoms.

Arginine   Histidine

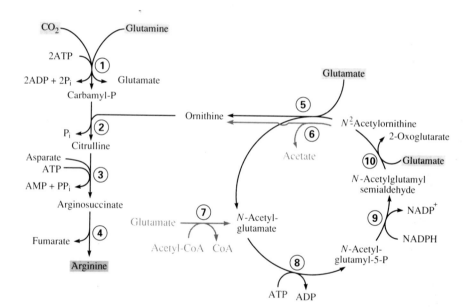

Fig. 13.15 Summary of the pathway of arginine biosynthesis. Substrates which contribute nitrogen or carbon atoms to the final product are shaded light grey. The relative importance of the alternative mechanisms for net synthesis of ornithine from glutamate, i.e. (5) versus (6) and (7), is uncertain. Enzymes are: (1) carbamyl-P synthetase; (2) ornithine carbamyl transferase; (3) arginosuccinate synthetase; (4) arginosuccinate lyase; (5) acetylornithine:glutamate acetyltransferase; (6) acetylornithine amidohydrolyase; (7) acetyl-CoA:glutamate acetyltransferase; (8) acetylglutamyl kinase; (9) acetylglutamyl semialdehyde dehydrogenase; (10) acetylornithine:2-oxoglutarate aminotransferase.

The sulphonylureas (R'-SO$_2$-NH-CO-NH-R'') are very slow acting herbicides of great potency, e.g. chlorsulphuron.

Addition of all three branched chain amino acids (valine, isoleucine and leucine) reverses chlorsulphuron toxicity in various experimental systems (e.g. growth of excised pea roots). Individual branched chain amino acids are not effective indicating that chlorsulphuron inhibits one or more of the enzymes common to the synthesis of all three (see Fig. 13.5). Acetohydroxy acid synthase, the first enzyme of the common pathway, is especially sensitive (about 50% inhibition at 20 nM). Inhibition of the enzyme by the sulphonylureas is reportedly competitive with respect to pyruvate.

13.8 Some organisms use gaseous nitrogen as a nitrogen source by the process of nitrogen fixation

N$_2$ fixation is the process by which gaseous N$_2$ is used as a nitrogen source for growth. N$_2$ is reduced to NH$_3$ which is assimilated into the nitrogen-containing compounds of cells. Eukaryotic plants *per se* cannot fix N$_2$ but some plants form symbiotic relationships with various heterotrophic prokaryotic organisms which can. The N$_2$-fixing organism supplies NH$_3$ or glutamine, which is in excess of its needs, to the host. The plant, in return, provides a niche and a continuous supply of organic carbon. In legumes (quantitatively the most important group of N$_2$-fixing plants), bacteria of the genus *Rhizobium* induce the production of root nodules and adopt a morphology known as bacteroids. Other associations involving various N$_2$-fixing organisms and non-leguminous plants (e.g. grasses and *Casuarina*) account for comparatively very little N$_2$ fixation. Except for a brief mention of the 'actinorhizal plants', they will not be discussed here.

13.8.1 *Nitrogenase catalyses the reduction of nitrogen to ammonia*

The reduction of N$_2$ to NH$_3$ in nodulated legumes is catalysed by nitrogenase within the *Rhizobium* bacteroids. The enzyme is irreversibly inactivated by O$_2$ so

that it is necessary to extract, purify and assay it anaerobically. The enzyme consists of two proteins, neither of which catalyses N$_2$ reduction without the other. The larger of the two proteins consists of four subunits, two identical, and a substantial number of iron atoms bound to an equal number of atoms of acid-labile sulphur (i.e. sulphur which is displaced as H$_2$S by weak mineral acids). The large protein also contains two atoms of molybdenum. The small protein consists of two identical subunits, each containing two atoms of iron.

Nitrogenase activity can be monitored using sodium dithionite (Na$_2$S$_2$O$_4$) as reductant. Mg^{2+} and ATP are required for N$_2$ reduction *in vitro*. The ADP produced inhibits nitrogenase activity so that a system must be used to phosphorylate ADP to ATP (e.g. addition of creatine-P and creatine kinase). Given these conditions, nitrogenase catalyses reduction of N$_2$ (N$\equiv$N) to NH$_3$, and also acetylene (CH$\equiv$CH) and HCN (CH$\equiv$N). Since acetylene is reduced to ethylene which can be readily separated from acetylene by gas chromatography, this reduction is often used to estimate nitrogenase activity. Nitrogenase also supports the reduction of H$^+$ to H$_2$ (hydrogenase activity).

Mossbauer spectroscopy shows that the iron in the large nitrogenase protein exists in several oxidation states; the most reduced state only occurs with the small protein plus Na$_2$S$_2$O$_4$, ATP and Mg^{2+} and is presumed to be the functional form of the enzyme. The reduced form also has a characteristic electron paramagnetic resonance (e.p.r.) absorption, slightly altered upon addition of acetylene which also suggests that it is the active form, possibly reacting through the molybdenum atoms.

The iron in the small protein exists in oxidized and reduced forms, the latter also having a characteristic e.p.r. spectrum. ATP and Mg^{2+} cause a change in the e.p.r. spectrum and an increase in chemical reactivity. It is proposed that the reduced (excited) form of the small protein maintains the iron in the large protein in its reduced state. *In vivo*, low molecular mass proteins of the ferredoxin or flavodoxin type (Fig. 13.16) act as the reductant in place of Na$_2$S$_2$O$_4$.

13.8.2 *Nitrogen fixation entails a heavy expenditure of ATP which is exacerbated by hydrogenase activity*

About 12–15 ATP are hydrolysed per N$_2$ reduced to NH$_3$ by isolated nitrogenase. This, together with the requirement for Fd$_{red}$ or reduced flavodoxin (Fld$_{red}$) makes great demands on the energy-generating capacity of a cell. Moreover, any diversion of reducing equivalents into H$^+$ (hydrogenase activity) further increases the energy expended per N$_2$ fixed, although in some strains the H$_2$ produced is oxidized to

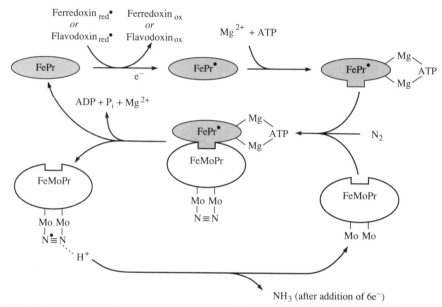

Fig. 13.16 Representation of the components of nitrogenase and their role in N_2 fixation. Reducing equivalents (denoted as $^\bullet$) are supplied from reduced forms of ferredoxin or flavodoxin to the small Fe-containing protein (FePr). In the presence of ATP and Mg^{2+} the reduced form of FePr binds to the large Fe/Mo-protein (FeMoPr) with the concomitant binding of N_2 to the molybdenum atoms of FeMoPr. The reducing equivalents associated with FePr$^\bullet$ are transferred to the N_2 bound to FeMoPr, effecting its partial reduction. Successive additions of reducing equivalents from FePr$^\bullet$ by repeated cycling of FeMoPr leads to eventual reduction of N_2 to NH_3.

H_2O in an exergonic reaction with the phosphorylation of ADP. Strains exhibiting this respiratory-like activity (sometimes incorrectly referred to as hydrogenase activity) use the ATP for N_2 fixation, offsetting direct loss of reducing equivalents from N_2 reduction, but organisms or strains of organisms which minimize leakage of reducing equivalents into H^+ are energetically most efficient. Much more energy is expended assimilating nitrogen from N_2 gas than from inorganic nitrogen salts.

Control of N_2 fixation at the post-transcriptional level occurs through the inhibitory effect of ADP on nitrogenase; restoration of activity depends on phosphorylation of ADP which is dependent on other aspects of cellular metabolism. Control also occurs at the level of gene expression since high concentrations of NH_3 suppress nodulation and the establishment of an active N_2-fixing symbiosis.

13.8.3 Supply and exchange of nitrogen, carbon and oxygen in nodules

The nitrogenase in *Rhizobium* bacteroids only functions in the relative absence of O_2 but the reducing equivalents required for N_2 reduction are formed by aerobic oxidation of a carbon source within the bacteroid. Controlled entry of O_2 for respiration is achieved by the O_2-carrying protein, leghaemoglobin, which has a high affinity for O_2 and supplies the bacteroids with O_2 in a form which protects nitrogenase from damage. Leghaemoglobin is produced by the host, not the bacteroids, and gives nodules their characteristic pink colour. The host also supplies C_4 dicarboxylates (e.g. succinate) which are used as respiratory substrates. They are metabolized to malate, decarboxylated to pyruvate via malic enzyme and then oxidized via the TCA cycle (Fig. 13.17) to generate the ATP and reducing equivalents required for N_2 fixation. Sugars are not used as an energy source for N_2 fixation by *Rhizobium* bacteroids.

Most of the NH_3 produced by bacteroids is excreted into the host cells where it is assimilated into glutamine. Table 13.2 shows that asparagine and glutamine are the main transport compounds in the xylem sap of most temperate legumes but that the ureides, allantoin and allantoate are the most important transport compounds in many tropical legumes.

Asparagine   Allantoin   Citrulline

Glutamine is the primary source of nitrogen for the synthesis of these transport compounds, all of which have high nitrogen:carbon ratios. The synthesis and metabolism of ureides is described in Section 15.2.2.

Some plants (actinorhizal plants, e.g. *Alnus*) establish N_2-fixing symbioses in their roots with members of the Actinomycete genus, *Frankia*. In many of these plants N_2 is incorporated into amides and transported but some (and also some legumes) transport assimilated N_2 principally as the ureide, citrulline (Table 13.2).

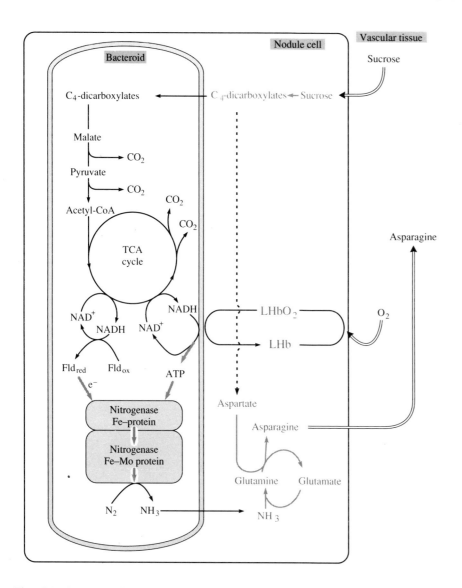

Fig. 13.17 Summary of the respiratory mechanisms of *Rhizobium* bacteroids and the flow of carbon, nitrogen and oxygen between bacteroids and host cells in the N_2-fixing *Rhizobium*/legume symbiosis. Abbreviations: $LHbO_2$ and LHb = oxygenated and unoxygenated forms of leghaemoglobin; Fld_{ox} and Fld_{red} = oxidized and reduced forms of flavodoxin.

Table 13.2 Principal nitrogenous compounds in the xylem sap of some N_2-fixing plants. After Schubert, K.R. (1986) *Annu. Rev. Plant Physiol.*, **37**, 539–74.

Plant species	Principal nitrogenous constituents
Temperate legumes	
Lathyrus sativa	Asparagine, glutamine
Lupinus albus (white lupin)	Asparagine, glutamine
Pisum sativum (pea)	Asparagine, glutamine
Trifolium repens (white clover)	Asparagine
Vicia faba (broad bean)	Asparagine, glutamine
Tropical legumes	
Arachis hypogaea (peanut)	4-Methyleneglutamine, allantoin, allantoate, (asparagine)
Glycine max (soya bean)	Allantoin, allantoate asparagine
Phaseolus vulgaris (French bean)	Allantoin, allantoate
Vigna radiata (mung bean)	Allantoin, allantoate
Vigna unguiculata (cowpea)	Allantoin, allantoate
Actinorhizal plants	
Alnus sp. (alder)	Citrulline
Myrica gale (sweet gale)	Asparagine

13.8.4 Heterocystous cyanobacteria fix nitrogen gas when inorganic nitrogen salts are unavailable

Certain filamentous cyanobacteria (blue-green algae) fix N_2 without forming symbioses. These prokaryotic organisms (e.g. *Anabaena*) form large specialized cells, called heterocysts, at regular intervals along the filament when grown in medium lacking a soluble nitrogen source. N_2 fixation is restricted to the heterocysts which lack an active PSII, the O_2 evolving component of the light reaction, maintaining an internal environment low in O_2 and avoiding inactivation of nitrogenase. The ATP and reductant needed for N_2 fixation are produced by light-dependent electron transport. Heterocystous cyanobacteria also exhibit light-dependent reduction of H^+ to H_2. Certain eukaryotic plants form symbiotic associations with cyanobacteria in which conditions conducive to N_2 fixation are established. Associations of this type include *Azolla* (a water fern) with *Anabaena azollae* and *Macrozamia* (a cycad) with *Nostoc punctiforme*.

13.9 Some plants contain toxic non-protein amino acids

Various plants contain amino acids which are not found in protein and are not essential intermediates in metabolism. These compounds, of which several hundred are known, are usually restricted to a few closely related species or genera (see Section 2.8). Some have structures very similar to the protein amino acids. For example, L-azetidine-2-carboxylic acid, found in some members of the family Liliaceae, is structurally similar to L-proline but contains one less carbon atom in the ring.

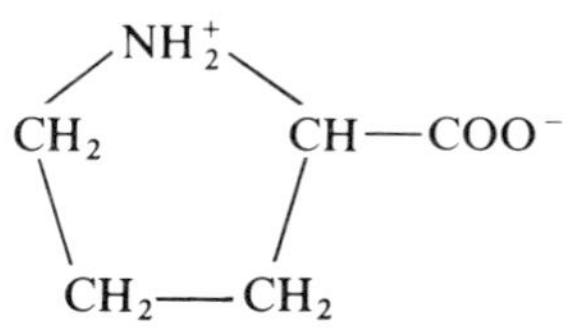

L-Proline

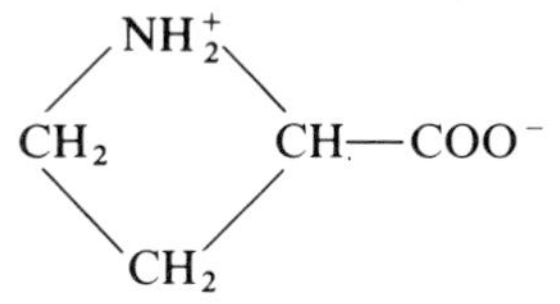

L-Azetidine-2-carboxylic acid

Some protein amino acids are so similar to their structurally related non-protein amino acids that enzymes from plants lacking the non-protein amino acids cannot distinguish between them (i.e. the amino acids are structural analogues). Thus, azetidine-2-carboxylic acid is an analogue of proline and since it interferes with proline metabolism it is rather toxic to some plants but not to those species which synthesize it.

Further reading

Monographs and treatises: Läuchli & Bieleski (1983); Miflin (1980).
Nitrogen fixation: Postgate (1982).
Photorespiratory nitrogen cycle: Singh *et al.* (1985).
Aromatic amino acid synthesis: Jensen (1985).
Non-protein amino acids: Bell & Charlwood (1980) Chapter 7; Conn (1981) Chapter 8.
Herbicide inhibition of amino acid synthesis: Fedtke (1982); Jensen (1985); Kishore & Shah (1988).
Transport of nitrogen in plants: Pate (1980); Schubert (1986).

Other light-coupled assimilatory
and reductive mechanisms

14.1 Light supports various processes in chloroplasts and protoplasts

Plants use ATP and reducing equivalents formed in the light for various processes, in addition to the assimilation of CO_2 into triose-P. These processes include the reassimilation of photorespired NH_3, the assimilation of NO_2^- into glutamate (Sections 11.6 & 13.5) and synthesis of many of the 20 protein amino acids (Section 13.7). This chapter is concerned with some other processes which consume light-generated ATP and/or NADPH or Fd_{red}, and includes shuttle mechanisms for the export of reducing equivalents and phosphorylation potential from chloroplasts to support various light-enhanced processes in the cytosol. C_1 metabolism, allied to methionine synthesis, is also described.

14.2 Assimilation of inorganic sulphur

14.2.1 Inorganic sulphate is assimilated into cysteine by the 'bound pathway' and involves a carrier

Many compounds, essential for cell function, contain sulphur; these include the protein amino acids cysteine and methionine, glutathione, S-adenosylmethionine, biotin, coenzyme A and thiamine pyrophosphate. Sulphur in aerobic soils is available principally as sulphate (SO_4^{2-}) and is taken up via an active mechanism in root tip parenchyma cells. Selenate (SeO_4^{2-}) and various anions of group VI transition elements (e.g. chromate, CrO_4^{2-}) competitively inhibit SO_4^{2-} uptake and are toxic to most plants. In cultured tobacco cells low internal concentrations of SO_4^{2-} promote uptake. When $^{35}SO_4^{2-}$ is supplied to root systems, 99% of the ^{35}S-label in xylem sap occurs as SO_4^{2-}, demonstrating that SO_4^{2-} assimilation in roots is minor.

The reductive assimilation of inorganic sulphur and nitrogen have several common features. They are normally acquired in oxidized forms and undergo an eight electron reduction before incorporation into organic molecules, in contrast to carbon which is assimilated first and then reduced. Inorganic sulphur and nitrogen are assimilated to varying extents in the chloroplasts of photosynthetic

cells since assimilation of NO_3^- in roots is important in some species (see Chapter 13). Assimilation of the two elements differs in that, for nitrogen assimilation, ATP is required after reduction but for sulphur, ATP is expended before reduction (SO_4^{2-} activation) and also involves binding to a carrier.

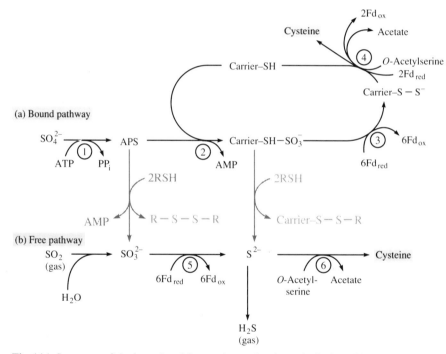

Fig. 14.1 Summary of the bound and free pathways for the assimilation of inorganic sulphur into cysteine in chloroplasts. Sulphur can be lost from intermediates of the bound pathway to the free pathway by side reactions with thiols (shown in grey). Both pathways utilize reducing equivalents (and in the case of the bound pathway, ATP) from the light reactions. Enzymes are: (1) ATP sulphurylase; (2) APS sulphotransferase; (3) thiosulphonate reductase; (4) cysteine synthase; (5) SO_3^{2-} reductase; (6) cysteine synthase. Abbreviation: APS = adenosine 5'-sulphatophosphate.

The incorporation of sulphur as SO_4^{2-} into cysteine, the first organic product of SO_4^{2-} assimilation, proceeds via the bound pathway (Fig. 14.1A), in which SO_4^{2-} is bound to a carrier during reduction. SO_4^{2-} is first activated by ATP to form adenosine 5'-sulphatophosphate, better known as adenosine phosphosulphate (APS), catalysed by ATP sulphurylase, a chloroplast enzyme.

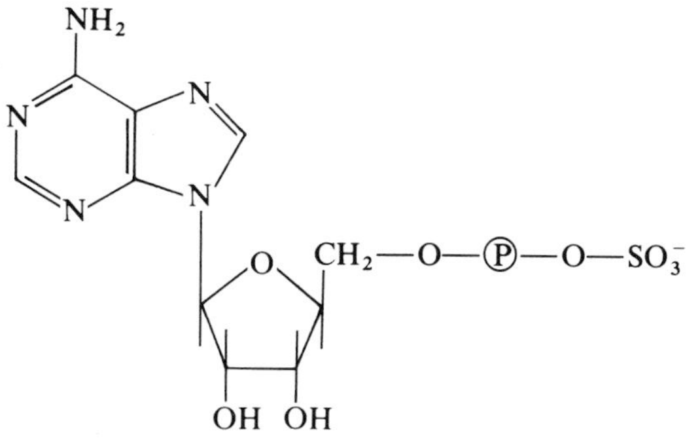

Adenosine 5'-phosphosulphate

$$ATP + SO_4^{2-} \rightleftharpoons APS + PP_i \qquad (14.1)$$

The $\Delta G^{\circ\prime}$ of this reaction is extremely unfavourable ($+80\,kJ\,mol^{-1}$) but PP_i is hydrolysed by pyrophosphatase in an exergonic reaction ($\Delta G^{\circ\prime} = -33.4\,kJ\,mol^{-1}$). When coupled to Eqn. 14.1, the $\Delta G^{\circ\prime}$ for the overall reaction (Eqn. 14.2) decreases to $+46.6\,kJ\,mol^{-1}$.

$$ATP + SO_4^{2-} \rightleftharpoons APS + 2P_i \qquad (14.2)$$

This supports the production of physiologically significant amounts of APS in illuminated chloroplasts.

SO_4^{2-} activation involves production of a sulphatophosphate anhydride bond which has a $\Delta G^{\circ\prime}$ of hydrolysis of $-18.5\,kJ\,mol^{-1}$. APS then has a standard redox potential ($E_o^{\prime}$) of $-0.06\,V$ which, unlike SO_4^{2-} with a $E_o^{\prime}$ of $-0.48\,V$, can be reduced by Fd_{red} and NADPH in an exergonic reaction.

Prior to reduction, the sulphonate moiety of APS is transferred to an enzyme-bound thiol carrier (carrier–SH) to form the thiosulphonate derivative (carrier–S—SO_3^-), catalysed by APS sulphotransferase which is specific for APS.

$$APS + carrier–SH \rightleftharpoons AMP + carrier–S–SO_3^- \qquad (14.3)$$

In chloroplasts, the carrier is most probably the reduced form of the tripeptide,

glutathione (GSH—see Section 14.5 for properties). The carrier–S–SO_3^- (G–S–SO_3^- in chloroplasts) is reduced by Fd_{red}, catalysed by thiosulphonate reductase (organic thiosulphate reductase).

$$Carrier–S–SO_3^- + 6Fd_{red} \rightleftharpoons carrier–S–S^- + 6Fd_{ox} \qquad (14.4)$$

The final step, synthesis of cysteine, involves reductive transfer of the sulphide group to O-acetylserine catalysed by cysteine synthase, another chloroplast enzyme.

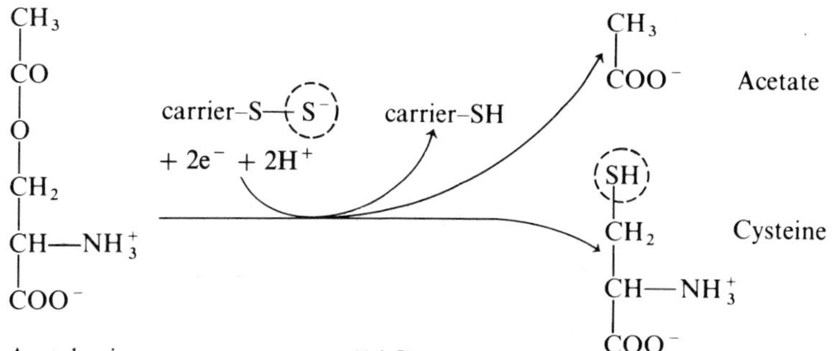

O-Acetylserine $\qquad (14.5)$

The source of the reducing equivalents ($2e^-$) is not clear but it is thought to be Fd_{red} in chloroplasts. Cysteine synthase also catalyses a reaction with H_2S in place of carrier-bound sulphide.

$$H_2S + O\text{-acetylserine} \rightleftharpoons cysteine + acetate \qquad (14.6)$$

Cysteine synthase has been extensively studied with H_2S as the donor (Eqn. 14.6) but not with bound sulphide (Eqn. 14.5). O-Acetylserine is produced from serine by reaction with acetyl-coenzyme A, catalysed by serine transacetylase.

$$Acetyl\text{-}CoA + serine \rightleftharpoons O\text{-acetylserine} + coenzyme\ A \qquad (14.7)$$

The free energy of hydrolysis of acetyl-coenzyme A (see Section 4.3) is effectively conserved in O-acetylserine, so its subsequent hydrolysis ensures that Eqns 14.5 and 14.6 favour product formation. Plant extracts also support the following reaction.

$$Serine + H_2S \rightleftharpoons cysteine + H_2O \qquad (14.8)$$

Although this activity has been attributed to serine sulph-hydrase, cysteine synthase may be involved. In any event, the reaction is probably not important *in vivo* since it is readily reversible and the rates are usually less than 10% of cysteine synthase (Eqn. 14.6).

Compounds containing sulphur in the oxidation state -2 are synthesized from cysteine. Thus, cysteine is a pivot in sulphur metabolism, like glutamate and glutamine are in nitrogen metabolism.

Chlorella mutants which lack thiosulphonate reductase cannot grow on SO_4^{2-}, demonstrating its assimilation via the bound pathway (Fig. 14.1A) since these mutants have sulphite reductase (see next Section). In theory, sulphur must flow through this pathway at about 0.4 μmol mg^{-1} chlorophyll h^{-1}.

In heterotrophic organisms which use SO_4^{2-} as a sulphur source (e.g. yeast), 3'-phosphate adenosine 5'-sulphatophosphate (PAPS) is the substrate for reduction. It is formed from APS with the expenditure of ATP, catalysed by APS kinase.

$$ATP + APS \rightleftharpoons PAPS + ADP \qquad (14.9)$$

This enzyme is present in chloroplasts but PAPS is not reduced since APS sulphotransferase is specific for APS. PAPS is thought to be involved in the synthesis of 6-sulphoquinovose, a component of sulpholipid found in chloroplasts (Section 6.1.2).

14.2.2 Sulphite is assimilated via the 'free pathway'

Plants have a separate mechanism to reduce free SO_3^{2-} (sulphite) to free S^{2-} (sulphide) without the formation of a bound intermediate (i.e. a 'free pathway'). In chloroplasts this reaction, catalysed by SO_3^{2-} reductase (Fig. 14.1b), uses reducing equivalents (Fd$_{red}$) from the light reactions making it light-dependent. The potential capacity for sulphur flow through this mechanism is well in excess of the bound pathway. The free pathway is thought to detoxify SO_3^{2-} formed from gaseous SO_2 by reaction with H_2O, and SO_3^{2-} formed by reactions between intermediates of the bound pathway and endogenous thiols (Fig. 14.1). The fate of free S^{2-} produced by SO_3^{2-} reductase *in vivo* is not clear. In theory, it could be incorporated into *O*-acetylserine (Eqn. 14.6) but this depends on the affinity of cysteine synthase for free S^{2-} (from SO_3^{2-}) and bound S^{2-} (from SO_4^{2-}). However, much of the free S^{2-} produced by SO_3^{2-} reductase is emitted into the air as H_2S in the light. An inverse correlation between this activity and SO_2 damage is consistent with this proposal. In some species (e.g. many cruciferous plants), emissions of volatile sulphur can reach astonishing proportions; consider the smell of a crop of maturing cabbages on a hot sunny afternoon. Nonetheless, the free pathway is thought to be important

in the acquisition of sulphur as SO_2 and its assimilation into cysteine, especially in plants on sulphur-deficient soils. Many lichens acquire their sulphur in this way and have efficient mechanisms to absorb it from the low concentrations present in unpolluted air. Lichens accumulate toxic amounts of H_2SO_3 and H_2SO_4 from air containing high concentrations of SO_2 and SO_3, making them very sensitive to atmospheric sulphur pollution.

14.2.3 Methionine is synthesized from cysteine via the trans-sulphuration pathway

Plants have two pathways of methionine synthesis (Fig. 14.2) but only the route via cystathionine (trans-sulphuration pathway) is physiologically important. Cystathionine is formed by condensation between cysteine and an α-aminobutyryl donor, catalysed by cystathionine γ-synthase. The plant enzyme is unique as homoserine-4-P acts as the α-aminobutyryl donor (Eqn. 14.10).

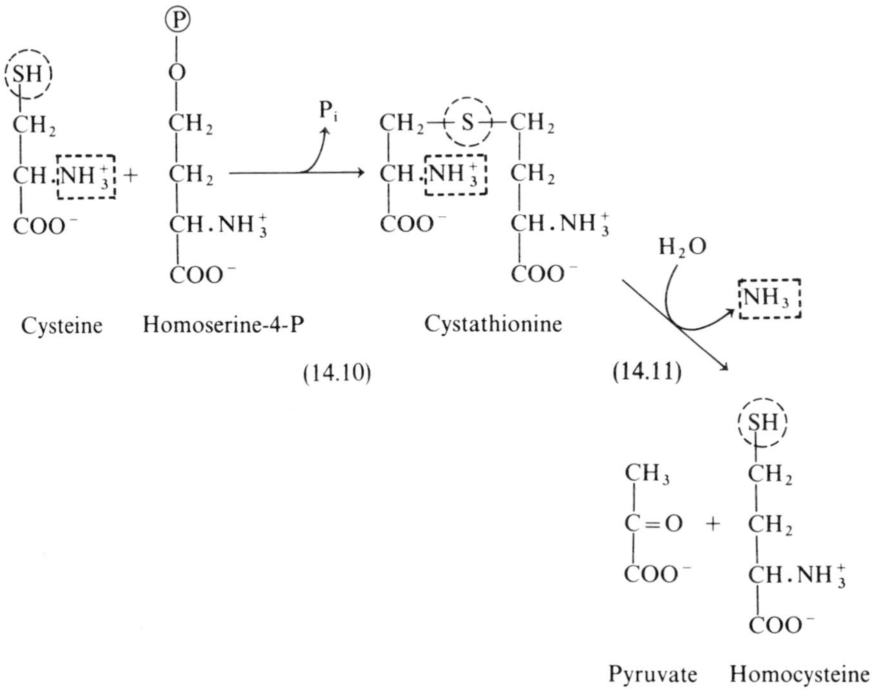

Cysteine   Homoserine-4-P   Cystathionine

(14.10)   (14.11)

Pyruvate   Homocysteine

The next reaction (Eqn. 14.11) catalysed by cystathionine β-lyase (β-cystathionase)

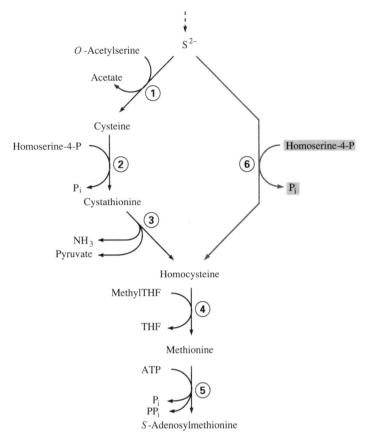

Fig. 14.2 Pathways for the incorporation of S^{2-} into methionine and S-adenosylmethionine. The trans-sulphuration pathway is shown in black; most methionine is made by this route. The alternative mechanism (shown in grey) is known as direct sulph-hydration. Note that, *in vivo*, a bound form of S^{2-} probably serves as the substrate for reaction (1) (see Eqn. 14.5). Enzymes are: (1) cysteine synthase; (2) cystathionine γ-synthase; (3) cystathionine β-lyase; (4) homocysteine:methylTHF methyltransferase; (5) S-adenosylmethionine synthetase; (6) homocysteine synthase. Abbreviation: methylTHF $= N^5$-methyltetrahydrofolate.

cleaves cystathionine to homocysteine, pyruvate and NH_3, effectively making the reaction sequence irreversible. Overall, the thiol group of cysteine is transferred to homoserine. In barley leaf protoplasts almost all the cystathionine γ-synthase and about 50% of the cystathionine β-lyase activity occurs in chloroplasts.

The final step is methylation of homocysteine catalysed by N^5-methyltetra-hydrofolate methyltransferase with N^5-methyltetrahydrofolate as methyl donor (Fig. 14.2—see also Section 14.3.4).

14.2.4 Cysteine and methionine fulfil many important functions

Cysteine and methionine are incorporated into proteins. Cysteine, in particular, determines the structure and catalytic activity of many enzymes (see Section 2.5.2). Cysteine and methionine are incorporated into other essential compounds and are used to support other metabolic pathways. Cysteine is used to synthesize coenzyme A and the tripeptide glutathione (see Sections 4.3 & 14.5 for structures and functions). Methionine is especially important in C_1 metabolism (Section 14.3.1) and smaller amounts are used in the synthesis of ethylene and polyamines.

14.2.5 Methionine biosynthesis is co-ordinated with other amino acids derived from aspartate

Homoserine-4-P is made from aspartate (Section 13.7.2). Since it is a direct precursor of both cystathionine and threonine (the starting point for the synthesis of isoleucine), mechanisms must exist for the regulated synthesis of the amino acids derived from it. Furthermore, the synthesis of lysine, also derived from aspartate, must be co-ordinated with methionine synthesis (see Section 13.7.2).

S-Adenosylmethionine, a product of methionine metabolism (see next Section) strongly enhances threonine synthase activity, diverting homoserine-4-P (see Fig. 13.5). Methionine synthesis is also controlled after the threonine branch point. For example, the flow of sulphur as SO_4^{2-} into the methionine residues of protein in the duckweed *Lemna paucicostata* is inhibited by very low concentrations of methionine ($< 1 \mu M$) but incorporation of sulphur as SO_4^{2-} into cysteine and the cysteine residues of protein is not. Examination of the enzymes involved in the synthesis of methionine from cysteine (Fig. 14.2) indicates that methionine (or a derivative) regulates cystathionine γ-synthase. This complements the regulation of the production of the α-aminobutyryl donor, homoserine-4-P.

14.3 C_1 fragments have various origins

C_1 fragments such as methyl, methenyl or formyl groups are used in the synthesis of many compounds in transfer reactions (e.g. transmethylation and transformylation). S-Adenosylmethionine and several C_1-derivatives of tetrahydrofolate (THF) are the main sources of methyl groups, and formylTHF is the principal donor of formyl groups. Some compounds with a carbon atom originating from these derivatives are listed in Table 14.1. This section is concerned with the production of C_1-donor compounds and their use in C_1 transfer reactions.

Table 14.1 Processes involving addition of C_1 fragments. Note that derivatives of tetrahydrofolate (THF) occur in plants as their polyglutamyl derivatives (see Section 5.2.1).

Process or Compounds synthesized	C_1-donor	Reference
S-adenosylmethionine; methionine	N^5-MethylTHF	Section 14.3.4
Formylmethionyl-tRNA	N^{10}-FormylTHF	Section 19.5
Purine ring and purine derivatives (e.g. purine nucleotides, cofactors and nucleic acids)	N^{10}-FormylTHF N^5, N^{10}-MethenylTHF	Section 15.2.1
Thymidine monophosphate	N^5, N^{10}-MethyleneTHF	Section 15.1.3
Serine	N^5, N^{10}-MethyleneTHF	Section 11.6
Methylation of tRNA	S-Adenosylmethionine	Section 2.6.3
Sinapate, ferulate and lignin	S-Adenosylmethionine	Section 17.11
Phosphocholine and phospholipids containing phosphocholine	S-Adenosylmethionine	Section 16.3.2
Methylated esters of polygalacturonic acid (pectin)	S-Adenosylmethionine	Section 17.9

14.3.1 S-Adenosylmethionine is a methyl group donor

When methionine labelled with ^{14}C in the S-methyl carbon atom is supplied to *Lemna paucicostata*, only about 20% of the label is incorporated into protein (Table 14.2). Of the remainder nearly all is transmethylated, 55% into methylated derivatives of ethanolamine (e.g. choline, phosphocholine and phosphatidyl choline). The methyl ester groups of pectin and chlorophyll are also strongly labelled.

Table 14.2 Estimates of the utilization of the methyl group of methionine for the synthesis of other metabolites by the duckweed *Lemna paucicostata*, as determined by the fate of exogenous additions of [S-methyl-^{14}C] methionine. From Mudd, S.H. & Datko, A.H. (1986) *Plant Physiol.*, **81**, 103–14.

Fraction	Methyl group equivalents (nEq frond^{-1})
Methylated ethanolamine derivatives	2.43
Methyl esters of pectin and other insoluble compounds	0.85
Methyl esters of chlorophyll	0.45
Neutral lipids	0.34
Nucleic acid derivatives	0.28
Methylated basic amino acids	0.09
Methionine residues in protein	[1.05]

The incorporation of the S-methyl carbon into these and other compounds involves formation of S-adenosylmethionine from ATP and methionine, catalysed by S-adenosylmethionine synthetase.

$$\text{Methionine} + \text{ATP} \rightleftharpoons \text{S-adenosylmethionine} + PP_i + P_i \qquad (14.12)$$

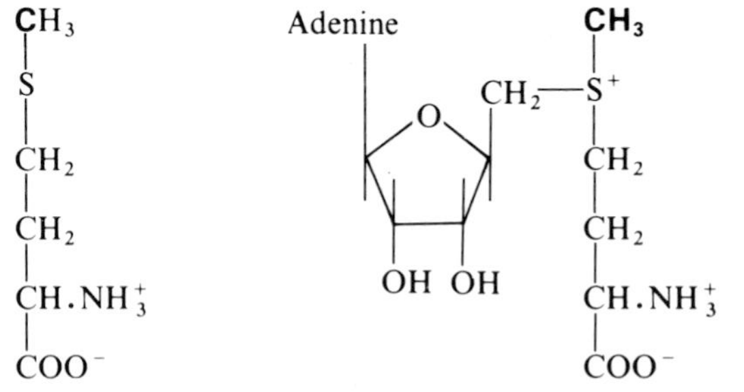

Methionine
(S-methyl carbon atom shown in bold)

S-Adenosylmethionine
(active methyl group shown in bold)

The methyl group on the sulphonium sulphur atom is transferred to appropriate acceptor molecules (R), catalysed by methyltransferase enzymes (Eqn. 14.14) and resulting in the production of S-adenosylhomocysteine which hydrolyses across the thioether bond (Eqn. 14.15) to form homocysteine.

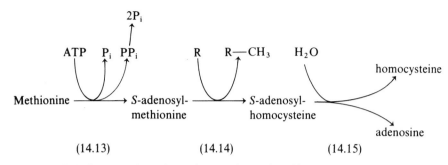

(14.13)   (14.14)   (14.15)

Transmethylation reactions involving S-adenosylmethionine are very energy demanding since all three of the phosphate ester bonds of ATP are hydrolysed. Table 14.1 lists some transmethylation reactions involving S-adenosylmethionine; the formation of sinapate and ferulate from 4-coumarate are especially important in lignin synthesis (see Section 17.11). Clearly, production of S-adenosylmethionine for use in C_1-transfer reactions depends on the production or turnover of methionine, which in turn depends on the production of C_1 derivatives of tetrahydrofolate for the methylation of homocysteine (Fig. 14.2).

14.3.2 Formation of C_1 derivatives of tetrahydrofolate

Plants have four common C_1 derivatives of tetrahydrofolate (THF); N^{10}-formyl-

THF, N^5-methylTHF, N^5,N^{10}-methyleneTHF and N^5,N^{10}-methenylTHF (see Section 5.2.1). These derivatives can be interconverted, although some component processes are irreversible (Fig. 14.3). FormylTHF and methyleneTHF are interconverted by N^5,N^{10}-methyleneTHF dehydrogenase (sensitive to aminopterin) and cyclohydrolase. Out of the total THF pool, formylTHF levels double at night, methyleneTHF declines to very low levels and methylTHF shows little change. The high proportion of formylTHF (i.e. oxidized form) formed at night reportedly reflects a decreased supply of reducing equivalents from light-driven electron transport (Fig. 14.3). Conversely, the enhanced levels of methyleneTHF in the light result from the reduction of formylTHF combined with the production of C_1 units by light-coupled metabolism (see Section 11.5).

THF is synthesized from 4-aminobenzoic acid, glutamate and 2-amino-4-hydroxy-6-hydroxymethyl dihydropteridine (formed from GTP). The synthesis of THF is inhibited by sulphanilamide, an antagonist of 4-aminobenzoic acid. Thus, all processes involving C_1 transfer via THF are also sensitive to this compound.

NH_2—⬡—COOH NH_2—⬡—SO_2—NH_2

4-Aminobenzoic acid   Sulphanilamide

Fig. 14.3 Interconversions of the C_1 derivatives of THF. With some exceptions, the pathway permits production of the various derivatives of THF regardless of the origin of the C_1 fragment, for use in the various reactions listed in Table 14.1. Reactions which can act as possible sources of C_1 units are shown in grey. Enzymes involved in the interconversions are: (1) N^5,N^{10}-methenylTHF cyclohydrolase; (2) N^5,N^{10}-methyleneTHF dehydrogenase; (3) N^5,N^{10}-methyleneTHF reductase. Other enzymes are: (4) formyl-THF synthetase; (5) glycine dehydrogenase; (6) serine hydroxymethyltransferase. Note that THF derivatives are shown in abbreviated form—in plants they occur as polyglutamyl derivatives (see Section 5.2.1).

14.3.3 Sources of C_1 units for production of C_1-THF derivatives

A considerable amount of carbon must pass through the THF pool since all C_1 fragments, including those from S-adenosylmethionine, are derived from C_1-THF derivatives. Lignin can comprise up to 30% of the dry weight of the walls of lignified cells and up to 18% of the carbon in the sinapyl residues is derived from C_1 fragments (see Section 17.11). A substrate closely linked to light-coupled carbon assimilation would seem a logical source of C_1 units for the synthesis of C_1-THF and, in fact, irreversible oxidation of glycine in leaf mitochondria catalysed by glycine dehydrogenase is closely linked to light-coupled CO_2 assimilation with the production of N^5, N^{10}-methyleneTHF.

$$\text{Glycine} + \text{THF} \xrightarrow{\;\;\overset{\text{NAD}^+ \quad \text{NADH} + \text{H}^+}{\curvearrowright}\;\;} N^5, N^{10}\text{-methyleneTHF} + CO_2 + NH_3 \quad (14.16)$$

The C_1 fragment of N^5, N^{10}-methyleneTHF is derived from C-2 of glycine and CO_2 from C-1 of glycine. However, it is not clear whether this is the major route for C_1-THF synthesis since the C_1 fragments produced in this way are assumed to regenerate serine (see Fig. 11.3). However, germinating peas, which do not photo-respire, incorporate ^{14}C-label from [2-^{14}C]glycine into N^{10}-formylTHF and N^5-methylTHF and subsequently into some of the products listed in Table 14.1. Perhaps this occurs in C_4 plants which also do not photorespire.

The reaction catalysed by serine hydroxymethyltransferase, also from the C_2-PR cycle, involves metabolism of a C_1 derivative of THF as well.

$$\text{Glycine} + N^5, N^{10}\text{-methyleneTHF} + H_2O \rightleftharpoons \text{serine} + \text{THF} \quad (14.17)$$

This reaction, unlike the previous one, is readily reversible so that serine can yield N^5, N^{10}-methyleneTHF under suitable conditions.

C_1 derivatives of THF can also be formed from formate, catalysed by N^{10}-formylTHF synthetase.

$$\text{Formate} + \text{ATP} + \text{THF} \rightleftharpoons N^{10}\text{-formylTHF} + \text{ADP} + P_i \quad (14.18)$$

This enzyme is very active in leaves, and extracts incorporate ^{14}C-label from [^{14}C]-formate into formylTHF and methylTHF (Fig. 14.3) and methionine (see Eqn. 14.19). In spinach leaves formate is incorporated at rates of 60 μmol mg^{-1} chlorophyll h^{-1}, more than fast enough for the requirement for C_1 fragments. However,

a mechanism for producing formate at this rate is not known *in vivo*. Envelope-free chloroplasts can form formate at rapid rates by decarboxylating glyoxylate in the presence of H_2O_2 but there is no evidence for this *in vivo*. Other oxidative systems involving H_2O_2 for the production of formate have been proposed. Another possibility not yet ruled out is that formate is formed directly from CO_2.

14.3.4 N^5-MethylTHF is the methyl donor for methionine synthesis

S-Adenosylmethionine is the methyl donor for the synthesis of many compounds (Table 14.1). The synthesis of methionine, in turn, involves the methylation of its precursor, homocysteine, by N^5-methylTHF.

$$\text{Homocysteine} + N^5\text{-methylTHF} \rightleftharpoons \text{methionine} + \text{THF} \quad (14.19)$$

The enzyme in animals and some bacteria (including *Rhizobium*) involves vitamin B_{12}, a cofactor containing cobalt. This explains the requirement of nodulated legumes for cobalt for N_2 fixation.

Plants also support the methylation of homocysteine by S-adenosylmethionine with S-methylmethionine as intermediate. This involves reactions catalysed by S-adenosylmethionine:methionine S-methyltransferase (Eqn. 14.20) and S-methyl-methionine:homocysteine S-methyltransferase (Eqn. 14.21).

$$\begin{aligned}\text{Methionine} + S\text{-adenosylmethionine} &\rightleftharpoons \\ S\text{-methylmethionine} + S\text{-adenosylhomocysteine}\end{aligned} \quad (14.20)$$

$$S\text{-Methylmethionine} + \text{homocysteine} \rightleftharpoons 2 \text{ methionine} \quad (14.21)$$

However, there is no net synthesis of methionine from homocysteine since S-adenosylmethionine is itself synthesized from methionine (Eqn. 14.12).

14.4 Reducing equivalents and phosphorylation potential are exported from illuminated chloroplasts by shuttle mechanisms

14.4.1 Shuttle mechanisms involve the counter-exchange of metabolites which comprise redox pairs

Some reductive anabolic processes in the cytosol are enhanced by light. However, direct transport of NADPH, Fd$_{\text{red}}$ and ATP from chloroplasts is impossible as the envelope is either impermeable or slowly permeable to these compounds (see Table 10.1).

Certain metabolites which cross the chloroplast envelope (see Table 10.1) constitute redox pairs which counter-exchange at rapid rates on specific translocators (Section 10.8). The oxidized forms are reduced by illuminated chloroplasts and counter-exchanged on the appropriate translocator with the oxidized form. This mechanism transports reducing equivalents and, in some cases, phosphorylation potential from illuminated chloroplasts to the cytosol. The exported metabolite is oxidized in the cytosol and the reducing equivalents are recovered in NADH and phosphorylation potential in ATP, which are used to support anabolic processes.

14.4.2 Triose-P/glycerate-3-P shuttle

This shuttle entails reduction of glycerate-3-P to triose-P in chloroplasts in reactions involving glycerate-3-P kinase, glyceraldehyde-3-P dehydrogenase and triose-P isomerase (Fig. 14.4) using ATP and NADPH supplied by the light reactions. The triose-P is exported on the phosphate translocator to the cytosol where it is oxidized (Fig. 14.4). One mechanism involves glycerate-3-P kinase and NAD-specific glyceraldehyde-3-P dehydrogenase, producing both ATP and NADPH. Another reaction involves non-phosphorylating glyceraldehyde-3-P dehydrogenase in which triose-P is a carrier of reducing equivalents only.

$$\text{Glyceraldehyde-3-P} + NAD^+ \rightleftharpoons \text{glycerate-3-P} + NADH + H^+ \qquad (14.22)$$

This reaction, unlike the mechanism involving phosphorylation of ADP, is highly exergonic and essentially irreversible. The glycerate-3-P formed exchanges with triose-P and conserves the carbon skeleton (cf. exchange of triose-P for P_i in sucrose synthesis in the cytosol—see Section 10.8).

Under steady state conditions an equilibrium is established among all the reactants of the shuttle (Fig. 14.4) on both sides of the membrane—a change in the concentration of a reactant will change the concentrations of others on both sides of the membrane. Although the shuttle is shown exporting reducing equivalents and phosphorylation potential from the chloroplast, in theory, it can reverse in the dark with the cytosol supplying these materials to the stroma.

The export of triose-P from illuminated chloroplasts is one of the most important transport processes in plants so that the theoretical capacity of the triose-P/glycerate-3-P shuttle is very high. Experimentally, isolated chloroplasts support a triose-P/glycerate-3-P shuttle which reduces various substrates outside the envelope in the light at rapid rates. The triose-P/glycerate-3-P shuttle might be

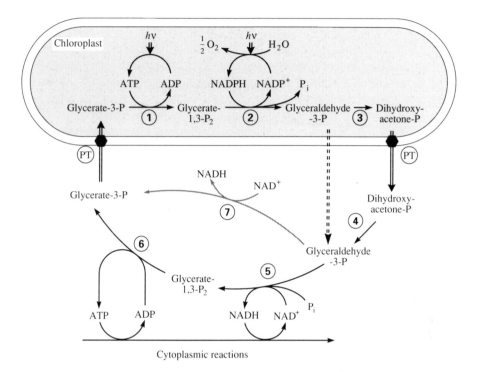

Fig. 14.4 Triose-P/glycerate-3-P shuttle for the export of light-generated reducing equivalents and phosphorylation potential from illuminated chloroplasts. Dihydroxyacetone-P is thought to be the principal form of triose-P exported from chloroplasts; this exchanges with glycerate-3-P on the phosphate translocator, (PT). Triose-P exported from chloroplasts is oxidized in the cytosol, either via the glycolytic sequence (shown in black) or via an irreversible non-phosphorylating mechanism (shown in grey). Note that, according to this model, reductive processes in the cytosol can be linked to O_2 evolution in the chloroplast without involving CO_2 assimilation. Enzymes are: (1) chloroplast glycerate-3-P kinase; (2) NADP glyceraldehyde-3-P dehydrogenase; (3) chloroplast triose-P isomerase; (4) cytosolic triose-P isomerase; (5) NAD glyceraldehyde-3-P dehydrogenase; (6) cytosolic glycerate-3-P kinase; (7) glyceraldehyde-3-P dehydrogenase (non-phosphorylating).

quantitatively important in photosynthetic cells to support sucrose synthesis in the cytosol but there is little data to support this. It is possible that relatively little of the triose-P exported from illuminated chloroplasts *in vivo* is metabolized via the triose-P/glycerate-3-P shuttle.

14.4.3 Dicarboxylate shuttles

Various C_4 and C_5 dicarboxylates can exchange at rapid rates across the chloroplast envelope (see Table 10.1) via one or more antiport translocators. Several functions for the dicarboxylate translocators have been proposed; one is that the redox pair malate/oxaloacetate (and/or aspartate) acts as a shuttle for the transport of light-generated reducing equivalents from stroma to the cytosol (Fig. 14.5). Supporting this, isolated chloroplasts can reduce oxaloacetate to malate in the light at rapid rates (e.g. 150 μmol mg^{-1} chlorophyll h^{-1} is typical), evolving 0.5 O_2 per oxaloacetate reduced and demonstrating that the reducing equivalents are derived from H_2O by electron transport. Further, in experimental conditions, isolated chloroplasts support light-dependent reduction of various substrates outside the chloroplast via a malate/oxaloacetate shuttle. Some processes possibly linked to light in this way include NO_3^- reduction in the cytosol (see Section 13.2.3) and hydroxypyruvate reduction in peroxisomes. However, no definitive demonstrations of the role of a dicarboxylate shuttle in these processes *in vivo* have been made.

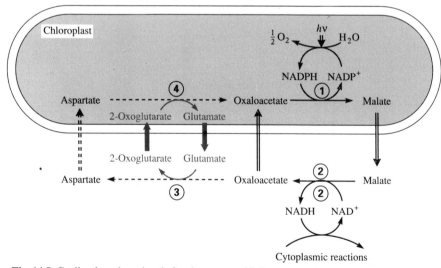

Fig. 14.5 C_4 dicarboxylate shuttle for the export of light-generated reducing equivalents from illuminated chloroplasts. Exchange with aspartate requires concomitant transport of another oxo acid and amino acid to maintain nitrogen balance; this is thought to involve 2-oxoglutarate and glutamate (shown in grey). Enzymes are: (1) NADP malate dehydrogenase; (2) NAD malate dehydrogenase; (3) cytosolic glutamate:oxaloacetate aminotransferase; (4) chloroplast glutamate:oxaloacetate aminotransferase.

Another proposed function of these dicarboxylate translocators is the counter-exchange of the C_5 dicarboxylates, glutamate and 2-oxoglutarate. Chloroplasts have very active mechanisms for the assimilation of NH_3 produced at rapid rates, partly by reduction of NO_2^- (Section 13.5) but mainly through glycine dehydrogenase in the C_2-PR cycle (see Section 11.5). The NH_3 is assimilated via glutamine into glutamate in a light-coupled process (see Section 13.5). The reassimilation of NH_3 derived from the C_2-PR cycle requires a constant input of 2-oxoglutarate into chloroplasts and an associated export of glutamate, i.e. the photorespiratory nitrogen cycle (see Section 11.6). The transport of 2-oxoglutarate, glutamate and NH_3 across the chloroplast envelope must be consistent with the observed rates of NH_3 production via the C_2-PR cycle in C_3 plants (about 30–40 μmol mg^{-1} chlorophyll h^{-1} at 20 °C), as are the rates in Table 10.1. Studies with photorespiratory mutants of *Arabidopsis* having a deficient dicarboxylate translocator also support the proposed role of a translocator in the photorespiratory nitrogen cycle—they only survive at low O_2 and high CO_2 concentrations which do not support the C_2-PR cycle and the production of NH_3.

Preincubation of isolated chloroplasts with low concentrations of malate or various other dicarboxylates increases the incorporation of NH_3 into glutamate and the associated O_2 evolution several-fold. The model shown in Fig. 14.6 proposes that two antiport dicarboxylate translocators are involved in glutamate/ 2-oxoglutarate exchange; one supporting inward transport of 2-oxoglutarate in exchange for malate and the other export of glutamate in exchange for another dicarboxylate (not 2-oxoglutarate). Figure 14.6 shows malate as the counter metabolite for both translocators but *in vivo* malate export could exchange with 2-oxoglutarate and glutamate with oxaloacetate, resulting in an associated malate/ oxaloacetate shuttle as set out in Fig. 14.5.

In theory, glutamate exported from illuminated chloroplasts can be oxidized by mechanisms other than the photorespiratory nitrogen cycle, e.g. glutamate dehydrogenase activity.

$$\text{Glutamate} + \text{NAD}^+ \rightleftharpoons \text{2-oxoglutarate} + \text{NH}_3 + \text{NADH} + \text{H}^+ \qquad (14.23)$$

NH_3 and 2-oxoglutarate could be recycled via the chloroplast with the production of more glutamate but there is little evidence *in vivo* for this proposal.

14.4.4 Other shuttle mechanisms in photosynthetic cells

A shuttle between chloroplasts and peroxisomes involving the redox pair glycollate/glyoxylate has been proposed. Glyoxylate is reduced to glycollate in

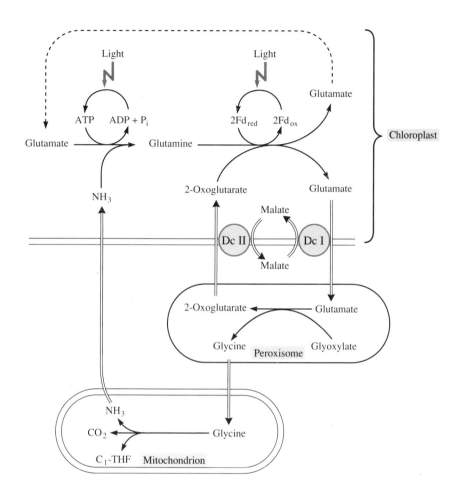

Fig. 14.6 Proposed role for the involvement of two dicarboxylate translocators in the photorespiratory nitrogen cycle for counter-exchange of glutamate and 2-oxoglutarate across the chloroplast envelope. Glutamate is exported from chloroplasts on one translocator (Dc I) in exchange for another dicarboxylate (e.g. malate) and metabolized in mitochondria and peroxisomes to 2-oxoglutarate and NH_3 (see Section 11.6). 2-Oxoglutarate is returned to chloroplasts via another translocator (Dc II) in exchange for another dicarboxylate (e.g. malate). Within the chloroplasts, 2-oxoglutarate and NH_3 are metabolized to glutamate using Fd_{red} and ATP, formed in the light. The cyclic exchange of glutamate for 2-oxoglutarate and NH_3 in this way constitutes a C_5 dicarboxylate shuttle. Enzymes are: (1) glutamine synthetase; (2) glutamate synthase; (3) aminotransferase; (4) glycine dehydrogenase. Abbreviation: C_1-THF = methylenetetrahydrofolate.

illuminated chloroplasts, catalysed by NADP-specific glycollate dehydrogenase.

$$\text{Glyoxylate} + \text{NADPH} + \text{H}^+ \rightleftharpoons \text{glycollate} + \text{NADP}^+ \qquad (14.24)$$

Glycollate is exchanged with glyoxylate and oxidized in the peroxisome, catalysed by glycollate oxidase.

$$\text{Glycollate} + \text{O}_2 \rightleftharpoons \text{glyoxylate} + \text{H}_2\text{O}_2 \qquad (14.25)$$

This sequence achieves no useful purpose and chloroplasts show little glycollate dehydrogenase activity, so it is not held to be important.

Various exchange mechanisms in photosynthetic cells constitute shuttles. In NADP M-E variants of C_4 plants malate acts as a carrier for the transfer of reducing equivalents and CO_2 from mesophyll to bundle sheath chloroplasts with malate/pyruvate forming the redox pair (see Section 12.7). However, PEP/pyruvate exchange between chloroplast and cytosol in the mesophyll cells of several C_4 variants (see Section 12.10) is not a shuttle since the phosphorylation potential of PEP is not transferred to ADP nor to any other compound in the cytosol.

14.5 Illuminated chloroplasts reduce oxidized glutathione

Reduced glutathione (GSH) is a tripeptide of γ-glutamyl-cysteinyl-glycine.

The concentration in chloroplasts is about 2 mM. Like most thiols, the —SH group oxidizes to form a disulphide bond which links two molecules together to produce a dimer of oxidized glutathione (GSSG). Chloroplasts contain a very active GSSG reductase which catalyses the reduction of GSSG by NADPH.

$$\text{GSSG} + \text{NADPH} + \text{H}^+ \rightleftharpoons 2\text{GSH} + \text{NADP}^+ \qquad (14.26)$$

The chloroplast envelope is impermeable to GSSG/GSH but osmotically shocked chloroplasts in the light support O_2 evolution in the presence of GSSG at rates up to $25\,\mu\text{mol mg}^{-1}$ chlorophyll h^{-1} with the production of four GSH per O_2 evolved. Thus, light-generated reductant (NADPH) supports reduction of GSSG *in vitro* in accordance with the theory given by the boxed area in Fig. 14.7.

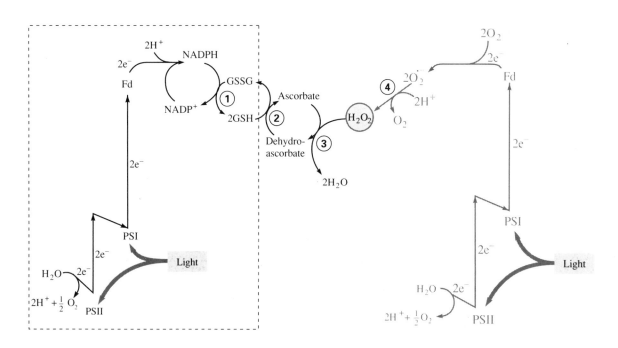

Fig. 14.7 Mechanism of light-dependent O_2 consumption and H_2O_2 production by chloroplasts (shown in grey) and the ascorbate/glutathione pathway for the light-coupled detoxification of H_2O_2 (shown in black). The boxed area denotes the theory of GSSG-dependent O_2 evolution by ruptured chloroplasts. Enzymes are: (1) GSSG-reductase; (2) glutathione dehydrogenase; (3) ascorbate peroxidase; (4) superoxide dismutase. Abbreviations: GSSG = oxidized glutathione; GSH = reduced glutathione; PSI and PSII = photosystems I and II.

In chloroplasts GSH might act as the carrier for the bound pathway of SO_4^{2-} reduction (Section 14.2.1) but this accounts for only a small proportion of the potential rate of GSSG reduction.

A major proposed role for GSH concerns the detoxification of H_2O_2, a powerful inhibitor of chloroplast activity, e.g. 0.6 mM H_2O_2 inhibits CO_2 assimilation by 90% in spinach chloroplasts. H_2O_2 is thought to be produced at about 15 μmol mg^{-1} chlorophyll h^{-1} due to channelling of electrons from PSI via ferredoxin into O_2 instead of NADPH (see Section 9.18). Light-dependent O_2 reduction is promoted by high concentrations of O_2, low levels of CO_2, addition of bipyridilium herbicides (e.g. methyl viologen) and, in reconstituted chloroplast systems, high concentrations of ferredoxin. Addition of an electron to O_2 (Eqn. 14.27) produces superoxide ($O_2^{\cdot-}$) which, because of its unpaired electron, is extremely reactive, especially with lipids. However, superoxide dismutase, a copper-containing enzyme, catalyses the dismutation of $O_2^{\cdot-}$ in chloroplasts to form H_2O_2 and O_2 (Eqn. 14.28).

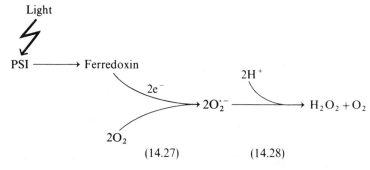

$$(14.27) \qquad (14.28)$$

The H_2O_2 produced in this light-dependent reaction (with net consumption of O_2) inhibits chloroplast activity, but is far less deleterious to chloroplasts than the superoxide from which it is derived.

H_2O_2 is detoxified by a light-dependent mechanism with GSH as the intermediary reductant (Fig. 14.7). H_2O_2 is reduced by ascorbate, present at high concentrations in chloroplasts, catalysed by ascorbate peroxidase. Dehydroascorbate, produced in this reaction, is reduced by GSH, catalysed by GSH dehydrogenase (dehydroascorbate reductase). In the light, the GSSG formed is reduced through the action of GSSG reductase (Eqn. 14.26) using reducing equivalents derived from H_2O by light-driven electron transport (boxed area in Fig. 14.7). As predicted in Fig. 14.7, addition of H_2O_2 to isolated chloroplasts causes oxidation of endogenous ascorbate and GSH to dehydroascorbate and GSSG. In the light, GSSG initiates O_2 evolution through GSSG-reductase. When all the H_2O_2 is reduced the endogenous pools of GSH and ascorbate are restored. In the dark, however, the endogenous pools of GSH and ascorbate are not regenerated and O_2 is not evolved. Spinach chloroplasts detoxify very low concentrations of H_2O_2 at rapid rates of up to $100\,\mu mol$ reduced mg^{-1} chlorophyll h^{-1}, well in excess of the expected rate of production *in vivo*.

In animals and most aerobic heterotrophs, H_2O_2 is detoxified by a high affinity mechanism involving GSH peroxidase, an enzyme containing selenium.

$$2\,GSH + H_2O_2 \rightleftharpoons 2H_2O + GSSG \qquad (14.29)$$

Although present in some unicellular algae, this enzyme has not been found in higher plants. This could explain why selenium is essential for animal nutrition but probably not essential for higher plants.

14.6 Illuminated chloroplasts synthesize fatty acids, proteins and other metabolites

Most of the lipid in leaves is in the thylakoid membranes of chloroplasts. Chloroplasts are important sites for the synthesis of saturated fatty acids from acetate using ATP and NADPH from the light reactions. For example, the synthesis of palmitic acid (16:0) from eight acetate molecules requires 15 ATP and 14 NADPH. Illuminated chloroplasts incorporate acetate into fatty acids at about $1.5\,\mu mol\,mg^{-1}$ chlorophyll h^{-1} which, per mol of carbon, is some 30–50 times faster than with CO_2. Acetate, or an acetate precursor, for fatty acid synthesis must come from outside the chloroplast, presumably from mitochondria (see Section 16.1.1).

Many, but by no means all, chloroplast proteins are synthesized within the organelle on 70 S ribosomes using chloroplast mRNA and light as an energy source. Other biosynthetic processes also occur in chloroplasts, although details of some of these are not clear (e.g. purine and pyrimidine synthesis).

Table 14.3 Estimates of the rates of processes catalysed *in vitro* by isolated chloroplasts in C_3 plants and their projected requirements for ATP and reducing equivalents (expressed as NADPH or its $2e^-$ equivalent).

Process	Estimated rate ($\mu mol\,mg^{-1}$ chlorophyll h^{-1})	Requirements	
		ATP	NADPH
CO_2 assimilation	200	600	400
O_2 reduction	15	0	15
H_2O_2 reduction	(200)*	0	(200)
SO_4^{2-} assimilation (bound pathway)	0.4	0.4	1.6
SO_3^{2-} reduction (free pathway)	(100)†	0	(300)
Photorespiration			
reassimilation of NH_3	35	35	35
reassimilation of glycerate-3-P produced by RuP_2-Oase activity	70	105	70
reassimilation of glycerate produced from glycollate via serine	35	70	35
Dicarboxylate shuttle	15	0	15
Incorporation of acetate into fatty acids	1.5	1.5	0
Reductive assimilation of NO_2^- into glutamine	6	6	18

* The rate *in vivo* is thought to be about $15\,\mu mol\,mg^{-1}$ chlorophyll h^{-1}.
† The rate *in vivo* is thought to be less than that for SO_4^{2-} assimilation by the bound pathway.

14.7 Rates of processes associated with chloroplasts

Table 14.3 provides a very approximate idea of the potential demand of processes in chloroplasts for ATP and reducing equivalents based on rates observed *in vitro*. The demands of many of these processes *in vivo* remain to be established. It is also not known whether they exceed the capacity of the light reactions to make them and, if so, how the various competing demands are controlled.

Further reading

Sulphur metabolism: Läuchli & Bieleski (1983) Chapter III; Miflin (1980) Chapters 5 and 12.
C_1 *metabolism*: Davies (1980) Chapter 9; Davies (1987) Chapter 9.
Light-dependent oxygen reduction: Badger (1985); Davies (1987) Chapter 8.
Shuttles and metabolite transfer between chloroplasts and other subcellular compartments: Davies (1987) Chapter 2; Edwards & Walker (1983) Chapter 8.

PART 4
SYNTHESIS OF NEW CELLS
AND CELL STRUCTURES

Synthesis of nitrogenous compounds
from amino acids

Amino acids provide the starting materials for the synthesis of a large number of compounds of low molecular weight, some of which are building blocks for the formation of other macromolecules. This chapter considers the synthesis of purines and pyrimidines (essential components of nucleotides and nucleic acids), and of chlorophyll. Glutamate, glycine and aspartate are particularly important sources of carbon and nitrogen and the amide group of glutamine is an important source of nitrogen.

15.1 Synthesis of pyrimidines and pyrimidine nucleotides

15.1.1 Uridine and cytidine nucleotides are synthesized de novo *from aspartate, glutamine and carbon dioxide*

The structures and associated nomenclature of the three pyrimidines commonly found in plants (uracil, cytosine and thymine) are given in Section 2.6. These occur mainly in the nucleic acids and nucleotides; only small amounts occur in free forms.

Pyrimidines are constructed from a few simple starting molecules: CO_2, aspartate and the amide nitrogen atom from one or more molecules of glutamine (Fig. 15.1a). This is known as synthesis *de novo* to distinguish it from the synthesis of specific pyrimidines from pre-existing pyrimidine ring structures by recycling— see Section 15.1.4).

The three major pyrimidines are synthesized as their nucleotides along a common route to UMP (Fig. 15.2). Cytosine and thymine are formed as their nucleotides (e.g. cytosine as CMP, CDP and CTP) from the appropriate uridine nucleotide (UMP, UDP or UTP). The first two reactions of UMP synthesis largely determine the framework of the pyrimidine ring. In the first reaction, carbamyl-P is synthesized from CO_2 and the amide group of glutamine using two ATP as the energy source (see also Section 13.7.5) and making the reaction essentially irreversible.

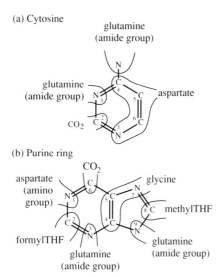

Fig. 15.1 Origin of the carbon and nitrogen atoms in the pyrimidine, cytosine (a) and in the purine ring in animals (b). (Purines of plants are thought to be formed in a similar way). Abbreviation: THF = tetrahydrofolate.

$$CO.NH_2 \quad CH_2 \quad CH_2 \quad CH.NH_3^+ \quad COO^- \ + \ CO_2 \xrightarrow[\ \ 2ADP+P_i\ \ ]{\ \ H_2O+2ATP\ \ } C=O \ + \ \text{(Glutamate)}$$

Glutamine   Carbamyl-P   Glutamate

(15.1)

Carbamyl-P synthetase, which catalyses this reaction, is a regulatory enzyme; its activity is enhanced by ornithine and inhibited by UMP (Section 15.1.2). In the second reaction, carbamyl-P is hydrolysed in a highly exergonic reaction and the free energy is used to join the carbamyl group to aspartate, eliminating P_i, in an essentially irreversible reaction catalysed by another regulatory enzyme, aspartate

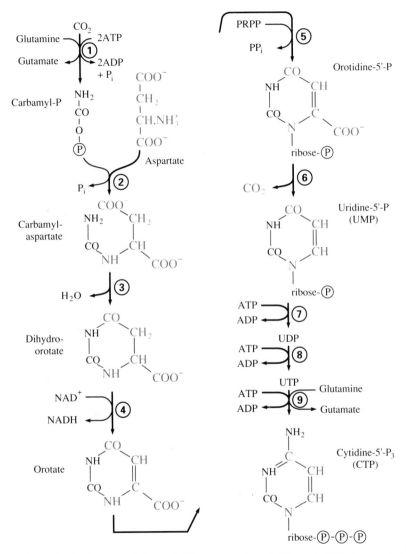

Fig. 15.2 Synthesis *de novo* of the pyrimidine nucleotides, UTP and CTP. Atoms derived from aspartate are shown in grey. Enzymes are: (1) carbamyl-P synthetase; (2) aspartate transcarbamylase; (3) dihydro-orotase; (4) dihydro-orotate dehydrogenase; (5) orotate phosphoribosyl transferase; (6) orotodine monophosphate decarboxylase; (7) UMP kinase; (8) UDP kinase; (9) CTP synthetase. Abbreviation: PRPP = phosphoribosyl-pyrophosphate.

transcarbamylase which, in plants, is inhibited by UMP. Both carbamyl-P synthetase and aspartate transcarbamylase occur in the stroma of chloroplasts. The latter enzyme is especially active but even carbamyl-P synthetase exhibits rates of about 1.5 μmol mg^{-1} chlorophyll h^{-1} (compare with rates of other processes in chloroplasts—see Table 14.3). In radish cotyledons, about 65% of the activity of each enzyme is found in chloroplasts and, in pea leaf tissue, all of the aspartate transcarbamylase occurs in chloroplasts. This implies that the production of carbamyl-P and carbamyl aspartate is light-coupled *in vivo*.

Water is removed from carbamyl aspartate in a freely reversible reaction. This results in ring closure with the formation of dihydro-orotate, catalysed by dihydro-orotase. Orotate, the first product with an entire pyrimidine ring, is formed by oxidation of dihydro-orotate. In the next step, ribose-5-P is attached to form the nucleotide, orotidine-5-P. However, ribose-5-P is not added directly to orotate, but first undergoes a reaction with ATP (Eqn. 15.2) to make the activated derivative, 5-phosphoribosyl-1-pyrophosphate (PRPP).

$$(15.2)$$

This reaction, catalysed by PRPP synthetase, transfers PP$_i$ rather than P$_i$ to the hemiacetal carbon (C-1). The reaction between PRPP and orotate, catalysed by orotate phosphoribosyl transferase, involves attachment of C-1 of PRPP to N-1 of the pyrimidine ring with the elimination of PP$_i$. The enzyme has very high affinity for both substrates ($K_m < 5 \mu$M) and its activity is enhanced by Mg^{2+}. The product, orotidine-5-P, is very difficult to detect in plants due to the very high substrate affinity of the next enzyme, orotidine-5-P decarboxylase ($K_m \sim 1 \mu$M), which catalyses the production of UMP. UDP and UTP are synthesized by successive phosphorylations of UMP and UDP using ATP as the phosphoryl donor, catalysed by two independent kinases found in chloroplasts and cytosol. Dihydro-orotase, orotate phosphoribosyl transferase and orotidine-5-P decarboxylase reportedly occur in chloroplasts but dihydro-orotate dehydrogenase is found in mitochondria. With the exception of the oxidative step, chloroplasts contain the necessary mechanisms for the complete synthesis of the pyrimidine nucleotides from simple assimilation products of inorganic carbon and nitrogen.

The cytidine nucleotides are synthesized from the uridine nucleotide (CTP from UTP and CMP from UMP). This entails addition of an amino group in the pyrimidine ring at C-4. The amide group of glutamine is thought to be the source of the amino group of CTP in a reaction involving CTP synthetase (Eqn. 15.3), since the incorporation of [^{14}C]orotate and [^{14}C]uracil into the cytosine moieties of RNA is very sensitive to azaserine, an inhibitor of amide group transfer reactions involving glutamine (see Section 13.3).

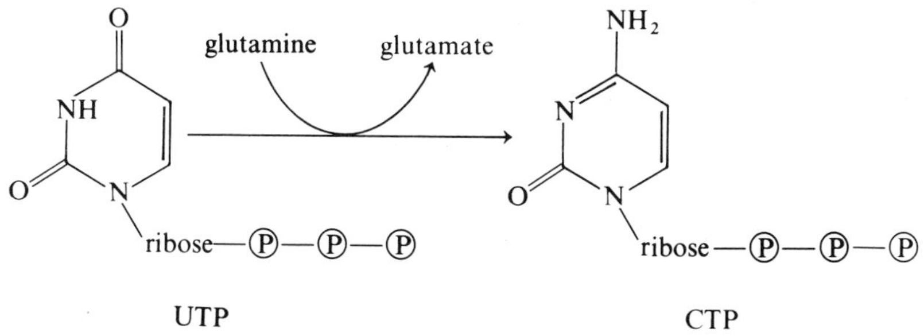

(15.3)

15.1.2 Formation of pyrimidine nucleotides is controlled by carbamyl-P synthetase activity

The first enzyme of pyrimidine synthesis, carbamyl-P synthetase, is inhibited by UMP (Table 15.1). The pyrimidine trinucleotides UTP and CTP are less inhibitory. On the other hand, several purine nucleotides (especially inosine monophosphate (IMP), the first of the purines synthesized) enhance activity, effectively keeping pyrimidine synthesis in step with purine synthesis. Overproduction of pyrimidine nucleotides under these conditions is controlled by UMP.

Carbamyl-P is a substrate for the synthesis of arginine (see Section 13.7.5) as well as pyrimidine nucleotides (Fig. 15.3). UMP inhibits aspartate transcarbamylase so that carbamyl-P is diverted into arginine, but carbamyl-P synthetase is stimulated by ornithine (Table 15.1) which is the carbamyl-P receptor molecule for the synthesis of citrulline, a precursor of arginine. Thus, if UMP levels rise, both

Table 15.1 Effect of ornithine and various nucleotides on the activity of carbamyl-P synthetase from pea shoots. Activity values are expressed as % of control without effector. From O'Neal, T.D. & Naylor, A.W. (1976) *Plant Physiol.*, **57**, 23–8.

		Activity (%)			
		Ornithine (mM)			
Expt	Negative effector	0	0.25	2	10
1*	nil	100	782	798	643
	50 μM UMP	7	15	50	85
	150 μM UDP	14	25	52	75
2*	nil	100	361	417	337
	300 μM TMP	6	19	54	89
	1000 μM UTP	21	44	66	97
		IMP (mM)			
		0	0.25	2	
3	nil	100	185	184	
	50 μM UMP	9	25	78	
	150 μM UDP	11	45	75	

*Separate enzyme preparations were used in experiments 1 and 2.

carbamyl-P synthetase and aspartate transcarbamylase are inhibited, causing ornithine levels to rise. This, in turn, ameliorates the UMP inhibition of carbamyl-P synthetase (but not aspartate transcarbamylase), and therefore controls the flow of carbamyl-P for arginine synthesis (Fig. 15.3).

15.1.3 Deoxyribonucleotides are formed by reduction of ribonucleotides

Cytosine and thymine occur in DNA as deoxyribonucleotides, i.e. deoxy-CMP and deoxy-TMP (also known as TMP or thymidilate since the ribonucleotides of thymine are not known in cells). The synthesis of the pyrimidine deoxyribonucleotides involves reduction of the ribosyl moieties, CDP and UDP, to form deoxy-CDP and deoxy-UDP. Deoxy-TMP is synthesized from deoxy-UDP.

The deoxyribonucleotide diphosphates (deoxy-NDP) are formed by reduction of their NDP counterparts with thioredoxin (Td), catalysed by NDP reductase (Eqn. 15.5). Several plant thioredoxins (see Section 10.13.1) support reduction of the

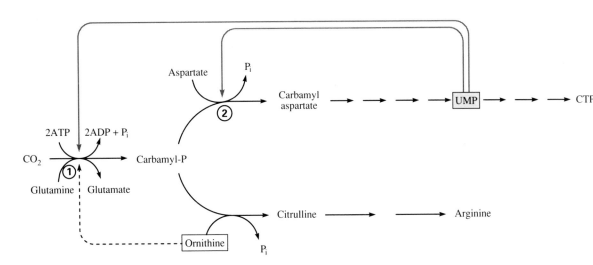

Fig. 15.3 Metabolic regulation of UMP and arginine synthesis via control of (1) carbamyl-P synthetase and (2) aspartate trans-carbamylase. Inhibition by UMP is shown in grey. Inhibition of carbamyl-P synthetase by UMP is alleviated by ornithine (dashed line). Carbamyl-P synthetase is also promoted by purine nucleotides, especially IMP (see Table 15.1).

ribonucleotides by the NDP reductase from the bacterium *Escherichia coli* (Eqn. 15.5). NDP reductase is assumed to occur in plants where Td is maintained in its reduced form by NADPH (Eqn. 15.4).

$$NADPH + H^+ + Td \overset{S}{\underset{S}{\big|}} \rightleftharpoons NADP^+ + Td \overset{SH}{\underset{SH}{}} \tag{15.4}$$

$$NDP + Td \overset{SH}{\underset{SH}{}} \rightleftharpoons deoxy\text{-}NDP + Td \overset{S}{\underset{S}{\big|}} + H_2O \tag{15.5}$$

The reaction shown in Eqn. 15.4 is thought to be catalysed by an enzyme containing a flavin prosthetic group and not the Fd:Td oxidoreductase of chloroplasts (see Section 10.13.1). Separate NDP reductases exist for the reduction of UDP and CDP. The purine deoxyribonucleotides, deoxy-ADP and deoxy-GDP, are synthesized from ADP and GDP in reactions analogous to Eqn. 15.5.

Deoxy-TMP synthesis is presumed to involve metabolism of deoxy-UDP to deoxy-UMP and methylation of the uracil moiety at C-5 in a reaction involving N^5,N^{10}-methyleneTHF as methyl donor, catalysed by thymidilate synthase.

$$\text{Deoxy-UMP} + N^5,N^{10}\text{-methyleneTHF} \rightleftharpoons \text{deoxy-TMP} + \text{dihydrofolate} \tag{15.6}$$

15.1.4 Nucleotides are reformed from free pyrimidines and purines

RNA and nucleotides undergo metabolic turnover resulting in the production of nucleosides and free pyrimidines and purines. These degradation products can be converted to nucleotides (Fig. 15.4). Free pyrimidines and purines react with PRPP, catalysed by phosphoribosyl phosphotransferases to form ribonucleoside monophosphates in reactions analogous to PRPP and orotate (Fig. 15.2, reaction (5)).

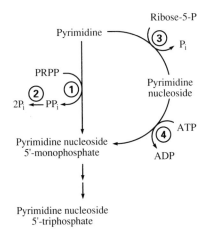

Fig. 15.4 Mechanisms for the reformation of pyrimidine nucleotides from free pyrimidines. Enzymes are: (1) phosphoribosyl phosphotransferase; (2) pyrophosphatase; (3) nucleoside phosphorylase; (4) nucleoside kinase. Abbreviation: PRPP = phosphoribosyl-pyrophosphate.

The PP$_i$ produced in these reactions is hydrolysed by pyrophosphatase activity, making the coupled reaction highly exergonic. Alternatively, free pyrimidines and purines can react with ribose-1-P to form the corresponding nucleoside with the elimination of P$_i$. Phosphorylation of the nucleoside at the 5' position is achieved at the expense of ATP in an irreversible reaction.

15.2 Synthesis and metabolism of purines and purine nucleotides

15.2.1 *Purine nucleotides are synthesized* de novo *from simple metabolites*

Despite the obvious importance of purines in the synthesis of the nucleic acids and various nucleotides, relatively little is known about their synthesis in plants. The available information suggests that the pathways are similar to those in other organisms. Like the pyrimidine ring, the purine ring is built up from simple molecules (Fig. 15.1), most of which are short-term products of carbon and nitrogen assimilation.

Pyrimidine synthesis involves formation of the ring first with subsequent attachment of ribose-5-P, whilst purine synthesis entails the successive addition of other structures to PRPP (Fig. 15.5). Purines are synthesized as their nucleotides by a common pathway with IMP as the first product with an entire purine ring. Nucleotides containing adenine and guanine are synthesized from IMP. IMP synthesis begins with the addition of the amide group of glutamine to C-1 of PRPP with the elimination of PP$_i$ and the formation of ribosylamide-5-P. Amidophosphoribosyl transferase, which catalyses this reaction, is inhibited by various purine nucleotides. This appears to be the main regulatory step. Subsequent steps are not known to be subject to metabolic control. The synthesis of ribosylamide-5-P involves a change in the configuration at C-1 from α to β, which persists in all the purine nucleotides.

The following reactions involve addition of glycine to the amino group of ribosylamine-5-P, addition of a formyl group by hydrolysis of N^5, N^{10}-methenylTHF, addition of a further amide group from glutamine, and closure of the imidazole ring (Fig. 15.5). CO_2 and aspartate are then added, fumarate is

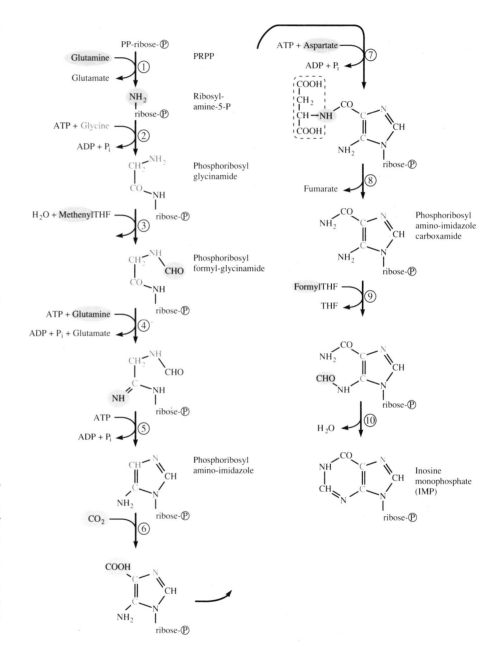

Fig. 15.5 Pathway for the synthesis of IMP, precursor of the purine nucleotides AMP and GMP. Atoms derived from glycine are shown in grey. PRPP is denoted as PP-ribose-P, the pyrophosphate moiety being attached at C-1 and the phosphate group at C-5. Abbreviations: PRPP = phosphoribosyl-pyrophosphate; THF = tetrahydrofolate.

eliminated, a further formyl group is added, this time from N^{10}-formylTHF, and finally the ring is closed by removal of water. The synthesis of GMP entails oxidation of IMP and addition of yet another amide group of glutamine (Fig. 15.6). AMP synthesis involves addition of aspartate to IMP and removal of fumarate, leaving an amino group attached at C-6. Large inputs of energy are required to synthesize GMP and AMP. For example, the synthesis *de novo* of a complete molecule of ATP from glycine, glutamine, aspartate, etc. requires four ATP and one GTP; a further two ATP are required for the synthesis of PRPP from ribose-5-P (Eqn. 15.1).

Studies of the synthesis of the purine, caffeine, provide the strongest evidence that purine synthesis in plants proceeds via the pathway shown in Fig. 15.5.

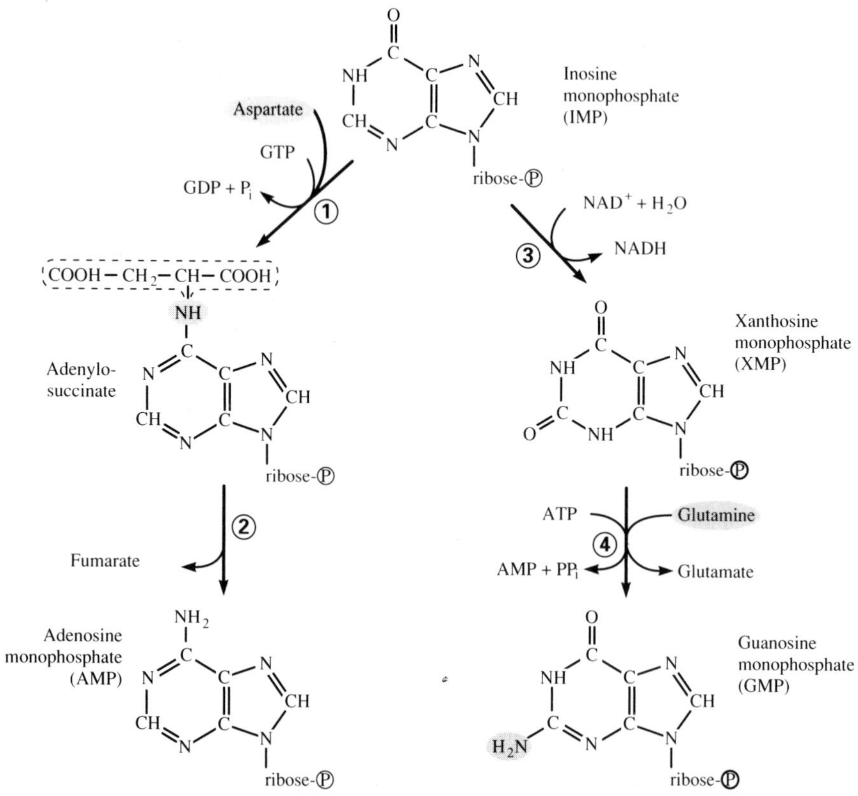

Fig. 15.6 Formation of AMP and GMP from IMP. Enzymes are: (1) adenylosuccinate synthetase; (2) adenylosuccinate lyase; (3) IMP dehydrogenase; (4) GMP synthetase.

Caffeine, a secondary compound, is not essential for growth; it is the addictive stimulant of tea and coffee and is of considerable commercial importance.

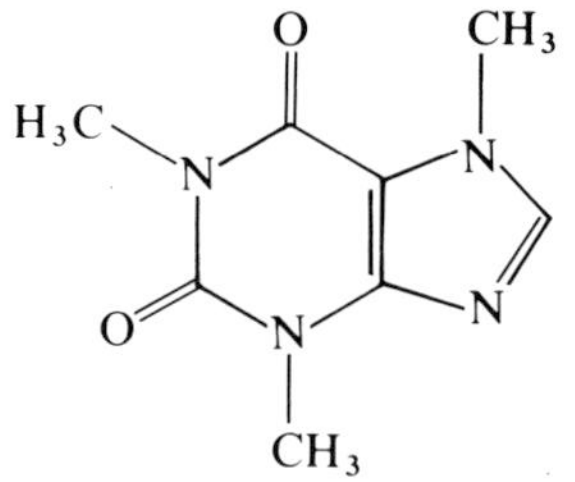

Caffeine

Caffeine synthesis involves incorporation of CO_2, formate and glycine into the specific members of the purine ring predicted by Fig. 15.1b. Other evidence is also in agreement with Fig. 15.5. Plants treated with sulphanilamide, an inhibitor of THF synthesis (see Section 14.3.2), accumulate phosphoribosyl glycinamide and phosphoribosyl amino-imidazole carboximide, intermediates of the pathway whose further metabolism involves C_1 derivatives of THF (Fig. 15.5, reactions (3) and (8)). Further, several enzymes of the pathway have been reported in plants, including enzymes which catalyse the phosphorylation of AMP and GMP, e.g. adenylate kinase (Eqn. 15.7).

$$ATP + AMP \rightleftharpoons 2ADP. \tag{15.7}$$

Mechanisms for the phosphorylation of ADP to ATP include substrate level phosphorylation, oxidative phosphorylation in mitochondria and photophosphorylation.

Purine nucleotides can be reformed from free purines and purine nucleosides by mechanisms analogous to those described for pyrimidines (Section 15.1.4). The synthesis of purine deoxyribonucleotides is similar to that of pyrimidine deoxyribonucleotides (Section 15.1.3).

15.2.2 Purines are oxidized to ureides which are important to nitrogen transport in some species

Ureides are substituted ureas with the general structure NH_2—CO—NH—R and R—NH—CO—NH—R^1, e.g. allantoin and allantoate (Fig. 15.7). Ureides are quite common in plants, although they are not involved in the synthesis of essential macromolecules. Several ureides are readily synthesized from IMP and the purines, adenine and guanine, by oxidative purine catabolism (Fig. 15.7). Ureides are

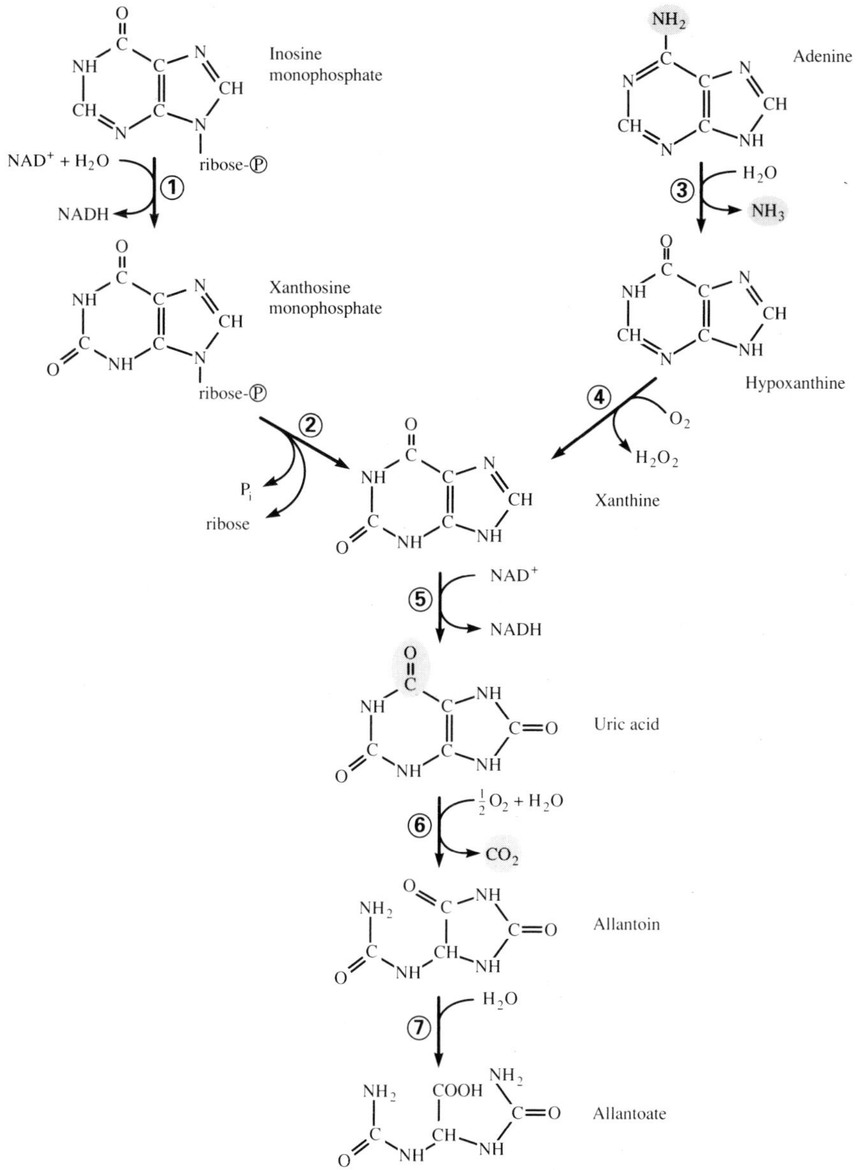

Fig. 15.7 Pathway for the synthesis of the ureides, allantoin and allantoate, from IMP and adenine. Enzymes are: (1) IMP dehydrogenase; (2) nucleotidase; (3) adenase; (4) xanthine dehydrogenase; (5) xanthine oxidase; (6) uricase; (7) allantoinase.

important transport compounds of nitrogen in some plants, especially nodulated tropical legumes (see Section 13.8.3). The nodules of these species (see Table 13.2) have very active systems for the synthesis *de novo* of purines, with enzyme levels many times greater than in temperate legumes, which transport nitrogen as amides. Also, the activities of the enzymes of purine and ureide synthesis in tropical legumes closely parallel the onset of N_2 fixation and the export of ureides from nodules.

In nodulated tropical legumes gaseous nitrogen is reduced to NH_3 by *Rhizobium* bacteroids, which is then exported to the host nodule cells where it is incorporated into glutamine as in temperate legumes (Section 13.8.3). The amide nitrogen of glutamine is used directly or indirectly for the synthesis of purines and ureides involving production *de novo* of IMP (Fig. 15.5). Of the four nitrogen atoms in IMP and the ureides derived from it, two come directly from glutamine, one from aspartate and one from glycine. Non-photosynthetic plastids are reportedly the site of IMP synthesis in the nodule cells of the host. There is some uncertainty about the pathway of ureide synthesis from IMP (Fig. 15.7). Allopurinol, which irreversibly blocks the oxidation of xanthine, catalysed by xanthine dehydrogenase, blocks ureide synthesis by isolated nodules and causes xanthine to accumulate, indicating that xanthine is an intermediate. However, the route of xanthine synthesis is uncertain since dehydrogenases acting on both hypoxanthine and IMP (Fig. 15.7) occur in nodules, although accumulation of hypoxanthine in nodules treated with allopurinol has not been reported.

Nodulation, and hence N_2 fixation, does not occur in legumes supplied with inorganic nitrogen salts (e.g. NO_3^-). In amide-transporting temperate legumes, the relative proportions of the nitrogenous compounds in the xylem sap are not greatly affected by the NO_3^- supply. However, in tropical legumes, the concentration of ureides in the xylem sap in un-nodulated plants (i.e. with NO_3^-) is very low indeed, clearly demonstrating the importance of nodulation in ureide synthesis.

In ureide-transporting legumes, nitrogen must be recovered in forms which can be used for the synthesis of other molecules. Tropical legumes, especially the shoots and pods, contain very active allantoicase and urease activities suggesting that allantoin and allantoate are metabolized via urea, from which the nitrogen is released as NH_3 (Fig. 15.8). According to this mechanism, allantoin labelled with ^{14}C in carbon atoms C-2 and C-7 should give rise to [^{14}C]urea and, ultimately, [14]CO_2 through the action of urease. However, although leaf discs of soya bean produce [14]CO_2 from [2,7-^{14}C]allantoin, this is not associated with the production of [^{14}C]urea. Further, the production of [14]CO_2 is not inhibited by phenylphosphordiamidate, a potent inhibitor of urease. However, leaf discs incubated with

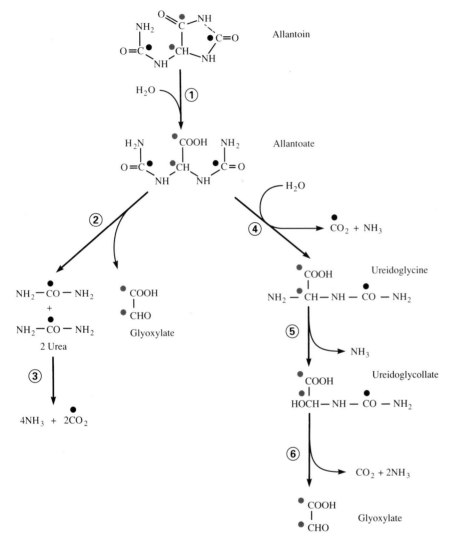

Fig. 15.8 Postulated pathways for the release of NH_3 from the nitrogen-transport compounds, allantoin and allantoate. Experiments to establish the importance of the two pathways are described in the text. According to theory, carbon atoms marked with black circles are derived from C-2 and C-7 of allantoin and those marked with grey circles are derived from C-4 and C-5. NH_3 released is reassimilated into glutamine. Enzymes are: (1) allantoinase; (2) allantoicase; (3) urease; (4) allantoate amidohydrolase; (5) ureidoglycine aminohydrolase; (6) ureidoglycollate amidohydrolase.

[4,5-^{14}C]allantoin readily produce ^{14}C-labelled allantoate, CO_2, glyoxylate, glycine and serine. The data are consistent with metabolism of allantoate via two amidohydrolase reactions (not involving urease) to release CO_2, NH_3 and glyoxylate (Fig. 15.8), the latter being further metabolized via the C_2-PR cycle to form glycine and serine.

The NH_3 released in the shoots and pods is reassimilated into glutamine and glutamate. Since purine synthesis involves the consumption of ATP, glutamine, glycine and aspartate (Fig. 15.5), and the release of nitrogen as NH_3 is associated with the loss of carbon (as CO_2) during ureide catabolism (Figs 15.7 & 15.8), the transport of an atom of nitrogen via a ureide is more energy-demanding than transport via glutamine and/or asparagine (see Section 13.7.1).

15.3 Synthesis of chlorophyll

Chlorophyll is a tetrapyrrole and so consists of four pyrrole molecules linked together.

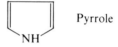

Pyrrole

Plants contain several important tetrapyrroles: chlorophyll, the haems present in various cytochromes and sirohaem. Plants also contain open-chain tetrapyrroles formed by oxidative opening of the tetrapyrrole ring, e.g. phytochrome (a light-absorbing molecule important in determining certain morphogenetic responses of plants to light) and the phycobilins. Tetrapyrroles are synthesized by a common pathway involving the polymerization of eight molecules of 5-aminolevulinate.

15.3.1 Plants synthesize 5-aminolevulinate from glutamate

Plants and animals differ in their synthesis of 5-aminolevulinate. In animals the most important pathway involves condensation of glycine with succinyl-coenzyme A, catalysed by aminolevulinate synthetase. In plants, however, glutamate and 2-oxoglutarate are far more effective sources of 5-aminolevulinate. Labelling patterns of the carbon atoms of the 5-aminolevulinate formed from [^{14}C]glutamate and [^{14}C]2-oxoglutarate indicate that the carbon skeletons of these C_5-dicarboxylates are not degraded. Plants contain an aminotransferase that catalyses the synthesis of 5-aminolevulinate from 4,5-dioxovalerate (Eqn. 15.9) which, in theory, could be formed by reduction of 2-oxoglutarate (Eqn. 15.8).

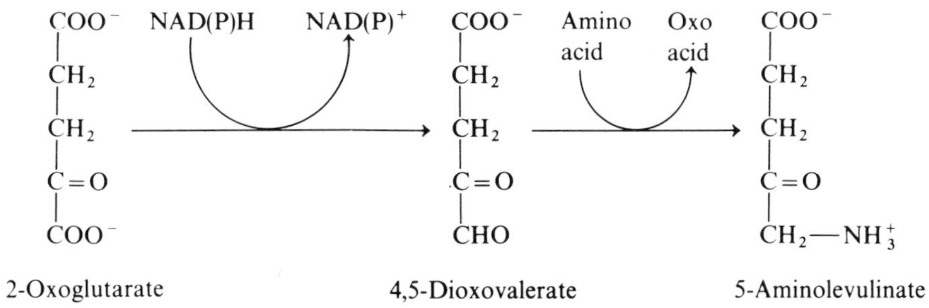

2-Oxoglutarate (15.8)   4,5-Dioxovalerate (15.9)   5-Aminolevulinate

However, the aminotransferase (Eqn. 15.9) is not very active and 4,5-dioxovalerate has not been reported in plants, casting doubt on any suggestion that this pathway contributes significantly to chlorophyll synthesis *in vivo*.

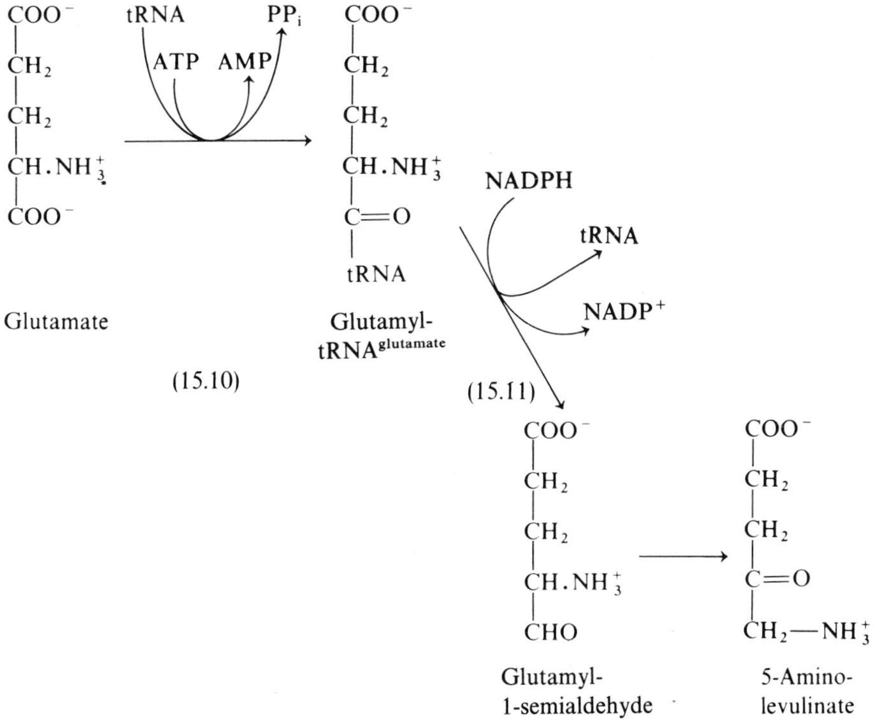

Glutamate (15.10)   Glutamyl-tRNA^{glutamate} (15.11)

Glutamyl-1-semialdehyde   5-Amino-levulinate (15.12)

Plants contain another more active aminotransferase which catalyses the internal rearrangement of the amino group of glutamyl-1-semialdehyde to 5-aminolevulinate in a reaction that does not require an external amino donor (Eqn. 15.12). This enzyme has an extremely high affinity for its substrate and is inhibited by gabaculine. It is now evident that this is a major route of 5-aminolevulinate synthesis. Glutamyl-1-semialdehyde is formed by attachment of glutamate to the chloroplast species of glutamyl-tRNA, catalysed by glutamyl-tRNA ligase (Eqn. 15.10), and reduction of glutamyl-tRNA^{glutamate} by a NADPH-specific dehydrogenase (Eqn. 15.11).

All three enzymes are located in the stroma of greening tissue.

15.3.2 Protochlorophyllide is synthesized from 5-aminolevulinate

Aminolevulinate dehydratase (porphobilinogen synthase) catalyses the formation of porphobilinogen from two molecules of 5-aminolevulinate (Fig. 15.9). In plants, this enzyme requires Mg^{2+} or K^+ and is inhibited by levulinate; the enzyme from heterotrophs requires Zn^{2+}. Production of porphobilinogen is probably not regulated *in vivo* since only small changes in aminolevulinate dehydratase occur during greening.

The formation of a tetrapyrrole ring structure involves the condensation of four molecules of porphobilinogen to yield uroporphyrinogen III (Fig. 15.9). Three porphobilinogen units (A, B and C) have the acetyl side chains to the left of the propionyl side chains while the fourth unit (D) is the other way about (hence uroporphyrinogen III). Formation of uroporphyrinogen III requires the concurrent involvement of two enzymes, porphobilinogen deaminase and uroporphyrinogen III cosynthase. The acetate side-chains on all four rings of uroporphyrinogen III are successively decarboxylated by uroporphyrinogen decarboxylase, leaving methyl groups, thus forming coproporphyrinogen III. This is then oxidized by O_2 in two oxidase reactions to form protoporphyrin IX with protoporphyrinogen IX as an intermediate. The latter reaction transforms the unstable protoporphyrinogen ring system into a stable protoporphyrin which, because of its aromatic structure, is extremely stable and absorbs light very strongly. Further, it is planar and readily chelates with many metals.

Protoporphyrin IX is the branch point for the formation of the haems and the chlorophylls, the synthesis of the latter beginning with addition of Mg catalysed by magnesium chelatase. Mg-protoporphyrin IX is then esterified by transfer of a methyl group from *S*-adenosylmethionine to the —COOH group of propionate on ring C, catalysed by a methyltransferase. A series of three reactions of uncertain

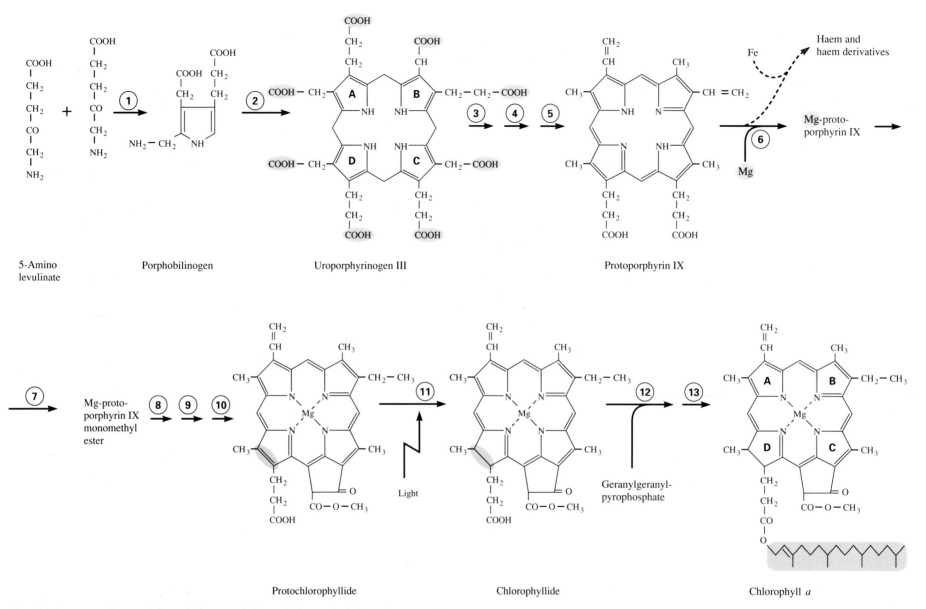

Fig. 15.9 Summary of some of the main features of chlorophyll *a* synthesis from eight molecules of 5-aminolevulinate. The nomenclature for the pyrrole rings is shown for uroporphyrinogen III and chlorophyll *a*. Enzymes are: (1) 5-aminolevulinate dehydratase; (2) porphobilinogen deaminase and uroporphyrinogen III cosynthase; (3) uroporphyrinogen decarboxylase; (4) coproporphyrinogen oxidase; (5) protoporphyrinogen oxidase; (6) magnesium chelatase; (7) *S*-adenosylmethionine:Mg-protoporphyrin methyltransferase; (8), (9) and (10) are poorly defined reactions of uncertain order, involving closure of ring C and reduction of the vinyl group on ring B; (11) protochlorophyllide reductase; (12) and (13) represent attachment of geranylgeranyl pyrophosphate and reduction of the C_{20} moiety to phytol. Chlorophyll *b* is synthesized from chlorophyll *a* by oxidation of the methyl group on ring B to a formyl group.

order follows this step. Two of the reactions involve modification of the esterified propionate residue to form a cyclopentanone ring and the third concerns reduction of the vinyl group on ring B to an ethyl group. Collectively, they result in the production of protochlorophyllide.

15.3.3 Light is required for the synthesis of chlorophyll from protochlorophyllide

The first of the three steps in the formation of chlorophyll from protochlorophyllide involves reduction of ring D (Fig. 15.9). In etiolated tissue this is initiated by light and involves reduction of protochlorophyllide by NADPH catalysed by NADPH:protochlorophyllide reductase. The photo-receptor for initiating this reaction is protochlorophyllide itself—action and absorption spectra are indistinguishable. Etioplasts contain extremely high activity of the reductase but the activity declines rapidly after illumination. Thus, although the enzyme is important in initiating protochloropohyllide reduction in etiolated tissue, it is not clear whether it is also involved in the sustained synthesis of chlorophyllide in the light.

Photo-reduction of protochlorophyllide following dark pretreatment is partially temperature sensitive, e.g. 60% inhibition at $-60\,°C$. Plants contain two pools of protochlorophyllide; one can be transformed to chlorophyllide ('active' or P_{650} with an absorption maximum at 650 nm), and the other cannot ('inactive' or P_{630}). 5-Aminolevulinate supplied to etiolated leaves promotes accumulation of the P_{630} form which can be converted to the P_{650} form at physiological temperatures in the dark, indicating that an enzyme is involved. Since the level of P_{650} in etioplasts depleted in P_{650} is enhanced by NADPH, it is envisaged that non-transformable P_{630} is free unbound protochlorophyllide and that P_{650} represents protochlorophyllide bound to an apoprotein to form a holoprotein (or, more correctly, a holochrome because P_{650} is a pigment). Binding of protochlorophyllide (PChde) to form the holoprotein (H) requires the presence of NADPH.

The reactions shown in Eqns 15.13 and 15.15 are temperature sensitive. The production of chlorophyllide following illumination of etiolated plants is extremely rapid.

The ensuing steps of chlorophyll synthesis are relatively slow; they involve esterification of the propionyl residue of ring D with geranylgeranyl pyrophosphate and its subsequent reduction of the geranylgeranyl moiety to form the phytol side-chain which typifies chlorophylls. Geranylgeranyl pyrophosphate (C_{20}), is synthesized via the terpenoid pathway from the C_5 precursors, Δ^2-isopentenyl pyrophosphate and Δ^3-isopentenyl pyrophosphate (see Section 16.4). In etiolated plants, the geranylgeranyl side-chain is reduced after its attachment to chlorophyllide to form chlorophyll a; in green plants the C_{20} chain of the pyrophosphate is reduced prior to attachment. Chlorophyll b is formed from chlorophyll a by oxidation of the methyl group of ring B.

15.3.4 Regulation of chlorophyll synthesis

Metabolic regulation of chlorophyll synthesis involves inhibition of magnesium

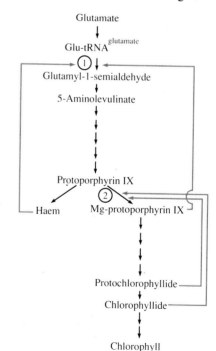

Fig. 15.10 Metabolic regulation of chlorophyll synthesis from glutamate. Grey arrows indicate inhibition of specific enzymes. The principal sites of regulation are (1) glutamyl-tRNA[glutamate] dehydrogenase and (2) magnesium chelatase.

$$\text{PChde}_{630} \longrightarrow \text{H} \longrightarrow \text{H} \longrightarrow \text{Chlorophyllide}_{672}$$

(15.13)   (15.14)   (15.15)

chelatase by protochlorophyllide and chlorophyllide (Fig. 15.10), causing diversion of protoporphyrin IX into haem. This mechanism is complemented by regulation of glutamyl-tRNA^{glutamate} dehydrogenase which is extremely sensitive to haem and, to a lesser extent, to Mg-protoporphyrin IX. This in effect controls the availability of 5-aminolevulinate.

In addition to these effects the concentration of protochlorophyllide reductase is extremely sensitive to light which causes decreased translation and enhanced degradation of the enzyme.

Other sites in the pathway are also regulated by light. Administration of 5-aminolevulinate to etiolated plants elicits rapid accumulation of proto-chlorophyllide implying that light *in vivo* controls the synthesis of 5-amino-levulinate. Further evidence comes from the rapid and parallel increase in the synthesis of chlorophyll and 5-aminolevulinate when etiolated seedlings are illuminated. However, synthesis of 5-aminolevulinate declines rapidly when the seedlings are returned to the dark, without similar changes occurring in the extractable activity of all the enzymes involved (Eqns 15.10–15.12). The site(s) of control by light are not understood but they could involve both translational control (synthesis of glutamyl-tRNA?) and post-translational control (e.g. dark inhibition of internal aminotransferase?).

Further reading

Purine and pyrimidine synthesis: Marcus (1981) Chapter 4; Parthier & Boulter (1982) Chapter 9.

Chlorophyll synthesis: Castelfranco & Beale (1983); Hatch & Boardman (1981) Chapter 9; Kannangara *et al.* (1988).

Nitrogen transport compounds: Miflin (1980) Chapter 16; Schubert (1986).

16
Synthesis of lipids

Lipids are important constituents of plant cells; they act as storage reserves in some tissues (see Section 2.4.2), as important structural constituents of membranes (see Chapter 6) and are involved in energy transduction (see Chapters 7 & 9). Furthermore, plants (but not animals) synthesize linoleic acid which is essential to the human diet as a precursor for the synthesis of the prostaglandin group of mammalian hormones.

This chapter describes the pathways of biosynthesis of structural, reserve, and metabolically active plant lipids and their assembly, with protein, to form membranes.

16.1 Fatty acids are synthesized from acetyl-CoA

Fatty acids are common components of all acyl lipids, so the first step to consider in acyl lipid biosynthesis is the route of synthesis of the fatty acids. As palmitate is the most common saturated fatty acid in plants and because other fatty acids derive from it, its synthesis from acetyl-CoA is described first.

16.1.1 Palmitate is synthesized in plastids from acetyl-CoA

Biosynthesis of fatty acids occurs in the plastids; in leaves, this means in the chloroplasts whilst in other tissues, such as developing seeds, proplastids are involved. Preparations of intact chloroplasts from spinach incorporate [^{14}C]acetate into fatty acids at rates greater than $1.5\,\mu mol\,mg^{-1}$ chlorophyll h^{-1}, at least as fast as synthesis in intact leaves as determined by rates of $^{14}CO_2$ incorporation into leaf lipids. Furthermore, all enzymes necessary for the synthesis of palmitic acid from acetyl-CoA are found only in plastids. In experiments to investigate the time course of incorporation of carbon from acetate or bicarbonate into fatty acids in isolated chloroplasts, acetate is a far better source of carbon for fatty acid synthesis than is CO_2. Chloroplasts do not incorporate CO_2 into acetyl-CoA at significant rates.

It was originally believed that the acetyl-CoA for fatty acid synthesis in chloro-plasts originated solely in mitochondria as shown in the scheme in Fig. 16.1. Since illuminated leaves incorporate CO_2 at rates similar to those for the incorporation of acetate by chloroplasts, it has been proposed that dihydroxyacetone-P, synthesized by the C_3-CR cycle in the chloroplast (see Section 10.3), is exchanged for P_i from the cytosol (see Section 10.8) where it is oxidized to pyruvate by the process of glycolysis (see Chapter 7). Pyruvate is converted within mitochondria to acetyl-CoA by the action of the pyruvate dehydrogenase complex (see Section 7.6.1). Acetyl-CoA does not diffuse readily across membranes, but mitochondria contain an acetyl-CoA hydrolase which releases acetate. Acetate then diffuses from the mitochondria to the chloroplast where the action of an acetyl-CoA synthetase supports the synthesis of acetyl-CoA, a process known as 'acetate activation' since it involves the expenditure of ATP to form a high energy compound (Eqn. 16.1).

$$ATP + acetate + coenzyme\ A \xrightleftharpoons{Mg^{2+}} acetyl\text{-}CoA + AMP + PP_i \qquad (16.1)$$

More recently, however, pyruvate dehydrogenase activity (see Eqn. 7.25) has been found in pea and spinach chloroplasts and in non-photosynthetic plastids from castor bean endosperm. There is evidence that plastids can synthesize pyruvate from triose-P via glycolysis (see Section 7.4) so, theoretically, triose-P synthesized by the C_3-CR cycle could be used to support the synthesis of acetyl-CoA within plastids. However, given the low rates of CO_2 incorporation into fatty acids and the low activities, within plastids, of certain enzymes of the glycolytic sequence (Section 7.4), intra-plastid synthesis of acetyl-CoA from CO_2 is unlikely to be quantitatively important. It is possible, however, that pyruvate synthesized in the cytosol is transported into plastids and subsequently oxidized to acetyl-CoA.

Whatever the source of acetyl-CoA, the pathway responsible for converting it to palmitate in chloroplasts is well-characterized. The first step in this process is the carboxylation of acetyl-CoA to form malonyl-CoA, carried out in two stages and catalysed by the acetyl-CoA carboxylase complex. This complex has three different functions; a biotin carboxylase activity, a biotin carboxyl carrier protein (BCCP), and carboxyl transferase activity. CO_2 is first bound to BCCP in an ATP-dependent reaction, catalysed by biotin carboxylase (Eqn. 16.2).

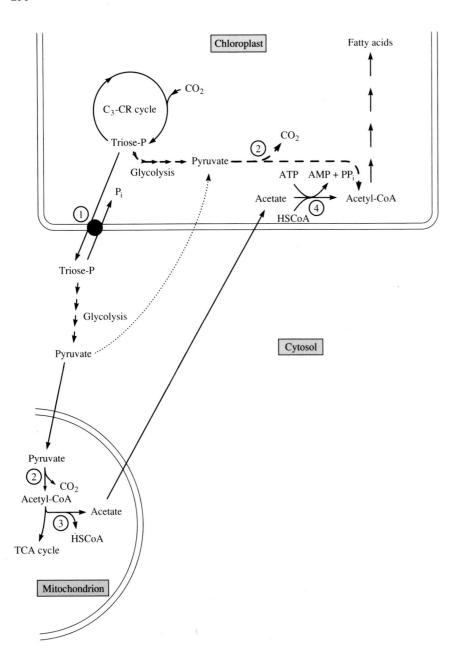

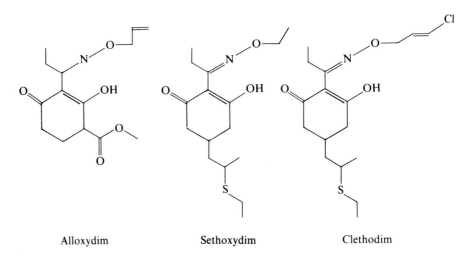

$$ATP + CO_2 + BCCP \rightleftharpoons CO_2{-}BCCP + ADP + P_i \qquad (16.2)$$

Subsequently, the CO_2 moiety of carboxybiotin ($CO_2{-}BCCP$) is transferred to acetyl-CoA by carboxyl transferase (Eqn. 16.3) to form malonyl-CoA. (For the structure and functions of biotin see Section 5.2.2.)

$$CO_2{-}BCCP \;+\; CH_3.CO.S.CoA \;\rightleftharpoons\; {}^-OOC.CH_2.CO.S.CoA \;+\; BCCP \qquad (16.3)$$

Carboxybiotin   Acetyl-CoA   Malonyl-CoA

The acetyl-CoA carboxylase complex is the site of action of the cyclohexane-dione herbicides, of which three are shown below. They inhibit the acetyl-CoA carboxylase of grasses at much lower concentrations than are needed for broad-leaf species such as spinach, and are therefore very useful for controlling grass weeds in broad-leaved crops.

Alloxydim   Sethoxydim   Clethodim

Fig. 16.1 Origins of the acetate for fatty acid synthesis in chloroplasts. Triose-P synthesized in the chloroplast by the C_3-CR cycle is exported by the phosphate translocator and is oxidized to pyruvate in the cytosol. It is theoretically possible that additional acetyl-CoA could also arise from triose-P via glycolytic reactions and pyruvate dehydrogenase activity within the chloroplast; this is shown by the heavy dashed line. It is also possible that pyruvate formed in the cytosol is transported into the chloroplast (dotted line) and then oxidized to acetyl-CoA. Enzymes and proteins are: (1) triose-P:P_i translocator; (2) pyruvate dehydrogenase complex; (3) acetyl-CoA hydrolase; (4) acetyl-CoA synthetase.

The inhibition constants (K_i, see Chapter 5) for the acetyl-CoA carboxylase from wheat are 0.96 μM and 0.14 μM for sethoxydim and clethodim respectively, whilst for spinach the equivalent values are 2160 μM and 1240 μM.

The second major step in the formation of palmitate is the elongation of acetyl-CoA by the successive addition of C_2 units from malonyl-CoA, catalysed by a complex of enzymes known as fatty acid synthetase (FAS).

The FAS in plants is found in chloroplasts and non-photosynthetic plastids and consists of six enzymes which, unlike the complexes found in mammalian liver or in yeast, are readily separated from one another (as found in prokaryotic organisms).

All the reactions catalysed by FAS involve an acyl carrier protein (ACP) with a molecular mass of about 9.5–11.5 kDa. In spinach protoplasts all the ACP is in the chloroplasts, a fact which supports the idea that synthesis of palmitate, and probably other fatty acids, occurs solely in these organelles. The six enzymes comprising the FAS complex are: acetyl transferase, malonyl transferase, β-ketoacyl ACP synthetase I, β-ketoacyl ACP reductase, β-ketoacyl ACP dehydratase, and enoyl ACP reductase. The sequence (Fig. 16.2) begins by attachment of acetyl-CoA and malonyl-CoA, derived from acetyl-CoA as in Eqns 16.2 and 16.3, to two separate ACP molecules. These reactions, catalysed by acetyl transferase and malonyl transferase respectively, involve the formation of a thioester bond between the acyl moiety of the substrate and the single thiol group of ACP. The ensuing reaction, catalysed by β-ketoacyl ACP synthetase, involves a head-to-tail addition of malonyl-ACP to acetyl-ACP with the elimination of CO_2 (Eqn. 16.4).

$$CH_3-\overset{O}{\overset{\|}{C}}-S-ACP + (^-OOC)-CH_2-\overset{O}{\overset{\|}{C}}-S-ACP$$

Acetyl-ACP   Malonyl-ACP   (16.4)

$$CH_3-\overset{O}{\overset{\|}{C}}-CH_2-\overset{O}{\overset{\|}{C}}-S-ACP + ACP-SH + (CO_2)$$

Acetoacetyl-ACP

Thus, the acetyl moiety gains two carbon atoms from malonyl-ACP to form acetoacetyl-ACP. The remaining reactions shown in Fig. 16.2 involve the reduction,

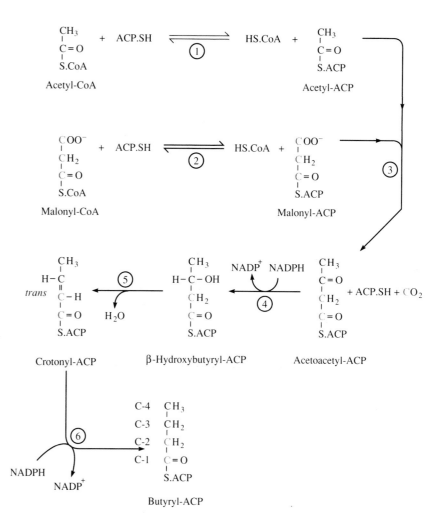

Fig. 16.2 Reactions catalysed by the FAS system of plants. This system catalyses the successive addition of C_2 moieties from malonyl-CoA to a growing acyl chain, the first 'round' of which is shown here. Each C_2 addition consumes two NADPH and releases one CO_2. The two terminal C atoms (C_3 and C_4 of butyrate in this example) are always derived from acetate, regardless of chain length; all other carbon atoms (shown in grey) are derived from malonate. Enzymes are: (1) acetyl transferase; (2) malonyl transferase; (3) β-ketoacyl ACP synthetase; (4) β-ketoacyl ACP reductase; (5) β-ketoacyl ACP dehydratase; (6) enoyl ACP reductase. Abbreviations: FAS = fatty acid synthetase; ACP = acyl carrier protein.

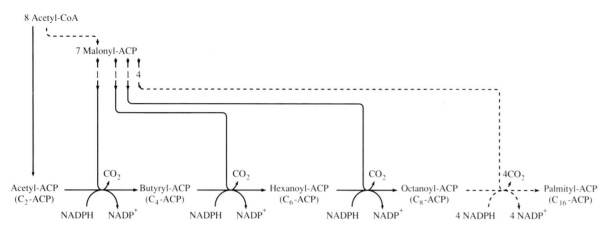

Fig. 16.3 Synthesis of palmityl-ACP by seven successive rounds of fatty acid synthetase activity; each adds a C_2 moiety from malonyl-ACP to a pre-existing acyl-ACP containing an even number of carbon atoms. Acetyl-CoA acts as the initial C_2 'primer'.

dehydration and further reduction of the acetoacetyl-ACP moiety. This results in the production of butyryl-ACP.

The butyryl-ACP then reacts with a second molecule of malonyl-ACP in another reaction, also catalysed by β-ketoacyl ACP synthetase (Fig. 16.3). This ultimately results in the formation of hexanoyl-ACP via reduction, dehydration and reduction (as above), catalysed by the enzymes β-ketoacyl ACP reductase, β-hydroxyacyl ACP dehydratase and enoyl ACP reductase. This process repeats to successively add a further five C_2 moieties from malonyl-CoA to the growing acyl-ACP. This 'elongation' usually proceeds until the acyl moiety contains 16 carbon atoms, i.e. palmityl-ACP has been formed. The enzymes involved are non-specific; the same set supports the synthesis of butyryl-, hexanoyl-, octanoyl-etc. derivatives of ACP. Most palmityl-ACP is metabolized further. However, termination of the growing acyl-ACP chain can be brought about by highly specific acyl-ACP hydrolases which cleave particular acyl-ACP molecules to release free fatty acids. In the case of free palmitate the overall reaction involves the input of one acetyl-CoA (referred to as a 'primer') and seven malonyl-CoA.

Note that in these reactions the original acetyl-CoA primer contributes the two terminal carbon atoms at the ω (methyl) end of the palmitate chain (boxed). All the remaining carbon atoms are derived from malonyl-CoA. With the exception of C_{15} and C_{16}, the odd-numbered carbon atoms of palmitate are derived from the carbonyl carbon atom of malonyl-CoA and the even-numbered atoms from the methylene group. The CO_2 displaced during the reaction (bold) is derived from the free carboxyl group of malonyl-CoA.

Alternatively, acyl-ACP chain termination can involve a specific acyl transferase to catalyse the transfer of the palmityl moiety of palmityl-ACP to a suitable acceptor such as 1-monoacylglycerol-P (lysophosphatidate).

16.1.2 Palmitate is attached to ACP and metabolized in plastids to stearyl- and oleyl-ACP

Most of the palmityl-ACP synthesized by the reactions above is lengthened by addition of another C_2 moiety from malonyl-ACP to the C_{16} acyl chain to form the C_{18} product stearyl-ACP. This elongation is also catalysed by the FAS system, except that the β-ketoacyl ACP synthetase involved is highly specific to palmityl-ACP (stearyl-ACP is inert as a substrate). Thus, in chloroplasts, only C_{16} and C_{18} fatty acids are formed, as elongation cannot proceed further than stearyl-ACP. This enzyme, β-ketoacyl ACP synthetase II, has, in spinach at least, a slightly greater molecular mass (57.5 kDa) than its equivalent, synthetase I (56 kDa) involved in the addition of C_2 moieties to C_2–C_{14} acyl-ACP molecules. Synthetase II is more sensitive to inhibition by arsenite but less sensitive to cerulenin than is synthetase I.

$$\underset{\text{Acetyl-CoA}}{\boxed{CH_3}-\overset{\overset{\displaystyle O}{\|}}{C}\text{-}S.CoA} + 7^-OO\mathbf{C}-CH_2-\overset{\overset{\displaystyle O}{\|}}{C}.S.CoA + 14\,NADPH + H_2O$$

$$\underset{\text{Malonyl-CoA}}{}$$

$$\downarrow\downarrow$$

$$\underset{\text{Palmitate}}{\boxed{CH_3-CH_2}-(CH_2)_{13}-COO^- + 7\,\mathbf{CO_2} + 14\,NADP^+ + 8\,HS.CoA} \qquad (16.5)$$

Plants contain relatively few saturated fatty acids such as stearate or palmitate. Most of the C_{16} and C_{18} fatty acids have one or more double bonds (see Table 2.4). Stearyl-ACP can be desaturated (oxidized) by stearyl-ACP desaturase to form oleyl-ACP. This introduces a double bond between the C-9 and C-10 atoms (oleate is an 18:1 ($9c$) fatty acid). There are three proteins required for desaturation: NADPH:Fd oxidoreductase (N:F-ORase), Fd and the desaturase (Eqn. 16.6).

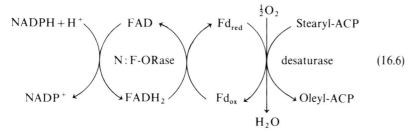

$$(16.6)$$

The desaturase system is found in plastids of both photosynthetic and non-photosynthetic cells. In both non-photosynthetic plastids and in chloroplasts the enzyme is in the soluble phase and is sensitive to cyanide at high levels (1 mM).

Oleyl-ACP is rapidly converted, by hydrolysis of its thioester bond, to oleate by oleyl-ACP thioesterase. Oleate is the major fatty acid exported from plastids. It is then converted to the acyl-CoA derivative which is used in the synthesis of acyl lipids. In seed tissue, free fatty acids are exported from proplastids to the cytosol where they are further modified.

16.2 Waxes are derived from palmitate and stearate

Waxes are a heterogeneous collection of compounds (see Section 2.4.3) derived from the saturated fatty acids synthesized in the plastids of epidermal cells. The structures of some of these compounds, and their derivation from palmitate, are shown in Fig. 16.4. Synthesis of waxes involves the activation of free palmitate (or

Fig. 16.4 Synthesis of various components of waxes. Palmitate synthesized in the plastids of epidermal cells is released into the cytosol and, after being activated to its coenzyme A derivative, is elongated and modified to generate the components of waxes. In higher plants the major components are odd-numbered alkanes and wax esters formed by esterification of long-chain fatty acids with long-chain alcohols. Structural formulae are given in a shorthand form with hydrogen atoms omitted; k and h followed by a number represent the position of carbonyl and hydroxyl groups respectively. Enzymes are: (1) NAD-specific fatty acyl-CoA reductase; (2) NADP-specific fatty aldehyde reductase; (3) fatty acyl-CoA:fatty alcohol acyl transferase; (4) decarbonylase; (5) hydroxylase; (6) oxidoreductase.

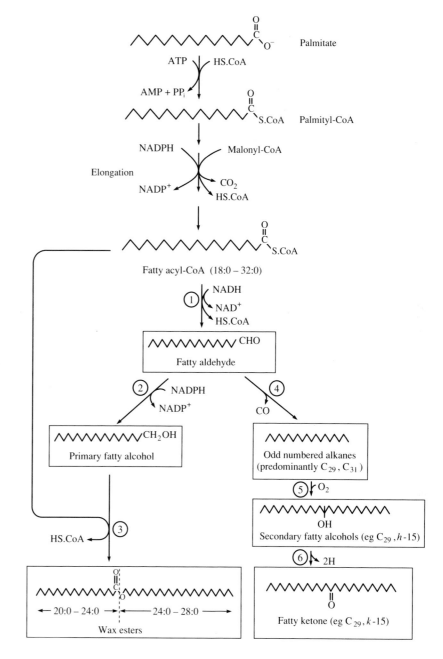

stearate), derived from plastids, by ATP to its coenzyme A derivative (see Eqn. 16.1). This derivative is then elongated by addition of C_2 units from malonyl-CoA, in a process requiring NADPH. The elongation process, in contrast to that of plastids, is poorly characterized. In this way a range of fatty acyl-CoA molecules are synthesized with even-numbered acyl moieties of C_{18}–C_{32}. These are then modified to form the various components of wax (as shown in the boxes in Fig. 16.4). Thus, the fatty aldehydes are synthesized by reduction of fatty acyl-CoA molecules in reactions catalysed by a NAD-specific fatty acyl-CoA reductase. Primary alcohols of the fatty acids are produced by reduction of the corresponding aldehyde in a reaction catalysed by a NADP-specific fatty aldehyde reductase. The alcohols thus formed can then react with fatty acyl-CoA molecules to produce wax esters by a condensation reaction, catalysed by fatty acyl-CoA:fatty alcohol acyl transferase. Generally, the fatty alcohol moiety is C_{20}–C_{24} in length and the fatty acid moiety is somewhat longer, C_{24}–C_{28}.

The unstable long-chain fatty aldehydes, formed by the action of acyl-CoA reductase on long-chain fatty acyl-CoA molecules, undergo a decarbonylation reaction (Eqn. 16.7), which involves a decarbonylase associated with the endoplasmic reticulum.

$$RCH_2CHO \rightarrow RCH_2H + CO \qquad (16.7)$$

This reaction accounts for the formation of alkanes with an odd number of carbon atoms. In higher plants these are predominantly C_{29} or C_{31} in length. Oxygen can be introduced into the alkanes by hydroxylases to yield secondary alcohols which, in waxes, tend to have the hydroxy (—OH) group near the centre of the hydrocarbon chain. For example, the C_{29} alkane, nonacosane, is converted by broccoli leaves to alcohols, with about 40% of the products having the hydroxyl group at C-14 and the remainder at C-15. Further oxidation of the secondary alcohols yields ketones in reactions catalysed by oxidoreductases which are highly specific to the position of the hydroxyl group. Thus, although broccoli leaves contain both nonacosan-14-ol and nonacosan-15-ol (with —OH at C-14 and C-15 respectively), only the latter is a substrate for the oxidoreductases and only the C-15 ketone is formed.

Cutin is a polymer of hydroxy-fatty acids cross-linked through ester bonds between the hydroxy and carboxy residues (see Section 2.4.3). The starting point

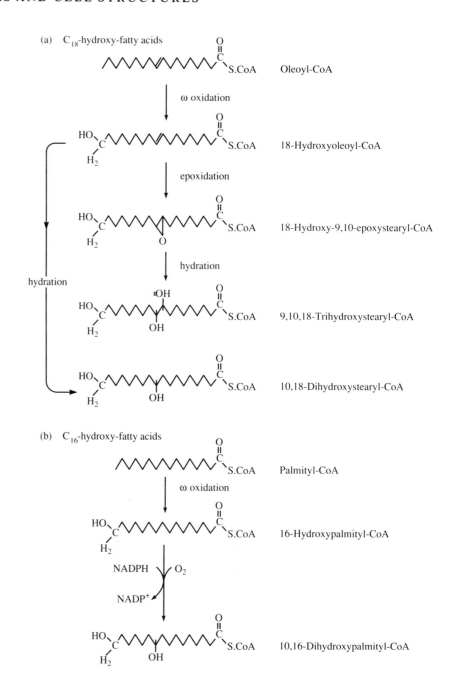

Fig. 16.5 Biosynthesis of cutin precursors. The hydroxy fatty acids involved vary between species. (a) In apple and grape hydroxy fatty acids are derived from the unsaturated C_{18} fatty acyl-CoA, oleyl-CoA. (b) In broad beans C_{16} hydroxy fatty acids are derived from palmityl-CoA. Polymerization of the hydroxy fatty acids is shown in Fig. 16.6.

for cutin synthesis is species specific. In apples and grapes, for example, cutin is derived from the C_{18} compound, oleyl-CoA, as shown in Fig. 16.5a. This unsaturated fatty acid (18:1, 9c) is subjected to oxidation at the ω or terminal methyl end of the acyl chain (ω oxidation) to yield 18-hydroxyoleyl-CoA. This is then either subjected to a hydration to introduce another hydroxyl group at C-10 to form 10,18-dihydroxystearyl-CoA, or is converted by epoxidation to 18-hydroxy-9,10-epoxystearyl-CoA. This, in turn, is converted to the trihydroxy fatty acid, 9,10,18-trihydroxystearate.

In other plants, such as *Vicia faba* (broad bean), the starting point is palmityl-CoA. This is also subjected to ω oxidation, giving rise to 16-hydroxypalmityl-CoA (Fig. 16.5b) which can undergo further oxidation to introduce a second hydroxyl group at C-10, forming 10,16-dihydroxypalmityl-CoA.

These hydroxy fatty acids are transferred to a developing cutin polymer—a cutin primer. For example, the addition of the acyl moiety of 10,16-dihydroxypalmityl-CoA to the primer (the acyl-acceptor) is shown in Fig. 16.6.

16.3 Phosphatidate is an important intermediate in the synthesis of acyl lipids other than waxes

The biosyntheses of triglycerides and the polar acyl lipids all proceed with phosphatidate as a common intermediate (Fig. 16.7). Phosphatidate is derived from

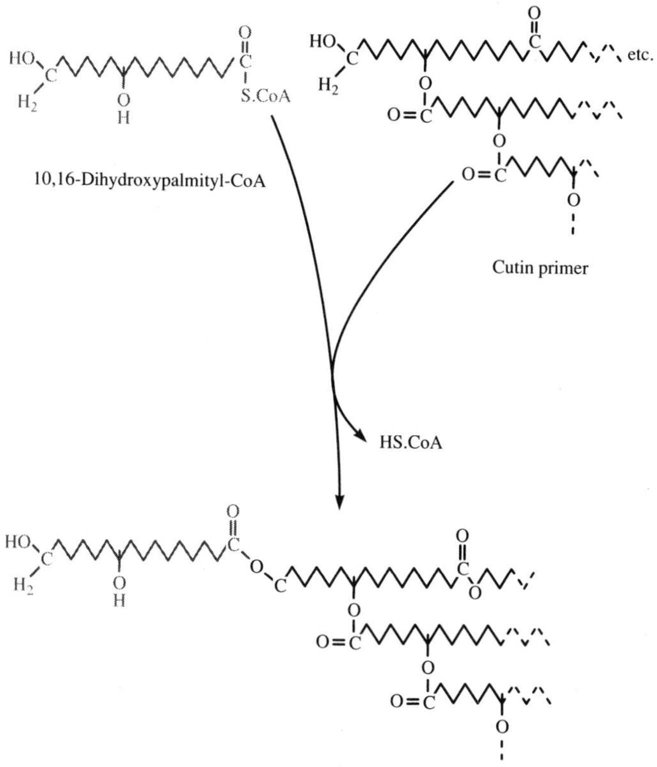

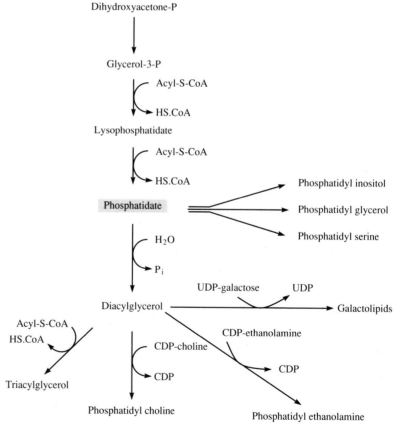

Fig. 16.6 Polymerization of hydroxy fatty acids to form cutin. Hydroxy fatty acyl-CoAs such as 10,16-dihydroxypalmityl-CoA are joined to the growing cutin polymer by ester linkages.

Fig. 16.7 The central position of phosphatidate in the synthesis of glycerolipids. The glycerol residue of phosphatidate is synthesized from dihydroxyacetone-P. Phosphatidate is an important branch point for the synthesis of phospholipids, triacylglycerides and galactolipids.

the coenzyme A derivatives of two fatty acids and from dihydroxyacetone-P, the latter being a triose-phosphate generated by the C_3-CR cycle in chloroplasts (see Section 10.3), as well as an intermediate in glycolysis (see Section 7.3) in the cytosol. Dihydroxyacetone-P is reduced to glycerol-3-P by NADH in a reaction involving glycerol-3-P dehydrogenase (sometimes known as dihydroxyacetone-P reductase) (Eqn. 16.8).

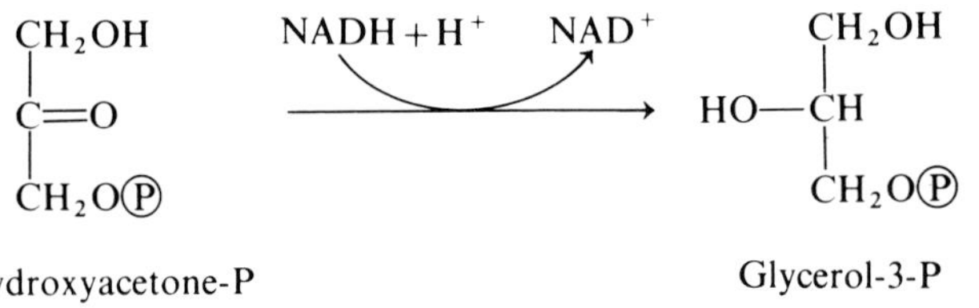

$$(16.8)$$

Dihydroxyacetone-P   Glycerol-3-P

Although plants contain both cytosolic and chloroplastic forms of glycerol-3-P dehydrogenase, in a range of plants (spinach, soybean, pea and corn) *both* forms use NADH. This is unusual as most other chloroplastic dehydrogenases (reductases) use NADPH.

In two subsequent reactions, two acyl residues are added to glycerol-3-P to give phosphatidate; the first reaction is catalysed by glycerol-P-1-acyl transferase (Eqn. 16.9) and involves the addition of an acyl chain (R^1—CO—, shown below as the boxed area) to the C-1 of glycerol-3-P.

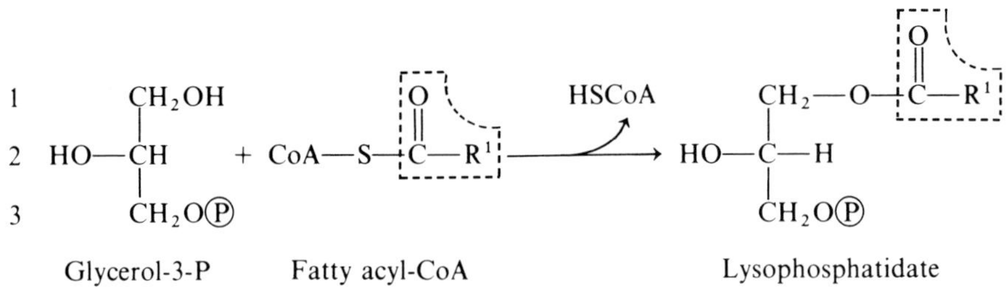

Glycerol-3-P   Fatty acyl-CoA   Lysophosphatidate

$$(16.9)$$

The products of this reaction are coenzyme A (HSCoA) and the monoacyl glycerol, lysophosphatidate.

The second acyl residue (R^2—CO—) is added, also via acyl transferase, to lysophosphatidate, which is thereby converted to phosphatidate (Eqn. 16.10).

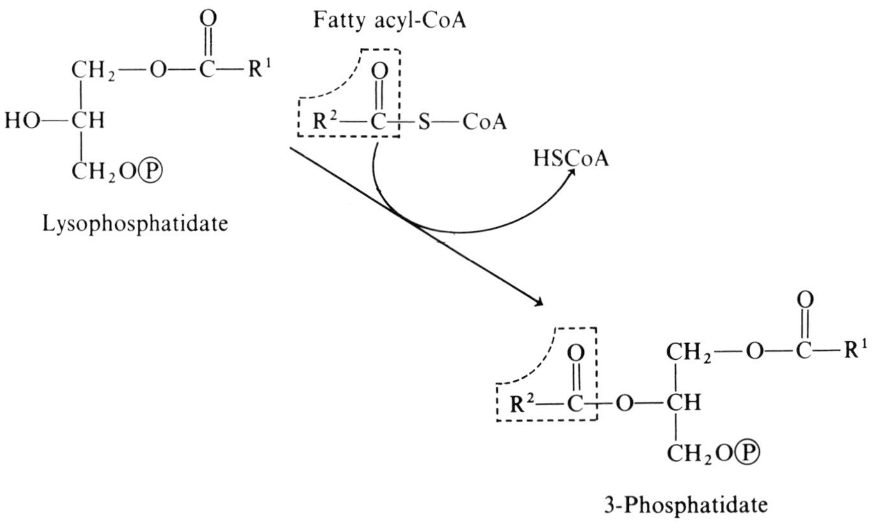

Lysophosphatidate

3-Phosphatidate

$$(16.10)$$

The C-1 of phosphatidate tends to bear saturated fatty-acyl residues whilst unsaturated residues are often attached to C-2. The phosphatidate is then used in the synthesis of the range of phospholipids and galactolipids found in cell membranes, as well as the triglycerides found as storage reserves in certain seeds (Fig. 16.7). Most of the synthesis of phosphatidate from glycerol-3-P and fatty acids occurs within the endoplasmic reticulum, but plastids do have at least some capacity for its synthesis.

16.3.1 Triacylglycerols are synthesized from fatty acids and phosphatidate in storage tissues of some plants

Triacylglycerols are a major storage product in some plant tissues, such as castor bean endosperm, the cotyledons of plants like sunflower and peanuts, and the mesocarp tissue of avocado, oil palm and coconut fruits. These triacylglycerols are stored in oleosomes which take up most of the space in the cells of such tissues when fully mature.

Triacylglycerol synthesis involves the formation of fatty acids, the synthesis of glycerol-3-P and the subsequent linking of the fatty acids to the glycerol moiety to form phosphatidate, as described in the previous section. Various studies have shown that the enzymes required for these syntheses are not found in plastids in storage tissues but are bound to the endoplasmic reticulum which gives rise to the membrane of the oleosome.

The action of phosphatidate phosphatase on the phosphate ester bond of phosphatidate results in the formation of 1,2-diacylglycerol (Eqn. 16.11) which, in turn, is converted into triacylglyceride by the action of diacylglycerol acyltransferase (Eqn. 16.12).

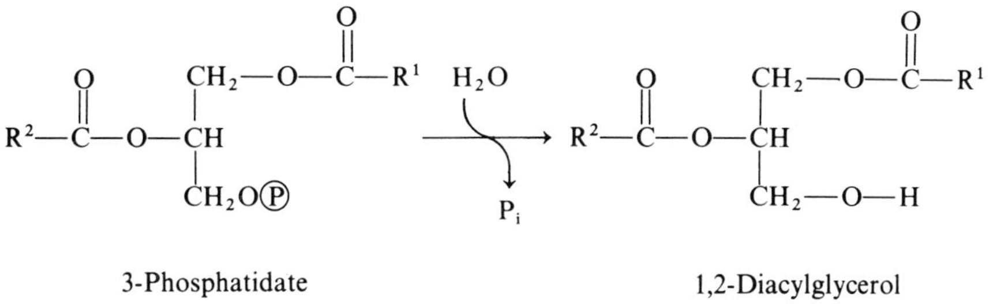

3-Phosphatidate

1,2-Diacylglycerol

(16.11)

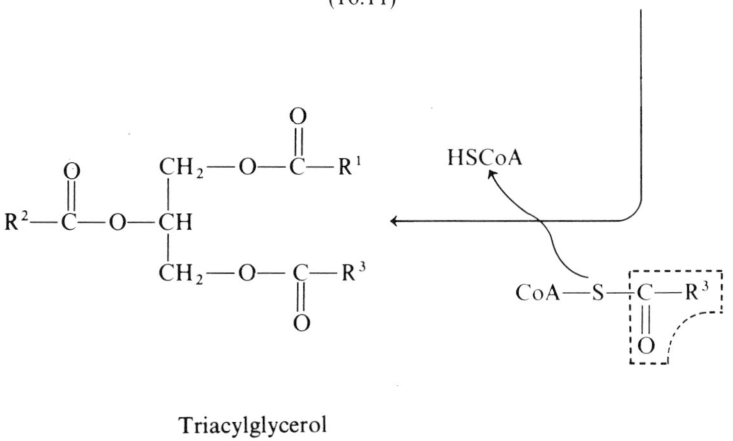

Triacylglycerol

(16.12)

The precise fatty acid composition of triacylglycerols from oil-storing seeds is species-dependent. Indeed, there are also considerable differences in the relative proportions of the various fatty acids attached to the three carbon atoms of the glycerol moiety. The oil from edible oil seed crops, for example, contains triacylglycerol with mainly the unsaturated C_{18} fatty acids such as oleate (18:1), linoleate (18:2) and linolenate (18:3) together with palmitate (16:0) and stearate (18:0) (see Table 2.5). In oils from the seeds of plants such as oilseed rape (*Brassica napus*), triacylglycerols contain mainly the unsaturated fatty acids eicosenoate and erucate (20:1 and 22:1).

There is considerable evidence that the fatty acids of polyunsaturated triacylglycerol (i.e. where all three of the fatty acid moieties in the lipid contain more than one double bond) in seeds are derived from phosphatidyl choline (PC; see Section 16.3.2). Thus, if radioactively labelled free fatty acids such as oleate are supplied to developing cotyledons of flax, label is incorporated rapidly into PC and diacylglycerol but only at a slow rate into triacylglycerols. However, if a pulse of labelled oleate is followed by a chase of unlabelled free fatty acids, radioactivity is lost from oleyl moieties of PC and appears in the polyunsaturated C_{18} fatty acid moieties (linoleyl and linolenyl) of triacylglycerols. Furthermore, desaturase enzymes isolated from microsomal fractions of developing linseed and safflower cotyledons convert supplied oleyl-PC (i.e. PC bearing only oleate) and linoleyl-PC to linoleyl-PC and linolenyl-PC respectively; both products have one more double bond than their respective substrates. Thus, PC serves both as a donor of fatty acids for triacylglycerol synthesis and as the substrate for the desaturases.

16.3.2 Phospholipids are also synthesized from phosphatidate

In contrast to the triacylglycerols, which are mainly storage products, the phospholipids, together with the glycolipids (Section 16.3.3), are important membrane constituents. The phospholipids are all derived from phosphatidate and are made by the addition of hydrophilic side-chains (e.g. choline, inositol, serine, ethanolamine or glycerol—see Fig. 2.11) to the phosphate group of this compound.

A scheme for the biosynthesis of the phospholipids is given in Fig. 16.8. Phosphatidyl glycerol (PG), phosphatidyl inositol (PI) and phosphatidyl serine (PS) are formed from cytidyldiphosphate diacylglycerol (CDP-diacylglycerol), which is a product of the reaction between cytidine triphosphate (CTP) and phosphatidate (PA), catalysed by phosphatidate cytidyl transferase as shown in Eqn. 16.13.

Phosphatidate

CTP

(16.13)

CDP-diacylglycerol   pyrophosphate

The CDP-diacylglycerol acts as a donor of the phosphatidyl moiety to the various alcohols, glycerol-3-P, *myo*-inositol or serine, as shown in the generalized scheme in Eqn. 16.14 in which the alcohols are represented as R—OH.

CDP-diacylglycerol

R—OH

(16.14)

+ CMP

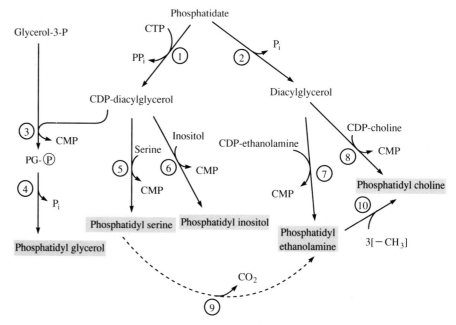

Fig. 16.8 Biosynthesis of phospholipids. Phosphatidate is converted to either CDP-diacylglycerol or to diacylglycerol and these two compounds then accept various hydrophilic side-chains to yield the phospholipids. Enzymes are: (1) phosphatidate cytidyl transferase; (2) phosphatidate phosphatase; (3) glycerol-P phosphatidyl transferase; (4) phosphatidyl glycerol-P phosphatase; (5) phosphatidyl serine synthase; (6) phosphatidyl inositol synthase; (7) CDP-ethanolamine:diacylglycerol ethanolamine phosphotransferase; (8) CDP-choline:diacylglycerol choline phosphotransferase; (9) phosphatidyl serine decarboxylase; (10) phosphatidyl ethanolamine *N*-methyl transferase.

Since the phosphatidyl residue forming the substrate for the synthesis of CDP-diacylglycerol (Eqn. 16.13) is also the group attached to the alcohol (Eqn. 16.14), the biosynthetic origin would be clearer if CDP-diacylglycerol was referred to as CMP-phosphatidate. However, the former term is more prevalent in the literature and is used in this text. In the reaction between CDP-diacylglycerol and glycerol-3-P, the immediate product is phosphatidyl glycerol-P. This is subsequently dephosphorylated to give PG (Fig. 16.8, reactions (3) and (4)).

These lipids are synthesized by expending CTP to activate phosphatidate and its CDP derivatives, whereas those containing choline and ethanolamine are formed by expending CTP to activate the amino alcohol, which then reacts with phosphatidate. Thus, phosphatidyl choline (PC) and phosphatidyl ethanolamine (PE) are formed by the addition of the respective CDP-amino alcohol (or CMP-phosphoamino alcohols) to diacylglycerol (Fig. 16.8, reactions (7) and (8)). This involves transfer of the phosphocholine and phosphoethanolamine residues of CDP-choline and CDP-ethanolamine to the 3-hydroxy group of diacylglycerol. The CDP-amino alcohols are formed by reactions between CTP and phosphocholine or phosphoethanolamine; the latter compounds, in turn, are synthesized by the phosphorylation of choline and ethanolamine by ATP in the presence of choline kinase and ethanolamine kinase respectively.

PE and PC can be formed from other phospholipids. For example, PE can be made by decarboxylation of PS, a process catalysed by phosphatidyl serine decarboxylase (Fig. 16.8, reaction (9)). PC can then be formed by methylation of PE (Fig. 16.8, reaction (10)). This has been demonstrated in fungi and in the alga *Ochromonas malhamensis* but probably also occurs in higher plant tissues. Three methyl groups are added to the amino nitrogen of PE with *S*-adenosylmethionine as donor. The first methylation is catalysed by PE methyltransferase but the subsequent ones probably involve separate enzymes.

With the exception of PC (and possibly PI) all the phospholipids can be synthesized within chloroplasts of leaves and the non-photosynthetic plastids from roots, as well as in the endoplasmic reticulum. As will become apparent below, this self-sufficiency of plastids extends to the synthesis of glycolipids.

Very little is known about the synthesis of the other group of phospholipids, the sphingolipids.

16.3.3 Glycolipids are synthesized by two pathways

Glycolipids are also polar acyl lipids. Structurally, they are *O*-glycosides, i.e. a monosaccharide or disaccharide linked to lipid via an oxygen atom (see Section 2.4.5). The most important glycolipids in plants are the galactosyl diglycerides, monogalactosyl diglyceride (MGDG) and digalactosyl diglyceride (DGDG) which consist of galactose attached to diacylglycerol. These two galactolipids are not only the main glycolipids but are also the predominant lipid of chloroplast membranes. Only small amounts of MGDG and DGDG are present in mitochondrial membranes (see Section 6.1.2). The chloroplasts of higher plants also contain a sulpholipid called sulphoquinovosyl diacylglycerol (SQDG). This is comprised of 6-sulpho-6-deoxy-D-glucose (i.e. sulphoquinovose) attached to diacylglycerol.

Analyses of the fatty acyl moieties of the leaf galactolipids in a large number of plant species show that two types of organism can be distinguished according to the dominant acyl moiety at *sn*-2; those with galactolipids containing 16:3 acyl (hexadecatrienoate) moieties and those with 18:3 (linolenate). Species with 16:3 (which is usually found predominantly in the *sn*-2 position of MGDG) are called 16:3-galactolipid plants (*sn*-2 refers to the C-2 of the glycerol moiety of MGDG according to the stereospecific *n*umbering system from the IUPAC–IUB Commission on Biochemical Nomenclature—see Section 2.4). These species include many bryophytes, pteridophytes and gymnosperms as well as angiosperms such as members of the Chenopodiaceae (e.g. spinach), the Solanaceae and Brassicaceae. Other more advanced angiosperms, such as the Fabaceae, Asteraceae and Poaceae, show a virtual absence of 16:3 at *sn*-2 in MGDG and DGDG, but have instead high levels of 18:3 (linolenate) residues. These are 18:3-galactolipid plants. Since galactolipids in prokaryotic photosynthetic organisms, such as the cyanobacteria, usually show C_{16} acyl moieties at the *sn*-2 position of the glycerol residue, compounds with this feature are often referred to as 'prokaryotic galactolipids'. Galactolipids with C_{18} acyl moieties at *sn*-2 of the glycerol residue are consequently 'eukaryotic galactolipids'. Some other glycerolipids also contain 16:3 or 18:3 fatty acids at the *sn*-2 position but these are present in relatively small amounts compared to the galactolipids. Some examples of the distribution of fatty acids at the *sn*-2 position of the galactolipids, PG and PC of a cyanobacterium (*Nostoc*), a 16:3-galactolipid plant (spinach) and an 18:3-galactolipid plant (rye) are shown in Fig. 16.9.

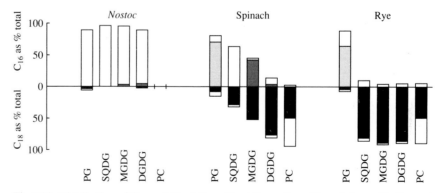

Fig. 16.9 Distribution of C_{16} and C_{18} fatty acids at the *sn*-2 position of glycerol in the major glycerolipids of a cyanobacterium (*Nostoc*), a 16:3-galactolipid plant (spinach) and an 18:3-galactolipid plant (rye). Hexadecenoate 16:1(3*t*) is given by the light grey area, hexadecatrienoate (16:3) by the dark grey area and linolenate (18:3) by the black area. After Heemskerk, J.W.M. & Wintermans, J.F.G.M. (1987) *Physiol. Plantarum*, **70**, 558–68.

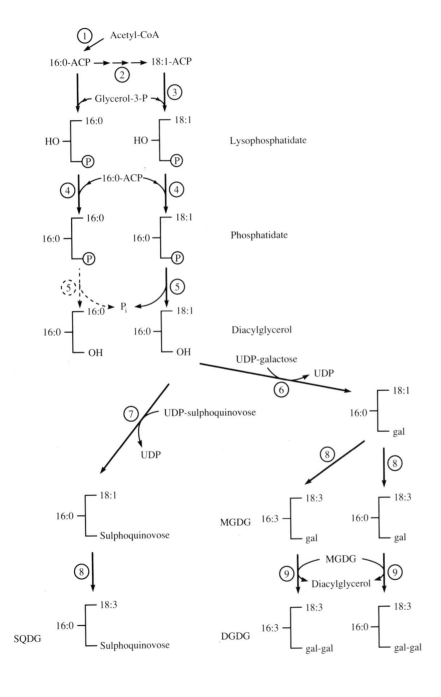

The routes of biosynthesis of the galactolipids in 16:3- and 18:3-galactolipid plants are rather different (Figs 16.10 & 16.11). In 16:3-galactolipid plants, galactolipids are synthesized entirely within the chloroplasts of the leaf cell (Fig. 16.10). Palmityl-ACP and oleyl-ACP (16:0-ACP and 18:1-ACP respectively) are synthesized as described in Section 16.1.1 and 16.1.2 and the acyl moieties are transferred to glycerol-3-P to yield lysophosphatidate bearing 16:0 or 18:1 at C-1 (*sn*-1). Either of these forms of lysophosphatidate can accept a palmityl (16:0) residue at C-2 (*sn*-2) to yield phosphatidate (PA). Unlike the PA bearing 16:0 at both positions, the PA bearing 16:0 at *sn*-2 and 18:1 at *sn*-1 is selectively dephosphorylated to give diacylglycerol. Glycosyl residues are transferred to diacylglycerol from UDP-galactose or UDP-sulphoquinovose to yield glycosylated derivatives bearing 18:1 and 16:0 fatty acids. These, in turn, are subject to desaturation to give a mixture of MGDG species with a *sn*-1 fatty acid residue of 18:3 but bearing either 16:0 or 16:3 at *sn*-2. Sulphoquinovosyl diacylglycerol (SQDG) bearing 18:1 and 16:0 fatty acyl residues is also subject to desaturation of the 18:1 moiety to 18:3. DGDG is synthesized from two MGDG molecules by transfer of the galactosyl residue from one MGDG to the galactosyl residue of the other, forming diacylglycerol and the corresponding DGDG. This reaction is carried out by galactolipid: galactolipid galactosyltrasferase located in the chloroplast envelope.

In contrast, 18:3-galactolipid plants synthesize their galactolipids by the 'eukaryotic pathway' (Fig. 16.11). Here, palmitate and oleate synthesized in chloroplasts are exported to the cytosol and activated to the corresponding thioester of coenzyme A. After transfer to the endoplasmic reticulum, the acyl-CoA molecules are converted, by the steps shown in Fig. 16.11, to phosphatidate bearing oleyl (18:1) residues at *sn*-2 and with 18:1 or 16:0 residues at *sn*-1. Phosphatidate with 18:1 at both *sn*-1 and *sn*-2 is dephosphorylated to diacylglycerol in the presence of

Fig. 16.10 Synthesis of glycolipids containing C_{16} fatty acids at the *sn*-2 position. In 16:3-galactolipid plants the glycolipids MGDG, DGDG and SQDG contain C_{16} fatty acids at the *sn*-2 position of glycerol. The 'backbone' of the glycerol moiety is represented as ⌐. Enzymes are: (1) fatty acid synthetase; (2) those responsible for elongation and desaturation of palmityl-ACP; (3) acyl-ACP:*sn*-glycerol-3-P acyl transferase; (4) acyl-ACP:1-acyl-*sn*-glycerol-3-P acyl transferase; (5) phosphatidate phosphatase; (6) UDP-galactose:*sn*-1,2-diacylglycerol galactosyltransferase; (7) UDP-sulphoquinovose:*sn*-1,2-diacylglycerol sulphoquinovosyltransferase; (8) desaturase enzymes; (9) galactolipid:galactolipid galactosyltransferase. After Heemskerk, J.W.M. & Wintermans, J.F.G.M. (1987) *Physiol. Plantarum*, **70**, 558–68.

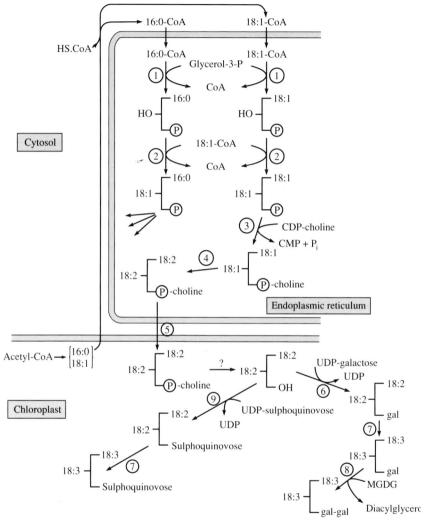

Fig. 16.11 Synthesis of glycolipids containing C_{18} fatty acids at the *sn*-2 position. In 18:3-galactolipid plants the glycolipids contain C_{18} fatty acids at the *sn*-2 position. The backbone of the glycerol moiety is depicted as ⌐. Enzymes are: (1) acyl-CoA:*sn*-glycerol-3-P acyl transferase; (2) acyl-CoA:1-acyl-*sn*-glycerol-3-P acyl transferase; (3) phosphatidate phosphatase and CDP-choline:*sn*-1,2-diacylglycerol choline phosphotransferase; (4) phosphatidyl choline desaturase; (5) phospholipid transfer protein; (6) UDP-galactose:*sn*-1,2-diacylglycerol galactosyltransferase; (7) desaturase enzymes; (8) galactolipid:galactolipid galactosyltransferase; (9) UDP-sulphoquinovose:*sn*-1,2-diacylglycerol sulphoquinovosyl transferase. After Heemskerk, J.W.M. & Wintermans, J.F.G.M. (1987) *Physiol. Plantarum*, **70**, 558–68.

phosphatidate phosphatase. The phosphocholine residue of CDP-choline is then attached to diacylglycerol to give phosphatidyl choline (PC) in a reaction involving CDP-choline:*sn*-1,2-diacylglycerol choline phosphotransferase. The 18:1 acyl groups of PC are then desaturated to 18:2 and finally to 18:3.

At this stage the PC is transferred from the endoplasmic reticulum to the chloroplast by a phospholipid transfer protein (PLTP). A series of experiments with isolated spinach chloroplasts have shown that, alone, these organelles are capable of synthesizing free fatty acids, acyl-CoA thioesters, diacylglycerol and a trace of MGDG when supplied with substrates such as acetate, glycerol-3-P and UDP-galactose. Adding a microsomal fraction (i.e. fragments of endoplasmic reticulum—see Chapter 3) results in the incorporation of ^{14}C-label from [^{14}C]acetate into PC, other phospholipids and triacylglycerol. Addition of PLTP isolated from spinach leaves causes transfer of the radioactively labelled PC from the microsomal fraction to the chloroplasts where the label is incorporated into MGDG containing 18:3 fatty acyl residues ('eukaryotic galactolipid').

The synthesis of MGDG from PC transported into the chloroplast takes place via diacylglycerol, although the precise mechanism is unknown. The diacylglycerol bearing 18:2 acyl residues at both *sn*-1 and *sn*-2 is converted to MGDG, DGDG and SQDG in much the same manner as in plants which synthesize 16:3-galactolipid (Fig. 16.10). Glycolipids bearing linolenyl (18:3) moieties are formed by further desaturation of the 18:2 acyl residues.

In any one species neither of the two pathways for the synthesis of galactolipid operates exclusively, although the presence of one of the two fatty acids with three double bonds tends to predominate. For example, 16:3-galactolipid plants such as *Arabidopsis thaliana* and spinach synthesize MGDG, DGDG and SQDG with different proportions of the 16:3 and 18:3 forms of galactolipid. Spinach contains only about 50% of its MGDG in the 16:3 configuration (see Fig. 16.9).

16.4 Biosynthesis of prenyllipids

Most of the terpenes and terpenoids, derived from terpenes by modification of the carbon framework, present in plants are not considered to be essential to growth and are thus regarded as *secondary plant products*. They are therefore regarded as being outside the scope of the present text, with its emphasis on the mechanism of plant growth. However, there are some prenyllipids which are absolutely essential to the functioning of plants. Clearly, the phytol chain of chlorophyll and various other compounds derived from terpenoids (e.g. carotenoids, certain plant growth regulator compounds and the redox compounds ubiquinone and plastoquinone)

Table 16.1 Some terpenoids and terpenoid precursors which are important to plant growth.

Compound	Number of carbon atoms	Structure	Function
Isoprene	5	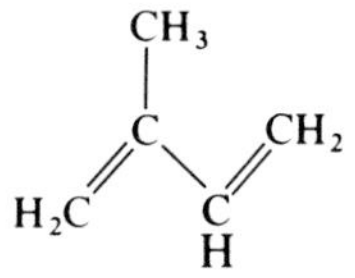	Fundamental building block of all terpenoids
Farnesol	15		Farnesyl-PP$_i$ and geranyl-geranyl-PP$_i$ are important branch points in terpenoid biosynthesis
Geranylgeraniol	20	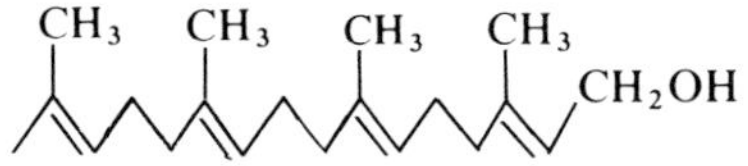	
Phytoene	40	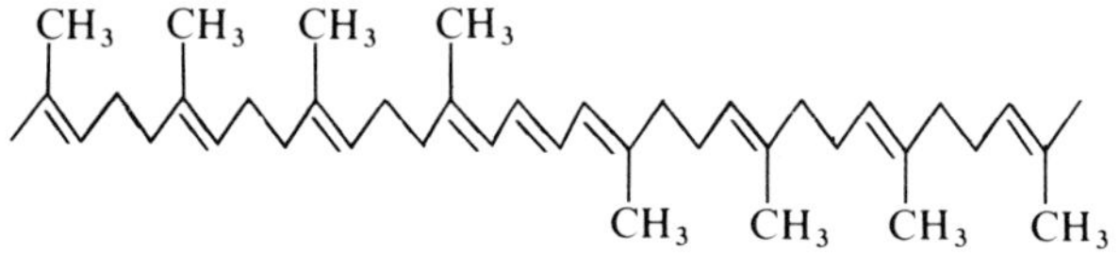	Carotenoid precursor (light absorption)
Phytol	20	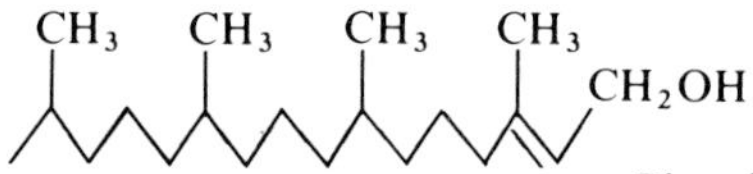	Phytol chain of chlorophyll (light-harvesting)
Polyprenol	45–50	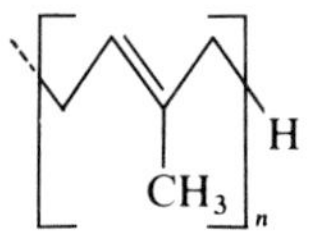	Polyprenol chain of plasto-quinone ($n=9$) and ubiqui-none ($n=10$) (electron transport)
Campesterol	28	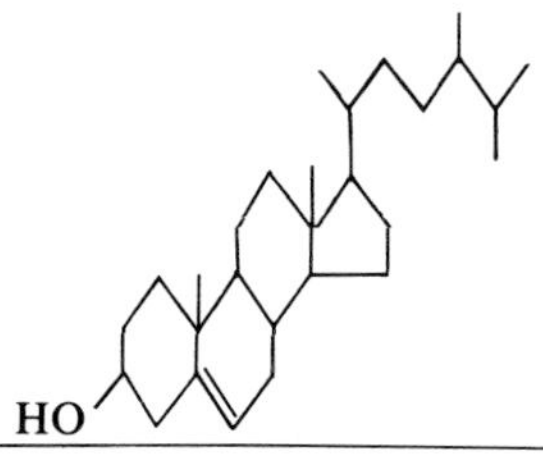	Widespread sterol in plants (membrane component)

are of fundamental importance to growth. Some of these compounds are shown in Table 16.1.

16.4.1 Terpenoids are synthesized from isoprenoid units

The terpenoids are all based on the basic C_5 unit, isoprene.

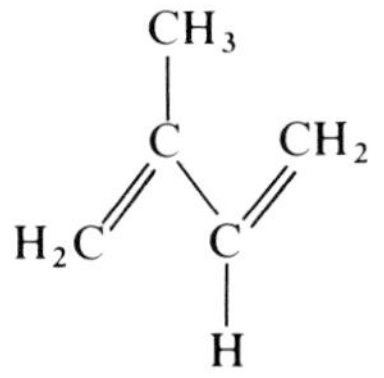

Two isoprenoid units attached head-to-tail form a monoterpene (C_{10}), such as geraniol.

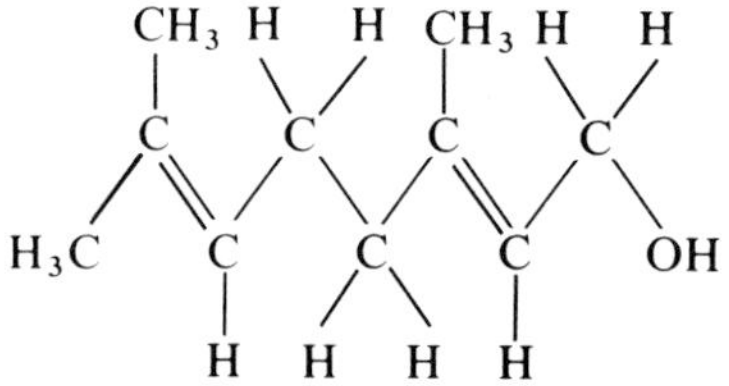

Geraniol

This, and other terpenoid structures, are often shown in a shorthand fashion as shown in Table 16.1. Alternatively, the presence of methyl or methylene moieties can be implied; thus geraniol can be represented as

Further head-to-tail attachments yield C_{15} *sesquiterpenes*, C_{20} *diterpenes* and the *polyprenols* ($C_{>40}$). Two sesquiterpenes joined together form a *triterpene* whilst two diterpenes produce a *tetraterpene* (Table 16.1). Mixed terpenoids contain a terpenoid (generally the diterpene phytol or a related compound) and a non-terpenoid component. The exceptions are those which have polyprenol side-chains—usually with nine (C_{45}) or ten (C_{50}) isoprenoid units. The terpenoids discussed in this chapter are various diterpenes (e.g. phytol), triterpenes such as the sterols (e.g. campestrol) and tetraterpenes like β-carotene.

All the terpenoids have a common biosynthetic route involving the synthesis

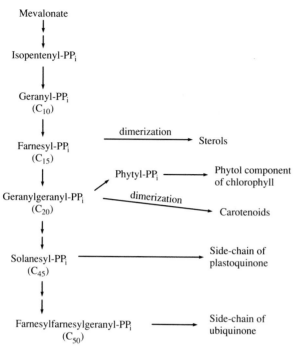

Fig. 16.12 A generalized scheme for the biosynthesis of terpenoids from mevalonate, which is synthesized from acetyl-CoA.

of isopentenyl pyrophosphate (IPP) from acetyl-CoA and its conversion to the C_{15} intermediate, farnesyl pyrophosphate. The subsequent biosynthetic routes to the various terpenoids are different and are summarized in Fig. 16.12.

The principal starting compound is the C_6 branched chain hydroxy acid, mevalonate, synthesized from three molecules of acetyl-CoA (Fig. 16.13). Two acetyl-CoA molecules react to form acetoacetyl-CoA; a third molecule is added, with water, to form 3-hydroxymethylglutaryl-CoA (HMG-CoA). Reduction of HMG-CoA and the elimination of coenzyme A yields free mevalonate. Production of Δ^3-isopentenyl pyrophosphate (Δ^3-IPP) requires the consumption of two ATP to form mevalonate pyrophosphate and finally, the elimination of CO_2 in a complex reaction involves the expenditure of a third molecule of ATP. Terpenoids are constructed by condensation of varying numbers of the activated C_5 compounds, Δ^3-IPP and Δ^2-IPP. Δ^2-IPP (sometimes known as dimethylallyl pyrophosphate) is formed from Δ^3-IPP by isomerization.

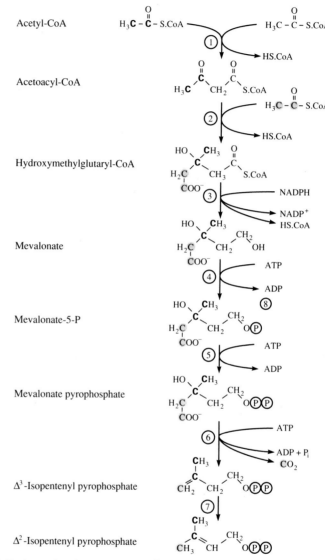

Fig. 16.13 Pathway for the synthesis of Δ^3-isopentenyl pyrophosphate and Δ^2-isopentenyl pyrophosphate from three molecules of acetyl-CoA. Enzymes are: (1) acetyl-CoA:acetyl-CoA acetyltransferase; (2) hydroxymethylglutaryl-CoA synthase; (3) hydroxymethylglutaryl-CoA reductase; (4) mevalonate kinase; (5) phosphomevalonate kinase; (6) pyrophosphomevalonate decarboxylase; (7) IPP isomerase.

16.4.2 Enzymes for the synthesis of isopentenyl pyrophosphate from acetyl-CoA are found in the cytosol

Although there have been reports that mitochondria and plastids are capable of IPP synthesis, a considerable amount of recent data suggest that IPP is synthesized solely in the cytosol and is exported to the organelles. Very pure preparations of chloroplasts assimilate ^{14}C-labelled IPP into plastid prenyllipids at rapid rates but do not incorporate intermediates of IPP biosynthesis (mevalonate, mevalonate-5-P and mevalonate-5-pyrophosphate—see Fig. 16.13) unless ATP and the enzymes listed in Fig. 16.13, derived from the cytosol, are supplied. Similarly, mitochondria from various species do not incorporate mevalonate or mevalonate-5-P into IPP, although they do assimilate IPP. Furthermore, recent evidence suggests that HMG-CoA reductase, in radish seedlings at least, is located solely on the endoplasmic reticulum. It is still possible that synthesis of IPP from acetyl-CoA within the chloroplast occurs but the balance of evidence suggests that IPP is synthesized in the cytosol and is transported to the mitochondria or chloroplasts for the biosynthesis of the prenyllipids found in these organelles.

16.4.3 Isopentenyl pyrophosphate is used in the synthesis of simple and mixed terpenoids

IPP, formed as described above, is used to synthesize simple and mixed terpenoids (Fig. 16.12). Δ^2-IPP reacts with Δ^3-IPP to form a C_{10} product, geranyl pyrophosphate (Fig. 16.14), which can react with another Δ^3-IPP to form farnesyl pyrophosphate (C_{15}). Yet another Δ^3-IPP can be added to form geranylgeranyl pyrophosphate (C_{20}). Each addition is catalysed by prenyl transferase and is accompanied by the elimination of PP_i. Farnesyl pyrophosphate and geranylgeranyl pyrophosphate are major 'stepping-off' points for the biosynthesis of different terpenoids, most of them secondary compounds.

Sterols are terpenoids which are important to the properties of membranes (see Section 6.1.3). They are derived from the intermediate, squalene, synthesized from two farnesyl pyrophosphate residues (Fig. 16.15). Squalene is converted via squalene-2,3-epoxide to the cyclic product, cycloartenol, from which all other sterols in higher plants are derived (Fig. 16.15). The formation of cycloartenol from squalene-2,3-epoxide is a distinguishing feature of photoautotrophic organisms since, in fungi and animals, the cyclization product is lanosterol.

Geranylgeranyl pyrophosphate provides the carbon backbone for the synthesis

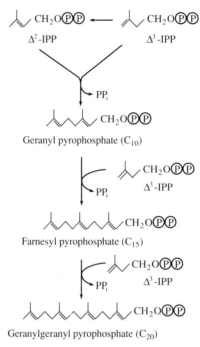

Geranyl pyrophosphate (C_{10})

Farnesyl pyrophosphate (C_{15})

Geranylgeranyl pyrophosphate (C_{20})

Fig. 16.14 Pathway for the formation of geranylgeranyl pyrophosphate from Δ^3-IPP and Δ^2-IPP. Pyrophosphates containing 10, 15 or 20 carbon atoms serve as parent molecules for the synthesis of monoterpenes, sesquiterpenes and diterpenes respectively. The reactions shown here are catalysed by prenyl transferase.

of the carotenoids, and for the phytol chains of the antioxidants tocopherol and vitamin K, and of chlorophyll (see Table 16.1). The phytol chain is formed by reduction of geranylgeranyl pyrophosphate and removal of pyrophosphate.

Carotenoid formation proceeds by the combination of two geranylgeranyl pyrophosphate residues, in a manner analogous to the formation of squalene from farnesyl pyrophosphate, to give the C_{40} compound phytoene which is the direct precursor of the carotenoids.

Phytoene

Farnesyl-⑫⑫

Presqualene-⑫⑫

Squalene

Squalene-2,3-epoxide

Cycloartenol

Phytoene is oxidized in a number of steps to form lycopene. One or both ends of the lycopene molecule is (are) then subjected to a cyclization process to form a group of carotenoids known as carotenes (e.g. see Eqn. 16.15).

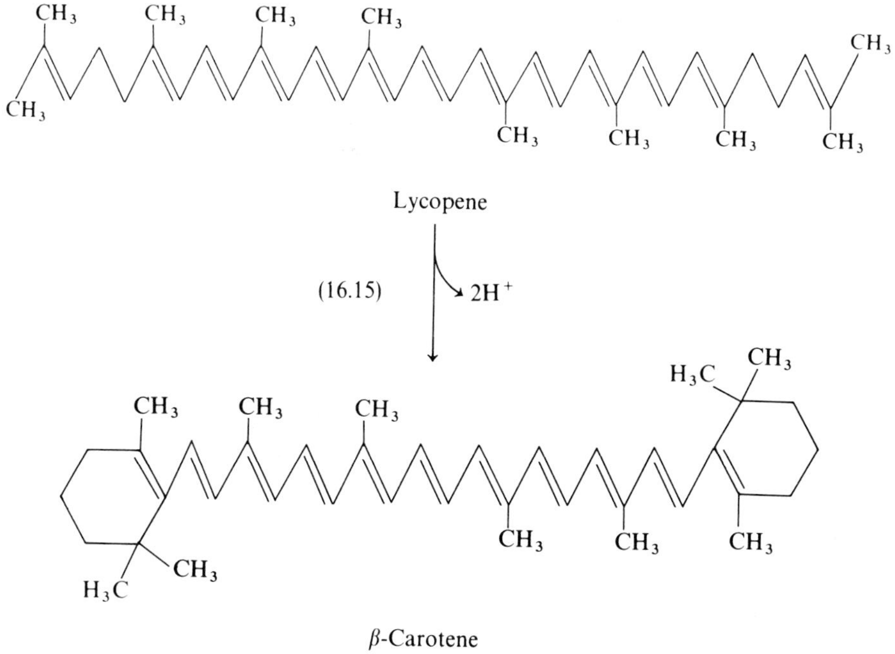

Lycopene

(16.15)

β-Carotene

Other carotenoids, such as the xanthophylls, are formed by the introduction of one or more oxygen-containing moieties (e.g. hydroxyl groups) but little is known about the mechanism involved.

The redox compounds plastoquinone and ubiquinone are terpenoid derivatives. They are of central importance in the energy-generating processes of chloroplasts (see Chapter 9) and mitochondria (see Chapter 8).

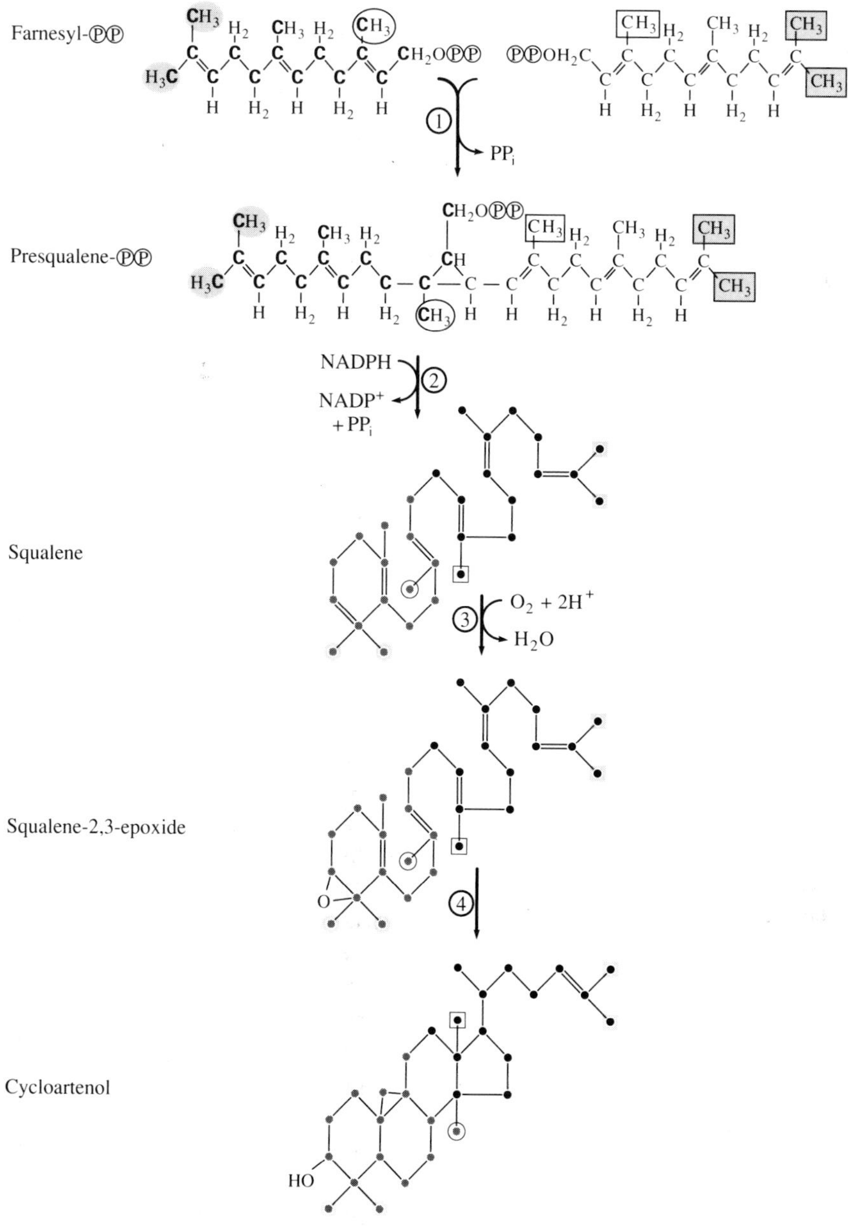

Fig. 16.15 Synthesis of cycloartenol, the precursor of plant sterols, from farnesyl pyrophosphate. Enzymes are: (1) presqualene synthase; (2) squalene synthase; (3) squalene epoxidase; (4) squalene epoxide cycloartenol cyclase. The structures of squalene and its derivatives are shown in a shorthand form, the grey or black circles representing carbon atoms from the two farnesyl pyrophosphate molecules. Terminal methyl groups derived from farnesyl pyrophosphate are shaded.

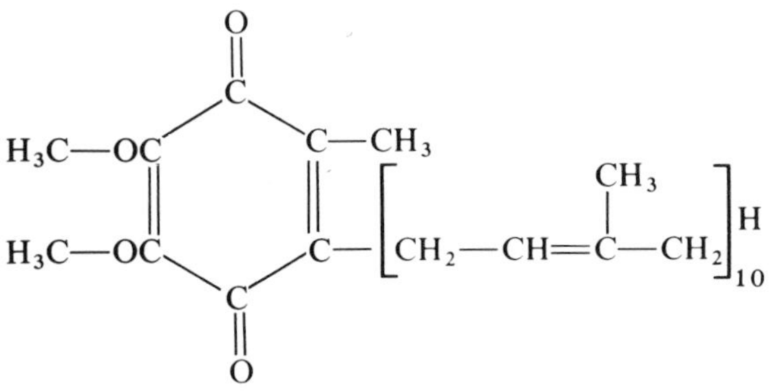

Ubiquinone
(oxidized form)

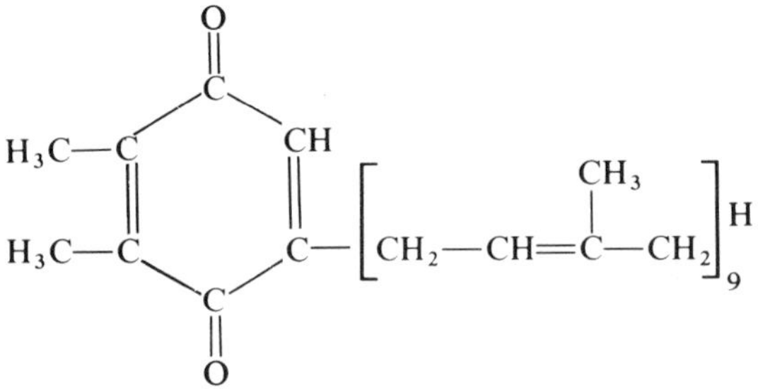

Plastoquinone
(oxidized form)

Both molecules consist of a quinone moiety which, in the case of plastoquinone, is attached to a C_{45} solanesyl moiety derived from solanesyl pyrophosphate. In ubiquinone the quinone is bound to a C_{50} residue derived from farnesylfarnesylgeranyl pyrophosphate. The quinones in plastoquinone and ubiquinone are derived from homogentistic acid and 4-hydroxybenzoic acid respectively.

16.5 Membranes have a capacity for self-assembly

Phospholipids are amphipathic lipids which comprise a large proportion of cell membranes (see Section 6.1.2). When mixed with water, amphipathic lipids spontaneously form vesicles (liposomes) which consist of a lipid bilayer surrounding an aqueous interior phase. The lipids have a natural tendency to form the membrane-like structures of bilayers and liposomes. However, the internal face of a membrane often has a very different lipid composition to the exterior face. How this asymmetry is established and maintained is unknown but, possibly, molecules such as phospholipid transfer protein (see Section 16.3.3) are important.

Liposomes have a natural tendency to 'round up' into spherical vesicles and thereby achieve the most stable conformation. However, many membranes, such as chloroplast thylakoids or the cristae of mitochondria, are much more flattened. Biological membranes contain various proteins in addition to amphipathic lipids and it is believed that the interaction of the proteins with the lipids prevents 'rounding up'. In thylakoids, galactolipids are also believed to be important in stabilizing the flattened organization of the membranes.

The interactions between lipids and proteins are extremely important in thylakoid membranes—the lipids cannot form uniform lamellar bilayers without protein. The arrangement of lipid and protein must therefore proceed simultaneously for correct assembly of the membrane. The synchronized onset of the synthesis of chlorophylls, polyunsaturated polar lipids and apoproteins of the light-harvesting complexes of chloroplast thylakoids by etiolated plants, following exposure to light, is consistent with this. The majority of proteins found in chloroplasts are coded for in the nucleus and synthesized on 80 S ribosomes in the cytosol. They are guided through and/or to membranes by specific amino acid sequences attached to the polypeptide. These 'transit' or 'signal' peptides are removed as the polypeptide is inserted into the membrane (see Chapter 21). However, many chloroplast proteins are synthesized without such sequences; their insertion into a developing thylakoid membrane is likely to occur by self-assembly as the membrane develops. These, and other aspects of the biogenesis of chloroplasts and mitochondria, are discussed further in Chapter 21.

Further reading

Monographs and treatises: Stumpf (1980); Stumpf (1987).
Fatty acid metabolism: Harwood (1988).
Galactolipid biosynthesis: Heemskerk & Wintermans (1987).
Compartmentation of lipid biosynthesis: Heemskerk & Wintermans (1987).

17

Synthesis of plant cell walls

17.1 Plant cell walls are important biologically and economically

The size, shape and form of plant cells are determined by the cell wall, which strongly influences the structural and mechanical properties of the cells. Plant cell walls comprise the structural elements of paper, certain building materials and cotton goods and they are important sources of domestic and industrial energy (e.g. wood and fossil fuels). Cell wall carbohydrates are an important food source for herbivorous animals.

The form and composition of cell walls varies during cell development and there are differences between tissues and species. This chapter is concerned with the molecular and supermolecular structure of cell walls and with current concepts of wall synthesis. The structure of primary and secondary walls is described in Section 3.10 and forms important background reading to this chapter.

17.2 Molecular composition of plant cell walls

Cellulose microfibrils are the structural elements or fibrillar components of cell walls, analogous to reinforcing rods. The matrix between fibrils consists of various compounds with different properties which differ quantitatively and qualitatively between cells and provide 'packing' in the cell wall. Some components (e.g. pectic substances) are gel-like so that the microfibrils can be displaced relative to each other as, for example, during axial cell growth, while other components (e.g. lignin) stiffen the matrix and strengthen the wall.

17.2.1 Fibrillar polysaccharides

The structure and properties of cellulose, a polymer of β-glucose residues joined in $\beta(1\rightarrow4)$ linkage, and its arrangement in microfibrils, are described in Sections 2.3.6 and 3.10. In higher plants the microfibrils are about 5–8 nm in cross-section with up to 70 cellulose molecules, but those of certain green algae are larger with cross-sections of about 12–19 nm and contain about 500 cellulose molecules.

X-ray analysis reveals that the cellulose microfibrils are in precise crystalline arrangements but those towards the outer edge are less ordered because of the presence of water and traces of non-cellulosic polysaccharides.

In the past there has been uncertainty as to whether the cellulose molecules of microfibrils are 'parallel' (i.e. with all the free C-4 hydroxy groups at one end and with the free C-1 hydroxy groups at the other end) or 'anti-parallel' (i.e. head-to-tail). Until recently opinion based on X-ray diffraction patterns supported an anti-parallel arrangement; however, current understanding of polysaccharide synthesis (see Section 17.8) suggests that they are parallel and this is supported by recent interpretations of the X-ray data.

The walls of various lower plants and fungi contain polysaccharides other than cellulose as fibrillar components (see Section 2.3.6).

The arrangement of microfibrils in higher plants is discussed in Sections 3.10 and 17.4.

17.2.2 Matrix polysaccharides

The matrix or non-cellulosic polysaccharides of cell walls form solutions of high viscosity; the most important of these are the pectic substances and hemicelluloses (see Section 2.3.6). The pectic substances are water-soluble and form viscous gels in water. Most cell wall pectic substances are rich in galacturonate which binds Ca^{2+} and Mg^{2+} between the carboxyl groups on adjacent polygalacturonyl chains. Thus, the chelating agent EDTA is commonly used for their extraction. The hemicelluloses are distinguishable as they are insoluble in water but soluble in alkali.

When pectic substances and hemicelluloses are extracted from cell walls with EDTA and alkali, various uncontrolled reactions can rupture internal bonds. Analysis of fractions prepared in this way only provides information about the sugars present. Uncontrolled reactions can be minimized by enzymic hydrolysis of specific glycosidic bonds to produce soluble fragments which can be separated. The isolated fractions are methylated and the polymeric components are hydrolysed to expose the hydroxy groups involved in glycosidic linkage. The hydrolysis products are then identified and, since the hydroxy groups involved in glycosidic linkage remain unmethylated, the structure of segments of the polysaccharides can be deduced.

Table 17.1 lists some of the more important classes of non-cellulosic polysac-

Table 17.1 Representative classes of structural cell wall polysaccharides.

Polymer	Major features
Cellulose	Linear chains of $\beta(1\rightarrow4)$ glucose which aggregate into fibrils
Pectic substances	
Galacturonans	Chains of $\beta(1\rightarrow4)$ galacturonate methylated to various extents on the —COO⁻ groups at C-6; uncommon
Rhamnogalacturonans	Chains as for galacturonans but with (a) occasional residues of β-L-rhamnose in $1\rightarrow2$ linkage inserted between galacturonan blocks and/or (b) attachment of short or extended side-chains containing xylose, galactose, L-arabinose or other neutral sugars. Most of the pectic material in the matrix is of this type
Galactans	Chains of $\beta(1\rightarrow4)$ galactose; uncommon
Arabinogalactans I*	As for galactans but with short side-chains of L-arabinose attached
Arabinans	Chains of $\alpha(1\rightarrow5)$L-arabinose with further α-L-arabinose units attached from C-3 of the residues in the main chain to C-1 of residues in the side-chain.
Hemicelluloses	
Xylans	Chains of $\beta(1\rightarrow4)$ xylose with short side-chains of other sugars attached to the main chain at C-2 or C-3. Those attached at C-2 are typically (4-*O*-methyl) α-glucuronate and those at C-3, α-L-arabinose. Attachment is either as single residues or as short chains of $\alpha(1\rightarrow3)$ L-arabinose (i.e. arabinoxylans), sometimes with still more substituents attached. Arabinoxylans are the most abundant of the non-cellulosic polysaccharides in cell walls of grasses
Glucomannans	Chains of $\beta(1\rightarrow4)$ mannose containing glucose residues, also in $\beta(1\rightarrow4)$ linkage, within the main chain. Ratio of mannose to glucose is commonly about 3:1. Especially important in gymnosperms

Table 17.1 (cont.)

Polymer	Major features
Galactoglucomannans	As for glucomannans but with single α-galactose residues forming side-chains attached via C-1 to C-6 of glucose or mannose residues in the main chain
Xyloglucans	Chains of $\alpha(1\rightarrow4)$ glucose with short side-chains of α-xylose linked via C-1 to C-6 of alternate glucose residues in the main chain. Other substituent sugars also occur in the side-chains. Principal hemicelluloses in dicotyledons
Mixed glucans	Linear chains of β-glucose containing both $\beta(1\rightarrow3)$ and $\beta(1\rightarrow4)$ linkages in the ratio of 1:2 to 1:3
Callose†	Essentially unbranched polymer of $\beta(1\rightarrow3)$ glucose deposited in granular form around the sieve plates of sieve tubes (main vascular cells for long-distance transport of organic materials in plants)

* Arabinogalactans II, distinguished from arabinogalactans I by having a structure based on a main chain of $\beta(1\rightarrow3)$ galactose, do not occur in cell walls.
† Callose is a non-structural polysaccharide.

charides and their properties. Individual pectic substances and hemicelluloses differ in their component monosaccharides, their sequences, glycosidic linkages and branching (Figs 17.1 & 17.2). With a few exceptions, most matrix pectic substances contain a high proportion of galacturonate—the rhamnogalacturonans being the most important. The hemicelluloses as a group have similar macromolecular conformations and fulfil similar functions in the matrix.

The ability of pectic substances to form gels is important in determining the plasticity of cell walls, particularly in the early phases of cell growth. Pectic substances are major components of the cell plate, formed when cells divide, and they are continually added as the middle lamella is formed. Indeed, some plant tissues separate into protoplasts when treated with pectinase, a fungal enzyme which hydrolyses $\beta(1\rightarrow4)$ galacturonate linkages. Hemicelluloses contribute to the plasticity and porosity of cell walls, which is important to the exchange of metabolites and water between cells. Some classes of hemicellulose form an extensive network of hydrogen bonds with the outer, non-crystalline regions of cellulose microfibrils; this may be important in cross-linking the microfibrils to the matrix (see Section 17.2.4).

(a) Rhamnogalacturonan

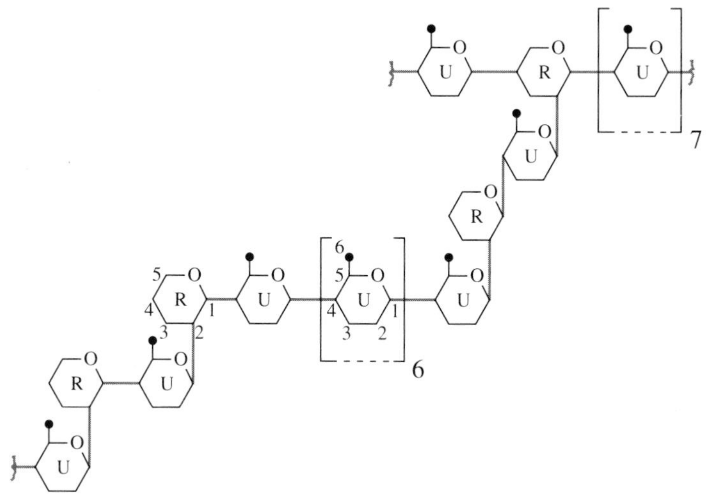

(b) Arabinogalactan (type I)

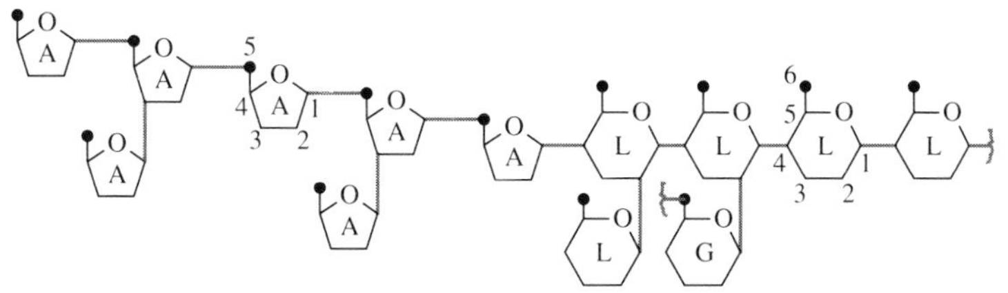

Fig. 17.1 Proposed structures of some pectic substances from sycamore callus cells; rhamnogalacturonan, an acidic pectin-containing galacturonate and arabinogalactan (type I), a neutral pectin. Glycosidic bonds linking the glycosyl residues are shown in grey. Furanose forms are shown as pentagons and pyranose forms as hexagons. Details of the structure of the individual glycosyl residues have been omitted except for the oxygen atoms in the ring. Carbon atoms lying outside the ring are represented as (—●). Numbering of the carbon atoms of some representative residues is shown. Abbreviations: A = arabinose; G = glucose; L = galactose; R = rhamnose; U = galacturonate.

17.2.3 Other components

Plant cell walls contain small amounts of protein, including various glycoprotein enzymes involved with the final processing of cell wall substances (e.g. peroxidases for the oxidative polymerization of lignin precursors) or with various hydrolytic functions. However, most of the cell wall protein has a structural function and contains a high proportion of the amino acid, hydroxyproline, which is absent from most proteins. It is combined with carbohydrate to form several types of hydroxyproline rich glycoprotein (HRGP). The most common of these are the extensins (some species contain several types, e.g. four in tomato) which are insoluble, basic and contain about 35% protein. Glycoproteins can account for up to 10% of the dry weight of the cell walls in some dicotyledons. Hydroxyproline accounts for about 46% of the amino acid residues of the principal extensin from carrot root cell walls (Table 17.2). The gene coding for this protein contains 25 repeated sequences for the pentapeptide, serine–(proline)$_4$. The proline in the peptide synthesized from this sequence is oxidized, post-translationally, to hydroxyproline which accounts for its high level in the completed gene product *in vivo*. Specific sugars are attached to particular residues within the peptide chain; chains of three or four arabinose residues are attached to each hydroxyproline residue and galactose is attached to serine.

Extensins occur in tight association with cellulose but they are not covalently bonded to it. However, it is not clear as to whether they are essential structural components. They are synthesized as monomers but within the matrix they are extensively cross-linked, possibly via isodityrosine linkages (see Section 17.5), to form a rigid network. The large number of short repeating gene sequences coding for polyproline in extensin is characteristic of structural proteins, such as collagen in animals.

Another group of compounds rich in hydroxyproline are the arabinogalactan proteins (AGPs). They differ from HRGPs in that the main framework consists of a polysaccharide, they contain only a small amount of protein (commonly 2–10%) and they are very small and soluble. They are therefore regarded as proteoglycans rather than as glycoproteins. The main backbone of the AGPs is a polymer of β-galactose in $\beta(1 \rightarrow 3)$ linkage with short galactosyl side branches attached to C-6; these are, in turn, substituted with other sugars, arabinose being the most common. The protein component is especially rich in hydroxyproline, serine, alanine and glycine. Their solubility makes the subcellular location of the AGPs difficult to establish, but they commonly occur as secretions and at the surfaces of protoplasts.

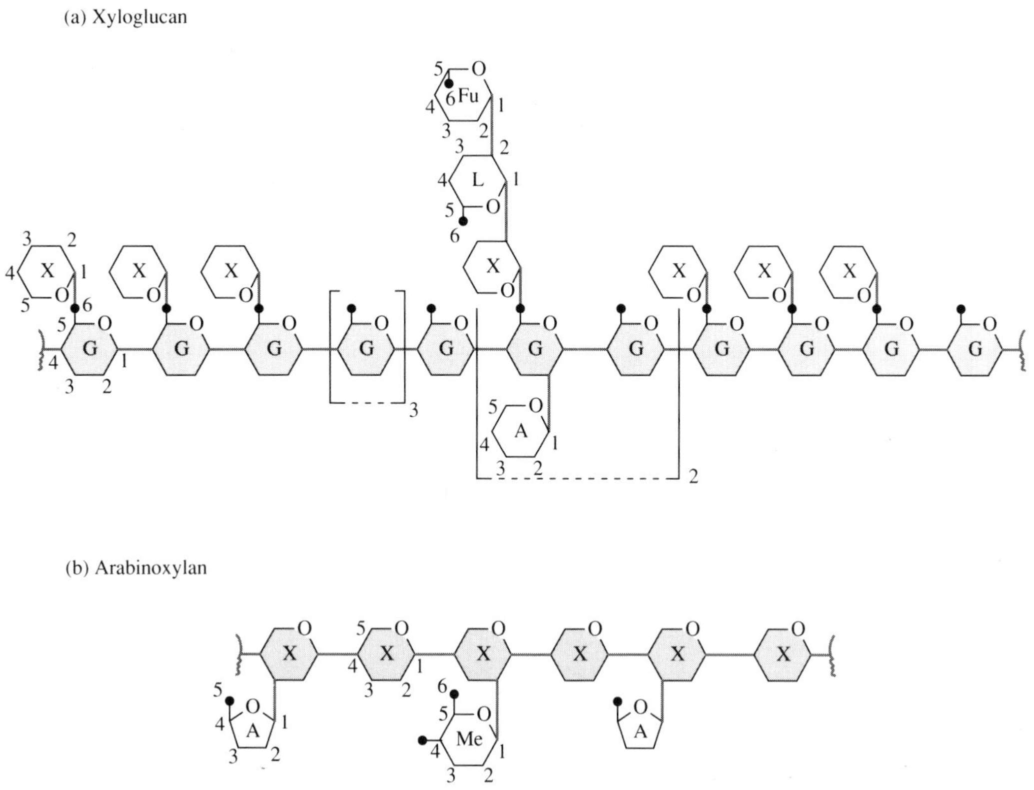

(a) Xyloglucan

(b) Arabinoxylan

Fig. 17.2 Proposed structure of some hemicelluloses from cultured sycamore cells; (a) fragment of xyloglucan with the main glucan chain shaded; (b) fragment of arabinoxylan with the main xylan chain shaded. Bonds linking the glycosyl residues are shown in grey (for other details see legend to Fig. 17.1). Abbreviations: A = arabinose; Fu = fucose; G = glucose; L = galactose; Me = 4-O-methylglucuronate; X = xylose.

They appear to be associated with the peripheral interface(s) between cytoplasm, plasma membrane and cell wall.

The other important, non-carbohydrate component is lignin, a collective term for various complex cross-linked phenylpropanoid compounds of particular plant cell walls (see Section 2.7). The form and phenylpropanoid composition varies between species and tissues. Lignin, when present, is deposited during secondary thickening in both the primary and secondary walls (but is not essential to it) and contributes significantly to the high rigidity of lignified cells, rendering them more or less impermeable to water. Many types of linkages have been found in isolated fragments, as shown by the extensive cross-linking of coniferyl and sinapyl residues in beechwood lignin (Fig. 17.3).

17.2.4 Matrix components are cross-linked to each other and to cellulose microfibrils

Matrix polysaccharides do not form close-packed structures like the cellulose molecules of microfibrils because of the presence of side-chains, non-repeating sequences of monosaccharide residues in the main chain and 'kinking' of the main chain by mixed glycosidic linkages (e.g. the mixed β-glucan of many cereals with both $\beta(1\rightarrow3)$ and $\beta(1\rightarrow4)$ linkages). Rather, the individual matrix polysaccharides form hydrated gels of varying 'texture'. Although they do not pack in a regular way, some form networks at two levels of organization. One is an open, highly hydrated region in which individual chains do not form close associations with others. The

Table 17.2 Amino acid and sugar composition of the major extensin (hydroxyproline rich glycoprotein) from carrot root cell walls. From Van Holst, G-J. & Varner, J.E. (1986) *Plant Physiol.*, **74**, 247–51.

Residue	Residues per mol of glycoprotein
Sugars	
Arabinose	407
Galactose	14
Amino Acids	
Hydroxyproline	116
Serine	36
Histidine	30
Tyrosine	28
Lysine	17
Valine	15
Threonine	3
Tryptophan	3
Proline	2
Aspartate/asparagine	1
Glutamate/glutamine	1
Glycine	1
Alanine	1
Isoleucine	1
Leucine	1

17.3 Plant cells vary in the composition of their walls

In the woody stems of dicotyledons the sequence cambium→sapwood→heartwood reflects changes during cell growth and development (Table 17.3). However, it is very difficult to obtain sufficient cambial cells for analysis without contamination by non-cambial cells; this accounts for the significant levels of lignin (generally absent from cambial cells) in the 'cambium' reported in Table 17.3. Nevertheless, the data show that the composition varies between species and during development. The principal matrix components of primary walls (cambium) are pectic substances and hemicelluloses, whereas lignin and hemicelluloses are more important in secondarily thickened walls (sapwood and heartwood). Adjusting for cell growth as the cambial cells grow, production of pectic substances ceases and lignin synthesis increases. The walls of secondarily thickened cells also contain a slightly higher proportion of cellulose than do primary walls. These differences explain the greater plasticity of primary walls compared to secondarily thickened walls.

The primary walls of dicotyledons commonly have more pectic material than monocotyledons and also lower levels of hemicellulose. Xyloglucan is the principal hemicellulose in dicotyledons whilst arabinoxylan is the principal hemicellulose in monocotyledons. Dicotyledons contain about 10 times more HRGP than do monocotyledons. The lignin of dicotyledons (hardwoods) is composed mainly of coniferyl alcohol and sinapyl alcohol; that of monocotyledons also has a considerable amount of 4-coumaryl alcohol and that of softwoods (gymnosperms) is principally composed of coniferyl alcohol.

other, known as a 'junction zone', contains close complementary associations of segments of separate molecules; these are, in general, linear unbranched sections of the main frameworks, free from 'kinks'. These are stabilized by hydrogen bonds, and possibly ionic bonds, between complementary segments. The greater the size and number of the junction zones, the greater the rigidity of the gel.

In addition to the internal non-covalent bonds between the same type of polysaccharide, there may be non-covalent cross-linking between different matrix components, e.g. unbranched sections of arabinoxylan with unbranched segments of xyloglucan. Thus, the cell wall matrix can be regarded as a multi-component cross-linked network. Moreover, the junction zones of certain matrix components not only link to each other but also to the cellulose microfibrils (see Section 17.5).

Covalent cross-linkages also occur between cell wall components. These could involve oxidative dimerization of ferulate, tyrosine and/or cysteine residues between various components (see Section 17.5).

17.4 The orientation of cellulose microfibrils is normal to the direction of cell growth

Alterations in the bonding of matrix components during growth permits movement between cellulose microfibrils. The cell expands, reducing turgor and increasing water uptake into vacuoles. In a cylinder, the tensile stress is twice as great tangentially as longitudinally. In cylindrical cells extending longitudinally, the fibrils are oriented around the circumference of the cell as are the strong covalent bonds (i.e. glycosidic bonds) linking glucose residues of the cellulose. The hydrogen bonds linking adjacent parallel cellulose molecules and the matrix are weaker. This distribution of the primary and secondary bonds provides the best arrangement for resisting the stresses of a growing cylindrical cell.

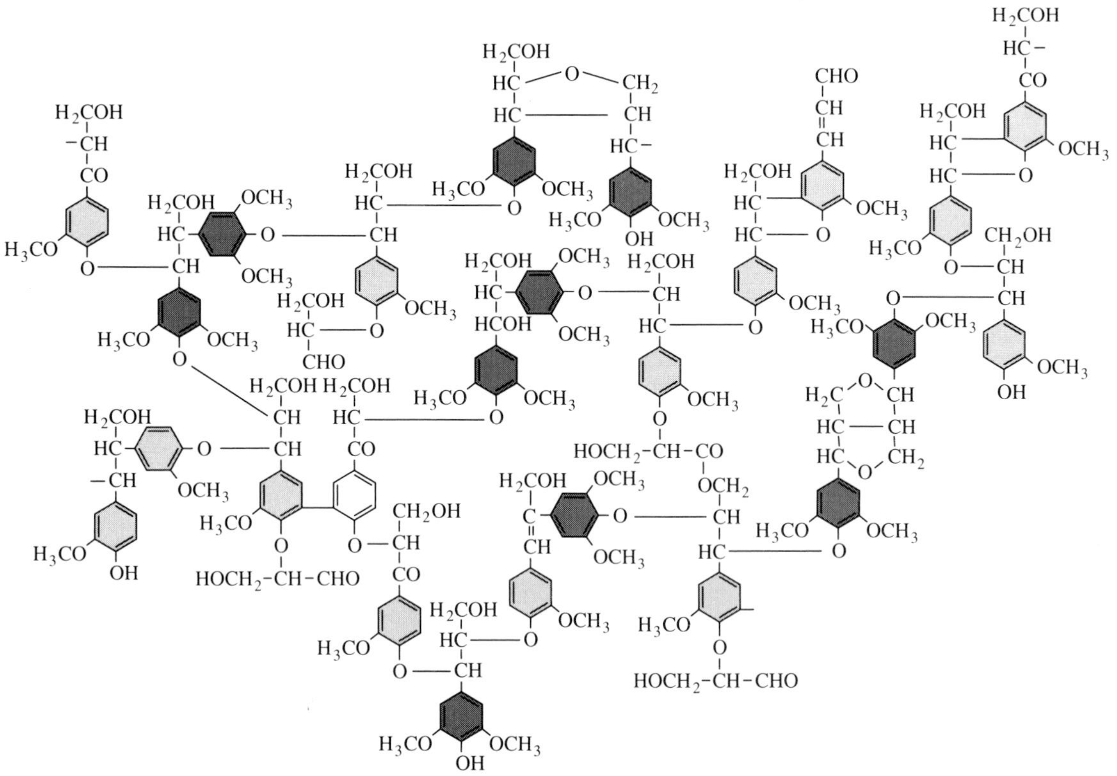

Fig. 17.3 Structure of a fragment of beechwood lignin, formed mostly by condensation of coniferyl and sinapyl alcohol. Rings of coniferyl residues (and rings derived from coniferyl residues) are shown in light grey and those derived from sinapyl residues in dark grey. After Nimz, H. (1974) *Angew. Chem. Internat. Edn.*, **13**, 313–21.

17.5 Model for the organization of the primary cell wall

A model incorporating the common features of primary cell walls is shown in Fig. 17.4. It is based on the isolation of polysaccharide fragments containing remnants of xyloglucan (a hemicellulose) attached to arabinogalactan (a neutral pectin), the identification of fragments of arabinogalactan linked to rhamnogalacturonan (an acidic pectin) and the demonstration that certain hemicelluloses form stable hydrogen-bonded complexes with cellulose. Small amounts of xylose and other sugars found in hemicellulose in the alkali-insoluble fraction of cell walls suggest that some hemicellulose is strongly bonded to cellulose, perhaps as attachment points between microfibrils and the matrix. The main structural components are the cellulose microfibrils, with the outermost molecules hydrogen-bonded to the glucan spine of xyloglucan. Binding between the two can be demonstrated *in vitro* by mixing purified xyloglucan with powdered cellulose; they are dissociated by 8 M urea, which disrupts hydrogen bonds. Most models, including the one in Fig. 17.4, show hemicellulose covalently linked to a neutral pectin of the arabinogalactan type, which is covalently bonded to an acidic pectin (rhamnogalacturonan). However, some recent models suggest that rhamnogalacturonan is bonded directly to cellulose. Several matrix polysaccharides are thought to be cross-linked via diferulate bridges (Fig. 17.5), possibly through the action of a cell wall peroxidase between ferulate residues present in polysaccharide chains. Some of the bonds linking components are shown in a hypothetical model in Fig. 17.6. (Note, however, that Fig. 17.6 is not a model of a cell wall and that it shows rhamnogalacturonan bonded directly to cellulose.)

Table 17.3 Composition of cell walls from cambium, sapwood and heartwood in sycamore and ash. The absolute amount of each component is obtained from the product of the relative amount of the component and the cell wall weight.* From Thornber, J.P. & Northcote, D.H. (1961) *Biochem. J.*, **81**, 449–55.

Species	Component	Composition (% of wall dry wt)		
		Cambium	Sapwood	Heartwood
Sycamore	Pectic substances	15.0	3.8	1.3
	Hemicellulose	45.0	31.5	31.4
	Cellulose	36.7	40.0	43.0
	Lignin	2.5	25.3	24.3
Ash	Pectic substances	6.6	1.4	0.5
	Hemicellulose	40.8	32.9	29.8
	Cellulose	42.6	42.4	44.4
	Lignin	10.0	22.8	22.5

* For sycamore, wall weights (ng cell^{-1}) are: cambium = 15.3; sapwood = 157; heartwood = 171. For ash, wall weights (ng cell^{-1}) are: cambium = 21.3; sapwood = 155; heartwood = 173.

The model of the primary cell wall (see Fig. 17.4) allows for growth by 'loosening' of the hydrogen bonds attaching xyloglucan to the microfibrils. The acid-growth hypothesis postulates that wall extension is accompanied by secretion of protons from the cytoplasm into the wall, weakening the hydrogen bonds between the matrix components, including those between the microfibrils and xyloglucan. Alternatively, specific covalent bonds cross-linking key components could be hydrolysed by wall enzymes.

The model illustrated in Fig. 17.4 does not include glycoproteins of the extensin type. These could be linked by ionic bonds to the acidic pectins (Fig. 17.6) or to some independent network. The former proposal implies that extensins are a key component in cell wall structure but this is not universally accepted.

17.6 Structural studies indicate that microtubules and Golgi bodies are involved in cell wall synthesis

A mechanism must exist on or near the plasma membrane to determine the orientation of microfibrils during synthesis. Several observations suggest that microtubules are involved in this process. Microtubules occur in the cytoplasm of interphase cells parallel to, and approximately 20 nm distant from, the plasma

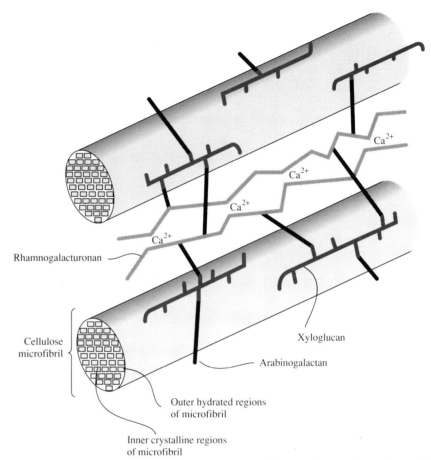

Fig. 17.4 Model for the structure of primary cell walls of dicotyledonous plants. See text for explanation of interaction between components.

membrane. They are arranged around the longitudinal axis of the cell, spaced at about 40 nm intervals, with each tubule extending about halfway round the cell so that they appear in their tubular dimension in sections taken parallel to the direction of growth. The similarity of the spacing and orientation of the microtubules in the cytoplasm to that of the microfibrils, especially during primary and secondary thickening in xylem elements, suggests that microtubules determine microfibril orientation.

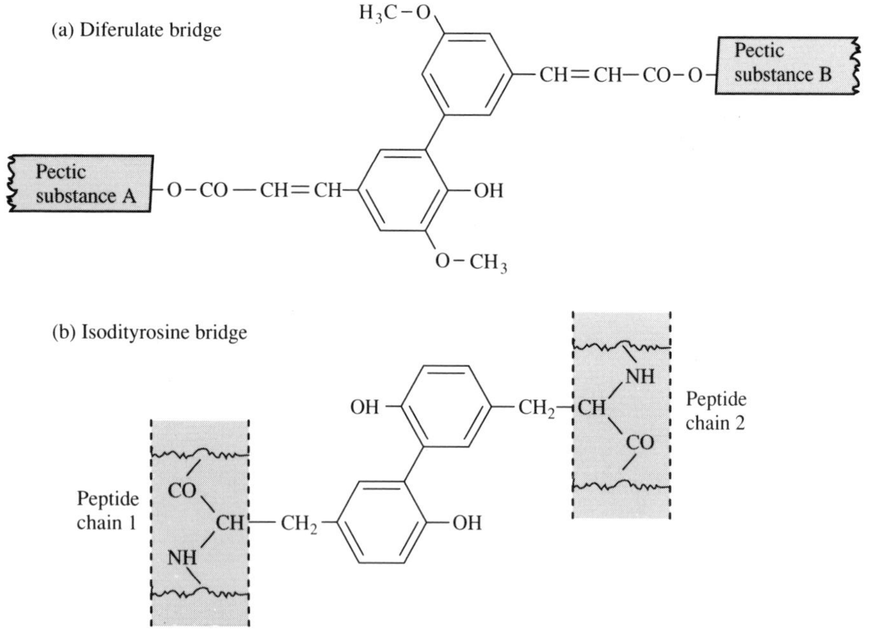

(a) Diferulate bridge

(b) Isodityrosine bridge

Fig. 17.5 Some examples of covalent cross-linking between matrix components involving oxidative coupling of residues: (a) linkage of two pectic substances via a diferulate bridge; (b) linkage of two extensin molecules by an isodityrosine bridge.

Colchicine, which causes depolymerization of microtubules to their constituent protomer, tubulin, causes loss of control of orientation in cells which have not commenced formation of microfibrils. However, it has little or no inhibitory effect in cells already producing microfibrils.

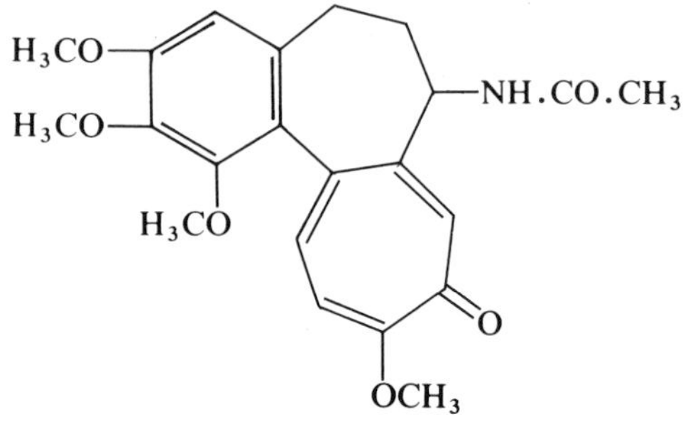

Colchicine

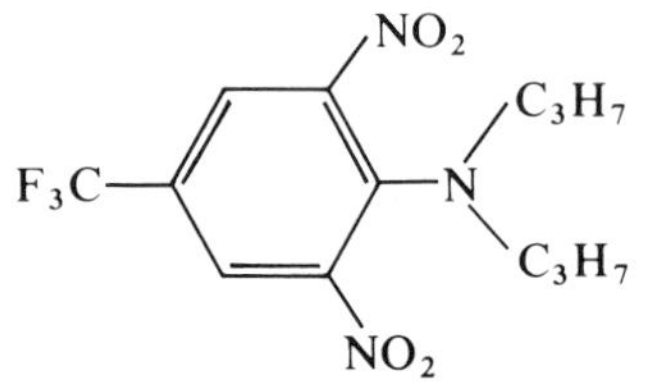

Trifluralin

The dinitroaniline herbicides (e.g. trifluralin) are even more effective in de-polymerizing microtubules and they also cause various wall abnormalities. Thus, microtubules appear to orient microfibrils, although they are not essential for the synthesis of cellulose *per se* or for the aggregation of cellulose molecules into microfibrils. The formation of plasmodesmata and pit fields and, in secondary walls, of pits, implies that there is control of the distribution of microfibrils and other wall materials and co-ordination between adjacent cells.

Structural evidence suggests that Golgi bodies also participate in cell wall synthesis. They are present in great numbers in cells with rapid tip growth (e.g. pollen tubes). Golgi bodies are active in the production of matrix polysaccharides which are transported in Golgi vesicles; these fuse with the plasma membrane and release their contents into the periplasmic space from where the released material is incorporated into the wall (see Section 17.9).

17.7 Monosaccharides are incorporated into cell wall polysaccharides as their nucleoside diphosphate derivatives

The formation of nucleoside diphosphate (NDP) derivatives of sugars and their importance as activated forms for the synthesis of glycosidic bonds is described in Section 10.9. Monosaccharide residues are incorporated into the polysaccharides of cell walls either as a NDP-derivative or as a NDP-derivative of an appropriate precursor.

Sucrose is the principal source of carbon for cell wall synthesis. UDP-glucose can be formed directly through the action of sucrose synthase.

$$\text{Sucrose} + \text{UDP} \rightleftharpoons \text{UDP-glucose} + \text{fructose} \qquad (17.1)$$

Alternatively, sucrose can be hydrolysed by invertase to glucose and fructose, which can be metabolized to glucose-1-P and, in the presence of specific NDP pyrophos-phorylases, to UDP-glucose, GDP-glucose and ADP-glucose. Glucose and fructose can also be metabolized to mannose-1-P and so to GDP-mannose by yet another specific pyrophosphorylase. Plants contain other pyrophosphorylases which catalyse the synthesis of UDP-galactose, UDP-glucuronate, UDP-galacturonate, UDP-arabinose and UDP-*N*-acetylglucosamine from the corresponding monosac-charide-1-phosphates. Various NDP-sugars and their derivatives are also syn-thesized by modification of related NDP-sugars, usually by direct or indirect metabolism of UDP-glucose (Fig. 17.7). Thus, UDP-galactose is formed by epimer-ization of UDP-glucose, and UDP-glucuronate by oxidation of UDP-glucose. UDP-galacturonate, in turn, is synthesized by epimerization of UDP-glucuronate. GDP-mannuronate is thought to be synthesized from GDP-mannose in a similar

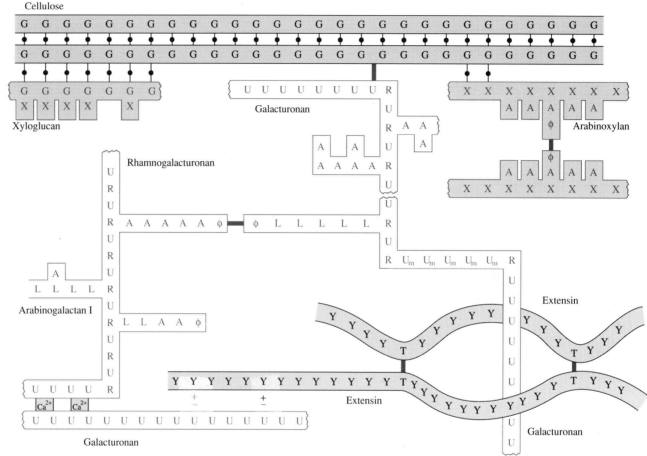

Fig. 17.6 Hypothetical representation of the various bonds between components within the primary cell wall of dicotyledons. Note that some features shown do not constitute essential features of primary cell wall structure (e.g. binding between hydroxyproline rich glycoprotein and various pectic substances). Also, orientation, spacing and chain lengths of various components are not intended to reflect these features *in vivo*. Abbreviations: A = arabinose; G = glucose; L = galactose; R = rhamnose; T = tyrosine; U = galacturonate; U_m = galacturonic acid methyl ester; X = xylose; Y = amino acid other than tyrosine; ϕ = ferulate. Notation of bonds: grey bar = covalent bonds; —●— = hydrogen bonds; Ca^{2+} = calcium bridges; $\pm$ = ionic bonds other than Ca^{2+}. (Bonds within components are not shown.) After Fry, S.C. (1986) *Annu. Rev. Plant Physiol.*, **37**, 165–86.

way to synthesis of UDP-glucuronate from UDP-glucose. UDP-pentoses can be formed from UDP-glucose. For example, UDP-xylose is formed by decarboxylation of UDP-glucuronate, and UDP-arabinose by epimerization of UDP-xylose. UDP-rhamnose is synthesized from UDP-glucose in an even more complex process involving reduction at C-6 and epimerizations at C-3, C-4 and C-5.

Another, quantitatively minor, pathway is the synthesis of UDP-glucuronate from glucose-6-P and inositol (Fig. 17.8). The first reaction, the synthesis of inositol-1-P from glucose-6-P, is exceedingly complex and involves oxidation by NAD^+ to form a transitory intermediate which is subsequently reduced by NADH,

with no net change of NAD. Inositol is thought to be an important source for the synthesis of some matrix polysaccharides, especially in primary cell walls.

17.8 Cellulose microfibrils are thought to be formed by enzyme assemblies associated with the plasma membrane

Cellulose synthesis refers to the production of the polymer of $\beta(1\rightarrow4)$ glucose, regardless of its physical organization. The formation of microfibrils means the

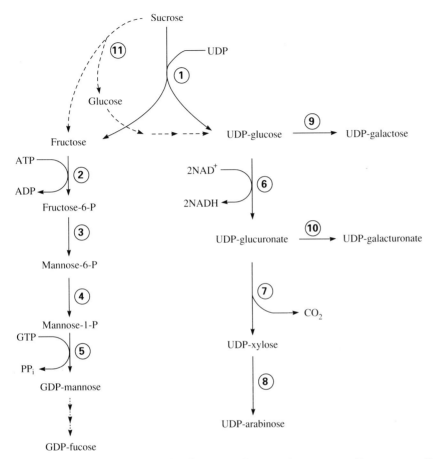

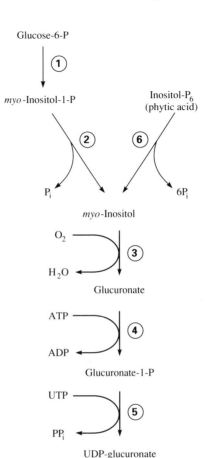

Fig. 17.7 Pathways for the synthesis of some NDP-sugars from sucrose. Enzymes are: (1) sucrose synthase; (2) hexokinase; (3) mannose-6-P isomerase; (4) mannose-P mutase; (5) GDP-mannose pyrophosphorylase; (6) UDP-glucose dehydrogenase; (7) UDP-glucuronate decarboxylase; (8) UDP-xylose 4-epimerase; (9) UDP-glucose 4-epimerase; (10) UDP-glucuronate 4-epimerase; (11) invertase. Production of UDP-glucose from free glucose and of GDP-fucose from GDP-mannose are not shown.

Fig. 17.8 Pathway for the metabolism of inositol to NDP-sugars, which can be incorporated into plant cell walls. Enzymes are: (1) glucose-6-P cycloaldolase; (2) *myo*-inositol-1-P phosphatase; (3) *myo*-inositol oxygenase; (4) glucuronate kinase; (5) UDP-glucuronate pyrophosphorylase; (6) phytase. (UDP-glucuronate can be further metabolized to UDP-xylose, UDP-arabinose and UDP-galacturonate as shown in Fig. 17.7.)

organization of the constituent cellulose molecules. It is thought that the two events are integrated *in vivo*.

Cellulose synthesis and microfibril formation must occur within or outside the plasma membrane since there is no evidence for them in the cytoplasm. Cellulose synthesis is thought to occur within the plasma membrane. The enzyme, cellulose synthase, is supplied with NDP-glucose from the cytoplasm and adds glucosyl residues in $\beta(1\rightarrow4)$ linkage to a growing cellulose chain, which emerges from the other side of the membrane.

This proposal is consistent with the enzymic synthesis of other macromolecules. It is unlikely that one molecule of cellulose synthase systematically adds glucosyl residues to 70 or so growing cellulose molecules to form a growing microfibril. It has been proposed that oligomeric assemblies of cellulose synthase support the simultaneous synthesis of several cellulose molecules which aggregate into microfibrils (Fig. 17.9). This hypothesis predicts that the cellulose molecules in the resulting

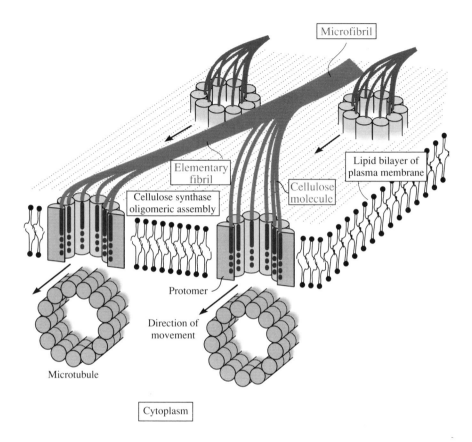

Microfibril

Elementary
fibril

Cellulose synthase
oligomeric assembly

Cellulose
molecule

Lipid bilayer of
plasma membrane

Protomer

Direction of
movement

Microtubule

Cytoplasm

Fig. 17.9 Speculative model for cellulose synthesis in the plasma membrane of plants. Cellulose synthase is shown as an oligomeric assembly of protomers, each of which synthesizes a cellulose molecule that aggregates with others to form an elementary fibril. Fibrils associate with one another to form a native microfibril containing 50–70 cellulose molecules. In a related model (not illustrated here), cellulose synthase is depicted as an assembly of six particles arranged in a rosette, each of which synthesizes six molecules of cellulose (i.e. 36 cellulose molecules per assembly). This model could produce sufficient cellulose molecules to form a native microfibril without involvement of elementary microfibrils. The direction of microfibril elongation is thought to be determined by the orientation of cytoplasmic microtubules, located about 20 nm from the plasma membrane and spaced at 40 nm intervals. It is envisaged that the oligomeric assembly moves in the plane of the plasma membrane causing microfibril formation in the direction shown by the arrows. There is little direct evidence to support the proposed location of functional cellulose synthase oligomers in the plasma membrane.

fibril are parallel. The hypothetical assemblies of cellulose synthase ('terminal complexes') may give rise to elementary fibrils which aggregate to form native microfibrils or they may contain sufficient enzyme molecules to simultaneously produce the 50–70 cellulose molecules present in a microfibril.

The synthesis of cellulose is best understood in the bacterium *Acetobacter xylinum*, which produces an extracellular pellicle of cellulose microfibrils when in liquid culture. An enzyme which catalyses the synthesis of a $\beta(1\rightarrow4)$glucose polymer from UDP-glucose can be extracted by the detergent digitonin from the membrane fraction of cells ruptured in a French press. In the presence of GTP and a separate protein factor from the membrane, the extract incorporates UDP-glucose (but not GDP-glucose) into a product similar to $\beta(1\rightarrow4)$glucan at rates approximating those of cellulose synthesis in intact cells. In the absence of either GTP or the membrane protein the rate is very low. The protein is a specific guanyl cyclase which converts two molecules of GTP to bis($3'\rightarrow5'$)-cyclic diguanylic acid and activates the cellulose synthase.

bis ($3'\rightarrow5'$)-cyclic diguanylic acid

Control of cellulose production in *Acetobacter* must involve regulation of guanyl cyclase and a membrane-bound enzyme to degrade diguanylic acid, inhibited by Ca^{2+}.

Cellulose, made by the extracted enzyme from *Acetobacter*, has fibrils about 1.7 nm in diameter, considerably less than the diameter of microfibrils *in vivo*. However, intact cells produce fibrils of similar size in the presence of calcofluor and Congo red. The smaller diameter fibrils are thought to represent a lower degree of organization. Perhaps cellulose molecules spontaneously aggregate, dehydrate and crystallize to form this rudimentary fibrillar structure which, in the right conditions, interacts laterally with others to form a native microfibril. Aggregation of the rudimentary fibrils may be inhibited by calcofluor and Congo red.

Little is known about cellulose synthase in plants. Reports dating from the 1960s, describing cellulose synthesis from GDP-glucose *in vitro*, are now thought to have been describing formation of a polymer of $\beta(1\rightarrow3)$glucose (callose) and various mixed linkage polysaccharides. Furthermore, the rates are too slow to account for synthesis *in vivo* and recent attempts to demonstrate cellulose synthase using the procedures employed in *Acetobacter* have been unsuccessful. The technique widely used to determine cellulose synthase activity involves measuring the incorporation of ^{14}C-label from UDP-[^{14}C]glucose into $\beta(1\rightarrow4)$glucan by membrane extracts. The rates obtained with this and other NDP-[^{14}C]glucose compounds have been negligible, perhaps because the enzyme is unstable in solution and readily inactivated or because cofactors or regulatory enzymes (analogous to guanyl cyclase in *Acetobacter*) are lost or inactivated during extraction.

Production of cellulose by a plant cell stops when the cell is stressed (e.g. following physical injury) and synthesis of callose, a $\beta(1\rightarrow3)$glucan, increases correspondingly. This inverse correlation of the two structurally related polymers occurs under a variety of conditions. Fractions prepared from the plasma membrane with digitonin catalyse the synthesis of callose from UDP-glucose at rapid rates. Perhaps cellulose and callose are synthesized by a common protein and the extraction procedures result in expression of callose synthesis only.

The proposed assemblies of cellulose synthase in the plasma membrane should be detectable with the electron microscope. There is good evidence for their existence in cellulosic algae (e.g. *Chaetomorpha*) with large microfibrils. The evidence in higher plants with smaller microfibrils is indefinite, although recent reports of rosette-shaped impressions in freeze-fractured plasma membranes may support the idea. It is not clear whether these rosette structures could simultaneously produce sufficient cellulose molecules to form entire microfibrils.

The orientation of cellulose microfibrils is thought to involve microtubules in the cytoplasm (see Section 17.6). Cellulose synthase assemblies could move within the membrane as they synthesize cellulose in a direction determined by the underlying microtubules—the microtubules in the cytoplasm act as tracks guiding the assemblies as they move within the membrane (Fig. 17.9). Perhaps this movement of the assemblies is determined by the growth of the microfibrils themselves, although other explanations are possible. Since microtubules are polymers of tubulin (see Section 3.9), then logically microfibril orientation could be controlled by the pattern of tubulin polymerization in the cytoplasm.

An alternative hypothesis for formation of microfibrils is that they arise by random crystallization of independently synthesized cellulose molecules ('self-assembly'). Although this model does not require cellulose synthase assemblies, it does not explain microfibril orientation; for this reason, the enzyme assembly hypothesis is more popular.

17.9 Matrix polysaccharides are synthesized in Golgi bodies and the endoplasmic reticulum

Several systems incorporate NDP-sugars and NDP-sugar derivatives into matrix polysaccharides. Particulate preparations incorporate ^{14}C-label from UDP-[^{14}C]-galacturonate into pectic polysaccharides, and sugar residues from UDP-xylose and UDP-glucose into xyloglucan. Control of the sequence and arrangement of the monosaccharides in the matrix polymers is not understood. The arabinose and galactose sequences of neutral pectins (e.g. Fig. 17.1b) are synthesized independently of the rhamnogalacturonan backbone and are subsequently joined together. Also, some monosaccharide residues in some matrix polysaccharides are modified after formation of the polymer. For example, the carboxyl groups of galacturonate and glucuronate residues in acidic pectins form *O*-methyl esters using *S*-adenosylmethionine as the methyl group donor (see Section 14.3.1).

The enzymes responsible for the synthesis of matrix polysaccharides occur in a membrane fraction derived from Golgi bodies. The data in Fig. 17.10 show that when extracts of etiolated pea seedlings are equilibrated on a sucrose density gradient, two turbid (paticulate) fractions are resolved; one equilibrates at a density of $1.15\,\mathrm{g\,cm^{-3}}$ and the other at $1.18–1.19\,\mathrm{g\,cm^{-3}}$. Because of its density, and the presence of cytochrome oxidase, the latter fraction is attributed to mitochondria (see Table 3.5). The lighter fraction incorporates UDP-[^{14}C]glucose and GDP-[^{14}C] glucose into a product with the characteristics of a hemicellulosic glucan; examination in the electron microscope shows that it consists primarily of Golgi bodies.

Golgi bodies contain various polysaccharides which, when hydrolysed, yield monosaccharides characteristic of pectic substances and hemicelluloses. Also, fractions of peas containing Golgi-derived vesicles support the synthesis of xyloglucan from UDP-xylose and UDP-glucose. Thus, cytoplasmic material is incorporated into matrix polysaccharides by an enzyme complex in Golgi bodies which accumulates in Golgi vesicles. The mechanism for transfer of the contents of Golgi vesicles to the cell wall is unclear, but electron microscope evidence shows that they fuse with the plasma membrane. Homology of these membranes is consistent with this. A mechanism must exist to determine the sites where Golgi vesicles fuse since cells maintain strict control over the synthesis of their walls.

The endoplasmic reticulum is also thought to contribute to the synthesis of matrix polysaccharides and their incorporation into walls as it is closely associated with the sites of cell plate formation, callose deposition and newly forming walls. Some electron micrographs show connections between the endoplasmic reticulum

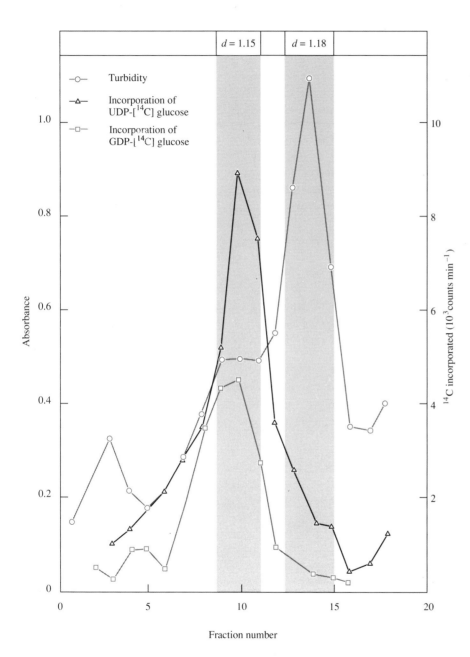

and plasma membrane and there is some indirect evidence for the presence of material within the lumen. It may be that enzymes involved in matrix polysaccharide synthesis are transported by the endoplasmic reticulum by membrane flow from their site of synthesis on the rough endoplasmic reticulum to the plasma membrane. Certain matrix polysaccharides could be transported in the same way.

17.10 Extensin synthesis involves post-translational oxidation of proline and attachment of sugars

Little is known about the synthesis of the extensins which constitute the bulk of the protein in plant cell walls. Separate genes have been found for individual HRGPs: one in tomatoes is only expressed after wounding. Studies of amino acid incorporation and of the DNA sequences encoding these proteins show that hydroxyproline residues in the completed proteins are incorporated as proline on the rough endoplasmic reticulum; hydroxylation of specific prolyl residues occurs subsequent to peptide synthesis (i.e. 'post-translational modification'). Hydroxylation involves a prolyl hydroxylase in the endoplasmic reticulum in a reaction requiring O_2, Fe^{2+}, ascorbate and 2-oxoglutarate. One atom of oxygen is incorporated into the hydroxyproline residue and the other is reduced to H_2O by ascorbate. Glycosylation of the peptide to form a soluble glycoprotein is presumed to occur in Golgi vesicles. In the extensins, three or four arabinose residues are attached to the hydroxy group of hydroxyproline and galactose is added to the hydroxy group of serine. In HRGPs associated with animal tissues, N-acetylglucosamine is added to the amide nitrogen of asparagine. This implies the existence of some sort of glycosylation recognition code, especially since not all hydroxyproline, serine and asparagine residues in proteins are glycosylated. A specific sequence of amino acids in a peptide ('sequon') appears to provide the signal for glycosylation. For glycosylation of asparagine by N-acetylglucosamine, the sequence is thought to be asparagine–X–serine/threonine, where X is variable.

Fig. 17.10 Association of hemicellulose synthesis with Golgi bodies. Extracts of etiolated pea seedlings were fractionated by isopycnic centrifugation in a sucrose density gradient (20–50% w/v) and analysed for turbidity (particulate material) and cytochrome oxidase (mitochondrial marker). Turbid fractions with a density of $1.15 \, \text{g cm}^{-3}$ contain principally Golgi bodies and support the incorporation of UDP-[^{14}C]glucose and GDP-[^{14}C]glucose into hemicellulosic glucan. After Ray, P.M., Shininger, T.L. & Ray, M.M. (1969) *Proc. Natl. Acad. Sci. Wash.*, **64**, 602–12.

Presumably, all or part of the pentapeptide, serine–(hydroxyproline)$_4$, provides the necessary message for the specific glycosylation of the extensins. Presumably a different sequence is involved in AGPs in which certain hydroxyproline residues are glycosylated with galactose, rather than with arabinose.

The addition of extensin to the matrix follows the mechanism described for the matrix polysaccharides. This involves transport of a soluble precursor in vesicles to the plasma membrane and their release into the cell matrix. The final production of the insoluble form of extensin involves oxidative dimerization between tyrosine residues to form isodityrosine linkages (see Fig. 17.5), probably through the action of a cell wall peroxidase.

17.11 Lignin is synthesized from phenylalanine

Lignin, the stiffening material of most secondary cell walls, is synthesized from phenylalanine produced via the shikimate pathway (see Section 13.7.4). Phenylalanine and its precursors are readily incorporated into lignin. In grasses, tyrosine also serves as a lignin precursor. The aromatic ring of shikimate is incorporated into lignin without modification and is not then turned over significantly.

Incorporation of phenylalanine into lignin first involves deamination to form cinnamate, catalysed by phenylalanine ammonia lyase (Fig. 17.11) and resulting in formation of a *trans* double bond between C-2 and C-3 in the C$_3$ side-chain. Cinnamate is metabolized to 4-coumarate by a mixed function oxygenase called cinnamate 4-hydroxylase which uses both O$_2$ and NADPH as substrates. (A mixed function oxygenase supports the concomitant oxidation by O$_2$ of two substrates). Both cinnamate and 4-coumarate are readily incorporated into lignin. In grasses, 4-coumarate is also formed directly from tyrosine in a reaction catalysed by tyrosine ammonia lyase (Fig. 17.11).

The phenylpropanoid compounds comprising lignin are formed from 4-coumarate. Modifications to the ring structure are made prior to reduction of the carboxyl group of the C$_3$ side-chain. Hydroxy groups are introduced at the C-3 and, less frequently, C-5 positions of the ring and are subsequently methylated with *S*-adenosylmethionine as methyl donor to form ferulate and sinapate (Fig. 17.12). Coumarate and its methoxy derivatives are reduced to their corresponding alcohols via the mechanism shown for 4-coumarate (Fig. 17.11). 4-Coumarate (and ferulate and sinapate) is activated to give the coenzyme A thioester derivative, involving coumarate:coenzyme A ligase and the expenditure of ATP. The CoA derivatives are thought to be reduced to their free aldehydes by NADPH and then reduced by

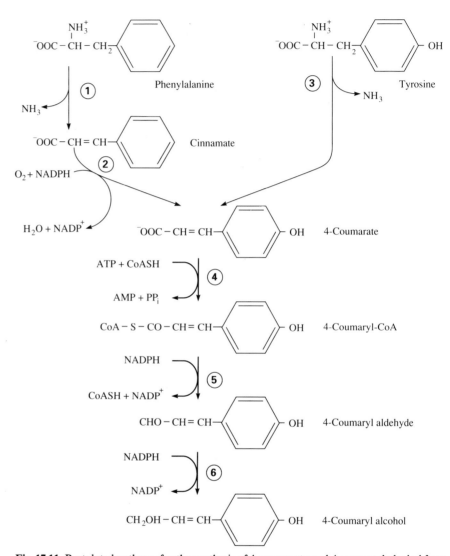

Fig. 17.11 Postulated pathway for the synthesis of 4-coumarate and 4-coumaryl alcohol from phenylalanine and, in grasses, tyrosine. All compounds with a double bond in the C$_3$ side-chain have the *trans* configuration. 4-Coumarate is thought to be the principal substrate for the synthesis of coniferyl alcohol and sinapyl alcohol. Enzymes are: (1) phenylalanine ammonia lyase; (2) cinnamate 4-hydroxylase; (3) tyrosine ammonia lyase; (4) 4-coumarate:coenzyme A ligase; (5) 4-coumaryl-CoA:NADPH oxidoreductase; (6) 4-coumaryl alcohol:NADP oxidoreductase. Few details for reactions (4)–(6) are available.

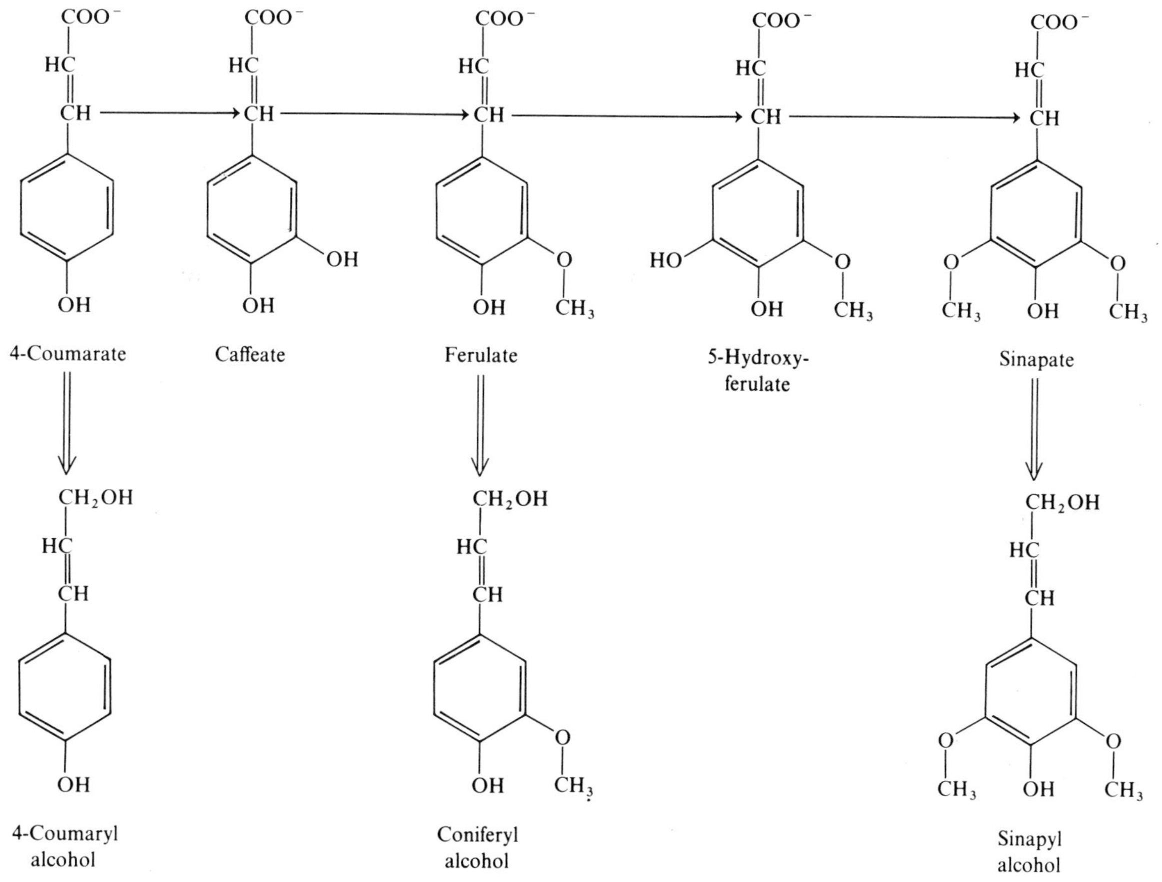

Fig. 17.12 Presumed pathway for the production of 4-coumaryl alcohol, coniferyl alcohol and sinapyl alcohol from 4-coumarate.

another NADPH to form alcohols. However, the nature of these reactions and the enzymes involved are very uncertain. The alcohols found in lignin also form β-1-glucoside derivatives through the hydroxy group at C-4 on the aromatic ring; their function is uncertain.

The phenylpropanoid alcohols are formed within the cell, possibly in the endoplasmic reticulum and Golgi bodies. Oxidation and polymerization of the phenylpropanoid alcohols (PH) occur outside the plasma membrane but the mechanisms are unclear; oxidation at the double bond in the side-chain by H_2O_2, catalysed by a cell wall peroxidase is the most popular proposal.

$$PH + H_2O_2 \rightleftharpoons POH + H_2O \qquad (17.2)$$

H_2O_2 is found in plant cell walls, although its origin is not clear. In theory, the oxidation product (POH) could polymerize, either enzymically or non-enzymically, in a non-specific way by forming links from the modified side-chain with the side-chains or rings of unoxidized phenylpropanoid compounds. Alternatively, free radicals could form, catalysed by the copper-containing enzyme, monophenol oxidase (denoted as Enz in the reactions below), and reducing the copper, which is then reoxidized by molecular O_2.

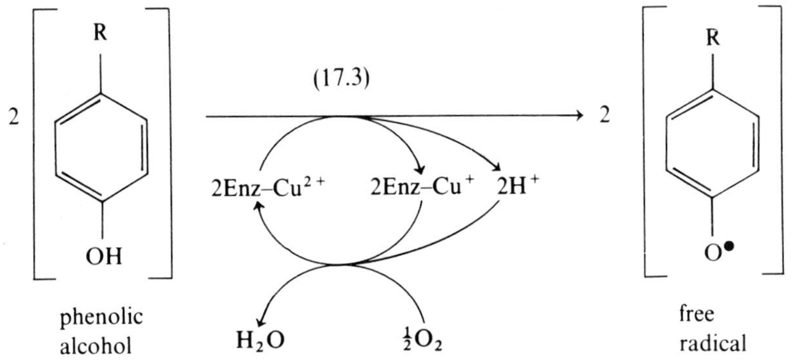

Theoretically, the free radicals can react non-enzymically with other molecules to form addition products (Fig. 17.13). Consecutive reactions could lead to the

formation of the spreading net-like polymer found *in vivo* (see Fig. 17.3). Supporting this, monophenol oxidase *in vitro* catalyses the formation of a lignin-like polymer from coniferyl alcohol under aerobic conditions.

Significant deposition of lignin in cell walls does not begin until primary growth is completed. Control of the initiation of lignin synthesis is not understood. The activity of phenylalanine ammonia lyase, the first enzyme of phenylalanine metabolism (Fig. 17.11), is strongly influenced by day length and other environmental events via a specialized pigment, phytochrome, and so, for lack of an alternative, this could regulate lignin synthesis.

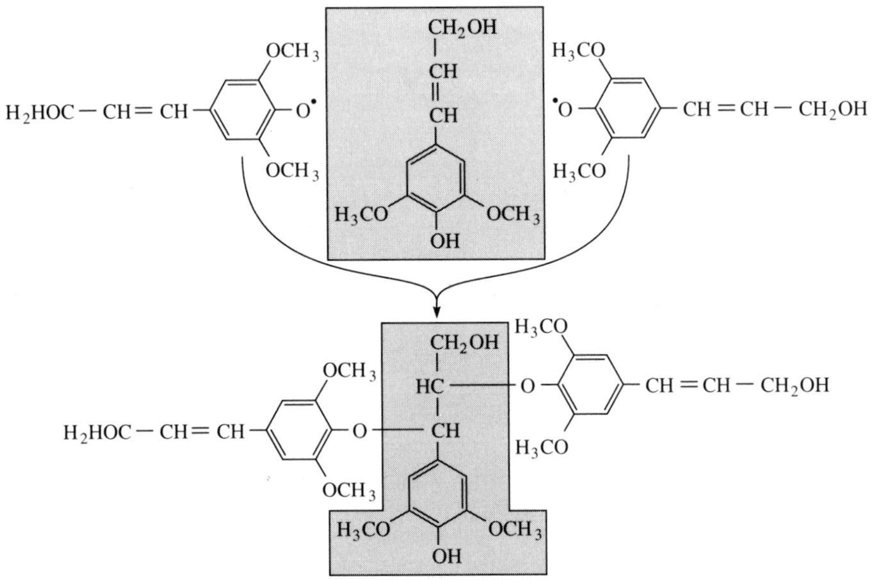

Further reading

General texts on carbohydrates and their roles in cell walls: Preiss (1980); Preiss (1988); Tanner & Loewus (1981); Tolbert (1980) Chapter 3.

Composition and synthesis of matrix polysaccharides and proteins: Cassab & Varner (1988); Fry (1986); Hayashi (1989); Marcus (1989) Chapter 12.

Microtubules and Golgi apparatus in wall synthesis: Gunning & Hardham (1982); Robinson (1985); Tolbert (1980) Chapter 12.

Fig. 17.13 Suggested mechanism for the non-enzymic coupling of two free radicals of sinapyl alcohol to an unmodified molecule of sinapyl alcohol (in shaded box), a process which could account for the formation of lignin. Double bonds of the trimeric product can react with further free radicals of various phenylpropanoid compounds to increase the size and cross-linking of the product.

18

The plant genome and its replication

All the information required for the formation of plant material and the biochemical activity of plant cells is referred to as the genome of the cell. The largest portion of this information is contained within the nucleus but at least two other organelles are repositories of some genetic information. This chapter examines how this information is stored and how it is replicated prior to its transmission from 'parent' to 'progeny' cells during cell division.

18.1 Organization of the plant genome

18.1.1 DNA is the genetic material

One of the most fundamental ideas in biology is that genetic information is carried in the nucleic acid of the cells of living organisms. As discussed in Chapter 19, this information is stored as the sequence of bases present in the nucleic acids. Sets of three bases code for particular amino acids. Thus, the sequence in which these bases occur in nucleic acids dictates the sequence in which amino acids are put together to form polypeptides. The portion of the DNA molecule coding for an individual polypeptide is known as a *gene* and the polypeptide for which it codes is the *gene product*. All the genes contained within a particular cell or organelle are referred to collectively as a *genome*, e.g. nuclear or chloroplastic genomes or the genome of *Pisum*.

The DNA molecules comprising the genome in the nucleus are organized into rod-like structures termed *chromosomes*. Most cells are diploid, i.e. there are two sets of each sort of chromosome. Germ cells, however, have only one of each chromosome and are termed haploid. The number of chromosomes per diploid cell is constant for a given species or variety of plant but is highly variable between species. Barley (*Hordeum vulgare*), for example, has 14 chromosomes and tobacco (*Nicotiana tabacum*) has 48, whilst for comparison, humans, the fungus *Penicillium* and chickens have 46, 2 and 78 chromosomes respectively. In many higher plants, polyploid forms exist in which the cells contain more than two sets of each chromosome.

18.1.2 Plant cells contain large amounts of DNA

The amount of DNA contained in plant nuclei has been estimated for many species by direct chemical analysis and by micro-densitometry of individual nuclei, stained specifically for DNA. These values are usually expressed as the amount of DNA present in a haploid cell. Values for a range of plants are given in Table 18.1, which also includes, for comparison, data for human, viral and bacterial (*Escherichia coli*) genomes. From Table 18.1, it can be seen that plants contain relatively large amounts of DNA even in comparison to the human genome. Values for plants range from only

Table 18.1 The range of genome size in plants. For comparison, equivalent values are given for the bacterium *Escherichia coli*, cauliflower mosaic virus and *Homo sapiens*.

Species	DNA content* (pg/haploid genome)	Single copy DNA (% of total DNA)
Cauliflower mosaic virus	0.84×10^{-5}	?
Escherichia coli	4.0×10^{-3}	>99.9
Marchantia polymorpha chloroplast	1.2×10^{-4}	83
Arabidopsis thaliana	0.5	?
Liriodendron tulipifera	0.8	52.5
Raphanus sativus	1.55	82
Nicotiana tabacum (tobacco)	2.0	45
Avena sativa (oats)	4.3	25
Pisum sativum (pea)	4.8	30
Triticum aestivum (wheat)	5.0	25
Hordeum vulgare (barley)	5.5	30
Vicia faba (broad bean)	14.5	20
Allium cepa (onion)	17.8	5
Lilium longiflorum	36.0	?
Viscum album (mistletoe)	107.0	?
Homo sapiens	6.0	?

*1 pg of DNA contains 10^9 base pairs and corresponds to a molecular mass of about 6.4×10^8 kDa.

0.5 pg DNA/haploid nucleus in *Arabidopsis* to 107 pg DNA/haploid nucleus in *Viscum album* (mistletoe), almost twenty times as much DNA as human cells! Even the smallest plant genome contains about five times the DNA content of the fruit fly, *Drosophila melanogaster*. One picogram of DNA contains approximately 10^9 pairs of the N-containing bases of DNA (referred to as 'base pairs'); the average gene is equivalent to 1000 base pairs. This means that a plant such as *Lilium longiflorum*, with about 36 pg DNA/haploid nucleus, contains enough DNA for 36 000 000 genes. A realistic estimate, based on biochemical and genetic studies, for the number of genes necessary to encode the proteins required for plant growth and functioning is only some 40 000–100 000, i.e. only 0.05–0.14% of the *L. longiflorum* genome!

18.1.3 *Plant nuclear DNA consists of many repeated sequences*

There is evidently a far greater amount of DNA present than is required to store *single copies* of the information required for forming each of the proteins essential for the development of plant structure and function. Many investigations have attempted to ascribe a function to the additional DNA. The best overall impression of genome organization comes from studies of DNA renaturation kinetics.

Double-stranded DNA can be denatured into two single strands by heating to above its melting temperature or by disrupting hydrogen bonds with alkali. Under suitable conditions, complementary strands cross-link via hydrogen bonds to form a double strand, a process known as reannealing. The rate of this reassociation is dependent on factors such as temperature, ionic concentration and the length of the DNA strands involved, as well as the concentration of complementary sequences. If the other parameters are standardized, then the rate of reassociation is dependent only on the concentration of complementary sequences. Thus, if a sequence is present a large number of times in a given population of DNA fragments, it reanneals more quickly than a sequence that is present only once. DNA is commonly examined by shearing it into double-stranded fragments of 200–400 base pairs which are then separated into single-stranded fragments. The mixture of sequences is then allowed to reanneal and the fraction of DNA reforming double strands is determined as a function of time. The proportion of the total DNA which is present as double-stranded DNA is usually determined by treating with S1 nuclease (which degrades only single-stranded DNA) or by the degree of binding of the DNA to hydroxyapatite (which does not bind single-stranded DNA). This data can be used to compare the complexity of the DNA of different species by plotting the fraction of double-stranded DNA as a function of the initial quantity of DNA (C_0) multiplied by the time (t), i.e. C_0t. Half the DNA is reannealed at $C_0t_{1/2}$. Comparison of $C_0t_{1/2}$ values for different species allows comparison of the complexity of their respective DNAs (Fig. 18.1). DNA with sequences which are often repeated reanneal faster and consequently have a lower $C_0t_{1/2}$ value than a similar amount of DNA composed of unique sequences.

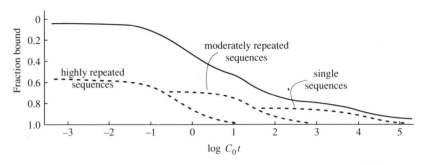

Fig. 18.1 The use of renaturation kinetics to determine the degree to which repeated sequences are represented in a sample of DNA. DNA from peas was denatured and sheared to lengths of about 300 nucleotides. The fragments were allowed to renature and the proportion of fragments able to bind to hydroxyapatite (double-stranded fragments) was plotted as a function of C_0t. The curve obtained (solid line) can be described mathematically by three derived curves (dashed lines). About 15% of the genome renatures with the slowest kinetics and represents single DNA sequences or sequences with few copies. The remaining 85% (in two classes) represents DNA which renatures faster and contains repeated sequences. Although the solid curve is resolved into three derivatives, this is probably an oversimplification and the range of repetition frequency can be more than 1000-fold. After Murray, M.G., Cuellar, R.E. & Thompson, W.F. (1978) *Biochemistry*, **17**, 5781–90.

Analysis of reassociation kinetics of plant DNA shows that much of the genome reanneals more rapidly than would be expected if it were composed of single copy or unique sequences. In fact, plant DNA appears to comprise:

1 Unique sequences (single copy DNA) of bases which mostly code for proteins with specialized functions (i.e. the structural genes).

2 Low and moderately repeated sequences which are reiterated from several to thousands of times. These are often interspersed with single copy DNA in an orderly arrangement along the chromosomes. Some of the DNA in this category codes for histones (Section 18.1.5) and some for ribosomal RNA.

3 Highly repeated sequences of DNA consisting of short sequences of some six to eight base pairs which are repeated from 10^5–10^7 times per genome. They have a different base composition from the rest of the DNA and are often associated at cell division with the region of the chromosome known as the centromere.

For onion, only about 5% of the DNA behaves as single copy or 'low repeat' sequences, although for most plants this value ranges from 20–40%, depending on genome size. Greater genome size is associated with a greater proportion of repetitive DNA.

18.1.4 Many plant genes have coding sequences interrupted by non-coding, intervening sequences

By a combination of gene cloning (see Section 20.5) and DNA sequencing, detailed analysis has been carried out on many plant genes. A number of these genes have their coding sequences interrupted by intervening, non-coding sequences called *introns*. Introns have been found, for example, in the phaseolin gene from French beans (see Fig. 20.8) and in the leghaemoglobin gene from soybean. In the yeast cytochrome *b* gene, introns play a vital role in the processing of primary RNA transcripts of DNA by coding for a 'maturase' or splicing enzyme (see Section 19.2). Their function in higher plant genes, however, remains obscure. Since introns separate parts of a gene, coding for different structural or functional regions (domains) of polypeptide chains, it has been suggested that introns could be important in evolution. Separating and spreading out the DNA sequences (known as exons) that code for the various domains of the gene increases the possibility of recombination between one gene sequence and another, to increase the probability of forming proteins with altered amino acid composition. However, no examples have been found of the exons from one gene corresponding to exons taken from other, different genes.

18.1.5 DNA is associated with proteins to form chromatin

The composition of DNA and its organization into a double helix has been discussed in Section 2.6. In common with other eukaryotes, the nuclei of plant cells contain components in addition to DNA; these include various enzymes involved in DNA replication, as well as structural proteins and RNA. In particular, nuclear DNA is associated with basic proteins termed histones and with acidic regulatory proteins in a ratio of DNA:histones:acidic proteins of approximately 1:1:0.6.

There are five histone proteins, termed H1, H2A, H2B, H3 and H4. All contain a high proportion of the amino acid arginine. H1 and, to a lesser extent, H2A and H2B have very variable amino acid sequences, in contrast to H4 in which the sequence is very similar in different organisms (e.g. the amino acid sequence of H4 is almost identical in peas and cows!). DNA combines with the histones to form structures known as nucleosomes (Fig. 18.2). Each nucleosome consists of two molecules of each of H2A, H2B, H3 and H4. These aggregate to form a flattened disc of 10–11 nm diameter and 6 nm depth, around which the DNA helix is wound approximately twice (equivalent to about 146 base pairs). H1 is attached to the

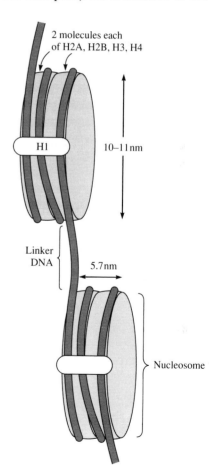

Fig. 18.2 Double-stranded DNA in the nucleus is associated with histones to form nucleosomes.

DNA strand where it joins the nucleosome. Adjacent nucleosomes are joined by regions of the DNA strand known as 'linker' DNA, the length of which varies between species and tissues. The DNA helix is thus associated into nucleosomes to form a structure rather analogous to a string of beads.

The major result of the binding of DNA into nucleosomes is that the DNA molecule is condensed into a structure known as a chromatin fibre with a length approximately one-seventh that of the naked DNA helix and a diameter of about 10 nm. The DNA in this structure is still several orders of magnitude less condensed than in a chromosome during cell division. It is believed that the nucleosome chain is twisted into a 'solenoid' (spring) structure and that this might be coiled and condensed even further. Although phosphorylation of H1 might be involved in the condensation of chromatin during mitosis it is not clear how the other histones could be involved in determining or modifying chromatin organization.

18.1.6 *Plant cell organelles have their own genomes*

The suggestion that plastids and mitochondria possess genetic information dates back to the concept of hereditary continuity, first proposed in the late 19th century. This theory suggested that new cells, and the organelles contained within them, could only arise from pre-existing cells and organelles. Thus, hereditary continuity demands that new plastids and mitochondria arise from pre-existing plastids and mitochondria. The simplest mechanism to satisfy this is a straightforward binary fission of these organelles and such a phenomenon has now been widely observed in both algae and higher plants (see Chapter 21).

The hereditary continuity of plastids and mitochondria, together with observations of 'non-nuclear inheritance', support the notion that these organelles possess their own genetic systems. For example, in *Mirabilis* it is possible to cross mutants which are deficient in a factor affecting chloroplast development (and hence having very pale plastids) with the green wild type. These crosses show that the progeny resemble the parent donating the ovum (i.e. the 'female' parent— whether wild type or mutant). That is to say, there is strict maternal inheritance of chloroplast characteristics which would not be expected from normal Mendelian inheritance involving the nuclear genome. Since the female parent contributes most of the cytoplasm and the subcellular organelles within it (in addition to a haploid nucleus), but the male parent (donating the pollen) contributes mainly nuclear material, this 'maternal inheritance' is often, though rather inaccurately, referred to as 'cytoplasmic inheritance'. In green plants, many of the features transmitted in this way are associated with chloroplast and mitochondrial development and function (see Section 21.4). This led to the idea that the genes concerned might be wholly or partly located in these organelles. Parallel experiments in fungi suggest that non-nuclear genes affecting mitochondrial development are actually located in mitochondria. It is now recognized that genetic systems of organelles are a fundamental feature of the organization of eukaryotic cells, although these systems do not contain all the information required for the complete development of the organelles.

The existence of DNA and RNA in chloroplasts has now been established without doubt by methods including electron microscopy and light microscopy using DNA-binding fluorescent dyes. Chloroplast DNA (ctDNA) accounts for 10–20% of the total leaf DNA in higher plants, although the corresponding value for root plastids is only about 1% of cellular DNA. With careful preparation techniques it has been shown that ctDNA is a circular molecule (the plastid genome or plastome) with a helix between 37–55 μm in length, depending on the species. There are no histones associated with ctDNA, nor does it contain methylcytosine which is a feature of nuclear DNA. The ctDNA is envisaged as being attached to chloroplast membranes but extending into the stroma. Several chloroplast DNA molecules are arranged together into regions termed nucleoids.

Between 10–30% of the ctDNA is further twisted about itself to form a 'supercoiled' structure. The molecular weight of ctDNA is, on average, about 90×10^6, corresponding to 135 000 base pairs. Values of plastome size for a range of species are given in Table 18.2. The entire chloroplast genomes from tobacco and the liverwort *Marchantia polymorpha* have recently been mapped and sequenced. The chloroplast genome of *Marchantia*, for example, consists of 121 024 base pairs with a set of large inverted repeat sequences, each of 10 058 base pairs, that encode chloroplast ribosomal RNA (rRNA) and some chloroplast transfer RNA species (tRNAs—see Section 2.6.3). The repeat sequences are separated by a small single copy region (SSC) of 19 813 base pairs and a large single copy region (LSC) of 81 095 base pairs (Fig. 18.3). In spinach, and some other higher plants, a sequence of 24 500 base pairs (coding for chloroplast rRNA) is repeated twice. Plastomes from most species contain sufficient information to code for about 100 proteins as well as for chloroplast rRNA and tRNAs. The coding functions of the plastome and the capacity of chloroplasts for protein synthesis are described in Chapter 21. Each chloroplast contains a single chromosome, present as multiple copies. The number of copies per chloroplast varies between species and during development; thus, pea chloroplasts from mature leaves normally contain about 14

Table 18.2 The range of plastome size in various species. From Bohnert, H.J., Crouse, E.J. & Schmitt, J.M. (1982) In Parthier, B. & Boulter, D. (eds) *Encyclopedia of Plant Physiology*, Vol. 14B, pp. 475–530. Springer Verlag, Berlin.

Species	Length of genome (µm)*	Number of base pairs (kb.p.)[†]
Euglena gracilis	40–44	125–143
Chlorella pyrenoidosa	27	85
Marchantia polymorpha (liverwort)	–	121
Avena sativa (oats)	37	–
Zea mays (maize)	43	–
Secale cereale (rye)	–	136–144
Triticum aestivum (wheat)	–	135
Atriplex triangularis	–	152
Lactuca sativa (lettuce)	44–46	–
Petunia hybrida	46	153
Pisum sativum (pea)	39	136
Spinacia oleracea (spinach)	46	145
Vicia faba (broad bean)	39	121

* Determined by electron microscopy.

[†] Determined by restriction endonuclease analysis.

copies of the plastome whilst *Euglena* chloroplasts have 67–80 copies. In very young leaves there can be up to 200 or more copies of the plastome per chloroplast.

The mitochondrial genome is also believed to be a circular, double-stranded DNA molecule. Although there are a number of reports of linear mitochondrial DNA molecules these are generally believed to arise from the degradation, during preparation, of a circular mitochondrial genome. Like chloroplasts, mitochondria contain their own transcription and translation apparatus, used to synthesize some of the polypeptides involved in ATP production. The size of the mitochondrial genome is highly variable and some values are given in Table 18.3. In general, the mitochondrial genomes of plants are larger than those from other organisms. Also, plant mitochondria contain more copies of the genome per organelle. Maize, for instance, contains seven DNA molecules per mitochondrion whilst humans contain only a single copy. In maize, the main mitochondrial DNA molecule or 'chromosome' consists of 570 000 base pairs containing six repeated sequences. The remaining six DNA molecules range from 47 000–503 000 base pairs in size and are thought to be derived from the main molecule. A similar situation occurs in *Brassica campestris*.

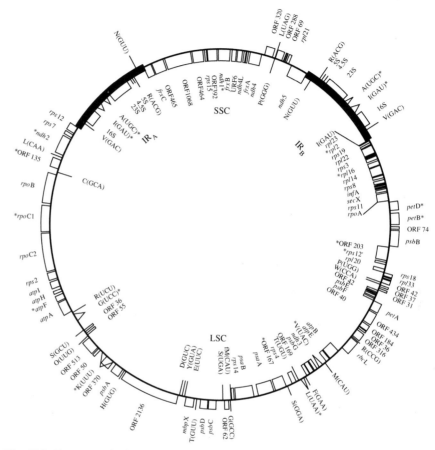

Fig. 18.3 Gene organization of the chloroplast genome from the liverwort, *Marchantia polymorpha*. The genome is organized into two inverted repeat sequences, IR_A and IR_B, which are separated by a short single copy region (SSC) and a large single copy region (LSC). There were 128 possible genes detected, including coding sequences for four rRNAs, 32 tRNAs and 55 identified proteins. Genes identified included: *rpl* and *rps*, 50 S and 30 S ribosomal proteins respectively; *atp*, subunits of H^+-ATPase; *psa*, proteins in PSI; *psb*, proteins in PSII; *pet*, cytochrome b_6/f complex; *rbc*L, the RuP_2-carboxylase large subunit; and *rpo*, genes for RNA polymerase. tRNA genes are indicated by their one letter amino acid code (see Table 19.1) with anticodons in brackets. Other abbreviations are: ORF, open reading frame (unidentified protein); *frx*, open reading frame for proteins homologous to 4Fe–4S proteins; *inf* A, initiation factor; *mbp*, sequences homologous to that for an inner membrane permease from *E. coli* From Ohyama, K. *et al.* (1986) *Nature*, **322**, 572–4.

Table 18.3 Size of the mitochondrial genome in some plants, with those from *Homo sapiens* and the yeast *Saccharomyces cerevisiae* given for comparison. The mitochondrial genome is larger than that of the chloroplast, although it is more variable in size. From Grierson, D. & Covey, S. (1984) *Plant Molecular Biology*. Blackie, Glasgow.

Species	Size of genome (kb.p.)	Copies per organelle
Brassica campestris	218	3
Zea mays	570	7
Cucumis melo	2400	n.d.*
Saccharomyces cerevisiae	75	1
Homo sapiens	16	1

* n.d. = not determined.

The mitochondrial genome codes for a range of functions which are summarized in Table 18.4. It is worth noting that these coding sequences account for only a small proportion of the mitochondrial genome. The maize mitochondrial genome has been shown to contain sequences corresponding to those coding for some chloroplast genes. Although it is known that movement of DNA sequences between organelles (promiscuous DNA) occurs, it is not known how such transfer,

Table 18.4 Some of the main coding functions of the mitochondrial genome. The genome also codes for tRNA, a ribosomal protein and, in maize, contains sequences for chloroplast 16 S rRNA, chloroplast tRNAs and the large subunit of ribulose-1,5-P_2 carboxylase/oxygenase.

Gene product	Mol. mass (kDa)
Ribosomal RNAs	
26 S	1120–1160
18 S	690–780
5 S	39
α-subunit of F_1-ATPase	58
DCCD binding protein of F_0	8
Cytochrome *b* apoprotein	42.9
Cytochrome *c* oxidase	
subunit I	38
subunit II	34

Abbreviation: DCCD = dicyclohexylcarbodiimide.

which would account for the common sequences in chloroplasts and mitochondria, takes place.

18.2 Replication of the plant genome

18.2.1 DNA is replicated semi-conservatively

The DNA present in the chromosomes of living organisms must be duplicated during cell division to ensure that 'progeny' nuclei each receive the full complement of genetic information. The Watson–Crick hypothesis proposes that each strand of DNA in the double helix acts as a template on which complementary new strands are synthesized. According to this hypothesis, DNA replication gives rise to two 'daughter' molecules which are identical to the parent DNA but with each daughter molecule containing one intact strand from the parent molecule and one newly synthesized strand (i.e. semi-conservative replication of DNA). Evidence in support of this proposal was obtained in an elegant experiment by Meselson and Stahl who grew *E. coli* for several generations in media containing NH_4Cl as the sole nitrogen source. The NH_4Cl was labelled with ^{15}N (a 'heavy', stable isotope of nitrogen) and all the compounds containing nitrogen in the cell, including DNA, became labelled with ^{15}N. As a result, the DNA of such cells had a density about 1% greater than that of cells grown with unlabelled NH_4Cl (i.e. ^{14}N). DNA molecules labelled with ^{15}N and ^{14}N can be separated by high-speed centrifugation. Upon transfer of ^{15}N-labelled cells to fresh medium containing ^{14}N only, the DNA after one generation's growth was a hybrid of 'heavy' and 'light' DNA strands. After a second generation, double-stranded 'light' DNA also appeared. This is precisely what would be predicted if the 'parental' DNA molecule passed on its component DNA strands to its 'progeny' and this provides direct evidence for the semi-conservative replication of DNA.

18.2.2 Initiation of DNA synthesis occurs at many specific sites on the DNA molecule

Separation of the two strands of the DNA double helix takes place at a rate of about 10 μm h^{-1}. Most actively dividing plant cells complete a cycle of growth and division within 15–40 h, with DNA replication confined to a period of 7–11 h (see Table 3.4). However, the genome of *Vicia faba*, for instance, contains almost 9 m of DNA. If replication of this DNA started at one end and moved along the molecule at 10 μm h⁻¹, replication of the genome would take over a century! That this is not the case is perfectly evident and is due to the fact that, as with all organisms, replication begins at numerous points along the DNA molecule. This can be demonstrated by growing plant cells with radioactive precursors of DNA, extracting the DNA at intervals and examining it by autoradiography. Each molecule shows replication at hundreds of points with separation of DNA strands and replication occurring from the origins in opposite directions on each of the two strands. There can be from 5000–60 000 replication points ('forks') per diploid genome. Each segment between two replication forks is termed a replicon. Replicons are generally 20–30 μm in length, containing 60 000–90 000 base pairs. There are distinct 'families' of replicons with anything from 2–25 'families' in a given plant. Replicons in each family undergo DNA replication simultaneously but different families show different times of DNA replication. For example, in *Arabidopsis* one of the two families of replicons begins DNA replication 36 min before the other family.

18.2.3 DNA synthesis is initiated by nicking and unwinding of the double helix

Our understanding of the processes involved in the replication of higher plant DNA is somewhat scanty. Studies on animal systems and lower eukaryotes are more complete and the following discussion is based on these investigations. It is unlikely, however, that the replication of higher plant DNA differs drastically.

The first step in the process of DNA replication is the cutting of one of the DNA strands by endonuclease activity. This 'nicking' of the DNA is accompanied by the unwinding of the supercoiled double helix. This is an energy-dependent process, the separation of each base pair involving the expenditure of two molecules of ATP. The nicking and unwinding processes are believed to both be catalysed by the same unwinding enzyme called a 'helicase'. The relationship between DNA and the nucleosome during unwinding is not fully understood but since there is some data suggesting that newly synthesized DNA is more susceptible to nuclease attack than the more 'mature' molecule, it is possible that nucleosome configuration during replication is abnormal. It has been suggested that the nucleosomes in fact dissociate into two half-nucleosomes during DNA replication.

The separated strands of DNA are stabilized by interaction with several single-stranded DNA binding proteins (SS-binding protein), which bind with higher affinity to single-stranded DNA than to double-stranded DNA. Thus, the nicking

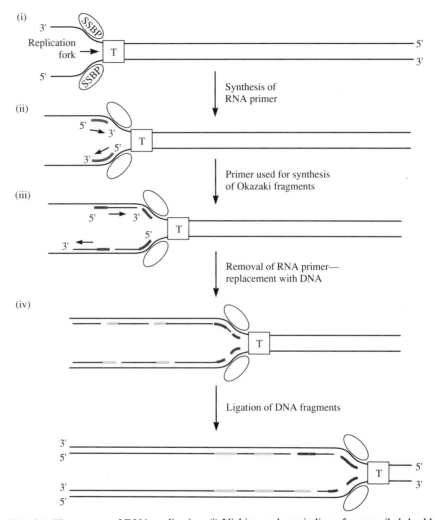

Fig. 18.4 The process of DNA replication. (i) Nicking and unwinding of supercoiled double helix by helicases. The two single strands of DNA are stabilized by single-stranded binding proteins (SSBP). These processes give rise to the replication fork. (ii) RNA primer (shown in dark grey) is synthesized by the action of primase. (iii) RNA primers are used as primers for DNA synthesis, catalysed by DNA polymerase (see Fig. 18.5), to form mixed DNA/RNA chains (Okazaki fragments). (iv) Primers are removed by nuclease action and gaps in the DNA are filled by DNA polymerase. (v) Adjacent DNA fragments are linked by the action of DNA ligase (Fig. 18.6).

and unwinding catalysed by DNA helicase gives rise to the replication fork (Fig. 18.4).

18.2.4 Separated single strands of DNA are used as a template for synthesis of new DNA

The two strands of DNA in a double helix run anti-parallel (see Section 2.6.3). Thus, as the two strands are separated, one of the strands is separated from the direction of the 5′-phosphate end in the direction of the 3′-hydroxyl end (5′→3′), whilst the other strand is separated from the 3′ end to the 5′ end (Fig. 18.4). The two strands are used as templates for the synthesis of new DNA.

The formation of a new DNA strand occurs under the direction of the single strand and involves the linking together of deoxyribonucleotides, catalysed by DNA polymerase. There are three DNA polymerase enzymes in higher plants. α-DNA polymerase is a soluble enzyme of high molecular mass (100–200 kDa), whilst β-DNA polymerase is a smaller enzyme (50 kDa) bound to chromatin. The third enzyme, γ-DNA polymerase, of molecular mass 100–150 kDa, is possibly associated with organelle DNA replication rather than that of nuclear DNA. The activities of α- and γ-DNA polymerases are highest during periods of DNA replication, suggesting that they play an important role in this process. The function of β-DNA polymerase is uncertain. Mutants lacking this enzyme appear to function normally. It has been suggested that β-DNA polymerase is associated

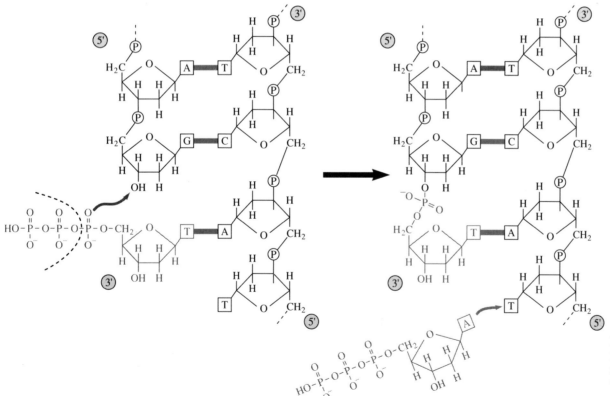

Fig. 18.5 Addition of deoxyribonucleotides to a growing DNA chain. Bases of incoming deoxyribonucleotides (grey) form hydrogen bonds with the corresponding bases of the 'parental' DNA strand. Abbreviations: A = adenine; T = thymine; G = guanine; C = cytosine.

with DNA repair. The α-, β- and γ-DNA polymerases are sometimes called DNA polymerases I, II and III, respectively. DNA polymerases act by adding successive deoxyribonucleotide units (dNTP) to the end of a DNA strand, the latter being denoted as (dNMP)$_n$ in Eqn. 18.1.

$$(dNMP)_n + dNTP \rightleftharpoons [dNMP]_{n+1} + PP_i. \qquad (18.1)$$

Thus, the end result of DNA polymerase action is a chain lengthened by one nucleotide unit with the elimination of pyrophosphate. The order of attachment of the nucleotides in a newly synthesized strand of DNA is determined by the base sequence in the parental strand which acts as a template. Incoming bases form hydrogen-bonded pairs such that adenine is matched with thymine and guanine is matched with cytosine. Thus, where the template has, say, a guanine residue, a cytosine residue will be inserted in the daughter strand (Fig. 18.5). DNA polymerase cannot initiate synthesis of a new DNA strand without a template of 'parental' single-stranded DNA. The DNA polymerase catalyses the covalent attachment of new deoxyribonucleotide units by the phosphate groups at the 5′ position to the terminal unsubstituted 3′-hydroxyl end of the pre-existing DNA chain (see Fig. 18.5). As a result, DNA polymerase catalyses the elongation of a new DNA chain in the 5′→3′ direction. DNA polymerase requires Mg^{2+} ions for activity and the active site contains tightly bound Zn^{2+}.

18.2.5 Synthesis is discontinuous and requires an RNA primer

DNA polymerase requires not only a single-stranded DNA template but also a 'priming' molecule. As noted above, DNA polymerase acts by the successive addition of nucleotides to a growing DNA chain by esterification of the 5′-phosphate group of the incoming nucleotide to a free 3′-hydroxyl group on the developing strand (Fig. 18.5). However, when the two strands of the 'parental' DNA molecule are first separated there is no pre-existing strand of growing DNA with

Fig. 18.6 The action of DNA ligase in linking the 5′-phosphate of one DNA fragment to the 3′-hydroxyl of an adjacent fragment. The reaction involves formation of an intermediate between an amino group of lysine in the DNA ligase (E) and an AMP moiety (derived from ATP with the elimination of PP$_i$). The AMP moiety is transferred to the 5′-phosphate group of one DNA fragment to form an ADP derivative (shaded), which reacts with the 3′-hydroxyl group of the adjacent fragment. Bases are indicated ①–⑥ and linking phosphate groups as ℗. Deoxyribose moieties in the DNA fragments are indicated by heavy black lines.

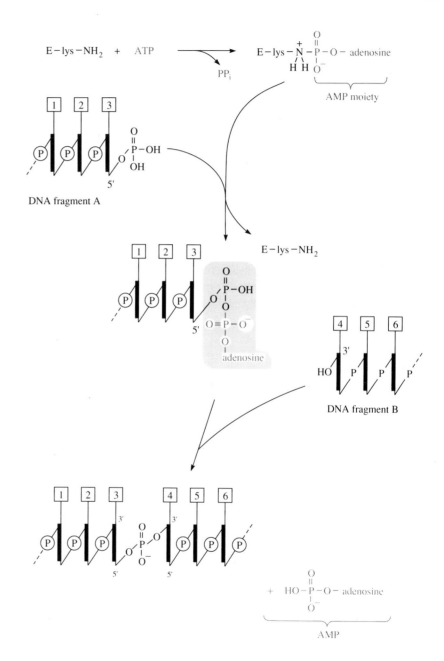

an available 3'-hydroxyl group to which additional nucleotides can be added. This difficulty is overcome by the synthesis, on both of the single-stranded DNA chains, of a short RNA primer of about 10 nucleotides in length. This primer is produced by base pairing of *ribonucleotides* with a specific sequence of bases on the single-stranded DNA at a particular site known as a primer region. Synthesis of the RNA primer is catalysed by a RNA polymerase (primase; see Fig. 18.4). DNA polymerase then acts to extend the primer in the 5'→3' direction. The end result of this activity is the production of Okazaki fragments (attached to the parent DNA strand) comprising both ribonucleotides and deoxyribonucleotides (Fig. 18.4). In soybean, for example, these fragments are about 200 nucleotides long, although only about 10 of these consist of ribonucleotides.

After synthesis of Okazaki fragments the RNA primers are removed by nuclease action and the resulting 'gaps' in the newly synthesized DNA strand are filled by DNA polymerase activity. This leaves a number of DNA fragments attached to the DNA template. These are finally joined by links between the 5'-phosphate of one fragment and the 3'-hydroxyl end of the adjacent segment. This reaction is catalysed by DNA ligase and involves the formation of an ADP derivative at the terminal 5'-phosphate of one fragment, which reacts with the 3'-hydroxyl group of the adjacent fragment (Fig. 18.6).

As the replication fork moves along the parental DNA molecule, so new RNA primers and Okazaki fragments are synthesized. Eventually, adjacent replicons are joined up and when all the replicons are ligated together, the whole DNA molecule is duplicated.

Following DNA replication, *S*-adenosylmethionine is used as a donor for the methylation of up to 25% of cytosine residues. This methylation is catalysed by DNA methylase. It is more common for cytosine to be methylated when it occurs in the sequences GC, CG, CAG or GTC. Methylation might be important in the regulation of the expression of the genes encoded in the DNA (see Chapter 20).

Further reading

Monographs and treatises: Alberts *et al.* (1989); Boulter & Parthier (1982); Grierson & Covey (1988); Marcus (1989); Parthier & Boulter (1982).
Nuclear genome organization: Heidecker & Messing (1986); Spiker (1985).
Organelle genome organization: Mullet (1988); Newton (1988).

19
Processes involved in protein synthesis

Growth and development of plants involves many different processes, each depending ultimately on the expression of a particular gene or genes. The end-products of gene expression are proteins used in a whole range of structural and catalytic roles in the cell. The central dogma of gene expression can be represented at its simplest as

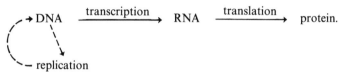

The information in the DNA of the genome is transcribed into RNA and then translated to protein using the ribosomal system. This chapter describes the biochemical processes involved in transcription and translation. The signals causing expression of particular genes and the way in which gene expression is controlled are dealt with in Chapter 20.

19.1 The information required to build proteins is contained in a genetic code

As described in Section 2.5.2, individual proteins are composed of a unique sequence of amino acids. Although most proteins consist of the same 20 amino acids, the sequence in which these are arranged in the polypeptide chain is crucial to the function of the protein. Any given amino acid may have a different function in different proteins depending upon its position and interaction with other amino acids or molecules. It is, therefore, imperative that the amino acids are laid down in the correct order. This is achieved using DNA as a template for the synthesis of messenger RNA (mRNA). mRNA acts as a template upon which ribosomes synthesize polypeptides. Studies on the amino acid sequences of proteins and the base sequences of mRNA have shown that each amino acid is coded for by a sequence of three bases (known as triplets) on the DNA or RNA chain. With four bases, adenine (A), guanine (G), cytosine (C) and thymine (T) (or uracil (U) in RNA), there are 64 possible nucleotide triplets or codons. The 'dictionary' of nucleotide triplets in mRNA and the amino acids for which they code are shown in Table 19.1.

Table 19.1 The genetic code used in plant nuclei and plastids. The code given here is that found in mRNA. Amino acids are denoted by their three letter codes and their single letter abbreviations (bold).

Base in first position (5′ end)	Base in second position				Base in third position (3′ end)
	U	C	A	G	
U	Phe (F)	Ser (S)	Tyr (Y)	Cys (C)	U
	Phe	Ser	Tyr	Cys	C
	Leu (L)	Ser	TERM*	TERM	A
	Leu	Ser	TERM	Trp (W)	G
C	Leu (L)	Pro (P)	His (H)	Arg (R)	U
	Leu	Pro	His	Arg	C
	Leu	Pro	Gln (Q)	Arg	A
	Leu	Pro	Gln	Arg	G
A	Ile (I)	Thr (T)	Asn (N)	Ser (S)	U
	Ile	Thr	Asn	Ser	C
	Ile	Thr	Lys (K)	Arg (R)	A
	Met (M)	Thr	Lys	Arg	G
G	Val (V)	Ala (A)	Asp (D)	Gly (G)	U
	Val	Ala	Asp	Gly	C
	Val	Ala	Glu (E)	Gly	A
	Val	Ala	Glu	Gly	G

*TERM refers to a translation terminator codon.

For example, the triplet UUU specifies phenylalanine whilst AAA codes for lysine. These codons are in turn recognized by a complementary group of bases on small RNA molecules known as transfer RNA (tRNA—see Section 2.6.3 for structural details). Each tRNA has its own characteristic base sequence and binds a specific amino acid. Whilst the first two bases in any given codon are highly specific, the third base is more flexible so, for example, phenylalanine can be specified by UUU, UUC, UUA or UUG. This flexibility or 'wobble' means that cells do not require the full complement of 64 tRNA molecules corresponding to all the possible triplets—many tRNAs recognize more than one of the 64 codons. Three of the codons (UAA, UAG and UGA) halt translation. Translation is initiated with a methionine residue but there is only one codon (AUG) to code both for initiator methionine and methionine within polypeptides. AUG sequences used as initiator codons are preceded by purine-rich sequences such as AGGA that might help to position the initiating AUG codon in the correct way to the ribosome. In the DNA of the gene, the base sequences corresponding to particular codons on mRNA are composed of the complementary bases. Thus, the base sequence AUGCCAG in mRNA is transcribed from a DNA base sequence of TACGGTC. Since one codon in mRNA would signify a different amino acid from that coded by the same sequence in DNA, it is important to specify whether the codon described is in DNA or in mRNA (e.g. GCC codes for alanine in the mRNA code but GCC in DNA would give rise to CGG in the corresponding mRNA and would be translated as arginine).

19.2 Transcription of DNA to RNA

19.2.1 RNA polymerase binds to DNA at a specific site

The first step in the expression of a gene is the transcription of DNA into RNA. RNA comprises ribonucleotides (see Section 2.6.2) joined by phosphodiester linkages between the 3′-hydroxyl of a ribose residue in one nucleotide and the 5′-hydroxyl of the ribose residue in another nucleotide (i.e. 3′→5′ linkage; Fig. 19.1). The DNA in the genome serves as a template for RNA synthesis as the bases of incoming ribonucleotides form hydrogen-bonded pairs with complementary bases in the DNA, i.e. 'base pairing'. The RNA molecules formed consequently have a base sequence complementary to that of the DNA so that C, G, A and T in the DNA give rise to G, C, U and A respectively in RNA. The enzymes responsible for RNA synthesis are RNA polymerases (or DNA-dependent ribonucleoside triphosphate:RNA nucleotidyl transferases). They catalyse the sequential addition of ribonucleotide units from ribonucleoside triphosphates (NTP) to the growing RNA chain, $(NMP)_n$ in Eqn. 19.1.

$$(NMP)_n + NTP \xrightarrow[\text{Mg}^{2+} \text{ or } \text{Mn}^{2+}]{\text{DNA template}} (NMP)_{n+1} \qquad (19.1)$$

The enzymes require a DNA template and Mg^{2+} or Mn^{2+} ions, as well as the ribonucleoside triphosphates.

RNA polymerases consist of a number of subunits (Section 19.2.3). In bacteria,

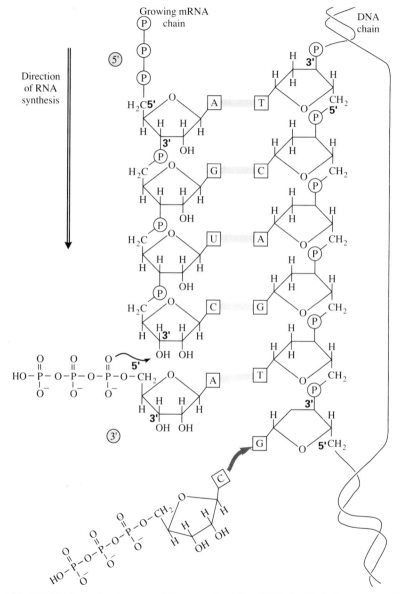

Fig. 19.1 RNA is formed using one of the strands of the DNA double helix as a template. The 5′→3′ phosphodiester bonds are shown as ℗. Note that the first base in the RNA chain retains its triphosphate residue.

one of these subunits, σ, acts as a recognition factor to select the correct site on the DNA to start transcription, although no analogous role for a subunit of RNA polymerase in plants has yet been demonstrated. The region of the DNA recognized as the site of initiation of transcription is the promotor site or 'TATA box'. In higher plants this is a sequence of bases represented as $T^C_G TATA^T_A A_{1-3}{}^C_T A$ (where

^C_G, ^T_A and ^C_T denote alternative base arrangements and the subscripts are the number of residues of adenine). The TATA box is situated about 16–54 nucleotides closer to the 3' end of the DNA transcribed chain, i.e. upstream from the 5' end of the transcribed RNA (Fig. 19.2). The rate of transcription can be increased by the CAAT box, some distance upstream from the TATA box. In plants, the CAAT

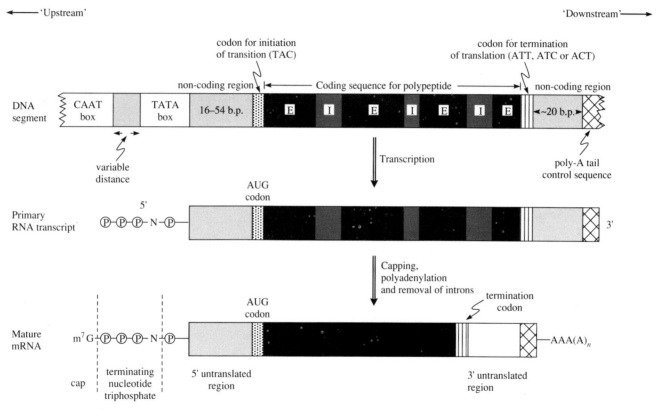

Fig. 19.2 A generalized scheme for the structure of a plant nuclear gene and its associated control sequences and for the corresponding primary transcripts and mature mRNA. The coding sequence for the polypeptide is preceded by a codon (TAC) for the initiation of translation. Upstream are a promoter site (TATA box) and also a CAAT box which controls the rate of transcription of RNA. After the coding sequence for the polypeptide has been transcribed into RNA, one of three codons indicates the signal for translation to stop. The site of addition of the poly-A tail, added to many mRNAs after transcription, is marked by a controlling sequence on the DNA, usually $^C_G ATAA_{1-3}$. Exons (E) may be interrupted by non-coding regions termed introns (I), depending on the gene in question, or may contain additional sequences for transport of the polypeptide to its correct site within the cell. Other non-coding regions are present. Following transcription the primary mRNA transcript is modified by: addition of a 7-methylguanosine (m^7G) cap, the addition of a poly-A tail and removal of introns (I). In the mature mRNA, the cap is separated from the polypeptide coding region by a short sequence of bases (5' untranslated region) and the poly-A tail is separated from the coding sequence by the 3' untranslated region.

sequence is variable but follows the general formula $^C_TA_{2-5}^C_TNGA_{2-4}^{CC}_{TT}$ (where N is any base).

19.2.2 RNA polymerase catalyses initiation of RNA synthesis and elongation of the RNA strand

Following binding of the RNA polymerase to the DNA molecule, which is still in the double helical configuration, a conformational change occurs, causing localized unwinding and separation of the DNA strands. The resultant single-stranded DNA is then available to pair with incoming ribonucleoside triphosphates.

Initiation of the RNA chain occurs when ATP attaches to this DNA by complementary base pairing between the adenine moiety and a specific thymine residue situated towards the 5' end, downstream from the promoter site. The ATP then reacts with the next incoming nucleotide as specified by the template. The initiating ATP retains the triphosphate residue and thus, the newly transcribed RNA chain starts with an entire ATP moiety (Fig. 19.1). RNA polymerase then catalyses the successive addition of further nucleoside-5'-monophosphate residues from their triphosphates according to the template.

Two points must be made about RNA synthesis.

1 Only one of the two single strands of double-stranded DNA is transcribed into RNA.

2 The RNA molecule grows from its 5' end in the 5'→3' direction. Synthesis proceeds from the 3' end of the template DNA or in an 'anti-parallel' direction. The DNA chain transcribed runs 3'→5' from the promotor site.

RNA synthesis stops when the RNA polymerase encounters a specific termination sequence (Fig. 19.2).

19.2.3 Three major classes of RNA are synthesized by three classes of RNA polymerase

There are three forms of RNA polymerase associated with eukaryotic nuclei: I, II and III. RNA polymerase I is situated in the nucleolus and consists of 6–10 polypeptide subunits of molecular masses ranging from 8–185 kDa. Unlike the other RNA polymerases it retains activity at low ionic strength and is resistant to the fungal toxin α-amanitin. RNA polymerase II, on the other hand, is sensitive to low concentrations of α-amanitin (50% inhibition of activity can be achieved by 0.01–0.05 μg ml^{-1}). It is located in the nucleoplasm and consists of 8–14 poly-

peptide subunits with molecular masses of 14–220 kDa. RNA polymerase III is also located in the nucleoplasm and consists of 8–14 polypeptide subunits with molecular masses of 16–155 kDa. It is sensitive to high concentrations of α-amanitin, requiring 50–1000 μg ml^{-1} for 50% inhibition of enzyme activity.

RNA polymerases I, II and III catalyse the synthesis of most ribosomal RNA (rRNA), all messenger RNA (mRNA) and all transfer RNA (tRNA), respectively. RNA polymerase III also catalyses the synthesis of the low molecular weight, 5 S rRNA.

19.2.4 Primary RNA transcripts are modified

The RNA species transcribed by RNA polymerases are not in the final state used for synthesis of protein; they undergo several modifications after transcription.

Many of the RNA molecules present in the nucleus are primary transcription products of nuclear genes and are precursors of messenger RNA (pre-mRNA). Each type of mRNA in the primary transcripts codes for only one species of polypeptide (i.e. they are monocistronic—a cistron being the sequence of codons for a particular protein) and ranges from several hundred to several thousand nucleotides in length. The primary transcripts are about 50% longer than the mRNA involved in protein synthesis because the genes for many proteins contain non-coding intervening sequences called introns (Section 18.1.4), which are also present in the pre-mRNA. The introns must be removed before the mRNA moves from the nucleus to the cytoplasm. How this is achieved in plants is uncertain. In yeast mitochondria, the gene for cytochrome *b* comprises six coding sequences interspersed with five introns. When transcribed, this gives a pre-mRNA containing 8000 bases, which is then modified to 7000 bases by excision of one intron by a nuclear-coded protein. This apparently permits a region of the rest of the precursor, including part of a remaining intron, to code for an enzyme termed a 'maturase'. This enzyme splices some of the coding regions (exons) to give an intermediate RNA molecule with 5000 bases. Further splicing removes the remaining introns. Thus, the introns are involved actively in the processing of pre-mRNA's. In mammalian cells, removal of introns and splicing of pre-mRNA involves four complexes of RNA and protein in the nucleus, termed small nuclear ribonucleoproteins (snRNPs). Each consists of a protein component and one or two of five different types of small nuclear RNA. The snRNPs bring the two ends of the intron into proximity, excise it and join up the then adjacent ends of the exons. There is, however, as yet no evidence for snRNPs in plants.

Messenger RNA is characteristically modified in two other ways. Most plant mRNAs possess a 7-methylguanosine residue joined to the end of the RNA molecule bearing a 5′ phosphate (the 5′ terminal). The methylguanosine (m⁷G) 'cap' is joined to the 5′ terminal of the RNA through an unusual 5′→5′ linkage, involving three phosphate groups (see below), and is added after transcription (Fig. 19.2). The

cap does not appear to be essential to the translation of mRNA but it might serve to bind a specific protein which has a role in regulating the initiation of protein synthesis. It might also protect the 5′ end of the mRNA from attack by various enzymes, such as phosphatases and nucleases. The other modification concerns the presence of a polyadenylate (poly-A) region, up to 200 nucleotides long, attached to the 3′ ends of many plant mRNAs. The poly-A 'tail' is also added after transcription, but before excision of introns and entry of the completed molecule into the cytoplasm. About 50% of mRNA molecules found in the cytosol contain these long poly-A tails, whilst the remainder possess shorter ones. The poly-A sequence confers stability on the mRNA, although stability might be increased by interaction between the mRNA and specific stabilizing proteins.

Plant tissues contain several million ribosomes per cell. These structures, found either free in the cytoplasm or attached to the membranes of the endoplasmic reticulum, consist of various rRNA components with an equal proportion of protein (see Section 3.3 and Table 3.3). In non-photosynthetic cells, the majority of ribosomes are responsible for the translation of nuclear DNA coded proteins and are of the typical eukaryotic 80 S type. They contain RNA subunits of 25 S, 5.8 S and 18 S assembled, with ribosomal proteins, to form the complete ribosome in the nucleolus (see Section 3.3). Mitochondrial and plastid ribosomes are considered in Section 19.5.

Ribosomal RNA accounts for up to 70% of the total cellular RNA and is synthesized at very high rates. Approximately 0.1–1% of plant DNA hybridizes (i.e. pairs with complementary bases) to the 18 S and 25 S rRNA extracted from 80 S ribosomes, implying that a significant amount of plant DNA is concerned with the synthesis of rRNA. The gene sequences controlling the formation of 18 S and 25 S subunits occur in equal numbers. If the cell division time is 15 h and there are several million ribosomes per cell, several hundred thousand ribosomes must be made every hour. Since the transcription of a rRNA molecule from DNA takes several minutes, the cell must have multiple sites involved in rRNA synthesis, achieved partly by multiple rRNA genes in the nucleoli. These genes (up to several hundred copies) are situated one after the other in the genome, arranged as tandem arrays. Such genes contain sequences for more than one product (i.e. they are polycistronic) and code for 18 S, 5.8 S and 25 S rRNA. They also have repeating transcribed and non-transcribed 'spacer' regions. Each DNA coding sequence is transcribed simultaneously by several molecules of RNA polymerase I. Plants also contain multiple copies of 5 S rRNA genes in tandem arrays, not linked to the 25 S, 5.8 S and 18 S rRNA genes. The structure and organization of the 5 S RNA gene has been investigated in wheat and two genes have been found with a size difference of about 410–500 base pairs due to non-coding or spacer DNA. The larger gene also has a small (15 base pair) region of duplication in the coding sequence which appears to render it inoperative. Immediately upstream from this sequence is a region, reading ATTAAG in the 3′→5′ direction, believed to play a role in the regulation of transcription (see Section 20.2.3). There is also a region following the 5′ end of the coding sequence enriched in AT residues. This is believed to be a termination sequence (see Fig. 19.2).

The transcripts of the polycistronic rRNA genes are precursors, containing one each of the 25 S, 5.8 S (not to be confused with the 5 S) and 18 S rRNA sequences, which together account for about 5660 nucleotides. The entire transcripts, however, contain 6500–10 000 nucleotides, the difference in length being accounted for in part by variability in length of the transcribed spacer. It is also possible that the nucleolar rRNA coding sequences contain introns, as is the case for rRNA from *Chlamydomonas* chloroplasts.

Each rRNA precursor molecule undergoes a series of maturation processes. Methylation, mainly of ribose, occurs at about 100 sites, present in regions of rRNA that are very similar in different species. This suggests that methylation is very important to fundamental features of rRNA function and ribosomal structure. Successive cleavage of the rRNA precursor yields intermediates of molecular mass 900–1000 and 1400–1450kDa. These are processed to yield the 18 S, 25 S and 5.8 S subunits. The 5.8 S subunit is derived from the same precursor as the 25 S rRNA; in fact they remain attached by hydrogen bonds even after the intervening linking sequences have been removed. The rRNAs, together with cytoplasmically synthesized proteins, are then assembled within the nucleolus into the 60 S and 40 S ribosomal subunits (Fig. 19.3).

Transfer RNA is also transcribed, by RNA polymerase III, as a precursor from multiple tRNA genes. Some tRNA precursors contain introns which are removed during processing. In many cases, the CCA sequence, characteristic of the 3'-hydroxyl end of all functional tRNA molecules, is added after tRNA synthesis. Individual tRNA molecules are also extensively modified to produce the rare bases which characterize tRNA (see Sections 2.6.3 & 19.3.1) and which improve binding of the tRNA molecule to ribosomes during protein synthesis.

The processing of RNA transcripts occurs in the nucleus. Processed RNA molecules (or, in the case of rRNA, the ribosomal subunits) are subsequently exported to the cytoplasm.

19.3 Translation of mRNA into protein occurs in the cytoplasm

19.3.1 Transfer RNA provides the recognition needed for translation of mRNA into polypeptides

Each amino acid in a polypeptide is encoded by one or more codons in the DNA molecule, mediated by the mRNA transcript. There is, however, no direct specific affinity between amino acids and the bases on the mRNA; this gap is filled by the tRNAs which 'line up' the amino acids at the mRNA template (see Section 19.1). Transfer RNA molecules have a region known as the anticodon, which contains the bases complementary to particular codons on mRNA. Very often the anticodon region contains modified nucleosides (e.g. inosine or 5-*N*-methylamino-methyl

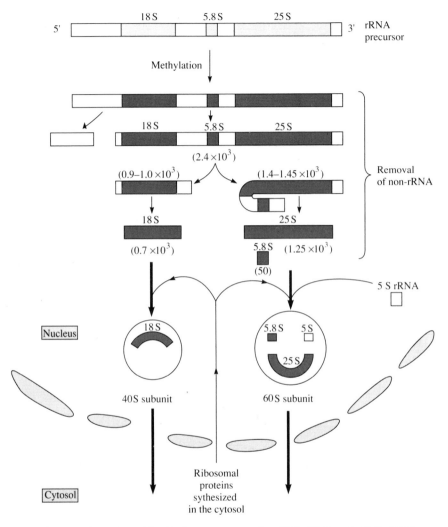

Fig. 19.3 Processing of rRNA. Coding sequences for 18 S, 5.8 S and 25 S rRNA are interrupted by various introns, removed during processing. 5 S rRNA is synthesized from genes elsewhere on the genome and is incorporated with other rRNAs and ribosomal proteins to form ribosomal subunits, which are exported to the cytosol. Values in brackets indicate molecular masses (kDa) of the rRNA components and precursors. After Grierson, D. (1982) In Parthier, B. & Boulter, D. (eds) *Encyclopedia of Plant Physiology*, Vol. 14B, pp. 192–233. Springer Verlag, Berlin.

2-thiouridine) in the position corresponding to the wobble position in the codon. Adenine and cytosine in the anticodon pair only with uracil and guanine respectively in the codon but inosine can pair with adenine, uracil or cytosine.

19.3.2 Aminoacyl-tRNA synthetases bind the specific amino acid to the relevant tRNA

The first step in protein synthesis in the cytoplasm is activation of the amino acids. This occurs when the relevant amino acids are esterified to their specific tRNAs by specific aminoacyl-tRNA synthetases, which catalyse the overall reaction shown in Eqn. 19.2.

$$\text{Amino acid} + \text{tRNA} + \text{ATP} \underset{}{\overset{\text{Mg}^{2+}}{\rightleftharpoons}} \text{aminoacyl} - \text{tRNA} + \text{AMP} + \text{PP}_i \quad (19.2)$$

The formation of an aminoacyl-adenylate complex as an enzyme-bound intermediate (Fig. 19.4) is a component step in this reaction. A reaction takes place between ATP and the amino acid, in which the carboxyl group of the amino acid is bound in an anhydride linkage to the 5'-phosphate of AMP, displacing PP_i. This is shown in Fig. 19.4a. Subsequently, the aminoacyl moiety is transferred from the enzyme to the specific tRNA (Fig. 19.4b).

The bond between tRNA and its amino acid is a high energy bond with a free energy of hydrolysis of $-29.2 \text{ kJ mol}^{-1}$. The pyrophosphate formed during activation (Fig. 19.4a) is hydrolysed to orthophosphate by the action of pyrophosphatase. As a result, the activation of every amino acid involves the expenditure of two high energy phosphate bonds. The overall reaction is essentially irreversible (Eqn. 19.3).

$$\text{Amino acid}^a + \text{tRNA}^a + \text{ATP} \underset{}{\overset{\text{Mg}^{2+}}{\rightleftharpoons}} \text{aminoacyl}^a - \text{tRNA}^a + \text{AMP} + 2P_i \quad (19.3)$$

The aminoacyl-tRNA synthetases, like DNA polymerases, have a 'proofreading' capacity. They are highly specific for both the tRNA and the amino acid. If the incorrect aminoacyl-adenylate is formed, this is detected and hydrolysed by the enzyme whilst it is still on the active site (Eqn. 19.4) so that the correct amino acid can be incorporated.

$$\text{Enzyme}^a\text{-[aminoacyl}^b\text{-adenylate]} + \text{H}_2\text{O} \rightleftharpoons \text{amino acid}^b + \text{AMP} + \text{enzyme}^a \quad (19.4)$$

The proofreading ability comes about because there are three specific sites, which function by recognizing the amino acid, the tRNA and ATP respectively. A fourth site permits the entry of water for hydrolysis of incorrect aminoacyl-adenylates. The

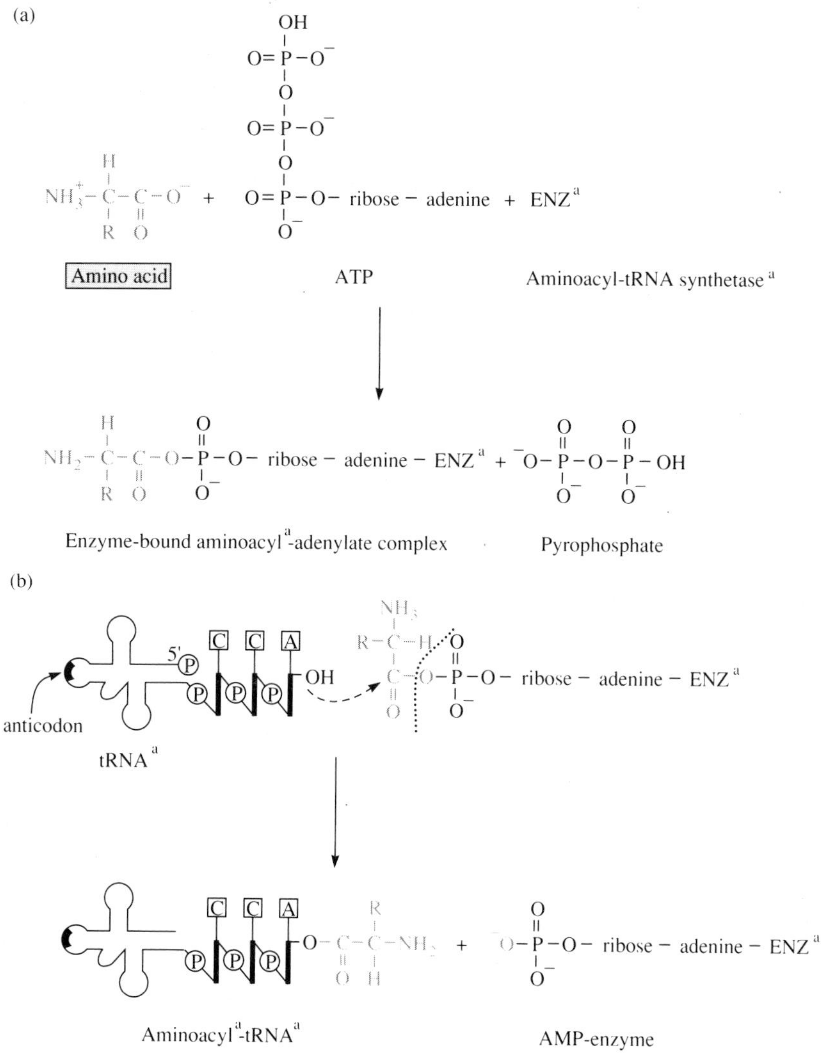

Fig. 19.4 Reactions involved in amino acid activation and attachment to tRNA. (a) The amino acid (shown in grey) is bound to the specific aminoacyl-tRNA synthetase (ENZ) as an aminoacyl-adenylate complex. (b) The aminoacyl moiety is transferred from the aminoacyl-tRNA synthetase to the specific tRNA for that amino acid (AMP and ENZ are released). Attachment of the amino acid is to the terminal adenine at the 3'-hydroxyl terminal of the tRNA. Ribose moieties in terminal nucleotides of tRNA are shown as heavy black lines.

specificity of the introduction of amino acids into a growing polypeptide is governed by the tRNA alone, with recognition between the codon on the mRNA and the anticodon loop (see Section 2.6.3, Fig. 2.21) on the tRNA paramount. This has been shown by an elegant experiment in which the aminoacyl-tRNA of cysteine (cysteinyl-tRNA^{cys}) was enzymically synthesized and the cysteine residue was then converted chemically to alanine. The aminoacyl-tRNA thus possessed the anticodon region corresponding to cysteine but was attached to alanine. A polypeptide, synthesized from this in a cell-free system, contained alanine in place of cysteine, illustrating that the aminoacyl group of an aminoacyl-tRNA is not recognized by the mRNA or by the ribosomes, despite the specificity of the codon/anticodon binding.

19.3.3 Initiation of a polypeptide chain involves a number of factors and occurs in several steps

Initiation of polypeptide chain synthesis in the cytoplasm requires: (i) the presence of the smaller 40 S ribosomal subunit (containing 18 S rRNA); (ii) the particular mRNA to code for the polypeptide; (iii) an initiating aminoacyl-tRNA (in the case of cytoplasmic protein synthesis in eukaryotes, methionyl-tRNA$_i^{met}$ where the met and i indicate that the tRNA involved is specific for an initiating methionine); (iv) a number of initiating factors (IF—up to six of which are involved in translation in wheatgerm); and (v) the trinucleotide, GTP.

The first step in the formation of the initiation complex is binding between the 40 S subunit and initiation factor 3 (IF3). This ensures that the small, 40 S ribosomal subunit does not combine with the large subunit at this stage. The 40 S subunit then binds to the mRNA so that a specific site on the 40 S subunit attaches to the initiating codon on the mRNA (AUG for methionine—Fig. 19.5). A special

Fig. 19.5 Initiation of synthesis of a polypeptide. (i) Binding of the 40 S subunit to initiation factor 3 (IF3). (ii) The 18 S RNA (shown in dark grey) in the 40 S ribosomal subunit binds to the initiating signal (IS) in the mRNA, aligning the 40 S subunit correctly. (iii) The complex [40 S subunit–IF3–mRNA] binds a second complex, [IF2–GTP–methionyl-tRNA]. The resulting methionyl-tRNA$_i^{met}$ complex then binds to the 60 S large ribosomal subunit, to give an intact and fully functional ribosome with an initiating methionyl-tRNA$_i^{met}$ attached to the initiating codon (AUG) at site P (peptidyl). (Other initiating factors may be required for optimal rates of translation; up to six such factors have been identified from wheatgerm, although the precise function of most of these is uncertain.)

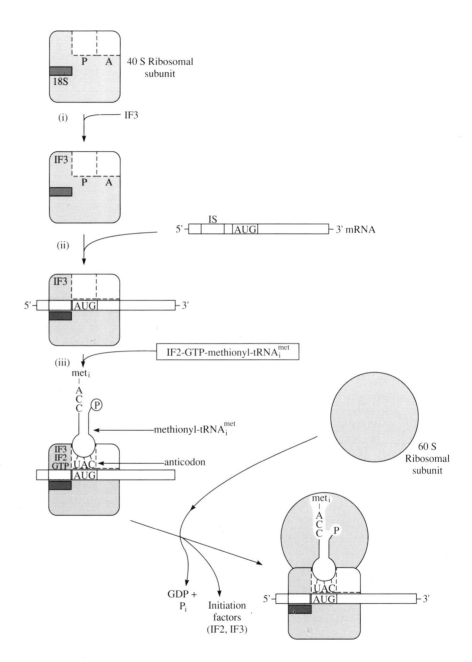

initiation signal in the mRNA helps to position the mRNA and ribosomal subunit correctly. In plants, this corresponds to a base sequence of $_G^C$AANNATGG in the genome (where N can be any base). The corresponding sequence in mRNA binds to a complementary sequence in the 18 S rRNA of the 40 S subunit.

In the next stage of initiation the complex of [40 S ribosomal subunit–IF3–mRNA] binds a complex of IF2, to which GTP and methionyl-tRNA$_i^{met}$ are attached (Fig. 19.5). The resulting structure combines with the large ribosomal subunit (60 S). During binding between ribosomal subunits and complexes, the GTP bound to IF2 is hydrolysed to GDP and P$_i$, which are released. IF2 and IF3 are also released at this stage, leaving an intact functional ribosome attached to the mRNA and an initiating methionyl-tRNA$_i^{met}$ attached both to the initiating codon and to the P site on the ribosome. (Ribosomes have two sites to which aminoacyl-tRNA can bind. These are formed from sections of both 60 S and 40 S subunits and are termed the P (*peptidyl*) site and the A (*aminoacyl*) site.) With the exception of an initiating methionyl-tRNA$_i^{met}$, all new incoming aminoacyl-tRNAs awaiting addition to the growing peptidyl chain bind to the A site. As will be shown later, the P site is the region of attachment for the growing peptidyl chain (peptidyl-tRNA) and of exit of tRNA molecules after they have transferred amino acids to the growing polypeptide chains.

19.3.4 Elongation of the peptide chain is a cyclic process

Elongation of a growing polypeptide chain involves three steps, repeated for each new amino acid (Fig. 19.6). The following must be present: the initiating complex (consisting of the assembled ribosome bound to the initiating methionyl-tRNA$_i^{met}$ at the first codon, 'codon a'), GTP and the next aminoacyl-tRNA (of a type to match the next codon) and three soluble proteins termed elongation factors (EF1, EFTs and EF2).

In the first step a molecule of EF1, bound to GTP (EF1–GTP), reacts with the incoming aminoacyl-tRNA (aminoacyl^b-tRNA^b) and then attaches to the initiation complex in a reaction driven by the hydrolysis of GTP. This binds the aminoacyl^b-tRNA^b to the A site on the ribosome/mRNA complex. EF1–GTP is regenerated from EF1–GDP and GTP, in a reaction catalysed by EFTs (Fig. 19.6). (EF1 binds each aminoacyl-tRNA to the A site on the ribosome. It binds all aminoacyl-tRNA species except methionyl-tRNA$_i^{met}$.) The methionyl-tRNA$_i^{met}$ is now attached to the ribosomal P site and the aminoacyl^b-tRNA^b to the A site. Attachment of aminoacyl^b-tRNA^b is brought about both by binding at the A site and by

recognition between the anticodon loop on the tRNA^b and the corresponding codon on the mRNA (codon b). The next step in the elongation process will only occur if the binding is correct in both of these positions.

The second step is the formation of a peptide bond between the carboxyl group on the methionine residue of the methionyl-tRNA$_i^{met}$ on the P site and the amino terminal group of aminoacyl^b-tRNA^b. This is catalysed by peptidyl transferase, one of the proteins associated with the large ribosomal subunit. The net result is a dipeptide of methionine and amino acid^b linked to tRNA^b which, in turn, is attached to the A site. The now redundant tRNA$_i^{met}$, with methionine removed, occupies the P site.

In the third stage of elongation (translocation), the ribosome moves one codon (three bases) along the mRNA in the 5'→3' direction. The dipeptidyl-tRNA^b remains bound to the second codon of the mRNA (codon b) and is thus displaced from the A site to the P site of the ribosome as the ribosome moves relative to the mRNA. The initiating tRNA$_i^{met}$ is released from the P site into the cytoplasm and the A site is now situated at codon c on the mRNA. Translocation is believed to involve a change in ribosomal conformation. It requires energy and involves the action of EF2 (sometimes termed translocase) and the hydrolysis of an additional molecule of GTP.

Addition of a third amino acid can now proceed in the same way as for the second. This process can be repeated until all the required amino acids are added to the growing polypeptide chain. This chain is bound to the ribosome–mRNA complex by attachment to whichever tRNA occupies the P site. In doing this, the ribosome continues to move towards the 3' end of the mRNA and the growing polypeptide chain remains attached to the tRNA of the last amino acid to be inserted. It is worth reiterating that the polypeptide chains are started at the amino-terminal end and elongate by successive additions of amino acids to the carboxyl-terminal end.

19.3.5 Termination of peptide chain synthesis requires a terminator codon and a release factor

The final step in the formation of a polypeptide occurs when the A site of a ribosome contacts one of the three possible terminator codons in the mRNA (Table 19.1). Three termination proteins (release factors), R1, R2 and S, hydrolyse the polypeptide from the terminal tRNA, releasing both into the cytoplasm. They also

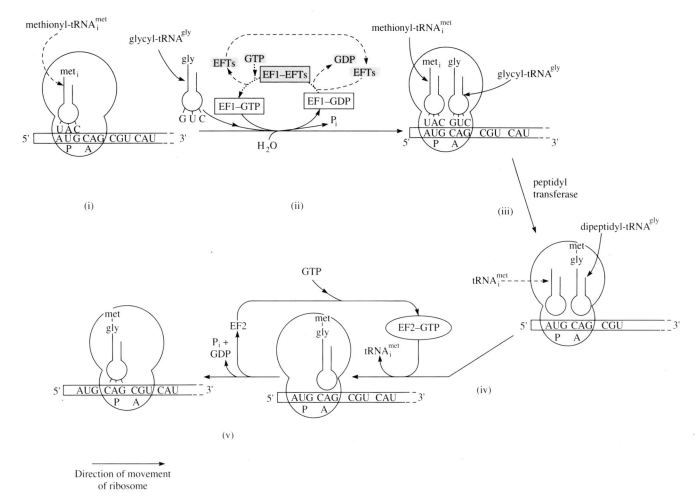

Fig. 19.6 Steps in the formation of a peptide bond during elongation of a nascent polypeptide chain. (i) Newly formed initiation complex (see Fig. 19.5) with met-tRNA$_i^{met}$ at the P site. (ii) A second aminoacyl-tRNA reacts with EF1–GTP to form a complex which attaches to the A site on the initiation complex. Since, in this example, the second codon on the mRNA is CAG, this is glycyl-tRNA^{gly}. Binding of glycyl-tRNA^{gly} to mRNA requires energy, obtained by hydrolysis of GTP. EF1–GTP is regenerated by the elongation factor, EFTs and GTP in a series of reactions. (iii) A peptide bond is formed between the carboxyl group of the methionine residue of methionyl-tRNA$_i^{met}$ on the P site and the amino group of the glycine residue of glycyl-tRNA^{gly} to form dipeptidy-tRNA^{gly} at the A site. This reaction is catalysed by peptidyl transferase. (iv) The tRNA$_i^{met}$ is displaced from the P site on the ribosome in a reaction involving the complex EF2-GTP which subsequently binds to the ribosome. (v) The ribosome moves one codon along the mRNA in the 5′→3′ direction, in an energy-requiring reaction that involves hydrolysis of GTP complexed with EF2 and attached to the ribosome. The dipeptidyl-tRNA^{gly} remains bound to the second codon (CAG) of the mRNA and is consequently displaced from the A site to the P site of the ribosome. Deacylated tRNA$_i^{met}$ is released and the A site, now at the third codon (CGU) on the mRNA, is read to bind another aminoacyl-tRNA (arginyl-tRNA^{arg} in this example).

catalyse the dissociation of the ribosome into the 60 S and 40 S subunits which are subsequently reused to synthesize another polypeptide chain.

19.3.6 Many ribosomes can be involved in translation of a mRNA strand

In most cells, ribosomes are often found both in clusters and singly. The clusters are polyribosomes or polysomes (see Section 3.3) and consist of numbers of ribosomes bound to a single strand of mRNA. This mRNA is translated by all the ribosomes along its length, permitting the message on a single strand of mRNA to be copied many times simultaneously and increasing efficiency of the use of the mRNA template.

19.3.7 Energy is required for protein synthesis

The processes involved in amino acid activation and initiation and elongation of the polypeptide chain require energy. As shown in Eqn. 19.3, activation of an amino acid involves the cleavage of two high energy phosphate bonds. Proofreading by the aminoacyl-tRNA synthetases might consume additional ATP. Two molecules of GTP are hydrolysed to GDP and P_i for every amino acid added to the polypeptide; one with the addition of each succeeding aminoacyl-tRNA to the A site on the ribosome and one during translocation. This totals at least four high energy phosphate bonds consumed for every amino acid added to the polypeptide chain. The energy required for the synthesis of each peptide bond in a protein is, therefore, at least $4 \times 30.5\, \mathrm{kJ\, mol^{-1}}$. Since the standard free energy of hydrolysis of the peptide bond is only $-20.9\, \mathrm{kJ\, mol^{-1}}$, the net free energy expended in the synthesis of peptide bonds is $101.1\, \mathrm{kJ\, mol^{-1}}$.

19.4 Many proteins are modified after mRNA translation

Many proteins are not translated from mRNA directly into their biologically active form but are subject to a range of *post-translational modifications*. For example, methylated lysine residues occur in certain ribosomal proteins found in yeast and in cytochrome *c* in higher plants. The histone H3 also contains methyl-lysine and a protein, 'methylase', has been isolated from plants which catalyses the methylation of lysine residues in H3 and H2B (for details of histones see Section 18.1.5). Methylation occurs only when lysine is present at particular locations in the polypeptide chain.

Many proteins are phosphorylated, particularly at serine residues and also sometimes at threonine and, more rarely, tyrosine residues. Phosphorylation is important in the regulation of enzyme activity (see Section 5.5.4) and occurs on proteins as diverse as histones and the light-harvesting chlorophyll *a/b* binding proteins of thylakoids. Histone proteins are also subject to acetylation at terminal serine residues and, in some species, at internal lysine residues. The 32 kDa ('D1') protein of PSII is modified by palmitoylation after translation.

Glycoproteins are widespread in plants and are involved in a range of functions such as recognition (lectins) and cell wall structure (e.g. extensin—see Section 17.2.3). The side-chains of glycoproteins contain carbohydrate moieties added after synthesis of the polypeptide chain (see also Section 17.10).

The glycosyl component of glycoproteins can contain a range of sugars such as hexoses (e.g. mannose, glucose or galactose), hexosamines (*N*-acetylglucosamine and *N*-acetylgalactosamine) and pentoses (e.g. xylose and arabinose). Mannose, galactose, *N*-acetylglucosamine and arabinose are involved in carbohydrate-peptide linkages, some of which are shown below (amino acid residues are boxed).

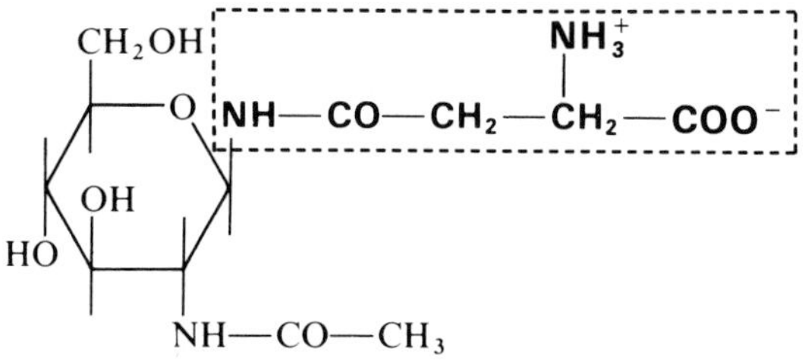

N-Acetylglucosamine–L-asparagine

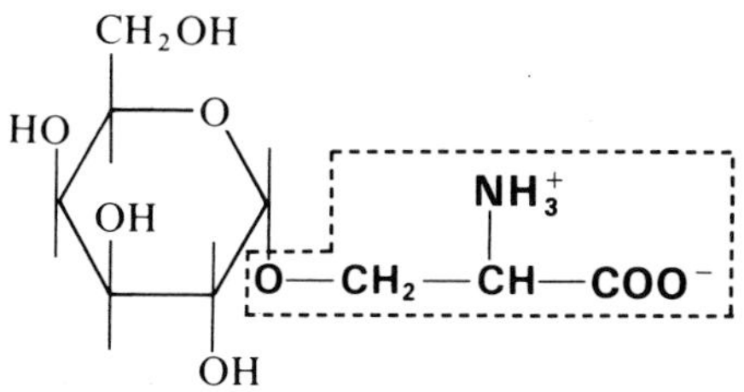

Galactosyl–serine

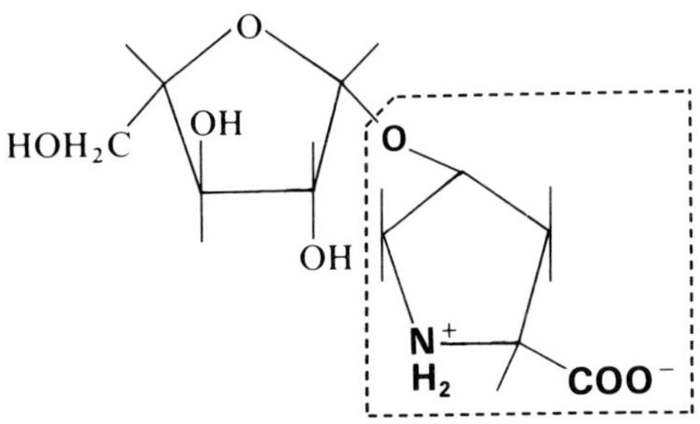

Arabinosyl–hydroxyproline

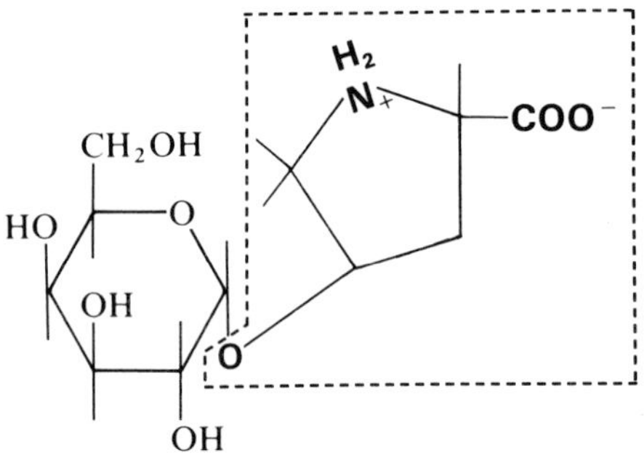

Galactosyl–hydroxyproline

In some glycoproteins (particularly structural glycoproteins found in cell walls), proline residues in the polypeptide are hydroxylated and the resulting hydroxyproline residues are linked to the glycosyl moieties in O-glycosidic linkages. For example, in extensin, arabinose residues in the furanose configuration (Ara$_f$) are joined to hydroxyproline (Hyp) residues in the polypeptide chain as shown below (the linkage involved is shown in bold).

$$[\text{Ara}_f\,\beta(1\rightarrow2)\text{Ara}_f\,\beta(1\rightarrow2)\text{Ara}_f\,\beta(1-4)\text{Hyp}] \quad \text{or}$$

$$[\text{Ara}_f\,\beta(1\rightarrow3)\text{Ara}_f\,\beta(1\rightarrow2)\text{Ara}_f\,\beta(1\rightarrow2)\text{Ara}_f\,\beta(1\rightarrow4)\text{Hyp}]$$

In some glycoproteins, the glycosyl residues are bound by N-glycosidic linkages between the amido-N atom of asparagine (Asn) in the protein component and N-acetylglucosamine (N-AcGln) in the carbohydrate component. Such linkages are found in a range of proteins, including horseradish peroxidase, lectins from lima beans and *Onobrychis viciifolia* (sainfoin) and seed storage proteins. These glycoproteins have a 'core oligosaccharide' linked to an asparagine residue in the polypeptide chain; the core oligosaccharide consists of mannose (Man) linked to N-AcGln as depicted below (the linkage involved is shown in bold).

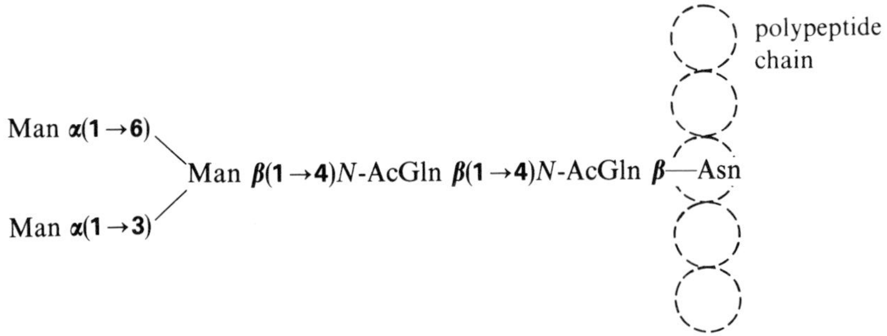

The oligosaccharide core is transferred to the polypeptide by the action of glycosyl transferases. The final steps in glycoprotein synthesis are the addition of a range of mannose, xylose or fucose residues to this core. Thus, the core not only links the polysaccharide component of a glycoprotein to the polypeptide chain but is also the structure to which other glycosidic components are added. Because of the variety of monosaccharide components which can be added to the core, there is a large variation in the type of glycoproteins produced.

Assembly of the oligosaccharide component of glycoproteins occurs on the endoplasmic reticulum and involves an intermediate linked to a lipid.

Many enzymes contain prosthetic groups, which are necessary for activity (see Section 5.2.2). One such protein is cytochrome c with a haem prosthetic group that is added to the polypeptide chain after translation. Although many other types of prosthetic groups are added to polypeptides post-translationally, little is known about the biochemical mechanisms involved.

In recent years, a number of genes have been found to contain a nucleotide sequence coding for some 18–100 amino acids at the amino-terminus of the polypeptide chain, which are not present in the biologically active or 'mature' protein. In some cases the additional amino acids comprise a sequence (synthesized prior to the structural polypeptide) to bind the ribosomes on which it is being translated to a receptor on the membrane of the endoplasmic reticulum. This results in attachment of polyribosomes to the endoplasmic reticulum so that the polypept-

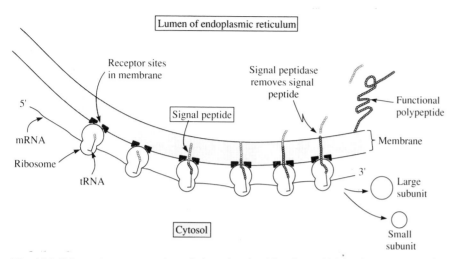

Fig. 19.7 Schematic representation of the role of a 'signal peptide' in the transport of a polypeptide across the membrane of the endoplasmic reticulum. Coding sequences for some polypeptides contain, immediately after the codon for initiation of translation, sequences which code for about 20 additional amino acids. These, when translated, bind the ribosomes to which they are attached to the membrane of the endoplasmic reticulum and, as a result, the polypeptide is translated across the membrane into the lumen of the endoplasmic reticulum. These short sequences (signal peptides) are removed by signal peptidases, once attachment of the ribosome to the membrane of the endoplasmic reticulum is complete.

ide passes into the lumen of the endoplasmic reticulum, or vesicles derived from the endoplasmic reticulum, as it is translated (Fig. 19.7) in a process termed co-translational transport. Amino acid sequences which cause this binding are termed signal peptides. They are removed from the growing polypeptide by the action of signal peptidases associated with the membrane of the endoplasmic reticulum (Fig. 19.7). The signal peptide for soya bean actin, for example, has 32 amino acids, whilst one of 18–21 amino acids is involved in translation of the maize (*Zea mays*) storage protein, zein. Many of the proteins transported across the endoplasmic reticulum membrane are glycosylated once the signal peptide is removed.

In contrast, other polypeptides are transported to their correct site within the cell as the native protein with an additional amino acid sequence (transit peptide) attached. Such polypeptides are translated in this form and the transit peptide controls the transport and final localization of the polypeptide. The best-studied

system is that associated with the chloroplast, for which about 100 nuclear-coded polypeptides are synthesized in the cytoplasm and transported across the double membrane system of the chloroplast envelope. One such polypeptide is the small subunit of ribulose-1,5-P_2 carboxylase/oxygenase. Each gene in the multigene family coding for this polypeptide contains two introns, one of which separates the sequence coding for the transit peptide from that for the rest of the polypeptide. The initial translation product has a molecular mass of 20 kDa, whereas that of the 'mature' subunit is 14 kDa. The transit peptide for this subunit is of variable length, depending on species, but is generally 40–60 amino acid residues long. The particular transit peptide involved is also specific to the protein transported; the transport process requires ATP. The role of transit peptides in chloroplast biogenesis is discussed in more detail in Section 21.4. Transit peptides are also believed to be involved in controlling the intracellular location of different forms of the same enzyme (isozymes).

19.5 Protein synthesis in mitochondria and chloroplasts

Organelles such as mitochondria and chloroplasts have their own genome and can code for and synthesize certain of their component proteins (see Tables 18.4 & 21.3). Although the basic process of translation is similar to that in the cytosol, there are certain features unique to the organelles.

The ribosomes of mitochondria and chloroplasts are somewhat different but they are more alike than those found in the cytoplasm. Chloroplastic and mitochondrial ribosomes have sedimentation coefficients of 70 S and 77–78 S respectively, compared to the 80 S of cytosolic ribosomes. The organelle ribosomes have a different subunit and rRNA composition, summarized in Table 3.3. The protein components also differ. It is possible to isolate tRNAs and aminoacyl-tRNA synthetases from chloroplasts. These are often, but not always, distinguishable from those found in the cytosol.

Chloroplast and mitochondrial mRNAs lack both the 5' cap and (usually) the 3' poly-A tail, characteristic of nuclear transcribed mRNAs. There also seems to be some specificity in recognition of chloroplast and cytoplasmic mRNA by 70 S and 80 S ribosomal systems respectively. Thus, translation of chloroplast mRNA by an *in vitro* system based on 80 S ribosomes from wheatgerm is severely inhibited if any cytoplasmic mRNA is present. Similarly, chloroplast mRNA competitively inhibits translation of cytoplasmic mRNA by a system based on 70 S ribosomes from *E. coli*. Initiation of protein synthesis in chloroplasts occurs with formyl methionine, as in bacterial systems, instead of with methionine as occurs in the cytoplasm.

Less is known of protein synthesis by plant mitochondria, although the coding capacity of the genome, and transcription have been studied in some detail. Plant mitochondria synthesize 20–30 proteins (see Table 18.4), almost double that of animal and fungal systems. It is interesting to note that mitochondria use a slightly different genetic code; for example, codon CUA in mRNA derived from the nuclear genome generally specifies leucine but, in yeast mitochondria, it codes for threonine. Similarly the 'stop' codon UGA codes, in mitochondria, for tryptophan.

Further reading

Monographs and treatises: Alberts *et al.* (1989),; Boulter & Parthier (1982); Grierson & Covey (1988); Marcus (1989); Parthier & Boulter (1982).

20
Regulation of gene expression

During the course of development, plant cells undergo changes at the ultrastructural and biochemical levels. In higher plants, cell division occurs mainly in the regions known as meristems, where cell lines go through a number of division cycles before they cease dividing and go on to differentiate. As cells differentiate, they tend to lose their capacity for division and assume specialized functions.

In Chapter 18 it was shown that the process of cell division and development is often accompanied by the appearance of various enzymes. Many of these are involved directly in cell division, but others are not; these include enzymes involved in respiration, CO_2 assimilation or phosphate metabolism. In multicellular organisms the appearance of enzymes or other gene products is also tissue- and/or organ-specific. Other products are formed only in response to specific environmental stimuli. Although plant cells contain all the genetic information necessary for the functioning of the plant only some of this information is required at any one time.

Table 20.1 Examples of proteins found in specific tissues in plants. After Trewavas, A.J. (1982) In Smith, H. & Grierson, D. (eds) *Molecular Biology of Plant Development*, pp. 7–27. Blackwell Scientific Publications, Oxford.

Developmental stage	Tissue/organ	Enzyme/protein	Function
Seed germination	Cotyledons of fatty seeds	Isocitrate lyase	Fat metabolism
	Endosperm of cereals	α-Amylase	Carbohydrate metabolism
Leaf development	Leaf	Light-harvesting chlorophyll-binding protein	Light harvesting
	Leaf	Ferredoxin	Electron transport
Cell differentiation	Epidermis	Palmitate hydroxylase	Cutin formation

The correct proteins must be produced at the right time in the right cells for normal development and growth to occur. Some examples of tissue- or organ-specific proteins are given in Table 20.1.

20.1 Control of gene expression can occur at several points

The formation of a biologically functional gene product from the genetic information contained in DNA requires the co-ordinated activity of transcription and translation (see Chapter 19) (Fig. 20.1). Although the first step is transcription of DNA into RNA, the signal causing the expression of a particular gene does not necessarily act at this level. Control can also be exercised at a post-transcriptional, translational, or even post-translational stage. At each stage, there are a number of possible control points (Table 20.2).

The absolute criteria for establishing whether the appearance of a particular gene product is controlled transcriptionally, translationally or post-translationally are straightforward.

1 Transcriptional control: the appearance of the protein must be preceded by the appearance and stabilization of a species of mRNA previously degraded or not synthesized.

2 Translational control: the synthesis of a protein, not previously found in the cell, without the synthesis of any 'new' species of mRNA (i.e. pre-existing mRNA is translated into protein).

3 Post-translational control: the appearance of the protein in an active form without the synthesis of any new polypeptides.

Many early investigations into the control of gene expression relied on the use of 'specific' inhibitors of RNA or protein synthesis. Actinomycin and 6-methylpurine have been used as inhibitors of RNA synthesis, whilst puromycin, cycloheximide or chloramphenicol inhibit protein synthesis. Therefore, if synthesis of a particular enzyme is blocked, for example, by cycloheximide but not by 6-methylpurine, this suggests that formation of that protein is under translational control. If, however, 6-methylpurine inhibits protein synthesis, transcriptional control is

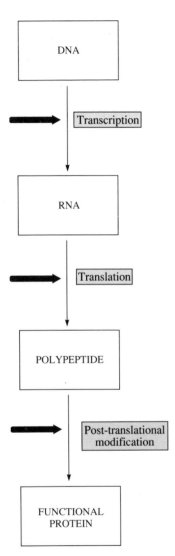

Fig. 20.1 The major steps involved in the formation of a functional protein from the information contained in the DNA of the plant genome. Regulation of gene expression can occur at any of the steps and in response to a range of environmental or developmental signals (bold arrows).

Table 20.2 Possible control points in gene expression.

Transcription
   Availability of DNA as a template for RNA synthesis
   Levels and/or activity of RNA polymerases
   Availability of transcriptional factors and their
      interaction with regulatory sequences associated with
      the gene

Post-transcription and translation
   Processing of mRNA precursors
   Transport of mRNA from the nucleus
   Availability of, and competition for, initiation factors
   Rate of turnover of mRNA

Post-translation
   Post translational cleavage
   Methylation/phosphorylation/acetylation of certain amino acid
      residues
   Addition and/or cleavage of prosthetic groups or carbohydrate
      side chains
   Transport of polypeptides to site of action e.g. organelles,
      cell walls

indirect and represent secondary effects. It is now possible in many cases to monitor levels of specific mRNAs directly to determine whether control is exercised at transcription or translation, thus avoiding the problems of non-specificity of inhibitors. Some examples are discussed in the following sections.

20.2 Regulation of some genes occurs at transcription

20.2.1 Availability of the DNA template influences transcription rates

For the DNA sequence of a gene to be transcribed to RNA, the DNA must be available to the action of RNA polymerase and other factors involved in transcription. In animal systems, switching on genes involves changes in the conformation of chromatin (see Chapter 18) from an inactive to a transcriptionally active state. Similar effects are also found in plant tissues. For example, the state of chromatin organization can be assayed by measuring the sensitivity of DNA to treatment by the DNA-degrading enzyme DNase 1. Transcriptionally active chromatin is sensitive to this enzyme but inactive regions are relatively insensitive. In barley, for

implied. Although the use of inhibitors has, in the past, provided some useful insights into regulation of gene expression, it must be stressed that many do not possess ideal specificity and such data must be carefully interpreted. Observed changes in gene product formation after inhibitor treatment might, for instance, be

example, there is good agreement between the level of mRNAs transcribed from several genes and the sensitivity of the corresponding regions of the genome to DNase 1 treatment (Fig. 20.2). The specific mRNA for the seed storage protein, hordein, is transcribed only in endosperm tissue. In the endosperm, but not in leaves, a family of component genes encoding the synthesis of hordein (a 'multigene family') is susceptible to DNase 1 treatment and most of the hordein gene components involved are degraded rapidly. In leaf tissue, in which the hordein genes are not expressed, DNase 1 treatment has little effect.

This suggests that the chromatin must be in a particular configuration for a gene to be transcribed. However, any conformational change in chromatin only

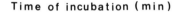

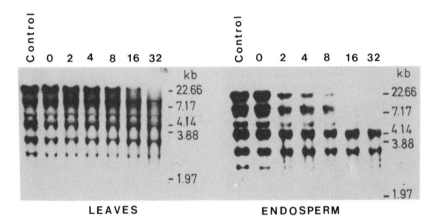

Fig. 20.2 Transcription of DNA can be affected by the conformation of chromatin. Effects of the DNA digesting enzyme DNase 1 on the DNA of the multigene family encoding the storage protein hordein in barley are shown. Chromatin samples from leaves and endosperm were treated with DNase 1 for various times and subsequently cut into a number of small fragments with the restriction enzyme *Eco* RI (see Section 20.5.1). Fragments were separated by electrophoresis on agarose gel, transferred to nitrocellulose by Southern blotting and hordein gene fragments were detected using a cDNA probe (see Section 20.5.3). The hordein-encoding DNA of endosperm tissue is degraded rapidly by DNase 1 whilst the corresponding DNA sequences in leaf tissue are unaffected. This implies that for the hordein DNA to be transcribed, there must be a conformational change in the chromatin such that the DNA is more accessible to DNase 1 attack. After Steinmüller, K., Batschauer, A. & Apel, K. (1986) *Eur. J. Biochem.*, **158**, 519–25.

converts genes from an inactive state to a *potentially* active state. For example, synthesis of the light-harvesting chlorophyll-binding protein (LHCP) is light-dependent. The formation of transcripts of LHCP genes is more marked in plants maintained under a light/dark cycle than in dark-grown, etiolated plants. However, the genes for LHCP and NADPH:protochlorophyllide oxidoreductase (involved in the synthesis of chlorophyll inserted into LHCP) are sensitive to DNase 1 in leaves of both light- and dark-grown plants, despite very marked differences in the rate of appearance of transcripts. Full control over the activity of a particular gene must rely on some other feature(s) of transcription or stability of transcripts. Conformational change in chromatin represents a first (coarse) control of gene expression, with a second (finer) level of control involving RNA polymerase and transcription factors.

20.2.2 Rates of transcription are affected by the levels and/or activity of RNA polymerase

RNA polymerase catalyses the transcription of DNA to RNA and factors influencing this enzyme might be expected to affect the rate of transcription. The rate of RNA synthesis can be influenced by the number of enzyme molecules ('level' of enzyme) and the catalytic activity per molecule. Measured *activity* can be a function of either or both of these factors. Plants contain three RNA polymerases which transcribe different types of RNA (see Section 19.2.3) and in some plant systems changes in RNA polymerase correlate with changes in transcriptional activity during development. During the germination of wheat, for example, RNA polymerase I is essentially absent in dormant embryos but its level increases dramatically a few hours after imbibition, at a stage associated with increased transcriptional activity. RNA polymerase II levels remain steady. Greatly increased levels (approximately 10-fold) of both RNA polymerase I and II can be induced by auxin treatment of soya bean hypocotyls.

However, in most cases, the level of the RNA polymerases is more or less constant despite changes in transcription rates. During germination of soya bean a 25-fold increase in RNA polymerase activity occurs without a significant change in its level. There is some evidence that modulation of polymerase activity is due to changes in binding between RNA polymerase and chromatin; in actively transcribed chromatin the polymerases are bound, whilst in inactive tissues they are unbound. This might be related to chromatin conformation.

Studies on soya bean hypocotyls, wheatgerm and parsley cell tissue culture show that RNA polymerase II exists in two forms: A and B. Form A from soya bean axes has a large subunit of molecular mass 215 kDa, and during germination it is converted to a polypeptide with a molecular mass of 180 kDa—form B. Similar changes occur in germinating wheat seeds. Form B is more active, with transcription rates in wheat increasing some 13-fold in the first 36 h of germination.

The activity and specificity of the polymerase can also be regulated by protein or other factors, interacting with either the template or the polymerase molecule. The role of initiation, elongation and termination factors have been discussed elsewhere (see Section 19.3).

20.2.3 Specific sequences in the plant genome regulate transcription

The regulation of gene expression is highly specific. It is debatable whether the more general effects on, say, RNA polymerase could exert specific control over expression of a single gene or small group of genes as occurs in vivo. In prokaryotes control is exercised through certain regulatory sequences in the microbial genome in an operon. This concept was originally proposed by Jacob and Monod, who were working on the regulation of the β-galactosidase gene in E. coli. In prokaryotic systems, genes with related functions are often grouped together and are subject to the same control system. The transcription product is a mRNA molecule containing the coding sequences for all the encoded proteins. For instance, the operon coding for the eleven proteins required for histidine synthesis produces a mRNA of 10 000 base pairs. In some cases, in the absence of substrate, a repressor protein is synthesized from a regulatory gene and binds to a region of the genome (the operator), effectively blocking transcription. Inducer compounds bind to the repressor releasing the repressor–inducer complex from the operator so that transcription occurs. A similar mechanism occurs during the repression of enzyme activity by the end-product of a biosynthetic pathway. In this case, however, the end-product is thought to bind to an inactive repressor protein to form a repressor complex which attaches to the operator and inhibits transcription. RNA polymerase normally binds to the promoter region situated between the regulatory gene and the operator. The action of the repressor at the operator prevents this binding.

The eukaryote genome is far more complex, of course, than that of a bacterium like E. coli and, with the exception of a group of ribosomal proteins encoded in the plastid genome, there is little evidence for the existence of 'operon' systems in plants. However, there are regions on plant genes that are implicated in the regulation of transcription. Close to the transcription start-point is a sequence called the 'TATA box' (see Section 19.2.1), which determines the site of initiation of transcription (promoter region). This sequence is generally some 16–54 base pairs upstream of the cap site, although in rare cases some genes have a second TATA box some 900 base pairs further upstream. The actual nucleotide sequence in the TATA box varies slightly in different plant genes. Another sequence called the 'CAAT box' is also implicated in control of promoter activity (Section 19.2.1). Furthermore, transcription is also influenced by other regulatory sequences referred to as enhancers; these are characteristically found 1000 base pairs or more upstream of the TATA and CAAT boxes and activate gene expression in response to environmental stimuli, but only in specific types of cell.

It has been possible to investigate the regulation of the small subunit of ribulose-1,5-bisphosphate carboxylase (sRuP$_2$-Case) and the light-harvesting chlorophyll-binding protein (LHCP) using gene manipulation techniques to construct transgenic plant systems (see Fig. 20.3) in which presumed regulatory regions of the sRuP$_2$-Case or the LHCP genes are attached to coding sequences for enzymes such as neomycin phosphotransferase (N-PTase) or chloramphenicol acetyltransferase (C-AcTase) (which confer antibiotic resistance and are easily assayed). The role of promoter and enhancer sequences can be followed by examining the degree of expression of C-AcTase or N-PTase activity. For example, a 247 base pair upstream element (U.E.) from the LHCP gene, termed AB80, fused to the N-PTase coding sequence is responsible for the tissue specificity and light regulation of the construct in transgenic tobacco plants (Fig. 20.3). Addition of the 247 base pair element from the LHCP gene not only caused light-dependent induction of enzyme activity but also tissue specificity, in that activity in the roots was not expressed (i.e. 'silenced'). Thus, this particular sequence possesses both enhancer and 'silencer' properties. Curiously, the orientation of the upstream element does not markedly affect its enhancing action. Addition of a second (repeated) enhancing region stimulated enzyme formation to an even greater extent. This effect is the result of transcriptional or post-transcriptional regulation as mRNA levels correlate well with enzyme activities.

20.3 Control of gene expression also occurs post-transcriptionally or at translation

The primary site of regulation of gene expression appears, in most cases, to be at transcription. There are, however, some situations in which a change in the rate of production of a protein, or a group of proteins, in response to a developmental or environmental signal cannot be ascribed wholly to changes in the rate of synthesis

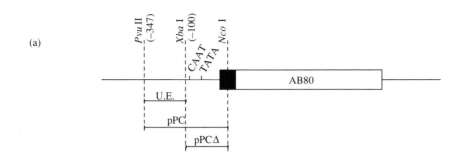

(a)

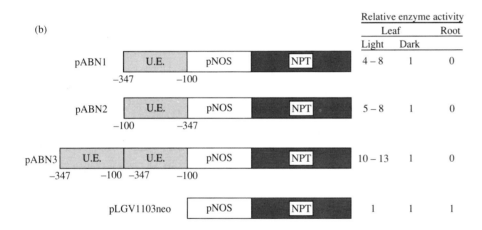

(b)

				Relative enzyme activity		
				Leaf		Root
				Light	Dark	
pABN1	U.E.	pNOS	NPT	4 – 8	1	0
	−347 −100					
pABN2	U.E.	pNOS	NPT	5 – 8	1	0
	−100 −347					
pABN3	U.E. U.E.	pNOS	NPT	10 – 13	1	0
	−347 −100 −347 −100					
pLGV1103neo	pNOS	NPT		1	1	1

(c)

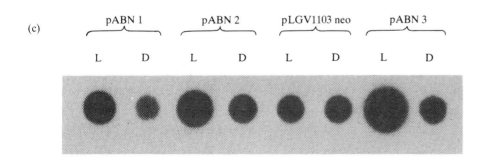

of the relevant mRNA. In such cases, control of gene expression must be exercised at a post-transcriptional level. In chloroplasts, genes are transcribed continually and the stability of mRNA controls the abundance of the particular mRNA and appearance of the gene product.

Potential candidates for post-transcriptional control are the rate of processing of nuclear mRNA precursors, and transport of mRNA from the nucleus. For instance, during recovery of peas from sulphur deficiency, transcription of mRNA from the gene encoding the sulphur-rich seed storage protein legumin doubles during the first 48 h but levels of legumin mRNA increase 20-fold. Since legumin mRNA levels accumulate in seeds of plants recovering from sulphur deficiency faster than can be accounted for by the relative level of transcription it has been proposed that the legumin mRNA transcripts are less stable in the developing seeds of sulphur-stressed plants than in developing seeds during recovery.

In studies on tobacco the polyribosomal mRNA in leaf tissue contained transcripts of about 27 000 different genes, representing approximately 5% of the single copy DNA in the genome. Some 8000 transcripts were common to all organs of the plant but many were unique to the leaf. The total number of transcripts in the mRNA populations from all organs was around 60 000. Pre-mRNA primary transcripts in nuclei from the leaf, however, contained all these transcripts. Therefore, a large number of RNA transcripts found in the pre-mRNA in the

Fig. 20.3 Effects of upstream elements on gene expression in transgenic tobacco plants. A 247 base pair upstream element (U.E.) from the gene encoding the light-harvesting chlorophyll-binding protein was attached in various configurations to the structural gene (NPT) for neomycin phosphotransferase (N-PTase) and the nopaline synthase promoter region (pNOS). These constructs were introduced to tobacco plants using *Agrobacterium* (see Section 20.6.2) and the level of N-PTase mRNA and N-PTase activity were determined in leaves and roots of the transgenic plants. (a) The AB80 gene, encoding LHCP, showing the position of a 400 base pair region (pPC) involved in the regulation of expression of the gene. Part of this region (pPCΔ) is not essential for gene expression and can be deleted to leave the active U.E. of 247 base pairs. The position of relevant restriction nuclease sites is shown. (b) Effects of the U.E. in various constructs designated pABN1, pABN2 and pABN3 on N-PTase activity. Activity in root tissue of transgenic plants without the U.E. (construct pLGV1103 neo) is ascribed a value of 1. Note that the presence of the U.E. stimulates light-induced expression of N-PTase activity and suppression of this activity in root tissue. Orientation of the U.E. does not affect activity but two U.E. in tandem increase N-PTase activity in leaf tissue in the light. (c) Levels of N-PTase mRNA in leaves of transgenic plants containing the various constructs shown in (b) (L and D indicate light and dark treatments respectively). Activities of N-PTase are paralleled by changes in the amounts of N-PTase mRNA. After Simpson, J., Schell, J., Van Montagu, M. & Herrera-Estrella, L. (1986) *Nature*, **323**, 551–4.

nucleus are not expressed as polyribosomal mRNA in specific tissues. This again suggests an important level of control between transcription of nuclear DNA into a primary transcript and the translation of a finished (processed) transcript by the polyribosome.

Some other very clear examples of control after transcription come from studies of the unicellular alga *Acetabularia* which develops as a giant cell from a zygote into a stalk of up to 5 cm in length, with rhizoids containing the single nucleus. After several weeks a structure called a cap is formed which eventually plays a role in reproduction. Cap formation is a developmental process involving the sequential appearance and disappearance of a number of enzymes and the synthesis of several proteins and specific polysaccharides. The large cell size of *Acetabularia* allows the nucleus to be removed without major damage to the rest of the cell and RNA synthesis stops immediately. However, cells enucleated several weeks prior to cap formation form a cap identical to that of a normal cell and show perfectly normal patterns of appearance and disappearance of specific enzymes and synthesis of polysaccharides. Since the genes controlling cap formation are not encoded in the organelle genomes (chloroplasts and mitochondria), regulation must occur at a post-transcriptional stage. The mRNA for the enzyme UDP-glucose pyrophosphorylase is exported into the cytosol well before cap formation, yet this enzyme remains at very low levels until cap formation begins. In this, and some other systems, the mRNAs are 'long-lived' and are translated into protein at a stage much later than their transcription. The precise mode of regulation at the translational step is not known.

The availability of mRNA for translation is also influenced by competition between different mRNAs for initiation factors. There is some suggestion that in the wheatgerm translation system the affinity of particular mRNAs for one of the initiation factors governs which mRNA is translated. During anaerobiosis, mRNAs for alcohol dehydrogenase and a few other proteins dominate the translation machinery. Similarly, in aleurone layers from barley seeds treated with gibberellic acid (see Section 20.4), mRNA for α-amylase competes strongly with mRNAs for other proteins. In prokaryotic systems the concentration of ribosomal proteins can regulate the translation of their mRNAs but there are no reports of a similar mechanism in plants.

Polypeptides are often subject to post-translational modifications before they appear in an active form (see Section 19.4). This includes, for instance, addition of carbohydrate side-chains to glycoproteins or the introduction of prosthetic groups to enzymes.

20.4 Gene expression is regulated in response to a range of environmental and developmental signals

Although some doubts may exist over precise mechanisms, there is no shortage of examples of regulation of gene expression at transcriptional or translational levels in response to various signals.

Light stimulation of the appearance of LHCP mRNA has already been described (Section 20.2.1). Synthesis of LHCP, like many other proteins involved in chloroplast function, is regulated via the phytochrome system (see Section 21.3.2). Figure 20.4 shows the appearance of LHCP mRNA in dark-grown barley seedlings after exposure to periods of red and far-red light. As expected for a phytochrome-based system, accumulation of mRNA is stimulated by red light and inhibited by far-red light.

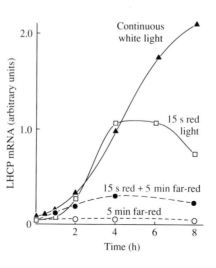

Fig. 20.4 Phytochrome-mediated regulation of transcription of LHCP mRNA in barley seedlings. Seedlings were grown in the dark and illuminated as shown on the figure. Poly-A-containing mRNA was extracted from the plants and translated *in vitro* using a wheatgerm system. Antibodies raised to the precursor of the LHCP were used to immunoprecipitate this translation product, which was solubilized and separated by polyacrylamide gel electrophoresis. The amount of radioactive label incorporated into the LHCP precursor polypeptide during translation was used as an indirect measurement of the amount of LHCP mRNA. Note that exposure to continuous white light induces formation of LHCP mRNA. Induction is also achieved with 15 s exposure of seedlings to red light but inhibited if the red light treatment is followed by 5 min of far-red light. Far-red light treatment for 5 min results in no increase in LHCP mRNA level. This indicates that regulation of LHCP mRNA synthesis in response to light is achieved via the phytochrome system. After Apel, K. (1979) *Eur. J. Biochem.*, **97**, 183–8.

Although light regulation of RuP$_2$-Case might involve transcriptional regulation in the majority of species, in *Amaranthus hypochondriacus* both the large and small subunits are regulated at both translation and transcription. Upon exposure of dark-grown cotyledons of *A. hypochondriacus* to light, synthesis of RuP$_2$-Case subunits begins rapidly and within 4 h the levels are similar to those of light-grown plants, representing a rapid translational control of gene expression since mRNA levels do not increase until 5 h after illumination. Once mRNA levels start to increase, the rate of increase in subunit synthesis decreases. These longer-term changes involve transcriptional control. The role of light in regulating the development of chloroplasts is discussed further in Chapter 21.

Enzymes involved in the uptake and assimilation of mineral nutrients are often inducible by the presence of the substrate, or removal of a repressor. For instance, nitrate reductase is induced in the roots of higher plants when nitrate is supplied. In many unicellular algae, formation of nitrate reductase is induced by nitrogen starvation as well as by the presence of NO_3^- (in the absence of NH_4^+, a repressor). In *Dunaliella*, a unicellular marine alga, the inhibitor of RNA synthesis, 6-methylpurine, has little effect on the induction of nitrate reductase activity whilst cycloheximide strongly inhibits formation of the enzyme. This suggests that nitrate induction of nitrate reductase is controlled at the level of translation or post-translational processing. By contrast, in the green alga *Ankistrodesmus*, formation of urea amido lyase, after removal of NH_4^+ from the growth medium, is inhibited by 6-methylpurine, indicating transcriptional regulation. Once again, however, the limitations and problems associated with the use of inhibitors (Section 20.1) must be emphasized.

The enzyme phenylalanine ammonia-lyase (P-ALase) is involved in the metabolism of phenylpropanoid compounds. These are normal metabolites in many species, being especially important in lignin synthesis, but they are also produced in response to stress, such as damage by ultraviolet light, wounding and pathogen attack. P-ALase activity is inducible in parsley cell cultures by exposure to ultraviolet light. Increase in activity is preceded by a marked increase in P-ALase mRNA synthesis (Fig. 20.5a), suggesting regulation at the level of transcription. Fungal elicitors also induce rapid formation of the mRNA for P-ALase and other enzymes of phenylpropanoid metabolism (Fig. 20.5b).

The role of plant growth substances in development is complex. Five major groups of 'phytohormones' (auxins, gibberellins, cytokinins, abscisic acid and ethylene) are known but their mechanism of action is poorly understood. One system more fully understood is the mobilization of reserve carbohydrate during

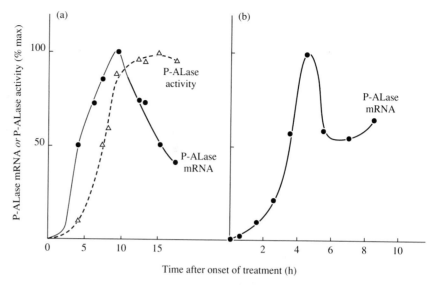

Fig. 20.5 Appearance of P-ALase and its corresponding mRNA in cell cultures in response to stress by ultraviolet light or the application of a fungal elicitor. (a) The appearance of P-ALase and P-ALase mRNA in parsley cell cultures following exposure to ultraviolet light. (b) P-ALase mRNA is also formed rapidly by a suspension of cultured bean cells in response to application of an elicitor from the fungus *Colletotrichum lindemuthianum*. Levels of mRNA specific to chalcone synthase and chalcone isomerase (also involved in phenylpropanoid biosynthesis) showed similar responses. After (a) Schröder, J., Kreuzaler, F., Schäfer, E. & Hahlbrock, K. (1979) *J. Biol. Chem.*, **254**, 57–65 and (b) Cramer, C.L., Ryder, J.B., Bell, J. & Lamb, C.J. (1985) *Science*, **227**, 1240–3. (Copyright 1985 by the AAAS.)

germination of cereals. When grains of plants such as barley imbibe water there is a rapid rise in metabolic activity of the embryo and aleurone. This is followed, after a few days, by mobilization of the grain reserves, especially starch and protein stored in the endosperm tissue, and is related to the formation of proteases and α-amylases in aleurone and scutellum cells and their release into the endosperm. If, however, the embryo is removed at an early stage, α-amylases and proteases are formed at a slower rate and mobilization of starch and protein is decreased. This is related to production, by the embryo, of the plant growth substance gibberellic acid (GA$_3$), one of more than 50 naturally occurring gibberellins which influence development. GA$_3$ has two important effects on the aleurone layer: it induces the synthesis *de novo* of proteases and α-amylase; and induces release of newly synthesized and pre-existing enzymes from the aleurone. Other hydrolytic enzymes are synthesized during imbibition which also depend on GA_3 for induction. GA_3 induction of expression of α-amylase genes is controlled at the level of transcription. Thus, α-amylase mRNA appears within 1 h of GA_3 application to isolated aleurone layers, the level of this mRNA increasing with time. The appearance of α-amylase mRNA is inhibited if GA_3 is added with the protein synthesis inhibitor, cyclo-heximide, indicating that protein synthesis is required for the transcription of α-amylase mRNA. The protein synthesized could act as a receptor for GA_3, to promote transcription or to stabilize the mRNA and prevent its degradation.

Cytokinins, as well as phytochrome, are involved in regulation of chloroplast development in detached leaves and cotyledons and in cultured cell lines of tobacco and petunia. For instance, the cytokinin benzyladenine stimulates red-light induced synthesis of RuP_2-Case and LHCP in greening tissues of duckweed (*Lemna gibba*), and also preferentially retards the decrease in the levels of the mRNAs encoding both the small and large subunits of RuP_2-Case in the darkness. However, when the levels of LHCP mRNA present in the nucleus, rather than in the cytosol, are examined, benzyladenine has no effect, suggesting that cytokinins stabilize specific mRNA species, although the mechanism involved is not understood.

20.5 Recombinant DNA technology is a powerful tool for investigating gene organization and function

Investigations into the regulation of gene expression and the organization of genes rely heavily on techniques based on recombinant DNA technology. This section considers some of the basic techniques used.

20.5.1 Restriction enzymes are a major tool used in recombinant DNA technology

Studies on the characterization and manipulation of DNA rely very heavily on restriction enzymes. These enzymes are found in bacterial cells and protect them from infection by degrading invading foreign DNA molecules. The enzymes, also termed restriction nucleases, are able to recognize specific sequences of 4–6 nucleotides in DNA and hydrolyse the phosphodiester bonds of that DNA in specific ways. Corresponding sequences in the host bacteria are protected by methylation of adenine or cytosine residues.) To date, nearly 100 different and specific cleavage sites have been identified and restriction enzymes have been

Table 20.3 Restriction enzymes isolated from different bacterial strains, from which they derive their abbreviated name. Base sequences in each of the two strands of double-stranded DNA, which are cleaved by the different enzymes are also shown. In some cases this results in short, single-stranded regions known as cohesive ends.

Source of enzyme	Enzyme abbreviation	Sequence recognized and cleavage position $5' \rightarrow 3'$ $3' \leftarrow 5'$
Bacillus amyloliquefaciens H	*Bam* H1	G|GATC C C CTAG|G
Escherichia coli RY13	*Eco* RI	G|AATT C C TTAA|G
Haemophilus aegyptius	*Hae* III	GG|CC CC|GG
Haemophilus influenzae Rd	*Hin* dIII	A|AGCT T T TCGA|A
Haemophilus parainfluenzae	*Hpa* II	C|CG G G GC|C
Providencia stuartii 164	*Pst* I	C TGCA|G G|ACGT C
Streptomyces albus	*Sal* I	G|TCGA C C AGCT|G

isolated from over twice that number of bacterial strains. Some examples of these nucleases and the sequences they recognize are given in Table 20.3.

A particular restriction nuclease cuts any long length of DNA double helix into a series of fragments known as restriction fragments. These fragments can be separated by size using electrophoresis on agarose gels (Fig. 20.6). Comparison of restriction fragments from a particular region of the genome using different nucleases allows construction of a restriction map for gene mapping and DNA sequencing (see Sections 20.5.2 & 20.5.8).

Many restriction nucleases produce staggered cuts which leave short, single-stranded ends on both fragments of the double-stranded DNA (Table 20.3). These are usually called cohesive (or sticky) ends and can form complementary base pairs with any other end produced by the same enzyme. Thus, a circular DNA molecule (such as a plasmid—see Section 20.5.4) cut at any single site by a restriction nuclease will tend to reform a circle by basepairing (reannealing) of its cohesive

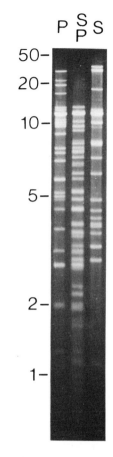

Fig. 20.6 Separation of DNA fragments, produced by restriction nuclease digestion, by electrophoresis on agarose gels. Smaller fragments migrate faster through the gel than do the larger fragments. In this example mitochondrial DNA from *Brassica campestris* was digested with *Pst* I (P), *Sal* I (S) or both *Pst* I and *Sal* I together (SP). Fragments were then separated by electrophoresis on 0.7% agarose gels. The size of the fragments in thousands of base pairs is given on the left and was established using λ phage DNA restriction fragments generated with *Sal* I, *Hin* dIII and *Eco* RI as standards. Fragments were visible under ultraviolet light after treatment with ethidium bromide. From Palmer, J.D. & Shields, C.R. (1984) *Nature*, **307**, 437–40.

ends. Once the two ends have joined by complementary base pairing they can be sealed by the enzyme, DNA ligase. The combined use of restriction enzymes and DNA ligase makes it possible to graft fragments of any DNA into other genetic elements (see Fig. 20.10).

20.5.2 Restriction enzymes can be used to map genes

Different restriction enzymes cut DNA at different specific sites, determined by particular four-, five- or six-nucleotide sequences. Cutting a DNA sequence with one or several restriction enzymes generates a series of fragments of differing size.

The order of restriction sites along the original DNA molecule can be determined by comparing the pattern of fragmentation obtained with different enzymes (Fig. 20.7). This allows the relative position of restriction sites on a DNA molecule to be mapped. Once these are known, restriction enzymes can be used to make highly

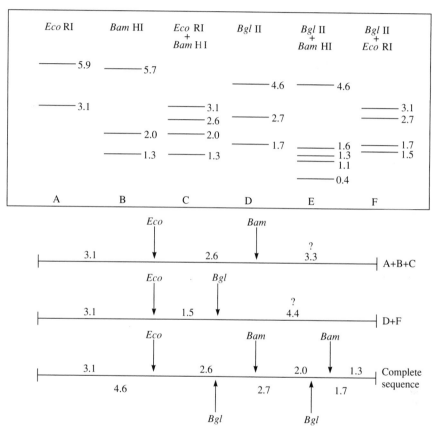

Fig. 20.7 Mapping of restriction nuclease sites in a segment of DNA. The DNA is cleaved using a number of restriction nucleases, singly and in combination, and fragments are separated by electrophoresis in agarose gels. A simple map can be drawn up from the data. For instance, data in lanes A, B and C permit identification of the *Eco* RI site and one of the two *Bam* HI sites. Lanes D and F contain data relating to one of the *Bgl* II sites and its position relative to *Eco* RI. Lanes B, D and E permit deduction of the position of the second *Bgl* II and *Bam* HI sites. From Grierson, D. & Covey, S. (1984) *Plant Molecular Biology*. Blackie, Glasgow.

specific cuts to allow manipulation of the genome or to facilitate further analysis such as sequencing. A knowledge of which fragments contain sequences coding for particular genes allows maps to be drawn up to show the relative position of those genes on the genome (see, for example, Fig. 18.3). Comparison of the size and sequence of fragments obtained from native DNA and a DNA 'clone' made by reverse transcription from mRNA (see Section 20.5.5) can provide information on the existence and size of introns within the DNA sequence (Fig. 20.8).

20.5.3 Hybridization of restriction fragments with specific probes allows determination of the site of particular sequences on the genome

The position of particular nucleotide sequences on a restriction map can be worked out by using the ability of nucleic acids to hybridize by complementary base pairing. In this procedure, called 'Southern blotting', the DNA restriction fragments separated by agarose gel electrophoresis are denatured by high pH to give two single strands of DNA. These single-stranded DNA fragments are transferred (blotted) from the agarose gel to a sheet of nitrocellulose paper (filter) or similar support and are fixed to it by heat treatment (usually by baking at 80 °C). The DNA fragments are then incubated with a range of radioactively labelled oligonucleotides of known sequence. These are known as 'probes' since they hybridize with complementary sequences in the restriction fragments. After washing and autoradiography of the filter those fragments which have combined with the radioactive probe, and therefore contain the sequence of interest to be identified from the usually large number produced by the action of the restriction enzyme(s), can be identified (Fig. 20.9). In a similar way, fragments of RNA molecules, produced by ribonuclease activity on RNA, can be transferred from gels to nitrocellulose paper and tested to detect sequences which bind to various radioactively labelled oligodeoxynucleotide probes ('Northern blotting'). It is thus possible to determine RNA sequences homologous to the DNA 'probes'. It is also possible to use cDNA (Section 20.5.5) as a probe.

20.5.4 DNA sequences of interest can be propagated using a suitable cloning vector

DNA cloning is the process by which portions of DNA are isolated and reproduced in large amounts. This permits the isolated and purified segments of DNA to be obtained in large numbers (i.e. 'amplified') by growth of a cloning vector containing

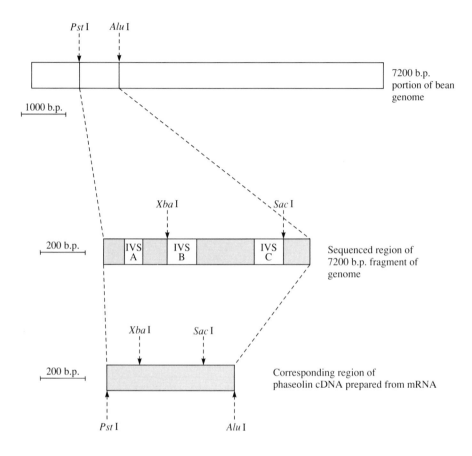

Fig. 20.8 Comparison of the restriction nuclease maps for a portion of the phaseolin gene from french bean and the cDNA prepared from phaseolin mRNA. *Pst* I and *Alu* I cut the bean genome as indicated to produce a fragment of ∼840 base pairs (b.p.). This fragment contains sites for *Xba* I and *Sac* I. The cDNA prepared from phaseolin mRNA and cloned in the *E. coli* plasmid pBR322 also contains sites for *Xba* I and *Sac* I in the region bounded by *Pst* I and *Alu* I sites. However, in this case the distances between *Pst* I and *Xba* I and *Sac* I are much smaller than in the case of the genomic DNA. This indicates the presence of non-coding intervening sequences (introns) in the gene for phaseolin. Sequencing of the region between *Pst* I and *Alu* I sites has revealed the presence of three intervening sequences (IVS A, IVS B and IVS C). After Sun, S.M., Slightom, J. & Hall, T.C. (1981) *Nature*, **289**, 37–41.

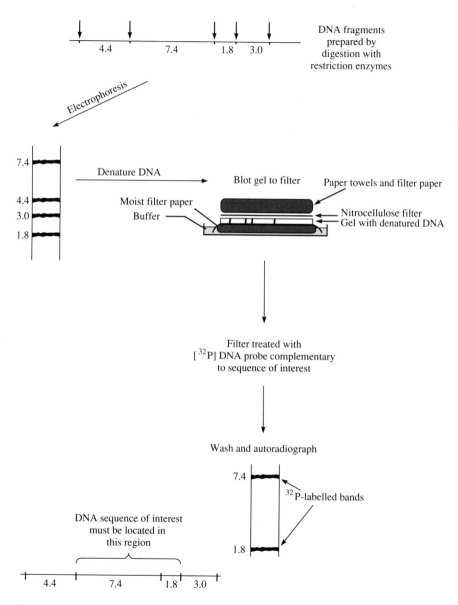

Fig. 20.9 The process of 'Southern blotting'. See text for full explanation of this process.

the desired DNA sequence in a suitable host. Although a number of bacteria, and even yeasts, have been used as hosts, the most common cloning vectors are those associated with the bacterium *Escherichia coli*, namely bacteriophage λ and various plasmids and cosmids.

The genome of bacteriophage λ is a double-stranded DNA molecule of about 50 000 base pairs that is linear within bacteriophage particles. The two 5′ ends of the molecule, at opposite ends of the DNA double strand, have single-stranded, 'sticky' ends of 12 bases, complementary with each other. The phage DNA can anneal these ends to form a circle after infection of the host. Since just under half the genome of phage λ is not essential for phage growth, such sequences can be excised with restriction enzymes and replaced with foreign DNA sequences of similar length, without affecting normal growth of the virus. This limits the amount of foreign DNA that can be inserted into the viral DNA to about 18 000–21 000 base pairs. This DNA is introduced into *E. coli* and amplified very efficiently by the normal processes of bacteriophage infection and reproduction within the bacterial cell. The DNA introduced in this way comprises vector (bacteriophage) DNA into which 'foreign' sequences are inserted and is known as *recombinant DNA*.

Plasmids are small, circular molecules of double-stranded DNA that exist outside the chromosome and occur naturally in bacteria and yeast. They replicate independently of the host cells. They generally account for only a small fraction of the host cell DNA but often carry important genes such as those for resistance to antibiotics or heavy metals. In *Rhizobium*, the N_2-fixing bacterial symbiont of legumes (see Section 13.8), the genes for N_2 fixation are carried on a plasmid. Plasmids are introduced into host bacterial cells by transformation. In order for bacterial cells to take up plasmids from the surrounding medium, the cells must first be rendered permeable to macromolecules (or made 'competent') by treatment in cold $CaCl_2$, heat shock and subsequent incubation with $CaCl_2$. Even so, only about 20% of the surviving cells will take up the added DNA and fewer do so if the plasmids are large. Therefore, smaller plasmids with small amounts of introduced DNA are better, and phage λ or cosmids are the preferred vector for introducing long lengths of DNA. All three types of vector are (or contain) much smaller DNA molecules than the *E. coli* host genome, so their DNA can readily be separated and purified.

Cosmids are segments of DNA which are hybrids of plasmids to which the cohesive ends of the phage λ have been added. They do not contain the phage λ

genes necessary for replication but inside the host they replicate as a plasmid. Foreign DNA of 28 000 – 42 000 base pairs can be inserted into these vectors which, provided the DNA is the correct length, can be 'packaged' into phage λ particles and introduced into *E. coli* by the normal processes of bacteriophage infection (transfection).

A basic procedure for introducing DNA sequences into a cloning vector is outlined in Fig. 20.10. Restriction fragments with sticky ends are created from the foreign DNA to be cloned by treating it with a suitable restriction enzyme. The cloning vector DNA is treated with the same enzyme, allowing the restriction fragment to anneal to it. The ends are joined by T4 DNA ligase in a reaction requiring ATP. The hybrid (recombinant) vector can then be introduced into *E. coli* by transformation or transfection. As the bacteria divide, so does the cloning vector, to produce up to 200 copies of the original DNA fragment. The hybrid plasmid, cosmid or phage DNA can be harvested, purified and the original DNA fragments excised by retreatment with the original restriction nuclease.

This essentially provides a 'gene library' of many fragments of the foreign genome. Only some of the bacteria will contain cloning vectors bearing part or all of the gene in question; some will contain fragments of other genes and others will contain non-coding foreign DNA. The real problem with recombinant DNA technology and gene cloning is to sort out which bacteria contain a cloning vector with the DNA sequences of interest.

20.5.5 *The synthesis of cDNA from mRNA can offer a quicker approach to gene isolation*

Plant genomes are large and complex, which causes considerable problems in isolating a single gene from the large numbers of DNA fragments generated by the methods described above. It is often quicker to use an indirect approach through the synthesis of highly specific clones of DNA complementary to mRNA sequences. First, mRNA is isolated. In some instances this can be achieved by isolating cytosolic polyribosomes (as in leaf tissue) or membrane-bound ribosomes (as in seed endosperm), although many mRNA species can be lost by this procedure. In many cases, mRNA can be isolated by oligo deoxythymidine (oligo-dT) cellulose affinity chromatography because most mRNA molecules possess a poly-A tail which binds to the oligo-dT cellulose. Alternatively, mRNA can be isolated by size fractionation using electrophoresis. This allows purification of mRNA from the other nucleic acids in the sample.

Having isolated and purified a mRNA fraction, there are two basic procedures

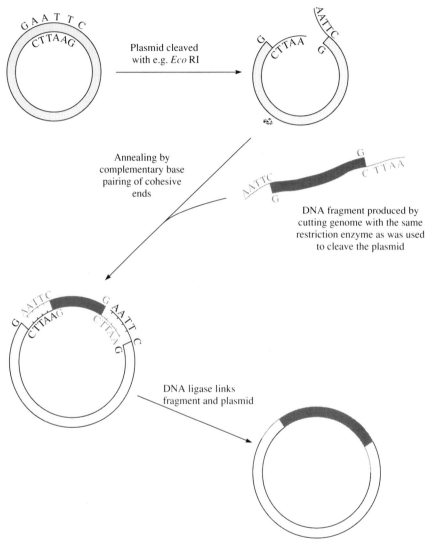

Fig. 20.10 The basic process by which a 'foreign' DNA fragment is introduced into a plasmid. Both plasmid and foreign DNA are cut with the same restriction nuclease so that foreign DNA fragments can anneal with the plasmid DNA. Fragments are then linked by DNA ligase (usually from T4 phage) and the hybrid plasmid (cloning vector) is re-introduced into *E. coli*, for example, in which it is produced in large numbers as the bacteria divide.

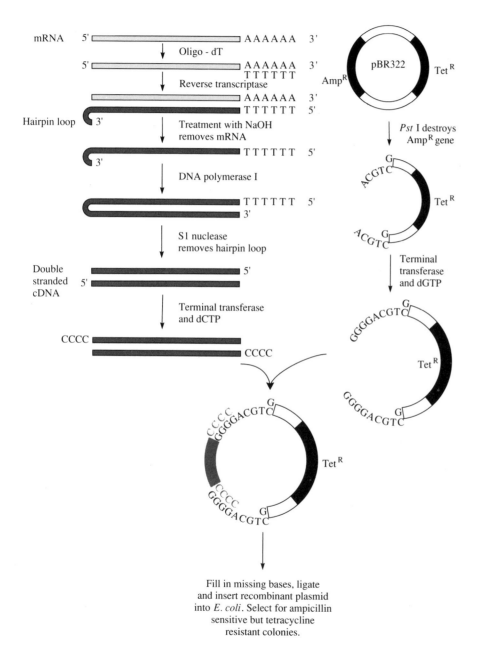

Fill in missing bases, ligate
and insert recombinant plasmid
into *E. coli*. Select for ampicillin
sensitive but tetracycline
resistant colonies.

that can be used to create 'copy' DNA (cDNA). The first is illustrated in Fig. 20.11. Short chains of oligo-dT, mixed with the purified mRNA, hybridize to the poly-A tail to provide a primer for the action of the enzyme, reverse transcriptase. This enzyme, isolated from certain RNA tumour viruses, uses RNA as a template to synthesize a complementary DNA strand to form a DNA–RNA hybrid. The RNA is removed by alkaline hydrolysis, leaving DNA with a 'hairpin loop' at one end. The loop acts as a primer for the synthesis, by the action of DNA polymerase I, of a second strand of DNA and is then removed by S1 nuclease (a single-strand-specific nuclease) to yield a double-stranded cDNA molecule. To incorporate cDNA into the DNA of the cloning vector it is treated with oligodeoxynucleotides and terminal transferase to add short base sequences or 'linkers' at the 3′ ends of the DNA strands. Cutting the cloning vector DNA with a restriction enzyme and adding a linker of the complementary oligonucleotide (i.e. oligo-dG) permits the cDNA to be annealed into the cloning vector.

There are disadvantages in this system, the most important being that the yield of full length cDNA sequences is very low. S1 nuclease digestion of the hairpin loop usually removes some of the coding region of the cDNA, leaving it truncated. These cDNAs are still useful as hybridization probes but they cannot direct the synthesis of complete proteins after their introduction into host cells.

A more widely used technique devised by Okayama and Berg is illustrated in

Fig. 20.11 A basic procedure for producing cDNA from specific mRNA sequences and the introduction of this cDNA into a cloning vector such as pBR322. The first step is the isolation of mRNA. Short chains of oligo-dT hybridize to the poly-A tail of the mRNA and provide a primer for the action of reverse transcriptase, which forms a DNA chain with a base sequence complementary to that of the mRNA. Alkaline hydrolysis removes the mRNA leaving the single-stranded DNA molecule. Reverse transcriptase produces a 'hairpin loop' at the 3′ end of the DNA strand; this acts as a primer for synthesis of a second strand of DNA by DNA polymerase I. The double-stranded DNA chain thus formed has the hairpin loop removed by treatment with S1 nuclease. Treatment with terminal transferase and dCTP produces short oligo-dC sequences (linkers) at the 3′ ends of the DNA strands. A vector (e.g. pBR322) is used to introduce the cDNA into a bacterial cell so that it can be cloned. This particular vector contains genes for ampicillin resistance (Amp^R) and tetracycline resistance (Tet^R). *Pst* I cuts pBR322 in the region of the Amp^R gene, thus destroying that function (Tet^R is unaffected). Treatment with terminal transferase and dGTP produces a short-chain oligo-dG sequence, which anneals with the oligo-dC region on the cDNA. After the missing bases in the recombinant DNA molecule are filled in, the plasmid is inserted into *E. coli*. Colonies of *E. coli* containing the recombinant plasmid are identified by having ampicillin sensitivity whilst retaining tetracycline resistance.

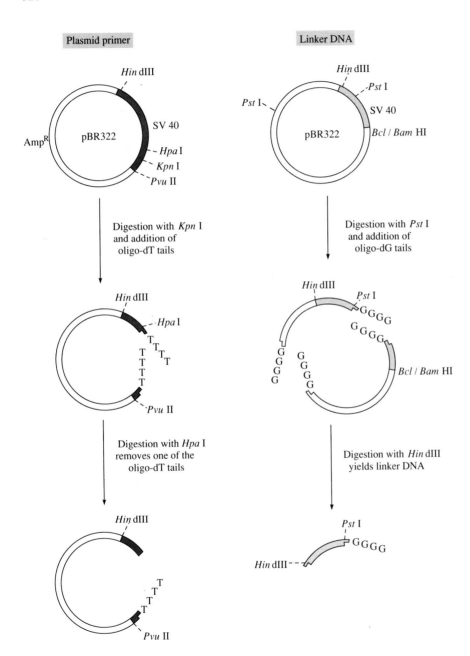

Figs 20.12 and 20.13. The first step involves the synthesis of a plasmid primer and an oligo-dG tailed linker DNA (Fig. 20.12). The system shown uses a construct of the plasmid pBR322 and a segment of a simian virus (SV 40). This recombinant plasmid is useful because it has two unique restriction sites: *Kpn* I where single strand ends can be generated for addition of oligo-dT tails; and *Hpa* I, near one end to allow one of the oligo-dT tails to be removed. Treatment of the recombinant plasmid with *Kpn* I and *Hpa* I therefore results in a primer with an oligo-dT tail. Linker DNA is also prepared from another, similar, recombinant pBR322–SV 40 construct, in this case by digestion with *Pst* I and addition of oligo-dG tails, followed by digestion with *Hin* dIII. The primer plasmid and oligo-dT tailed linker are then used in the synthesis of cDNA as shown in Fig. 20.13, by annealing the poly-A tail of mRNA to the oligo-dT tail on the primer plasmid. Reverse transcriptase is then used to catalyse the synthesis of a cDNA strand. Oligo-dC tails are added by the action of deoxynucleotidyl transferase. Digestion with *Hin* dIII allows a portion of the primer to be replaced with the oligo-dG tailed linker. This forms complementary base pairs with the oligo-dC cohesive end on the cDNA and the circular recombinant plasmid is completed by annealing the linker with DNA ligase. This yields a recombinant plasmid containing an RNA strand and a complementary cDNA strand. The RNA strand is replaced by a complementary DNA sequence through the action of RNase H, DNA polymerase and DNA ligase from *E. coli*. The end result is a recombinant plasmid containing a cDNA sequence. This plasmid can be used as a cloning vector by transformation of *E. coli*, yielding large quantities of the cDNA sequence.

20.5.6 Clones containing cDNA must be identified

Even if cDNA is successfully incorporated into a bacterium it might not contain the desired sequence. Except for certain cases, such as seed storage proteins, where

Fig. 20.12 Synthesis of a plasmid primer and linker DNA for the Okayama and Berg method of cDNA cloning. Both primer and linker are derived from plasmid pBR322 containing different segments of SV 40 DNA (light grey and dark grey shaded regions). pBR322 also contains a gene for ampicillin resistance (Amp^R). For primer preparation the pBR322–SV 40 recombinant is first digested with *Kpn* I and oligo-dT tails are added. Digestion with *Hpa* I leaves a primer plasmid with one oligo-dT tail of ~60 residues. The linker is prepared by digestion of the pBR322–SV 40 recombinant with *Pst* I and addition of oligo-dG tails. Digestion with *Hin* dIII releases an oligo-dG tailed linker derived from SV 40 DNA. After Okayama, H. & Berg, P. (1982) *Mol. Cell. Biol.*, **2**, 161–70.

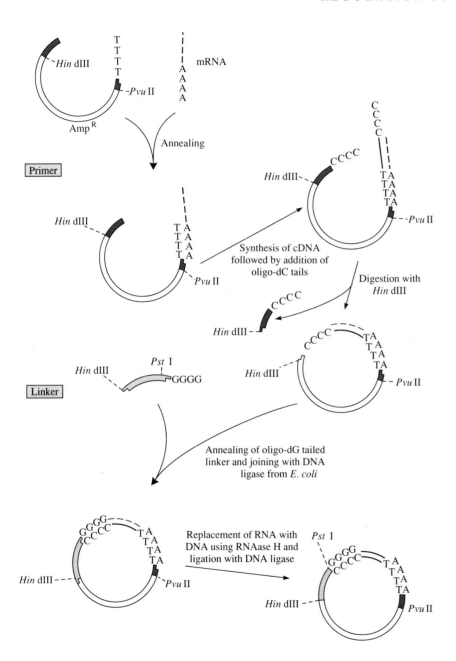

Fig. 20.13 Construction of cDNA by the Okayama and Berg method. Poly(A)$^+$ RNA (mRNA) is isolated from plant tissue and the poly-A tail hybridizes to the oligo-dT tail of the primer plasmid. cDNA synthesis is carried out by reverse transcription of the mRNA. Oligo-dC tails are added and *Hin* dIII digestion yields primer, to which an mRNA sequence and a corresponding cDNA with an oligo-dC tail are attached. The oligo-dG tailed linker hybridizes to the oligo-dC tail of the modified primer and the circular recombinant molecule is joined using DNA ligase. The RNA strand of this molecule is replaced through the combined action of RNase H, DNA polymerase 1 and DNA ligase to yield a plasmid containing a double-stranded cDNA. After Okayama, H. & Berg, P. (1982) *Mol. Cell. Biol.*, **2**, 161–70.

the relevant mRNA is copious in developing seeds, most mRNA preparations are a heterogeneous collection. The first step is to identify clones containing any cDNA. Most plasmids and other cloning vectors contain genes for antibiotic resistance. It is easy, therefore, to plate out cultures on a medium containing antibiotic and to screen for surviving colonies. Some plasmids are particularly useful because they contain genes for resistance to two antibiotics. For example, the plasmid pBR322 contains genes for ampicillin resistance and for tetracycline resistance. Inserting cDNA into the plasmid using the restriction enzyme *Pst* I interferes with the function of the gene coding for ampicillin resistance but leaves the gene for tetracycline resistance intact. As a result, those *E. coli* cells containing a *recombinant* plasmid are resistant to tetracycline but they do not appear on replica plates containing ampicillin. This makes it relatively easy to identify the original colonies of *E. coli* that contain cDNA.

Alternatively, a plasmid such as pUC9 can be used. The pUC plasmids contain the ampicillin resistance gene and the DNA replication origin (i.e. the base sequence at which DNA replication is initiated) from pBR322 attached to a portion of the *lac* Z gene of *E. coli* which codes for β-galactosidase. The *lac* region of the plasmid contains a DNA insert with an array of unique sites that are susceptible to particular restriction enzymes. When the plasmids are introduced into *E. coli* lacking a fully functional β-galactosidase, the incorporated *lac* Z region results in formation of the functional enzyme which can be detected chemically. Colonies in which the plasmid is present appear blue when grown on agar plates containing 5-bromo-4-chloro-3-indolyl-β-D-galactoside, an analogue of the β-galactosidase enzyme substrate. If, however, DNA fragments are cloned into any of the multiple restriction sites, the *lac* gene is inactivated and the colonies appear white. Although there are a great many other plasmids which can be used as cloning vectors, pUC9 is often the plasmid of choice.

20.5.7 *Clones containing the DNA sequence of interest must be identified from the many that comprise the 'gene library'*

The procedures described so far allow the identification of clones of, for example, *E. coli* containing cloning vectors with any of a large number of DNA or cDNA fragments. There are a number of approaches that lead to identification of those clones containing the sequence of interest.

If a suitable DNA or RNA hybridization probe is available the various clonal populations can be plated out and allowed to develop into colonies. These colonies are replica-plated onto nitrocellulose filters, which are placed on top of nutrient agar to allow the colonies to grow. The filters are removed, the cells are lysed and the double-stranded DNA is converted to single strands ('denatured') by treatment with alkali. The DNA is then firmly attached to the filter by baking and the whole is washed thoroughly. Finally, the DNA is exposed to a solution containing a radioactive probe (see Section 20.5.3), which binds only to DNA with a complementary sequence and identifies the colonies containing that sequence. Alternatively, colonies can be screened for DNA sequences complementary to particular mRNAs by hybrid arrest and hybrid release translation (HART and HRT respectively; Fig. 20.14). In HART, DNA from the clones (i.e. host cells containing modified (recombinant) DNA) to be tested is denatured and allowed to hybridize with free mRNA in solution. The information contained in the mRNA is then translated into protein using an *in vitro* protein synthesis system. (This is usually a system from wheatgerm or rabbit reticulocytes which is free of mRNA but contains ribosomes and the various enzymes and protein factors needed for translation; sometimes a prokaryotic system from *E. coli* is used.) The proteins formed are separated by polyacrylamide gel electrophoresis. The mRNA bound to the complementary cDNA sequence will not be available for translation and therefore no corresponding protein will appear on the gel. Control and cDNA-treated systems are compared in order to identify clones containing the particular mRNA sequence.

The other approach (HRT) firstly involves denaturation of the cDNA isolated from the colonies selected. The single-stranded cDNA is bound to nitrocellulose filters and exposed to mRNA. Complementary mRNA hybridizes to the cDNA and remains bound to the filter; other mRNA species are washed off. The bound mRNA is then released, translated *in vitro* and identified. In this case, only the protein coded for by the mRNA bound to the cloned DNA is translated. This latter technique is also used to isolate specific mRNA species from the general mRNA population.

The various probes can either be a closely related gene sequence (e.g. the DNA

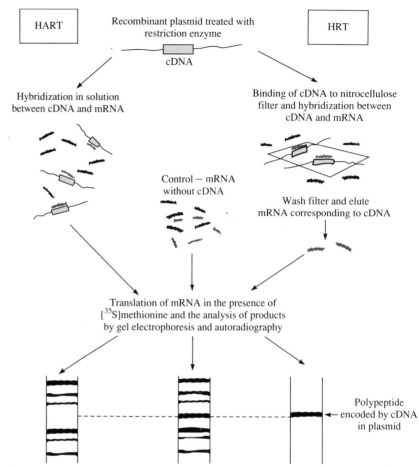

Fig. 20.14 The processes of hybrid arrest translation (HART) and hybrid release translation (HRT) can be used to screen for the presence of particular DNA sequences in recombinant plasmids. The two processes rely on the hybridization, by complementary base pairing, between cDNA sequences and mRNA. In HART, cDNA sequences are mixed in solution with mRNAs extracted from plant tissue. Hybridization between the cDNA and the complementary mRNA prevent that mRNA from being translated in an *in vitro* translation system. As a result, that particular gene product will be absent when the translation products are analysed. In contrast, in HRT the DNA is first bound to a nitrocellulose filter. Complementary mRNA hybridizes to the cDNA on the filter whilst other mRNAs are washed off. The mRNA can then be eluted in a pure form and translated. In this case a single translation product is formed corresponding to the cDNA.

sequence coding for a protein in one species can be used as a probe in another) or, if the protein sequence of the product is known, an oligonucleotide sequence of about 15 nucleotides, specific to a particular region of the protein, can be constructed. This oligonucleotide probe hybridizes only to DNA sequences containing the complementary base sequence. Alternatively, a differential screening process must be carried out. Total mRNA is isolated and purified from a tissue expressing the required gene, and is copied as radioactively labelled cDNA. This can serve as a probe since it hybridizes to the DNA from clones carrying complementary sequences to any of the messages in the original mRNA fraction. By repeating the process with mRNA from the same species, but by not expressing the required gene (i.e. a negative probe), positively selected colonies of cells, which also hybridize to the negative probe, can be discarded.

20.5.8 Cloned DNA fragments can be sequenced to provide information on gene organization

Having isolated and cloned a specific DNA fragment or cDNA, determination of the base sequence can provide considerable additional information regarding the organization of the gene. The sequencing of nucleic acids was, at one time, an extremely difficult and time-consuming process. However, DNA cloning processes which permit the production of large amounts of pure DNA and techniques based on the enzymic manipulation of DNA have dramatically improved DNA sequencing methods.

There are two basic methods used to sequence DNA. These are the chemical degradation method developed by Maxam and Gilbert, and the 'chain termination' method of Sanger and his co-workers. In the chemical degradation process, segments of DNA produced by restriction enzymes are labelled at one end of each strand with $[^{32}P]ATP$ using polynucleotide kinase. The two strands of the DNA chain are separated by denaturation and purified by gel electrophoresis. The labelled DNA is then divided into four aliquots and each is subjected to a chemical treatment to destroy one or two of the bases in the DNA. Subsequent treatment with piperidine breaks the DNA strand at this site. Cleavage of the DNA strand is base-specific and occurs at G, A + G, C + T or C. Cleavage occurs randomly at any of the appropriate bases in the chain. The reaction conditions are adjusted so that only some of the molecules have their bases destroyed at any one position, giving rise to a series of fragments of different size which can be separated by polyacryl-

amide gel electrophoresis. Labelled fragments are detected by autoradiography and a sequence of bases is built up (Fig. 20.15a).

The Sanger sequencing procedure relies on the synthesis of DNA chains and depends on the incorporation of 2',3'-dideoxynucleotide analogues of each of the four deoxynucleoside triphosphates. The dideoxynucleotides are incorporated into a growing DNA chain, but they cannot form phosphodiester bonds with the next incoming deoxynucleotide.

General structure of dideoxynucleotides

The Sanger technique also involves the isolation of single-stranded (but unlabelled) DNA from restriction fragments. A ^{32}P-labelled primer is annealed to the single-stranded DNA fragment and DNA polymerase 1 is used to copy the DNA strand. Normal polymerization occurs until a base is reached where a dideoxynucleotide is incorporated and elongation ceases at that point. Four reactions are carried out with DNA synthesis inhibited by each of the four dideoxynucleotides. By manipulating the ratio of the 'normal' deoxynucleotide and dideoxynucleotide, not every fragment in a sample is stopped at the same position and so, a series of labelled fragments of different lengths results. These can be separated by polyacrylamide gel electrophoresis and the sequence of bases in the original strand determined (Fig. 20.15b).

In order to obtain sufficient material for sequencing, single-stranded DNA is often cloned using the bacteriophage M13 as a vector. This filamentous bacteriophage contains a circular DNA genome, single-stranded when the phage is mature but double-stranded during replication within E. coli. The double-stranded form of the phage DNA is isolated and the DNA to be sequenced is inserted. The recombinant DNA is re-introduced into E. coli by transfection. Mature M13 bacteriophage particles, containing single-stranded DNA, are subsequently released from E. coli cells. The M13 genome has been modified to include a portion of the E. coli β-galactosidase gene. As with pUC9 plasmids described in Section 20.5.6,

(a) Maxam–Gilbert chemical degradation

(b) Sanger chain termination

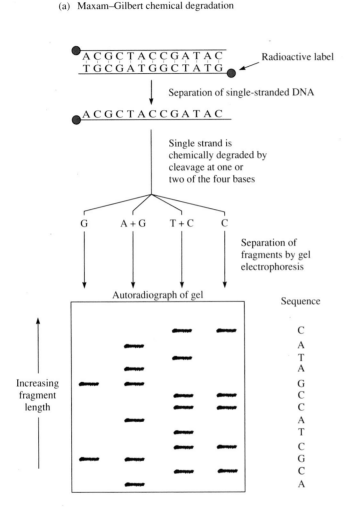

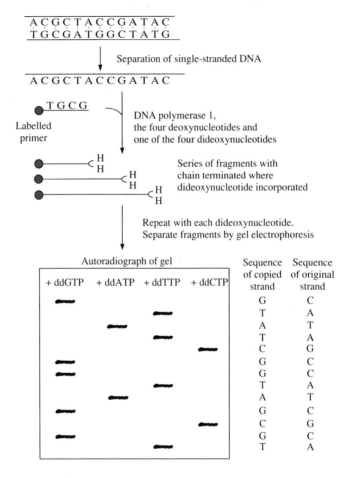

Fig. 20.15 DNA sequencing by the Maxam–Gilbert or Sanger methods. (a) The Maxam–Gilbert method relies on the chemical degradation of single-stranded DNA labelled at the 5′ end. This breaks the DNA strand at one of the four bases and produces a number of fragments which can be separated by gel electrophoresis. Different chemical treatments are used to break the strand. From the resultant sizing pattern of radioactively labelled fragments the sequence of bases in the chain can be inferred. (b) The chain termination technique of Sanger relies on the synthesis of short DNA strands from single-stranded DNA. In the presence of radioactively labelled primer DNA, DNA polymerase I directs the formation of a complementary DNA strand. If the reaction mixtures contain all four deoxynucleotides and one of the four possible dideoxynucleotides, the growing DNA strand is terminated when the chain reaches a base complementary to the dideoxynucleotide. Not all the developing chains are terminated at this position however, and a series of fragments is produced each being terminated by the same dideoxynucleotide. By comparison of the fractionation pattern the sequence of bases in the copied strands, and hence that in the original strand, can be inferred.

this section of the β-galactosidase gene complements sequences in a mutant *E. coli* host strain so that it can produce a functional β-galactosidase. If 'foreign' DNA fragments are cloned into M13, these are inserted into the β-galactosidase sequences so that the mutant *E. coli* host cannot make this enzyme. With a suitable indicator, mutant *E. coli* cells containing recombinant M13 phage show up as

clear/white areas against a background of blue (formed by mutant *E. coli* with non-recombinant M13 and, hence, a functional β-galactosidase).

It is now possible to obtain a number of plasmids for use as cloning vectors for single-stranded DNA for dideoxynucleotide sequencing. These, and a large range of other cloning vectors, are available commercially.

DNA base sequences for restriction fragments can be used to build up sequences for much larger portions of the plant genome by comparison of base ordering in overlapping fragments. These DNA base sequences can be used to provide a considerable amount of information, such as data on putative promoter or enhancer sites. Comparison of DNA base sequences for fragments of the DNA genome and the corresponding cDNA can yield information about the nature of introns in a gene (e.g. the phaseolin gene, see Fig. 20.8).

20.5.9 Plants pose specific problems for recombinant DNA work

There are particular problems associated with recombinant DNA work in higher plants. Plant cells have tough walls which are difficult to break, and they contain high levels of nucleases (especially ribonuclease) which make it difficult to obtain high yields of nucleic acid per gram of starting tissue, especially in mature seeds and leaves. It is generally easier to make RNA preparations from young seedlings. Thirdly, plant cells contain large amounts of polysaccharide which sometimes co-purifies with RNA and leads to purification problems. Isolation of nuclear DNA sometimes requires the absence or removal of both chloroplastic and mitochondrial DNA, achievable to some extent by choice of tissue or careful isolation and treatment of intact nuclei. Finally, the typical plant nucleus contains a very high proportion of repeated (non-coding) DNA from which the gene(s) of interest have to be selected.

20.6 Genetic engineering: manipulation of the genetic information of plant cells

The last half century has seen a revolution in agricultural practice, technology and plant breeding, producing a marked increase in the global capacity for food production. This development has been rather one-sided, however, and has been restricted mainly to the developed countries that can afford the pesticides, fertilizers and research effort to develop new varieties. In many 'Third World' countries this intensive, high-cost approach is not practical and climatic conditions are often unsuitable for many of the new varieties. There is, therefore, considerable pressure to modify existing crops to suit needs. Furthermore, the energy- and management-intensive agricultural practices of developed countries are becoming less economical as reserves of fossil fuels dwindle. It is an attractive proposition therefore to manipulate crops so that productivity is maintained (or increased) for a lower energy input.

20.6.1 Some objectives for the alteration of plant performance

There are many features of the biology of plants that are possible candidates for alteration.

Many crop plants are C_3 species and are susceptible to photorespiration (see Chapter 11). Since the basis of photorespiration is the oxygenase activity of RuP_2-C/Oase, it may be possible to manipulate the gene for this enzyme to produce an enzyme active in CO_2 fixation but with a reduced oxygenase function. Since relevant information about the nature and functioning of the RuP_2-C/Oase active site is still sparse, such suggestions are probably premature.

There are many cultivars of crops and species of non-crop plants that possess resistance to diseases and pests that reduce crop production but there is a strong need for more resistant varieties. Other resistance mechanisms could be introduced to plants. For instance, the bacterium *Bacillus thuringiensis* var. *berliner* synthesizes a glycoprotein that is highly toxic to some Lepidoptera (moths and butterflies), and other varieties of *B. thuringiensis* produce toxins that act specifically on Diptera, such as mosquitoes and blackflies, or on Coleoptera (beetles). The genes for these polypeptides have been characterized and their introduction into brassicas, for example, could have dramatic effects on the ravages of predators such as the caterpillar of the cabbage white butterfly. The gene for the toxin from *B. thuringiensis* var. *berliner* has been introduced into tobacco and confers resistance to tobacco hornworm infestations. Similarly, resistance to microbial (viral, fungal and bacterial) infections could be introduced.

A very great proportion of the world's population depends, to a large extent, on seed storage proteins and carbohydrate for their nutrition. However, cereals and legumes contain limited amounts of some of the amino acids essential to humans. For instance, many cereal storage proteins are deficient in lysine and tryptophan, while legumes have a low proportion of methionine and cysteine which contain sulphur. The possibility of introducing more of these amino acids into various proteins is, therefore, attractive, although changing the primary structure of proteins in this way could dramatically affect their properties. Alternatively, genes for new polypeptides, rich in these amino acids, could be introduced so that they are expressed in seeds or the levels of existing storage proteins, rich in the relevant amino acids, could be amplified.

Modern agricultural practice relies heavily on the use of nitrogenous fertilizers. This is costly and uses a lot of energy, both for synthesis and transport to the site of application. Since one group of crop plants, nodulated legumes, can fix and assimilate molecular dinitrogen (N_2) from the atmosphere, much interest has been expressed in, and research effort devoted to, transferring the nitrogen fixation genes of the legume–*Rhizobium* symbiosis into non-leguminous crop plants such as maize, which normally rely on applied fertilizers. However, the N_2 fixation process and the interactions between host and symbiotic bacteria (see Section 13.8) is highly complex and it is likely to be some time before the molecular biology of the system is fully understood and practical advances are made. Nevertheless, significant advances have been achieved and it may yet prove possible to extend the host range of *Rhizobium* spp. to other, non-leguminous, crop plants or at least to improve the efficiency of N_2 fixation in existing host species.

However much agricultural practice is improved, crop plants will always be subject to a range of environmental extremes with losses resulting from, for example, drought, temperature extremes and salt accumulation in soil. The introduction of cultivars with increased tolerance to such stresses would be of great importance. Plant breeding has achieved some notable success in this area but the most dramatic stress-tolerance capabilities are shown by plants of low economic importance (weeds). Such plants are not genetically compatible with most crops and cannot be considered useful for conventional plant breeding techniques. It is tempting to consider genetic engineering as a method of overcoming this obstacle, although many adaptations to stress or pathogen resistance involve morphological features controlled by many, rather than a single, easily manipulated gene.

Many of these manipulations of gene expression could be (and have been) achieved with conventional breeding techniques where a large number of plants are screened for desirable characteristics. In addition, the techniques of protoplast fusion to fuse cells from different species and regenerate hybrids could produce plants with useful new properties. However, these techniques are essentially random and do not identify particular genetic targets for modification. (They are also very time consuming!) This specificity *is* possible, however, with the recombinant DNA techniques now available to plant molecular biologists. Not only do these techniques, developed during the last two decades, allow identification, mapping and purification of some simple genes (Section 20.5) but they also permit the isolation of genes from one genome and their introduction into the genome of another organism. This section is designed to provide an introduction to how these manipulations can be carried out in plant systems.

20.6.2 Introduction of cDNA into plant genomes using Agrobacterium tumefaciens

A. tumefaciens is a common soil bacterium which can enter wounds on the stems of plants, inducing the plant cells to proliferate and form a tumour or gall. This usually happens at the junction of root and stem (the crown) and so the infection is known as crown gall disease. The tumour tissue, in contrast to normal plant cells, proliferates in culture without addition of auxin and cytokinin. The gall cells are induced to synthesize opines, derived from amino acids, keto acids or sugars (Table 20.4), which are used as carbon and nitrogen sources by the bacteria. Opine synthesis is due to the introduction of bacterial genes into the host cell genome.

Tumour induction, the induction of opine synthesis and suppression of host cell differentiation all depend on the presence of plasmids in the bacterium, termed T_i plasmids, which are circular DNA molecules with a molecular mass of about 1.2×10^5 kDa. *Agrobacterium* without a T_i plasmid is unable to infect susceptible plants. The T_i plasmids are classified according to the type of opine they induce the host cell to make. Most are octopine or nopaline plasmids and *Agrobacterium* harbours only one sort of T_i plasmid per cell. Genes for tumour induction, opine synthesis and the suppression of differentiation map close together in one segment of the plasmid, called the transforming DNA region (T-DNA) and this portion of the plasmid DNA is transferred to the host genome during infection. Transfer and integration can be shown by Southern blotting and copies of T-DNA of about 20 000 base pairs are found in tumour cells. The T-DNA in octopine tumour cells has seven genes that specify particular RNA transcripts in the plant. Most of these genes are controlled by separate promoters.

Because of their wide host range and integration of T-DNA into host cells, *Agrobacterium* T_i plasmids are excellent vectors for introducing foreign genes into plants (transfer vectors). The simplest way to introduce T-DNA into plant cells is to infect them with *A. tumefaciens* containing the appropriate T_i plasmid and then let the process take place naturally. It is necessary, therefore, to insert the derived DNA sequence(s) into the T region of the T_i plasmid. Since the entire T_i plasmid is larger than DNA molecules normally used for recombinant DNA studies, it is modified by the following procedure (Fig. 20.16). The T region is excised from the T_i plasmid and is introduced into one of the standard cloning vector plasmids used with *E. coli* (e.g. pBR322). This provides large amounts of vector carrying T-DNA. The relevant foreign gene is inserted into the T-DNA region of the modified *E. coli* plasmid and the hybrid plasmid is grown in large quantities by re-introducing it

Table 20.4 Opines synthesized in plant cells that have been transformd by *Agrobacterium tumefaciens*.

Octopine family

General Structure

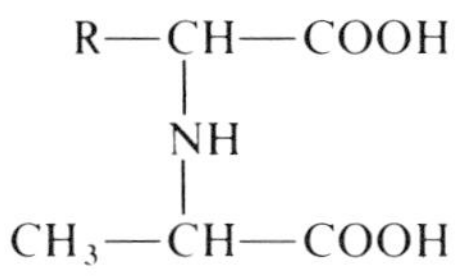

$$R-CH-COOH$$
$$|$$
$$NH$$
$$|$$
$$CH_3-CH-COOH$$

Octopine   R = $NH_2-\overset{\overset{NH}{\|}}{C}-NH-(CH_2)_3-$

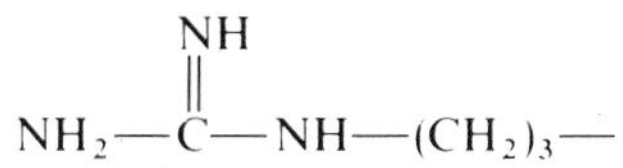

Lysopine   R = $NH_2-(CH_2)_4-$

Histopine   R =

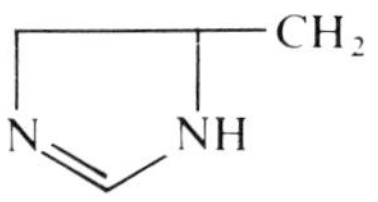

Nopaline family

General structure

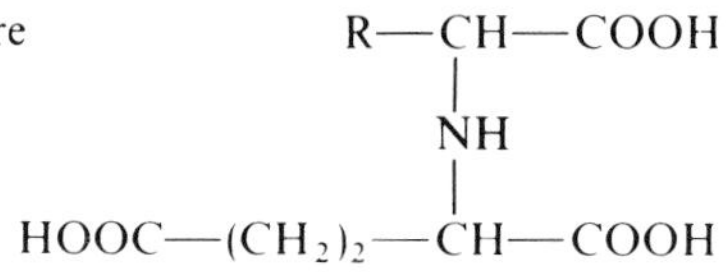

$$R-CH-COOH$$
$$|$$
$$NH$$
$$|$$
$$HOOC-(CH_2)_2-CH-COOH$$

Nopaline   R = $NH_2-\overset{\overset{NH}{\|}}{C}-NH(CH_2)_2-$

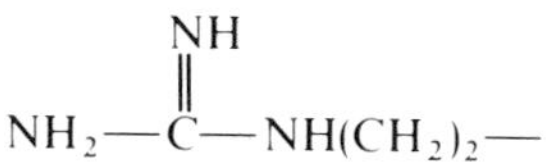

Ornaline   R = $NH_2-(CH_2)_3-$

Agropine

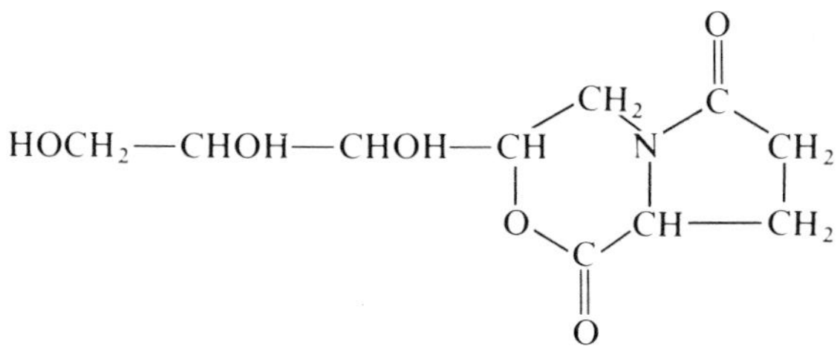

$HOCH_2-CHOH-CHOH-CH$ ...

into *E. coli* and growing quantities of this host. The cloned hybrid plasmid is isolated and introduced into *Agrobacterium* cells containing a normal T_i plasmid. Homologous genetic recombination (cross over) between the T-DNA segment of the native T_i plasmid and the cloned modified T-DNA (containing the foreign gene) results in transfer of the engineered T-DNA to the T_i plasmid and the displacement of its normal T-DNA. The end result is an *Agrobacterium* strain containing a T_i plasmid with the desired foreign gene inserted in the T region. Such strains are then used to infect host plants which will be transformed by the T-DNA, resulting in plant cells containing the foreign genes.

Only the T-DNA region and another region called '*vir*' (virulence) are essential in a plasmid for infection and bacterial proliferation to occur. To regenerate whole transformed plants, the T region must be modified to remove the factor which causes cells to remain undifferentiated (i.e. the oncogenicity factor). There is also a need to incorporate a working promoter for the foreign gene into the modified T_i plasmid in the correct position. This is often a promoter for the opine synthesis genes but it can be a promoter for a plant gene. A marker must also be introduced to show that the DNA is inserted into the plant tissue. This is usually a gene(s) for synthesis of opines or for antibiotic resistance (e.g. that coding for kanamycin resistance).

More frequently, binary (hybrid) vectors are used for plant transformations based on the *Agrobacterium* system. For example, the binary vector Bin 19 is a plasmid of 10 000 bases, containing short repeated sequences (25 base pairs) from the region bordering the T-DNA region of the *Agrobacterium* genome (border repeats). Within the borders of the T-DNA region are a hybrid nopaline synthase/neomycin phosphotransferase gene (which confers kanamycin resistance) and the β-galactosidase region from M13 bacteriophage. Bacteria containing plasmids with DNA inserts can then be identified following growth on agar plates containing kanamycin and the β-galactosidase substrate analogue 5-bromo-4-chloro-3-indolyl-β-galactoside (Section 20.5.6). When introduced into a suitable strain of *A. tumefaciens*, sequences contained within the T-DNA borders are transferred into the nuclear genome of infected plants provided that the bacteria contain a plasmid with the *vir* region.

20.6.3 Other vectors can be used to introduce foreign genes into plants

One drawback with the *A. tumefaciens* system is that the bacterium is restricted largely to dicotyledonous plants. For monocotyledons other mechanisms are normally used for gene transfer.

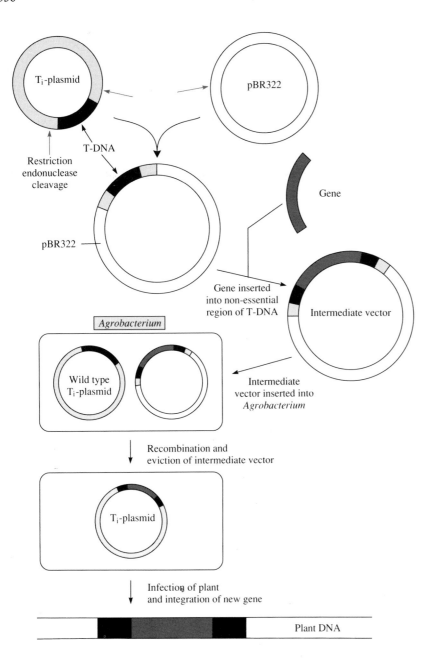

Alternative methods for introducing foreign DNA sequences into plant genomes include: (i) mechanical methods, such as microinjection of cDNA into single cells or protoplasts or shooting DNA-coated tungsten or gold spheres into cells or protoplasts ('biolistics'); (ii) introducing DNA by making protoplasts permeable to macromolecules through polyethylene glycol treatment or subjection to high voltage electrical pulses (electroporation); or (iii) the use of liposomes (containing cDNA) which fuse to protoplasts. However, it is difficult to regenerate plants from protoplasts and the cDNA rarely integrates with the host genome by this method.

Nonetheless, electroporation and microinjection are commonly used, especially in monocotyledons not subject to infection by *Agrobacterium*. Other natural routes involving plant pathogens have been investigated. Initially, considerable interest was shown in viral vectors such as cauliflower mosaic virus (CaMV). However, this double-stranded DNA virus has a host range limited primarily to the Brassicaceae (cabbage, cauliflower, turnips, Brussels sprouts, etc.) and has a relatively small genome size which restricts insertion of DNA to only about 250 base pairs (most genes are 1500–6000 base pairs). Also, the DNA does not integrate with the host genome. Investigations of CaMV have, nevertheless, proved valuable as the CaMV promoter works extremely well in a number of chimaeric systems (i.e. those involving recombinant DNA from different species). More recently, the tomato golden mosaic virus (TGMV) has been used as the basis of a transfer vector system. This virus has two advantages over CaMV as it proliferates rapidly within the cells

Fig. 20.16 The introduction of foreign genes into a plant genome using an *Agrobacterium* T$_i$ plasmid as a vector. The T$_i$ plasmid and one of the standard cloning vectors, such as pBR322, are both treated with restriction nucleases to cut them as shown by the grey arrows. This excises the T-DNA region of the T$_i$ plasmid which is inserted into the cloning vector. The foreign gene of interest is inserted into the non-essential region of T-DNA and the resulting plasmid (intermediate vector) is grown in large quantities in a host, such as *E. coli*. The cloned intermediate vector is then introduced into *Agrobacterium* cells which contain wild type T$_i$ plasmid. Homologous recombination between the intermediate vector and the T$_i$ plasmid results in exchange of the manipulated T-DNA region from the intermediate vector and the wild type T-DNA. This produces a T$_i$ plasmid containing a recombinant T-DNA region with the foreign gene inserted. The intermediate vector (now containing a 'normal' T-DNA region) is evicted by introduction of a third plasmid. The recombinant *Agrobacterium* is used to infect a host plant and the 'foreign' gene becomes incorporated into the host genome during the normal course of the infection process.

and tissues of a wide range of plants including monocotyledons and dicotyledons and the viral genome (plus the introduced 'foreign' plant gene) integrates into the host DNA.

Many plant genes or cDNAs have now been cloned and a number of them have been transferred between species. Although many of these gene transfers have served only to clarify aspects of technique (e.g. moving pea RuP_2-Case to tobacco), they have been invaluable as the experimental basis for more applied aspects of engineering the constitution of plant genomes. It must also be appreciated that the ability to manipulate plant genomes by the techniques of genetic engineering provides an extremely powerful tool for understanding the processes by which gene expression is regulated.

Further reading

Monographs and treatises: Alberts *et al.* (1989); Grierson & Covey (1988); Mantell *et al.* (1985); Marcus (1989); Parthier & Boulter (1982).

Regulation of gene expression: Kuhlemeier *et al.* (1987); Mullet (1988); Tobin & Silverthorne (1985).

Genetic engineering of plants: Klee *et al.* (1987).

21

Biogenesis of organelles

In the preceding chapters, the biochemical mechanisms for synthesis of the various molecules which make up cellular material and their assembly to form cellular structures were discussed. New organelles must be synthesized to keep pace with cell growth and division and to maintain an active complement of organelles in non-dividing cells to offset losses from normal cellular turnover. This final chapter is concerned with organelle biogenesis and the regulation of these processes. Mitochondria and, especially, chloroplasts have been extensively studied but little is known about the biogenesis of other organelles. Any bias in the ensuing pages towards chloroplast biogenesis reflects the available information and not personal prejudice. However, this emphasis is not altogether misplaced since the presence of chloroplasts and other plastids is one of the most distinguishing features of plant cells, both morphologically and functionally.

21.1 Mitochondria and chloroplasts are not formed *de novo*

When a plant cell divides, each daughter cell must synthesize a full complement of the requisite organelles in order to function normally. Organelles are not made *de novo* but are formed from pre-existing structures. Even chloroplasts formed during seed germination originate from colourless proplastids in seed tissues.

Mitochondria and plastids must, on average, double their size and divide at least once in each cell generation, but the number of chloroplasts per cell is very variable, ranging in peas from ~50 in young cells to 200–300 in mature tissue. Some algae, however, contain only a few chloroplasts, or a single chloroplast, which divide just before cell division occurs. In higher plants, in contrast, organelle division appears in most cases to occur throughout the cell cycle.

The replication of organelle DNA (see Chapter 18) is not usually correlated with nuclear DNA synthesis and in fact the number of copies of the plastome per chloroplast in dividing cells declines markedly with cell age, although the number of chloroplasts per cell increases (Fig. 21.1). This suggests that synthesis of chloroplast DNA is not an essential requirement for chloroplast division.

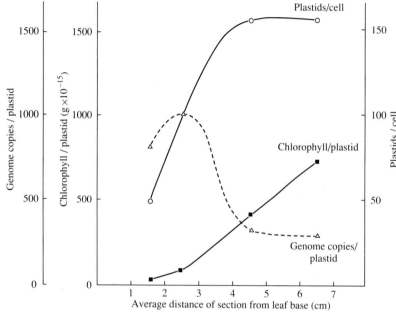

Fig. 21.1 Changes, associated with cell development, in the number of plastids per cell and number of genome copies per plastid in leaves of wheat. Wheat seedlings were grown for 7 days and the leaves cut into 1 cm sections. Because leaves of monocotyledons grow from a basal meristem, a series of sections along the leaf reflects a developmental sequence with older cells further from the leaf base (see Fig. 21.3). This developmental sequence is reflected in the increased chlorophyll content per plastid with distance from the leaf base. Plastids were isolated from leaf sections, counted, and the DNA extracted to estimate DNA/plastid and, hence, the number of genome copies/plastid, assuming a genome size of 1.5×10^{-16}g. In the basal 2–3 cm of leaf, increments in plastid DNA synthesis occurred in much the same ratio as plastid division so that the DNA per plastid remained relatively constant. At about 3 cm from the base, however, replication of plastid DNA ceased, whereas plastid division continued up to 4–5 cm from the base so that the number of genome copies per plastid decreased markedly. In the oldest tissue both plastid genome and plastids stopped dividing and genome copies per plastid became constant. After Boffey, S.A. & Leech, R.M. (1982) *Plant Physiol.*, **69**, 1387–91.

21.2 Chloroplasts develop from proplastids

In the very young cells of leaf meristems the progenitors of chloroplasts are small organelles known as proplastids, about 1 μm in diameter and bounded by a double membrane but with little internal membrane structure. Proplastids divide and in the course of development differentiate into chloroplasts (Fig. 21.2). Young chloroplasts can also divide. This combination of organelle division and subsequent development is referred to as organelle biogenesis. Much of the discussion that follows is restricted to the acquisition of functional competence by organelles that have undergone division, i.e. organelle differentiation.

Most studies of the development of chloroplasts from relatively undifferentiated proplastids have used one or other of two basic types of leaf development processes. In monocotyledons, leaves grow from a basal meristem so that cells from the base of the leaf are youngest and those at the tip are oldest. In such a system, the development of chloroplasts from proplastids is easily followed (Fig. 21.3). This developmental sequence is representative of the normal course of chloroplast development in higher plants.

In the absence of light, proplastids undergo a different sequence of development and become etioplasts (Figs 21.2 & 21.4). When light is supplied to etiolated plants, etioplasts differentiate into chloroplasts in the process known as 'greening'. Although many seedlings spend the early stages of growth in the dark under a layer of soil, chloroplast development during greening is *not* typical of normal differentiation from proplastids. Nonetheless, the use of greening tissues has helped to elucidate a number of problems involving the regulation of chloroplast development and gene expression and, therefore, both the 'normal' proplastid→chloroplast and the proplastid→etioplast→chloroplast developmental sequences are discussed below.

21.2.1 Chloroplast development involves quantitative and qualitative changes in many molecular components

As chloroplasts develop from proplastids the internal membrane structure proliferates as thylakoids are formed. In developing plastids of monocots the first thylakoids are formed by an invagination of the inner membrane of the proplastid envelope. As the chloroplasts develop, additional thylakoids proliferate from existing ones (Fig. 21.2). Changes in membrane structure also occur during greening. For instance, in dark-grown bean plants transferred to light, the tubular

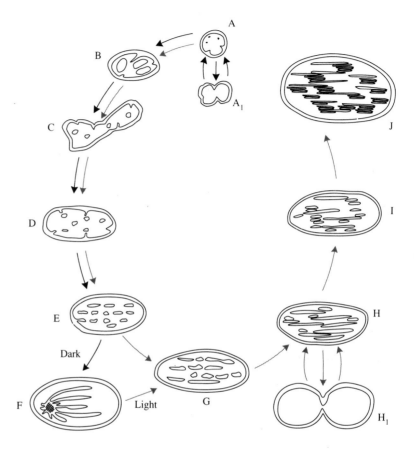

Fig. 21.2 Development of a chloroplast from a proplastid. Proplastids (A) are small (~1 μm) organelles which repeatedly divide in meristematic cells (A₁). They increase in size and synthesize starch grains (B), which are lost before the organelle passes through an amoeboid phase (C). Subsequently membrane proliferation occurs (D, E and G) and eventually granal stacks appear (H). The young chloroplasts at this stage are still capable of division (H₁). The complexity of granal stacking increases until the fully mature chloroplast is formed (I, J). In wheat the differentiation from A to J takes about 6 h. If leaves are held in the dark, development from A is arrested at E and the developing organelle forms an etioplast (F), containing an internal membranous structure called a prolamellar body. Under subsequent illumination etioplasts can differentiate to chloroplasts. Grey lines indicate events occurring in the light, black lines indicate events occurring in the dark. After Leech, R.M. & Baker, N.R. (1983). In Dale, J.E. & Milthorpe, F.L. (eds) *Growth and Functioning of Leaves*, pp. 271–307. Cambridge University Press, Cambridge.

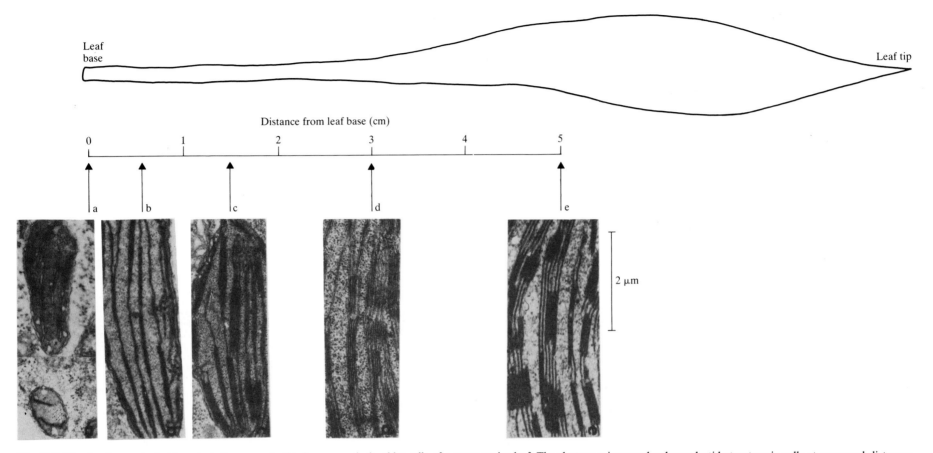

Fig. 21.3 The development of chloroplasts from proplastids in progressively older cells of a green maize leaf. The electron micrographs show plastid structure in cells at measured distances from the leaf base: (a) 0 cm; (b) 0.5 cm; (c) 1.5 cm; (d) 3 cm; (e) 5 cm. After Baker, N.R. & Leech, R.M. (1977) *Plant Physiol.*, **60**, 640–44.

membranes of the prolamellar body (Fig. 21.5) disperse as thylakoids are formed. The increase in the area of thylakoid membrane during the first 10 h of illumination approximates to the loss of membrane material from the prolamellar body, suggesting that the one gives rise to the other. However, on prolonged illumination, the thylakoid area per plastid increases ninefold, the additional area arising from the synthesis of new membrane components (Fig. 21.5).

This net membrane proliferation of necessity involves lipid synthesis but there are also striking changes in the lipid composition of thylakoids as they develop. In leaves of maize, for instance, chloroplast development is accompanied by a relatively much greater increase in galactolipid content, particularly that of monogalactosyl diacylglycerol (Table 21.1), compared to other lipids. Phospholipid content, in contrast, remains relatively constant except for phosphatidyl glycerol which increases in parallel with chlorophyll content. During chloroplast development there is also a marked change in fatty acid composition with a large increase in the amount of linolenic acid (18:3) in all lipids but particularly in galactolipids, where it makes up more than 75% of the fatty acids (Table 21.1). These changes are mainly quantitative; the only new species of lipid that appears as chloroplasts differentiate is phosphatidyl glycerol containing *trans*-3-hexadecanoyl (16:1(3*t*)) residues. Overall, the content of membrane lipids in plastids increases three to four-fold.

The other major constituents of plastids also show major changes during chloroplast biogenesis. In maize, for instance, the chlorophyll content of plastids increases with age whilst the DNA content drops due to a decrease in the number of copies of the genome per plastid (Fig. 21.1). In developing wheat, increasing cell age correlates with a massive increase in synthesis of chloroplast ribosomal RNA and in the number of 70S ribosomes per cell and per plastid (Fig. 21.6). The number of 80S ribosomes in the cytosol also increases but not to the same extent. The major protein in wheat leaves, ribulose-1,5-P_2 carboxylase (RuP$_2$-Case), increases at the same time as 70S ribosome numbers, with the maximum rate of increase in RuP$_2$-Case levels at the stage of maximum numbers of 70S ribosomes per plastid (Fig. 21.7). In wheat cells, rates of accumulation of chlorophyll, 70S ribosomes and RuP$_2$-Case during plastid development have been estimated at 4 pg cell^{-1}h^{-1}, 4.5×10^5 ribosomes cell^{-1}h^{-1} and 8 pg cell^{-1}h^{-1} respectively.

21.2.2 Chloroplasts are normally fully functional early in their development

Although the capacity of developing chloroplasts to carry out normal functions, such as CO_2 assimilation, O_2 evolution and ADP phosphorylation, is less than in mature cells, it is clear that this capacity is acquired at a very early stage of differentiation. For instance, plastids in all stages of development in 4-day-old wheat leaves exhibit significant rates of light-dependent O_2 evolution, indicating the presence of a functional non-cyclic electron transport chain. This is consistent with observations of the fluorescence characteristics of leaf sections (Fig. 21.8). Values of 0.5–0.6 for F_v/F_m at 695 nm (see Section 9.15.2) indicate that cells in even the basal 0.5 cm of 4-day-old wheat leaves have a photochemically competent photosystem II (PSII). A similar analysis can be made using fluorescence at 735 nm (emanating from photosystem I—PSI). The ratio $(F_v/F_m)_{735}/(F_v/F_m)_{695}$, termed β_N, reflects the proportion of excitation energy reaching PSI which originates from PSII and is unaffected by the extent of plastid maturity (Fig. 21.8). Furthermore, all but the most immature plastids in the base of these wheat leaves are capable of State 1–State 2 transitions (see Section 9.16). Thus, the capacity to modify the

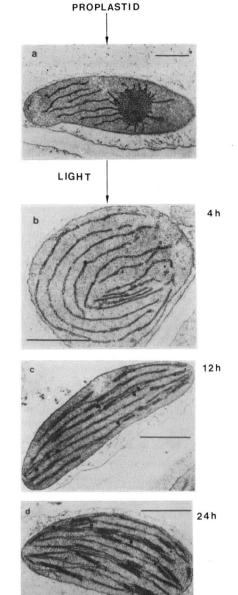

PROPLASTID

LIGHT

a

b 4h

c 12h

d 24h

Fig. 21.4 The development of chloroplasts from etioplasts in barley. When plants are grown in the dark, proplastids develop into etioplasts (a) which characteristically contain one or more prolamellar bodies that develop from invaginations of the inner plastid membrane. Upon illumination (in the examples shown, blue light was used), the paracrystalline structure of the pro-lamellar body is lost and there is a gradual organization of primary thylakoids and development of normal chloroplast structure (b–d). These ultrastructural changes are accompanied by the development of chloroplast function (see text). Bars represent 1 μm. After Tevini, M. (1982). In Parthier, B. & Boulter, D. (eds), *Encyclopedia of Plant Physiology*, Vol. 14B, p.p. 121–45. Springer Verlag, Berlin.

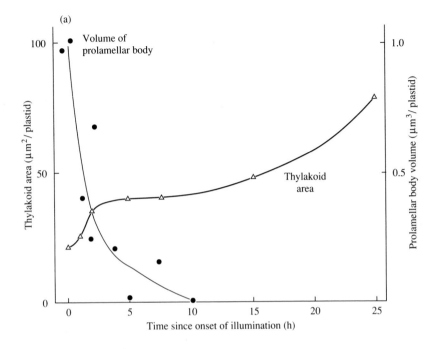

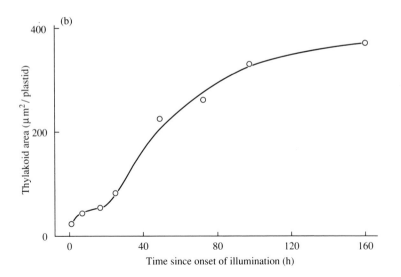

distribution of excitation energy between the two photosystems also develops early in differentiation, but only after the establishment of fully competent and linked PSII and PSI.

In contrast, when chloroplasts develop from etioplasts in greening tissue, the functional components of chloroplasts are not all formed simultaneously, but are added in a series of steps. Almost all chloroplast polypeptides are present, although in many cases at very low levels, in etioplasts. Upon illumination, the level of these polypeptides increases dramatically. In the case of many reaction centre proteins, however, light causes an absolute change in gene expression giving rise to polypeptides not previously found.

As greening proceeds, chlorophyll accumulates steadily for at least 48 h. In contrast, development of the photosystems occurs over the first 8 h of greening of etiolated oat seedlings (Fig. 21.9). In oats, PSI activity was first detected as measurable electron flow from an ascorbate/diaminodurane couple to methyl viologen after only 15 min illumination. At this stage, synthesis of chlorophyll from glutamate (see Section 15.3) had not occurred. Chlorophyll required for PSI activity must have arisen, therefore, from conversion of protochlorophyllide pre-existing in the etioplast. The development of some PSII activity occurred after 2 h illumination, and only after 8 h was complete electron flow from water to NADP established. Although the absolute timing varies, the development of PSI in advance of PSII has been observed in a number of different, greening, plants.

As mentioned earlier, chlorophyll synthesis and accumulation in greening leaves continues for at least 48 h. Since about 50% of the photochemical reaction centres found in fully greened leaves are formed over the first 8 h of illumination, the accumulation of chlorophyll over longer periods is believed to be due to increases in the size of the light-harvesting antenna serving each reaction centre.

In some cases, enhanced levels of certain polypeptides during greening are a result of transcriptional or post-transcriptional changes in gene expression. Light

Fig. 21.5 Changes in the volume of the prolamellar body and thylakoid area when leaves of dark-grown bean plants are illuminated. Bean plants germinated and grown in the dark for 14 days were illuminated ($t=0$) at an irradiance of $1.6\,mW\,cm^{-2}$. (a) Over the first 10 h of illumination increases in thylakoid area could be accounted for by breakdown and re-organization of the membrane material associated with the prolamellar body, the volume of which decreased dramatically. (b) Further increases in the thylakoid area (15–160 h) could not be accounted for by re-organization of the prolamellar body; this implies that membrane formation during this period involves synthesis *de novo*. After Bradbeer, J.W., Glydenholm, A.O., Ireland, H.M.M., Smith, J.W., Rest, J. & Edge, M.J.W. (1974) *New Phytol.*, **73**, 271–9.

Table 21.1 Changes in the lipid and fatty acid composition of chloroplast lipids during development of maize leaves. Seedlings of *Zea mays* were grown for 6 days in the light and the developing leaves cut into sections (A–E), corresponding to different distances from the basal meristem. Sections A and E represented the youngest and oldest sections respectively, this developmental sequence being reflected in the chlorophyll content. Chloroplasts were extracted from these sections and their lipids analysed. From Leech, R.M. & Walton, C.A. (1983) In Thomson, W.J., Mudd, J.B. & Gibbs, M. (eds) *Biosynthesis and Function of Plant Lipids*, pp. 56–80. Am. Soc. Plant Physiologists, Rockville.

	Lipid content (nmol/10^6 plastids)							
Section	MGDG	DGDG	SQDG	PG	PC	PI	PE	Chl
A	0.34	0.31	0.20	0.09	0.26	0.04	0.05	0.22
B	0.44	0.35	0.18	0.12	0.13	0.06	0.04	0.37
C	1.03	0.75	0.29	0.19	0.09	0.04	0.01	0.52
D	2.12	1.10	0.57	0.34	0.10	0.02	0.01	0.97
E	3.54	1.81	0.69	0.41	0.13	0.03	0.02	1.53

		Fatty acid composition (mol as % of total)						
Section	Lipid	16:0	16:1	18:0	18:1	18:2	18:3	20:0
A	MGDG	13	0	6	8	15	52	4
E		5	trace	2	2	6	85	0
A	DGDG	23	0	3	7	28	34	3
E		12	0	3	3	5	77	0
A	SQDG	35	0	17	13	16	16	trace
E		34	0	8	10	7	30	11
A	PG	53	0	3	6	32	3	1
E		35	13.0	5	6	8	27	6

Abbreviations: MGDG = monogalactosyl diacylglycerol; DGDG = digalactosyl diacylglycerol; SQDG = sulphoquinovosyl diacyglycerol; PG = phosphatidyl glycerol; PC = phosphatidyl choline; PI = phosphatidyl inositol; PE = phosphatidyl ethanolamine; Chl-chlorophyll.

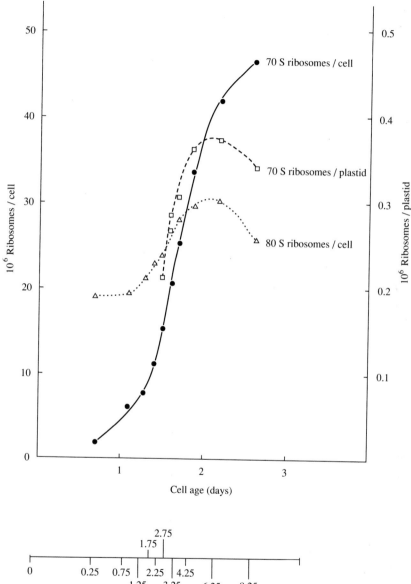

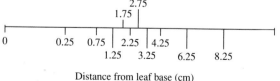

Fig. 21.6 Changes in the number of ribosomes per cell and per plastid as a function of cell age in the first leaf of wheat (*Triticum aestivum*). Sections of leaf at different distances from the basal meristem (equivalent to different ages) were taken and examined for the number of 80 S and 70 S ribosomes in the cytosol and plastids respectively. The number of 70 S ribosomes increased dramatically with the age of the cell. From Dean, C. & Leech, R.M. (1982) *Plant Physiol.*, **69**, 904–10.

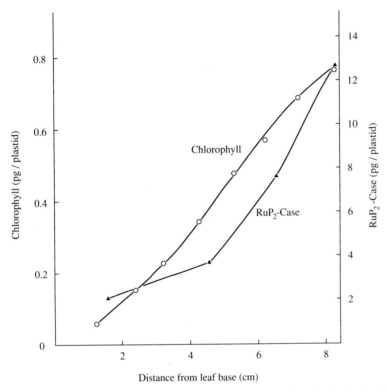

Fig. 21.7 Changes in chlorophyll and ribulose-1,5-P_2 carboxylase levels in developing plastids of wheat (*Triticum aestivum*). Both parameters increase steadily with distance from the leaf base, which corresponds to increasing cell age. From Leech, R.M. (1986) In *Plasticity in Plants*, Society for Experimental Biology Symposium 40, pp. 121–153. Company of Biologists Limited, Cambridge.

affects the expression of genes coding for a number of chloroplast components. Some of these 'photogenes' are in the plastid genome (plastome) whilst others are present in nuclear DNA. 'Photogenes' include those coding for the 32 kDa thylakoid protein involved in photosystem II (32 kDa or 'D1' protein) and the small subunit of ribulose-1,5-P_2 carboxylase (sRuP$_2$-Case). For these genes, illumination results in accumulation of the respective mRNA prior to synthesis of the polypeptide. In the case of the light-harvesting chlorophyll *a/b*-binding protein (LHCP), the apoprotein is transcribed to the same extent in the light and dark but degrades rapidly without concurrent synthesis of chlorophyll *a* and *b* for binding to LHCP.

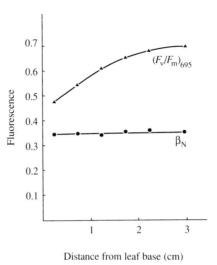

Fig. 21.8 Changes in the fluorescence parameters (F_v/F_m) measured at 695 nm and β_N (see text) along the length of 4-day-old wheat leaves. F_v/F_m is a measure of the ability of PSII to function normally, whilst β_N represents an indication of the fraction of energy reaching PSI that is passed on from PSII. Clearly, PSII is fully functional at a very early stage in development and even in the youngest plastids, energy transfer from PSII to PSI is occurring to the same extent as in older tissue. This is in contrast to the situation in greening tissue (see, for example, Fig. 21.9) in which development of the two photosystems is not tightly integrated. From Webber, A.N., Baker, N.R., Platt-Aloia, K. & Thomson, W.W. (1984) *Physiol. Plant.*, **60**, 171–9.

Until the developing plastid can synthesize ATP, the energy requirements for protein and lipid biosynthesis are supplied by ATP imported from mitochondrial respiration (see Chapter 7). This transfer of ATP is achieved by the glycerate-3-P/triose-P shuttle (see Chapter 14) and by the adenine nucleotide translocator (see, for example, Chapter 7), although the capacity of the chloroplast envelope to transport adenine nucleotides directly is limited (see Table 10.1). The translocators must be present at an early stage of greening; this seems to be the case. Furthermore, when dark-grown oat seedlings are illuminated, the capacity for ATP formation by subsequently isolated mitochondria rises markedly but drops to preillumination levels as the capacity of plastids for photophosphorylation develops (Fig. 21.10).

Although the greening of etiolated tissue is convenient for studies of the effect of light on chloroplast development, it must again be stressed that this system is not representative of normal chloroplast differentiation.

21.3 Perception of the light stimulus in chloroplast development involves a number of receptors

A number of photoreceptors have been implicated in chloroplast (and other organelle) development; in higher plants three are thought to be important. These

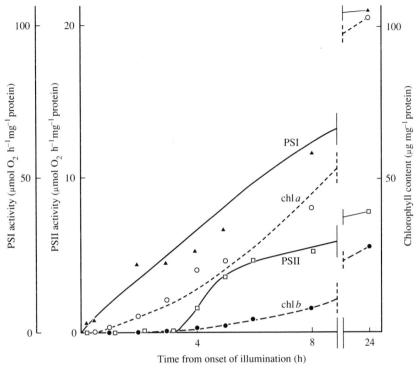

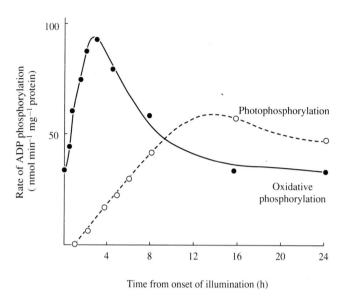

Fig. 21.10 Capacity for ADP phosphorylation by isolated mitochondria (solid line) and plastid thylakoid preparations (dashed line) during greening of 8-day-old dark-grown oat seedlings exposed to light (12 000 lux, ~130 μmol m^{-2} s^{-1}). Until the capacity of plastids for photophosphorylation is fully developed, the ATP requirement for protein synthesis in plastids is met by increased capacity for ADP phosphorylation by oxidative phosphorylation in mitochondria. After Wellburn, A.R. (1984) In Baker, N.R. & Barber, J. (eds) *Chloroplast Biogenesis*, pp. 253–303. Elsevier Sciences Publishers BV., Amsterdam.

Fig. 21.9 Development of photochemical activities during greening of etiolated levels of *Avena sativa* (oats). Seedlings of *Avena* were grown in the dark and illuminated at $t = 0$. The levels of chlorophylls *a* and *b* and activities of PSI and PSII were determined over a 24 h period following illumination. PSI activity of plastid internal membranes was measured by the rate of O$_2$ consumption coupled to electron transport from an ascorbate/diaminodurane couple, via PSI, to methyl viologen. PSII activity was determined by the rate of O$_2$ consumption coupled to electron transport from H$_2$O→PSII→dimethylmethyleneioxybenzoquinone. Clearly, plastid membranes possess an active PSI before PSII activity develops. At the time at which PSI is first active, chlorophyll synthesis had not commenced and any chlorophyll active in light harvesting for PSI activity must have arisen from photoconversion of pre-existing protochlorophyllide in the etioplast. After Wellburn, A.R. & Hampp, R. (1979) *Biochim. Biophys. Acta*, **547**, 380–97.

which lack phytochrome, a pigment called cryptochrome, which absorbs blue light, might be important in photomorphogenesis.

21.3.1 Light absorption by protochlorophyll(ide) results in stimulation of chlorophyll synthesis

The synthesis of chlorophyll via 5-aminolevulinate and protochlorophyll(ide) is described in Section 15.3. Protochlorophyll(ide) inhibits 5-aminolevulinate synthesis but this inhibition is relieved in the light by the conversion of protochlorophyll(ide) to chlorophyll(ide) *a* via NADPH:protochlorophyll(ide) oxidoreductase. However, this enzyme is present at high activity only in etiolated tissue and during the early stages of light-induced chlorophyll synthesis. The levels of this

are phytochrome, protochlorophyll(ide)* (and in a sense chlorophyll itself) and the 'blue-light receptor' (at present rather poorly characterized). In algae and fungi,

*Since both protochlorophyllide and its esterified form protochlorophyll can be converted to chlorophyllide *a* and chlorophyll *a* respectively, the notation protochlorophyll(ide)/chlorophyll(ide) *a* is used to denote both forms.

enzyme and its corresponding mRNA decrease markedly with prolonged illumination.

21.3.2 Many light-regulated steps in organelle development are regulated by phytochrome

Strictly speaking, protochlorophyll(ide) does not act as a sensory pigment in the true sense of a receptor which converts an environmental (light) signal to a developmental response. Rather, its light-dependent reduction is simply an essential step in initiating the formation of chlorophyll and it does not mediate other biochemical and developmental processes. Phytochrome, on the other hand, is a true light receptor and has been implicated in a whole range of developmental processes. Some of the genes known to be regulated by phytochrome are shown in Table 21.2.

Table 21.2 Some genes that have been shown to be regulated by phytochrome. The list is not intended to be comprehensive, but illustrates the range of genes and processes concerned. In all cases regulation is brought about at the transcriptional or post-transcriptional level. After Tobin, E.M. & Silverthorne, J. (1985) *Ann. Rev. Pl. Physiol.*, **36**, 569–93.

Protein encoded	Function of protein
NADPH:protochlorophyllide oxidoreductase	Chlorophyll synthesis
32 kDa thylakoid protein of PSII	O_2 evolution
Light-harvesting chlorophyll *a/b*-binding protein	Light harvesting
RuP$_2$-Case, small subunit	CO_2 assimilation
RuP$_2$-Case, large subunit	CO_2 assimilation
rRNA (cytoplasmic and plastid)	Protein synthesis
Phytochrome	Light regulation
Nitrate reductase	Nutrient assimilation

Phytochrome is a chromoprotein (i.e. a protein with a light-absorbing chromophore attached) with a molecular mass of 120–127 kDa, dependent on the plant species from which it is isolated. The chromophore is an open-chain tetrapyrrole and exists in two, reversibly photoconvertible, states (Eqn. 21.1).

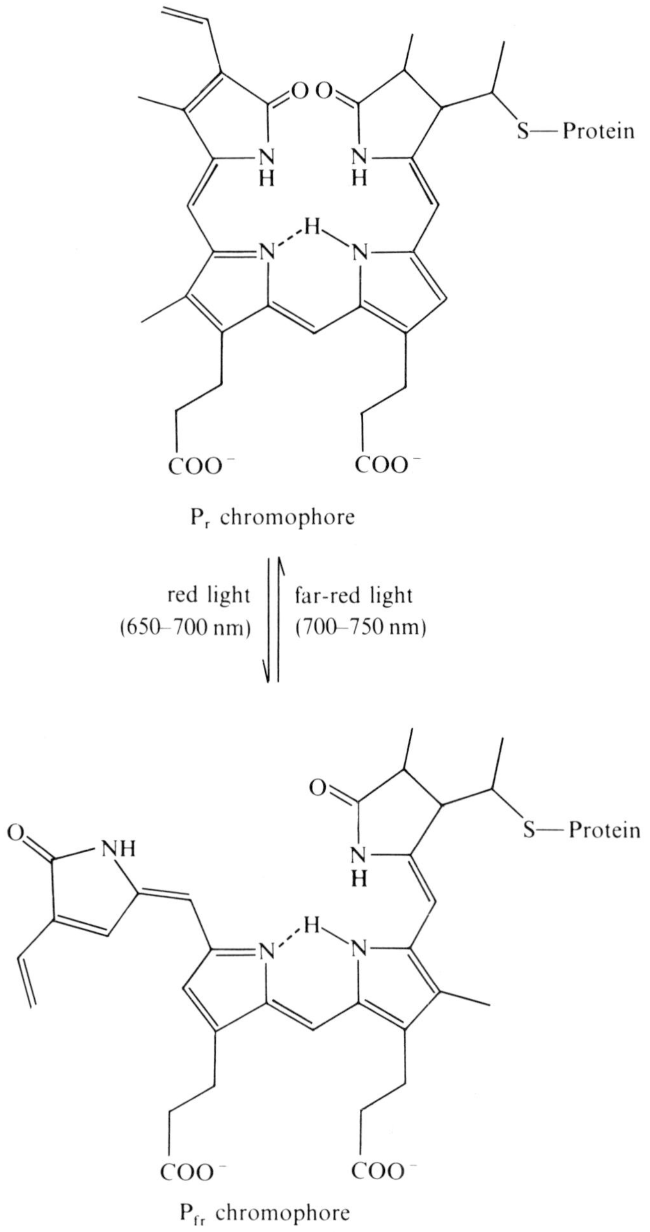

P$_r$ chromophore

red light (650–700 nm) ⇌ far-red light (700–750 nm)

(21.1)

P$_{fr}$ chromophore

In dark-grown plants, only the form of phytochrome known as P_r, which absorbs red light and has an absorption maximum at 666 nm, exists. The P_r form is converted by red light to P_{fr}, a form which absorbs light of longer wavelengths (far-red light, absorption maximum at 730 nm) and is the biologically active molecule, the effector molecule of the phytochrome system. Absorption of far-red light converts P_{fr} back to P_r. It is a characteristic feature of phytochrome regulated phenomena that a short pulse of far-red light reverses the effect of a pulse of red light (see, for example, the regulation of LHCP mRNA by phytochrome in Fig. 20.4).

It is now some 50 years since phytochrome-mediated phenomena were first recognized and more than 20 years since phytochrome was first purified, but the precise mechanism by which it effects a biochemical or development change is not clear. Immunocytochemical and cell fractionation experiments suggest that the P_f form of phytochrome is soluble and occurs in the cytosol whilst P_{fr} is rapidly converted from a soluble form to one which is membrane bound, possibly to specific receptors. Following conversion to P_{fr}, phytochrome is subject to degradation due to ATP-dependent proteolysis mediated via a protein termed ubiquitin.

It is worth pointing out that phytochrome influences aspects of mitochondrial structure and function including stimulation of cytochrome oxidase, fumarase and succinate dehydrogenase activities.

21.4 Biosynthesis of organelles involves interaction between organelles and the nucleus

21.4.1 The RNA and protein components of chloroplasts and mitochondria are encoded by both nuclear and organelle genomes

Many organelle proteins are coded and synthesized outside the organelle, although the organelles also contain their own genomes (see Chapter 18). This raises the questions of which organelle components are coded by the organelle genome and which by the nuclear genome, and how their synthesis is co-ordinated and controlled. These issues are intriguing, particularly as, although organelles such as chloroplasts are thought to have originated by endosymbiosis of free-living prokaryotic organisms, the coding capacity of the chloroplast genome is much less than that of the putative prokaryotic ancestors, implying that much coding information has been transferred to the nucleus during evolution.

Extensive studies, using a range of techniques, of the coding and synthesis of

plastid protein and RNA suggest that most of the steps of the various biochemical pathways in plastids are controlled by nuclear genes (Table 21.3). However, genetic analysis shows several chloroplast proteins are usually inherited through the female parent, suggesting involvement of the chloroplast genome (see Section 18.1.6). This is supported by studies on the origin of specific mRNA molecules coding for

Table 21.3 Location of the genes for some plastid polypeptides. (For the full complement of plastid genes see, for example, Fig. 18.3.)

Polypeptide	Gene location*
Thylakoid proteins	
P_{700}-chlorophyll *a* binding protein	P
P_{680}-chlorophyll *a* binding protein (32 kDa protein, 'D1')	P
'D2' protein of PSII	P
43 and 47 kDa chlorophyll binding proteins	P
Cytochrome b_{559} apoprotein	P
Light-harvesting chlorophyll *a/b*-binding protein	N
Cytochrome *f* apoprotein	P
Cytochrome b_{563} apoprotein	P
α, β and ε subunits of ATP synthetase, CF_1	P
γ and δ subunits of ATP synthetase, CF_1	N
Subunits I, III and IV of ATP synthetase, CF_1	P
Plastocyanin	N
Ferredoxin:NADP oxidoreductase	N
Stromal proteins	
RNA polymerase	N
DNA polymerase	N
Elongation factors EF1, EF2	N
11–19 proteins of 70 S ribosomes	P
Aminoacyl-tRNA synthetases	N
RuP_2-Case, large subunit	P
RuP_2-Case, small subunit	N
Ferredoxin	N
5-Aminolevulinate dehydrase	N
Enzymes of starch formation	P
Enzymes of the C_3-CR cycle (other than RuP_2-Case)	N

*P = plastid coded; N = nuclear coded.
Abbreviations: RuP_2-Case = ribulose-1,5-P_2 carboxylase; CF_1 and CF_0 = coupling factors of ATP synthetase.

chloroplast proteins, which hybridize with plastid DNA. With the relatively new recombinant DNA techniques, specific probes can be constructed to map the genes coding for particular polypeptides in organelles. Indeed, the complete chloroplast genome of tobacco and the liverwort *Marchantia* have now been mapped (see Fig. 18.3).

The chloroplast genome contains sufficient DNA to code for the chloroplast ribosomal RNA (rRNA) and transfer RNA (tRNA) and about 100 other polypeptides, assuming an average gene size of 1000 base pairs. Isolated chloroplasts can synthesize about this number of proteins. The polypeptides present in thylakoids can be grouped into five distinct complexes; namely PSI, PSII, and the LHCP, cytochrome *b/f* and ATP synthetase complexes. With the exception of LHCP, all of these complexes have at least some polypeptides coded in the chloroplast DNA (Table 21.3).

Most of the proteins found in the stroma of chloroplasts are encoded on the nuclear genome. In tobacco the coding sequences for 12 of the proteins of the small subunit of 70 S ribosomes and eight of the 34 large subunit proteins are contained in chloroplast DNA, although in the liverwort *Marchantia* the corresponding numbers are 11 small subunit proteins and nine large subunit proteins. The large subunit of ribulose-1,5-P_2-carboxylase is also encoded in chloroplast DNA whilst the multigene family for the small subunit is found in the nuclear genome. All of the aminoacyl-tRNA synthetases of the chloroplast are coded for in the nucleus. RNA polymerase of chloroplasts is of interest, as probes for the α, β and β' subunits show homology with *both* chloroplastic and nuclear DNA in tobacco, pea and spinach, although binding of the probes to nuclear DNA is stronger. This implies that the chloroplast RNA polymerase is encoded in both genomes. However, since the enzyme is found in chloroplasts under conditions where no protein is being synthesized within these organelles, it is believed that the nuclear gene is the functional one and the chloroplastic sequence could be a non-functional relic from the ancestral endosymbiont. Similarly, the proton translocation subunit of the mitochondrial ATP synthetase of the fungus *Neurospora crassa* is represented by coding sequences in both nuclear and mitochondrial genomes.

The locations of the coding sequences for mitochondrial proteins of plants are less well characterized than those of chloroplast components. Isolated plant mitochondria synthesize 18–20 polypeptides of molecular mass of 8–54 kDa, all but two being membrane bound (see Section 18.1.6). At least eight of these polypeptides have been identified (see Table 18.4) but virtually nothing is known about the way in which the nuclear and mitochondrial genomes interact in mitochondrial biogenesis.

21.4.2 Nuclear-coded organelle components are synthesized on cytosolic ribosomes

There is, as discussed above, good evidence that coding sequences for some organelle polypeptides are in the organelle genome, although coding for the majority of proteins is within nuclear DNA. In all cases studied, proteins and RNA encoded in organelle DNA are synthesized within that organelle. Organelle polypeptides encoded by the nuclear genome are synthesized in the cytoplasm and transported to the organelle. Evidence for this comes from a number of approaches including *in vivo* labelling of proteins in the presence of inhibitors of protein synthesis (e.g. cycloheximide which inhibits synthesis on 80 S (cytoplasmic) ribosomes, and D-*threo*-chloramphenicol which is inhibitory to synthesis on 70 S (plastid) ribosomes), the examination of proteins synthesized by isolated organelles and *in vitro* protein synthesis using cytosolic or organelle RNA or DNA as templates for translation or coupled transcription/translation respectively (Table 21.4).

21.4.3 Transport of proteins synthesized in the cytosol into organelles involves a transit peptide

Many polypeptides necessary for organelle function are synthesized in the cytosol and are then transported across the organelle peripheral membranes to their site of action. Such polypeptides are synthesized as higher molecular mass precursors containing an additional sequence termed a transit peptide. For example, chloroplast proteins synthesized in the cytosol are translated with a transit peptide of 34–100 amino acid residues attached to the amino terminus of the functional polypeptide.

The first step in the transport of precursor proteins into organelles involves binding to the organelle peripheral membranes. In both mitochondria and chloroplasts this involves specific receptors in the outer envelope; for example, there are 1500–3000 binding sites for the small subunit of RuP$_2$-Case per chloroplast. Binding of precursors to chloroplasts requires ATP in the space between the envelope membranes. The subsequent transport step in both mitochondria and chloroplasts also requires ATP—in chloroplasts ATP from the stroma is used whereas cytosolic ATP is used for protein import into mitochondria. In addition, import into mitochondria requires a proton motive force across the peripheral membranes. The exact mechanism by which transport across membranes is brought about is, however, not known.

Within the organelle, the transit peptide is cleaved by proteases (transit

Table 21.4 Sites of synthesis of some chloroplast proteins. Polypeptides synthesized in the chloroplast are coded on the plastid genome whilst nuclear-coded chloroplast polypeptides are synthesized in the cytosol. After Bottomley, W. & Bohnert, J.H. (1982) In Parthier, B. & Boulter, D. (eds) *Encyclopedia of Plant Physiology*, Vol. 14B, pp. 531–95. Springer Verlag, Berlin.

Polypeptide	Site of synthesis	Methods used*
Ribulose-1,5-P_2 carboxylase large subunit	chloroplast	I, O, IV
Ribulose-1,5-P_2 carboxylase small subunit	cytosol	I, IV
Most other enzymes of the C_3-CR cycle	cytosol	I
Ferredoxin	cytosol	I, IV
Ferredoxin:NADP oxidoreductase	cytosol	I
Cytochrome f	chloroplast	I, O, IV
Cytochrome b_{559}	chloroplast	I, O, IV
P_{700}-chlorophyll a binding protein	chloroplast	O, IV
Light-harvesting chlorophyll a/b-binding protein	cytosol	IV
32 kDa protein of PSII	chloroplast	O, IV
Plastocyanin α, β, ε subunits of ATP synthetase, CF_1.	cytosol	I, IV
	chloroplast	O, IV
γ and δ subunits of ATP synthetase, CF_1.	cytosol	I, IV
Aminoacyl-tRNA synthetases	cytosol	I
RNA polymerase	cytosol	I
5-Aminolevulinate dehydratase	cytosol	I
Some 70 S ribosomal proteins	cytosol	I, IV
Phosphate translocator	cytosol	I, IV

*Method used to demonstrate the location of the site of synthesis: I = inhibitor studies; O = synthesis by isolated organelles; IV = *in vitro* transcription/translation using RNA or cloned nuclear or chloroplastic DNA fragments.
Abbreviations: CF_1 = coupling factor of ATP synthetase; PSII = photosystem II.

peptidases) to release the native protein. The enzyme which removes the transit peptide from the sRuP$_2$-Case precursor in pea chloroplasts is a soluble, metallo-protease of about 180 kDa molecular mass. This protease also cleaves transit peptides from the precursors of wheat and barley plastocyanin and ferredoxin:NADP oxidoreductase. The corresponding transit peptidase from *Chlamydomonas* is also a high molecular mass protein but, unlike the pea enzyme, it is not active against precursors of other proteins and does not remove the transit peptide from the pea sRuP$_2$-Case precursor.

Transit peptides are also important in controlling the location of the protein within the organelle. Those chloroplast proteins synthesized in the cytosol but finally located in the stroma or the stromal side of the thylakoid complex have to cross only the chloroplast envelope. However, plastocyanin is located within the intra-thylakoid space and has to cross not only the chloroplast envelope but also the thylakoid membrane. In pea chloroplasts, there are two transit peptidases, one in the stroma and one in the intra-thylakoid space. These decrease the size of a precursor plastocyanin in two discrete steps. The initial translation product has a molecular mass of about 20.6 kDa. Incubation with the stromal transit peptidase produces an intermediate sized precursor of about 16.2 kDa. If this intermediate plastocyanin precursor is incubated with the thylakoid transit peptidase then native plastocyanin is released. This has a slightly lower apparent molecular mass of about 14.1 kDa. When the plastocyanin precursor is supplied to intact chloroplasts both the intermediate plastocyanin precursor and native plastocyanin are formed. The importance of the transit peptide in localizing polypeptides is shown by experiments in which recombinant DNA techniques were used to produce a polypeptide consisting of the transit peptide for plastocyanin attached to ferredoxin, or vice versa. If the transit peptide for ferredoxin (a stromal protein) is attached to native plastocyanin, the resulting complex is transported across the chloroplast envelope and native plastocyanin is released into the stroma by peptidase action. Conversely, by attaching plastocyanin transit peptides to native ferredoxin, native ferredoxin is not found in the stroma but is embedded in thylakoids (albeit with a very low efficiency of transport).

The 32 kDa protein in the core of PSII is synthesized within the chloroplast, at least in maize and peas, as a slightly larger precursor protein of ~34.5 kDa. This is made on ribosomes attached to stromal thylakoids and is subsequently transported to the appressed regions of granal thylakoids. The additional amino acid residues are probably involved in this transport and insertion of the native protein into the grana thylakoids. The cytochrome f apoprotein and subunits I and IV of the CF_0 component of the chloroplast ATP synthetase, also found within thylakoids, are also made as precursors.

21.5 Regulation of the interaction between nuclear and organelle genomes

It is evident from the above that organelles are able to encode and synthesize many polypeptides necessary for their functioning. However, very few of the functions are dependent only on polypeptides synthesized within the organelle; an excellent example is RuP₂-Case which requires both the large and small subunits to function, these being encoded and synthesized in the plastid and nucleocytosolic systems respectively. A second example is the electron transport chain in the chloroplast which requires polypeptides synthesized in the cytosol (e.g. ferredoxin) and in the chloroplast (e.g. cytochromes). Without both internally and externally synthesized proteins the organelle cannot function properly. Furthermore, the machinery required to synthesize proteins within the organelle is dependent on nuclear genome products. Transcription of nuclear DNA must influence the rates of transcription of chloroplast genes, since the RNA polymerases required in the chloroplast are encoded in the nucleus and synthesized in the cytosol.

There is a gross imbalance in the number of copies of organelle and nuclear-encoded genes within plant cells. If photosynthetic cells of higher plants contain an average of 100 chloroplasts per cell, each with 20–50 copies of the chloroplast genome, then each chloroplast encoded gene is repeated 2000–5000 times per cell. In contrast, there might be only a few copies of the gene for a particular chloroplast protein on the single nuclear genome (e.g. petunia has eight genes in the sRuP₂-Case multigene family). Similar arguments apply for mitochondrial genes. The rate of transcription of nuclear genes is therefore far more likely to limit the development of chloroplast functions than that of chloroplast genes.

Clearly, the activities of the nuclear/cytosolic and organelle systems must be integrated and regulated. The synthesis of the two subunits of RuP₂-Case is an excellent example of co-ordination between nuclear and cytosolic genomes. During the course of development of chloroplasts in barley seedlings, for instance, the rates of synthesis of small and large subunits of RuP₂-Case change dramatically (Fig. 21.11a). However, they change in unison and the ratio of the rates of synthesis remains the same throughout development (Fig. 21.11b). Synthesis of RuP₂-Case appears to be under control of the nucleus, as in cultivars of wheat with different ploidy levels (diploid, tetraploid or hexaploid) the amount of RuP₂-Case synthesized reflects the number of copies of the nuclear genome present (two, four or six respectively). This 'nucleotypic control' also occurs in other species such as *Festuca* and lucerne. It is not known how co-ordinated expression of the organelle and nuclear genomes is achieved. In the case of RuP₂-Case, it has been suggested that the small subunit, synthesized in the cytosol, acts as a positive control signal for synthesis of the large subunit in the chloroplast. The tightly controlled synthesis of the two subunits and their nucleotypic control is consistent with this. There is, however, mounting evidence that chloroplast products control the synthesis of

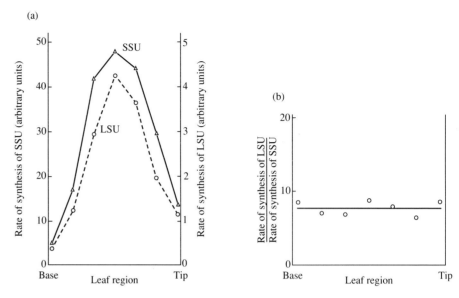

Fig. 21.11 (a) Changes in the rates of synthesis of the small (SSU) and large (LSU) subunits of RuP₂-Case in successively older leaf sections of 7-day-old barley seedlings. Rates of translation of the two subunits were determined using the incorporation of [¹⁴C]leucine into the relevant polypeptides in an *in vitro* translation system; the two subunits were separated by two-dimensional gel electrophoresis. (b) Ratio of the rates of synthesis of the two subunits in the same samples as in (a). Clearly, both subunits are synthesized in a co-ordinated fashion, despite very large changes in absolute rates of synthesis. From Nivison, H.T. & Stocking, C.R. (1983) *Plant Physiol.*, **73**, 906–11.

chloroplast proteins made in the cytosol. Nuclear-encoded genes for chloroplast proteins are not expressed in root tissue unless the roots are exposed to light and turn green. When this occurs, nuclear genes for chloroplast proteins are transcribed as the plastids develop. Futhermore, in certain mutants of barley unable to make plastid ribosomes, the synthesis of plastid enzymes formed in the cytosol (e.g. ribulose-5-P kinase and glyceraldehyde-3-P dehydrogenase) is inhibited. In the cotyledons of developing mustard seedlings, the activities of a number of enzymes found in plastids, but synthesized in the cytosol, are decreased by treating the seedlings with chloramphenicol in the early stages of development (up to 30 h after sowing). Conversely, proteins synthesized in the cytosol but which are not constituent components of organelles (e.g. the glycolytic enzymes of the cytosol) are mostly unaffected. In the case of sRuP$_2$-Case, regulation is achieved at the level of transcription as levels of sRuP$_2$-Case mRNA are lowered dramatically by chloramphenicol treatment. However, there are other cases which argue against a role for chloroplast protein synthesis in regulating the expression of nuclear-encoded chloroplast polypeptides. Rye seedlings, for example, have no ribosomes in chloroplasts when grown at moderately high temperatures (e.g. 25 °C) but show almost

normal synthesis of sRuP$_2$-Case and other stromal proteins. Furthermore, one maize mutant lacks chloroplast ribosomes but, unlike the barley mutant described above, *does* transcribe nuclear genes for chloroplast components. Thus, the nature of the message controlling the transcription of nuclear genes for chloroplast components is uncertain at present. An attempt to summarize the integration of expression of plastid and nuclear genomes is given in Fig. 21.12.

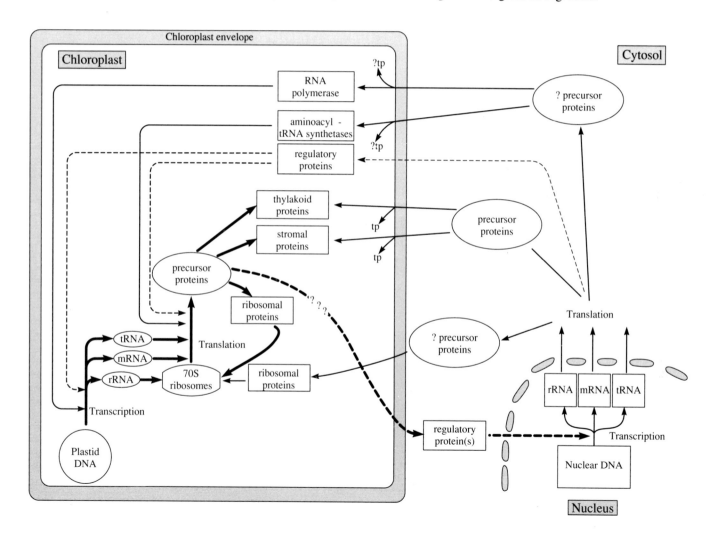

Fig. 21.12 Summary of the integration between the nuclear and plastid genomes in the co-ordinated synthesis of plastid proteins. Proteins involved in plastid function are shown by continuous lines. Postulated regulatory proteins are indicated by the dotted lines. Abbreviation: tp = transit peptide.

There is little data available on the co-ordination between mitochondrial and nuclear genomes during development. However, in the yeast *Saccharomyces cerevisiae*, mutations in the mitochondrial genome cause changes in the level of nuclear-encoded transcripts. As in the plastid system, an organelle product might be responsible for regulating nuclear genes encoding organelle proteins.

Further reading

Monographs and treatises: Baker & Barber (1984).

Light and chloroplast biogenesis: Castelfranco & Beale (1983); Leech (1986); Tobin & Silverthorne (1985).

Gene expression during organelle biogenesis: Leaver & Gray (1982); Mullet (1988); Newton (1988).

Cell development: Burgess (1989).

Further reading

Alberts, B., Bray, D., Lewis, J., Raff, M., Roberts, K. & Watson, J.D. (1989) *Molecular Biology of the Cell*, 2nd edn. Garland Publishing, New York.

Anderson, J.M. (1986) Photoregulation of the composition, function and structure of thylakoid membranes. *Annual Review of Plant Physiology*, **37**, 93–136.

Badger, M.R. (1985) Photosynthetic oxygen exchange. *Annual Review of Plant Physiology*, **36**, 27–53.

Baker, N.R. & Barbar, J. (1984) *Chloroplast Biogenesis*. Elsevier, Amsterdam.

Bell, E.A. & Charlwood, B.V. (eds) (1980) *Encyclopedia of Plant Physiology* new series, Vol. 8, *Secondary Plant Products*. Springer Verlag, Berlin.

Boulter, D. & Parthier, B. (eds) (1982) *Encyclopedia of Plant Physiology* new series, Vol. 14A., *Nucleic Acids and Proteins in Plants: I Structure, Biochemistry and Physiology of Proteins*. Springer Verlag, Berlin.

Burgess, J. (1985) *Plant Cell Development*. Cambridge University Press, Cambridge.

Cassab, G.I. & Varner, J.E. (1988) Cell wall proteins. *Annual Review of Plant Physiology and Plant Molecular Biology*, **39**, 321–53.

Castelfranco, P.A. & Beale, S.I. (1983) Chlorophyll biosynthesis: recent advances and areas of current interest. *Annual Review of Plant Physiology*, **34**, 241–78.

Chothia, C. (1984) Principles that determine the structure of proteins. *Annual Review of Biochemistry*, **53**, 537–72.

Conn, E.E. (ed.) (1981) *The Biochemistry of Plants*, Vol. 7, *Secondary Plant Products*. Academic Press, New York.

Cornish-Brown, A. (1979) *Fundamentals of Enzyme Kinetics*. Butterworths, London.

Davies, D.D. (ed.) (1980) *The Biochemistry of Plants*, Vol. 2, *Metabolism and Respiration*. Academic Press, New York.

Davies, D.D. (ed.) (1987) *The Biochemistry of Plants*, Vol. 11, *Biochemistry of Metabolism*. Academic Press, San Diego.

Davies, D.D. (ed.) (1987) *The Biochemistry of Plants*, Vol. 12, *Physiology of Metabolism*. Academic Press, San Diego.

Dixon, M. & Webb, E.C. (1979) *Enzymes*, 3rd edn. Longmans, London.

Douce, R. & Day, D.A. (eds) (1985) *Encyclopedia of Plant Physiology* new series, Vol. 18, *Higher Plant Cell Respiration*. Springer Verlag, Berlin.

Edwards, G. & Walker, D.A. (1983) C_3, C_4: *Mechanisms, and Cellular and Environmental Regulation of, Photosynthesis*. Blackwell Scientific Publications, Oxford.

Eisenberg, D. (1984) Three-dimensional structure of membrane and surface proteins. *Annual Review of Biochemistry*, **53**, 595–623.

Fedtke, C. (1982) *Biochemistry and Physiology of Herbicide Action*. Springer Verlag, Berlin.

Findlay, J.B.C. & Evans, W.H. (eds) (1987) *Biological Membranes: a practical approach*. IRL Press, Oxford.

Finean, J.B., Coleman, R. & Michell, R.H. (1984) *Membranes and their Cellular Functions*, 3rd edn. Blackwell Scientific Publications, Oxford.

Fork, D.C. & Satoh, K. (1986) The control by state transitions of the distribution of excitation energy in photosynthesis. *Annual Review of Plant Physiology*, **37**, 335–61.

Fry, S.C. (1986) Cross-linking of matrix polymers in the growing cell walls of angiosperms. *Annual Review of Plant Physiology*, **37**, 165–86.

Giaquinta, R.T. (1983) Phloem loading of sucrose. *Annual Review of Plant Physiology*, **34**, 347–87.

Gibbs, M. & Latzko, E. (eds) (1987) *Encyclopedia of Plant Physiology* new series, Vol. 6, *Photosynthesis: II Photosynthetic Carbon Metabolism and Related Processes*. Springer Verlag, Berlin.

Glazer, A.N. & Melis, A. (1987) Photochemical reaction centres: structure, organization and function, *Annual Review of Plant Physiology*, **38**, 11–45.

Grierson, D. & Covey, S. (1988) *Plant Molecular Biology*, 2nd edn. Blackie, Glasgow.

Gunning, B.E.S. & Hardham, A.R. (1982) Microtubules. *Annual Review of Plant Physiology*, **33**, 651–98.

Gunning, B.E.S. & Steer, M.W. (1975) *Ultrastructure and the Biology of Plant Cells*. Arnold, London.

Hall, J.L., Flowers, T.J. & Roberts, R.M. (1981) *Plant Cell Structure and Metabolism*, 2nd edn. Longman, London.

Harold, F.M. (1986) *The Vital Force: A Study of Bioenergetics*. W.H. Freeman & Co., New York.

Harwood, J.L. (1988) Fatty acid metabolism. *Annual Review of Plant Physiology and Plant Molecular Biology*, **39**, 101–38.

Hatch, M.D. & Boardman, N.K. (eds) (1981) *The Biochemistry of Plants*, Vol. 8, *Photosynthesis*. Academic Press, New York.

Hatch, M.D. & Boardman, N.K. (eds) (1987) *The Biochemistry of Plants*, Vol. 10, *Photosynthesis*. Academic Press, San Diego.

Hayashi, T. (1989) Xyloglucans in the primary cell wall. *Annual Review of Plant Physiology and Plant Molecular Biology*, **40**, 139–68.

Heemskerk, J.W.M. & Wintermans, J.F.G.M. (1987) Role of the chloroplast in leaf acyl-lipid synthesis. *Physiologia Plantarum*, **70**, 558–68.

Heidecker, G. & Messing, J. (1986) Structural analysis of plant genes. *Annual Review of Plant Physiology*, **37**, 439–66.

Hipkins, M.F. & Baker, N.R. (1986) *Photosynthesis: energy transduction: a practical approach*. IRL Press, Oxford.

Ho, L.C. (1988) Metabolism and compartmentation of imported sugars in sink organs in relation to sink strength. *Annual Review of Plant Physiology and Plant Molecular Biology*, **39**, 355–68.

International Union of Biochemistry (1979) *Enzyme Nomenclature 1978*. Academic Press, New York.

Jensen, R.A. (1985) The shikimate/arogenate pathway: link between carbohydrate metabolism and secondary metabolism. *Physiologia Plantarum*, **66**, 164–8.

Kannangara, C.G., Gough, S.P., Bruyant, P., Hoober, J.K., Kahn, A. & von Wettstein, D. (1988) tRNA^{Glu} as a cofactor in γ-aminolevulinate biosynthesis: steps that regulate chlorophyll synthesis. *Trends in Biochemical Sciences*, **13**, 139–43.

Kirk, J.T.O. & Tilney-Basset, R.A.E. (1978) *The Plastids: Their Chemistry, Structure Growth and Inheritance*, 2nd edn. Elsevier/North Holland, Amsterdam.

Kishore, G.M. & Shah, D.M. (1988) Amino acid biosynthesis inhibitors as herbicides. *Annual Review of Biochemistry*, **57**, 627–63.

Klee, H., Horach, R. & Rogers, S. (1987) *Agrobacterium*-mediated plant transformation and its further applications to plant biology. *Annual Review of Plant Physiology*, **38**, 467–86.

Kuhlemeier, C., Green, P.J. & Chua, N.H. (1987) Regulation of gene expression in higher plants. *Annual Review of Plant Physiology*, **38**, 221–57.

Lange, O.L., Nobel, P.S., Osmond, C.B. & Ziegler, H. (eds) (1982) *Encyclopedia of Plant Physiology* new series, Vol. 12B, *Physiological Plant Ecology: II Water Relations and Carbon Assimilation*. Spring Verlag, Berlin.

Läuchli, A. & Bieleski, R.L. (eds) (1983) *Encyclopedia of Plant Physiology*, Vol. 15A, *Inorganic Plant Nutrition*. Springer Verlag, Berlin.

Lawlor, D.W. (1987) *Photosynthesis: Metabolism, Control and Physiology*. Longman Scientific & Technical, Harlow.

Leaver, C.J. & Gray, M.W. (1982) Mitochondrial genome organization and expression in higher plants. *Annual Review of Plant Physiology*, **33**, 373–402.

Leech, R.M. (1986) Stability and plasticity during chloroplast development. In Jennings, D.H. & Trewavas, A.J. (eds) *Plasticity in plants*, pp. 121–53. *Society for Experimental Biology Symposium*, No. 40. Society for Experimental Biology, Cambridge.

Leopold, A.C. & Kriedemann, P.E. (1975) *Plant Growth and Development*, 2nd edn. McGraw-Hill, New York.

Loewus, F.A. & Tanner, W. (eds) (1982) *Encyclopedia of Plant Physiology*, Vol. 13A, *Plant Carbohydrates: II Intracellular Carbohydrates*. Springer Verlag, Berlin.

Lorimer, G.H. (1981) The carboxylation and oxygenation of ribulose-1,5-bisphosphate: the primary events in photosynthesis and photorespiration. *Annual Review of Plant Physiology*, **32**, 349–83.

Mantell, S.H., Matthews, J.A. & McKee, R.A. (1985) *Principles of Plant Biotechnology*. Blackwell Scientific Publications, Oxford.

Marcus, A. (ed.) (1981) *The Biochemistry of Plants*, Vol. 6, *Proteins and Nucleic Acids*. Academic Press, New York.

Marcus, A. (ed.) (1989) *The Biochemistry of Plants*, Vol. 15, *Molecular Biology*. Academic Press, San Diego.

Mengel, K. & Kirkby, E.A. (1982) *Principles of Plant Nutrition*, 3rd edn. International Potash Institute, Bern.

Miflin, B.J. (ed.) (1980) *The Biochemistry of Plants*, Vol. 5, *Amino Acids and Derivatives*. Academic Press, New York.

Morris, J.G. (1974) *A Biologist's Physical Chemistry*, 2nd edn. Arnold, London.

Mullet, J.E. (1988) Chloroplast development and gene expression. *Annual Review of Plant Physiology and Plant Molecular Biology*, **39**, 475–502.

Murphy, D.J. (1986) The molecular organization of the photosynthetic membranes of higher plants. *Biochemica et Biophysica Acta*, **865**, 33–94.

Newton, K.J. (1988) Plant mitochondrial genomes: organization, expression and variation. *Annual Review of Plant Physiology and Plant Molecular Biology*, **39**, 503–32.

Nicholls, D.G. (1982) *Bioenergetics: An Introduction to the Chemiosmotic Theory*. Academic Press, London.

Parthier, B. & Boulter, D. (eds) (1982) *Encyclopedia of Plant Physiology* new series, Vol. 14B, *Nucleic Acids and Proteins in Plants: II Structure, Biochemistry and Physiology of Nucleic Acids*. Springer Verlag, Berlin.

Pate, J.S. (1980) Transport and partitioning of nitrogenous solutes. *Annual Review of Plant Physiology*, **31**, 313–40.

Postgate, J.R. (1982) *The Fundamentals of Nitrogen Fixation*. Cambridge University Press, Cambridge.

Preiss, J. (ed.) (1980) *The Biochemistry of Plants*, Vol. 3, *Carbohydrates: Structure and Function*. Academic Press, New York.

Preiss, J. (ed.) (1988) *The Biochemistry of Plants*, Vol. 14, *Carbohydrates*. Academic Press, San Diego.

Ranjeva, R. & Boudet, A.M. (1987) Phosphorylation of proteins in plants: regulatory effects and potential involvement in stimulus/response coupling. *Annual Review of Plant Physiology*, **38**, 73–93.

Reinhold, L. & Kaplan, A. (1984) Membrane transport of sugars and amino acids. *Annual Review of Plant Physiology*, 35, 45–83.

Robertson, R.N. (1983) *The Lively Membranes*. Cambridge University Press, Cambridge.

Robinson, D.G. (1985) *Plant Membranes: Endo- and Plasma Membranes of Plant Cells*. John Wiley & Sons, New York.

Salisbury, F.B. & Ross, C.W. (1985) *Plant Physiology*, 3rd edn. Wadsworth Publishing, Belmont.

Schubert, K.R. (1986) Products of biological nitrogen fixation in higher plants: synthesis, transport and metabolism. *Annual Review of Plant Physiology*, 37, 539–74.

Singh, P., Kumar, P.A., Abrol, Y.P. & Naik, M.S. (1985) Photorespiratory nitrogen cycle—A critical evaluation. *Physiologia Plantarum*, **66**, 169–76.

Somerville, C.R. (1986) Analysis of photosynthesis with mutants of higher plants and algae. *Annual Review of Plant Physiology*, 37, 467–507.

Spanswick, R.M. (1987) Electrogenic ion pumps. *Annual Review of Plant Physiology*, **23**, 267–89.

Spiker, S. (1985) Plant chromatin structure. *Annual Review of Plant Physiology*, **36**, 235–53.

Staehlin, L.A. & Arntzen, C.J. (eds) (1986) *Encyclopedia of Plant Physiology* new series, Vol. 19, *Photosynthesis: III Photosynthetic Membranes and Light Harvesting Systems*. Springer Verlag, Berlin.

Stumpf, P.K. (ed.) (1980) *The Biochemistry of Plants*, Vol. 4, *Lipids: Structure and Metabolism*. Academic Press, New York.

Stumpf, P.K. (ed.) (1987) *The Biochemistry of Plants*, Vol. 9, *Lipids: Structure and Function*. Academic Press, Orlando.

Tanner, W. & Loewus, F.A. (eds) (1981) *Encyclopedia of Plant Physiology* new series, Vol. 13B, *Plant Carbohydrates: II Extracellular Carbohydrates*. Springer Verlag, Berlin.

Thorne, J.H. (1985) Phloem unloading of C and N assimilates in developing seeds. *Annual Review of Plant Physiology*, **36**, 317–43.

Tobin, E.M. & Silverthorne, J. (1985) Light regulation of gene expression in higher plants. *Annual Review of Plant Physiology*, **36**, 569–93.

Tolbert, N.E. (ed.) (1980) *The Biochemistry of Plants*, Vol. 1, *The Plant Cell*. Academic Press, New York.

Walker, D. (1987) *The Use of the Oxygen Electrode and Fluorescence Probes in Simple Measurements of Photosynthesis*. Research Institute for Photosynthesis, University of Sheffield.

Index

Note: Page numbers shown in *italic* refer to illustrations and structural formulae.

A site, *see* aminoacyl site
Abscisic acid 321
Absorption spectrum, of chloroplasts 144
Acacia georginae 110
Acetabularia 20, 320
Acetaldehyde *103*
Acetate
  conversion to sucrose 141
  metabolism in chloroplasts 241
  metabolism via glyoxylate pathway 138
Acetoacetyl-ACP *259*
Acetoacetyl-coenzyme A 271
Acetobacter xylinum 285
Acetohydroxy acid reductoisomerase 220
Acetohydroxy acid synthase 218–19
  inhibition by chlorsulphuron 226
Acetohydroxy acids, metabolism 218–19
Acetolactate, metabolism 219
Acetyl transferase 259
Acetyl-coenzyme A *52*, 109, *110*, 138
  in *O*-acetylserine synthesis 231
  elongation 259
  in fatty acid synthesis 257–60
  formation from fatty acyl-CoA 135
  $\Delta G^{\circ\prime}$ of hydrolysis 52–4
  in prenyllipid synthesis 269–71
Acetyl-coenzyme A carboxylase 257
Acetyl-coenzyme A hydrolase 258
Acetyl-coenzyme A synthetase 257–8
Acetyl-coenzyme A:acetyl-CoA
  acetyltransferase 271
Acetyl-coenzyme A:glutamate
  acetyltransferase 225
6-Acetyl-dihydrolipoate *110*
Acetylation, of proteins 311, 316
Acetylene, reduction by nitrogenase 226
N-Acetylgalactosamine *15*, 311–12
  in chitin 19
  in HRGP synthesis in animals 287
N-Acetylglucosamine-*L*-asparagine *311*

Acetylglutamyl kinase 225
Acetylglutamyl semialdehyde dehydrogenase
  225
Acetylornithine amidohydrolyase 225
Acetylornithine:2-oxoglutarate
  aminotransferase 225
Acetylornithine: glutamate acetyltransferase
  225
O-Acetylserine *231*
Acid-growth hypothesis 281
Acid lipase 133
Acidic regulatory proteins 293
Aconitase 109–10, 139
cis-Aconitate *110*
ACP (acyl carrier protein) 259
Acrylyl-coenzyme A 136
Actin 29, 44
  signal peptide from soya bean 313
Actinomycetes 227
Actinomycin, inhibition of RNA
  synthesis 315
Actinorhizal plants 226–7
Action spectra
  in chloroplasts 144
  in dinoflagellates 144
  of PSI 161
  of PSII 161
Activase, of RuP$_2$-Case 179
Activated sugars 54
Activation energy 69
Activators 62–3
Active transport 94
Acyl carrier protein, *see* ACP
Acyl lipid, importance in thylakoids 156
Acyl transferase 260
Acyl-ACP:1-acyl-*sn*-glycerol-3-P acyl
  transferase 268–9
Acyl-ACP:*sn*-glycerol-3-P acyl
  transferase 268–9
Acyl-coenzyme A oxidase 134, 136

Acylated sterylglycoside, in plasma
  membrane 84
Acyldiaminopimelate aminotransferase 219
Acyldiaminopimelate deacylase 219
Adenase 251
Adenine *30*, 302
  base pairing with thymine 32–3
  catabolism 251
  synthesis 250
Adenine nucleotide translocator 344
  of mitochondria 113, 123–4
Adenine nucleotides
  effects on specific enzymes 114
  in TCA cycle regulation 114
Adenosine 31
Adenosine 5′-diphosphate, *see* ADP
Adenosine 5′-monophosphate, *see* AMP
Adenosine 5′-phosphosulphate, *see* APS
Adenosine 5′-sulphatophosphate, *see* APS
Adenosine 5′-triphosphate, *see* ATP
Adenosine diphosphate glucose, *see* ADP-
  glucose
S-Adenosylhomocysteine, metabolism 234
S-Adenosylmethionine *234*, 300
  in chlorophyll synthesis 253
  metabolism 234–6
  methyl group donor 234–6
  methylation of acidic pectins 286
  in phenylpropanoid metabolism 288–9
  in phospholipid biosynthesis 267
  in regulation of methionine
    synthesis 233
  regulation of aspartate kinase 217
  regulation of aspartate metabolism 218
  regulation of threonine synthase 217
  synthesis 233–4
S-Adenosylmethionine:methionine
  S-methyltransferase 236
S-Adenosylmethionine:Mg-protoporphyrin
  methyltransferase 254

S-Adenosylmethionine synthetase 233–4
Adenylate kinase 134, 250
Adenylic acid 31
Adenylosuccinate lyase 250
Adenylosuccinate synthetase 250
ADP (adenosine 5′-diphosphate) 31
  $\Delta G^{\circ\prime}$ of hydrolysis to AMP 53
  phosphorylation
    in chloroplasts 157
    comparison with free energy of proton
      gradient 121
    in conservation of energy of hexose
      oxidation 125
    efficiency of respiration 124
    in glycolysis 103
    linked to mitochondrial electron
      transport 119
    in mitochondria 121
    proton requirements 124
    stoichiometry with proton
      transport 159
    in TCA cycle 111
  regulation of pyruvate, P$_i$ dikinase 199
  theory of substrate level
    phosphorylation 54
ADP-glucose (adenosine diphosphate
  glucose) 175
  metabolism 282
  in starch synthesis 175–6
ADP-glucose pyrophosphorylase 175–6,
  181
  in C$_4$ plants 202
  regulation 179–81
ADP-glucose starch transglucosylase 176
  in C$_4$ plants 202
ADP/ATP translocator 93
AGP (arabinogalactan protein) 277–8
Agrobacterium 334–5
  as transfer vector 335–6
Agropine *335*

Alanine *27*
  in AGPs 277
  formation from pyruvate 214
  metabolism in C$_4$ plants 200
  recycling in C$_4$ plants 201–2
Alcohol dehydrogenase 103
  mRNA 320
Aldo-sugars 10
Aldohexoses 11–12, 14
Aldolase(s) 61
  in C$_3$-CR cycle 170–1, 174, 177, 180
  in glycolysis 101–2, 104
  synthesis of sedoheptulose-1,7-P$_2$ 171
Aldonic acids (aldonates) 14
Aldopentoses 11–12, 14
Aldoses 10
Aleurone 320–1
  and GA$_3$ 321–2
Alkaline lipase 133
Alkaloids, examples 34
n-Alkane hydrocarbons, in cuticles 22
Allantoate amidohydrolase 252
Allantoate
  catabolism 252
  in long-distance N transport 227
  synthesis 251
Allantoicase 251–2
Allantoin *227*
  catabolism pathway 252
  in long-distance N transport 227
  sytnesis 251
Allantoinase 251–2
Allopurinol 251
Allose *12*
Allosteric enzymes 79, 80
Allosteric inhibition, of TCA cycle
      enzymes 114
Alloxydim *258*
Alnus 227
Alternative oxidase 117
Altrose *12*
Amaranthus 321
Amide-N transfer 216
Amidohydrolase, in allantoate
      metabolism 252
Amidophosphoribosyl transferase 249

Amino acids
  abbreviations 39
  activation 307
  classification 26
  concentration in stressed plants 26
  enantiomers 26
  essential 25, 209
  in foodstuffs 209
  formation from oxo acids 214
  general synthetic routes 215–16
  linkage in proteins 27
  in proteins 25
  in purine synthesis 249–50
  in pyrimidine synthesis 245–7
  nomenclature 26
  non-essential 26
  non-protein 229
  properties in solution 26
  structures 27
  synthesis 209, 211–23, 230–3
2-Amino-2-deoxygalactose 16
2-Amino-2-deoxyglucose 16
2-Amino-4-hydroxy-6-hydroxymethyl
      dihydropteridine 235
Amino-oxyacetate 188, *215*
Aminoacyl site 309–10
Aminoacyl-adenylate complex 307
Aminoacyl-tRNA(s) 308–10
Aminoacyl-tRNA synthetase(s) 307
  from chloroplasts 313
  equilibrium exchange 75
  location of genes 348
  proofreading capacity 307
4-Aminobenzoic acid *235*
Aminohydroxy acid synthase 220
5-Aminolevulinate *253*
  in chlorophyll synthesis 254
  effect on chlorophyllide synthesis
      255
  metabolism 253–4
  synthesis 252–3
    regulation of 256
Aminolevulinate dehydratase 253–4
Aminolevulinate synthetase, in
      animals 253
Aminopterin 235

Aminotransferase(s) 61, 188, 214–15, 220,
    239
  in amino acid synthesis 216, 219
  in C$_2$-PR cycle 188–9
  in C$_4$ plants 200
  4,5-dioxovalerate as substrate 253–4
  examples 214
  glutamyl-1-semialdehyde as
      substrate 253
  in peroxisomes 188–9
  kinetics of reaction 75
  pyridoxal phosphate cofactor 215
  specificity 215
Ammonia
  accumulation in photorespiratory
      mutants 191
  assimilation 211–14
    in chloroplasts 213
      effect of dicarboxylates 235
  formation from glycine 189, 213
  formation from N$_2$ 226–7
  formation in C$_2$-PR cycle 187
  oxidation by chemoautotrophs 5
  rate of production in leaf
      mitochondria 189
  reassimilation 213, 238
    energy requirements 191
    rates 241
  release from ureides 252
  repression of NO$_3^-$ reductase 211
  toxicity 212
  uncoupling of
      photophosphorylation 212
AMP (adenosine 5′-monophosphate) 31
  $\Delta G^{\circ\prime}$ of hydrolysis to adenosine 53–4
  synthesis 250
Amphipathic proteins, in membranes 37
Ampicillin resistance 327–9
  genes 329
α-Amylase(s) 100, 321
  gene expression 322
β-Amylase 100
Amylases, in chloroplasts 177
Amylo(1,4 → 1,6) transglucosylase 176
Amylopectin 100
  structure 18–19

synthesis 176
Amylose 100
  structure 18
  synthesis 176
Amytal, inhibitor of mitochondrial electron
    transport 115
Anabaena 228
Analogues, of amino acids 229
Analytical centrifugation 48–9
Anaphase 45–6
Anaplerotic reactions 125
Ankistrodesmus 321
Anomers, of sugars 11–12
Antenna complexes,
  of PSI 149–50
  properties 149
Anthocyanins 34
Anthranilate, metabolism 221–3
Anthranilate phosphoribosylisomerase 223
Anthranilate phosphoribosyltransferase 223
Anthranilate synthase 223–4
  regulation 222
Antibiotic resistance 329, 335
Anticodon 33, 306, 308
Antimycin A, as inhibitor of mitochondrial
    electron transport 115
Antiport 93–4, 174
Apoenzyme 63
Apoprotein 56
Apple, hydroxy fatty acids in cutin 262
APS (adenosine 5′-phosphosulphate) *231*
  metabolism 231–2
APS kinase 232
APS sulphotransferase 230–1
  light modulation 178
Aqueous polymer two-phase partition 155
Arabidopsis 179
  DNA content 292
  photorespiratory mutants 189–90, 238
  replicons 297
Arabinans, structure 276
Arabinogalactan
  in cell wall organization 280–1
  structure 277
Arabinogalactan protein, *see* AGP
Arabinogalactans I, structure 276

Arabinose *12*, 311
in AGPs 277
in extensin 312
in HRGP synthesis 287
Arabinosyl–hydroxyproline *312*
Arabinoxylan
cross linking to xyloglucan 279
structure 278
Arachidic acid 20
Arachidonic acid 20
Arbutin 34
Arginine 27
in proline synthesis 222
synthesis 222–3
pathway 225
regulation 247–8
Arginosuccinate lyase 225
Arginosuccinate synthetase 225
Arogenate
metabolism 214, 221–3
regulation 222
regulation of DAHP synthase 221–2
Arogenate dehydratase 223–4
regulation 222
Arogenate dehydrogenase 223–4
regulation 222
Aromatic amino acids
general metabolism 221
synthesis 221–3
regulation 224
synthetic pathway 222–3
Aromatic pathway 221–3
Ascorbate peroxidase 240–1
Ascorbate *15*
as electron donor to chloroplast electron
transport 148
in H_2O_2 reduction 240–1
in proline hydroxylation 287
metabolism 240
Ascorbate/glutathione pathway 240
Ash (*Fraxinus*) 281
Asparagine 27
glycoslyation by *N*-
acetylglucosamine 287
synthesis 215–16
in vascular transport of N 215

Asparagine synthetase 216
Asparagine: glyoxylate
aminotransferase 189
Aspartate 27
in amino group transfer 214
in arginine synthesis 223
in asparagine synthesis 215–16
in CO_2 assimilation in C_4 plants 197–8
decarboxylation in C_4 plants 199–202
formation from oxaloacetate 214
metabolism 216–21
regulation 217–18, 233
metabolism in C_4 plants 200–2
in purine synthesis 249
in pyrimidine synthesis 245–6
in threonine synthesis 216–17
in ureide synthesis 251
Aspartate kinase 216–19
isoenzymes 217
regulation 217
subcellular location 217
Aspartate transcarbamylase 245–6, 248
regulation 247
Aspartyl-4-P, metabolism 216
Aspartyl-4-semialdehyde, metabolism 216–
17
Aspartyl-4-semialdehyde
dehydrogenase 217, 219
Asteraceae 19, 100
ATP (adenosine 5′-triphosphate) 31, *52*
in assimilation of inorganic N 213–14
consumption, associated changes in
chlorophyll fluorescence 183–4
expenditure in glycolysis 102
export from mitochondria 112
hydrolysis 4
coupling to other reactions 53
energetics of coupled reactions 53
$\Delta G^{o\prime}$ for ADP production 52–3
$\Delta G^{o\prime}$ for AMP production 52–3
in glutamine synthesis 211, 213
in glycolysis 100, 102, 104
light-enhanced production in
cytosol 237
phosphorylation of glycerate-3-P 169
rates of consumption in

chloroplasts 241
requirements for C_3-CR cycle 172
requirements for N and S
assimilation 230
requirements of C_2-PR cycle 191–3
requirements of protein synthesis 311
supply during plastid development 345
synthesis in mitochondria, proton:ATP
ratio 123
ATP sulphurylase 230–1
ATP synthetase 121
of chloroplasts
CF_0 158–9
coupling factor (CF_1) 158
distribution in thylakoids 155–6
structure and composition 157, 159
effect of monogalactosyl
diacylglycerol 86
of mitochondria 122–4
location in inner membrane 118
location of genes 348
mechanism of ADP
phosphorylation 122–3
structure 122
Atrazine *153*
Autocatalysis, in chloroplasts 172–3
associated chlorophyll fluorescence
182–3
Autotrophs, characteristics 5
Auxins 321
Avogadro number 59
Azaserine *212*, 247
L-Azetidine-2-carboxylic acid 34, *229*
Azide
inhibition of mitochondrial electron
transport 115
inhibition of NO_3^- reductase 210
Azolla 228

Bacillus thuringiensis var. *berliner*,
toxin 333
Bacterial toxins, effects on plant pests 333
Bacteriophage M13 331–2
as cloning vector 331
Bacteriophage λ 325
as cloning vector 325

genome 325
Barley, chromosomes 291
Benzyladenine 322
Bicarbonate, permeability of
membranes 88
Bicarbonate pump, in algae 170
Binary fission, of organelles 294
Binary vectors 335
Biotin *65*, 67
as prosthetic group 64–5
Biotin carboxyl carrier protein 257
Biotin carboxylase 257
Biotinyl lysine 65
Bipyridylium herbicides 154, 240
interaction with electron flow in
PSI 153
Bis-($3′ \to 5′$)-cyclic diguanylic acid *285*
Bisubstrate reactions 68, 73, 75, 78
Blue-green algae, *see* cyanobacteria
Blue-light receptor 345
Branching
in glycosidic chains 17
in matrix polysaccharides 276
Branching enzyme 176
Brassica campestris, vegetable oil 20
Brewing 103–4
Broad beans, hydroxy fatty acids in
cutin 262
Bundle sheath 196
Bundle sheath cells in C_4 plants
CO_2 assimilation 198–9
enzyme complement 200–2
function 198–9
micrograph 196
Butyryl-ACP 260

C_1 fragments 233–6
sources 234–5
C_1 metabolism, reaction products 234
C_2-photorespiratory cycle (C_2-PR
cycle) 187–90
carbon recovery 191–2
C_1 metabolism 236
energy requirements 191, 193
location of enzymes 187–90
metabolite transport 190–1

pathway 188
stoichiometry 191–2
C$_3$-carbon reduction cycle (C$_3$-CR
        cycle) 167–75
associated ion movements 179
in C$_4$ plants 202
in CAM plants 206
coupling to O$_2$ evolution 169
effect of light-induced ion changes 179
energy requirements 172
inhibition by glyceraldehyde 198
light-modulated enzymes 177–9
membrane permeability to
        intermediates 172–4
methods 167
occurrence in organisms 167
pathway 171
perturbation and changes in chlorophyll
        fluorescence 183–4
regulation 177–80
stoichiometry 170, 172
C$_3$ plants
energy requirements 204
quantum efficiency 205
rate of CO$_2$ assimilation 7
C$_4$ dicarboxylates
in CO$_2$ assimilation in C$_4$ plants 197–8
in C$_4$ plants 195
decarboxylation in C$_4$ plants 200–2
in N$_2$ fixation 227
turnover in C$_4$ and CAM plants 207
C$_4$ dicarboxylate shuttle 238–9
C$_4$-CO$_2$ concentrating cycle (C$_4$-CC
        cycle) 198, 205
inhibitors 208
C$_4$ plants 195
characteristics 196, 204–5
decarboxylation mechanisms 200–2
energy requirements 202–4
enzymes of CO$_2$ assimilation 199–200
herbicides 208
O$_2$ tolerance 205
photorespiration 196
phylogenetic distribution 207
quantum efficiency 205
weeds 195, 208

C$_4$ variants 200–2
energy requirements 202–4
NAD malic enzyme type 200–2, 204
NADP malic enzyme type 200, 202–3
PEP carboxykinase type 200–3
C$_5$-ammonia assimilation cycle 212–13
co-ordination with C$_2$-PR cycle 214
C$_5$ dicarboxylate shuttle 139
CAAT box 303–4, 318
Caffeine 250
synthesis 250
Calcium
binding by pectic substances 275
control of cellulose synthesis 285
in enzyme regulation 81–2
levels of free Ca^{2+} in cytosol 81–2
release fron endoplasmic reticulum 82
as secondary messenger 82
Calcofluor 285
Callose 285–6
role of endoplasmic reticulum in
        synthesis 286
structure 276
synthesis, inverse correlation with
        cellulose synthesis 286
Calmodulin 81–2
Calvin cycle, *see* C$_3$-CR cycle
Calvin–Benson cycle, *see* C$_3$-CR cycle
CAM (Crassulacean acid metabolism)
        plants 205–7
definition 195
energy requirements 207
metabolite recycling 206
photorespiration 207
phylogenetic distribution 205, 207
properties 205, 207
rate of CO$_2$ assimilation 7
CAM variants 206
energy requirements 204
Cambium, cell wall composition 279, 281
Campestrol 270
Camphor 34
Capric acid 20
Caproic acid 20
Caprylic acid 20
Carbamyl-P 245

metabolism 223, 245–6
control of synthesis 247
Carbamyl-P synthetase 223, 225, 245–6,
        248
regulation 247
Carbohydrates 10–20
content in germinating seeds 139
mobilization 99
oxidation 99
storage 99
Carbohydrate side-chains 316
Carbon
asymmetric bonding of substituents
        9–10
inorganic, transport in
        chloroplasts 169–170
metabolism
        co-ordination by phosphate 181–2
        diurnal levels in leaves 6
        export from chloroplasts 173–4
        fluorescence methods 182–4
        regulation in photosynthetic cells
                177–82
        routes in mesophyll cells 181
        ratio with respect to other elements 3
Carbon dioxide
accumulation in C$_4$ plants 205
assimilation
        associated changes in chlorophyll
                fluorescence 183
        associated metabolite transport 204
        in CAM plants 205
        in cells pre-adapted to carbon 193
        in chemoautotrophs 5
        in C$_3$ plants 167–75
        in C$_4$ plants 195–204
        into C$_2$ products 187
        during chloroplast development 341
        energy requirements 4, 172, 203–5
        light energy requirements 59–60
        methods 167
        pathway in C$_3$ plants 171
        pathway in CAM plants 206
        pathway in C$_4$ NAD M-E variant 201
        pathway in C$_4$ NADP M-E
                variant 198

pathway in C$_4$ PEP variant 102
products 167
pulse-chase experiment 197
rates 7–8, 173, 241
regulation by P$_i$ 182
steady state 170
        into sugar phospnates 170
        into sucrose 173
compensation point(s) 186
        in C$_4$ plants 193, 205
        in algae 193
carbon source for plant growth 4
effect on CO$_2$ assimilation 185–6
effect on RuP$_2$-C/Oase 186
equilibrium with HCO$_3$$^-$ in
        chloroplasts 169–70
evolution, in light 185
evolution, via OPP pathway 128
exchange, by leaves 185–6
formation from glycine 189
inhibition of photorespiration 185
permeability of membranes 88
plant requirements 3
production in C$_2$-PR cycle 187
        in purine synthesis 249
        in pyrimidine synthesis 245
reassimilation, in CAM plants 205, 207
uptake 170
        in C$_4$ plants 205
Carbonic anhydrase 170
in algae 194
Carbonyl cyanide *m*-chlorophenyl
        hydrazone (CCCP) 121
Carbonyl cyanide-4-
        trifluoromethoxyphenylhydrazone, *see*
        FCCP
3-Carboxy-2-hydroxyisocaproate,
        metabolism 219
2-Carboxyarabinitol-1,5-P$_2$ 168
2-Carboxyarabinitol-1-P 179
Carboxyl transferase 257
β-Carotene(s) 144, 270, *273*
Carotenoids 25, 34, 142
biosynthesis 272–3
in light harvesting 144
Carrier proteins

characteristics 92
kinetics 92–3
Cassuarina 226
Castor bean
glycolysis in plastids 104
β-oxidation 140
Catalase 43, 64, 134, 137, 140
Cauliflower mosaic virus, as transfer
vector 336
cDNA (copy DNA) 324, 326, 334
from phaseolin mRNA 324
identification 330–1
insertion into cloning vector 327
synthesis 326–9
by Okayama and Berg method 328–9
by S1 nuclease method 327
CDP-choline:diacylglycerol choline
phosphotransferase 266, 269
CDP-diacylglycerol 266
CDP-ethanolamine:diacylglycerol
ethanolamine phospho-
transferase 266
Cell cycle 45–6
Cell differentiation, and gene
expression 315
Cell extraction, methods 46
Cell fractionation
by centrifugation 47
example 48
Cell plate 45, 276
role of endoplasmic reticulum in
synthesis 286
Cell wall
bonding between components 283
composition 275–9
growth 279, 281
importance 275
matrix 275–9
methods for disruption 46
organization 280–1
plasticity 276, 279
polysaccharides 276
structure 45
synthesis 275–90
Cellobiose 17
Cells

diversity 36
structural organization 36–46
Cellulase 19
in protoplast preparation 46
Cellulose
biological degradation 19
in cell wall organization 280
in cell walls 45
complexes with hemicelluloses 280
cross-linking to matrix 279
crystalline regions 19
intermolecular hydrogen bonding 18–19
intramolecular hydrogen bonding 18–19
microfibrils 275
stability 19
structure 18, *19*, 276
synthesis 283–6
in *Acetobacter* 285
inverse correlation with callose
synthesis 286
model 285
Cellulose synthase 284–6
in *Acetobacter* 285
methods 286
movement in plasma membrane 286
oligomeric assemblies 284
in plants 285
in plasma membrane 285
Centrifugation 47–9
conditions for separating organelles 47
sedimentation coefficient 48–9
sedimentation rate 47–9
Centromere 293
Ceramide monohexoside
in plasma membrane 84
in tonoplasts 84
CF$_1$, *see* coupling factor CF$_1$
Chaetomorpha 286
Charge separation, at reaction centres of
photosystems I and II 150
Chemiosmosis 119
Chemoautotrophic bacteria 5, 209
C$_3$-CR cycle 167
Chitin 19–20
Chlamydomonas reinhardtii, CO$_2$
assimilation 193

Chloramphenicol 315, 348
effect on synthesis of plastid
enzymes 350
Chlorate 210
Chlorella 232
Chlorobium 167
3-(*p*-Chlorophenyl)-1,1-dimethyl urea,
see CMU
Chlorophyll *a* 142
absorption spectrum 143
different forms 143–4
structure 143
synthesis 254–5
Chlorophyll *b* 142
absorption spectrum 143
synthesis from chlorophyll *a* 254–5
Chlorophyll
absorption of light 4, 59, 142–3
accumulation during 'greening' 342
in antennae serving reaction centres 149
basis for expressing growth 7–8
content of plastids during
development 338, 341, 344
electron orbitals 143
excitation 144
excitation states 145
first singlet state 145
first triplet state 146
fluorescence 145–6, 182–4
changes, 'fast' 160
changes, 'slow' 160
changes during perturbation of C$_3$-CR
cycle 183
during plastid development 341, 344
emission maximum 146
heat-loss quenching 182
induction 160
induction curves 159–160, 182–3
induction kinetics 160
Kautsky curves 160
in monitoring activity of
photosystems 159
non-photochemical quenching 182
photochemical quenching 182
photosystem II 182
quenching 146, 182–4

quenching, in monitoring activity of
photosystems 159
state transitions 161
transients during autocatalysis 182–3
number of molecules serving reaction
centres 149
second singlet state 144–5
second singlet state 144–5
synthesis 234, 252–5
regulation 255–6
in thylakoids 41
Chlorophyll-binding protein, *see* LHCP
Chlorophyll:cytochrome *f* ratio, values in
thylakoids 156
Chlorophyllide
regulation of magnesium chelatase
255–6
synthesis 255
Chlorophylls 142–3
absorption spectra 143
electronic state of porphyrin ring 143
phylogenetic distribution 142
Chloroplast DNA 294
Chloroplast envelope
permeability to glycerate-3-P 169
permeability to GSH 239
rates of metabolite transport 173
Chloroplast extracts, enzymes 167
Chloroplast function, location of
genes 347
Chloroplast genome, *see* plastome
Chloroplast proteins
control by nuclear genes 347–8
control by plastid genome 348
regulation of synthesis 350
sites of synthesis 349
Chloroplast ribosomal RNA, genes in
spinach 294
Chloroplasts
absorption spectrum 144
acetate metabolism 241
amino acid synthesis 217, 221
ammonia reassimilation 238
aromatic amino acid synthesis 221
aspartate metabolism 217, 221
assimilation of inorganic N 213–14

assimilation rates 241
autocatalysis 172–3
biogenesis 338–52
centrifugation characteristics 48
from C_4 plants, CO_2 assimilation 198
CO_2 assimilation 167
CO_2 uptake 170
cysteine synthesis 231
decarboxylation mechanisms 200–2
development
    changes in fatty acid composition 343
    changes in lipid composition 343
    from etioplasts 341
    at molecular level 339–47
    in monocots 339
    from proplastids 339–40
dicarboxylate shuttles 238
enzymes of C_3-CR cycle 167–9
enzymes of methionine synthesis 233
enzymes of pyrimidine synthesis 246–7
enzymes of starch synthesis 175–6
fatty acid synthesis 241
fluorescence quenching 182–4
lipids 25
lysis 41
membrane asymmetry 42
membrane lipids 85
    fatty acid composition 84, 86
methionine synthesis 233
micrographs 197
in NADP C_4 M-E variant 202
number per cell 41, 338
OPP pathway 130–1
O_2 evolution 172
particles 41
permeability
    of envelope 172–4
    to inorganic carbon 170
    to metabolites 173
    methods 172
pH of stroma 169
preparation 46
protein synthesis 313
pyruvate, P_i dikinase 199
rates of light-coupled processes 241
reconstituted system 172

ribosomes 40
shrinking 170
shuttle mechanisms 236–9
starch synthesis 175–6
structure 41–2
sulphate assimilation 231
sulphite assimilation 232
triose-P/glycerate-3-P shuttle 237
Chlorsulphuron *226*
    inhibition of acetohydroxy acid
        synthase 226
Choline
    in phospholipids *23*
    synthesis 234
Choline kinase 267
Chorismate
    metabolism 221–3
        regulation 221, 224
Chorismate mutase 221, 223–4
    regulation 221
Chorismate synthase 222
Chromate 230
Chromatids 45
Chromatin 40, 43, 293–4, 316
    acidic regulatory proteins 293
    conformation 317
    conformational changes 316
    DNA 293
    histones 293
Chromatin fibre 294
Chromosome(s) 45, 291, 294
    polyploids 291
Cinchona 35
Cinnamate, metabolism 288
Cinnamate 4-hydroxylase 288
Cisternae 43
Cistron 304
Citrate *110*
    efflux from mitochondria 112
Citrate synthase
    in glyoxylate cycle 139–40
    inhibition by ATP 114
    location 140
    in TCA cycle 109–10, 112
Citrulline *227*
    in long-distance N transport 227

Clethodim *258*
Cloning vector(s) 325
    bacteriophage M13 331
    bacteriophage λ 325
    cosmids 325
    introduction of DNA 326
    plasmids 325
Cloning, of DNA 324–6
CMP (cytidine 5′-monophosphate),
    synthesis 247
CMU (3-(*p*-chlorophenyl)-1,1-
    dimethyl urea) 153
Co-translational transport 313
Cobalt, in vitamin B_{12} 236
Cocaine 34
Codeine 34
Codium 19
Codon(s) 34, 302, 304
Coenzyme A 63
    in fatty acid synthesis 257–60
    in phenylpropanoid synthesis 288
Coenzymes 63–4
Cofactors 63
    as vitamins in animal nutrition 67
Cohesive ends 322
Colchicine *282*
Coleoptera, susceptibility to bacterial
    toxins 333
Collenchyma cell, structure 37, 44
Competitive inhibition 76–7
Complex 1, of mitochondrial electron
    transport chain 114, 116
    composition 114–15
Complex II, in mitochondrial electron
    transport 116
    composition 116
Complex III, in mitochondrial electron
    transport 116
    composition 116
Complex IV, in mitochondrial electron
    transport 116
    composition 116–17
Congo red 285
Coniferyl alcohol *34*
    polymerization 290
    synthesis 289

Coniferyl residues, in lignin 278–9
Consumer cells, definition 6
Copper
    in monophenol oxidase 289–90
    plant requirements 3
    in superoxide dismutase 240
Coproporphyrinogen III, metabolism 253
Coproporphyrinogen oxidase 254
Copy DNA, *see* cDNA
Corn, sucrose transport 175
Cosmids 325
    as cloning vectors 326
4-Coumarate, metabolism 288–9
4-Coumarate:coenzyme A ligase 288
Coumarins 34
4-Coumaryl alcohol 34
    in monocotyledon lignin 279
    synthesis 288–9
4-Coumaryl alcohol:NADP
    oxidoreductase 288
4-Coumaryl-coenzyme A:NADPH
    oxidoreductase 288
Coupling factor CF_1 (*see also* ATP
    synthetase of chloroplasts) 158
Crassulacean acid metabolism, *see* CAM
Cristae 42
Crotonase, *see* enoyl-CoA hydratase
Crown gall disease 334
Cryptochrome 345
ctDNA, *see* plastome
CTP (cytidine 5′-triphosphate) 265
    $\Delta G^{\circ\prime}$ of hydrolysis 53
    synthesis 246–7
CTP synthetase 246–7
Cuticle
    lipid composition 22
    permeability 22
Cutin 22
    biosynthesis 262–3
Cyanide
    inhibition of mitochondrial electron
        transport 115
    inhibition of NO_3^- reductase 210
    metabolism 216
    reduction by nitrogenase 226
β-Cyanoalanine, in asparagine

synthesis 216
Cyanobacteria 5, 210
  C_2 metabolism 194
  N_2 fixation 228
Cyanogenic glycosides 35
Cyclic electron transport 161–2
  in purple sulphur bacteria 148
Cyclic photophosphorylation 147, 161–2
  efficiency of energy conversion 162
  importance 162
  in NADP M-E variant 202
  rates 162
Cycloartenol 272
  biosynthesis 273
Cylohexanedione herbicides 258
  differential effect on grasses and broad-
    leafed species 259
Cycloheximide 315, 321–2, 348
Cyclohydrolase 235
Cyperus rotundus 208
β-Cystathionase 232
Cystathionine 26, 232
  metabolism 232–3
    regulation 217, 233
Cystathionine β-lyase 232–3
Cystathionine γ-synthase 232–3
  regulation 233
Cysteine 27
  in asparagine synthesis 216
  in ferredoxins 55–6
  in functions 233
  in methionine synthesis 232
  origin of sulphur atom 231
  role in S metabolism 232
  in S–S bonds in proteins 28
  synthesis in chloroplasts 230–1
  synthesis via bound pathway 231
Cysteine synthase 230–1, 233
  in assimilation of SO_3^{2-} 232
Cytidine 31
Cytidine 5'-monophosphate, see CMP
Cytidine 5'-triphosphate, see CTP
Cytidine nucleotides, synthesis 245–7
Cytidyldiphosphate diacylglycerol 265
Cytochrome a 117
  location in inner mitochondrial

membrane 118
Cytochrome a_3 117
  location in inner mitochondrial
    membrane 118
Cytochrome b
  introns in gene 293, 304
  in NO_3^- reductase 210
Cytochrome b/c_1 complex, in mitochondrial
  electron transport 116
Cytochrome b_6/f complex, location in
  thylakoids 155–6
Cytochrome c 117, 312
  inhibition 118
  location in inner mitochondrial
    membrane 118
Cytochrome c oxidase, subcellular
  distribution 48
Cytochrome c reductase 118
Cytochrome f, in electron transport 151
Cytochrome oxidase 116–18, 137
  affinity for O_2 188
  conformation pump 121
  in mitochondria 47
  mitochondrial marker 286
Cytochromes 56, 116
Cytokinesis 46
Cytokinins 321–2
Cytoplasm 36
  definition 37
  structure 44–5
Cytoplasmic inclusions, definition 37
Cytoplasmic inheritance, see maternal
  inheritance 294
Cytoplasmic ribosomes 38, 40
Cytoplasmic streaming 45
Cytosine 30, 302
  base pairing with guanine 32–3
  synthesis 245
Cytoskeleton 45
Cytosol
  definition 36–7
  preparation 44

Dahlias, fructans 100
DAHP, (3-deoxy-arabinoheptulosonate-7-P)
  221

DAHP synthase 221–2, 224
Dalton, definition 27
Dark respiration, in NO_3^- reduction 210
DBMIB (2,3-dibromo-3-methyl-6-
  isopropyl-p-benzoquinone),
  inhibition of cytochrome f
  reduction 151
DCMU, (3-(3',4'-dichlorophenyl)-1,1-
  dimethylurea) 148, 151, 153
Decarbonylase 261–2
Decarboxylases 61
Dehydroascorbate 241
Dehydroascorbate reductase 241
trans-2,3-Dehydro-fatty acyl-CoA, see trans-
  2-enoyl-CoA
Dehydrogenase(s) 61, 127
3-Dehydroquinate dehydratase 222
Dehydroquinate dehydrogenase 222
Density gradient centrifugation 47–8
3-Deoxy-arabinoheptulosonate-7-P, see
  DAHP
Deoxy-CDP, synthesis 247
Deoxy-TMP, synthesis 248
Deoxy-UDP, synthesis 247
Deoxyadenosine 31
Deoxycholate 40
6-Deoxygalactose 15, 16
6-Deoxymannose 15, 16
Deoxynucleotidyl transferase 328
Deoxyoligonucleotide, synthetic 32
Deoxyribonucleic acid, see DNA
Deoxyribonucleosides 30
Deoxyribonucleotide diphosphates,
  synthesis 247
Deoxyribonucleotides, synthesis 247–8
2-Deoxyribose 15, 16, 30
Deoxythymidine 31
Desaturase 268–9
Desaturation, of fatty acids in phosphatidyl
  choline 265
Dextrorotation 10
Diacylglycerol 265, 266, 269
  formation from phosphatidate 265
Diacylglycerol acyltransferase 265
Diacylglycerol choline 269
Diacylsulphoquinovosylglycerol, in

chloroplasts 84
Diaminopimelate decarboxylase 219
Diaminopimelate epimerase 219
Diaphorase 210
Diastereoisomers, of sugars 11
2,3-Dibromo-3-methyl-6-isopropyl-p-
  benzoquinone, see DBMIB
Dicarboxylate shuttle(s) 238–9
  rate in chloroplasts 241
Dicarboxylate translocators 238–9
  in C_4 plants 204
  in mitochondria 112–13
  mutant 238
  specificity 238
Dichapetalum, fluoroacetate
  accumulation 110
2,4-Dichlorophenoxyacetic acid 88
3-(3',4'-Dichlorophenyl)-1,1-dimethylurea, see
  DCMU
Dicotyledons, cell wall composition 279
Dictyosomes 43
Dideoxynucleotides 331
  in DNA sequencing 331–2
Diethyldithiocarbamate 47
Diferulate bridges 280–2
Differential centrifugation 47
Differential screening 331
Differentiation, specific proteins 315
Diffusion 88–90
  across membranes 88–90
  of ions through 'gated channels' 93
Diffusion potential 89
Digalactosyl diglyceride 24–5, 85–6, 267
  in chloroplast lamellae 84
  structure 24
Digitaria sanguinalis, chloroplast structure
  41, 197
Digitonin, preparation of membrane proteins
  285
Diguanylic acid 285
Dihydrolipoate 109, 110
Dihydrolipoate dehydrogenase 62, 110
Dihydrolipoyl transacetylase (lipoate
  acetyltransferase) 62
Dihydronicotinamide residue, in NAD and
  NADP 55

Dihydro-orotase 246
Dihydro-orotate, metabolism 246
Dihydro-orotate dehydrogenase 246
2,3-Dihydropicolinate 217
Dihydropicolinate reductase 219
Dihydropicolinate synthase 217–19
5,6-Dihydrouridine 33
Dihydroxyacetone *10*
Dihydroxyacetone-P *102*
  in lag phase in CO$_2$ assimilation 172–3
  from lipid oxidation 133–4
  metabolism in chloroplasts 169
  reduction to glycerol-3-P 264
  in triose-P shuttle 237
Dihydroxyacid dehydratase 220
10,16-Dihydroxypalmityl-CoA 263
10,18-Dihydroxystearyl-CoA 263
1,1-Dimethyl-4,4′-bipyridillium dichloride, *see*
    paraquat
Dimethylallyl pyrophosphate, *see* Δ^2-
    isopentenyl pyrophosphate
Dimethyloxazolidinedione, in determination
    of pH gradients 90
Dinitroanilines, as herbicides 282
Dinitrogen fixation, *see* nitrogen fixation
4,5-Dioxovalerate *254*
  metabolism 253–4
o-Diphenoloxidase
  inhibition 46
  properties 46
Diphenylcarbazide 156
Diphosphatidyl glycerol 23, 84–5
Diptera, susceptibility to bacterial
    toxins 333
Disaccharides 16–17
Disease resistance 333
Dissimilatory NO$_3^-$ reduction 209
Disulphide bond(s)
  in GSSG 239
  in proteins 28
Diterpenes 270
Dithiothreitol, activation of light-modulated
    enzymes 178
Diuron, *see* DCMU 153
DNA 31–4
  base sequencing 331–3

in chloroplasts, *see* plastome
chromatin 293
coding capacity 292–3
complementary base pairing 33
conformational change 304
content 45, 291–9, 341
  during development of plastids 338, 341
double helix 32
as genetic material 291
location in cell 32
methylation 300
nuclear 32, 40
in organelles 338
organization 294
renaturation kinetics 292
repair 299
repeated sequences 292–3
replication 297
satellite 32
semi-conservative replication 296–300
single copy 291–2, 294
structure 32
supercoiling 294
synthetic segments 32
synthesis 298–300
  elongation 298–9
  initiation 296–7
  ligation of DNA fragments 299
  rate 296
  replication points 297
  requirement for primer 299
  template
    availability for transcription 316–17
    in DNA synthesis 298
    in RNA synthesis 302
DNase 1 316–17
  susceptibility of expressed genes 316
DNA binding proteins 297
DNA ligase 297, 299–300, 323, 326, 328–9
DNA polymerase(s) 297–300, 327–9, 331
Double-displacement in enzymic
    reactions 74–5

Eadie–Hofstee plot 72–3
Effectors 62
  of allosteric enzymes 79–80

in chloroplasts, *see* plastome
Einstein, definition 59
Einstein's law of photochemical
    equivalence 59
Electrical transmembrane gradients, Δ*G* of
    58–9
Electrochemical potential gradient 94
Electrogenic transport 94
Electron carrier proteins 29
Electron donors
  to electron transport chain in
      chloroplasts 148
Electron orbitals, in chlorophyll 143
Electron transport
  in chloroplasts 150–2, 156–7
  in mitochondria 114–18
Electrophoresis, of DNA fragments 324
Electroporation 336
Elements, oxidation states in plants 3
Elicitors (fungal), induction of P-ALase 321
Elongation factors 309–10
Elongation, of polypeptide chain 309
Enantiomers 10
  of sugars 11
Endergonic reactions 4
  coupling to exergonic reactions 4, 52
Endoamylase (Ca^{2+}-independent) 177
Endonuclease activity, in DNA
    synthesis 297
Endoplasmic reticulum
  attachment of ribosomes 312
  marker protein 84
  in phenylpropanoid metabolism 289
  polypeptide transport 313
  structure 44
  synthesis of glycosyl residues of
      glycoproteins 312
  synthesis of matrix polysaccharides
      286–7
Endosperm 321
  hordein mRNA 317
Endosymbiosis, as origin of
    chloroplasts 347
Endothermic reactions 4, 51
Energetics
  principles 50–60
  of respiration 123–5

Energy, Einsteins required for CO$_2$
    assimilation 59–60
Energy change, principles 50–1
Energy conservation in ATP, efficiency of
    respiration 124
Energy distribution, between PSII and
    PSI 160–1
Energy requirements
  of C$_3$-CR cycle 172
  CAM plants 207
  CO$_2$ assimilation in C$_4$ plants 202–4
  for plant growth 4–5
  for protein synthesis 311
Energy-transducing membranes 58
Energy transfer
  from carotenoids 147
  use of fluorescence monitoring 159
Enhancement 148
Enhancer sequences, *see also* 'upstream
    elements'
  in expression of light-harvesting
      chlorophyll-binding protein 318
Enolase 101, 103–5, 107
  in chloroplasts 104
5-Enolpyruvyl-shikimate-3-P, *see* EPSP
Enoyl ACP reductase 259–60
trans-2-Enoyl-CoA 134
cis-Enoyl-CoA 136
Enoyl-CoA hydratase 134, 137
Enoyl-CoA hydrogenase 136
Enoyl-CoA isomerase 136–7
Entropy 50
Environmental signals, and gene
    expression 320–2
Enzyme kinetics 67–80
  allosteric enzymes 80
  bisubstrate reactions 73–5
    double-displacement 74–5
    single-displacement 73–4
  inhibition 77–8
  leading substrate 73
Enzymes 29, 61–82
  active site 69
  allosteric 79–80
  classification 61
  competition for substrate 75

covalent modification
  phosphorylation 81
  reduction 80
importance 4
inactivation by phenolics 46
inhibition 76–8
  applications 78
  competitive 77–8
  non-competitive 77–8
  uncompetitive 78
kinetics 69–80
light activation/deactivation 80
mode of action 69
pH effects 76
regulation 61–82
regulation of metabolism 79–82
structure 61–2
temperature effects 76
transition state complex 69
Epimerase(s) 61, 128
Epimers, of sugars 11
EPSP (5-enolpyruvyl-shikimate-3-P),
  metabolism 224
EPSP synthase 78, 222
  inhibition by glyphosate 224
Equilibrium, chemical 50
Equilibrium, transmembrane 59
Equilibrium constant 50
  relation to $\Delta G^{\circ\prime}$ 51
Equilibrium exchange 75
Erucic acid 20
Erythrose *11*
Erythrose-4-P *129*
  in amino acid synthesis 215
  in aromatic amino acid synthesis 221–2
  in chloroplasts 171
  in OPP pathway 129–30
Escherichia coli 325
Essential amino acids 25
Esterases 61
Ethanol *103–4*
Ethanolamine
  methylated derivatives 234
  in phospholipids *23*
Ethanolamine kinase 267
Ethylene 321

Ethylene, formation from acetylene 226
Ethylmaleimide, effects on stoichiometry of
  mitochondrial proton
  translocation 121
Etioplasts 339, 341
  chloroplast polypeptides 342
  enzymes of chlorophyll synthesis 255
  marker pigment 84
Etioporphyrins 64
Eukaryotic galactolipids 267
Euglena, plastome copy number 295
Excitation energy transfer 146
Excitation states
  of chlorophyll 145–5
    duration 146
    reversion to ground state 146
Exergonic reactions 4, 51–2
  coupling to endergonic reactions 4, 52
Exons (coding sequences) 293
Exothermic reactions 4
Extensin(s) 277, 312
  addition to cell wall matrix 288
  in cell wall organization 281
  composition 279
  structure of glycosyl moieties 312
  synthesis 287–8
Extrinsic proteins
  extraction 38
  in membranes 38

FAD (flavin adenine dinucleotide) *55*
  properties 55
  as prosthetic group 64
  in pyruvate dehydrogenase complex 110
  in succinate dehydrogenase 111
$FADH_2$ (reduced FAD) *55*
  oxidation by mitochondrial electron
    transport 114, 116
Far-red light, phytochrome and gene
  expression 321
Faraday constant 57
Farnesol *270*
Farnesyl pyrophosphate 271–2
Farnesylfarnesylgeranyl pyrophosphate 271
  as precursor of ubiquinone 274
Fats 21

Fatty acids 20–1
  activation 134
    energetics 134
  in chloroplasts during development 343
  composition
    in chloroplasts during
      development 340–1
    in membranes 86
  conformation 21
  conversion to sucrose 139
  distribution in vegetable oils 20
  elongation 260–1
  export from plastids 261
  in glycerides 21
  in lipids 133
  in membranes 84
  in phospholipids 22
  melting point 21
  nomenclature 21
  oxidation (dehydrogenation) 134
  oxidation of compounds with odd
    number of C-atoms 135
  oxidation of unsaturated fatty acids 136
  β-oxidation 134
  plant composition 21
  rates of synthesis 257
  secondary alcohols 262
    oxidation to ketones 262
  structures 20
  synthesis
    from acetyl-CoA 257–61
    intracellular location 257
Fatty acid synthetase 259–60, 268
  composition 259
  location 259
  reactions of complex 259
Fatty acid thiokinase, *see* fatty acyl-CoA
  synthetase
Fatty acyl-coenzyme A (fatty acyl-
  CoA) *134*
Fatty acyl-CoA oxidase 135
Fatty acyl-CoA reductase (NAD-specific)
  261–2
Fatty acyl-CoA synthetase 134
Fatty acyl-CoA:fatty alcohol acyl
  transferase 261–2

Fatty aldehydes 262
Fatty aldehyde reductase (NADP-
  specific) 261–2
FCCP (carbonyl cyanide-4-
  trifluoromethoxyphenylhydrazone)
  91, 210
  mode of action 91
Fe–S centres, *see* iron–sulphur centres
Ferredoxin(s) 151
  in assimilation of inorganic N 213–14
  as energy source 4
  in glutamate synthesis 212–13
  in light modulation of enzymes 178
  location in thylakoids 157
  molecular mass 28
  in nitrogenase reaction 226–7
  in NO_2^- reduction 211
  in O_2 reduction 240
  properties 55–6
  in reconstituted chloroplast system 172
  in SO_3^{2-} reduction 232
  standard redox potential 57
  structure 56
  in sulphate assimilation 231
Ferredoxin:$NADP^+$ oxidoreductase 151
  precursor polypeptide 349
Ferredoxin:thioredoxin reductase 80, 178
Ferulate
  metabolism 288
  residues in matrix polysaccharides 280
  synthesis 235
Fibrils
  aggregation 285
  elementary cellulose 285
  formation in *Acetobacter* 285
First-order reactions 67
  kinetics 68
First singlet state
  charge separation 150
  in chlorophyll 145
  dispersal of energy 145–6
  of oxygen, protective role of
    carotenoids 146
First triplet state in chlorophyll 146
Flash yield 152
Flavin adenine dinucleotide, *see* FAD

Flavin mononucleotide, *see* FMN

Flavin
in NDP reductase 248
in NO_3^- reductase 210
Flavodoxin, in nitrogenase reaction 226–8
Flavonoids 34–5
Flavoprotein 111
Fluid mosaic model 37–8
Fluorescence, *see* chlorophyll fluorescence
Fluoroacetate-accumulating plants 110
Fluoroacetyl-CoA 110
Fluorocitrate, as inhibitor of aconitase 110
Fluorophenylalanine 77
Flux
of charged solutes 89
across membranes via diffusion 88
of solutes via mediated transport 93
FMN (flavin mononucleotide) 55
in Complex I of mitochondrial electron
transport chain 115
as prosthetic group 64
properties 55
$FMNH_2$ (reduced FMN) 55
Folic acid (folate) 67
Formate, metabolism 236
Formyl groups, sources 233–6
Formyl methionine 313
N^{10}-FormylTHF *64*, 234–6
C_1 donor 234
metabolism 235–6
in purine synthesis 250
Formyl-THF synthetase 235–6
Fractionation, example of density gradient
centrifugation 48
Frankia 210, 227
Free energy change 50–1
oxidation/reduction reactions 57–8
at pH 7 51
relation to ratio of reactants 50
transmembrane electrical potential
difference 58–9
of transmembrane ion gradient 58–9
Free energy gradient as driving force for
diffusion 89
Free energy of activation 69
Free radicals, in phenylpropanoid

metabolism 289–90
French beans, phaseolin gene 293
Fructans 16, 19, 100
β-D Fructofuranose 100
β-Fructofuranosidases 100
Fructokinase 108
Fructose *13*
furanose form *14*
metabolism 175
phosphorylation 101
pyranose form *14*
Fructose-1,6-P_2 100, 101–2
effect on autocatalysis 172
CO_2 labelling pattern 170
metabolism in chloroplasts 170
regulation 180
in starch synthesis 175
synthesis in chloroplasts 171
Fructose-1,6-P_2 phosphatase, FP_2-Pase 106
in chloroplasts 171–2
light modulation 178–9
regulation in cytosol 180
regulation by fructose-2,6-P_2 107
in sucrose synthesis 174
Fructose-2,6-P_2
inhibition of fructose-1,6-P_2
phosphatase 180
metabolism 180
in regulation of carbon metabolism 180
in regulation of glycolysis 107
regulation of sucrose metabolism 180
stimulation of fructose-6-P:P_i
phosphotransferase 180
Fructose-2,6-P_2 kinase, regulation 107
Fructose-2,6-P_2 phosphatase 107, 180
Fructose-6-P *101*, 129–30
control of fructose-2,6-P_2
metabolism 180
$\Delta G^{o\prime}$ of hydrolysis 54
metabolism in chloroplasts 170–1
metabolism in cytosol 180
recycling through OPP pathway 131
in starch synthesis 175
in sucrose synthesis 174
Fructose-6-P kinase (FP-Kase) 81, 106–7,
180

Fructose-6-P:PP_i phosphotransferase,
FP-PTase 102, 104, 106–8
activities in different tissues 107
in C_3-CR cycle 171–2
in futile cycle 79
ingluconeogenesis 106
in glycolysis 102
light modulation 178–9
location 105
regulation 180–1
in sucrose cleavage to hexose-P 107–8
Fucose 15–16
Fucoxanthin 144
Fumarase 109, 111, 139–140
location 140
in synthesis of sucrose from lipid 139
Fumarate 76, 111
Functional groups 9
Fungi
cell walls 19
cellulose hydrolysis 19
Furan 13–14
Furanose sugar 11, 13–14
Futile cycling 79
ATP hydrolysis 177
between glycolysis and gluconeogenesis
106
of PEP 206
F_0 component of mitochondrial ATP
synthetase, structure 122
F_1 component of mitochondrial ATP
synthetase, structure 122

ΔG, *see* free energy change
ΔG°, *see* standard free energy change
Gabaculine 253
Galactans 16
structure 276
Galactoglucomannans, structure 276
Galactolipid:galactolipid
galactosyltransferase 268–9
Galactolipids 25
16:3 galactolipid plants 267
18:3 galactolipid plants 267
in membranes 86
routes of biosynthesis 268–9

stabilization of membranes 274
Galactomannans 100
Galactosamine *15*, 16
Galactose *12*, 311
in AGPs 277
in HRGP synthesis 287
in lipids 267
β-Galactosidase 335
in bacteriophage M13 331
in pUC plasmids 329
regulation in bacteria 318
Galactosyl–hydroxyproline *312*
Galactosyl–serine *311*
Galactosyl diglycerides 25
Galacturonans, structure 276
Galacturonic acid (galacturonate) *15*
6-methyl ester *15*
in pectic substances 275–6
incorporation into pectic
polysaccharides 286
Gastrolobium 110
Gastropods, cellulose hydrolysis 19
Gated channels 93
Gated pore, in mitochondria 93
GDP (guanosine 5′-diphosphate) 31
GDP-glucose
incorporation into hemicellulose 286–7
metabolism 282
GDP-mannose, metabolism 282
GDP-mannose pyrophosphorylase 284
GDP-mannuronate, metabolism 282
Gels
matrix polysaccharides 278
pectic substances 276
Gene expression 315–37
availability of DNA template 316–17
control points 315–16
control sequences in genome 318
in germinating seeds 321
integration of expression of plastid and
nuclear genomes 351
regulation by phytochrome 346
Gene library 326, 330
Gene mapping 322
Gene organization 294–5
base sequence 331

repeated sequences 292–3
single copy DNA 292
Gene probes, see oligonucleotide probes
Gene structure 303
Genes
    for extensin synthesis 277
    HRGPs 287
    ribosomal RNAs 305
Genetic code 301–2
    in mitochondria 314
Genetic engineering 333–7
    mechanical methods for introducing
        genes 336
    objectives for plants 333–4
    pest resistance 333
Genome 291–300
    organization 291–6
    organelles 294
    replication 296–300
    size 291–2
Geraniol 270
Geranyl pyrophosphate 272
Geranylgeraniol 270
Geranylgeranyl pyrophosphate 271–2
    biosynthesis 272
    in biosynthesis of terpenoids 272
    in chlorophyll synthesis 255
Germination, of cereals 321
Gibberellic acid 321
    effects on aleurone
Gibberellins 25, 321
Glucan synthetase II 84
β-Glucan, mixed β(1→3), β(1→4) 278
Glucan, β(1→3) 285–6
    in higher plants 20
    synthesis 286
Glucans, mixed 276
    terminology 16
Glucomannans 100
    structure 276
Gluconate-6-P 127
    regulation of ribulose-5-P kinase 179
Gluconate-6-P dehydrogenase 127–8, 130
Gluconate-6-P lactonase 127–8
Gluconeogenesis
    in conversion of lipids to sucrose

137–141
    as 'reversal' of glycolysis 105–6
    standard free energy changes 106
Gluconic acid (gluconate) 14
Gluconolactone-6-P 127
Glucopyranose
    three-dimensional structures 14
    anomers 12–13
Glucosamine 15–16
Glucose 12, 100, 311
    acyclic form 12
    anomers 11–12
    cyclic forms 12
    equilibrium between ring and acyclic
        forms 14
    furanose forms 13
    pyranose forms 13
    reduction of Cu^{2+} 14
Glucose hexokinase 101
Glucose-1-P 100–1
    in starch metabolism 176
    in starch synthesis 175
    in sucrose synthesis 174
Glucose-6-P
    branch point in glycolysis and OPP
        pathway 128
    $\Delta G^{\circ\prime}$ of hydrolysis 54
    in glycolysis 100–1
    in OPP pathway 127, 130
Glucose-6-P cycloaldolase 284
Glucose-6-P dehydrogenase 127–8, 130–1
    inhibition by NADPH 132
    inhibition by ribulose-1,5-P_2 132
    light modulation 177–8
    regulation of OPP pathway 132
Glucose-P isomerase 104
Glucose-P mutase
    in glycolysis 100–1, 104–5, 108
    in sucrose synthesis 174, 177
α-Glucosidase 100
Glucuronic acid (glucuronate) 15
Glucuronate kinase 284
Glufosinate, see phosphinothricin
Glutamate 27
    in 5-aminolevulinate synthesis 252–3
    in amino group transfer reactions 214

intracellular transport 190–1
    in ornithine synthesis 223
    oxidation in cytosol 238
    oxidation in photorespiratory N
        cycle 238
    in proline synthesis 222
    synthesis 211–14
    synthesis from ureides 252
    in THF synthesis 235
    transamination reactions 216
Glutamate dehydrogenase 211–12, 238
    covalent modification 81
Glutamate kinase 222, 225
Glutamate synthase 212, 239
    in chloroplasts 213
    inhibitors of 212
    in photorespiratory mutants 190
    subcellular location 213
Glutamate/2-oxoglutarate shuttle 238–9
Glutamate:oxaloacetate aminotransferase
    200–2, 214, 238
    specificity 215
Glutamate:prephenate aminotransferase 223
Glutamate:pyruvate aminotransferase 200–1
Glutamine 27
    in asparagine synthesis 215–16
    in carbamyl-P synthesis 223, 245–6
    in cytosine synthesis 247
    $\Delta G^{\circ\prime}$ of hydrolysis 54
    metabolism 212
    in purine synthesis 249
    in pyrimidine synthesis 245
    synthesis 211–14
        energetics 53
    in synthesis of N transport compounds
        227
    in ureide synthesis 251–2
Glutamine synthetase 211–12, 239
    in chloroplasts 213
    inhibition by phosphinothricin 191
    inhibitors 212
    light modulation 178
    in NH_3 reassimilation 213
    in photorespiratory mutants 190
    properties 212
    subcellular location 213

Glutamyl-1-semialdehyde 253
    in 5-aminolevulinate synthesis 253
Glutamyl-5-semialdehyde
    dehydrogenase 225
β-Glutamyl-cysteinyl-glycine
    (glutathione) 239
Glutamyl-tRNA ligase 253
Glutamyl-tRNA, in 5-aminolevulinate
    synthesis 253
Glutamyl-tRNA^{glutamate} dehydrogenase
    253, 255
    regulation 256
Glutathione (oxidized), see GSSG
Glutathione (reduced), see GSH
Glyceraldehyde 10, 26, 198
    optical isomers 10
Glyceraldehyde-3-P 102
    metabolism in chloroplasts 169, 171
    in OPP pathway 129–30
    oxidation in glycolysis 102–3
    production in C_4 plants 202
Glyceraldehyde-3-P dehydrogenase
    101, 104
    (NAD-specific) 102, 237
    (NADP-specific) 62, 171, 237
        in C_4 plants 202
        light activation 178
        properties 169
    (non-phosphorylating) 237
    in NO_3^- reduction 213
    in triose-P shuttle 237
Glycerate 189
    formation from serine 189
    intracellular transport 191
    metabolism in chloroplasts 190
    reassimilation rates 241
    synthesis from tartronyl semialdehyde
        194
Glycerate dehydrogenase 188–9
Glycerate kinase 188, 190
Glycerate-1,3-P_2 52, 102
    in glycolysis 102
    $\Delta G^{\circ\prime}$ of hydrolysis 52–4
    light-coupled metabolism 169
    metabolism in chloroplasts 169
    in substrate-level phosphorylation 54

Glycerate-2-P *103*, 107
Glycerate-3-P *103*, 107
  in C_3-CR cycle 167–72
  in C_4 plants 197–98
  CO_2 labelling pattern 167
  concentration in chloroplasts 169
  control of fructose-2, 6-P_2 metabolism
    180
  exchange with triose-P 237
  in glycolysis 103–4
  light-coupled metabolism 169
  metabolism in C_4 plants 202
  metabolism in NAD M-E variant 204
  metabolism in NADP M-E variant 202
  metabolism in PEP-CK variants 202–3
  permeability of chloroplasts 173
  reassimilation rates 241
  recycling in chloroplasts 170
  regulation of ADP-glucose
    pyrophosphorylase 179
  regulation of starch synthesis 179
  synthesis from CO_2 168
  synthesis from glycollate-2-P 187–90
  synthesis from serine 189
  transport across chloroplast envelope 174
  transport in C_4 plants 204
  trivalent anion in chloroplasts 169
  in triose-P shuttle 344
Glycerate-3-P kinase
  in C_3-CR cycle 171
  in C_4 plants 202
  in chloroplasts 169
  in cytosol 169
  in glycolysis 101, 103–4
  in substrate-level phosphorylation 54
  in triose-P shuttle 237
Glycerate-P mutase 101, 103–5
Glycerides
  enantiomers 22
  structure 21–2
Glycerol 14, *22–3*
  in glycerides 21–2
  in lipid oxidation 133–4
  stereospecific numbering 22
Glycerol kinase 133
Glycerol-3-P 22–*3*, 133–4, 264

in acyl lipid synthesis 264
  conversion to lysophosphatidate 264
  $\Delta G^{o\prime}$ of hydrolysis 54
  in oxidation of lipids 133
Glycerol-3-P dehydrogenase 133, 264
Glycerol-P 1-acyl transferase 264
Glycerol-P phosphatidyl transferase 266
Glycerolipids
  of 16:3 galactolipid plants 267
  of 18:3 galactolipid plants 267
Glycine 26, *27*
  in AGPs 277
  in C_2-PR cycle 187
  formation from CO_2 in leaves 187
  formation from glyoxylate 214
  intracellular transport 191
  metabolism in leaves 213
  oxidation 236
  oxidation in leaf mitochondria 189
  in photorespiration 213
  in purine synthesis 249
  in ureide synthesis 251
  source of C_1 fragments 236
  synthesis in peroxisomes 188
Glycine decarboxylase, *see* glycine
    dehydrogenase
Glycine dehydrogenase
  in C_2-PR cycle 188–9, 239
  in C_1 metabolism 235–6
Glycogen phosphorylase, covalent
    modification 81
Glycolipids 25
  biosynthesis 267–9
  content in membranes 85
  structures 24
  synthesis in 16:3 galactolipid plants 268
  synthesis in 18:3 galactolipid plants 269
Glycollate
  formation from CO_2 in leaves 187
  intracellular transport 191
  metabolism
    in algae 194
    in peroxisomes 188
  energy requirements 191
  oxidation in algae 194
  photorespiration 185

synthesis in leaves 185
Glycollate dehydrogenase 239
Glycollate oxidase 187–8
  affinity for O_2 188
  in glycollate shuttle 239
  in peroxisomes 47
  subcellular distribution 48
Glycollate-2-P *168*
  metabolism to CO_2 187–90
  metabolism via C_2-PR cycle 188
  synthesis 186
  synthesis from ribulose-1,5-P_2 168
  synthesis under physiological conditions
    169
Glycollate-2-P phosphatase 187–8
Glycollate/glyoxylate shuttle 238–9
Glycolysis 99–108
  ADP phosphorylation 102–3
  anaerobic stimulation of glucose
    consumption 105
  anaerobic yield of ATP 104
  ATP expenditure 102
  in chloroplasts 104
  efficiency of energy conservation 125
  energetics 104
  energy-conserving reactions 102
  equilibrium constants of reactions 105
  functions 125
  interaction with OPP pathway 128
  location 104
  primary reactions 102
  'priming reactions' 100–1
  products and energy yield 108
  rates under air and nitrogen 105
  regulation 107
  regulation of FP$_2$-Pase by fructose-2,6-P_2
    107
  regulation via PF-Kase and pyruvate
    kinase 105
  relationship to OPP pathway 131
  standard free energy changes of
    reactions 106
Glycolytic enzymes
  activity in plastids 104
  latency 104
Glycoproteins 311–12

in cell wall organization 281
  'core oligosaccharide' *312*
  role of endoplasmic reticulum 312
Glycoprotein enzymes, in cell walls 277
Glycosidases 61
Glycoside bond, $\Delta G^{o\prime}$ of hydrolysis 54
N-Glycoside glycoproteins, core
    oligosaccharide *312*
Glycosidic bonds 16
  determination of configuration 275–6
  in matrix polysaccharides 276
  principle 16
  synthesis in cell wall polysaccharides 282
  terminology 16
Glycosyl transferases 312
Glycosylation
  of proteins 313, 320
  recognition code 287
  signals 287
Glycylglycine, $\Delta G^{o\prime}$ of hydrolysis 54
Glyoxylate *138*
  as amino group acceptor 215
  in C_2-PR cycle 187–8
  decarboxylation 236
Glyoxylate pathway 137–40
  ^{14}C-labelling from acetate 141
  energetics 140
  evidence 137
  location 140
Glyoxysomes 133, 137, 140
  properties 43
  role in fatty acid conversion to
    sucrose 139
Glyphosate 78, *224*
  alleviation of toxicity 224
  herbicidal activity 224
  inhibition of EPSP synthase 224
GMP (guanosine 5′-monophosphate),
    synthesis 250
GMP synthetase 250
Golgi apparatus 43
  marker protein 84
Golgi bodies
  in cell wall synthesis 282
  enzymes 286
  hemicellulose synthesis 287

in matrix polysaccharide synthesis 282
in phenylpropanoid metabolism 289
polysaccharide content 286
structure 43
synthesis of matrix polysaccharides 286
Golgi body, model 43
Golgi vesicles
in formation of cell plate 46
in glycosylation of HRGPs 287
homology with plasma membrane 286
origin 43
Gramicidin 91
mode of action 91
structure and composition 91
Grana
in C_4 plants 42
structure 41–2
Granal sacs, structure 41–2
Granal thylakoid, structure 41–2
Grape, hydroxy fatty acids in cutin 262
Grasses 19
C_4 plants 207
Green algae, C_2 metabolism 194
Green sulphur bacteria
light-dependent electron transport 148
photosynthetic processes 146
reaction centre 148
Greening 339
development of photosystems 345
Growth, CO_2 assimilation rates 7–8
GSH (glutathione) 239
in APS metabolism 231
in detoxification of H_2O_2 240–1
functions 240
metabolism 240
GSH dehydrogenase 240–1
GSH peroxidase 241
GSSG (oxidized glutathione) 239
reduction in chloroplasts 239–41
GSSG reductase 239–41
GTP (guanosine 5'-triphosphate)
in cellulose synthesis 285
$\Delta G^{\circ\prime}$ of hydrolysis 53
requirements of protein synthesis 311
in translation 309
Guaiacyl alcohol 34

Guanine 30, 302
base pairing with cytosine 32–3
catabolism 251
synthesis 250
Guanosine 31
Guanosine 5'-diphosphate, see GDP
Guanosine 5'-monophosphate, see GMP
Guanosine 5'-triphosphate, see GTP
Guanyl cyclase 285
Gulose 12

H^+ reduction, by nitrogenase activity 226
Haem 56
as prosthetic group 64
in cytochromes 56
in NO_3^- reductase 210
regulation of glutamyl-tRNA^{glutamate}
dehydrogenase 256
synthesis 253
Hairpin loop, in cDNA synthesis 327
HART, see hybrid arrest translation
Heartwood, cell wall composition 279, 281
Helical coiling,
of DNA 32
in proteins 28
Helicase 297
Hemiacetal 11
in glycosidic linkages 16
Hemicellulose 20, 275–7
in cell wall organization 280
classification 276
complexes with cellulose 280
extraction 275
functions 276
structures 276
synthesis in Golgi bodies 287
Hemiketal 14
in glycosidic linkages 16
Henderson–Hasselbach equation 90
Herbicides
binding domain on thylakoids 153
bipyridiliums 153, 240
chlorsulphuron 226
CMU 153
cyclohexanediones 258–9
DCMU 148

dinitroaniline 282
glufosinate 191, 212
glyphosate 224
inhibition of amino acid synthesis
223–6
interaction with light-dependent electron
transport 153
mechanism of action 78
movement across membranes 88
phosphinothricin 191, 212
photo-oxidative damage 153
specificity towards C_4 plants 208
sulphonylureas 226
triazines 153
trifluralin 282
Heterocysts, N_2 fixation 228
Heterotrophs, characteristics 5
Heterotropic control 79
Hevea 25, 35
threo-Hex-2-enono-1,4-lactone 15
Hexadecatrienoate 267
Hexadecenoate 267
trans-3-Hexadecenoate, in chloroplasts
84–5, 341
Hexanoyl-ACP 260
Hexokinase(s) 100–1, 104–5, 284
in mitochondrial membrane 104
Hexosamines 311
Hexose monophosphate shunt, see OPP
pathway
Hexose-P 130
Hexose-P isomerase 101, 105, 108, 174,
177
in sucrose synthesis 174
Hexoses
conversion to triose-P 100
cyclic forms 11
oxidation via glycolysis 100, 103
High energy bonds 307
High energy compounds 52–4
$\Delta G^{\circ\prime}$ of hydrolysis 52–4
Hill reaction 147
Histidine 27
synthesis 222–3, 225
regulation in bacteria 318
Histones 40, 292–3

Histopine 335
Homoalanine 4-(methyl)phosphinite 212
Homocysteine 26, 232
metabolism 232–6
methylation 236
Homocysteine:methylTHF
methyltransferase 233
Homocysteine synthase 233
Homoserine 26
metabolism 216–17
Homoserine dehydrogenase 217–18
Homoserine kinase 216–18
subcellular location 217
Homoserine-4-P 232
metabolism 217, 232–3
regulation 217, 233
Homotropic control 79
Hordein 317
gene expression 317
mRNA 317
Horseradish peroxidase 312
HRGP (hydroxyproline rich glycoprotein—
see also extensin) 277
in dicotyledons 279
genes 287
HRT, see hybrid release translation
Humulus, collenchyma cell 37, 44
Hybrid arrest translation (HART) 330
Hybrid release translation (HRT) 330
Hybridization probes, see oligonucleotide
probes
Hydrocarbons, in cuticle 22
Hydrogen (H_2), requirement by
photosynthetic bacteria 148
Hydrogen bonds, in cell wall growth 281
Hydrogen electrode, standard 56–7
Hydrogen peroxide 134
in cell walls 289
detoxification in chloroplasts 240–1
metabolism in microbodies 43
oxidation of glyoxylate 236
production in illuminated
chloroplasts 240
rate of reduction in chloroplasts 241
Hydrogen sulphide (H_2S)
assimilation 231

emission by plants 232
requirement by photosynthetic
bacteria 148
Hydrogen, from H$^+$ reduction 228
Hydrogenase 226
Hydrolyases 61
Hydrolysis, $\Delta G^{\circ\prime}$ values 52–4
Hydroxylysine 26
Hydroxy fatty acids, in cutin 22
18-Hydroxy-9,10-epoxystearyl-CoA 263
β-Hydroxyacyl ACP dehydratase 260
3-Hydroxyacyl epimerase 136
L-3-Hydroxyacyl-CoA 136
D-3-Hydroxyacyl-CoA 137
β-Hydroxyacyl-CoA dehydrogenase 135–7
3-Hydroxyacyl-CoA epimerase 137
Hydroxyapatite, binding of double-stranded
DNA 292
Hydroxycinnamyl alcohol 34
α-Hydroxyethyl thiamine pyrophosphate
109, 218
Hydroxy fatty acids 262
β-Hydroxy-fatty acyl-CoA 135
oxidation 135
Hydroxylamine 215
Hydroxylase 261
3-Hydroxymethylglutaryl-CoA
reductase 81, 271
location 272
Hydroxymethylglutaryl-CoA synthase 271
18-Hydroxyoleyl-CoA 263
16-Hydroxypalmityl-CoA 263
Hydroxyproline 26
in AGPs 277
in cell wall proteins 277
in extensin 312
in glycoproteins, see HRGP
Hydroxyproline rich glycoprotein, see HRGP
3-Hydroxypropionate dehydrogenase 136
3-Hydroxypropionyl-CoA hydrolase 136
Hydroxypyruvate 189
as amino group acceptor 215
in C$_2$-PR cycle 189
Hydroxypyruvate reductase, see glycerate
dehydrogenase
Hyoscyamine 34

Hypoxanthine, metabolism 251

Idose 12
Imbibition, of cereal seeds 322
Immunological techniques, in study of
membranes 38
IMP (inosine monophosphate)
catabolism 251
metabolism 250
regulation of carbamyl-P synthetase 247
synthesis 249
in ureide synthesis 251
IMP dehydrogenase 250–1
Indole glycerol-P synthase 223
Indoleacetic acid, movement across
membranes 88
Inductive resonance, see resonance transfer
Inhibition constant (K_i) 76–7
Inhibitors
of mitochondrial electron transport 115,
118
of photosynthesis 148
of protein synthesis 315, 348
Initiation of translation
codon 301, 303, 308
complex 308–9
factors 308, 320
availability 316
signal, in mRNA 308–9
Inorganic carbon, different species in
solution 88
Inorganic salts, plant requirements 3
Inosine monophosphate, see IMP
Inositol(s) 14–15
in cell wall synthesis 283
in phospholipids 23
metabolism 283–4
Inositol oxygenase 284
Inositol hexaphosphate 15
Inositol triphosphate 82
myo-Inositol-1-P phosphatase 284
Inside-out vesicles from thylakoids 154
Integral proteins, see intrinsic proteins
Internal energy, of molecules 50
Interphase 45
Interphase cells, microtubule orientation 281

Intra-thylakoid space 41–2
Intrinsic protein 37–8
example of structure 39
Introns 293, 303–4, 324, 333
in cytochrome b gene 304
in evolution 293
in rRNA genes 306
in RuP$_2$-Case genes 313
Inulin 19, 100
degradation 100
Invertase 99, 282, 284
Inverted repeat sequences
chloroplast rRNA 294
in chloroplast tRNA 294
in liverwort plastome 295
Iodine, reaction with amylose 18
Ion channels 95–6
Ion electrochemical potential gradient
($\Delta \tilde{\mu}_x{}^m$) 58–9
Ion transmembrane gradients, ΔG of 58–9
Ionophores 91
mode of action 91
Ions
light-mediated movement in
chloroplasts 179
long-distance transport 6
IPP (isopentenyl pyrophosphate)
site of synthesis 272
Δ^2-IPP 271
biosynthesis 271
in biosynthesis of terpenoids 272
in phytol synthesis 255
Δ^3-IPP 271
biosynthesis 271
in phytol synthesis 255
in terpenoid synthesis 272
IPP isomerase 271
Iron
in cytochromes 56
in ferredoxins 55–6
in haem 56, 64
in nitrogenase 226–7
oxidation states in nitrogenase 226
in proline hydroxylation 287
Iron–sulphur centre(s)
in NO$_2^-$ reductase 211

in PSI 150
in succinate dehydrogenase 111
Iron–sulphur proteins 55–6, 116
in complex I of mitochondrial electron
transport chain 115
Irradiance 142
Irreversible inhibition, of enzymes 76
Isoalloxazine ring, in FMN 55
Isoamylase 100
Isocitrate 110, 138
in TCA cycle 110–11
Isocitrate dehydrogenase 109, 111
in mitochondria 47
Isocitrate lyase 137–40
in germinating seeds 139
location 140
Isodityrosine linkages 277, 282, 288
Isoelectric focusing 62
Isoenzymes 62
of glycolytic enzymes 104
intracellular location 313
of OPP pathway 130
Isoleucine 27
in regulation of threonine dehydratase 219
synthesis 218–221
regulation 217, 219
Isomerases 61
Isopentenyl pyrophosphate, see IPP
Isoprene 25, 270
Isoprenoid unit 25, 270
β-Isopropylmalate dehydrogenase 220
β-Isopropylmalate isomerase 220
Isopropylmalate synthase 218–20
Isopycnic centrifugation 47, 287
Isotopic equilibrium exchange, mechanism of
ADP phosphorylation 124

Jerusalem artichoke 19, 100
β-oxidation 137
Joule, definition 50
Junction zone, in matrix
polysaccharides 279

Kanamycin resistance 335
Kautsky curves 160
β-Ketoacyl ACP dehydratase 259

β-Ketoacyl ACP reductase 259–60
β-Ketoacyl ACP synthetase I 259
β-Ketoacyl ACP synthetase II 260
α-Ketoglutarate, *see* 2-oxoglutarate
Ketohexoses 11
    anomers 14
    ring structures 14
    structures 13
Ketopentoses 11
    structures 13
Ketoses 10
K_i, *see* inhibition constant
Kinetics (of chemical reactions) 67–9
    first-order reactions 67
    second-order reactions 67
    zero-order reactions 68
Klebsiella 210
Kranz anatomy 196

lac operon, use in gene cloning 329
Lactate *92, 104*
    in glycolysis 104
Lactate dehydrogenase 104
Lactonase, in regulation of OPP
        pathway 132
Lactose, $\Delta G^{o\prime}$ of hydrolysis 54
Laevorotation 10
Lamellae 41
Large single copy DNA region, in chloroplast
        genome 294–5
Latent activitoy, of glycolytic enzymes in
        chloroplasts 104
Lauric acid 20
Leading substrate 73
Leaf
    anatomy of maize 196
    carbon budget 6
    development, ribosome numbers in
        plastids 343
Leaf area ratio 7
Lectins 312
Leghaemoglobin 64, 227–8
    introns in gene 293
Legume–*Rhizobium* symbiosis 226, 334
    metabolite exchange 227–8
Legumes, cobalt requirement 236
Legumin, gene expression 319

Lemna 322
Lemna paucicostata 233–4
Lepidoptera, susceptibility to bacterial
        toxins 333
Leucine 27
    in regulation of isopropylmalate
        synthase 219
    synthesis 218–21
        pathway 220
        regulation 219
Levan (phlein) 100
    degradation 100
Levulinate 253
LHCP (light-harvesting chlorophyll-binding
        protein) 149
    complex 1 (LHCP 1) 150
    complex 2 (LHCP 2) 149
        associated electron transport
            components 149
        carotenoid content 149
        chlorophyll content 149
        composition 149
        location in thylakoids 155–6
    cytokinin stimulation of synthesis 322
    expression during greening 344
    gene expression 317–19
    migration following phosphorylation 161
    mRNA 320
    phosphorylation 81
    phytochrome control of gene
        expression 320
    in PSII 149
    structure and amino acid sequence 38–9
Lichens, sulphur nutrition 232
Ligases 61
Light
    absorption by pigments 59
    in assimilation of inorganic N 213–14
    in assimilatory processes 230
    changes in metabolite concentration 179
    characteristics 59
    in chlorophyll synthesis 255
    effect on pH in chloroplasts 169
    effect on THF metabolism 235
    energy required for CO_2 assimilation
        59–60
    energy sources for plant growth 5

enhancement of reactions in cytosol 237
    in fatty acid synthesis in chloroplasts 241
    and gene expression 320
    in ion movements in chloroplasts 179
    in reduction of GSSG 239–40
    in reduction of O_2 240
    requirement for C_3–CR cycle 169
    in sulphur assimilation 230
Light-coupled reactions, rates in
        chloroplasts 241
Light-dependent electron transport, in green
        sulphur bacteria 148
Light-harvesting chlorophyll-binding protein,
    see LHCP
Light harvesting, reaction centre 148
Light-modulated enzymes 177–9
Light modulation, pyruvate, P_i dikinase 199
Light reactions 142–63
    historical background 146
Lignin 278
    degradation products 34
    origin of C_1 units 236
    phenylpropanoid residues in hardwoods
        and softwoods 279
    phylogenetic distribution 34
    structure 280
    synthesis 235, 288–90
        control 290
        polymerization mechanism 290
Lilium, DNA content 292
Lima beans 312
Limit dextrin 100
Lineweaver–Burk equation 72
Lineweaver–Burk plot 72–3
Linker DNA
    in cDNA synthesis 328
    between nucleosomes 293
Linoleic acid 20–1
    conformation 21
    in human diet 257
    oxidation 136–7
Linolenic acid (linolenate) 20–1, 84, 267, 341
    oxidation 136
Lipases 133
    location 133
Lipid(s) 20–5, 340
    in chloroplasts during development 343

composition of thylakoids during
        development 340
    content in germinating seeds 139
    conversion to sucrose 138–41
    effect on enzyme activity 86–7
    energy content 133
    functions 257
    oxidation 133–41
    storage 133
    synthesis 257–74
    transformations to sugars 141
Lipid bilayer, in membranes 37
Lipid–protein interactions, in membranes 86
Lipid-soluble cations 90
Lipoate 109, *110*
    2-oxoglutarate dehydrogenase
        complex 111
Lipoate acetyl transferase 109–10
Liposomes 274, 336
Lipoxygenase 137
Liverwort 294
Lupinus albus, asparagine synthesis in 215
Lutein 144
Lyases 61
Lycopene *273*
Lysergic acid 34
Lysine 27
    in cereals 209
    in control of aspartate metabolism
        217–18
    effect on growth 218
    synthesis 217
        pathway 219
        regulation 217
Lysophosphatidate 260, *264*, 268
Lysopine *355*
Lyxose *12*

Macrozamia 228
Magnesium, plant requirements 3
Magnesium chelatase 253–5
    regulation 255–6
Magnesium ions
    activation of RuP_2-Case 179
    in chloroplasts 179
Magnesium-protoporphyrin IX
    metabolism 253

in regulation of chlorophyll synthesis 256
Maize
  elemental composition 3
  mitochondrial genome 295
  quantum efficiency 205
Malate *111*, 138–9
  in CO_2 assimilation in C_4 plants 197–8
  decarboxylation in CAM plants 205–6
  decarboxylation in C_4 plants 199–202
  formation in CAM plants 205
  metabolism in nodules 227
  in NO_3^- reduction 213
  oxidation by mitochondria 119
  oxidation in cytosol 238
  in TCA cycle 111
  transport in C_4 plants 204
Malate dehydrogenase 61
  in glyoxylate cycle 139–40
  inhibition 114
  kinetics 74
  in NO_3^- reduction 213
  subcellular distribution 48, 140
  in TCA cycle 109, 111
Malate dehydrogenase (NAD-specific) 201,
    238
  in CAM plants 205–6
  in C_4 plants 200–1
  subcellular distribution 48
Malate dehydrogenase (NADP-
    specific) 198, 238
  in C_4 plants 200
  light modulation 178
Malate shuttle 238
Malate synthase 137, 139–40
  location 140
Malate/oxaloacetate exchange 238
Malate/pyruvate shuttle 239
Malic enzyme
  anaplerotic function 125
  in N_2 fixation 227
Malic enzyme (NAD-specific) 200
  in CAM plants 205–6
  in C_4 plants 200–2
Malic enzyme (NADP-specific) 198, 202
  in CAM plants 205–6
  in C_4 plants 200

Malonate 76, 112
  inhibition of mitochondrial electron
    transport 115
Malonyl semialdehyde 136
Malonyl semialdehyde dehydrogenase 136
Malonyl transferase 259
Malonyl-ACP *259*
Malonyl-coenzyme A 257
  synthesis 258
Maltose 16, *17*, 100
  $\Delta G^{o'}$ of hydrolysis 54
Manganese–protein complex of PSII 152–3
  composition 152
  location in thylakoids 157
  redox states of Mn 152
  role in oxidation of H_2O 152
Mannan, $\beta(1\rightarrow4)$ in algae 20
Mannitol 14
Mannose *12*, 311
  phosphorylation 101
Mannose-6-P 101
Mannose-6-P isomerase 284
Mannose-P mutase 284
Margarine 21
Marker enzymes 47
Marker proteins 83–4
Mass-action ratios 132
  of glycolytic reactions 105
Maternal inheritance 294
Matrix (of cell wall) 275–9
Matrix (of mitochondria), structure 42
Matrix (M) face of inner mitochondrial
    membrane 118
Matrix polysaccharides (of cell walls) 20
  cross-linking 278–9
  synthesis 286–7
Maturase 293, 304
Maxam and Gilbert sequencing 331–2
Mediated transport 92
  characteristics of carrier proteins 92
  possible mechanisms 93
Mehler reaction 163
Membrane electrical gradients, ΔG of 58–9
Membrane flow, in transport of enzymes to
    plasma membrane 287
Membrane ion gradients, ΔG of 58–9

Membrane potential 89
  across illuminated thylakoids 158
  in mitochondria 119
Membrane proteins
  function 38
  interaction with lipids 274
  orientation 38
  techniques 38
Membrane transport
  carrier proteins 92
  driving forces 94–5
  electrogenic 93–4
  linking of transport of solutes (*see also*
    uniport, symport and antiport)
    93–4
Membrane vesicles, in determining transverse
    heterogeneity of thylakoids 156
Membrane(s) 36–9, 83–96
  asymmetry 38
  composition 83
    of granal thylakoids, stromal
      thylakoids 83
    inner mitochondrial 83
    plasma 83
  fluidity 86
  functions 36, 83
  galactolipids 86
  glycerolipid composition 85
  isolation 83
  lipid composition 84–5
  marker proteins 83–4
  model of structure 38
  permeability 36
    to gases 87
    to ions 87–8
    to metabolites 86–8
    to protons 87–8
  phospholipid composition 85
  properties 86
  protein extraction 38
  self assembly 274
  stability 86
  sterols 86
  structure 37–8
  transport 86–96
  unsaturated fatty acids 86

viscosity 86
Menthol 34
3-Mercaptopicolinic acid 200
Meristems 45
Mesembryanthemum crystallinum, CO_2
    assimilation 206
Mesophyll cells
  C_3-CR cycle enzymes in C_4 plants 202
  in C_4 plants 196
  chloroplast number 41
  function in C_4 plants 198–9
  sucrose content 175
Mesophyll chloroplasts, transport 204
Mesoporphyrins 64
Mesotome 196
Messenger RNA, *see* mRNA
Metabisulphite 46
Metabolic regulation
  role of enzymes 79–82
  of TCA cycle enzymes 113
Metabolite transport, *see* transport
Metal ions
  as enzyme activators 63
  in prosthetic groups 64
Metaphase 45
Methenyl groups, sources 233–6
N^5,N^{10}-methenylTHF *64*
  in C_1 donor 234
  in purine synthesis 249
  metabolism 235–6
N^5,N^{10}-MethenylTHF cyclohydrolase 235
Methionine *27*
  in control of aspartate metabolism 217
  functions 233
  inhibition of methionine synthesis 233
  initiation of translation 302
  metabolism 234–6
  methyl group metabolism 234
  origin of methyl group 236
  synthesis 222, 232–3
    co-ordination with lysine synthesis 233
    co-ordination with threonine synthesis
      233
    regulation 217, 233
Methionine sulphoximine *212*, 213
  effect on glycine metabolism 213

Methionyl-tRNA$_i^{met}$ 309
  in initiation of protein synthesis 308
Methyl groups, sources 233–6
Methyl viologen 153–4, 240
Methylammonium 90
Methylation 233–6, 311, 316
  of DNA 300
5-Methylcytosine *30*
N^5,N^{10}-MethyleneTHF *64*
  as C$_1$ donor 234
  in deoxy-TMP synthesis 248
  metabolism 235–6
  metabolism in leaf mitochondria 189
N^5,N^{10}-MethyleneTHF dehydrogenase 235
N^5,N^{10}-MethyleneTHF reductase 235
α-Methylglucoside 16
1-Methylguanine, in transfer-RNA 33
7-Methylguanine *30*
7-Methylguanosine 303, *305*
  'cap' in mRNA 303, 305
  in transfer RNA 33
1-Methylhypoxanthine, in transfer-RNA 33
S-Methylmethionine, metabolism 236
S-Methylmethionine:homocysteine S-
    methyltransferase 236
6-Methylpurine 315, 321
2′-Methylribosylguanosine, in transfer
    RNA 33
N^5-MethylTHF *64*
  as C$_1$ donor 234
  metabolism 235–6
  in methionine synthesis 233
  in methylation of homocysteine 236
N^5-MethylTHF methyltransferase 233
2-Methylthio-N^6-isopentenyladenosine, in
    transfer RNA 33
Methyltransferase 234
  in chlorophyll synthesis 253
Mevalonate, biosynthesis 271
Mevalonate kinase 271
Micelles 24–5
Michaelis–Menten constant (K_m) 71
Michaelis–Menten equation 71
  transformations 72
Michaelis–Menten kinetics 70
  transformations 73

Michaelis–Menten relationship 70–1
Microbodies 42–3
  enzyme complement 43
  enzymes of β-oxidation 137
  enzymes of glyoxylate pathway 140
  β-oxidation activity 137
  types 43
Microelectrodes 95
Microfibrils
  cellulose composition 19
  in cell wall organization 280
  cross-linking to matrix 279
  formation 281–6
    from fibrils 285
    by random crystallization 286
  non-crystalline regions 276
  orientation 45, 279
  properties 275
Microfilaments 44–5
Microinjection, of DNA 336
Microsomes 40
  fatty acid composition of membranes 86
Microtubules
  in cell wall synthesis 281–2
  control of cellulose synthase migration 286
  depolymerization 282
  in microfibril orientation 285–6
  orientation 281
  in phragmoplast 46
  in spindle fibres 46
  structure 44
Middle lamella 45
Mirabilis 294
Mistletoe, DNA content 292
Mitochondria
  aspartate decarboxylation in C$_4$
      plants 200–2
  biogenesis 338
  centrifugation characteristics 47
  electron transport 114–17
    Complex I 114–16
    Complex II 115–16
    Complex III 115–16
    Complex IV 115–17
    external NADH dehydrogenase 117
    FMN 115

  half times for oxidation of components
      118
  iron–sulphur proteins 115–16
  kinetics of redox changes 118
  organization of components in
      membrane 116, 118
  sequence of electron carriers 117–18
  stoichiometry with proton translocation
      121
  ubiquinone 114
  enzymes of C$_2$-PR cycle 189
  genome 295–6
    coding capacity 295–6
    copy number 295–6
    restriction nuclease digestion 323
    size 295–6
    structure 295–6
  marker protein 84
  membrane lipids 84–5
  membranes
    fatty acid composition 84, 86
    metabolite transport 113
  in photorespiration 189
  protein synthesis 313–14
  proteins, location of genes 348
  ribosomes 40
  role in sucrose synthesis in fatty seeds 139
  structure 42
  transport of metabolites across
      membranes 112
Mitosis 45–6
Mixed terpenoids 270
Mobilization, of storage products 99
Molybdenum
  in nitrogenase 226–7
  in NO$_3^-$ reductase 210
  plant requirements 3, 210
Monactin 91
Monocotyledons, cell wall composition 279
Monogalactosyl diacylglycerol 25, 85,
    267–8
  content in chloroplasts during
      development 340
  properties 86
  structure 24
Monogalactosyl diglycerides, *see*

  monogalactosyl diacylglycerol
Monoglycerides 22
Monomeric enzymes 62
Monomolecular reaction, kinetics 70
Monophenol oxidase 289–90
Monosaccharides 10–14
  in cell wall synthesis 282–3
  linking by glycosidic bonds 16
Monoterpene 270
Monuron, *see* CMU
Morphine 34
mRNA(s) (messenger RNA) 32–3, 302–4, 306,
    308
  for alcohol dehydrogenase 320
  for α-amylase 322
  availability for translation 320
  for hordein 317
  initiating signal 308
  isolation 330
  for legumin 319
  for LHCP 320, 322
  modifications of primary transcript 303,
      305
  in organelles 313
  for P-ALase 321
  in polyribosomes 311
  poly-A tail 305, 327
  precursors, processing in control of gene
      expression 319
  purification 326
  stabilization by cytokinins 322
  transport from nucleus 316, 319
  turnover 316
  for UDP-glucose pyrophosphorylase 320
Multi-enzyme complexes 62
Multigene family, for hordein 317
Mung bean, plasma membrane lipids 84
Mutarotation 12
Myo-inositol *15*
Myristic acid 20–1

NAD (nicotinamide adenine dinucleotide) 55
  absorption of u.v. light 55
  properties 55
  reduction
      by pyruvate dehydrogenase

complex 110
in TCA cycle 109, 111–12
NAD kinase 82
NADH (reduced NAD) 55
energetics of oxidation by O_2 57
inhibition of pyruvate dehydrogenase
complex 113
light-enhanced production in cytosol 237
oxidation by O_2 108, 114
energy conservation in ADP
phosphorylation 118
in mitochondria 118
potential to phosphorylate ADP 57
in NO_3^- reduction 210, 213
NADH cytochrome c reductase 86
NADH dehydrogenase 86
(external), in mitochondrial electron
transport 117
of mitochondrial electron transport
chain 114
NADP (nicotinamide adenine dinucleotide
phosphate) 55
absorption of u.v. light 55
in non-cyclic photophosphorylation 147
properties 55, 57
in reconstituted chloroplast system 172
reduction in OPP pathway 128, 130
$NADP^+$ (oxidized NADP), as a Hill
reagent 147
NADPH (reduced NADP)
as an energy source 4
in bundle sheath cells in C_4 plants 202
generation by OPP pathway 131–2
$\Delta G^{\circ\prime}$ of oxidation by O_2 57
inhibition of glucose-6-P
dehydrogenase 132
in NO_2^- reduction in fungi 211
rates of consumption in chloroplasts 241
requirement in C_3-CR cycle 169, 172
requirements of C_2-PR cycle 191–3
NADPH cytochrome c reductase 83
NADPH:Fd oxidoreductase 261
NDP reductase 248
NDP-sugars
in cell wall synthesis 282–3
energetics of synthesis 53–4

incorporation into matrix polysaccharides
286
synthesis 284
NDP-sugar pyrophosphorylases 174, 282
Negative co-operativity, of allosteric
enzymes 80
Neomycin phosphotransferase 319
Nernst equation 59, 90
Net assimilation rate 7
NH_3, see ammonia
Nicking, of DNA during replication 297
Nicotinamide adenine dinucleotide, see NAD
Nicotinamide adenine dinucleotide
phosphate, see NADP
Nicotinamide residue, in NAD and
NADP 55
Nicotine 34
Nigericin 91
as proton ionophore 91
structure 91
Nitrate
assimilation into amides 215–16
as nitrogen source for plants 4
reduction 209–11
as terminal electron acceptor 209
uptake 210
in xylem sap 210
Nitrate reductase 210–11
in cytosol 48
induction 321
Nitrite
assimilation in chloroplasts 213
formation from nitrate 210
Nitrate reductase
in chloroplasts 213
in C_4 plants 211
properties 211
subcellular location 211
Nitrobacter 209
Nitrogen
assimilation 209–14, 230
energy requirements 213–14
light requirement in leaves 213–14
rate in chloroplasts 241
rate in leaves 213–14
rate in plants 8

biological importance 209
human requirements 209
isotopes 210
organic transport compounds 251–2,
227–8
plant requirements 3
in soils 209
in xylem sap 210, 227
Nitrogen fixation 210, 226–8, 334
energy requirements 226–7
prospects for genetic engineering 334
requirement for cobalt 236
Nitrogenase 226–7
Nitrosomonas 5, 209
Nodules, ureide synthesis 251
Non-amphipathic proteins, in membranes 38
Non-competitive inhibition 77–8
Non-cyclic electron transport 150
efficiency of energy conversion 159
Non-cyclic photophosphorylation 147
in NADP M-E variant 202
Non-nuclear inheritance 294
Non-protein amino acids 229
Non-reducing sugars 17
Nonacosan-14-ol 262
Nonacosan-15-ol 262
Nonactin 91
Nopaline(s) *335*
Nopaline synthase 319
Nopaline synthase/neomycin
phosphotransferase 335
Northern blotting 324
Nostoc punctiforme 228
Nuclear envelope 40
changes during cell cycle 45–6
Nuclear membranes, fatty acid
composition 86
Nuclear pores 40–1
Nuclease(s) 297, 333
Nucleic acids, structure 31–4
Nucleolus 40
changes during cell cycle 45
Nucleoplasm 40
Nucleoside diphosphatase 83
Nucleoside diphosphate sugars, *see*
NDP-sugars

Nucleoside kinase 248
Nucleoside phosphates 30
Nucleoside phosphorylase 248
Nucleosides, nomenclature 31
Nucleosome(s)
configuration and DNA replication 297
Nucleotidase 251
Nucleotides
composition 30
nomenclature 30–1
occurrence 30
synthesis 245–50
role of OPP pathway 132
Nucleus
division 45–6
structure 40–1
N_2 fixation, *see* nitrogen fixation

cis-9-Octadecenoic acid (octadecenoate) 21
trans-9-Octadecenoic acid (octadecenoate),
conformation 21
Octopine *335*
Octopines, structure 335
Oil-bearing seeds, lipid oxidation during
germination 133
Oils, fatty acid composition 265
Okasaki fragments 297, 300
Okayama and Berg, method for cDNA
cloning 328
Oleic acid (oleate) 20, 21, 261
conformation 21
Oleosomes 133
Oleyl-ACP 261, 268
Oleyl-ACP thioesterase 261
Oleyl-CoA, as cutin precursor 263
Oligo-1,6-glucosidase 100
Oligomeric enzymes 62
Oligomycin 121–2, 124
Oligomycin-sensitivity-conferring-protein
(OSCP), of mitochondrial ATP
synthetase 122
Oligonucleotide probes 324, 330–1
Oligosaccharides 17–18
Onion, DNA content as single copy
sequences 293
Operon 318

Opines 334–5
OPP (oxidative pentose phosphate)
　　pathway 108, 127–3
　as a cycle 130, 132
　carbon flux 133
　in chloroplasts 131
　functions 125, 131–2
　glutathione reduction 132
　importance to nucleotide synthesis 132
　interaction with glycolysis 128
　labelling patterns 131, 133
　location 130–1
　NADP reduction 131–2
　non-oxidative phase 127–30
　oxidative phase 127–8, 130
　products 130
　rates 133
　regulation 131–2
　relationship to glycolysis 131
　stoichiometry 130
　as unidirectional process 130
Orchids, storage materials 100
Organelle function, need for products of both
　　nuclear and organelle genomes 349–
　　52
Organelle genomes, copies per cell 350
Organelle polypeptides
　binding to receptors during transport 348
　energy requirements for transport 348
　location of synthesis 348
Organelles
　definition 36–7
　division 294
　genome 294
　lysis during isolation 48
　methods of preparation 46–9
Organic thiosulphate reductase 231
Ornaline 335
Ornithine
　in arginine synthesis 223, 225
　in proline synthesis 222
　regulation of aspartate
　　transcarbamylase 247
　regulation of carbamyl-P synthetase 245,
　　247–8
Ornithine carbamyl transferase 225

Orotate, metabolism 246
Orotate phosphoribosyltransferase 246
Orotidine monophosphate
　decarboxylase 246
Orotidine-5-P, metabolism 246
Orotidine-5-P decarboxylase 246
Osmotic shock 41
Oxalate 200
Oxaloacetate 110
　as amino group acceptor 215
　in CO_2 assimilation 195
　decarboxylation 200–2
　decarboxylation to PEP 140
　formation in CAM plants 205
　in TCA cycle 110
Oxidases 61
　in chlorophyll synthesis 253
α-Oxidation 137
β-Oxidation 134–41
　energetics 140
　location 137, 140
ω-Oxidation 137, 263
Oxidation of lipids 133–41
Oxidation of water, in PSII 149
Oxidation reactions involving O_2, $\Delta G^{\circ\prime}$
　　values 57
Oxidation/reduction reactions
　characteristics 54–6
　free energy changes 56–8
　importance in plants 54
Oxidative decarboxylation
　in TCA cycle 109–11
　of malonyl semialdehyde 136
Oxidative pentose phosphate pathway, see
　　OPP pathway
Oxidative phosphorylation 114
　during leaf development 345
Oxidoreductases 61, 261
Oxo acids, in amino acid synthesis 214
2-Oxo-3-methylvalerate, metabolism 219
2-Oxobutyrate 218–19
　in isoleucine synthesis 218
2-Oxocaproate, metabolism 219
3-Oxofatty acyl-CoA 135
2-Oxoglutarate 111
　in glutamate synthesis 211–13

　in glutamine metabolism 212
　intracellular transport 190–1
　in NH_3 assimilation 212–13
　in proline hydroxylation 287
　in TCA cycle 111
　transporter of mitochondria 113
2-Oxoglutarate dehydrogenase 109, 111
　inhibition by succinyl-CoA 114
　regulation by adenine nucleotide
　　concentration 114
2-Oxoisovalerate metabolism,
　　regulation 214, 219
Oxygen
　effect on CO_2 assimilation 186
　effect on quantum efficiency 205
　effect on ribulose-1,5-P_2 carboxylase/
　　oxygenase 186–7
　effect on RuP_2-C/Oase 186
　evolution by chloroplasts with NH_3 238
　evolution by chloroplasts with
　　oxaloacetate 238
　inactivation of nitrogenase 226
　inhibition of CO_2 assimilation 185
　K_m values of oxidases/oxygenases 188
　light-dependent reduction 240
　production from H_2O by PSII 152
　reduction by ferredoxin 240
　requirement for photorespiration 185
　standard redox potential 57
Oxygen consumption
　in C_2-PR cycle 187
　light-dependent 162
　by Mehler reaction 163
　by mitochondria 119
　by pseudocyclic electron transport 163
Oxygen evolution
　action spectrum 144
　associated changes in chlorophyll
　　fluorescence 184
　by chloroplasts with GSSG 239–41
　coupling to C_3-CR cycle 169
　coupling to NH_3 assimilation 213
　coupling to NO_2 assimilation 213
　derivation from H_2O 146
　during chloroplast development 341
　evidence for location 156

　flash yields 149, 152
　stoichiometry associated with C_3-CR
　　cycle 172
Oxygen reduction, rate in chloroplasts 241
Oxygenase, mixed function 288

P_{430} 151
　in PSI 150
P_{680} 146
　location in thylakoids 157
P_{700} 146
　in PSI 149–50
　location in thylakoids 157
P site, see peptidyl site
Palmitate 20–1
　activation 261–2
　biosynthesis 257–61
　synthesis in chloroplasts 241
Palmitoleic acid 20
Palmitoylation 311
Palmityl-ACP 260, 268
　synthesis 260
Palmityl-CoA, as cutin precursor 263
Panicum miliaceum 198
Pantothenic acid 67
PAPS (3′-phosphate adenosine 5′-
　　sulphatophosphate) 232
Paracoccus denitrificans 209
Paraquat 153, 240
Parenchyma cells 36
Partition coefficient, between lipid and
　　aqueous phase 87, 89
Passive diffusion, across membranes
　　88–9
Pasteur effect 79
Patch clamping 95
pBR322 plasmid 328–9
Pea genome, plastome copy number 294
　DNA reassociation kinetics 292
Pectic substances 20, 275–7
　classification 276
　extraction with EDTA 275
　functions 276
　structures 276–7
Pectin(s)
　in cell walls 45, 286

methylation 286
synthesis 234
Pectinase 276
in protoplast preparation 46
Penicillium, chromosomes 291
Pentanyl-CoA *135*
oxidation 135
Pentose phosphates, metabolism in
chloroplasts 172
Pentoses, cyclic forms 11
PEP (phosphoenolpyruvate) *52*, 78
in amino acid synthesis 215
in aromatic amino acid synthesis 221–2
carboxylation in C$_4$ plants 199–200
formation from pyruvate in C$_4$
plants 199–200
$\Delta G^{o\prime}$ of hydrolysis 52–4
in gluconeogenesis 139–41
in glycolysis 103
metabolism in C$_4$ plants 200–2
recycling in C$_4$ plants 201–2
in substrate level phosphorylation 54
in synthesis of EPSP 224
synthesis in CAM plants 206
transport across chloroplast envelope 174
transport in C$_4$ plants 204
PEP carboxykinase (PEP-CKase) 139–41,
200–2, 206
equation 200
in CAM plants 206
PEP carboxylase 81, 195, 198, 201
anaplerotic function 125
in CAM plants 205–6
in C$_4$ plants 199–200, 205
in C$_3$ plants 200
equation 199
location in C$_4$ plants 198
properties 199–200
Peptidases 61
Peptide bond 27
formation 310
$\Delta G^{o\prime}$ of hydrolysis 54
Peptide linkages, example 28
Peptidyl site 308–9
Peptidyl transferase 310
Peridinin, in light harvesting 144

Perinuclear space 40
Peripheral proteins, in membranes 38
Permeability coefficient 89
of lipid bilayers
values for various solutes 88
Permeases 92
Peroxidase(s) 61, 64
in cell walls 277, 280, 288
isoenzymes 62
in phenylpropanoid cross-linking 289
Peroxisome
centrifugation characteristics 47
enzymes of C$_2$-PR cycle 188–90
fatty acid composition of membrane 86
marker protein 84
in photorespiration 188–9
properties 43
serine metabolism 189
Pest resistance 333
pH
optima for enzymes 76
light-induced changes in chloroplasts 179
of stroma in illuminated chloroplasts 157
pH gradient
across membranes 90
determination using weak acids 90
in mitochondria 119
measurement 90
Phaseolin
introns in gene 293
restriction map 324
Phenolic compounds
inactivation of enzymes 46
location in cells 46
Phenols, examples 34
Phenylalanine 27
in lignin synthesis 288
metabolism 288
regulation of synthesis 221–4
Phenylalanine ammonia lyase (P-ALase)
288–90, 321
gene regulation by fungal elicitor 321
induction by u.v. light 321
light modulation 178
Phenylalanyl transfer RNA, structure 33
Phenylphosphordiamidate 251

Phenylpropanoid complex, in lignin 278
Phenylpropanoids 34, 321
metabolism 288–90
polymerization 290
residues in hardwoods and softwoods 279
Pheophytin *a*
in PSII 149
role in light-dependent electron
transport 151
Phloem
loading 175
sucrose concentration 175
sucrose content 6
Phosphate
control of fructose-2,6-P$_2$ metabolism 180
co-ordination of carbon metabolism
181–2
in control of triose-P export from
chloroplasts 174
light-induced changes in
concentration 179
regulation of ADP-glucose
pyrophosphorylase 179
regulation of starch synthesis 179–80
in transport of mitochondrial
metabolites 112
transport across chloroplast envelope 174
3′-Phosphate adenosine 5′-
sulphatophosphate, *see* PAPS
Phosphate translocator 92, 173–4, 181
in C$_4$ plants 204
in mitochondrial envelope 112–13
in starch mobilization 177
in sucrose synthesis 174
in triose-P shuttle 237
in triose-P/glycerate-3-P exchange 237
Phosphatidic acid (phosphatidate) 22–3,
263–4, 268
content of different membranes 85
location of synthesis 264
as precursor of phospholipids 265
in synthesis of glycerolipids 263
Phosphatidate cytidyl transferase 265–6
Phosphatidate phosphatase 265–6, 268–9
Phosphatidyl choline 23, 84–5
content of different membranes 85

as donor of fatty acids for
triacylglycerols 265
fatty acid composition 84
synthesis 234
Phosphatidyl choline desaturase 269
Phosphatidyl ethanolamine 23, 84–5, 266–7
content of different membranes 85
fatty acid composition 84
Phosphatidyl ethanolamine *N*-methyl
transferase 266
Phosphatidyl glycerol 23, 85, 265–6
in chloroplasts 84
content of chloroplasts during
development 340
content of different membranes 85
Phosphatidyl glycerol-P, precursor to
phosphatidyl glycerol 266
Phosphatidyl glycerol-P phosphatase 266
Phosphatidyl inositol 23, 85, 265
content of different membranes 85
fatty acid composition 84
Phosphatidyl inositol synthase 266
Phosphatidyl inositol-4,5-P$_2$ 82
Phosphatidyl serine 24, 265–6
content of different membranes 85
decarboxylation 267
structure 24
Phosphatidyl serine decarboxylase 266–7
Phosphatidyl serine synthase 266
Phosphinothricin 191, 212
Phospholipids, biosynthesis 266
7-Phospho-2-oxo-3-deoxyheptanoate, *see* 3-
deoxy-arabinoheptulosonate-7-P
3′-Phospho-pantoyl-β-alanyl-cyteamine-
ADP, *see* coenzyme A
Phosphocholine, synthesis 234
Phosphoenolpyruvate, *see* PEP
Phosphoenolpyruvate carboxykinase, *see*
PEP-CKase
Phosphofructokinase (PF-Kase) 79, 101–2,
104–5, 108
in chloroplasts 182
in futile cycle 106
location 106
in regulation of carbon metabolism in
photosynthetic cells 180–1

in regulation of glycolysis 105, 107
regulation by glycolytic intermediates 107
in starch mobilization 177
Phosphoglycerides, structures 23
Phosphoglyceromutase see glycerate-P mutase
Phospholipase 36
Phospholipase C 81–2
Phospholipid transfer protein 269
Phospholipids 22–5, 265
amphipathic properties 24–5
biosynthesis 265–7
location of biosynthesis 267
in membranes 36, 84–5
Phosphomevalonate kinase 271
N-Phosphonomethylglycine, see glyphosate
Phosphoribosyl aminoimidazole carboximide, accumulation in plants 250
Phosphoribosyl phosphotransferases 248
5-Phosphoribosyl-1-pyrophosphate, see PRPP
in purine synthesis 249
metabolism 248
synthesis 246
Phosphoribosylglycinamide, accumulation in plants 250
Phosphorylases 61
Phosphorylation 311, 316
of ADP 54
in chloroplasts 157–9
in mitochondria 118–23
coupling to NADH oxidation 57
coupling to proton electrochemical gradient 59
of light-harvesting chlorophyll protein 161
of pyruvate dehydrogenase 113
substrate level 54
Phosphorylation/dephosphorylation, of enzymes 81
Phosphorylation potential, export from chloroplasts 236–7
Phosphosphingolipids 23–4
Phosphotransferase 269
Photoautotrophs 5

bacteria 210
Photogenes 344
Photon flux 142
Photons, characteristics 59
Photophosphorylation 157–9, 161–3
during chloroplast development 341
during leaf development 345
Photoreceptors, in organelle development 344–6
Photorespiration 185–94, 333
in algae 193–4
alternative routes 194
in C_4 plants 193, 196
CAM plants 207
definition 185
effect of temperature 187
inhibition by CO_2 185
light requirement 191
metabolite transport 190–1
O_2/CO_2 antagonism 185–7
rate 241
reassimilation of NH_3 213
source of respired CO_2 185, 187
Photorespiratory mutants 189–90
Photorespiratory nitrogen cycle 190–1, 238–9
Photosynthesis
amino acid synthesis 217, 221
in anoxygenic bacteria 148
assimilation of inorganic N 213–14
definition 5
Photosynthetic bacteria 5
C_3-CR cycle 167
Photosynthetic carbon reduction cycle, see C_3-CR cycle
Photosynthetic cells, requirements 5–6
Photosynthetic electron transport, reverse reactions 158
Photosynthetic unit 149
Photosynthetically active radiation 59, 142
Photosystem I (PSI) 148
activity during 'greening' 342–5
antenna complex of 149–50
chlorophyll content 149–50
during chloroplast development 341
composition 149–50

location in thylakoids 155–7
P_{700} 149–50
primary electron acceptor 151
in O_2 reduction 240
reduction of NADP 151
role in $NADP^+$ reduction 148
Photosystem II (PSII) 148
activity during greening 342, 345
composition 149
cytochrome b_{559} 149
during chloroplast development 341
deficiency in heterocysts 228
fluorescence 182
location in thylakoids 155–7
manganese 149
in NAD M-E variant 204
in NADP M-E variant 202
in oxidation of water 149
in PEP-CK variant 203
pheophytin a 149
quinones 149
secondary electron acceptor 151
Photosystems
co-operation between PSI and PSII 148
functions 148
Phragmoplast 46
Phycobilins 142
in algae and cyanobacteria 144
phylogenetic distribution 142
Phytase 284
Phytic acid 15
Phytin 15
Phytoalexins 35
Phytochrome 320, 345–6
in control of P-ALase 290
effect on LHCP gene expression 320
influence on mitochondrial function 347
intracellular location 347
in regulation of gene expression 346
structure of chromophore 347
Phytoene 270, 272
Phytoglycogen 100
Phytohormones 321
Phytol 25, 270
in chlorophyll 143
in chlorophyll synthesis 255

Phytosphingosine 23–4
Pigment–protein complexes
characteristics 149
major complexes in thylakoids 149
preparation 149
Pigments
absorption of light 59
in light harvesting 142
phylogenetic distribution 142
Ping-pong mechanism in enzymic reactions 74
Δ'-Piperidine dicarboxylate acylase 219
Planck's constant 59
Planck's quantum theorem 59
Plant pathogens, as transfer vectors for genes 336
Plant respiration, analysis of inputs and outputs 125
Plants
elemental composition 3
energy requirements 4–5
growth measurements 7–8
growth requirements 3–5
Plasma membrane 36
cellulose synthase assemblies 286
in cellulose synthesis 283–6
composition 83
fatty acid composition 84, 86
formation 45
lipids 84
marker protein 84
sucrose transport 175
Plasmodesmata 36
Plasmids 325
antibiotic resistance genes 329
as cloning vectors 325–6
introduction of DNA 326
pBR322 327
pUC 329
structure 325
T_i 334
Plastid genome, copies per plastid 338
Plastid polypeptides, location of genes 347
Plastids, number per cell during development 338
Plastocyanin 152

location in thylakoid 156–7
precursor polypeptide 349
reduction 152
Plastome
coding capacity 294, 348
copies per chloroplast 294, 338
DNA content 294
gene map 294
genes 295
of liverwort 295
size 294–5
structure 294
Plastoquinone 25, *151*, 273–4
abundance in thylakoids 151
in electron transport 151
half time for reoxidation 156
reduction 151
role as lateral shuttle for H^+ and e^- 156
Poaceae 100
C_4 plants 207
Polar acyl lipids 267
Polar lipids, content in membranes 85
Pollen tubes, Golgi bodies 282
Polyadenylate (poly-A) tail, on mRNA 303, 305, 326
controlling sequence in genes 303
functions 305
Polycistronic sequence 305
Polygalacturonyl chains, in pectic substances 275
cis-1,4-Polyisoprene 25
Polynucleotides 31–4
Polypeptides
post-translational modification 311
transport in control of gene expression 316
transport into organelles 348
Polyploids 291
Polyprenols *270*
Polyproline, in extensin 277
Polyribosomes 40, 311
Polyribosomal mRNA, composition 319
Polysaccharide
in AGPs 277
in cell wall matrix 20, 275–7
contamination of RNA preparations 333

fibrillar 275
non-cellulosic in cell walls 275–7
properties 18–20
storage 18–19
structural 19–20
Polysomes, *see* polyribosomes
Polyterpenes 35
Polyvinylpolypyrrolidone 47
Polyvinylpyrrolidone 46
Porine 112
Porphobilinogen, metabolism 253
Porphobilinogen deaminase 253–4
Porphobilinogen synthase 253
Porphyrin ring
chlorophyll 64
cytochromes 64
in haem 56
in prosthetic groups 64
Positive co-operativity, of allosteric enzymes 79
Post-transcriptional control, of gene expression 318–20
Post-transcriptional modification 304–6
Post-translational control, of gene expression 315
Post-translational modifications of proteins 311–12, 320
Post-illumination burst 185
Potassium, plant requirements 3
Potato, extraction of enzymes from 46
Potential difference (electrical)
across membranes 89
ionic concentrations 89
Pre-mRNA transcripts
processing 316
in control of gene expression 319
Precursor polypeptides, for chloroplast components 349
Prenyl transferase 272
Prenyllipids, biosynthesis 269–74
Prephenate, metabolism 214, 221–3
Presqualene synthase 273
Primary active transport 94
Primary alcohols, of fatty acids 262
Primary cell wall
organization 280–1

composition 279
Primary electron acceptors, oxidation following charge separation 151
Primary photochemistry, fluorescence characteristics 159–60
Primary transcripts of DNA 303–4
modification 304–6
Primase, *see also* RNA polymerase
Primer plasmid, in cDNA synthesis 328
Primer
in cDNA synthesis 328
in starch synthesis 176
Probes, *see* oligonucleotide probes
Prokaryotic galactolipids 267
Producer cells, definition 6
Prolamellar body 340–1
changes during chloroplast development 342
Proline 26, *27*
accumulation in stressed plants 222
post-translational hydroxylation 287
post-translational oxidation 277
in salt-stressed plants 26
structural analogue 229
synthesis 225
Proline dehydrogenase 225
Proline synthesis 222
Prolyl hydroxylase 287
Promiscuous DNA 296
Promoter sequences 303, 318, 335
see also 'TATA box'
base sequence 303
in expression of LHCP 318
Proofreading, by aminoacyl-tRNA synthetase 311
Prophase 45
Propionyl-CoA *135*
oxidation 136
Proplastids 338
development 339–40
differentiation into chloroplasts 339
Prosthetic groups 63–4, 316, 320
biotin 64–5
FAD 64
FMN 64
haem 64, 312

pyridoxal phosphate 65–6
TPP 65
Proteases 321
in vacuoles 44
Protein amino acids, classification 26
Protein kinases 81–2
Protein phosphatases 81
Protein synthesis 301–14
elongation 310
energy requirements 311
in vitro systems 330
initiation 308
initiation in chloroplasts 313
in organelles 313–14
Protein 25–30
amphipathic 29
β-pleated sheets 28
in cell walls 277
in chloroplasts 42
conformational changes 29
content in seeds 25–6
covalent bonds 27
dissociation of oligomers 29
disulphide bonds 28
electron carriers 29
enzymes 29
essential functions 25
functions in plants 29–30
helical coiling 28
helical destabilization 28
hydrogen bonding between chains 28
hydrogen bonding within peptide backbone 28
hydrophobic domains 29
in membranes 36
in mitochondria 42
mobilization 321
molecular mass 28
native configuration 28–9
in nuclei 40
oligomeric 29
primary structure 27–8
quarternary structure 29
redox 56
ribosomes 40
secondary structure 28

structural 29
tertiary structure 29
transport 29
Proteoglycans 29, 277
Protochlorophyll(ide) 342, 345
  binding protein 255
  inhibition of aminolevulinate
      synthesis 345
  photoreduction 255
  regulation of magnesium chelatase 255-6
  synthesis 253-5
Protochlorophyllide oxidoreductase: NADP-
      specific 254-5, 345
  gene expression 317
Protochlorphyllide reductase, see
      protochlorophyllide oxidoreductase
Protofilaments 44
Protomer 29
Proton-ATPase 95
  investigation with patch clamping 95-6
  role in solute transport 94
  stoichiometry 94
Proton:ATP ratio, in ATP synthetase of
      mitochondria 121, 123
Proton electrochemical gradient ($\Delta\tilde{\mu}_H^+$) 59
  typical values 59
  ΔG values 59
Proton gradient
  in fluorescence quenching 182
  generation across thylakoids 182
  generation across thylakoids 156-7
  generation using succinate 157
  magnitude in ulliminated chloroplasts 157
Proton motive force 95
  across illuminated thylakoids 158-9
  in mitochondria 119, 121
Proton motive Q-cycle 120-1, 162
  evidence 120
  in mitochondria 120
  mechanism 120
Proton pump, in sucrose transport 175
Proton translocation across thylakoids
  in cyclic electron transport 162
  in non-cyclic electron transport 156-7
Proton translocation in mitochondria 119
  mechanism of translocation 119-20

stoichiometry with electron transport 120
stoichiometry with ADP
    phosphorylation 159
Proton translocators 91
Protoplasm 37
  definition 37
Protoplasts 36
  from C_4 plants, CO_2 assimilation 198
  lysis 46
  methods of preparation 46, 198
Protoplast fusion 334
Protoporphyrins 64
Protoporphyrin IX
  metabolism 253
  regulation 256
Protoporphyrinogen IX, metabolism 253
Protoporphyrinogen oxidase 254
PRPP synthetase 246
PRPP (5-phosphoribosyl-1-
    pyrophosphate) 246
Pseudocyclic electron transport 163
Pseudocyclic photophosphorylation 163
Pseudouridine, in transfer RNA 33
PSI, see Photosystem I
Psicose 13
PSII, see Photosystem II
pUC plasmids 329
Purines 30
  in nucleic acids 32
  oxidative catabolism 250-2
  synthesis de novo 249-50
Purine nucleotides
  regulation of carbamyl-P synthetase 247
  synthesis from free purines 248-9
Purine synthesis
  energy requirements 250
  pathway 249
Puromycin 315
Purple sulphur bacteria 148
Pyran 12-13
Pyranose forms, of sugars 11-13
Pyridoxal phosphate 66
  in aminotransferase reactions 215
  as prosthetic group 65-6
Pyridoxamine phosphate 215
2-Pyridylhydroxymethanesulphonate 187-8

Pyrimidines 30
  metabolism 248
  in nucleic acids 32
  origin of ring atoms 245
  synthesis de novo 245-7
Pyrimidine nucleotides
  regulation of carbamyl-P synthetase 247
  synthesis 245-9
Pyrimidine synthesis, regulation 247
Pyrophosphate 102
  $\Delta G^{o'}$ of hydrolysis 53
  generation by FP-PTase 108
Pyrophosphate bond, in nucleoside
    diphosphate sugars 174
Pyrophosphatase 134
  in pyrimidine metabolism 248
  in sulphate assimilation 231
Pyrophosphomevalonate decarboxylase 271
Pyrophosphorylases 282
Pyrrole 252
Pyruvate 103
  as amino group acceptor 215
  anaerobic oxidation 103
  oxidation to acetyl-CoA and CO_2 108-10
  recycling in C_4 plants 202
  transport in C_4 plants 204
  transport into mitochondria 112
  in valine synthesis 218
Pyruvate transporter, of mitochondria 113
Pyruvate carboxylase, anaplerotic
    function 125-6
Pyruvate decarboxylase 62, 65, 108-9, 220
Pyruvate dehydrogenase complex 81, 104,
    108-9, 257-8
  in regulation of TCA cycle 113
  structure 62
Pyruvate dehydrogenase kinase 113-14
Pyruvate kinase 101, 103-5
  in chloroplasts 104
  in control of glycolysis 105
  inhibition during gluconeogenesis 106
  regulation by adenine nucleotides 107
  regulation by cations 107
  in regulation of glycolysis 105-7
  in substrate-level phosphorylation 54
Pyruvate, P_i dikinase 81, 198-9, 201-2

in CAM plants 206
regulation 199

Q_A, in fluorescence quenching 182
Q-cycle, see proton motive Q-cycle
Quantum efficiency 205
Quantum yield, in photosynthesis 147, 160
Quinate: NAD^+ oxidoreductase 81
Quinine 34-5
Quinones, in PSII 149

Racemases 61
Raffinose 18
Random bisubstrate reactions 73-4
Random walk 145-6
Rape, fatty acid composition of
    triacylglycerols 265
Reaction centres 146
  chlorophyll antennae 149
Reannealing, of cohesive ends 322-3
Reassociation kinetics, see renaturation
    kinetics
Recombinant DNA 322-37
  specific problems in plants 333
Red drop, in $NADP^+$ photoreduction 148
Red light, phytochrome and gene
    expression 320
Red rise, in quantum yield of $NADP^+$
    reduction 148
Redox loops, in mitochondria 120
Redox pairs 54
  in shuttle mechanisms 237
Redox potential 56
  of carriers in mitochondrial electron
      transport 118
Redox proteins 56
Reducing equivalents, export from
    chloroplasts 236-9
Reducing sugars 17
Reduction, of disulphide group enzymes 80
Reductive pentose phosphate pathway, see
    C_3-CR cycle
Regulation
  of glycolysis 107
  of OPP pathway 131
  of TCA cycle 112-13

Regulatory enzymes   80
Relative growth rate   7
Release factors, in translation   309
Renaturation kinetics   292
Repeated sequences (of DNA)   292, 333
Replication fork   297–8, 300
Replicons   297, 300
Repressor   318
Resonance transfer   145–6
Respiration   99–141
    analysis of energy inputs and outputs
        123–5
    functions   125
    importance in heterotrophs   5
    importance in plants   5
    rates in different tissues   99
    secondary oxidative mechanisms   127–41
Respiratory control ratio   119
Restriction enzymes, see restriction nucleases
Restriction fragments   322
Restriction mapping   322–4
Restriction nucleases   322–3
    abbreviations   322
    base sequences recognized   322
    in cloning   326
    use in gene mapping   323
Retranslocation   6
Reverse reactions
    of ATP synthetase   158
    of photosynthetic electron transport   158
Reverse transcriptase   324, 327–9
Reversible inhibition, of enzymes   76
Rhamnogalacturonans   16, 276
    in cell wall organization   280–1
    structure   276–7
    synthesis   286
Rhamnose   15, 16
Rhizobium   226, 236, 251, 334
    plasmids   325
Rhizobium–legume symbiosis   226
    metabolite exchange   227–8
Rhodopseudomonas sphaeroides, oxidation of
    H$_2$ and H$_2$S   148
Ribitol   14
Riboflavin   67
Ribonuclease   40, 333

Ribonucleic acid, see RNA
Ribonucleosides   30
Ribose   12
    in ribonucleotides and RNA   30
Ribose-5-P   129
    in amino acid synthesis   215
    metabolism inchloroplasts   171
    in OPP pathway   128–9
    in pyrimidine synthesis   246
Ribose-5-P isomerase   128, 130, 171–2
Ribosomal proteins
    gene expression   318
    location of genes   348
Ribosomal RNA, see rRNA
    content during chloroplast development
        341
    methylation of precursors   306
    modifications   306
    multiple genes   305
    post-transcriptional processing   306
Ribosome translocation   310
Ribosomes   38, 40, 341
    chloroplastic   313
    mitochondrial   313
    number per cell during leaf development
        343
    number per plastid during leaf
        development   343
    subunit composition   305
    subunits   40, 306, 308
Ribosylamide-5-P, metabolism   249
Ribulose   13
Ribulose-1,5-P$_2$ 168
    in activation of RuP$_2$-Case   179
    effect on autocatalysis   172
    in C$_3$-CR cycle   167–72
    inhibition of glucose-6-P dehydrogenase
        132
    light-induced changes in concentration
        179
    metabolism   185–7, 192
        effect of temperature   187
        energy requirements   191, 193
        stoichiometry   193
    pool size   172
    regeneration from triose-P   171

regulation of glucose-6-P
    dehydrogenase   132
synthesis   172
Ribulose-1,5-bisphosphate carboxylase
    (RuP$_2$-Case)   28, 81, 171
    activation   179
    affinity for CO$_2$ 168
    in CAM plants   206
    in chloroplants   48
    control by light   321
    cytokinin stimulation of synthesis   322
    in C$_4$ plants   195, 204–5
    $\Delta G^{o\prime}$ of reaction   168
    kinetics   186
    large subunit   350
        coding in plastid genome   348
        rate of synthesis   345, 350
    mRNA   321
    during plastid development   344
    properties   168–9
    reaction catalysed   168
    reaction with O$_2$ 168
    small subunit
        coding in nuclear genome   348
        gene expression   318
        gene expression during greening   344
        initial translation product   313
        precursor polypeptide   349
        rate of synthesis   345, 350
        regulation of synthesis   350
        transcriptional regulation   351
        transit peptide   313
    structure   62
        of large subunit   28
    subcellular distribution   48
    synthesis during chloroplast
        development   341
    transcriptional control   321
    translational control   321
Ribulose-1,5-P$_2$ carboxylase/oxygenase
    (RuP$_2$-C/Oase)   168, 333
    active site   70
    in anaerobic bacteria   192
    in chemoautotrophs   192
    in C$_4$ plants   202
    kinetics for O$_2$ and CO$_2$ 77–8, 186–7

in photorespiration   185–7
rates of oxygenase and carboxylase
    activities   168–9
Ribulose-1,5-P$_2$ oxygenase (RuP$_2$-Oase)   168,
    188
    in C$_4$ plants   204–5
    kinetics   186
    in photorespiration   185–7
    reaction catalysed   168
Ribulose-5-P   127, 128, 130
    in OPP pathway   128
    metabolism in chloroplasts   171
Ribulose-5-P 3-epimerase   128, 130, 171–2
Ribulose-5-P kinase   62, 171
    in chloroplasts   172
    light modulation   178
Ricinoleic acid   20
Ricinus communis, vegetable oil   20
Rieske centre   115–16
Rings, formation in sugars   11–12
RNA   31–4
    cytochrome b   304
    location in cells   32
    messenger, see mRNA
    modification of primary transcripts
        304–6
    in nuclei   40
    in ribosomes   40
    ribosomal, see rRNA
    structure   31–2
    synthesis using DNA template   302, 304
    transfer, see tRNA
    transport to cytoplasm   306
RNase, in vacuoles   44
RNase H   328–9
RNA polymerase(s)   300, 302, 316–18
    binding to chromatin   317
    binding to DNA   304
    control of gene expression   317
    in DNA synthesis   297
    forms   304, 317
    levels during gene expression   317
    location of genes   348
    regulatory proteins   318
    subunits   302–3
RNA primers   297

requirement for DNA synthesis 299
removal 300
Rotenone, inhibitor of mitochondrial electron transport 115
rRNA (ribosomal RNA) 33, 292, 304–5
Rubber 25, 34–5
Rubisco, see ribulose-1,5-P_2 carboxylase/oxygenase
RuP_2-C/Oase, see ribulose-1,5-P_2 carboxylase/oxygenase
RuP_2-Case, see ribulose-1,5-P_2 carboxylase
RuP_2-Oase, see ribulose-1,5-P_2 oxygenase

S1 nuclease 292, 327
Sainfoin 312
Salicylhydroxamic acid, see SHAM
Sanger sequencing 331–2
Sapwood, cell wall composition 279, 281
Satellite DNA 32
Scurvy 15
Scutellum 321
Second-order reactions 67–8
    kinetics 68
Second singlet state, of chlorophyll 144–5
Secondary active transport 94–5
Secondary alcohols of fatty acids 262
    oxidation to ketones 262
Secondary cell walls, composition 279
Secondary compounds 9, 34–5
Secondary messenger 82
Secondary oxidative mechanisms 127–41
Sedimentation coefficient 49
Sedimentation, rates for subcellular particles 47–9
Sedoheptulose-1,7-P_2 171
    synthesis in chloroplasts 171
Sedoheptulose-1,7-P_2 phosphatase 171
    in chloroplasts 171–2
    light modulation 178
    regulation 181
Sedoheptulose-7-P 129
    metabolism in chloroplasts 171
Seed storage proteins 312, 317, 319
    amino acid deficiencies 333
    human diet 333
    genetic engineering 333

mRNAs 328–9
Selenate 230
Selenium, in GSH peroxidase 241
Selenocysteine 77
Semicarbazide 215
Semiconservative replication 296
Sequon 287
Serine 27
    in AGPs 277
    enantiomer in protein 26
    formation from CO_2 in leaves 187
    metabolism in leaves 189
    synthesis from glycine 187
    synthesis in leaf mitochondria 189
Serine:glyoxylate aminotransferase 189
Serine hydroxymethyltransferase 188–9, 235–6
Serine:pyruvate aminotransferase 189
Serine sulph-hydrase 231
Serine transacetylase 231
Sesquiterpenes 270
Sethoxydim 258
SHAM (salicylhydroxamic acid) 115
Shikimate
    derivation from OPP pathway 132
    in lignin synthesis 288
    synthetic pathway 221–2
Shikimate dehydrogenase 222
Shikimate kinase 222
Shikimate-3-P 78
    metabolism 224
Shuttles
    in chloroplasts 236–9
    glutamate/2-oxoglutarate 238–9
    glycollate/glyoxylate 238–9
    malate/oxaloacetate 238
    malate/pyruvate 239
    mechanisms 140
    triose-P/glycerate-3-P 237
Signal peptides 313
    coding regions in genes 303
Signal peptidases 313
Simazine 153
Simian virus 40 (SV 40) DNA 328
Sinapate
    metabolism 288

synthesis 235
Sinapyl alcohol 34
    in lignin 278–9
    polymerization 290
    synthesis 289
Single copy DNA 292
Single-displacement reactions
    ordered binding 73
    random binding 73
Single-stranded binding proteins 297
Singlet state, of chlorophyll 145
Sinks, definition 6
Sirohaem, in NO_2^- reductase 211
Small nuclear ribonucleoproteins (snRNPs) 304
Small single copy DNA region, in chloroplast genome 294–5
Sodium dithionite, substrate for nitrogenase 226
Sodium dodecyl sulphate 38
Solanaceae, extraction of enzymes from 46
Solanesyl-pyrophosphate 271
Solar radiation 142
Sorbose 13
Southern blotting 324–5
Soya bean
    amino acid composition 209
    lipids 21
    Okasaki fragments 300
    protein content 25
    leghaemoglobin gene 293
SO_2, see sulphur dioxide
Spacer regions, in rRNA genes 305
Sphingosine 23
Spinach, glycolysis in chloroplasts 104
Spindle fibres 46
Squalene 34, 272–3
Squalene-2,3-epoxide 272–3
Squalene epoxidase 273
Squalene epoxide cycloartenol cyclase 273
Squalene synthase 273
Stachyose 17–18
Staggered cuts 322
Standard free energy change (ΔG°) 51
Standard free energy change at pH 7 ($\Delta G^{\circ\prime}$) 51

biological relevance 51
    of oxidation/reduction reactions 57
    relation to K_{eq} 51
    values for high energy compounds 52–5
Standard hydrogen electrode 56–7
Standard redox potential (E_0) 56
Standard redox potential at pH 7 (E_0') 57
Starch
    accumulation in producer cells 7
    in C_4 plants 196
    diurnal variation in leaves 6
    grains 18
    mobilization 100–1, 321
    mobilization in chloroplasts 177
    regulation 182
    mobilization to sucrose 99
    structure 18–19
    synthesis in chloroplasts 175–6
    regulation 179–80
    synthesis in C_4 plants 202
Starch phosphorylase 99, 101, 177
    in starch mobilization 177
State 1 – State 2 transitions 161
    during chloroplast development 341
    monitoring by fluorescence 161
Stearic acid 20–1
    conformation 21
Stearyl-ACP 260
    desaturation 261
Stearyl-ACP desaturase 261
    location 261
Stereoisomers, of sugars 10–11
Sterols 270
    biosynthesis 272
    in membranes 86
    in plasma membrane 84
Sterylglycoside, in plasma membrane 84
Sticky ends, see cohesive ends
Storage lipids 133
Storage polysaccharides 18–19
Storage proteins 25–6
Stress tolerance, prospects for genetic engineering 334
Stroma 41
    pH in light 157, 169
Stromal thylakoids 41–2

Structural isomers, of sugars 11
Structural polysaccharides 19–20, 276
Structural proteins 29
  in cell walls 277
Sub-chloroplast particles 41
Sub-mitochondrial particles 118, 121–2
  formation 118
Subcellular fractions, methods of
    preparation 46
Subcellular organelles, separation 48
Suberin 22
Substrate-level phosphorylation
  in glycolysis 103
  in TCA cycle 109, 111
  theory 54
Succinate 76
  for creating artificial proton gradients 157
  in glyoxylate cycle 138–40
  metabolism to sucrose, 140–1
  in N_2 fixation 227
  in TCA cycle 111
Succinate dehydrogenase 109, 111, 139–40
  location 118, 140
  in mitochondrial electron transport 111,
    116
Succinate thiokinase 109, 111, 114
Succinyl-coenzyme A 111
  inhibition of 2-oxoglutarate
    dehydrogenase 114
Sucrose 17, 18
  in cell wall synthesis 282
  export from photosynthetic cells 175
  metabolism to NDP-sugars 284
  phloem loading 175
  synthesis 141, 174–5
    in protoplasts 173
    regulation by fructose-2,6-P_2 180
    from starch 99, 177
    stoichiometry 174
  transport in phloem 6–7, 175
Sucrose synthase 99–100, 108, 175, 181–2,
    184
  inactivation by phenolics 46
Sucrose-6-P, metabolism 175
Sucrose-P phosphatase 174–5
Sucrose-P synthetase 174–5, 180–1

in C_4 plants 202
  regulation 180
Sucrose/proton co-transport 175
Sugar alcohols 14
Sugar beet 175
  carbon budget 6
Sugar cane 175
  CO_2 assimilation 197
Sugars
  activation 174–5
  aerobic oxidation 99
  in cell wall synthesis 282–3
  cyclic forms 11–12
  furanose forms 13–14
  Haworth representation 12
  L and D configuration 10
  linking by glycosidic bonds 16
  pyranose forms 12–13
Sulphanilamide 235, 250
Sulphate
  activation 230–1
  assimilation
    via bound pathway 231
    in chloroplasts 230, 241
    in heterotrophs 232
    in soils 230
  sulphur source for plants 4
  uptake 230
  in xylem sap 230
Sulphatophosphate anhydride bond 231
Sulph-hydration pathway 233
Sulphide
  assimilation 232
  bound 231
  formation from sulphite 232
  incorporation into cysteine 231
  incorporation into methionine 233
  oxidation by chemoautotrophs 5
Sulphite
  assimilation 232
    in chloroplasts 230, 232
  reduction rate in chloroplasts 241
Sulphite reductase 232
6-Sulpho-6-deoxyglucose, in sulpholipids 25
Sulpholipid 24–5, 85, 267
  synthesis from PAPS 232

Sulphonylureas 225–6
6-Sulphoquinovose
  in sulpholipids 25
  synthesis from PAPS 232
Sulphoquinovosyl diacylglycerol 267, 288
Sulphur
  acid-labile in nitrogenase 226
  assimilation 230, 232
  atmospheric pollution 232
  in ferredoxins 55–6
  plant requirements 3, 230
  volatile emissions 232
Sulphur dioxide 232
Sulphur trioxide 232
Supercoiling of DNA 294
Superoxide radical 154, 240
  generation by bipyridylium herbicides
    153–4
  metabolism 240
Superoxide dismutase 240
  in the Mehler reaction 163
Svedberg unit 49
Sycamore 281
Symport 93–4

Tagatose 13
Talose 12
Tartronyl semialdehyde, metabolism in
  algae 194
Tartronyl semialdehyde reductase 194
TATA box 303, 318 (see also promoter)
TCA cycle (tricarboxylic acid cycle) 108–14
  early evidence for the pathway 108
  efficiency of energy conservation 125
  evidence in plants 112
  functions 125
  labelling pattern 109, 112
  location 140
  reactions 109
  regulation 112–14
  regulation by mitochondrial membrane
    transport 112–13
TDNA, see transforming DNA
Telophase 45–6
Temperature optima, for enzymes 76
Terminal transferase 327

Termination sequence 304–5
Terminator codon(s) 301, 309
Terpenes 25, 269
Terpenoids 25, 269–70
  biosynthesis 271
  examples 34
  structure 270
Tetracycline resistance 327, 329
  genes 329
Tetrahydrofolate, see THF
Tetraphenylphosphonium 90
Tetrapyrrole(s)
  in chlorophyll 143
  occurrence 252
  in prosthetic groups 64
  synthesis 253
Tetraterpenes 270
Tetroses, structures 11
THF (tetrahydrofolate) 63, 64
  C_1 derivatives 235
  interconversions of C_1 derivatives 235
  metabolism 234–6
  in mitochondria 189
Thiamine (Vitamin B_1) 67
Thiamine pyrophosphate, see TPP
Thiobacillus 5
Thiocyanate 90
Thiol carrier, in APS metabolism 231
Thiolase 137, 140 (see also acetyl-CoA acyl
  transferase)
Thiols, in enzyme extraction 46
Thioredoxin(s) 80
  in light modulation of enzymes 178
  in synthesis of deoxyribonucleotides 247
Thiosulphonate reductase 230–1
  in Chlorella mutants 232
Threonine 27
  in control of aspartate metabolism 217–
    18
  effect on growth 218
  in regulation of homoserine dehydrogenase
    217
  synthesis 216–17
    regulation 217
    synthetic pathway 217
Threonine dehydratase 218–20

Threonine synthase 217–18
  regulation by *S*-adenosylmethionine 217
  subcellular location 217
Threose *11*
Thylakoids
  appressed regions 42
  in C_4 plants 196
  changes in area during chloroplast
     development 342
  components
    of appressed regions 154–5
    lateral distribution 154–6
    of non-appressed regions 154–5
    transverse distribution 157
  distribution of major components 156
  non-appressed regions 42
  proliferation during chloroplast
    development 339
  structure 41–2
  transverse heterogeneity of
    components 156
Thymidilate 247
Thymidilate synthase 248
Thymidine 31
Thymine *30*, 302
  base pairing with adenine 32–3
  in DNA 32
T_i plasmid(s) 334
  modification for genetic engineering 334
  as transfer vectors 334, 336
  vir region 335
TMP, *see* deoxy-TMP
Tobacco 175
  chromosomes 291
  extraction of enzymes from 46
  gene expression 319
  genetically engineered hornworm
    resistance 333
  isoenzymes of peroxidase 62
  regulation of LHCP 318
  upstream elements in gene expression 319
Tomato, extraction of enzymes from 46
Tomato golden mosaic virus, as transfer
    vector 336
Tonoplast 43
  lipids 84

marker protein 84
sucrose transport 175
TPP (thiamine pyrophosphate) 65
  in 2-oxoglutarate dehydrogenase 111
  as prosthetic group 65
  in pyruvate decarboxylase reaction 65,
    109
  in transketolase reaction 65, 129
Transaldolase 61, 128, 130
  in OPP pathway 129, 131
Transaminases, *see* aminotransferases
Transcription 302–6
  of chloroplast genes, effect of nuclear
    genome products 350
  effect of chromatin conformation 317
  initiation 304
  of nuclear DNA,
    effect on transcription of chloroplast
      DNA 350
    effect on organelle function 350
  regulatory sequences 303, 305
Transcriptional control 315, 318, 322
  of gene expression 315
  of P-ALase 321
  of RuP_2-Case 321
  of urea amido lyase 321
Transfer RNA, *see* tRNA
Transfer vectors 335–6
  Agrobacterium 335
  cauliflower mosaic virus 336
  tomato golden mosaic virus 336
Transferases 61
Transforming DNA (TDNA), of T_i
    plasmid 334
Transformylation reactions 233–6
Transit peptidases 348–9
  in intra-thylakoid space 349
  in stroma 349
Transit peptide(s) 313, 348
  coding regions in genes 303
  role in localization of organelle
    components 349
Transition state
  in chemical reactions 69
  of enzymes 69
Transketolase 61, 65–6, 128

in C_3-CR cycle 170–1
in OPP pathway 129–31
reactions catalysed 129
Translation 306–11
  activation of amino acids 307
  elongation 309
  initiation 308–9
  termination 309–11
  termination signal 303
  translocation of ribosomes 309
Translational control
  of gene expression 315
  of RuP_2-Case 321
Translocators 92, 237–8
  in C_4 plants 204
  in chloroplasts
    for dicarboxylates 238
    for phosphate 173–4
  in mitochondrial envelope 112
Transmethylation 233–6
  energy requirements 235
  methionine as donor 234
  reactions 235
Transport 83–96
  across membranes 83
  in C_2-PR cycle 190
  driving force 89
  of intermediates of C_3-CR cycle 173
  long distance 6–7
  of metabolites across chloroplast
    envelope 173
  of metabolites across mitochondrial
    membranes 112–13
  of metabolites in C_4 plants 204
  of polypeptides 316
  of proteins into chloroplasts 313
  of sucrose 175
Transporters 112
Trans-sulphuration pathway 232–3
Triacylglycerols *265*
  as storage products 264
  biosynthesis 264–5
  fatty acid composition 265
  location of synthesis 265
Triazines 153
Tricarboxylate transporter, of mitochondria

112–13
Tricarboxylic acid cycle, *see* TCA cycle
Trifluralin *282*
Triglycerides 21–2
9,10,18-Trihydroxystearate 263
Triose-P
  in C_3-CR cycle 167
  control of export from chloroplasts 174
  control of fructose-2,6-P_2 metabolism 180
  effects on fructose-2,6-P_2 kinase 107
  in NO_3^- reduction 213
  oxidation in cytosol 237
  in starch synthesis 175
  in sucrose synthesis 174
  transport across chloroplast envelope 174
Triose-P dehydrogenase (NADP-specific),
    subcellular distribution 48
Triose-P isomerase 101–2, 104–5, 171, 174,
    237
  in chloroplasts 169
  in triose-P shuttle 237
Triose-P/glycerate-3-P shuttle 237
Trioses, structure 10
Triphenylmethylphosphonium 90
Triplet code 302
Triterpene 270
tRNA (transfer RNA) 304, 306
  attachment of amino acids 307
  genes in chloroplast 294
  interactions with mRNA template 306
  precursors, modifications 306
  structure 33
Tryptophan *27*
  synthesis 221–3
    regulation 221–2, 224
Tryptophan synthase 223
Tubulin 29, 44
  polymerization, control of microfibril
    orientation 286
  depolymerization of microtubules 282
Tyrosine *27*
  dimerization in cell walls 288
  in lignin synthesis 288
  metabolism 288
  synthesis 221–3
    regulation 221–2, 224

Tyrosine ammonia lyase 288

Ubiquinol, *see* ubiquinone
Ubiquinone 25, 114, 273–4
    in mitochondrial electron transport 114,
        116
    oxidation of UQH$_2$ by alternative oxidase
        117
    properties 114
Ubiquitin 347
UDP (uridine 5'-diphosphate), synthesis 246
UDP kinase 246
UDP-arabinose, metabolism 282–3
UDP-galactose 268
    metabolism 282
UDP-galactose : *sn*-1,2-diacylglycerol
        galactosyltransferase 268–9
UDP-galacturonate, metabolism 282
UDP-glucose 52, 99, 174
    in cellulose synthesis 285
    $\Delta G^{\circ\prime}$ of hydrolysis 52–4
    in hemicellulose synthesis 256, 287, 296
    metabolism 282
    in sucrose synthesis 174–5
    synthesis from sucrose 282
    in xyloglucan synthesis 286
UDP-glucose 4-epimerase 284
UDP-glucose dehydrogenase 284
UDP-glucose pyrophosphorylase 100–2,
        108, 174
    mRNA 320
    in C$_4$ plants 202
UDP-glucuronate, metabolism 282
UDP-glucuronate 4-epimerase 284
UDP-glucuronate decarboxylase 284
UDP-glucuronate pyrophosphorylase 284
UDP-*N*-acetylglucosamine, metabolism 282
UDP-rhamnose, metabolism 283
UDP-sulphoquinovose 268
UDP-sulphoquinovose : *sn*-1,2-diacylglycerol

sulphoquinovosyl transferase 268–9
UDP-xylose
    incorporation into xyloglucan 286
    metabolism 283
UDP-xylose 4-epimerase 284
UMP (uridine 5'-monophosphate)
    regulation of aspartate transcarbamylase
        245, 247–8
    regulation of carbamyl-P synthetase 245,
        247–8
    synthesis 246
        regulation 248
UMP kinase 246
Uncompetitive inhibition 78
Uncouplers 91
    effects on mitochondrial proton gradient
        119
Uniport 93–4
Unsaturated fatty acids, oxidation 136–7
Unwinding, of DNA 297
Upstream element(s) 318–19
Uptake
    of inorganic carbon in algae 193–4
    of sucrose 175
Uracil *30*, 302
    in RNA 32
    synthesis 245
Urea
    disruption of H bonds in cell walls 280
    metabolism 251–2
Urea amido lyase 321
Urease 251–2
Ureides
    in long-distance N transport 227
    synthesis from purines 250–2
Ureidoglycine aminohydrolase 252
Ureidoglycollate amidohydrolase 252
Uricase 251
Uridine 31
Uridine 5'-diphosphate, *see* UDP

Uridine diphosphate sugars, *see* UDP-sugars
Uridine 5'-monophosphte, *see* UMP
Uridine 5'-triphosphate, *see* UTP
Uridine nucleotides, synthesis 245–7
Uronic acids (uronates) 14–15
    methyl esters 14
Uroporphvrinogen III, metabolism 253
Uroporphyrinogen decarboxylase 253–4
Uroporphyrinogen III cosynthase 253–4
UTP (uridine 5'-triphosphate)
    $\Delta G^{\circ\prime}$ of hydrolysis 53
    synthesis 246

Vacuoles
    in CAM plants 205
    properties 43–4, 46
Valine 27
    metabolism 214
    synthesis 218–21
        regulation 219
Valinomycin 91
    ionophore properties 91
    mode of action 91
Variable fluorescence of chlorophyll 160
Vegetable oils 21
    fatty acid composition 20
Verbascose 18
Vesicles, from thylakoids 154–5
Viologen dyes, in NO$_3^-$ reduction 210–11
Vir region of T$_i$ plasmid 335
Viscosity, of membranes 86
Vitamin B$_{12}$, function 236
Vitamin C 15
Vitamins 67

Wall, *see* cell wall
Water
    long distance transport 6
    membrane permeability 87

oxidation in light-dependent electron
        transport 152
Water-use efficiency 205, 207
Waxes 22, 261
    esters 262
    synthesis 261
Waxy corn 18
Weeds, C$_4$ plants 195
Wheat
    amino acid composition 209
    quantum efficiency 205
Wobble in genetic code 302, 307
Woolf plot 72–3

Xanthine, metabolism 251
Xanthine dehydrogenase 251
Xanthine oxidase 251
Xanthophylls 149, 273
    phylogenetic distribution 142
Xylans, structure 276
    β(1 → 3) in algae 20
Xylem, microfibril orientation 281
Xylem sap, N-transport compounds 227–8
Xyloglucan(s)
    in cell wall organization 280–1
    cross-linking to arabinoxylan 279
    structure 276, 278
    synthesis by Golgi bodies 286
Xylose *12*, 311
Xylulose *13*
Xylulose-5-P *129*, 130, 172
    metabolism in chloroplasts 170–1

Z scheme 150
Zea mays, leaf anatomy 196
Zein, signal peptide from maize 313
Zero-order reactions 68
    kinetics 68
Zwitterion 26